Edited by
Richard Dronskowski,
Shinichi Kikkawa, and
Andreas Stein

Handbook of
Solid State Chemistry

Edited by
Richard Dronskowski,
Shinichi Kikkawa, and
Andreas Stein

Handbook of Solid State Chemistry

Volume 2: Synthesis

WILEY-VCH

WILEY-VCH Verlag GmbH & Co. KGaA

Editors

Richard Dronskowski
RWTH Aachen
Institute of Inorganic Chemistry
Landoltweg 1
52056 Aachen
Germany

Shinichi Kikkawa
Hokkaido University
Faculty of Engineering
N13 W8, Kita-ku
060-8628 Sapporo
Japan

Andreas Stein
University of Minnesota
Department of Chemistry
207 Pleasant St. SE
Minneapolis, MN 55455
USA

Cover Credit: Sven Lidin, Arndt Simon and Franck Tessier

Library of Congress Card No.: applied for

British Library Cataloguing-in-Publication Data
A catalogue record for this book is available from the British Library.

Bibliographic information published by the Deutsche Nationalbibliothek
The Deutsche Nationalbibliothek lists this publication in the Deutsche Nationalbibliografie; detailed bibliographic data are available on the Internet at http://dnb.d-nb.de.

Print ISBN: 978-3-527-32587-0
oBook ISBN: 978-3-527-69103-6

Cover Design Formgeber
Typesetting Thomson Digital, Noida, India
Printing and Binding Markono Print Media Pte Ltd, Singapore

Printed on acid-free paper

Preface

When you do great science, you do not have to make a lot of fuss. This oft-forgotten saying from the twentieth century has served these editors pretty well, so the foreword to this definitive six-volume *Handbook of Solid-State Chemistry* in the early twenty-first century will be brief. After all, is there any real need to highlight the paramount successes of solid-state chemistry in the last half century? – Successes that have led to novel magnets, solid-state lighting, dielectrics, phase-change materials, batteries, superconducting compounds, and a lot more? Probably not, but we should stress that many of these exciting matters were derived from curiosity-driven research — work that many practitioners of our beloved branch of chemistry truly appreciate, and this is exactly why they do it. Our objects of study may be immensely important for various applications but, first of all, they are interesting to us; that is, how chemistry defines and challenges itself. Let us also not forget that solid-state chemistry is a neighbor to physics, crystallography, materials science, and other fields, so there is plenty of room at the border, to paraphrase another important quote from a courageous physicist.

Given the incredibly rich heritage of solid-state chemistry, it is probably hard for a newcomer (a young doctoral student, for example) to see the forest for all the trees. In other words, there is a real need to cover solid-state chemistry in its entirety, but only if it is conveniently grouped into digestible categories. Because such an endeavor is not possible in introductory textbooks, this is what we have tried to put together here. The compendium starts with an overview of materials and of the structure of solids. Not too surprisingly, the next volume deals with synthetic techniques, followed by another volume on various ways of (structural) characterization. Being a timely handbook, the fourth volume touches upon nano and hybrid materials, while volume V introduces the reader to the theoretical description of the solid state. Finally, the sixth volume reaches into the real world by focusing on functional materials. Should we have considered more volumes? Yes, probably, but life is short, dear friends.

This handbook would have been impossible to compile for three authors, let alone a single one. Instead, the editors take enormous pride in saying that they managed to motivate more than a hundred first-class scientists living across the globe, each of them specializing in (and sometimes even shaping) a subfield of solid-state chemistry or a related discipline, and all of these wonderful colleagues did their very best to make our dream come true. Thanks to all of you; we sincerely appreciate your contributions. Thank you, once again, on behalf of solid-state chemistry. The editors also would like to thank Wiley-VCH, in particular Dr. Waltraud Wüst and also Dr. Frank Otmar Weinreich, for spiritually (and practically) accompanying us over a few years, and for reminding us here and there that there must be a final deadline. That being said, it is up to the reader to judge whether the tremendous effort was justified. We sincerely hope that this is the case.

A toast to our wonderful science! Long live solid-state chemistry!

Richard Dronskowski
RWTH Aachen, Aachen, Germany

Shinichi Kikkawa
Hokkaido University, Sapporo, Japan

Andreas Stein
University of Minnesota, Minneapolis, USA

Contents

Volume 1: Materials and Structure of Solids

1 **Intermetallic Compounds and Alloy Bonding Theory Derived from Quantum Mechanical One-Electron Models** *1*
Stephen Lee and Daniel C. Fredrickson

2 **Quasicrystal Approximants** *73*
Sven Lidin

3 **Medium-Range Order in Oxide Glasses** *93*
Hellmut Eckert

4 **Suboxides and Other Low-Valent Species** *139*
Arndt Simon

5 **Introduction to the Crystal Chemistry of Transition Metal Oxides** *161*
J.E. Greedan

6 **Perovskite Structure Compounds** *221*
Yuichi Shimakawa

7 **Nitrides of Non-Main Group Elements** *251*
P. Höhn and R. Niewa

8 **Fluorite-Type Transition Metal Oxynitrides** *361*
Franck Tessier

9 **Mechanochemical Synthesis, Vacancy-Ordered Structures and Low-Dimensional Properties of Transition Metal Chalcogenides** *383*
Yutaka Ueda and Tsukio Ohtani

10 **Metal Borides: Versatile Structures and Properties** *435*
Barbara Albert and Kathrin Hofmann

11 **Metal Pnictides: Structures and Thermoelectric Properties** *455*
Abdeljalil Assoud and Holger Kleinke

12 **Metal Hydrides** *477*
Yaoqing Zhang, Maarten C. Verbraeken, Cédric Tassel, and Hiroshi Kageyama

13 **Local Atomic Order in Intermetallics and Alloys** *521*
Frank Haarmann

14 **Layered Double Hydroxides: Structure–Property Relationships** *541*
Shan He, Jingbin Han, Mingfei Shao, Ruizheng Liang, Min Wei, David G. Evans, and Xue Duan

15 **Structural Diversity in Complex Layered Oxides** *571*
S. Uma

16 **Magnetoresistance Materials** *595*
Ichiro Terasaki

17 **Magnetic Frustration in Spinels, Spin Ice Compounds, $A_3B_5O_{12}$ Garnet, and Multiferroic Materials** *617*
Hongyang Zhao, Hideo Kimura, Zhenxiang Cheng, and Tingting Jia

18 **Structures and Properties of Dielectrics and Ferroelectrics** *643*
Mitsuru Itoh

19 **Defect Chemistry and Its Relevance for Ionic Conduction and Reactivity** *665*
Joachim Maier

20 **Molecular Magnets** *703*
J.V. Yakhmi

21 **Ge–Sb–Te Phase-Change Materials** *735*
Volker L. Deringer and Matthias Wuttig

Index *751*

Volume 2: Synthesis

1 **High-Temperature Methods** *1*
Rainer Pöttgen and Oliver Janka
1.1 Introduction *1*
1.2 Resistance Furnace Types *2*
1.3 Crucible Materials *3*
1.4 The Ceramic Method and Precursor Decomposition Reactions *3*
1.5 The Plasma Burner *4*
1.6 Laser Heating and Laser-Assisted Synthesis *5*
1.7 Solar and Mirror Furnaces *6*
1.8 Arc Melting: Research Laboratory and Industrial Scale *7*
1.9 Induction Heating *9*
1.10 Flux-Assisted Synthesis: Metal and Salt Fluxes *10*
1.11 Microwave Heating *13*
1.12 Hot Isostatic Pressing and Reactive Sintering *14*
1.13 Current-Assisted Sintering *14*
1.14 Carbothermic Reductions *16*
1.15 Self-Propagating High-Temperature Synthesis *17*
References *18*

2 **High-Pressure Methods in Solid-State Chemistry** *23*
Hubert Huppertz, Gunter Heymann, Ulrich Schwarz, and Marcus R. Schwarz
2.1 Introduction *23*
2.2 Historical Aspects *24*
2.3 Static Methods *25*
2.3.1 Piston-Cylinder Apparatus *25*
2.3.2 Belt Module *26*
2.3.3 Toroid Type/Paris-Edinburgh Cell *28*
2.3.4 Multianvil Apparatus *29*
2.3.4.1 Multianvil Techniques *29*
2.3.4.2 *In situ* Multianvil Technique *31*
2.3.5 Diamond Anvil Cell *32*
2.3.5.1 Diamond Anvil Cell Technique *32*
2.4 Dynamic Methods *34*
2.4.1 Shock Wave Syntheses *34*
2.5 Principles of High-Pressure Chemistry *41*
2.6 Selected Examples of Chemistry Under High-Pressure Conditions *42*
2.7 Conclusion *44*
Acknowledgment *44*
References *44*

3 High-Pressure Perovskite: Synthesis, Structure, and Phase Relation *49*
Yoshiyuki Inaguma
3.1 Introduction *49*
3.2 Synthesis of Perovskite Aided by Pressure *50*
3.2.1 Synthesis under Oxygen Gas Pressure of ∼1 MPa *50*
3.2.2 Synthesis under Pressure of ∼100 MPa *50*
3.2.3 Synthesis under Pressure in the GPa Range *52*
3.2.4 Why Can Perovskites Be Stabilized by High-Pressure Synthesis? *53*
3.3 High-Pressure Perovskite ABX_3 *55*
3.3.1 Phase Stability of Perovskite *55*
3.3.2 Evolution from Layered Structure to Perovskite-Type Structure with Pressure *70*
3.3.3 $LiNbO_3$ as Derivatives of Perovskite *72*
3.4 Ordered High-Pressure Perovskites *73*
3.4.1 B-Site-Ordered High-Pressure Perovskite *74*
3.4.2 A-Site-Ordered High-Pressure Perovskite *77*
3.4.2.1 1: 3 A-Site-Ordered High-Pressure Perovskite $AA'_3B_4O_{12}$ *77*
3.4.2.2 Columnar-Type 1 : 1 A-Site-Ordered High-Pressure Perovskite $AA'B_2O_6$ *78*
3.5 Perovskite-Related Layered Compounds Synthesized under High Pressures *78*
3.6 Post-Perovskite: More Dense Than Perovskite *86*
3.7 Final Remarks *90*
References *90*

4 Solvothermal Methods *107*
Nobuhiro Kumada
4.1 Solvothermal Methods for Crystal Growth and Synthesis of New Compounds *107*
4.2 α-Quartz *107*
4.3 GaN *109*
4.4 ZnO *109*
4.5 YAG ($Y_3Al_5O_{12}$) *112*
4.6 $AlPO_4$ (Berlinite) and $GaPO_4$ *113*
4.7 $ABe_2BO_3F_2$ (ABBF, A = K, Rb, Cs, Tl) *114*
4.8 Synthesis of Zirconium Phosphates *115*
4.9 Synthesis of Bismuth Oxides *118*
References *119*

5 High-Throughput Synthesis Under Hydrothermal Conditions *123*
Nobuaki Aoki, Gimyeong Seong, Tsutomu Aida, Daisuke Hojo, Seiichi Takami, and Tadafumi Adschiri
5.1 Introduction *123*
5.2 Mechanism of Hydrothermal Nanoparticle Synthesis *124*
5.2.1 Properties of Water *124*

5.2.2 Reaction and Solubility in Water *125*
5.2.3 Nucleation and Particle Formation *127*
5.2.4 Mechanism of Nanoparticle Formation *129*
5.2.4.1 Hydrothermal Synthesis *129*
5.2.4.2 Metal Nanoparticle Formation *129*
5.2.4.3 Organic Modified Nanoparticles *134*
5.3 Flow Visualization Inside Supercritical Hydrothermal Process *138*
5.4 Process Design under a Reaction-controlled Condition Using Reynolds and Damköhler Numbers *140*
5.4.1 Reaction Rate of Hydrothermal Reaction *140*
5.4.2 Effect of Mixing Rate of Particle Size *141*
5.4.3 Determination of Reaction-Controlled Condition Using Dimensionless Numbers *143*
5.5 Applications of Nanoparticles *145*
5.5.1 Surface-Modified Nanoparticles by Organic Molecules *145*
5.5.2 Superhybrid Materials *146*
5.5.3 Catalysts *147*
5.5.4 Nanoink for 3D Printing *148*
References *149*

6 Particle-Mediated Crystal Growth *155*
R. Lee Penn
6.1 Introduction *155*
6.2 Classical Crystal Growth *157*
6.3 Nonclassical Crystal Growth *158*
6.4 Oriented Attachment *159*
6.4.1 Role of Attachment in the Development of Microstructure *162*
6.4.1.1 Twinned Attachment *162*
6.4.1.2 Nearly Oriented Attachment *163*
6.4.1.3 Oriented Attachment between Poorly Crystalline Particles *164*
6.4.2 Role of Attachment in Phase Transformation *166*
6.5 The Anatase to Rutile Phase Transformation *167*
6.6 Two Examples of Phase Transformations in the Iron Oxides *168*
6.7 Growth of Silicalite-1 from Amorphous Primary Particles *170*
6.8 Quantifying Crystal Growth *171*
6.9 Concluding Remarks *174*
Acknowledgments *175*
References *175*

7 Sol–Gel Synthesis of Solid-State Materials *179*
Guido Kickelbick and Patrick Wenderoth
7.1 Introduction *179*
7.1.1 Requirement for Low-Temperature Processes *180*

7.1.2 Inorganic–Organic Hybrid Materials *181*
7.2 Mechanisms *181*
7.2.1 Hydrolytic Sol–Gel Process *181*
7.2.2 Nonhydrolytic Sol–Gel Process *184*
7.2.3 Aging and Drying *185*
7.3 Precursors *186*
7.3.1 Silicon-Based Precursors *187*
7.3.2 Metal Precursors *188*
7.4 Materials Based on Sol–Gel Chemistry *189*
7.4.1 Bulk Materials *190*
7.4.2 Thin Films and Coatings *191*
7.4.3 Nanoparticles *191*
7.4.4 Porous Materials *193*
7.4.5 Hybrid Materials *194*
7.5 Applications *195*
7.6 Conclusion *196*
References *197*

8 Templated Synthesis for Nanostructured Materials *201*
Yoshiyuki Kuroda and Kazuyuki Kuroda
8.1 Introduction *201*
8.2 Variety of Templates *203*
8.2.1 Atomic-Scale Templates *203*
8.2.2 Microscale Templates *204*
8.2.2.1 Molecular Templates *205*
8.2.2.2 Molecular imprinting *205*
8.2.2.3 Microporous Solid Templates *206*
8.2.3 Mesoscale Templates *207*
8.2.3.1 Liquid Crystalline Templates *208*
8.2.3.2 Microphase-Separated Block Copolymer Templates *210*
8.2.3.3 Mesoporous Solid Templates *210*
8.2.4 Macroscale Templates *211*
8.2.4.1 Emulsion/Reverse Micelle Templates *211*
8.2.4.2 Water Droplets and Ice Templates *211*
8.2.4.3 Colloidal Templates *212*
8.2.4.4 Biotemplates *213*
8.3 Synthetic Methods *213*
8.3.1 Impregnation of Precursors *213*
8.3.2 Solidification of Wall Components *214*
8.3.2.1 Metal Oxides *214*
8.3.2.2 Polymers *214*
8.3.2.3 Carbon *214*
8.3.2.4 Metals *215*
8.3.3 Removal of Templates *215*
8.3.3.1 Calcination *215*

8.3.3.2 Solvent Extraction *215*
8.3.3.3 Chemical Etching *216*
8.3.4 Interfacial Control of Templated Synthesis *216*
8.4 Morphological and Hierarchical Control *217*
8.4.1 Nanoparticles *217*
8.4.2 Nanorods and Nanofibers *218*
8.4.3 Films and Membranes *219*
8.4.4 Hierarchical Structures by Combined Use of Templates *220*
8.4.4.1 Combination of Hard and Soft Templates *220*
8.4.4.2 Combination of Different Soft Templates *220*
8.4.4.3 Use of Spontaneous Phase Separation *221*
8.4.4.4 Use of Bifunctional Templates *221*
8.5 Applications of Nanostructured Materials Synthesized by Templating Methods *222*
8.6 Summary and Outlook *222*
References *224*

9 Bio-Inspired Synthesis and Application of Functional Inorganic Materials by Polymer-Controlled Crystallization *233*
Lei Liu and Shu-Hong Yu
9.1 Introduction *233*
9.2 Basic Principles of Crystal Growth and Aggregation *235*
9.2.1 Morphological Control in Classical Crystallization *235*
9.2.2 Nonclassical Crystallization *238*
9.2.2.1 Oriented Attachment *238*
9.2.2.2 Mesocrystal *240*
9.2.2.3 Amorphous Precursor *241*
9.3 Bio-Inspired Polymer-Controlled Crystallization *243*
9.3.1 Noble Metal *246*
9.3.2 Metal Oxides *248*
9.3.3 Metal Chalcogenides *253*
9.3.4 Other Functional Inorganic Materials *256*
9.4 Applications *257*
9.5 Summary and Outlook *260*
References *261*

10 Reactive Fluxes *275*
10.1 Introduction *275*
10.2 Reactive Metal Fluxes: Self-Flux Technique *275*
10.3 Reactive Salt Fluxes *277*
10.4 Polychalcogenide Fluxes *279*
10.5 Nitride Synthesis via Li_3N and NaN_3 *282*
10.6 Other Systems *283*
References *283*

11 Glass Formation and Crystallization *287*
T. Komatsu
11.1 Introduction *287*
11.2 Basic Scenario for Glass Formation *288*
11.3 Fragility Concept for Glass Formation *291*
11.4 Nanoscale Structure in Oxide Glasses *292*
11.5 Glass Formation and Crystallization in Oxide Glasses *295*
11.5.1 $BaO–TiO_2–SiO_2/GeO_2$ Glasses *296*
11.5.2 $ZnO–Bi_2O_3–B_2O_3$-Based Glasses *300*
11.5.3 $Gd_2O_3–MoO_3/WO_3–B_2O_3$-Based Glasses *302*
11.5.4 $Li_2O–Nb_2O_5–P_2O_5$ Glasses *306*
11.5.5 TeO_2-Based Glasses *308*
11.5.6 Oxyfluoride Glasses *313*
11.6 Conclusion *316*
Acknowledgments *317*
References *317*

12 Glass-Forming Ability, Recent Trends, and Synthesis Methods of Metallic Glasses *319*
Hidemi Kato, Takeshi Wada, Rui Yamada, and Junji Saida
12.1 Glass Formation and Glass-Forming Ability *319*
12.2 Improving the Glass-Forming Ability for Bulk Metallic Glasses *324*
12.3 Heterometallic Glasses *326*
12.3.1 High-Plasticity Zr–Al–Ni–Pd Bulk Metallic Glasses *326*
12.3.2 Fe-Based Heterometallic Glasses for Excellent Soft Magnetic Nanocrystalline Alloys *329*
12.4 Preparation Methods for Metallic Glasses *330*
12.4.1 Melt Spinning *330*
12.4.2 Sputtering *331*
12.4.3 Gas (Water) Atomization *331*
12.4.4 Pulsated Orifice Ejection Method *332*
12.4.5 Mechanical Alloying and Mechanical Grinding *334*
12.4.6 Water Quenching *335*
12.4.7 Metallic Mold Casting *336*
12.4.8 Arc-Melt Tilt Casting and Suction Casting *337*
12.4.9 Metallic Glass Powder Consolidation *339*
12.4.10 Severe Plastic Deformation *340*
12.5 Important Factors for Metallic Glass Preparation *341*
12.5.1 Impurities *342*
12.5.2 Atmosphere Gas and Its Pressure *344*
12.6 Conclusions *347*
References *347*

13 Crystal Growth Via the Gas Phase by Chemical Vapor Transport Reactions *351*
Michael Binnewies, Robert Glaum, Marcus Schmidt, and Peer Schmidt
13.1 Chemical Vapor Transport – Principles *351*
13.1.1 Transport via a Single Equilibrium Reaction *352*
13.1.2 The Gas Phase Solubility λ *355*
13.2 Working Techniques *358*
13.3 Examples *362*
13.3.1 Chemical Vapor Transport of Elements *362*
13.3.2 Chemical Vapor Transport of Intermetallics *363*
13.3.3 Halides *365*
13.3.4 Oxides *367*
13.3.5 Sulfides, Selenides, and Tellurides *369*
13.3.6 Pnictides *370*
13.4 Modeling of Chemical Vapor Transport Experiments *370*
References *373*

14 Thermodynamic and Kinetic Aspects of Crystal Growth *375*
Detlef Klimm
14.1 Introduction *375*
14.2 Equilibrium Thermodynamics *376*
14.2.1 Basic Terms *377*
14.2.1.1 System *377*
14.2.1.2 Component and Phase *379*
14.2.2 Phase Diagrams *383*
14.2.2.1 Solid–Liquid Equilibria *383*
14.2.2.2 Solid–Gas Equilibria *389*
14.3 Kinetic Aspects *392*
14.3.1 Segregation *392*
14.3.2 Lever Rule *392*
14.3.3 Constitutional Supercooling *395*
14.4 Conclusion *396*
References *397*

15 Chemical Vapor Deposition *399*
Takashi Goto and Hirokazu Katsui
15.1 Introduction *399*
15.2 Basics of CVD *399*
15.3 Kinetics of CVD *402*
15.4 Thermodynamics of CVD *405*
15.5 Rotary CVD *407*
15.6 Plasma CVD *415*
15.7 Laser CVD *417*
15.8 Summary *423*
References *423*

16 Growth of Wide Bandgap Semiconductors by Halide Vapor Phase Epitaxy *429*
Yuichi Oshima, Encarnación G. Víllora, and Kiyoshi Shimamura
16.1 Introduction *429*
16.2 Outline of HVPE Technique *430*
16.2.1 Principle of HVPE *430*
16.2.2 What Enables the High-Speed Growth? *431*
16.2.2.1 Determinant Factors of the Growth Rate *431*
16.2.2.2 Spontaneous Nucleation by a Parasitic Gas-Phase Reaction *432*
16.2.2.3 Comparison Between HVPE and MOCVD *433*
16.2.3 Technical Characteristics and Functionality of HVPE Equipment *435*
16.2.3.1 Outline of the Entire System *435*
16.2.3.2 Fundamentals of the Reactor Design *436*
16.2.3.3 Counter-Measures Against the Deposition of By-Products *437*
16.2.3.4 Counter-Measures Against the Corrosion *438*
16.3 Fabrication of GaN Wafers *438*
16.3.1 Features and Applications of GaN *439*
16.3.2 Necessity and Applications of GaN Wafers *440*
16.3.3 Basic Strategies and Technical Issues for the Fabrication of GaN Wafers *440*
16.3.3.1 Reduction of the Dislocation Density *441*
16.3.3.2 Removal of the Base Substrate *443*
16.3.3.3 Control of Electrical Properties *445*
16.3.3.4 Reduction of the Off-Angle Variation *446*
16.3.4 Concrete Examples of the Production Technology of GaN Wafers *450*
16.3.5 Future Prospectives *453*
16.4 HVPE of Ga_2O_3 *455*
16.4.1 Recent Progress in HVPE of β-Ga_2O_3 *457*
16.4.1.1 Technical Issues for the Epitaxial Growth of β-Ga_2O_3 *457*
16.4.1.2 Rapid Growth of β-Ga_2O_3 by HVPE *457*
16.4.1.3 Homoepitaxy of β-Ga_2O_3 by HVPE *458*
16.4.1.4 HVPE of β-Ga_2O_3 on Foreign Substrates *459*
16.4.2 Recent Progress in HVPE of α-Ga_2O_3 *460*
16.4.2.1 Technical Issues for the Epitaxial Growth of α-Ga_2O_3 *460*
16.4.2.2 Rapid Growth of α-Ga_2O_3 by HVPE *461*
16.4.3 Summary and Future Prospectives *463*
16.5 Conclusion *463*
References *464*

17 Growth of Silicon Nanowires *467*
Fengji Li and Sam Zhang
17.1 Growth Techniques for Silicon Nanowires *467*
17.1.1 Introduction *467*
17.1.2 Metal-Assisted Chemical Etching *469*

17.1.3 Vapor–Liquid–Solid Growth *470*
17.1.4 Vapor–Solid–Solid Growth *474*
17.1.5 Supercritical–Fluid–Liquid–Solid Growth *476*
17.1.6 Solid–Liquid–Solid Growth *478*
17.1.6.1 Molecular Beam Epitaxy *478*
17.1.6.2 Laser Ablation *479*
17.1.7 Summary *480*
17.2 Thermal Evaporation Growth of Silicon Nanowire *481*
17.2.1 Introduction *481*
17.2.2 Growth of Si Nanowires in Quartz Tube *481*
17.2.2.1 Metallic Al and Cu in Quartz Tube *481*
17.2.2.2 Source of Si: Ejected Micrometer-Sized Si Particles *487*
17.2.2.3 Alternative Growth Mechanism *490*
17.2.3 Growth of Si Nanowires in Alumina Tube *496*
17.2.3.1 Reduction of Alumina Tube by H_2 *496*
17.2.3.2 Si in the Ejected Micrometer-Sized SiO Particles *498*
17.2.3.3 Alternative Growth Mechanism: Al-Catalyzed Growth *498*
17.2.4 Conclusion *500*
17.3 Growth of Silicon Nanowires through Thermal Annealing *500*
17.3.1 Introduction *500*
17.3.2 Growth of the Si Nanowires under a-C *501*
17.3.2.1 Experimental Details *501*
17.3.2.2 Structural Evolution of the a-C/Ni Film *501*
17.3.2.3 Morphology of the Annealed a-C/Ni Film *502*
17.3.2.4 Composition of the Nanowire *504*
17.3.2.5 Structure of the Nanowire *507*
17.3.2.6 Conclusion *507*
17.3.3 Origin of the Growth Orientation *509*
17.3.3.1 Introduction *509*
17.3.3.2 Growth Orientation *510*
17.3.3.3 Interfacial Structure between Si Nanowire and Ni Catalyst *513*
17.3.3.4 Determinant Factors of Growth Orientation *515*
17.3.3.5 Conclusion *518*
17.3.4 Origin of the Morphology *518*
17.3.4.1 Introduction *518*
17.3.4.2 Coaxial Nanowire *520*
17.3.4.3 Side-by-Side Biaxial Nanowire *520*
17.3.4.4 Triple-Concentric Nanowire *521*
17.3.4.5 Ni-Nanosphere Entrapment in Nanowires *523*
17.3.4.6 Transitional Nanowire *525*
17.3.4.7 Growth Mechanisms of the Various Morphologies *527*

17.3.4.8 Conclusion *530*
Acknowledgment *531*
References *531*

18 Chemical Patterning on Surfaces and in Bulk Gels *539*
Olaf Karthaus
18.1 Background *539*
18.2 Examples of Chemical Processes that Lead to the Chemical Modification of Surfaces *540*
18.3 Surface Patterning *542*
18.4 Application of Surface Patterns and Combination with Other Methods *546*
18.5 In Bulk (Liesegang patterns) *547*
18.6 Summary *556*
References *556*

19 Microcontact Printing *563*
Kiyoshi Yase
19.1 Introduction *563*
19.2 Organic Thin Film Devices *563*
19.3 Printing Methods *564*
19.4 Microcontact Printing *565*
19.5 Printing of TFT Arrays on Substrate *568*
19.6 Liquid Crystal Display Packaging *569*
19.7 Conclusion *571*
Acknowledgments *572*
References *572*

20 Nanolithography Based on Surface Plasmon *573*
Kosei Ueno and Hiroaki Misawa
20.1 Introduction *573*
20.2 Surface Plasmon Interference Nanolithography *576*
20.3 Nanogap-Assisted Surface Plasmon Nanolithography *577*
20.3.1 Verification of Local Photochemical Reaction by LSPR *577*
20.3.2 Demonstration of Nanogap-Assisted Surface Plasmon Nanolithography *580*
20.4 Nanolithography Using Scattering Component of Higher order LSPRs as an Exposure Source *581*
20.5 Conclusion *586*
References *587*

Index *589*

Volume 3: Characterization

1 **Single-Crystal X-Ray Diffraction** *1*
Ulli Englert

2 **Laboratory and Synchrotron Powder Diffraction** *29*
R. E. Dinnebier, M. Etter, and T. Runcevski

3 **Neutron Diffraction** *77*
Martin Meven and Georg Roth

4 **Modulated Crystal Structures** *109*
Sander van Smaalen

5 **Characterization of Quasicrystals** *131*
Walter Steurer

6 **Transmission Electron Microscopy** *155*
Krumeich Frank

7 **Scanning Probe Microscopy** *183*
Marek Nowicki and Klaus Wandelt

8 **Solid-State NMR Spectroscopy: Introduction for Solid-State Chemists** *245*
Christoph S. Zehe, Renée Siegel, and Jürgen Senker

9 **Modern Electron Paramagnetic Resonance Techniques and Their Applications to Magnetic Systems** *279*
Andrej Zorko, Matej Pregelj, and Denis Arčon

10 **Photoelectron Spectroscopy** *311*
Stephan Breuer and Klaus Wandelt

11 **Recent Developments in Soft X-Ray Absorption Spectroscopy** *361*
Alexander Moewes

12 **Vibrational Spectroscopy** *393*
Götz Eckold and Helmut Schober

13 **Mößbauer Spectroscopy** *443*
Hermann Raphael

14 **Macroscopic Magnetic Behavior: Spontaneous Magnetic Ordering** *485*
Heiko Lueken and Manfred Speldrich

15 **Dielectric Properties** *523*
Rainer Waser and Susanne Hoffmann-Eifert

16 **Mechanical Properties** *561*
Volker Schnabel, Moritz to Baben, Denis Music, William J. Clegg, and Jochen M. Schneider

17 **Calorimetry** *589*
Hitoshi Kawaji

Index *615*

Volume 4: Nano and Hybrid Materials

1 **Self-Assembly of Molecular Metal Oxide Nanoclusters** *1*
Laia Vilà-Nadal and Leroy Cronin

2 **Inorganic Nanotubes and Fullerene–Like Nanoparticles from Layered (2D) Compounds** *21*
L. Yadgarov, R. Popovitz-Biro, and R. Tenne

3 **Layered Materials: Oxides and Hydroxides** *53*
Ida Shintaro

4 **Organoclays and Polymer-Clay Nanocomposites** *79*
M.A. Vicente and A. Gil

5 **Zeolite and Zeolite-Like Materials** *97*
Watcharop Chaikittisilp and Tatsuya Okubo

6 **Ordered Mesoporous Materials** *121*
Michal Kruk

7 **Porous Coordination Polymers/Metal–Organic Frameworks** *141*
Ohtani Ryo and Kitagawa Susumu

8 **Metal–Organic Frameworks: An Emerging Class of Solid-State Materials** *165*
Joseph E. Mondloch, Rachel C. Klet, Ashlee J. Howarth, Joseph T. Hupp, and Omar K. Farha

9 **Sol–Gel Processing of Porous Materials** *195*
Kazuki Nakanishi, Kazuyoshi Kanamori, Yasuaki Tokudome, George Hasegawa, and Yang Zhu

10 **Macroporous Materials Synthesized by Colloidal Crystal Templating** *243*
Jinbo Hu and Andreas Stein

11 **Optical Properties of Hybrid Organic–Inorganic Materials and their Applications – Part I: Luminescence and Photochromism** *275*
Stephane Parola, Beatriz Julián-López, Luís D. Carlos, and Clément Sanchez

12 **Optical Properties of Hybrid Organic–inorganic Materials and their Applications – Part II: Nonlinear Optics and Plasmonics** *317*
Stephane Parola, Beatriz Julián-López, Luís D. Carlos, and Clément Sanchez

13 **Bioactive Glasses** *357*
Hirotaka Maeda and Toshihiro Kasuga

14 **Materials for Tissue Engineering** *383*
María Vallet-Regí and Antonio J. Salinas

Index *411*

Volume 5: Theoretical Description

1 **Density Functional Theory** *1*
Michael Springborg and Yi Dong

2 **Eliminating Core Electrons in Electronic Structure Calculations: Pseudopotentials and PAW Potentials** *29*
Stefan Goedecker and Santanu Saha

3 **Periodic Local Møller–Plesset Perturbation Theory of Second Order for Solids** *59*
Denis Usvyat, Lorenzo Maschio, and Martin Schütz

4 **Resonating Valence Bonds in Chemistry and Solid State** *87*
Evgeny A. Plekhanov and Andrei L. Tchougréeff

5 **Many Body Perturbation Theory, Dynamical Mean Field Theory and All That** *119*
Silke Biermann and Alexander Lichtenstein

6 **Semiempirical Molecular Orbital Methods** *159*
Thomas Bredow and Karl Jug

7 **Tight-Binding Density Functional Theory: DFTB** *203*
Gotthard Seifert

8 **DFT Calculations for Real Solids** *227*
Karlheinz Schwarz and Peter Blaha

9 **Spin Polarization** *261*
Dong-Kyun Seo

10 **Magnetic Properties from the Perspectives of Electronic Hamiltonian: Spin Exchange Parameters, Spin Orientation, and Spin-Half Misconception** *285*
Myung-Hwan Whangbo and Hongjun Xiang

11 **Basic Properties of Well-Known Intermetallics and Some New Complex Magnetic Intermetallics** *345*
Peter Entel

12 **Chemical Bonding in Solids** *405*
Gordon J. Miller, Yuemei Zhang, and Frank R. Wagner

13 **Lattice Dynamics and Thermochemistry of Solid-State Materials from First-Principles Quantum-Chemical Calculations** *491*
Ralf Peter Stoffel and Richard Dronskowski

14 **Predicting the Structure and Chemistry of Low-Dimensional Materials** *527*
Xiaohu Yu, Artem R. Oganov, Zhenhai Wang, Gabriele Saleh, Vinit Sharma, Qiang Zhu, Qinggao Wang, Xiang-Feng Zhou, Ivan A. Popov, Alexander I. Boldyrev, Vladimir S. Baturin, and Sergey V. Lepeshkin

15 **The Pressing Role of Theory in Studies of Compressed Matter** *571*
Eva Zurek

16 **First-Principles Computation of NMR Parameters in Solid-State Chemistry** *607*
Jérôme Cuny, Régis Gautier, and Jean-François Halet

17 **Quantum Mechanical/Molecular Mechanical (QM/MM) Approaches** *647*
C. Richard A. Catlow, John Buckeridge, Matthew R. Farrow, Andrew J. Logsdail, and Alexey A. Sokol

18 **Modeling Crystal Nucleation and Growth and Polymorphic Transitions** *681*
Dirk Zahn

Index *701*

Volume 6: Functional Materials

1 **Electrical Energy Storage: Batteries** *1*
Eric McCalla

2 **Electrical Energy Storage: Supercapacitors** *25*
Enbo Zhao, Wentian Gu, and Gleb Yushin

3 **Dye-Sensitized Solar Cells** *61*
Anna Nikolskaia and Oleg Shevaleevskiy

4 **Electronics and Bioelectronic Interfaces** *75*
Seong-Min Kim, Sungjun Park, Won-June Lee, and Myung-Han Yoon

5 **Designing Thermoelectric Materials Using 2D Layers** *93*
Sage R. Bauers and David C. Johnson

6 **Magnetically Responsive Photonic Nanostructures for Display Applications** *123*
Mingsheng Wang and Yadong Yin

7 **Functional Materials: For Sensing/Diagnostics** *151*
Rujuta D. Munje, Shalini Prasad, and Edward Graef

8 **Superhard Materials** *175*
Ralf Riedel, Leonore Wiehl, Andreas Zerr, Pavel Zinin, and Peter Kroll

9 **Self-healing Materials** *201*
Martin D. Hager

10 **Functional Surfaces for Biomaterials** *227*
Akiko Nagai, Naohiro Horiuchi, Miho Nakamura, Norio Wada, and Kimihiro Yamashita

11 **Functional Materials for Gas Storage. Part I: Carbon Dioxide and Toxic Compounds** *249*
L. Reguera and E. Reguera

12 **Functional Materials for Gas Storage. Part II: Hydrogen and Methane** *281*
L. Reguera and E. Reguera

13 **Supported Catalysts** *313*
Isao Ogino, Pedro Serna, and Bruce C. Gates

14 **Hydrogenation by Metals** *339*
Xin Jin and Raghunath V. Chaudhari

15 **Catalysis/Selective Oxidation by Metal Oxides** *393*
Wataru Ueda

16 **Activity of Zeolitic Catalysts** *417*
Xiangju Meng, Liang Wang, and Feng-Shou Xiao

17 **Nanocatalysis: Catalysis with Nanoscale Materials** *443*
Tewodros Asefa and Xiaoxi Huang

18 **Heterogeneous Asymmetric Catalysis** *479*
Ágnes Mastalir and Mihály Bartók

19 **Catalysis by Metal Carbides and Nitrides** *511*
Connor Nash, Matt Yung, Yuan Chen, Sarah Carl, Levi Thompson, and Josh Schaidle

20 **Combinatorial Approaches for Bulk Solid-State Synthesis of Oxides** *553*
Paul J. McGinn

Index *573*

1
High-Temperature Methods

Rainer Pöttgen and Oliver Janka

Universität Münster, Institut für Anorganische und Analytische Chemie, Corrensstrasse 30, 48149 Münster, Germany

1.1
Introduction

The synthesis of ceramics and intermetallics, especially when starting from oxidic precursors and refractory metals, requires high temperatures. The decisive question then is "what is high temperature" and what do we mean by high-temperature methods.

One cannot answer this question by just mentioning a given temperature range. The answer should be a clear differentiation to typical solution chemistry that attains its limits when reaching the boiling point of the solvent. The same holds true for ionic liquids. Herein, when discussing high-temperature methods, we mean application to typical solid-state materials, metallic, semiconducting, and insulating ones. The respective methods depend on the form of the material, that is, large crystals for property investigations, small single crystals for structure determination, polycrystalline powdered samples, sintered powders, or thin films.

We will start with a standard laboratory course experiment for the synthesis of textbook examples in the field of oxides. A classical example is the synthesis of a spinell-type phase $MgAl_2O_4$, starting from the binary oxides MgO and Al_2O_3. Both precursor compounds are high-melting oxides and one needs sufficiently high diffusion to enable the reaction. According to the Tammann rule, at least two-third of the melting temperatures have to be reached to get a diffusion-controlled reaction in reasonable time. Usually powders of such starting compositions are ground and pressed to pellets. Diffusion takes place only at the contact areas between two adjacent particles. This requires repeated grinding and repressing of pellets. Which kind of furnace is suitable for such an experiment? We will first discuss standard commercially available and self-constructed furnace types.

Most of the high-temperature techniques described herein are well documented in literature. We understand this chapter as a concise and compact

Handbook of Solid State Chemistry, First Edition. Edited by Richard Dronskowski, Shinichi Kikkawa, and Andreas Stein.

summary of these techniques and we refer to the relevant textbooks, review articles, and/or original articles in order to avoid repetition.

1.2
Resistance Furnace Types

Many kinds of muffle, chamber, and tube furnaces are commercially available with variable tube size and length, multiple heating zones, and different volume. Meanwhile, such furnaces are routinely equipped with programmable electronic power devices that allow well-defined temperature programs. An Internet search for these three furnace types readily reveals a multitude of different models.

Laboratory courses and screening reactions (multiple annealings or tube reactions for phase analyses) often require a large number of standard furnaces. Solid-state research groups often use standard tube furnaces self-made by the mechanical workshops of the institutes using standard heating filaments. Such furnaces usually have permanent maximal working temperatures around 1170–1270 K. A summary of such self-made furnaces has recently been published [1]. This kind of tube furnace is usually equipped with an inner ceramic tube. Often lower temperatures are sufficient, especially in the case of reactions running in low-melting salt or metal fluxes. For such applications, aluminum block furnaces are suitable alternatives, although the low melting point of aluminum is the limiting parameter (the permanent operating temperature should not exceed 770 K). For details of the setups of both furnace types, we refer to a recent handbook on intermetallic materials [1].

Resistance furnaces for higher temperatures are also commercially available. The heating filaments can be high-melting transition metals; diverse standard alloys are on the market, or semiconducting materials that have increased conductivity at the desired temperatures. Typical heating sticks consist of $MoSi_2$ or silicon carbide. Formation of a silica surface layer protects the filaments from further oxidation.

Standard annealing experiments, especially in the case of oxides, for decomposition reactions or for annealing sealed silica tubes are performed in air. Tube and chamber furnaces can be run under defined conditions, that is, dynamic vacuum or a well-defined gas atmosphere, for example, argon, nitrogen, hydrogen, oxygen, or another gas mixture.

The resistance furnaces discussed above are usually of small size (active heating volume) and used for small samples for research purposes. Often such reactions are run on the milligram or gram scale. For industrial application, similar devices are available with much larger annealing chambers. Different setups can easily be found at the supplier's homepages using the keyword *resistance furnace*.

The temperature control for the chamber and tube furnaces mostly proceeds through commercially available thermocouples that are obtainable along with the programmable power supplies. For higher temperatures, special sets of thermocouples such as Pt/Re can be used, but meanwhile powerful small pyrometers find application.

1.3
Crucible Materials

The use of high-temperature synthesis techniques readily calls for suitable crucible materials that resist the synthesis conditions. Important parameters for a suitable crucible material are the melting temperature, the oxidation stability, the surface wetting, and the machinability. Widely used ampoule materials for solid-state synthesis are Duran® glass and silica, depending on the reaction temperature. Especially, silica has broad application up to around 1370 K due to its enormous thermal shock resistance and an easy workability. Silica ampoules are also used as gas-tight (mostly vacuum) reaction chamber for diverse ceramic and metallic crucibles. Some words of caution are appropriate at this point. Silica is not inert toward all reaction media. The ampoule walls can be attacked and a tiny attack might not be visible at first sight. Some compounds might incorporate oxygen, others silicon. Typically, during tin flux synthesis of metal-rich phosphides, one observes silicon occupancy on the crystallographic phosphorus sites after long annealing times.

The usual crucible materials can be regrouped into three types: (i) ceramic crucibles of MgO, Al_2O_3, ZrO_2, or CeO_2, (ii) metallic crucible materials of Fe, Cu, Ag, Pt, Au, Nb, Ta, Mo, W, or special alloys, and (iii) carbon-based crucible materials of graphite and glassy carbon. Hexagonal boron nitride is an additional inert crucible material with good workability. An important parameter for the ceramic crucibles is the content of sintering additives. This means that the entire crucible has not the simple formula as listed above. These additives can take part in the reaction. Under special redox conditions, the high-melting oxide ceramics might also decompose and lead to side products. A more detailed overview of these materials is given in Ref. [1–3]. Metallic crucible materials usually need an additional oxidation protection.

For large-scale synthesis in industry, one needs high-volume crucibles. Such reaction containers are mostly bricked with clinkers, similar to the classical cement kiln.

1.4
The Ceramic Method and Precursor Decomposition Reactions

The most frequently used technique in solid-state synthesis in classic laboratory courses as well as in basic research is the ceramic synthesis; diffusion controlled annealing of the starting constituents. Historically, this has widely been used for the synthesis of the many oxides known today [4]. As already mentioned for the classical textbook example of spinell synthesis, diffusion is the decisive reaction parameter and for sufficient mass transport, one needs close contacts of the particles. As emphasized in Figure 1.1, diffusion takes place only at the black points.

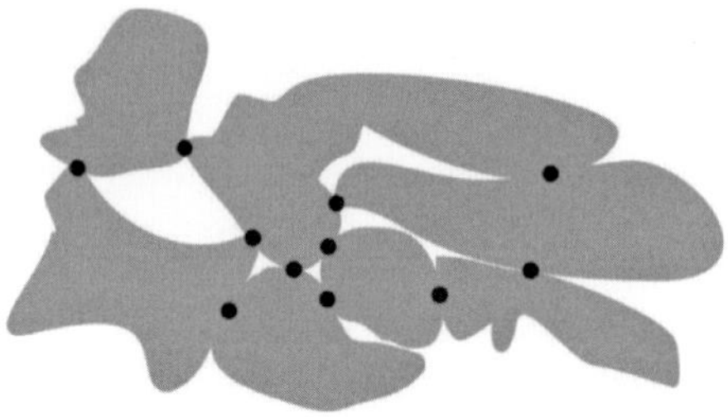

Figure 1.1 A sintering reaction of randomly arranged particles. Diffusion can occur only at the black points where the grains touch each other.

Thus, dense pressing and small size of the particles is a prerequisite for reasonable reaction times. This is independent of the class of materials, be it ceramics or intermetallics. Repeated grinding, repressing of the reaction intermediates, and increasing the temperature to the tolerable maximum (depending on the furnace and the crucible material) increase the product ratio.

These classical ceramic reactions (also applied to diverse intermetallic materials) often need several (up to a dozen) regrinding cycles. This is time-consuming, requires very high temperatures, and does not always lead to X-ray pure bulk samples that allow physical property studies. In the case of oxides, various precursor techniques have been developed. Instead of the high-melting binary oxides, mixtures of oxalates, carbonates, nitrates, or coprecipitated hydroxides can be used. Thermal decomposition of the precursors leads to fresh particles with large surface and thus much higher reactivity. This allows lower reaction temperatures and shorter annealing periods.

The Pechini method [5] is a similar approach using a polymeric precursor. Usually the metal nitrates are mixed in the desired stoichiometric portions along with an addition of equivalent molar amounts of citric acid. Such solutions are subsequently heated to moderate temperature (around 370 K) accompanied by addition of ethylene glycol until a viscous gel forms. Such a gel is then dried (usually 370 K are sufficient and finally fired at high temperature (900–1200 K, depending on the nature of the gel) until the desired complex oxide remains (complete removal of the organic material). Slightly modified syntheses use ethylenediaminetetraacetic acid (EDTA). Typical examples are reported in Refs [6,7]. For exact pH conditions and experimental details, we refer to the original literature. The decomposition reactions mostly take place in standard muffle furnaces.

1.5
The Plasma Burner

Oxide synthesis, especially with the high-melting rare earth oxides, often needs very high temperatures for the growth of small single crystals needed for X-ray structure determination. Simple resistance furnaces often do not achieve the melting temperature of the samples that are required for a crystallization from the melt. A useful technique for such crystal growth experiments is the low-pressure high-frequency plasma burner developed in the Müller-Buschbaum group. Such a furnace allows formation of plasma under reduced pressure of <1 Torr. The free electrons within the plasma have extremely high velocity, enabling the high heat transfer. The technical details are well described in Refs [8,9]. Although the plasma-assisted synthesis led to a huge number of new oxide phases, there is a severe limitation. The intrinsic brightness of the plasma torch disables a precise temperature control through a pyrometer. Furthermore, in larger samples, one might observe temperature gradients that may lead to crystals with different composition.

High-power plasma torches also exist for industrial applications with respect to large-scale synthesis, for example, a plasma torch ash-melting furnace [10].

1.6 Laser Heating and Laser-Assisted Synthesis

Within the field of high-temperature synthesis, not only the method of the energy transfer can be changed, but also the efficiency of the energy transfer varies significantly. Hot flames and argon plasmas exhibit an energy transfer of $0.2\,kW/cm^2$, arc melting of $16\,kW/cm^2$, while electron beams or lasers achieve up to $10^3\,kW/cm^2$. Different kinds of lasers (solid-state, semiconductor, chemical, and gas lasers) are known today. In the first report from 1979 [11], a CO_2 laser was used to initiate the solid-state reaction. Depending on the construction of the laser and the sample chamber, different sample geometries along with various atmospheres can be used. The first samples, which could be obtained as single-crystalline materials, were oxides, for example, $Ba_4FeTa_{10}O_{30}$ [12]. A number of oxide materials thermally decompose at temperatures $T < 1900\,K$; therefore, laser experiments were not available for these compounds. Sample chambers that allow the use of high oxygen partial pressure were successfully used to synthesize, for example, $SrNiO_2$ from SrO and NiO [13]. Due to the design of the laser hardware, the temperature on the surface of the sample is significantly higher, compared to the rest of the specimen. This temperature gradient can yield a number of inhomogeneous reaction products, therefore being a limitation for the preparation of phase pure samples. The temperature gradient, however, will produce a zone within the specimen that favors the formation of metastable samples. Rapid cooling will allow the crystallization of these metastable phases and a subsequent characterization, as shown for, for example, $SrCr_2O_4$ [14]. A number of examples exist in the literature in which the synthesis of high-temperature modifications was successful with the help of laser methods. One example is the series of rare earth titanates RE_2TiO_5, which was synthesized first for $RE = La$ [15]. Recently, these synthetic approaches yielded a number of new compounds in the field of alkaline earth and rare earth oxocobaltates [16]. For more examples and a detailed experimental setup, we refer to Ref. [17].

But not only different atmospheres and partial pressures can be utilized in combination with laser heating (LH) of solid samples. The combination of high pressures with a large number of physical measurements [18] along with, for example, laser-assisted high-temperature synthesis has been developed. Diamond anvil cells (DACs) are used to attain static pressures >500 GPa [19]; however, due to the size of the pressure cell (~100 μm) and the specimen (~30–50 μm), heating is quite challenging. Since the diamonds can serve as optical windows, the use of radiation from the IR to the UV as well as the use of X-rays is possible. The pressure within these cells is usually measured by the ruby pressure scale, which tracks the redshift of the R1 fluorescence line using an Ar laser. The use of an additional laser for sample heating therefore seems obvious. Laser heating within a DAC was first implemented by Ming and Bassett [20] who used continuous-wave Nd-YAG lasers for temperatures up to 2300 K and pulsed ruby lasers for $T > 3300$ K. The temperature was detected from the thermal radiation spectrum of the specimen using the Planck formula. The difficulties faced during the temperature detection have recently been assessed in Ref. [21]. The investigations of melting phenomena and the synthesis of new materials using synchrotron X-ray capabilities have been summarized in Ref. [22]. With the help of these LH-DAC experiments, a number of metastable high-temperature high-pressure phases have been identified. For $BaCO_3$, an orthorhombic HT–HP phase with space group *Pmmm* was found and spectroscopically investigated [23] and also $MgCO_3$ has been reinvestigated in the HT–HP regime between 2100 and 2650 K and pressures up to 84 GPa [24]. But not only oxide-containing materials can be investigated. The Fe–Si system has been in the focus of research from a geophysical point of view and a number of (p,T) phase diagrams for two different iron-to-silicon ratios (1 : 9 and 1 : 1) have been investigated [25]. For In_2S_3, a new high-pressure polymorph with a defect Th_3P_4-type structure has been found [26]. In the case of Hf_3N_4, a nanostructured precursor (tetragonally distorted CaF_2-type structure) prepared by solution-based methods has been crystallized (1500 K, 12 GPa) to yield a bulk sample with the same crystal structure as the starting material. At higher pressures and temperatures (2000 K and 19 GPa), a new orthorhombic polymorph with an anion-deficient TiO_2 (cottunite type) structure forms [27].

Laser-assisted synthesis has furthermore been used to prepare a number of different nanomaterials. The synthesis of ZnS and CdS [28], metal nanoparticles [29], or Fe-based nanoparticles [30] has been reported. For further information about nanosized materials, we refer to Chapter 11 of Volume 2 of this book.

1.7
Solar and Mirror Furnaces

Besides laser heating, focused light can also be utilized to achieve temperatures above 2500 K. While solar heating can be used for large-area materials processing, mirror furnaces are employed for selective heating, for example, crystal growth. Solar furnaces produce powers up to 1000 kW, which is large compared to CO_2 or Nd-YAG lasers with powers of 2–5 and 0.5 kW, respectively. The power flux densities of lasers (in W/cm^2), however, are two to three orders of magnitude larger than for solar systems. This is the result of the smaller affected area of lasers (<1 cm^2) compared to solar furnaces that can heat surface areas of ~1000 cm^2. In general, solar furnaces consist of a heliostat, which reflects the sunlight into a primary concentrator. The amount of sunlight will be controlled by an attenuator. A second concentrator, for example, a parabolic mirror, focuses the sunlight finally on the specimen. The parabolic mirror of the 2 kW furnace of IMP laboratory has a diameter of 2 m and a focal length of 85 cm with a peak flux up to 1400 W/cm^2. A number of different results, for example, surface hardening of steel or the formation of nanomaterials, have been achieved using this furnace [31]. The PROMES research unit of CNRS in France operates the megawatt solar furnace facility (MWSF), and consists of a facetted parabolic solar concentrator with an area of 1830 m^2. The focal length is 18 m and it creates an 80 cm spot with 1 MW energy flux [32].

For laboratory scale application, the use of mirror furnaces has been established, for example, for the growth of single crystals for various materials. One of the first mirror furnaces was described by de la Rue and Halden who used carbon arc theater projector lamps as light sources along with a double ellipsoidal mirror arrangement [33]. The powdered material gets dropped onto a pedestral just below the hot zone from where the material growth can occur. Due to the crucible free synthesis and the possibility to use different gas atmospheres, high-purity crystals of magnesium ferrite or sapphire have been obtained. Especially, refractory oxides can be grown under an O_2 atmosphere, thus yielding contamination-free samples with a length of several centimeters [34–36], when using the floating zone technique. Also, the growth of Si [37] or high-purity $CuGeO_3$ single crystals for property investigates [38] as well as the synthesis of superconducting Bi-containing oxides ($T_c > 85$ K) [39] could be achieved in mirror furnaces. Furthermore, the metal reduction by thermal dissociation of the oxides has been summarized in Ref. [40] (and references therein). In addition, mirror furnaces can be used to achieve high temperatures for experimental purposes, for example, neutron diffraction [41].

1.8
Arc Melting: Research Laboratory and Industrial Scale

The generation of very high temperatures (up to 4000 K) is possible by arc melting, where a plasma volume is heated by an electron beam between two electrodes, similar to the well-known welding techniques. Many homemade and commercial laboratory arc furnaces are available for research purposes [1,42–47]. Usually sample quantities from 300 mg to 5 g can be prepared as a single batch. Special setups allow upscaling up to 20 g (often used for larger quantities for neutron diffraction studies). Arc melting requires a certain metallic conductivity of the educts and the sample itself and is thus limited to the field of intermetallics. Typically diverse alloy systems, borides, gallides, indides, carbides, silicides, germanides, and stannides are synthesized via this high-temperature technique.

The two electrodes are a water-cooled copper crucible for the educt sample and a CeO_2-doped tungsten electrode (thoriated electrodes are no longer used). The size and shape of the copper crucibles can easily be adjusted in a mechanical workshop for the desired sample type, for example, a small spherical button or an elongated ingot. The strong water cooling of the copper hearths prevents a reaction of the sample with the crucible, allowing for a so-called crucible-free melting. The tungsten electrodes of the small devices are commercial ones with 1.5–2.4 mm diameter. More powerful devices use pure tungsten cylinders with up to 10 mm diameter; however, the mechanical machinability of such pieces is difficult.

The welding setup is placed in a gas-tight cylinder made of glass or stainless steel. These cylinders should be as small as possible in order to keep the gas volume (and the surfaces that allow adsorption of reactive rest gas) at minimum. Furthermore, the small volume allows fast evacuation and refilling with purified argon. Typically used purification sequences use titanium or magnesium sponge furnaces (870 K), silica gel, molecular sieves, or commercial BTX catalysts. Some setups have an additional small mold inside for melting titanium or zirconium sponge (getter material) as an additional trap for reactive gases (e.g., oxygen or nitrogen).

Usually the arc-melting chambers are operated under a reduced argon pressure of 700–800 mbar. A perfect melting process can be achieved only with a high-quality welding generator with high-frequency ignition. This avoids contact of the electrode with the sample and the crucible. Several generators are equipped with a foot pedal for better power dosage. For standard laboratory arc-melting furnaces, amperages of 180 A (standard welding generator) are largely sufficient. Larger sample quantities and samples with extremely high melting temperatures (e.g., transition metal carbides) might require a more powerful generator. A more technical documentation along with pictures of homemade arc-melting crucibles is published in a recent monograph on intermetallics [1].

Arc melting is a very efficient and fast reaction technique; however, some limitations and technical refinements need to be accounted for. To prevent arcing between the tungsten electrode and the crucible sidewalls, the diameter of the copper crucible is much larger than that of the electrode. Also, an optimal electrode gap (electrode–sample distance) is essential for a precise arc-melting process (optimal heat transfer). The operator needs to wear welding safety glasses with sufficient shading for eye protection. Especially for home-built devices, one needs to take care that all prerequisites for electrical safety are fulfilled.

Apart from these technical details, some chemical prerequisites are also important. Arc melting is possible only for elements with comparable melting and boiling points. Otherwise, severe evaporation problems for elements with low boiling point, for example, Mg, Cd, or Zn, occur, leading to significant deviations from the ideal starting composition. If very well-crystallized samples are remelted, microcracks can lead to strong shattering.

Besides powder melting and compound synthesis, the miniaturized arc-melting chambers also allow sealing high-melting metal tubes, for example, niobium, tantalum, or tungsten [46], and thermocouple junction welding.

Another special application that requires high temperatures is the synthesis of titanium suboxides; partial reduction of rutile with elemental titanium to TiO is possible through arc melting. One might think of many related reactions. Often one simply has to test a certain technique.

Tri-arc furnaces are an enhanced setup for crystal growth experiments on intermetallics with the Czochralski technique. The larger number of tungsten electrodes allows a more uniform melting of higher sample quantities. Tri-arc systems with large power supplies up to 650 A are commercially available. Detailed technical descriptions of such devices are readily available on the homepages of the suppliers.

On an industrial scale, arc melting is commonly used for melting processes in steel and alloy production with melting capacities up to 300 t. Other applications are the reduction of ilmenite ($FeTiO_3$) and phosphorite ($Ca_3(PO_4)_2$) with carbon for titanium and phosphorus production as well as for the synthesis of calcium and silicon carbide. The huge reaction containers are bricked with clinkers and large carbon cylinders are used as electrodes. Such technical setups can easily be found by an Internet search using the keywords *Lichtbogenofen* and *arc-melting furnace*. Also a technical handbook is available [48].

1.9 Induction Heating

The major field of applications of induction heating concerns diverse large-scale industrial processes in the field of steel, aluminum, copper, and zinc industry [49–51]. Several manufacturing processes are almost 100 years old. In contrast to classical flame or torch heating (which release combustion products in the inductive heating process!), the material to be heated is well isolated from the power supply.

In an induction furnace, an alternating current is passed through a coil with several windings in order to heat the metallic specimen inside the center of the coil. Eddy currents are generated within the sample and the resistance of the specific material leads to Joule heating.

One of the main advantages of induction heating is that it works quickly. The heat is developed directly inside the sample. The energy transfer is efficient and thus very attractive for cost-effective heating purposes on an industrial scale. Furthermore, a fixed setup leads to highly reproducible heating patterns.

Depending on the size of the volume to be melted or the shape of the material to be annealed, different inductors can be designed individually. Usually such inductors are made of copper tubing (with round or square diameter) since copper has both good electrical and good thermal conductivity and the tubing allows for an effective water-cooling. Many setups and technical details of such large-scale furnaces are well documented in Refs [49–51].

For laboratory applications, several small sample chambers have been developed for treatment of amounts from 200 to 300 mg up to the gram scale. The different devices are used either for direct reaction of elements or precursor compounds, for annealing purposes (sample homogenization or crystal growth) or for sample evaporation. Most of these devices are homemade ones and it is a prerequisite for the operator to fulfill the requirements of electromagnetic compatibility.

Since induction heating is directly coupled to electrical conductivity, most techniques have been developed in the field of intermetallic compounds [1]. Direct reactions of elements can be performed in open high-melting metal (e.g., Nb, Ta, Mo, W), graphite, or glassy carbon crucibles [52], provided that the elements have high boiling points. Synthesis with volatile elements (e.g., Mg, Zn, Cd, Eu, Yb) requires sealed metal tubes [53]. Such sample chambers are usually operated under a purified argon atmosphere for oxidation protection. Several techniques exist for the growth of small single crystals [54] since comparatively high diffusion is possible.

Nonconducting ceramic samples can be prepared only by indirect heating, using thick-walled molybdenum or tungsten crucibles that couple with the high-frequency field. This indirect heat transfer is fast and allows effective high-temperature reactions. Many ternary and quaternary nitridosilicates have been prepared via this high-frequency routine [55]. Another aspect concerns the synthesis of fullerenes. Evaporation of suitable carbon precursors followed by condensation leads to effective amounts of fullerenes [56,57]. Another interesting application is the use of a vertical high-frequency induction furnace for the growth of silicon nanowires [58].

For research purposes using small sample sizes, suitable high-frequency generators are commercially available. Such power supplies generally work with 1.5–10 kW and operating frequencies in the range of 30–300 kHz. In some special cases, more powerful generators up to 25 kW find application. The frequency range depends on the sample size. Generally, higher frequencies are required for smaller sample volumes. Many of the induction coils are homemade. The special design guarantees an optimal heat transfer to the sample. The design of the coil should be harmonized with the supplier in order to ensure optimal working conditions.

Induction levitation melting is a special field of induction heating, using a coil with counter-windings [59,60]. Levitation of the sample allows a quasi-crucible-free melting, thus effectively avoiding impurities. The cold-wall induction crucible is a special design as well as levitation boats. Setups of these techniques can easily be found through Internet search using these keywords.

1.10
Flux-Assisted Synthesis: Metal and Salt Fluxes

Application of high temperature is also a key parameter for high-temperature solution growth of crystals. In solid-state chemistry, mainly salt and metal fluxes find application and these two research fields are briefly summarized. Syntheses in ionic liquids proceed at much lower temperatures. Flux-assisted crystal growth has two main reasons: (i) the growth of small crystals (where only powders are available through conventional techniques) for structure determination; (ii) the growth of large single crystals for direction-dependent physical property studies. This allows measurement of the intrinsic behavior of the material in the absence of any grain boundaries.

Our short overview starts with salt fluxes [2,61–63]. A very old technical application of salt fluxes is the soldering process where borate-based salts are used for dissolution of surface impurities. Another very old application is sintering (with and without additives) for the synthesis of ceramic/oxidic materials. In basic research, salt fluxes meanwhile find broad application for explorative synthesis, mainly using simple salts and eutectic mixtures. The cheapest and easiest salt flux system is an equimolar NaCl/KCl mixture.

For an effective salt flux synthesis, the flux medium should have a sufficiently low melting point (large liquidus range) and dissolve appropriate amounts of the reagents. The solubility should vary over the temperature range in order to enable excellent recrystallization with low cooling rates. The flux should be nonvolatile and unreactive toward the crucible material and one should be able to easily dissolve the flux after the reaction. In many cases, dissolution of the flux with hot demineralized water is possible. When the product is moisture sensitive, one can use centrifugation techniques and organic solvents, sometime it is only possible to separate the crystals and the flux mechanically.

Besides simple halide fluxes, also hydroxides (simple binaries or hydroxide mixtures), carbonates, binary oxides such as PbO or Bi_2O_3, and complex vanadate, molybdate, and tungstate fluxes find application. Usual crucible materials can be noble metal tubes of silver, gold or platinum, or alumina tubes. A huge number of salt flux examples have been summarized in an excellent review by Bugaris and zur Loye [63]. Some of the reactions deserve maximum temperatures up to 1750 K, always followed by low cooling rates (generation of a few number of seeds that grow subsequently). As examples of salt-flux-grown specimens, we present YZnPO [64] and YZnAsO [65] crystals in Figure 1.2.

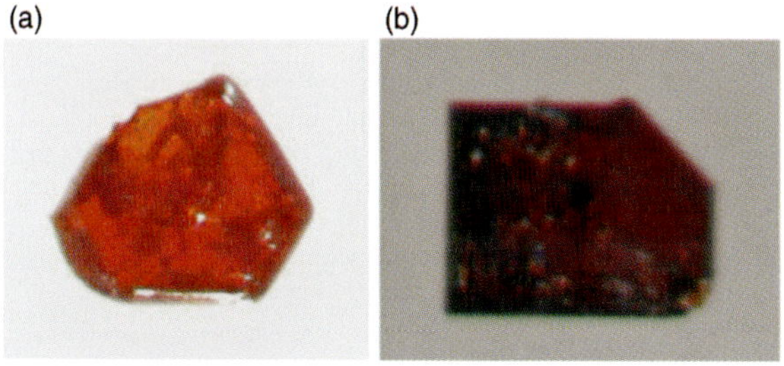

Figure 1.2 Single crystals of YZnPO (a) and YZnAsO (b), grown in NaCl/KCl fluxes. Edge lengths are approximately 300 μm.

Most of the salt flux experiments avoid the incorporation of the flux medium. This is best achieved if the flux medium is chemically similar to the desired crystals, although with differently large ions, suppressing substitution reactions just for geometrical reasons. Parts of the flux (either anions or cations, or even both components) can intentionally be included in the product. This technique is called the reactive flux [66–68] and reviewed in more detail in Chapter 12 of Volume 2. Halides and especially many complex sulfides have been effectively synthesized this way.

The second large family concerns the metal fluxes. Also, this field has extensively been reviewed and many examples have been presented [69–72]. Almost all low-melting metals find application for crystal growth experiments. The most frequently used flux media are tin, indium, and lead, but also gallium, bismuth, zinc, and the alkali metals have been introduced. The experimental requirements for metal flux synthesis are very similar to the salt fluxes. Alumina crucibles are the most used crucible material. The different flux media depend on the melting point and the liquidus range of the metal and another important parameter is the wetting capability. For a list of examples and required annealing sequences, we refer to Ref. [71].

The removal of the metal flux is more aggressive than the dissolution of simple salts. The alkali fluxes can be removed by ethanol, while tin and zinc deserve 1 M hydrochloric acid. At this point, it is important to note that not all products resist the acid. Thus, careful dissolution of the flux accompanied by observation of the product is essential. Other fluxes request even a more aggressive reagent, for example, the lead fluxes are dissolved with mixtures of acetic acid and hydrogen peroxide, since hydrochloric acid would lead to large amounts of lead chloride. As an example, we present $IrSb_2$ crystals grown in liquid bismuth in Figure 1.3.

A mild separation of the product crystals from the melt is possible through the melt-centrifugation technique [69,73], frequently used for tin and gallium fluxes. After the centrifugation process, often thin films of the flux remain on the crystal surface. A severe problem of metal flux synthesis concerns impurity inclusions. This is well known from the tin flux synthesis of metal-rich phosphides using silica ampoules as crucible material. DSC measurements showed the inclusion of tiny amounts of tin (signal at the melting point of 504 K) and the phosphorus

Figure 1.3 Single crystals of $IrSb_2$ in a residual bismuth matrix. Edge lengths are approximately 80–100 μm.

sites sometimes show silicon occupancy resulting from a reaction with the crucible material.

Similar to salts, metals can serve solely as solvent or play the role of a reactive flux.

1.11 Microwave Heating

Application of microwave heating for chemical reactions is already performed for more than 40 years. In the beginning, many solution-based reactions and decomposition reactions in Teflon-lined autoclaves have been developed. Today, especially the decomposition reactions find broad application in all fields of analytical chemistry. Synthesis of solid-state materials by microwave heating was developed later. The high potential and many examples of this technique are summarized in three extensive review articles [74–76]. The early work focused on the reproduction of already known materials; exclusively by heating with microwaves. These studies impressively showed that microwave heating works and leads to a drastic decrease of the reaction times compared to the conventional techniques. Meanwhile, microwave heating gives access to many metastable solids and further new materials and is rapidly growing.

The wavelengths of microwaves are in the range of 1 mm to 1 m. This corresponds to a frequency range from 0.3 to 300 GHz. The heart of a microwave device is the magnetron, a thermionic diode with a cathode as an electron source. The microwaves are generated through a stream of electrons with a magnetic field. Meanwhile, many setups for laboratory synthesis are commercially available. The technical details and several specifications of the experimental setups are summarized in Ref. [76]. An important prerequisite of the experimental setup is an efficient transfer of power from the generator to the sample.

Many families of solids can be synthesized by direct application of microwave heating. Typical applications concern binary, ternary, and even multinary oxides, related sulfides and selenides, complex phosphides, and nitridation reactions (binary transition metal nitrides as well as complex nitridometalates) as well as glass preparation. For all these applications, the requested reaction times are small and often the crystallinity of the product is better than in conventional solid-state synthesis. This has repeatedly been demonstrated by much sharper reflections in powder X-ray diffraction.

Even carbides and borides of the high-melting transition metals are available through microwave synthesis. Typical examples are TiC, WC grains, the superconductor MgB_2, or ZrB_2. Besides the enormous gain in reaction time, several of these reactions often proceed at lower temperature when applying microwave heating. Especially in the case of carbides, the strong coupling of carbon with microwaves is favorable. In addition to pure binary or ternary carbides, many composite materials have also been prepared, for example, ZrC-SiC or composites of the carbides with carbon nanotubes. Further classes of compounds are silicides (e.g., $Li_{21}Si_5$, Mg_2Si, or WSi_2), pnictides (e.g., Li_3Sb and Li_3Bi), oxynitrides, and carbonitrides.

Current research in the field of microwave heating concerns the understanding of the reaction mechanisms as well as the interaction of the microwave with the reactants. Further approaches focus on different *in situ* techniques for monitoring the steps within synthesis. Such *in situ* studies are possible with powder X-ray as well as neutron diffraction. Special sample chambers have been developed for these purposes.

1.12
Hot Isostatic Pressing and Reactive Sintering

During a sintering process, one obtains a dense material by application of heat (and pressure) without attaining the melting point of the material. This technique is typically applied to many of the very high-melting ceramics (typically oxides) and metals, for example, magnesium and aluminum oxide, molybdenum, and tungsten. In the case of metals, this is well known as powder metallurgy. In ceramics production, sintering is one of the key steps in the high-temperature firing process, leading to grain growth and densification of the material. The technical and theoretical details of ceramics and metal sintering are well documented in standard textbooks [77,78].

Coupling sintering with hydrostatic pressure is used in the hot isostatic pressing (HIP) process [79,80], which reduces the materials porosity and consequently leads to an increase of the density. The HIP process is used for metallic, ceramic, polymeric, and composite materials as well and improves the mechanical properties as well as the workability. The pressure medium usually is high-purity argon gas in order to avoid any side reactions. The temperature and pressure ranges for most applications extend to 1700 K and 200 MPa. Details on the experimental setups for many large-scale industrial machines as well as laboratory devices can easily be found by a routine Internet search. The theoretical background for the HIP process is well documented [81].

In the field of intermetallic materials, the HIP technique is used for aluminum- and magnesium-based lightweight alloys, cemented tungsten carbide, diverse magnetic materials, biomedical alloys for knee and hip joints, titanium alloys, metal–matrix composites, or several steel castings. On the ceramics side, oxide- and nitride-based ceramics as well as glasses are treated. The HIP technique is also used for high-quality plastic materials, however in these cases at much lower temperatures.

1.13
Current-Assisted Sintering

Reacting starting materials can be achieved by several methods, as already described. Sintering, with the goal to consolidate powders, is one form that has been used for millennia. To obtain high-density materials, the sintering is usually carried out at high temperatures to get a thermally activated material. Additionally, samples can be (uniaxially) compressed during sintering to achieve higher densities. This process is usually referred to as hot isostatic pressing. Another way, explored to activate the sintering process more efficiently, was the use of electrical current. In 1913, a patent was issued for a "process and apparatus for sintering refractory materials" [82]; in 1933, two patents were issued describing a "welding process" [83] and an "apparatus for making hard metal composites" [84] in which the sintering of powders with the help of an electric discharge/current is described. Subsequent investigations and development lead to the commercialization of these instruments. The terms pulsed electric current sintering (PECS), field-assisted sintering technology (FAST), plasma-assisted sintering (PAS), spark plasma sintering (SPS), or current-activated pressure-assisted densification (CAPAD) are used to describe related processes. Since the "plasma" in SPS has not yet been proven, the alternative term FAST is widely used instead [85,86]. The term PECS finally underlines the use of a noncontinuous, pulsed, electrical current. All terms, besides CAPAD, however, neglect the role of mechanical pressure used in these processes. The advantages proposed for these methods were lower sintering temperatures, shorter holding times, and improved material properties. Since these processes have become important for sintering as well as synthetic methods, a large number of reviews exist focusing on the various aspects such as electric field and pressure [87], the current activation of the sintering [88], metals and metal matrix nanocomposites [89], or mechanisms and materials [90]. Challenges and opportunities are discussed in Ref. [91].

In order to sinter a sample, the precursors or the final product get loaded into a die that is placed into a mechanical press supplying uniaxial pressure. The punchers of the press serve simultaneously as electrodes supplying voltages typically below 10 V and high currents (1–10 kA). Depending on the material loaded into the die and the material of the die itself (e.g., metal, carbides, or graphite can be used), Joule heating, either of the specimen or the die, can occur. Even in the case of low- or nonconductive samples, a rapid heat transfer into the sample is achieved by the close contact of die and specimen. The current is usually applied by a pulse program with defined pulse and pause phases. Typical pulse durations are on the order of a few milliseconds. With this setup, temperature ramps of 1000 °C/min can be run. Cooling rates of 150 °C/min are possible, actively cooled systems allow even rates up to 400 °C/min. During the heating step, mechanical pressure can be applied to obtain highly dense samples. Here, forces up to 250 kN can be used, resulting in pressures in the megapascal region, depending on the cross section of the die used. Sintering can take place under vacuum or in the case of volatile components also under protective gas atmosphere. The sample chamber and the electrode are water-cooled, which acts as an additional heat sink during the cooling step. The temperature can be tracked via thermocouples touching the die from the side or at the top/bottom; for high-temperature processes, pyrometers can also be used. During the electric pulse, the current flows through the die and the sample, where the heating power is dissipated at the contact points of the particles. This allows the sintering to happen, but in contrast to regular sintering, the grain growth can be suppressed due to the short times (on the minute scale) used. Depending on the pulse sequence, one can use a different denomination of the process. For plasma-assisted sintering, a pulsed current is applied in the beginning of the sintering followed by a direct current (DC) during the rest of the process. Spark plasma sintering in contrast uses a pulsed direct current throughout the whole process.

Due to the two possible heating processes, metallic as well as insulating materials can be sintered with these techniques. Refractory metals and intermetallics are one class of materials that have broad technical and industrial applications due to their physical properties. Fabrication of these compounds, however, requires high temperatures and pressures to obtain high densities and good mechanical properties. Furthermore, reducing conditions or vacuum has to be chosen for processing to avoid the oxidation of these materials. While tungsten powders with particle sizes of 2 μm usually are compacted at 2000 °C, the use of SPS/FAST leads to a significant reduction in the maximal temperature needed for densification. A reduction down to 1300 °C was observed [92]. Other refractory metals, for example, Ta [93], Mo [94], or Ru [95], exhibit similar behavior. Also, intermetallic systems, for example, superconducting Nb_3Al, can be compacted and sintered by SPS techniques. Sintered powders, prepared by a hydriding–dehydriding route, show superior properties compared to conventionally prepared Nb_3Al [96]. Besides bulk sintering, it has also been shown that nanostructured powders retain their grain size when SPSed using the right conditions [89]. Also ceramic oxide materials and so-called ultrahigh temperature ceramics (UHTC) of the boride, carbide, or nitride family were processed via current-assisted sintering techniques. They all share high melting temperatures, the latter in excess of 3000 °C, and high electrical and thermal conductivities, which are very welcome properties but at the same time can cause problems during sintering. To just give some examples, the refractory metal diborides ZrB_2 and HfB_2 [97] are well-studied UHTCs. The properties for ZrB_2, for example, the failure strength, could be significantly improved from ~300 MPa for pressureless sintered to ~460 MPa for hot-pressed to ~530 MPa for spark plasma sintered specimen [98]. Transparent ceramics such as yttria-stabilized zirconia (YSZ), Al_2O_3, MgO, or yttrium aluminum garnets (YAG) can also be densified by CAPAD [99]. Here, the role of the atmosphere used during sintering plays a significant role. Specimen sintered in vacuum and/or graphite dies exhibit coloring. The colors are results of oxygen vacancies and defects caused by the reduced atmospheric pressure and the reaction of the material with the carbon die. An extensive overview of a lot of different classes of materials and their consolidation and synthesis by electric current-activated sintering is given in Ref. [100].

1.14 Carbothermic Reductions

The carbothermic reduction is a typical high-temperature application using carbon as reducing agent. Usually metal oxides are reduced to produce the element.

Such carbon-based reductions are, however, possible only for elements that do not form stable carbides, for example, TiC, TaC, $Cr_{23}C_6$, or WC. Since the reactions run at very high temperatures (>1400 K), according to the Boudouard equilibrium, essentially carbon monoxide is formed as reaction product (thus, these reactions are entropy driven). Some of these reactions are carried out in arc-melting furnaces using consumable carbon electrodes.

The best-known example of a carbothermic reduction is the iron production in the blast furnace: $Fe_2O_3 + 3C \rightarrow 2Fe + 3CO$. Reduction of ternary oxides or oxide mixtures leads to the technically important ferromolybdenum and ferrotungsten alloys. Other important reactions are the production of raw silicon ($SiO_2 + 2C \rightarrow Si + 2CO$) and the direct reaction to silicon carbide ($SiO_2 + 3C \rightarrow SiC + 2CO$), one of the cheap abrasive materials. SiC can also be grown in the form of whiskers for use in composite materials [101]. Zinc and cadmium oxide can directly be reduced by carbon and distilled *in situ*, leading to the pure elements and carbon monoxide. Meanwhile, the preparation of magnesium and calcium metal via carbothermic reduction is also under discussion [102], since there is large demand for magnesium-based lightweight alloys.

The synthesis of white phosphorus either starts from fluoroapatite or phosphorite. Again, the carbon source is a consumable carbon electrode operated in an arc furnace:

$$4Ca_5(PO_4)_3F + 20SiO_2 + 30C \rightarrow P_4 + 18CaSiO_3 + 30CO + 2CaF_2$$

$$2Ca_3(PO_4)_2 + 6SiO_2 + 10C \rightarrow P_4 + 6CaSiO_3 + 10CO$$

Formation of stable carbides can be bypassed if the carbothermic reduction is coupled to direct chlorination, the so-called carbochlorination. The prominent examples are the formation of titanium tetrachloride from rutile or ilmenite. These two reactions run above 1300 K:

$$TiO_2 + 2Cl_2 + 2C \rightarrow TiCl_4 + 2CO$$

$$2FeTiO_3 + 7Cl_2 + 6C \rightarrow 2TiCl_4 + 2FeCl_3 + 6CO$$

Many other special examples and certain perspectives have been reported. The carbothermic reduction can be initiated by classical flame or torch heating, by a preliminary exothermic reaction (the classical thermite reaction), by arc melting, or via microwaves. An interesting new application focuses on a solar carbothermic reduction of ZnO with 3–8 kW solar furnaces, another example of a coupled reaction [103].

1.15
Self-Propagating High-Temperature Synthesis

Self-propagating high-temperature synthesis (SHS) is a synthetic approach to react inorganic compounds by an exothermic reaction. A variant of this process is called solid-state metathesis (SSM), which involves the formation of salts. For the reaction, intimately mixed, fine powdered reactants (elements or salts) are pressed into pellets. The reaction is initiated by point heating of the sample; the reaction propagates through the specimen due to its exothermic nature. In order to prevent, for example, the thermal decomposition of the product, the reaction can be moderated using an inert component. Alkali metal halides can be used, since melting or evaporation consumes parts of the generated heat [104–107]. The choice of the atmosphere plays a significant role as well. The reactions can be carried out in vacuum or in either reactive or inert gas atmosphere. Due to the temperatures generated in this reaction, the synthesis is especially suitable for refractory materials with unusual properties. First attempts were made reacting a transition metal halide with, for example, lithium oxide yielding the transition metal oxide and lithium halide as driving force [108]. But metastable oxides [109] and chalcogenides of lead and tin could also be synthesized by this approach [110]. Meanwhile, the classes of materials prepared are numerous. Carbides [111], silicides [112], nitrides [113], and phosphides [114] could be obtained by this approach as well. Besides the general interest in this synthesis technique, perovskite materials [115] or $CuCo_2O_4$ [116], a material for high-rate supercapacitors, can also be synthesized by solid-state metathesis reaction.

References

1 Pöttgen, R. and Johrendt, D. (2014) *Intermetallics: Synthesis, Structure, Function*, De Gruyter, Berlin.
2 Brauer, G. (1981) *Handbuch der Präparativen Anorganischen Chemie*, Band 1–3, Ferdinand, Enke Verlag, Stuttgart.
3 Herrmann, W.A. (1998–2002) *Synthetic Methods of Organometallic and Inorganic Chemistry*, vols. **1–10** (ed. G. Brauer), Georg Thieme Verlag, Stuttgart, Germany.
4 Villars, P. and Cenzual, K. (2013) Pearson's Crystal Data: Crystal Structure Database for Inorganic Compounds, ASM International®, Materials Park, OH.
5 Pecchini, M.P. (1967) U.S. Patent 3,330,697.
6 Mori, M. and Sammes, N.M. (2002) *Solid State Ionics*, **146**, 301.
7 Kodaira, C.A., Brito, H.F., Malta, O.L., and Serra, O.A. (2003) *J. Lumin.*, **101**, 11.
8 Müller-Buschbaum, Hk. (1967) *Z. Anorg. Allg. Chem.*, **355**, 30.
9 Müller-Buschbaum, Hk. (2006) *Z. Anorg. Allg. Chem.*, **632**, 369.
10 Kinoshita, K., Hayashi, A., Akahide, K., and Yamazaki, T. (1994) *Pure Appl. Chem.*, **66**, 1295.
11 Müller-Buschbaum, Hk. and Pausch, H. (1979) *Z. Naturforsch.*, **34b**, 371–374.
12 Brandt, R. and Müller-Buschbaum, Hk. (1986) *Z. Anorg. Allg. Chem.*, **542**, 18–24.
13 Pausch, H. and Müller-Buschbaum, Hk. (1976) *Z. Anorg. Allg. Chem.*, **426**, 184–188.
14 Pausch, H. and Müller-Buschbaum, Hk. (1974) *Z. Anorg. Allg. Chem.*, **405**, 1–7.
15 Bertaut, E.F. and Guillen, M. (1966) *Bull. Soc. Fr. Ceram.*, **72**, 57.
16 Müller-Buschbaum, Hk. (2013) *Z. Anorg. Allg. Chem.*, **639**, 2715–2735.
17 Möhr, S. and Müller-Buschbaum, Hk. (1995) *Angew. Chem., Int. Ed. Eng.*, **34**, 634–640.
18 Jayaraman, A. (1983) *Rev. Mod. Phys.*, **55**, 65–108.
19 Ruoff, A.L., Xia, H., and Xia, Q. (1992) *Rev. Sci. Instrum.*, **63**, 4342–4348.
20 Ming, L.C. and Bassett, W.A. (1974) *Rev. Sci. Instrum.*, **45**, 1115–1118.

21 Du, Z., Amulele, G., Robin Benedetti, L., and Lee, K.K.M. (2013) *Rev. Sci. Instrum.*, **84**, 075111–075119.

22 Salamat, A., Fischer, R.A., Briggs, R., McMahon, M.I., and Petitgirard, S. (2014) *Coord. Chem. Rev.*, **277–278**, 15–30.

23 Chaney, J., Santillán, J., Knittle, E., and Williams, Q. (2015) *Phys. Chem. Miner.*, **42**, 83–93.

24 Solopova, N.A., Dubrovinsky, L., Spivak, A.V., Litvin, Y.A., and Dubrovinskaia, N. (2015) *Phys. Chem. Miner.*, **42**, 73–81.

25 Fischer, R.A., Campbell, A.J., Caracas, R., Reaman, D.M., Heinz, D.L., Dera, P., and Prakapenka, V.B. (2014) *J. Geophys. Res.*, **119**, 2810–2827.

26 Lai, X., Zhu, F., Wu, Y., Huang, R., Wu, X., Zhang, Q., Yang, K., and Qin, S. (2014) *J. Solid State Chem.*, **210**, 155–159.

27 Salamat, A., Hector, A.L., Gray, B.M., Kimber, S.A.J., Bouvier, P., and McMillan, P.F. (2013) *J. Am. Chem. Soc.*, **135**, 9503–9511.

28 Onwudiwe, D.C., Krüger, T.P.J., and Strydom, C.A. (2014) *Mater. Lett.*, **116**, 154–159.

29 Kazakevich, P.V., Simakin, A.V., Voronov, V.V., and Shafeev, G.A. (2006) *Appl. Surf. Sci.*, **252**, 4373–4380.

30 Bi, X.-X. and Eklund, P.C. (1992) *MRS Proc.*, **286**, 161–166.

31 Flamant, G., Ferriere, A., Laplaze, D., and Monty, C. (1999) *Solar Energy*, **66**, 117–132.

32 http://www.promes.cnrs.fr/index.php?page=mega-watt-solar-furnace (accessed November 6, 2014).

33 de la Rue, R.E. and Halden, F.A. (1960) *Rev. Sci. Instrum.*, **31**, 35–38.

34 Kitazawa, K., Nagashima, K., Mizutani, T., Fueki, K., and Mukaibo, T. (1977) *J. Cryst. Growth*, **39**, 211–215.

35 Balbashov, A.M. and Egorov, S.K. (1981) *J. Cryst. Growth*, **52** (Part 2), 498–504.

36 Bednorz, J.G. and Arend, H. (1984) *J. Cryst. Growth*, **67**, 660–662.

37 Eyer, A., Kolbesen, B.O., and Nitsche, R. (1982) *J. Cryst. Growth*, **57**, 145–154.

38 Weiden, M. (1999) *Phys. Unserer Zeit*, **30**, 6–11.

39 Menken, M.J.V., Winkelman, A.J.M., and Menovsky, A.A. (1991) *J. Cryst. Growth*, **113**, 9–15.

40 Bjorndalen, N., Mustafiz, S., and Islam, M.R. (2003) *Energy Sources*, **25**, 153–159.

41 Lorenz, G., Neder, R.B., Marxreiter, J., Frey, F., and Schneider, J. (1993) *J. Appl. Crystallogr.*, **26**, 632–635.

42 Daane, A.H. (1952) *Rev. Sci. Instrum.*, **23**, 245.

43 Miller, A.E., Daane, A.H., Habermann, C.E., and Beaudry, B.J. (1963) *Rev. Sci. Instrum.*, **34**, 644.

44 Reed, T.B. (1967) *Mater. Res. Bull.*, **2**, 349.

45 Corbett, J.D. and Simon, A. (1983) *Inorg. Synth.*, **22**, 15.

46 Pöttgen, R., Gulden, T., and Simon, A. (1999) *GIT Labor-Fachzeitschrift*, **43**, 133.

47 Ferro, R. and Saccone, A. (2008) *Intermetallic Chemistry*, Pergamon Materials Science, Amsterdam.

48 Pfeifer, H., Nacke, B., and Beneke, F. (2011) *Praxishandbuch Thermoprozesstechnik, Band II: Anlagen-Komponenten-Sicherheit*, 2. Auflage, Vulkan-Verlag, Essen, Germany.

49 Zinn, S. and Semiatin, S.L. (1988) *Elements of Induction Heating: Design, Control, and Applications*, ASM International, OH.

50 Rudnev, V., Loveless, D., Cook, R., and Black, M. (2003) *Handbook of Induction Heating*, Marcel Dekker, Basel, Switzerland.

51 Dötsch, E. (2013) *Inductive Melting and Holding: Fundamentals, Plants and Furnaces, Process Engineering*, 2nd edn, Vulkan-Verlag, Essen, Germany.

52 Kußmann, D., Hoffmann, R.-D., and Pöttgen, R. (1998) *Z. Anorg. Allg. Chem.*, **624**, 1727.

53 Pöttgen, R., Lang, A., Hoffmann, R.-D., Künnen, B., Kotzyba, G., Müllmann, R., Mosel, B.D., and Rosenhahn, C. (1999) *Z. Kristallogr.*, **214**, 143.

54 Niepmann, D., Prots', Yu.M., Pöttgen, R., and Jeitschko, W. (2000) *J. Solid State Chem.*, **154**, 329.

55 Schnick, W. and Huppertz, H. (1997) *Chem. Eur. J.*, **3**, 679.

56 Peters, G. and Jansen, M. (1992) *Angew. Chem., Int. Ed. Engl.*, **31**, 223.
57 Müller, A., Ziegler, K., Amshoroa, K.Yu., and Jansen, M. (2011) *Eur. J. Inorg. Chem.*, 268.
58 Pang, C., Cui, H., Yang, G., and Wang, C. (2013) *Nano Lett.*, **13**, 4708.
59 Okress, E.C., Wroughton, D.M., Comenetz, G., Brace, P.H., and Kelly, J.C.R. (1952) *J. Appl. Phys.*, **23**, 545.
60 Morita, A., Fukui, H., Tadano, H., Hayashi, S., Hasegawa, J., and Niinomi, M. (2000) *Mater. Sci. Eng. A*, **280**, 208.
61 Scheel, H.J. (1974) *J. Crystal Growth*, **24/25**, 669.
62 Elwell, D. and Scheel, H.J. (1975) *Crystal Growth from High-Temperature Solutions*, Academic Press, London.
63 Bugaris, D.E. and zur Loye, H.-C. (2012) *Angew. Chem., Int. Ed.*, **51**, 3780.
64 Lincke, H., Glaum, R., Dittrich, V., Tegel, M., Johrendt, D., Hermes, W., Möller, M.H., Nilges, T., and Pöttgen, R. (2008) *Z. Anorg. Allg. Chem.*, **634**, 1339.
65 Lincke, H., Glaum, R., Dittrich, V., Möller, M.H., and Pöttgen, R. (2009) *Z. Anorg. Allg. Chem.*, **635**, 936.
66 Kanatzidis, M.G. (1990) *Chem. Mater.*, **2**, 353.
67 Pell, M.A. and Ibers, J.A. (1997) *Chem. Ber.*, **130**, 1.
68 Kanatzidis, M.G. and Sutorik, A.C. (1995) *Progr. Inorg. Chem.*, **43**, 151.
69 Fisk, Z. and Remeika, J.P. (1989) Growth of single crystals from molten metal fluxes, in *Handbook on the Physics and Chemistry of Rare Earths*, vol. **12** (eds K.A. Gschneidner, Jr., and L. Eyring), Elsevier Publishers B. V., Amsterdam, p. 53.
70 Canfield, P.C. and Fisk, Z. (1992) *Philos. Mag. B*, **65**, 1117.
71 Kanatzidis, M.G., Pöttgen, R., and Jeitschko, W. (2005) *Angew. Chem., Int. Ed.*, **44**, 6997.
72 Canfield, P.C. (2012) *Philos. Mag. B*, **92**, 2398, and review articles within this special issue.
73 Boström, M. and Hovmöller, S. (2000) *J. Solid State Chem.*, **153**, 398.
74 Mingos, D.M.P. and Baghurst, D.R. (1991) *Chem. Soc. Rev.*, **20**, 1.
75 Rao, K.J., Vaidhyanathan, B., Ganguli, M., and Ramakrishnan, P.A. (1999) *Chem. Mater.*, **11**, 882.
76 Kitchen, H.J., Vallance, S.R., Kennedy, J.L., Tapia-Ruiz, N., Carassiti, L., Harrison, A., Whittaker, A.G., Drysdale, T.D., Kingman, S.W., and Gregory, D.H. (2014) *Chem. Rev.*, **114**, 1170.
77 Kang, S.-J. (2005) *Sintering: Densification, Grain Growth and Microstructure*, Elsevier Butterworth-Heinemann, Oxford.
78 Fang, Z.Z. (ed.) (2010) *Sintering of Advanced Materials: Fundamentals and Processes*, Woodhead Publishing Limited, New Delhi.
79 Koizumi, M. (ed.) (1992) Hot isostatic pressing: theory and applications, in *Proceedings of the 3rd International Conference*, Osaka, Japan, June 10–14, 1991, Elsevier Applied Science, ISBN-10:185166744X.
80 Atkinson, H.V. and Rickinson, B.A. (1992) *Hot Isostatic Processing*, Springer, New Orleans.
81 Helle, A.S., Easterling, K.E., and Ashby, M.F. (1985) *Acta Metall.*, **33**, 2163.
82 Weintraub, G. and Rush, H. (1913) U.S. Patent 1,071,488A.
83 Taylor, G.F. (1933) U.S. Patent 1,896,853.
84 Taylor, G.F. (1933) U.S. Patent 1,896,854.
85 Hulbert, D.M., Anders, A., Andersson, J., Lavernia, E.J., and Mukherjee, A.K. (2009) *Scr. Mater.*, **60**, 835–838.
86 Hulbert, D.M., Anders, A., Dudina, D.V., Andersson, J., Jiang, D., Unuvar, C., Anselmi-Tamburini, U., Lavernia, E.J., and Mukherjee, A.K. (2008) *J. Appl. Phys.*, **104**, 033305.
87 Munir, Z.A., Anselmi-Tamburini, U., and Ohyanagi, M. (2006) *J. Mater. Sci.*, **41**, 763–777.
88 Munir, Z.A., Quach, D.V., and Ohyanagi, M. (2011) *J. Am. Ceram. Soc.*, **94**, 1–19.
89 Saheb, N., Iqbal, Z., Khalil, A., Hakeem, A.S., Al Aqeeli, N., Laoui, T., Al-Qutub, A., and Kirchner, R. (2012) *J. Nanomater.*, **2012**, 1–13.
90 Guillon, O., Gonzalez-Julian, J., Dargatz, B., Kessel, T., Schierning, G., Räthel, J., and Herrmann, M. (2014) *Adv. Eng. Mater.*, **16**, 830–849.

91 Suárez, M., Fernández, A., Menéndez, J.L., Torrecillas, R., Kessel, H.U., Hennicke, J., Kirchner, R., and Kessel, T. (2013) *Challenges and Opportunities for Spark Plasma Sintering: A Key Technology for a New Generation of Materials*, InTec., Croatia.

92 El-Atwani, O., Quach, D.V., Efe, M., Cantwell, P.R., Heim, B., Schultz, B., Stach, E.A., Groza, J.R., and Allain, J.P. (2011) *Mater. Sci. Eng.*, **528**, 5670–5677.

93 Angerer, P., Neubauer, E., Yu, L.G., and Khor, K.A. (2007) *Int. J. Refract. Metals Hard Mater.*, **25**, 280–285.

94 Ohser-Wiedemann, R., Martin, U., Seifert, H.J., and Müller, A. (2010) *Int. J. Refract. Metals Hard Mater.*, **28**, 550–557.

95 Angerer, P., Wosik, J., Neubauer, E., Yu, L.G., Nauer, G.E., and Khor, K.A. (2009) *Int. J. Refract. Metals Hard Mater.*, **27**, 105–110.

96 Li, X., Chiba, A., Sato, M., and Takashash, S. (2002) *J. Alloys Compd.*, **336**, 232–236.

97 Fahrenholtz, W.G., Hilmas, G.E., Talmy, I.G., and Zaykoski, J.A. (2007) *J. Am. Ceram. Soc.*, **90**, 1347–1364.

98 Thompson, M., Fahrenholtz, W.G., and Hilmas, G. (2011) *J. Am. Ceram. Soc.*, **94**, 429–435.

99 Kodera, Y., Hardin, C.L., and Garay, J.E. (2013) *Scr. Mater.*, **69**, 149–154.

100 Orrù, R., Licheri, R., Locci, A.M., Cincotti, A., and Cao, G. (2009) *Mater. Sci. Eng.*, **R63**, 127–287.

101 Wang, K., Cheng, Y.-B., and Wang, H. (2009) *J. Aust. Ceram. Soc.*, **45**, 10.

102 Hu, F.-P., Pan, J., Ma, X., Zhang, X., Chen, J., and Xie, W.-D. (2013) *J. Magnes. Alloys*, **1**, 263.

103 Kräupl, S., Frommherz, U., and Wieckert, C. (2005) *J. Sol. Energy Eng.*, **128**, 8.

104 Subrahmanyam, J. and Vijayakumar, M. (1992) *J. Mater. Sci.*, **27**, 6249–6273.

105 Wiley, J.B. and Kaner, R.B. (1992) *Science*, **255**, 1093–1097.

106 Rao, L., Gillan, E.G., and Kaner, R.B. (1995) *J. Mater. Res.*, **10**, 353–361.

107 Parkin, I.P. (1996) *Chem. Soc. Rev.*, **25**, 199–207.

108 Hector, A. and Parkin, I.P. (1993) *Polyhedron*, **12**, 1855–1862.

109 Gillan, E.G. and Kaner, R.B. (2001) *J. Mater. Chem.*, **11**, 1951–1956.

110 Parkin, I.P. and Rowley, A.T. (1993) *Polyhedron*, **12**, 2961–2964.

111 Nartowski, A.M., Parkin, I.P., Mackenzie, M., and Craven, A.J. (2001) *J. Mater. Chem.*, **11**, 3116–3119.

112 Nartowski, A.M. and Parkin, I.P. (2002) *Polyhedron*, **21**, 187–191.

113 Gibson, K., Ströbele, M., Blaschkowski, B., Glaser, J., Weisser, M., Srinivasan, R., Kolb, H.-J., and Meyer, H.-J. (2003) *Z. Anorg. Allg. Chem.*, **629**, 1863–1870.

114 Jarvis, R.F., Jacubinas, R.M., and Kaner, R.B. (2000) *Inorg. Chem.*, **39**, 3243–3246.

115 Mandal, T.K. and Gopalakrishnan, J. (2004) *J. Mater. Chem.*, **14**, 1273–1280.

116 Pendashteh, A., Rahmanifar, M.S., Kaner, R.B., and Mousavi, M.F. (2014) *Chem. Commun.*, **50**, 1972–1975.

2
High-Pressure Methods in Solid-State Chemistry

Hubert Huppertz,[1] Gunter Heymann,[1] Ulrich Schwarz,[2] and Marcus R. Schwarz[3]

[1]Leopold-Franzens-Universität Innsbruck, Institut für Allgemeine, Anorganische und Theoretische Chemie, Innrain 80-82, A-6020 Innsbruck, Austria
[2]Max-Planck-Institut für Chemische Physik fester Stoffe, Chemical Metals Science, Nöthnitzer Straße 40, D-01187 Dresden, Germany
[3]Technische Universität-Bergakademie Freiberg, Freiberg High Pressure Research Centre, Institut für Anorganische Chemie, Akademiestraße 6, D-09599 Freiberg, Germany

2.1
Introduction

The majority of syntheses in solid-state chemistry are carried out by variation of the two thermodynamic parameters "chemical composition" and "temperature" under ambient pressure conditions. Taking into account the variation of the third parameter "pressure," solid-state chemists can have access to a tremendous field of new compounds. High-pressure experiments usually take place between 0.1 and 300 GPa using a variety of experimental devices, which can be separated into two specific areas: "static" techniques, in which the extreme conditions can be maintained for an arbitrary length of time and "dynamic" techniques, creating high-pressure in the microsecond range. In both cases, the generated pressure can be classified as hydrostatic (directed everywhere), uniaxial (directed in a particular direction), and quasi-hydrostatic (both components). Currently, the highest static pressures in the multimegabar region (1 Mbar = 100 GPa) can be generated with a double stage diamond anvil cell (ds-DAC) [1] reaching values above 750 GPa [2]. Due to the fact that the sample volume under such extreme conditions is extremely small, solid-state chemists prefer devices with an appreciable cell volume for preparative chemistry. In the context of this handbook, an overview of suitable techniques for high-pressure/high-temperature chemical synthesis with reference to the most commonly used techniques will be given. For static high pressures, these are the piston-cylinder apparatus, the belt-module, the Paris-Edinburgh cell (toroidal-anvil apparatus), multianvil devices, and the diamond anvil cell. One chapter will be dedicated to dynamic methods, in

Handbook of Solid State Chemistry, First Edition. Edited by Richard Dronskowski, Shinichi Kikkawa, and Andreas Stein.

particular syntheses and chemical reactions in shock waves. Additionally, some basic principles of high-pressure chemistry are summarized. Finally, several selected examples of new discoveries from this highly interesting field of research accomplish this overview of high-pressure methods.

2.2
Historical Aspects

Before experiments simultaneously carried out at high pressures and high temperatures became interesting for chemists, geologists were interested in reproducing minerals and rocks. So it was not surprising that the geologist Sir James Hall (1761–1832) was the first who subjected limestone to pressures close to 0.1 GPa and temperatures around 600 °C converting it to marble [3]. Hall experimented in sealed gun barrels under addition of some water, thus inventing the first hydrothermal high-pressure autoclaves. His experiments were far ahead of his time and for almost one century following his work no significant progress was made in experimental mineralogy and petrology. In the middle of the nineteenth century, the interest in high-pressure and high-temperature phenomena increasingly inspired physicists and chemists. Percy W. Bridgman (1882–1961) was the most famous pioneer of high-pressure research, and his interest was focused on the effects of pressure on physical properties such as electric conductivity, thermal conductivity, viscosity, melting, and also on reaction kinetics and other material properties. Especially the finding of dramatic drops in the electrical resistance and volume of bismuth, lead, and other metals were carefully documented. These "fixed points" associated with respective phase transitions are still in use today for an internal calibration of high-pressure experimental devices [4]. Another remarkable discovery of Bridgman was the transformation of white phosphorus into black phosphorus using high-pressure as an additional parameter. It was probably the first example for the transformation of a nonmetallic element to its metallic polymorph and at the same time an example of the later formulated "pressure-homologue" rule. It states that at high pressures, elements (and compounds containing these) adopt the same structures as their heavier homologues below them in the periodic table at lower pressures (*vide infra* "Principles of High-Pressure Chemistry"). A major challenge during his work was the development of appropriate technical facilities for the realization of high pressures in combination with high temperatures. The so called Bridgman seals, a special type of seal, in which the pressure in the gasket always exceeds that in the pressurized material so that the closure is self-sealing, are still in use in modern laboratories. In 1946, Bridgman was awarded with the Nobel Prize for physics. Bridgman's work not only laid the foundation for the important artificial diamond synthesis (by General Electric Company in 1955 [5]) but also for other technological developments as they can be found in the following Chapters of this book.

2.3
Static Methods

2.3.1
Piston-Cylinder Apparatus

Markedly, higher volumes under high-pressure are attainable with piston-cylinder apparatuses developed by Boyd and England [6]. A typical assembly of an end-loaded piston-cylinder apparatus (Boyd & England type) is illustrated in Figure 2.1. It consists of a "bomb" with a cylindrical cavity, in which a piston is forced. To sustain the high mechanical stress, the inner parts of the "bomb" as well as the piston are fabricated from tungsten carbide and the outer part of the spherical bomb of prestressed steel. Performing a high-pressure synthesis, the bomb is compressed by an end load in the first step leading to a much higher stability of the bomb. In the second step, the main ram pushes the piston into the cylinder generating pressure on the sample. This arrangement leads to the accessibility of much higher pressures at the sample compared with a simplified

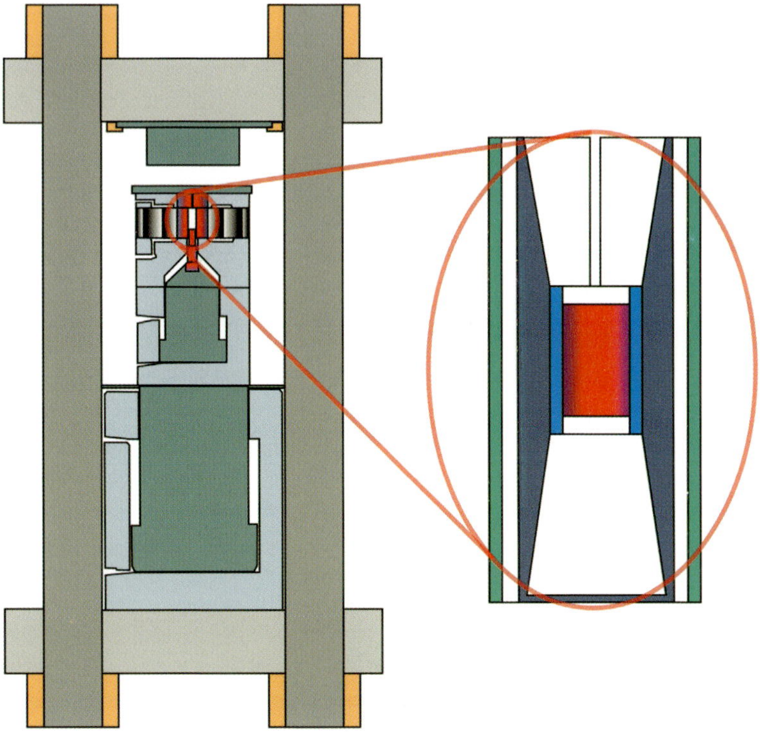

Figure 2.1 End-loaded piston-cylinder apparatus (left) and cross section of the sample assembly (right).

arrangement of a piston-cylinder apparatus without end load [7,8]. Due to the relatively large size of the piston (diameter: 1–2 cm), quite large samples can be reacted under high-pressure and high-temperature conditions up to 6 GPa and 1750 °C. Figure 2.1 (right) shows a cross section of the sample assembly. In the center, the capsule (Au, Pt, Mo, BN, etc.) is positioned in a pyrophyllite cylinder, which is surrounded by a tapered graphite heater. The pressure-transmitting medium inside of the heater is a soft form of alumina. The heater is surrounded by a Pyrex sleeve and additionally by a cylinder of salt (NaCl), which is an ideal pressure-transmitting medium. This arrangement can be adapted in manifold ways due to the requirements of the specific chemical system being investigated. To maintain the stability of the arrangement at high temperatures, the bomb is additionally cooled with water. The temperature is measured via a thermocouple and the pressure is calculated by dividing the force on the piston by the area of the piston taking into account a small reduction due to friction of the solid media. This way is acceptable for using salts as pressure transmitting media during the syntheses. Taking mechanically harder materials, the effect increases, so a calibration using standard methods (see Ref. [8] and Section 2.3.2) is necessary. For solid-state chemists, the piston cylinder apparatus is predestinated for high-pressure syntheses of large quantities of samples in a pressure range up to 6 GPa.

2.3.2
Belt Module

The desire to increase the sample size and pressure over that obtainable in Bundy's apparatus, led Tracy Hall to the development of the "belt" (Figure 2.2). This apparatus can maintain pressures of 10 GPa at temperatures more than 2000 °C for several hours [9,10]. The device makes use of two conical carboloy pistons that push into each end of a specially shaped carboloy chamber. Both, chamber and piston, receive lateral support from two stress-binding rings. In detail, the tungsten carbide pistons and the tungsten carbide core are cold-pressed into the hardened steel plates for radial preload. A safety ring made from mild steel completes each ring. Additionally, all three rings can be cooled with water between the last ring and the safety ring. The pressure is transmitted to the center of the assembly by pyrophyllite. In fact, the hydrous layered alumosilicate pyrophyllite ($Al_2(OH)_2(Si_2O_5)_2$) fulfills three functions. In the first instance, it acts as a pressure-transmitting medium. Second, it serves as a thermal, and third as an electrical insulator. The assembly inside of the rings consists of a boron nitride or a platinum capsule containing the sample, which is surrounded by a graphite heater centered in a pyrophyllite cylinder. Above and underneath, two soft iron rings with inserted corundum discs are placed on thin molybdenum plates to transfer the current from the pistons via the iron rings and the molybdenum plates to the graphite heater. To measure the temperature during the experiment, thermocouple wires can be inserted perpendicular to the graphite heater in a corundum tube. At this point, it has to be considered that the effect of pressure on the electromotive force of the thermocouple affects the

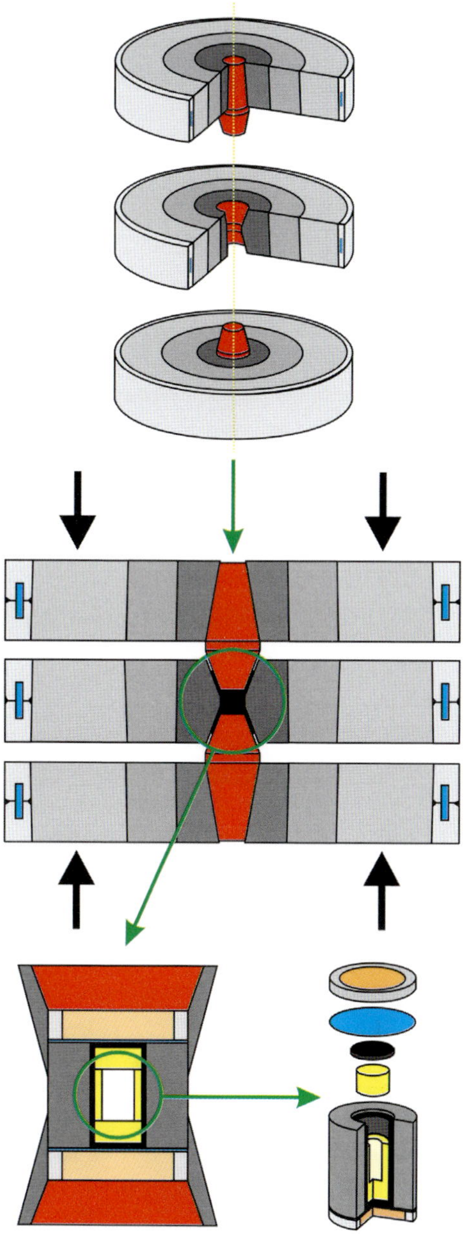

Figure 2.2 Exploded assembly drawing of a belt-apparatus. Top: the belt is built up from two tungsten carbide pistons and a tungsten carbide chamber merging together (center) via a standard press (not shown). Bottom: inner part of the belt with the specimen in a capsule (yellow) surrounded by a graphite heater (black), centered in a pyrophyllite cylinder (gray).

accuracy of the temperature measurement significantly. This is true for all temperature measurements under high-pressure conditions, so a general problem which can be corrected, for example, by the method of Getting and Kennedy [11]. Due to the fact that the applied pressure to the sample has to be corrected due to the typical friction loss of the apparatus, a pressure calibration is necessary, wherefore the phase transitions of Bi I-II (2.55 GPa), Tl II-III (3.68 GPa), Ba I-II (5.5 GPa), and Bi III-IV (7.7 GPa) are internationally accepted phase transformations, which can be used at room temperature. Additionally, a high-temperature calibration is necessary depending on the chemical systems the user is working in. Well-investigated examples for the belt-apparatus are the transformation quartz-coesite [12] or the melting curve of silver [13]. Modifications of the assembly, for example, a sodium chloride-based cell including the specimen, led to an extension of the pressure limit up to a pressure range of 10 GPa [14]. So far, the belt-module is one of the most used devices in preparative high-pressure research.

2.3.3 Toroid Type/Paris-Edinburgh Cell

The Toroid high-pressure device was developed at the Institute for High Pressure Physics of the Russian Academy of Sciences (IHPP RAS) a few decades ago [15]. Incentive was the demand of new synthetic techniques for the production of superhard materials, for example, synthetic diamond and c-BN products, in large-volume chambers being able to resist high pressures (>5 GPa) and high temperatures (>1500 °C) for a long time. As a first developmental stage, the so-called "Chechevitsa" arose from a modification of the original Bridgman anvils by introducing a small recess in the center of an anvil to increase the working volume. At the Toroid type device (Figure 2.3), the central recess was additionally concentrically surrounded by a circular groove, which reduced the pressure

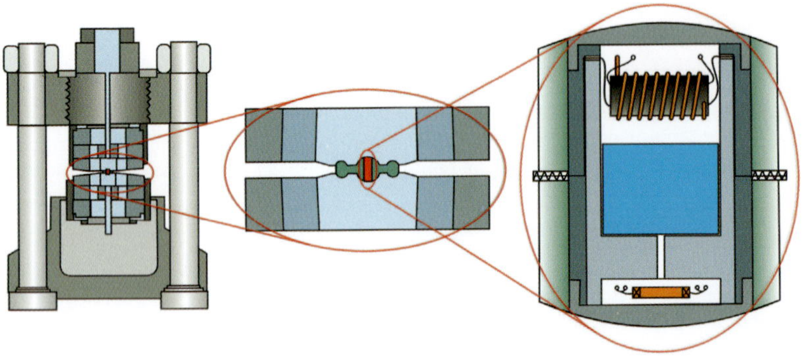

Figure 2.3 Toroid type/Paris-Edinburgh cell (left) with hard metal central region (center) [22]. The right part of the figure shows the central recess with a special sample arrangement containing a liquid and the sample simultaneously [16].

gradients in the working zone and improved the introduction of measuring lead-in wires into the high-pressure zone. Even at maximum pressures, the gap between the anvils remains large enough for various measuring wires or even *in situ* X-ray and neutron diffraction investigations. The maximum achievable pressure of the Toroid devices is depending on the central recess diameters. A chamber of 10 mm in diameter of the Toroid-10 device can be correlated to a maximum pressure of 14 GPa, for the Toroid-15, Toroid-25, Toroid-35, and Toroid-50 devices the corresponding maximum pressures are 12, 10, 9, and 7 GPa, respectively [16]. There exist numerous variations and modifications of Toroid devices, for example, central parts made of polycrystalline diamonds giving access to pressure regions >35 GPa in a volume of 0.1 mm^3 [17]. Pressures of around 25 GPa in volumes of ~10 mm^3 can be attained by using a two-stage device with inserts from diamond compacts [18]. It is also possible to work with a purely hydrostatic pressure up to 10 GPa using a heated capsule, which contains a liquid and the sample simultaneously, being placed in the central recess (Figure 2.3, right). In the Western countries, the Toroid chamber is better known as the so-called "Paris-Edinburgh" cell (Figure 2.3), which is a specific small-scale press machine with a toroidal high-pressure device developed for neutron and X-ray diffraction studies [19–21]. Although the Toroid system is cheap, simple, reliable, and can be installed in a conventional press machine, it was ignored in Western laboratories for a long time. The Toroid cell can also be a meaningful expansion in addition to the belt-apparatus and multianvil devices in existing high-pressure laboratories.

2.3.4 Multianvil Apparatus

2.3.4.1 Multianvil Techniques

Motivated by the advancement of hydrostatic pressure conditions inside of the sample cell, T. Hall constructed a three-dimensional tetrahedral anvil apparatus [9]. Instead of two opposing anvils with circular faces, he used four with triangular faces. Hexahedral devices with trihedral-pyramidal and cubic arrangements (DIA-geometry) of the anvils followed [23,24]. The first octahedral-anvil device using eight anvils segmented from a sphere was developed by Kawai [25]. The sphere was covered with a membrane, to which oil pressure was applied. A substantial improvement in the design was made by Kawai and Endo with double staging of the device, wherein eight cubic anvils were driven by six anvils, segmented from a sphere covered with a rubber membrane. The membrane was removed in a later modification, substituted by rigid driving units. As the two components approach each other, the polyhedral shrinks along the body diagonal. The anvils, contained in the space and in rigid contact with polyhedral surfaces, are thrust to slide and advance toward the center of the assembly. Further developments followed, which are impressively summarized by Onodera [26] and Liebermann [27].

In 1990, Walker *et al.* [28,29] described the development of a simplified multi-anvil device, starting from a previous design of Ohtani [30], who modified the nest for the anvils from a sphere to a long cylindrical cluster of wedges, contained by a pair of separable, massive rings. The modifications of Walker *et al.* led to an unanchored cylindrical cluster of removable tool steel wedges with the wedges clusters ratio of length to diameter less than unity, rather than being a few times greater. This cluster of wedges was completely free to float within the ring and substantial elastic strain of the supporting ring was induced during loading the module. The main aim was to construct a soft-shell cylindrical nest for wedges to drive the standard octahedron-within-cubes payload with a modular device that could be retrofitted into hydraulic presses of many existing piston-cylinder laboratories. These changes of established practices allowed a drastic reduction in hardware cost, bulk, and production difficulties relative to current installations, as well as being small and light enough to be inserted by hand in many standard hydraulic presses. This Walker-module and modified versions of it [31] have found their way in mineral physics, experimental petrology, and solid-state chemistry laboratories around the world, facilitating experimentation under high-pressure conditions up to pressures and temperatures of 26 GPa and 2500 °C. Figure 2.4 gives a view of the general assembly in a Walker-module: at the top (a), a cross section of the octahedral pressure cell is displayed. The sample is directly located in the center of the octahedron. As capsule material, different materials like hexagonal boron nitride, copper, molybdenum, platinum, or gold can be used. The capsule is surrounded by cylindrical resistance heaters, centered in an additional zirconia sleeve for thermal insulation of the furnace against the MgO octahedron. For heating, electrical contact is provided via molybdenum plates; the temperature is measured via a thermocouple. A cubic arrangement of eight inner anvils compress the octahedron (Figure 2.4b). Each anvil consists of a tungsten carbide cube with an edge length of 32 mm and a triangular truncation. Three wedges from the bottom (Figure 2.4c) and three from the top (Figure 2.4d) are located in a containment ring (Figure 2.4e). Two pressure distribution plates on top and bottom of the nest (Figure 2.4f) close the

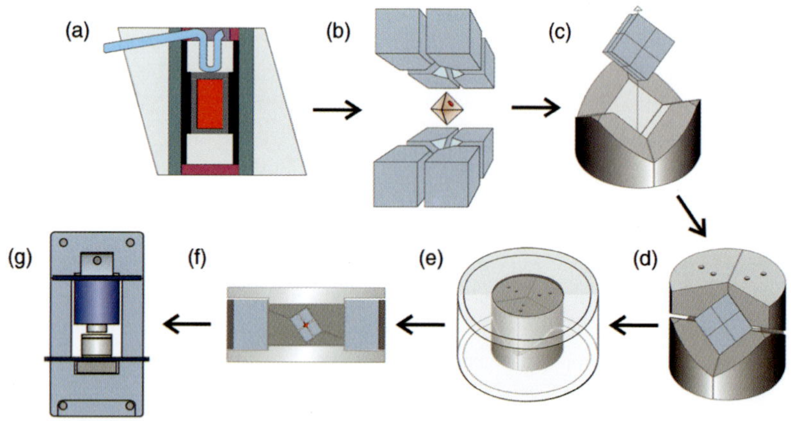

Figure 2.4 Schematic assemblage of a multianvil experiment.

module, which can be loaded with a uniaxial hydraulic press (Figure 2.4g). There exist several different assemblies, which are clearly defined by their octahedral edge length (OEL) and truncation edge length (TEL) of the corresponding tungsten carbide cube. For instance, an 18/11 assembly describes an octahedron with an edge length of 18 mm, including eight tungsten carbide cubes exhibiting truncated triangular faces with an edge length of 11 mm. Typical assemblies for the six–eight type multianvil apparatus are the 25/17 (OEL/TEL), 19/12, 18/11, 14/8, 10/5, 10/4, and 7/3. The sample volume and therewith the size of the octahedron determine the maximum achievable pressure. These values range typically from 12 mm^3 at 4–10 GPa (18/11), 5–8 mm^3 at 11–16 GPa (14/8), 3 mm^3 at 17–21 GPa (10/5), and 1 mm^3 at 22–26 GPa (10/4 or 7/3) giving access to new synthetic fields of solid-state chemistry [32].

2.3.4.2 *In situ* Multianvil Technique

Multianvil devices are mostly used for the synthesis of quenchable metastable phases and sample characterization is accomplished after synthesis. However, certain techniques are also realized for *in situ* measurements. Figure 2.5 gives a schematic view, how the incident X-ray beam (e.g., from a synchrotron source) passes through the gaps between the tungsten carbide cubes in the [110] direction of an eight-cube assembly and diffracts parallel to one of the planes through gaps between the anvils.

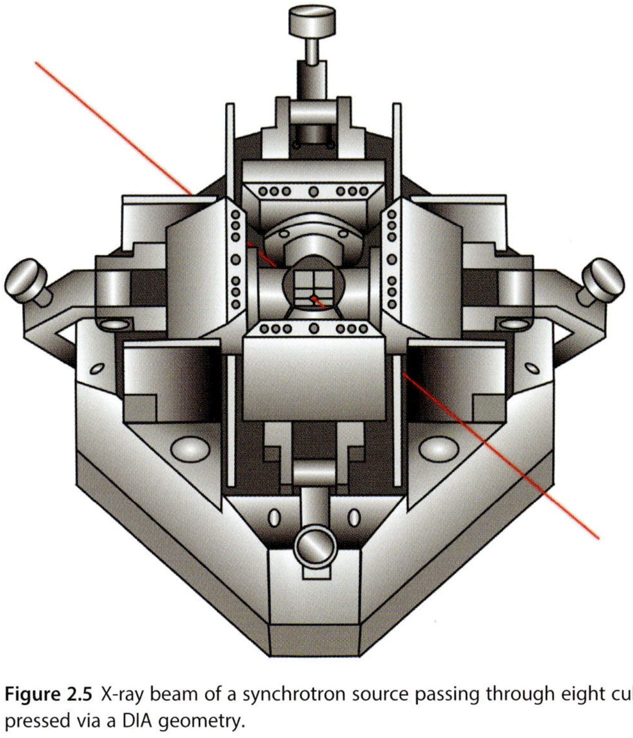

Figure 2.5 X-ray beam of a synchrotron source passing through eight cubes, which are compressed via a DIA geometry.

The majority of experiments concerns p,T phase diagrams of elements like iron [33] or oxide minerals [34,35] because of their relevance for the geochemistry and rheology of the Earth's upper mantle. These investigations are of special importance since the rock-forming minerals tend to deform by ductile flow rather than brittle fracture under the conditions of enhanced pressure and temperature. Thus, mainly seismic velocities, viscosities, and density measurements are valuable information in conjunction with *in situ* measured crystal structure parameters as input data for the theoretical modeling of geophysical properties [36]. Recently, special emphasis is directed to direct observation of uniaxial stress by diffraction techniques. For this purpose, the so-called D-DIA apparatus [37–39] has been developed with the aim to study the effects of controlled uniaxial stress on the mineral samples [40] by incorporating small hydraulic jacks within the upper and lower wedge blocks. The intense light of the synchrotron source is used to obtain X-ray diffraction data finally yielding stress data of the investigated sample.

Although the majority *in situ* multianvil work is presently done in the field of geoscience, *in situ* measurements are also of relevance for material science. In this field, interest is mainly focused on synthesis and crystal growth of new materials. Again, oxides are an important class of materials with often intriguing physical properties [37]. An example of suppressing decomposition of the target phase by application of high pressure is the crystal growth of GaN, which is of relevance as an optoelectronic material. At ambient pressure, the compound decomposes into gallium and nitrogen below the melting point that hampers crystal growth at ambient pressure. By melting the phase at 6 GPa and 2500 K, single crystals of the material were grown by cooling of the melt and can be recovered at ambient conditions [41].

2.3.5 Diamond Anvil Cell

2.3.5.1 Diamond Anvil Cell Technique

The *in situ* measurement of mostly statically compressed condensed phases by X-ray diffraction or spectroscopic methods has largely advanced after the introduction of the Diamond Anvil Cell (DAC) technique. Diamond windows combine the benefits of mechanical hardness with transparency for electromagnetic radiation in many frequency ranges [42]. The DAC is the most used high-pressure device for X-ray diffraction experiments at third generation synchrotron sources, which revealed a plethora of hitherto unknown phenomena, for example, for simple elemental solids like sodium [43,44]. The principle of the DAC is ingeniously simple (Figure 2.6).

A metallic seal is compressed between two opposing diamond anvils with typical diameters ranging from a few microns to roughly one millimeter. In the currently most utilized design, the anvils with a so-called conical crown shape are located in seats, which provide mechanical support and transmit the force from the cell body to the diamond tips. In order to guarantee optimal match of the

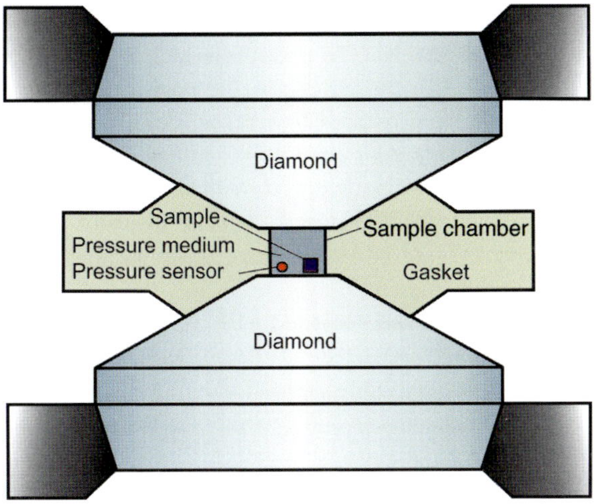

Figure 2.6 Section of the inner elements of the Diamond Anvil Cell (DAC). For pressure generation, force is applied to the seats of the diamonds (see text).

components, seats machined from steel or tungsten carbide and anvils are lapped together [45]. The design of support and anvils in combination with the option for large aperture is beneficial for various types of experiments like X-ray diffraction. Further modification of the anvils gives access to maximal pressures, which nowadays exceed those of the Earth's core [1,46].

For sample confinement, a metallic seal is positioned between the diamonds. Preindentation of the metal foil yields gaskets, which are tightly fitted to the diamond shape and provide lateral support for the diamond anvils. The metallic seal is manufactured from stainless steel or hard metals like rhenium or tungsten. The central hole serving as sample compartment is made by mechanical drilling or laser ablation. In setups, in which the X-ray beam is directed through the gasket, transparent materials with low atomic number like beryllium, boron [47], or boron nitride [48] are chosen.

Within the cylindrical hole serving as a sample chamber, a pressure-transmitting medium is used in order to provide hydrostatic pressure conditions. In some experiments, salts like CsCl or NaCl with low shear strength are employed. However, usually pressure media are selected, which are liquid at ambient conditions and solidify upon pressure increase. The frequently used mixture methanol/ethanol in a ratio 1 : 4 undergoes a transition into a glass at approximately 10 GPa [49] but even at pressures slightly above 6 GPa the samples are subject to some directed pressure components, that is, stress and strain. A reduction of these nonhydrostatic contributions can be achieved by using noble gases like neon or helium as pressure transmitting media. Even after solidification, their low shear strength impedes the transmission of significant strain forces so that they represent almost ideal media [50]. Helium is best suited with respect to its quasihydrostatic behavior even at low temperatures.

For pressure generation, force is applied to the opposing diamond culets, which effectively deform the gasket plastically. The pressure, which is created by the reduction of volume, is transferred to the sample by the pressure medium. The force to create the pressure may be generated by screws [51,52], gas membranes [53], or a suitable combination thereof.

The pressure within the sample chamber is measured by means of calibration substances, which are added together with the sample and the pressure-transmitting medium. The most frequently used pressure sensor is the luminescence of ruby [54,55]. The wavelength of the emission maximum is pressure dependent and the ruby signal is intense enough to be recorded with a standard optical spectrometer. Recently, also the temperature-independent wavelength of the fluorescence maximum of a samarium-doped borate was calibrated [56].

Another method for pressure calibration is to add calibration substances with precisely determined equations of state, for example, salts [57] or metals with simple crystal structure like copper, platinum, gold, or tantalum [58,59], and to determine the lattice parameters by diffraction experiments.

Heating in diamond-anvil cells is essential for material synthesis and the investigation of p,T diagrams. It can be achieved by heating the anvils involving resistive heaters around the diamonds or the pressure cell. Here, the temperature range is limited by the disintegration of diamond in air. The most commonly alternative method especially for higher temperatures is laser heating.

In these experiments, temperature measurement was initially done by means of an optical pyrometer to measure the intensity of the light, which was emitted by the sample. Later, determination of the black body radiation enabled more accurate measurements of the temperature. However, the small hot spot produced by the laser has the disadvantage of creating large thermal gradients within the sample. The solution to this problem is ongoing but advances have been made with the introduction of double-sided laser heating devices [60–62]. The simultaneous availability of the essential parameters temperature and pressure in combination with X-ray diffraction already opened new perspectives in material and geoscience [63,64].

2.4 Dynamic Methods

2.4.1 Shock Wave Syntheses

Application of dynamic pressure is a powerful tool to attain thermodynamic conditions and loading–unloading pathways not offered by any other technique. It thus can create exotic states of matter and modify materials in a unique way, but also produce "regular" high-pressure phases and even commercially applicable materials such as diamond and cubic boron nitride in much larger amounts then in a single static synthesis run.

While static pressure methods, such as those treated in this chapter so far, are ultimately based on the strength of their anvils, pistons, plungers, gaskets, autoclave walls, and so on, dynamic compression of matter is based on the "creation" and transfer of momentum – and often accomplished in the complete or partial absence of strength within the sample and its immediate environment (hydrodynamic regime). In dynamic experiments, the true backing "behind" the sample allowing for its rapid compression down to one half or an even smaller fraction of its original volume (V/V_0) is given by its own inertia, experienced by any (accelerated) massive object or particle.

The most prominent dynamic method is the so-called shock wave compression, where a sample material is subject to an almost instantaneous rise of intensive parameters such as pressure p, density ρ, and velocity U_p of its constituting particles as shown in Figure 2.7 [65,66]. The velocity of the frontline (shock front) of this discontinuity U_s is faster than the speed of sound c_0 of the undisturbed medium. As disturbances in an already compressed medium travel faster than in a dilute one, steep wave profiles (i.e., shocks) will eventually emerge in all materials within a broad range of dynamic loading conditions (see, e.g., Ref. [65], chapters 4 and 5 therein).

Shock waves therefore occur in gases, liquids, and solids. They are often accompanied with dramatic changes in the state of the shocked material, such as shock melting [67] or the conversion of a solid high explosive into a mixture of highly compressed and hot gases across the rapidly moving detonation front.

Figure 2.8 shows schematic sketches of typical setups for planar impact [68] (Figure 2.8a) and coaxial impact [66] (Figure 2.8b) shock recovery experiments on solids (and also on liquids [69]), using explosives for the creation and a

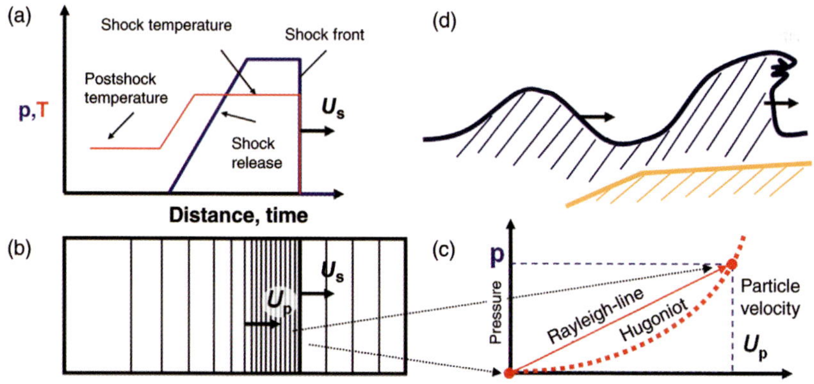

Figure 2.7 Schematic profiles of a shock wave: (a) pressure p and temperature T, (b) and density. The shock front moves with U_s, while the particle velocity U_p in the shocked material is smaller ($U_s \approx c_0 + b \cdot U_p$) [65,66]. (c) p-U_p diagram: initially resting particles are spontaneously accelerated to U_P (Rayleigh line). For a given material, each U_p leads to a unique corresponding value of p. The locus of all U_p–p pairs is called the Hugoniot (red dottet line). (d) Breaking of ocean waves as they enter into shallow water. Similar effects lead to steepening of wave profiles and emergence of shocks in all kinds of dynamically loaded matter.

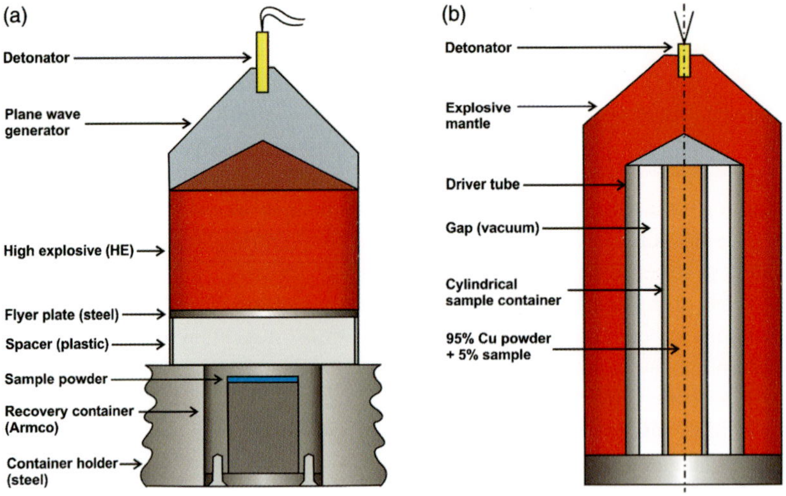

Figure 2.8 Typical setups for (a) planar impact (flyer plate) [68] and (b) coaxial impact [66] shock recovery experiments.

projectile (flyer plate, driver tube) for the transfer of momentum, respectively. A comprehensive list of techniques is presented further below.

For the solid-state chemist, crystalline–crystalline [67,70,71] and crystalline–amorphous [72] phase transitions, shock-induced defects [73], and chemical reactions between solids [74,75] should be the most relevant issues, with a few exemplary examples given here.

Diffusionless (topotactical, martensitic) phase transitions [70,71] and chemical decompositions, such as the decomposition of calcite ($CaCO_3 \rightarrow CaO + CO_2$) [76], are easily anticipated to proceed at high rates with respect to reaction time and completeness. However, it has been found that also the kinetics in reactive powder mixtures such as Ni + Ti, Mo + 2 Si [74], Zn + S [75], and at bulk interfaces [76] can be up to several orders of magnitude larger than would be expected from regular diffusion [75,76]. Reactions are found to be driven in "opposite" direction (e.g., $Ta + SiO_2 \rightarrow Si + Ta_2O_3$ [76]).

In "shock-driven" reactions, which proceed as fast as the shock itself (rather than after shock initiation in the superheated zone behind it), extensive microjetting is considered as one, if not *the* origin for these phenomena.

Similar to the macroscopic effect of an armor-piercing shaped-charge (Figure 2.9a)), millions of microjets may originate from inevitable inhomogeneities [74,75] and voids [77] (Figure 2.9c) within a prepressed powder mixture. Even in the absence of voids, a contrast between the acoustic impedance[1)] of the two materials will lead to different local particle speeds U_p and therefore extensive intermixing [75].

This also sheds light on the complex role of metal powder additives (Figure 2.9b and c) first introduced in the 1960s as cooling agent and impedance

1) Acoustic impedance $Z = c_0 \cdot \rho_0$, with ambient pressure sound velocity c_0 and density ρ_0, respectively. In acoustics and shock wave physics, it plays a similar role as the refractive index in optics.

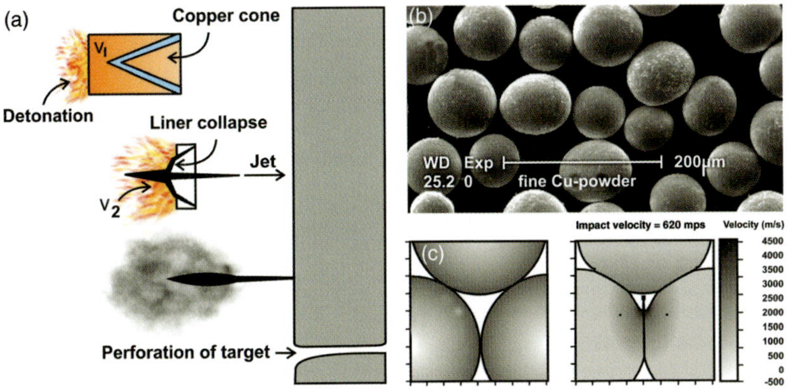

Figure 2.9 (a) Armor piercing action of a shaped charge [65]. The explosively driven collapse of a metal cone forms an up to 10 km/s fast material jet with extreme penetration capability. (b) Copper beads typically used as additive in shock wave synthesis [78]. (c) Microjetting in shock-loaded metal beads [77].

modifiers to enhance the yield of shock-synthesized diamond [79] and c-BN [70], which are successfully employed in the synthesis of high-pressure phases up to the present days [78,80].

The importance of microjetting in shock wave synthesis can thus barely be underestimated and needs intensive further investigation [81]. It becomes clear that a significant volume fraction of a shocked heterogeneous solid can experience much more severe conditions than the average bulk pressures and temperatures usually given in the experimental reports. These are calculated from the (tabellated) Hugoniot curve parameters of its constituents, as established from prior $U_s - U_p$ measurements (e.g., Ref. [65] chapter 5 and Ref. [66] chapter 1).

A rich variety of dynamic high-pressure techniques exists. In order to keep an overview, it is useful to classify them according to the following criteria.

Loading geometry:

a) planar (*uniaxial*) [65,82],
b) convergent [66] or
c) divergent,
d) shock wave [65–68,70–82] or
e) ramp wave (*quasi-isentropic*) compression [83–85].

Loading history:

f) single (*impedance method*) or
g) multistage (*reflection method*) compression [86].

Pulse wave creation (type of *momentum transfer*):

h) collision [65,66,82,86] or
i) direct contact [66,82,85].

Acceleration system:

j) high explosive [65,66,82,86],
k) gun [65,82,86],

l) laser pulse [83],
m) magnetic pulse [84,85],
n) electric discharge and other [86].

Experiment purpose:
o) sample recovery (synthesis) of solids [66,82,86] and liquids [69],
p) *in situ* diagnostics (e.g., *equation of state* (*EoS*) measurements) [66,82–86].

Figure 2.8a and b depicts realizations of *a-d-f/g-h-j-m* and *b-d-f-h-j-m*, respectively, and are two of the most popular setups for shock wave synthesis, where the sample is recovered after the treatment and can be subject to post-mortem investigations with conventional analytical methods [73,78,80]. Planar impact *a-d-h*, that is, uniaxial compression (Figure 2.8a) is the preferred scientific method with respect to the degrees of freedom in experimental design and simplicity of calculation of the average bulk values of p, (V), and T and their time history (i.e., loading path). As a starting point for this, Ref. [82] can be well-recommended to the novice. Coaxial impact *b-d-h* is suitable for (commercial) synthesis of larger amounts, but requires careful precautions against detrimental mach effects as the shock front converges along the cylinder axis [66]. Calculations are more complex and often carried out via numerical simulations using hydrocodes [87].

Figure 2.10 exemplifies the difference in loading paths and degree of energy deposition between *f* and *g* up to the same nominal maximum peak pressure according to Refs [86,88]. It signifies that care must be taken, when comparing different results from literature only in terms of a given pressure value if the method, and hence the full *p-T-t* history, are neglected or not disclosed [88].

In both cases, a flyer plate (yellow) strikes a sample recovery container (gray-blue) made of a material with high impedance Z_c (e.g., iron or stainless steel) surrounding a sample of sandstone, (i.e., porous quartz) with lower impedance Z_s. For the following, it is important to note that compressional shock waves (solid arrows in the right panel of a and b) are always transmitted forward, but can be (partially) reflected backwards at a boundary, when entering a material of higher impedance. Crossing a boundary from higher to lower impedance leads to a reflection of a sound wave (dotted arrows). In (a), the sample consists of a thin disc of sandstone with thickness smaller than that of the flyer, located in the upper part of the container. It is subject to a step-wise compression by consecutive shock reflections of the primary shock front with pressure P_0 emanating from the collision interface. The pressures in each step can be calculated from the intersections between the transmission and reflection Hugoniots of the sample and container according to the impedance-match technique [82], the graphic solution [86,88] of which is shown on the left hand side of Figure 2.10c. The authors in [88] model the initial compression of the sandstone sample to 4.7 GPa by its so-called principal Hugoniot (dashed line) and all following steps

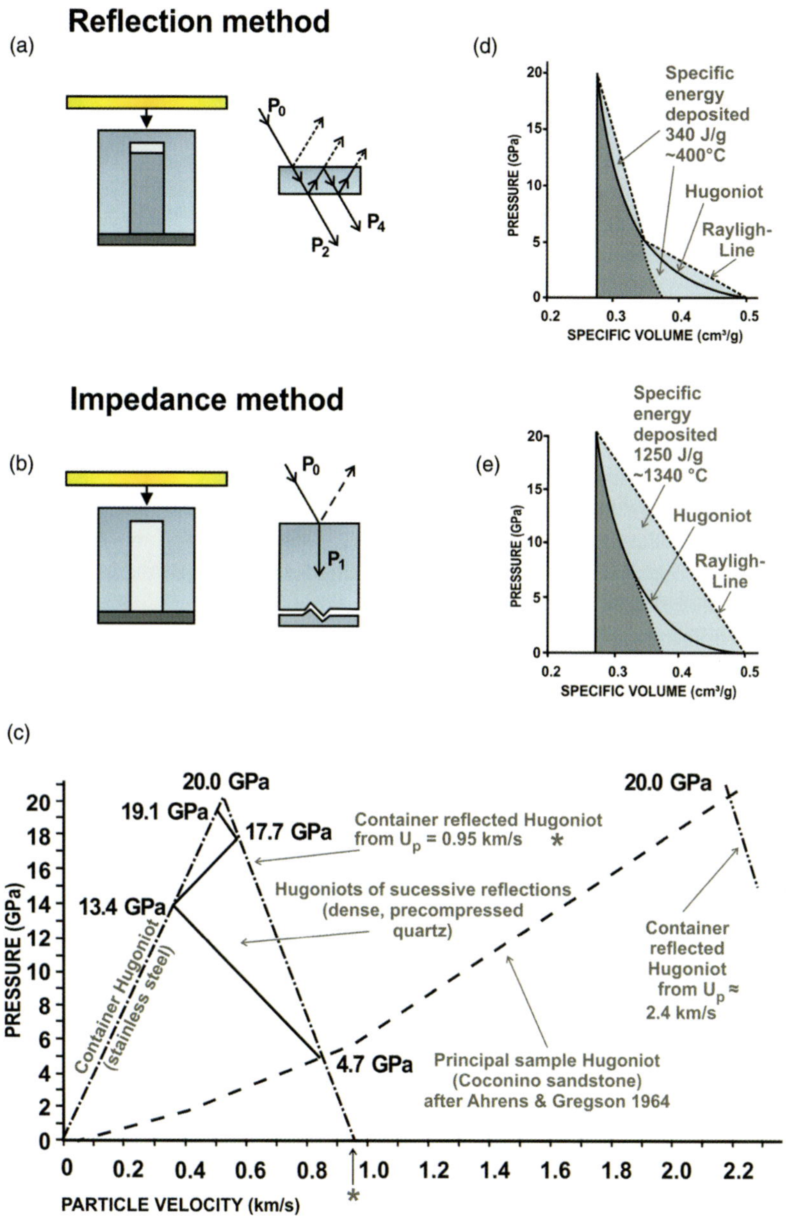

Figure 2.10 Example according to Ref. [88] for the different loading paths and degrees of energy deposition in planar impact shock wave recovery experiments with the same maximum peak pressure of 20 GPa, carried out by (a) reflection method (b) impedance method (see text).

by the Hugoniots for dense and precompressed quartz (solid lines), assuming that all porosity is essentially crushed in the first step. Due to the constructive interference of the multiple shock reverberations between the container lid above and the piston below the sample, an initial particle velocity of 0.95 km/s of the container lid (*-mark at the abscissa of Figure 2.10c) and a corresponding flyer velocity [82] of ~1.9 km/s is sufficient to achieve a terminal pressure of 20 GPa. The popularity of the reflection method stems from the fact that this terminal pressure is given by the intersection between transmission and reflected Hugoniot[2] of the container material alone, so that details about the shock compression behavior of the sample are apparently not required for its computation. However, this way of thinking grossly ignores the energy deposited in the sample in the form of mechanical work $\int \mathrm{d}p\,\mathrm{d}V$, and hence it's shock- and postshock temperature. From the p–V diagrams (d) and (e) it becomes obvious that the proportion of nonrecoverable mechanical work (light gray areas), which will be dissipated as thermal energy is significantly smaller in the reflection case with step-wise compression (here approximated by two steps) than if the same terminal pressure is achieved in a single compression step. This is because the intermediate shocked states have to lie on the respective Hugoniots (Figure 2.10d), so that the Rayleigh lines leading to them from the previous state cut out a smaller triangular area $\approx \frac{1}{2}\,\Delta p \cdot \Delta V$.

The conditions for the single step compression/impedance method (Figure 2.10b) are given for a sample with thickness in the same order or larger than the thickness of the flyer (here represented by a long sandstone cylinder filling the whole sample container), for which no backward shock reflection at high-impedance boundaries during passage of the whole shock pulse occurs. In order to know the pressure in experiments conducted with this method, the principal Hugoniot of the substance is either to be known beforehand, measured within the experiment, or at least approximated by mixing the sample with a large excess of a known pressure medium, such as the aforementioned metal powder additives. From the chosen sandstone example in Ref. [88], the difference in energy deposition between reflection and impedance method becomes now easily understandable from Figure 2.10c: For a pressure of 20 GPa in initially porous quartz by a single shock, its principal Hugoniot (dashed line) has to intersect with the reflected container Hugoniot at this pressure of 20 GPa, thus affording much higher particle velocities within the sample, the container cover plate (2.4 km/s), and hence for the flyer plate (4.8 km/s if made from the same material as the cover). The flyer therefore simply delivers more kinetic energy to the system. It becomes thus clear that samples treated by either the single shock or the reflection method could, for example, differ dramatically in the yield of high-pressure phase formed, although having been compressed to the same maximum pressure. High temperatures created by single shocks may promote reconstructive phase transitions, but will also lead to high postshock temperatures and thus to back-conversion. In order to make a

2) often used as an approximation for the release adiabat [82,88].

reasonable estimate of the loading path and temperature, knowledge about the principal (and higher order) Hugoniots are required for both methods. Moreover, as previously noted, the *p* and *T* values derived on this basis refer to the average bulk pressure and temperature, while the true local conditions in a heterogeneous solid can be much higher – a problem in shock wave chemistry and physics which is yet to be solved.

Finally, it remains to be pronounced that due to the steepening of the shock wave profile (cf. Figure 2.7), the conditions reached in any shock wave (c.f. Loading geometry *d*) experiment are more or less limited to the Hugoniot that is one discrete line within the *p*-*T*-*V* parameter. So called off-Hugoniot states (cf. Figure 2.7c) can be reached by so called ramp wave (*e*) experiments. Here, the momentum transfer to the sample is applied gradually in such a way that the steepening of the compression front into a single shock is avoided, allowing for a quasi-isentropic compression [85]. The majority of these experiments are of type *p*, where the dynamical response and not the recovered material (if any) are at the focus of interest. They are carried out with miniaturized samples, using high-amplitude lasers (i.e., *l*) [83] or pulsed magnetic fields (*m*) [84,85] and sophisticated *in situ* diagnostics with picosecond time resolution. Macroscopic ramp experiments can be performed by modifying conventional shock wave generators with buffers in front of the impactor or flyer plates with graded density [85]. A general review on time-resolved diagnostics for shock waves is given in Ref. [89].

2.5 Principles of High-Pressure Chemistry

All matter of the Universe, and in consequence its properties, are influenced by the parameters pressure and temperature. Normally, these two parameters interact in opposite directions, which means that increasing pressure at constant temperature leads to a transformation from the normal-pressure phase to a high-pressure phase, while the back-transformation to the normal-pressure phase is possible by increasing the temperature at constant pressure.

The above mentioned technologies permit the study of matter under pressures of several hundred gigapascal. By application of such harsh conditions, the properties of elements and compounds differ drastically from those known at ambient pressure. A hierarchy of responses to pressure can be specified starting with molecular crystals [90]. Primarily, a large volume decrease occurs at the early stages of squeezing (less than 10 GPa), when the so-called "van der Waals space" is compressed. In order of increasing energy (= increasing pressure), ionic and covalent structures (molecular or extended) respond to pressure by electron-density rearrangements leading to an extension of the coordination spheres of the atoms. For example, linear CO_2 molecules transform into a three-dimensional extended solid (CO_2-V; >35 GPa and 1800 K) with carbon close to tetrahedral coordination and quartz-like structure [91]. Neuhaus [92] and later

Prewitt and Downs [93] summarized the observations of different structural changes at high pressures in a set of useful thumb rules: for example, “Increasing pressure increases coordination number.” and “Elements behave at high pressures like the elements below them in the periodic table at lower pressures.” The result of an increasing coordination sphere of the atoms in a compound is a densification of the material. In the case of ionic crystals, the anions are more compressible than the cations. As a consequence, the coordination numbers of the cations grow distinctly with increasing pressure. But what happens if close packing is not close enough? Cubic and hexagonal closed packing of equally sized spheres show a space-filling of 74% and there is a lot of room to fill, if the spheres are sized unequally. So applying high pressure can cause an electronic disproportionation of the atoms A into sublattices of $(A^{\delta+})_m(A^{\delta-})_n$ of any dimensionality, or it causes nonclassical deformation of spherical electron densities [94]. Finally, both phenomena result in a densification of the material.

On the level of covalent bonding, the nuclei are moved closer to one another and the repulsive forces increase with growing external pressure. To counteract this, the electron density between the nuclei must be increased – the bond becomes more covalent [93]. Multicenter bonding, electron-rich or electron poor, is also a possible mechanism for compactification at elevated pressures. Under moderate up to extremely high-pressures, electrons may occupy orbitals, which are normally not involved. For example, EuO shows at pressures larger than 30 GPa an unexpected 4f → 5d electron transfer leading to a pressure-induced metallization [95]. This can be formulated as a general rule: “all materials become metallic under sufficiently high-pressure” [90]. All these rules are thumb rules and often exceptions can be found. For example, recent high-pressure experiments at the element sodium showed that at densities corresponding to a ~5.0-fold compression (at pressures exceeding 200 GPa) the pressure-metallization rule is no longer valid or rather has to be extended. Under these extreme conditions, an overlap of core electrons can be observed, which alters the electronic properties from those of a simple free-electron metal dramatically. Sodium becomes optically transparent and insulating at pressures of ~200 GPa [44]. So, chemistry under pressure is extraordinary but the effects are explainable, though not always predictable. For a more detailed view of the effects of the parameter pressure to matter, the reader is referred to the article of Grochala *et al.* [90] giving an excellent introduction into the field.

2.6
Selected Examples of Chemistry Under High-Pressure Conditions

The discovery of new polymorphs of known chemical compounds is an important field in solid-state chemistry mainly investigated via low- and high-temperature investigations under ambient pressure conditions. Taking into account the above mentioned methods, this field of research can be extended by the additional parameter pressure giving access to a variety of high-pressure polymorphs of elements (allotropes) or compounds. Certainly, the most famous high-pressure transformation is the synthesis of artificial diamond from graphite [5]. Based on this success, high-pressure investigations on elements and chemical compounds became more and more an important factor in solid-state chemistry. The recent publications on the transformation of hydrogen into a metal at 260–270 GPa [96], a transparent modification of the simple alkali metal sodium (~200 GPa) [44], or the finding of conventional superconductivity at 203 K at 90 GPa in the sulfur hydride system [97,98] represent remarkable results of high-pressure transformations, which were all obtained from investigations in the diamond anvil cell.

Also in the field of static high-pressure experiments with larger devices, a multitude of new synthetic achievements can be found in the literature. Some recent reviews cover several aspects of the production of new oxides, nitrides, intermetallics, and superhard materials under high pressures [99–105] like novel modifications of binary compounds with so-far unusual coordination numbers and significantly increased hardness (γ-Si_3N_4 [106] and γ-P_3N_5 [107]). Also, cage compounds and materials with framework structures are synthesized under high-pressure conditions, for example, nitridophosphates and phosphorus nitride imides like $P_4N_6(NH)$ [108] and HP_4N_7 [109] as well as superconducting silicides, germanides, and cuprates [110–114].

In the field of borates, high-pressure experiments in a standard multianvil equipment (Walker-type module) led to a remarkable extension of the compositional and structural possibilities. Especially in the series of trivalent rare earth borates, the original number of well-known ambient pressure phases could be increased by a multitude of new compounds with new compositions [115]. Additionally, several of these new high-pressure borates exhibited the structural motive of edge-sharing BO_4 tetrahedra, which was unknown in the structural chemistry of borates [116–118]. Interestingly, high-pressure conditions led also to the synthesis of crystalline borates in systems, which are well known as glass-forming systems. For example, experiments in the ternary glass-forming systems Zr–B–O, Hf–B–O, and Sn–B–O led to the pressure-induced crystallization of the new compounds ZrB_2O_5, HfB_2O_5, and β-SnB_4O_7, respectively [119–121]. Furthermore, high-pressure conditions opened an easy access into the new class of ammine borates with the compound $Cd(NH_3)_2[B_3O_5(NH_3)]_2$ being the first representative [122].

In the area of shock wave treatments, a new halide-based method was developed for the recovery of fluid-rich and decomposition-sensitive phases in the pressure range of 25–162 GPa [123]. The advantage of this method is the consequent avoidance of undesirable shock wave reflections and adiabatic decompression.

Another example is a new synthetic route to the high-pressure phase of aluminum nitride with rocksalt structure (rs-AlN) as a highly interesting ceramic material. In contrary to microcrystalline AlN, high yields of rs-AlN could be achieved by starting from (commercial) wurtzite-type AlN nanopowders subjected to multiple shock reflections [68]. One of the most striking advantages of shock wave experiments is the access to large quantities of high-pressure phases [124]. Köhler *et al.* [125] demonstrated the experimental shock wave synthesis of high-pressure spinel-type γ-$Si_{3-y}N_{4-x}O_x$ from amorphous precursors at peak shock pressures around 34 GPa yielding up to 8 g high pressure phase per shot. The employed recovery system allows for a complete sample recovery from experiments in the megabar range with quantities of synthesized high-pressure phases in the multigram scale – giving impressive prospects for an upscaled synthesis of material so far made via static high-pressure experiments.

2.7 Conclusion

High-pressure solid-state chemistry enjoys an increasing popularity due to the fact that more and more laboratories possess the experimental possibilities to perform syntheses under such extreme conditions. Nevertheless, the technical possibilities are huge. So this chapter should have given a brief impression of available high-pressure techniques to give a decision-making support to scientists, which technique could be the right one for the specific scientific task. Due to the fact that in most cases the parameter pressure opens up new synthetic conditions with an access to new allotropes, polymorphs, or compounds with unusual compositions, this synthetic way is one of the most promising approaches for the discovery of new materials.

Acknowledgment

The authors would like to thank Mr. Abraham Siedler (Universität Innsbruck) for the design or modification of the numerous figures of this chapter.

References

1 Dubrovinsky, L., Dubrovinskaia, N., Prakapenka, V.B., and Abakumov, A.M. (2012) *Nat. Commun.*, **3**, 1163.
2 Dubrovinsky, L., Dubrovinskaia, N., Bykova, E., Bykov, M., Prakapenka, V., Prescher, C., Glazyrin, K., Liermann, H.P., Hanfland, M., Ekholm, M. *et al.* (2015) *Nature*, **525**, 226–229.
3 Keppler, H. and Frost, D.J. (2004) Introduction to minerals under extreme conditions, in *EMU Notes in Mineralogy*, Mineralogical Society of America.
4 Bridgman, P.W. (1952) *The Physics of High Pressures*, G. Bell & Sons, London.
5 Bundy, F.P., Hall, H.T., Strong, H.M., and Wentorf, R.H. (1955) *Nature*, **176**, 51–55.
6 Boyd, F.R. and England, J.L. (1960) *J. Geophys. Res.*, **65**, 741–748.
7 Johannes, W. (1973) *Neues Jahrb. Miner. Monatsh.*, **1973**, 337–351.
8 Johannes, W., Bell, P.M., Mao, H.K., Boettcher, A.L., Chipman, D.W., Hays, J.F., Newton, R.C., and Seifert, F. (1971) *Contrib. Mineral. Petrol.*, **32**, 24–38.
9 Hall, H.T. (1958) *Rev. Sci. Instrum.*, **29**, 267–275.
10 Hall, H.T. (1960) *Rev. Sci. Instrum.*, **31**, 125–131.

11 Getting, I.C. and Kennedy, G.C. (1970) *J. Appl. Phys.*, **41**, 4552.

12 Brey, G.P., Weber, R., and Nickel, K.G. (1990) *J. Geophys. Res.*, **95**, 15603–15610.

13 Mirwald, P.W. and Kennedy, G.C. (1979) *J. Geophys. Res.*, **84**, 6750–6756.

14 Kanke, Y., Akaishi, M., Yamaoka, S., and Taniguchi, T. (2002) *Rev. Sci. Instrum.*, **73**, 3268–3270.

15 Brazhkin, V.V., Lyapin, A.G., Popova, S.V., Voloshin, R.N., Antonov, Y.V., Lyapin, S.G., Kluev, Y.A., Naletov, A.M., and Mel'nik, N.N. (1997) *Phys. Rev. B*, **56**, 11465–11471.

16 Khvostantsev, L.G., Slesarev, V.N., and Brazhkin, V.V. (2004) *High Press. Res.*, **24**, 371–383.

17 Evdokimova, V.V., Kuzemskaya, I.G., Pavlov, S.P., and Modenov, V.P. (1976) *High Temp. High Press.*, **8**, 705.

18 Bilyalov, Y.R., Kaurov, A.A., and Tsvyashchenko, A.V. (1992) *Rev. Sci. Instrum.*, **63**, 2311.

19 Besson, J.M., Nelmes, R.J., Loveday, J.S., Hamel, G., Pruzan, P., and Hull, S. (2006) *High Press. Res.*, **9**, 179–193.

20 Besson, J.M., Hamel, G., Grima, T., Nelmes, R.J., Loveday, J.S., Hull, S., and Häusermann, D. (1992) *High Press. Res.*, **8**, 625–630.

21 Besson, J.M., Nelmes, R.J., Hamel, G., Loveday, J.S., Weill, G., and Hull, S. (1992) *Phys. B*, **180–181**, 907–910.

22 Shen, G. and Wang, Y. (2014) *Rev. Mineral. Geochem.*, **78**, 745–777.

23 Contre, M. (1969) *High Temp. High Press.*, **1**, 339–356.

24 Yagi, T. and Akimoto, S.I. (1976) *J. Appl. Phys.*, **47**, 3350–3354.

25 Kawai, N. (1966) *Proc. Jpn. Acad.*, **42**, 385–388.

26 Onodera, A. (1987) *High Temp. High Press.*, **19**, 579–609.

27 Liebermann, R.C. (2011) *High Press. Res.*, **31**, 493–532.

28 Walker, D., Carpenter, M.A., and Hitch, C.M. (1990) *Am. Mineral.*, **75**, 1020–1028.

29 Walker, D. (1991) *Am. Mineral.*, **76**, 1092–1100.

30 Ohtani, E., Irifune, T., Hibberson, W.O., and Ringwood, A.E. (1987) *High Temp. High Press.*, **19**, 523–529.

31 Huppertz, H. (2004) *Z. Kristallogr.*, **219**, 330–338.

32 Rubie, D.C. (1999) *Phase Trans.*, **68**, 431–451.

33 Uchida, T., Wang, Y., Rivers, M.L., and Sutton, S.R. (2001) *J. Geophys. Res.*, **106**, 21799–21810.

34 Suzuki, A., Ohtani, E., Morishima, H., Kubo, T., Kanbe, Y., Kondo, T., Okada, T., Terasaki, H., Kato, T., and Kikegawa, T. (2000) *Geophys. Res. Lett.*, **27**, 803–806.

35 Inoue, T., Irifune, T., Higo, Y., Sanehira, T., Sueda, Y., Yamada, A., Shinmei, T., Yamazaki, D., Ando, J., Funakoshi, K. *et al.* (2006) *Phys. Chem. Miner.*, **33**, 106–114.

36 Liebermann, R.C., Chen, G., Li, B., Gwanmesia, G.D., Chen, J., Vaughan, M.T., and Weidner, D.J. (1998) *Rev. High Press. Sci. Technol.*, **7**, 75–78.

37 Durham, W.B., Weidner, D.J., Karato, S.-I., and Wang, Y. (2002) *Rev. Mineral. Geochem.*, **51**, 21–49.

38 Wang, Y., Durham, W.B., Getting, I.C., and Weidner, D.J. (2003) *Rev. Sci. Instrum.*, **74**, 3002–3011.

39 Guignard, J. and Crichton, W.A. (2015) *Rev. Sci. Instrum.*, **86**, 085112.

40 Li, L. (2009) *Prog. Nat. Sci.*, **19**, 1467–1475.

41 Utsumi, W., Saitoh, H., Kaneko, H., Watanuki, T., Aoki, K., and Shimomura, O. (2003) *Nat. Mater.*, **2**, 735–738.

42 van Valkenburg, A. (1964) *Diamond Res.*, **1**, 17–20.

43 Gregoryanz, E., Lundegaard, L.F., McMahon, M.I., Guillaume, C., Nelmes, R.J., and Mezouar, M. (2008) *Science*, **320**, 1054–1057.

44 Ma, Y., Eremets, M., Oganov, A.R., Xie, Y., Trojan, I., Medvedev, S., Lyakhov, A.O., Valle, M., and Prakapenka, V. (2009) *Nature*, **458**, 182–185.

45 Boehler, R. and de Hantsetters, K. (2004) *High Press. Res.*, **24**, 391–396.

46 Goettel, K.A., Mao, H.K., and Bell, P.M. (1985) *Rev. Sci. Instrum.*, **56**, 1420–1427.

47 Lin, J.-F., Shu, J., Mao, H.K., Hemley, R.J., and Shen, G. (2003) *Rev. Sci. Instrum.*, **74**, 4732–4736.

48 Funamori, N. and Sato, T. (2008) *Rev. Sci. Instrum.*, **79**, 053903.

49 Piermarini, G.J., Block, S., and Barnett, J.D. (1973) *J. Appl. Phys.*, **44**, 5377–5382.
50 Takemura, K. (2001) *J. Appl. Phys.*, **89**, 662–668.
51 Merrill, L. and Bassett, W.A. (1974) *Rev. Sci. Instrum.*, **45**, 290–294.
52 Adams, D.M., Christy, A.G., and Norman, A.J. (1993) *Meas. Sci. Technol.*, **4**, 422–430.
53 Letoullec, R., Pinceaux, J.P., and Loubeyre, P. (2006) *High Press. Res.*, **1**, 77–90.
54 Forman, R.A., Piermarini, G.J., Barnett, J.D., and Block, S. (1972) *Science*, **176**, 284–285.
55 Piermarini, G.J., Block, S., Barnett, J.D., and Forman, R.A. (1975) *J. Appl. Phys.*, **46**, 2774–2780.
56 Datchi, F., Letoullec, R., and Loubeyre, P. (1997) *J. Appl. Phys.*, **81**, 3333–3339.
57 Decker, D.L. (1971) *J. Appl. Phys.*, **42**, 3239.
58 Mao, H.K., Bell, P.M., Shaner, J.W., and Steinberg, D.J. (1978) *J. Appl. Phys.*, **49**, 3276.
59 Hanfland, M., Syassen, K., and Köhler, J. (2002) *J. Appl. Phys.*, **91**, 4143–4148.
60 Shen, G., Rivers, M.L., Wang, Y., and Sutton, S.R. (2001) *Rev. Sci. Instrum.*, **72**, 1273–1282.
61 ChandraShekar, N.V., Sahu, P.C., and Govinda Rajan, K. (2003) *J. Mater. Sci. Technol.*, **19**, 518–525.
62 Schultz, E., Mezouar, M., Crichton, W., Bauchau, S., Blattmann, G., Andrault, D., Fiquet, G., Boehler, R., Rambert, N., Sitaud, B. *et al.* (2005) *High Press. Res.*, **25**, 71–83.
63 Eremets, M.I., Gavriliuk, A.G., Trojan, I.A., Dzivenko, D.A., and Boehler, R. (2004) *Nat. Mater.*, **3**, 558–563.
64 Anzellini, S., Dewaele, A., Mezouar, M., Loubeyre, P., and Morard, G. (2013) *Science*, **340**, 464–466.
65 Meyers, M.A. (1994) *Dynamic Behavior of Materials*, John Wiley & Sons, Inc, New York.
66 Batsanov, S.S. (1994) *Effects of Explosions on Materials*, Springer, New York.
67 German, V.N., Mikhailov, A.L., Osipov, R.S., and Tsyganov, V.A. (2000) *AIP Conf. Proc.*, **505**, 247–250.
68 Keller, K., Schlothauer, T., Schwarz, M., Heide, G., and Kroke, E. (2012) *High Press. Res.*, **32**, 23–29.
69 Leighs, J.A., Appleby-Thomas, G.J., Hameed, S.C.A., Wilgeroth, J.M., and Hazell, P.J. (2012) *Rev. Sci. Instrum.*, **83**, 115113.
70 Sekine, T. (1998) *Eur. J. Solid State Inorg. Chem.*, **29**, 823–833.
71 Liu, X., Mashimo, T., Li, W., Zhou, X., and Sekine, T. (2015) *J. Appl. Phys.*, **117**, 095901.
72 Sikka, S.K. and Gupta, S.C. (1998) *AIP Conf. Proc.*, **429**, 145–150.
73 Atou, T. and Kikuchi, M. (2005) *Rev. High Press. Sci. Technol.*, **15**, 238–246.
74 Eakins, D.E. and Thadhani, N.N. (2009) *Int. Mater. Rev.*, **54**, 181–213.
75 Gordopolov, Y.A., Batsanov, S.S., and Trofimov, V.S. (2009) *Shock Wave Science and Technology Reference Library* (ed. F. Zhang), Springer, Berlin Heidelberg, pp. 287–314.
76 Ohno, S., Ishibashi, K., Sekine, T., Kurosawa, K., Kobayashi, T., Sugita, S., and Matsui, T. (2014) *J. Phys. Conf. Ser.*, **500**, 062001.
77 Cooper, S.R., Benson, D.J., and Nesterenko, V.F. (2000) *Int. J. Plast.*, **16**, 525–540.
78 Schlothauer, T., Schwarz, M.R., Ovidiu, M., Brendler, E., Moeckel, R., Kroke, E., and Heide, G. (2012) *Minerals as Advanced Materials II* (ed. S.V. Krivovichev), Springer, Berlin, Heidelberg, pp. 375–388.
79 Cowan, G.R., Dunnington, B.W., and Holtzman, A.H. (1968) Process for synthesizing diamond. US Patent 3, 401,019, Du Pont.
80 Keller, K., Brendler, E., Schmerler, S., Röder, C., Heide, G., Kortus, J., and Kroke, E. (2015) *J. Phys. Chem. C*, **119**, 12581–12588.
81 Schlothauer, T., Schimpf, C., Schwarz, M., Heide, G., and Kroke, E. (2016) The role of decompression and micro-jetting in shock wave synthesis experiments, *J. Phys. Conf. Series*, **774**, 012053.
82 DeCarli, P.S. and Meyers, M.A. (1981) *Shock Waves and High-Strain-Rate Phenomena in Metals* (eds M.A. Meyers

and L.E. Murr), Plenum Publishing Corp., Boston, MA, pp. 341–373.
83 Maddox, B.R., Park, H.S., Lu, C.H., Remington, B.A., Prisbrey, S., Kad, B., Luo, R., and Meyers, M.A. (2013) *J. Mater. Sci. Eng. A*, **578**, 354–361.
84 Ding, J.L. and Asay, J.R. (2007) *J. Appl. Phys.*, **101**, 073517.
85 Asay, J. and Knudson, M. (2005) *High-Pressure Shock Compression of Solids VIII* (eds L. Chhabildas, L. Davison, and Y. Horie), Springer, Berlin Heidelberg, Berlin/Heidelberg, pp. 329–380.
86 Thoma, K., Hornemann, U., Sauer, M., and Schneider, E. (2005) *Meteorit. Planet. Sci.*, **40**, 1283–1298.
87 Carton, E.P., Verbeek, H.J., Stuivinga, M., and Schoonman, J. (1997) *J. Appl. Phys.*, **81**, 3038–3045.
88 DeCarli, P.S., Bowden, E., Jones, A.P., and Price, G.D. (2002) *Catastrophic Events and Mass Extinctions: Impacts and Beyond* (eds C. Koeberl and K.G. McLeod), Geological Society of America, Boulder, Colorado, pp. 595–605.
89 Isbell, W.M. (2005) *High-Pressure Shock Compression of Solids VIII* (eds L. Chhabildas, L. Davison, and Y. Horie), Springer, pp. 311–328.
90 Grochala, W., Hoffmann, R., Feng, J., and Ashcroft, N.W. (2007) *Angew. Chem., Int. Ed.*, **46**, 3620–3642.
91 Iota, V. (1999) *Science*, **283**, 1510–1513.
92 Neuhaus, A. (1964) *Chimia*, **18**, 93–103.
93 Prewitt, C.T. and Downs, R.T. (1999) *Ultrahigh-Pressure Mineralogy: Physics and Chemistry of the Earth's Deep Interior*, Msapubs.org, pp. 283–318.
94 Donev, A., Stillinger, F.H., Chaikin, P.M., and Torquato, S. (2004) *Phys. Rev. Lett.*, **92**, 255506.
95 Jayaraman, A. (1972) *Phys. Rev. Lett.*, **29**, 1674–1676.
96 Eremets, M.I. and Troyan, I.A. (2011) *Nat. Mater.*, **10**, 927–931.
97 Drozdov, A.P., Eremets, M.I., Troyan, I.A., Ksenofontov, V., and Shylin, S.I. (2015) *Nature*, **525**, 73–76.
98 Gordon, E.E., Xu, K., Xiang, H., Bussmann-Holder, A., Kremer, R.K., Simon, A., Köhler, J., and Whangbo, M.H. (2016) *Angew. Chem.*, **128**, 3746–3748.
99 Demchyna, R., Leoni, S., Rosner, H., and Schwarz, U. (2006) *Z. Kristallogr.*, **221**, 420–434.
100 Rodgers, J.A., Williams, A.J., and Attfield, J.P. (2006) *Z. Naturforsch.*, **61b**, 1515–1526.
101 Inaguma, Y., Aimi, A., Katsumata, T., and Mori, D. (2014) *Rev. High Press. Sci. Technol.*, **24**, 212–222.
102 Yamada, I. (2013) *Rev. High Press. Sci. Technol.*, **23**, 167–173.
103 Horvath-Bordon, E., Riedel, R., Zerr, A., McMillan, P.F., Auffermann, G., Prots, Y., Bronger, W., Kniep, R., and Kroll, P. (2006) *Chem. Soc. Rev.*, **35**, 987–1014.
104 Salamat, A., Hector, A.L., Kroll, P., and McMillan, P.F. (2013) *Coord. Chem. Rev.*, **257**, 2063–2072.
105 Brazhkin, V.V. (2007) *High Press. Res.*, **27**, 333–351.
106 Riedel, R., Zerr, A., Miehe, G., Serghiou, G., Schwarz, M., Kroke, E., Fueß, H., Kroll, P., and Boehler, R. (1999) *Nature*, **400**, 340–342.
107 Landskron, K., Huppertz, H., Senker, J., and Schnick, W. (2001) *Angew. Chem.*, **40**, 2643–2645.
108 Baumann, D. and Schnick, W. (2014) *Angew. Chem.*, **126**, 14718–14721.
109 Baumann, D. and Schnick, W. (2014) *Inorg. Chem.*, **53**, 7977–7982.
110 Yamanaka, S., Komatsu, M., and Inumaru, K. (2012) *Rev. High Press. Sci. Technol.*, **22**, 9–16.
111 Schwarz, U., Wosylus, A., Rosner, H., Schnelle, W., Ormeci, A., Meier, K., Baranov, A., Nicklas, M., Leipe, S., Müller, C.J. *et al.* (2012) *J. Am. Chem. Soc.*, **134**, 13558–13561.
112 Fukuoka, H. and Yamanaka, S. (2003) *Phys. Rev. B*, **67**, 094501.
113 Schnelle, W., Ormeci, A., Wosylus, A., Meier, K., Grin, Y., and Schwarz, U. (2012) *Inorg. Chem.*, **51**, 5509–5511.
114 Er, G., Miyamoto, Y., Kanamaru, F., and Kikkawa, S. (1991) *Phys. C*, **181**, 206–208.
115 Huppertz, H. (2010) *Chem. Commun.*, **47**, 131–140.
116 Huppertz, H. and von der Eltz, B. (2002) *J. Am. Chem. Soc.*, **124**, 9376–9377.
117 Emme, H. and Huppertz, H. (2003) *Chem. Eur. J.*, **9**, 3623–3633.
118 Knyrim, J.S., Roessner, F., Jakob, S., Johrendt, D., Kinski, I., Glaum, R., and

Huppertz, H. (2007) *Angew. Chem., Int. Ed.*, **46**, 9097–9100.

119 Knyrim, J.S. and Huppertz, H. (2008) *Z. Naturforsch.*, **63**, 707–712.

120 Knyrim, J.S. and Huppertz, H. (2007) *J. Solid State Chem.*, **180**, 742–748.

121 Knyrim, J.S., Schappacher, F.M., Pöttgen, R., Schmedt auf der Günne, J., Johrendt, D., and Huppertz, H. (2007) *Chem. Mater.*, **19**, 254–262.

122 Sohr, G., Ciaghi, N., Schauperl, M., Wurst, K., Liedl, K.R., and Huppertz, H. (2015) *Angew. Chem., Int. Ed.*, **54**, 6360–6363.

123 Schlothauer, T., Schimpf, C., Brendler, E., Keller, K., Kroke, E., and Heide, G. (2015) *J. Phys. Conf. Ser.*, **653**, 012033.

124 Sekine, T. and Kobayashi, T. (2003) *New Diam. Front. Carbon Technol*, **13**, 153–160.

125 Köhler, A., Schlothauer, T., Schimpf, C., Klemm, V., Schwarz, M., Heide, G., Rafaja, D., and Kroke, E. (2015) *J. Eur. Ceram. Soc.*, **35**, 3283–3288.

3
High-Pressure Perovskite: Synthesis, Structure, and Phase Relation

Yoshiyuki Inaguma

Gakushuin University, Faculty of Science, Department of Chemistry, 1-5-1 Mejiro, Toshima-ku, 171-8588 Tokyo, Japan

3.1
Introduction

A huge number of compounds with chemical formula ABX_3 adopt the perovskite-type structure or its distorted derivatives [1]. The perovskite-type structure is characterized by three-dimensional corner-shared MX_6 octahedra and A ions placed in the interstice. The tilting of corner-shared MX_6 octahedra with several manners accommodates various A ion and permits distorted versions [2–10]. Perovskite and related compounds are of much interest in the field of materials science such as solid-state physics, solid-state chemistry, and materials engineering owing to their versatile and attractive physical and chemical properties. Perovskite is also closely related to the research of field “high-pressure earth science,” in which the researchers make efforts to clarify the constitution and state of the earth interior, as the deep part of earth under high-pressure condition is composed of perovskite-type phase. This is attributable to the fact that perovskite-type phase consists of close-packed ions and is dense, which allows perovskite-type compounds to be stable in a wide pressure range. Therefore, studies on *perovskite-type and related compounds synthesized under high pressures and high temperatures or stable under high pressure*, which *we named it high-pressure perovskite* in this chapter, have been developed by large contributions of earth science as well as materials science and engineering.

In the field of solid-state chemistry, the stabilization of perovskite-phase by high pressure, the phase relation under various pressure–temperature conditions, and the structure and properties of obtained metastable perovskite phases have been elucidated. Please refer to previous excellent review articles on high-pressure synthesis [1,11,12]. Among them, in the review articles by Goodenough *et al.* [11], the keyword “high-pressure perovskite” has already appeared and the review is still giving us the deep insight in the field of high-pressure perovskite.

Handbook of Solid State Chemistry, First Edition. Edited by Richard Dronskowski, Shinichi Kikkawa, and Andreas Stein.

Through synthesis aided by pressure, we could not only attain the synthesis of new compounds but also understand the thermodynamics as a function of pressure in addition to temperature and discuss the relationship between phase stability, structure, and chemical bonding. In this chapter, syntheses of high-pressure perovskite are reviewed in terms of synthetic methods, phase stability, structure, and chemical bonding. We hereafter abbreviate perovskite as Pv.

3.2
Synthesis of Perovskite Aided by Pressure

The development of pressure apparatus and materials used for the apparatus, and their availability have promoted solid-state chemists to engage in synthesis aided by pressure. Pressure apparatus and sample cell assembly, which make available to high-pressure synthesis, have been described in detail in many excellent review papers and books [13–15]. Therefore, in this section, synthetic methods of Pvs aided by pressure, from relative low gas pressure of ~1 MPa to hydrostatic pressure in solid medium on the order of 10 GPa and the examples, have been briefly introduced.

3.2.1
Synthesis under Oxygen Gas Pressure of ~1 MPa

Synthesis under oxygen pressure of ~1 MPa up to 1000 °C can be performed using quartz glass tube. Reagents such as CrO_3 and Ag_2O, which decompose at elevated temperatures to evolve oxygen gas, are used in order to control oxygen pressure. An example using Ag_2O as an oxidizer is shown in Figure 3.1 [16,17]. The pellets of starting materials and Ag_2O were sealed in an evacuated quartz glass tube. The Ag_2O pellet was separated from the pellets of starting materials. The tube was heated at elevated temperature. Ag_2O as an oxidizer was decomposed at approximately 160 °C into Ag and oxygen gas:

$$2Ag_2O \rightarrow 4Ag + O_2 \quad (3.1)$$

The oxygen pressure of ~1 MPa is available, and we have confirmed that quartz glass tube with a thickness of 1 mm could keep 1 MPa at 1000 °C, and one with a thickness of 1.5 mm could keep 1.5 MPa at 1000 °C if no reaction of sample with quartz glass tube occurs. The Pv-type oxides such as $Bi_{1/2}Ag_{1/2}TiO_3$ and $PrNiO_3$ were successfully synthesized using this method.

3.2.2
Synthesis under Pressure of ~100 MPa

The synthesis under more than 1 MPa oxygen gas pressure has been performed using high-pressure bomb or hot isostatic pressing apparatus (HIP), which realizes the synthesis under up to 300 MPa oxygen gas pressure. The oxygen

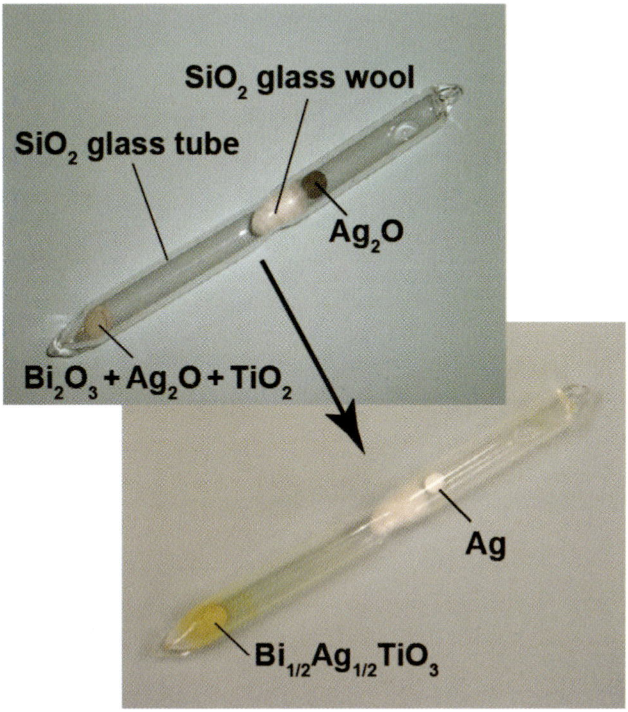

Figure 3.1 Photographs of sealed quartz glass tubes for synthesis under oxygen pressure. *Top*: before heating. *Bottom*: after heating.

pressure in this range makes it possible to attain the higher valance state of transition metal ions in perovskite-type oxides. Using high-pressure bomb, O_2 pressure of 58 MPa at 500 °C can be produced by decomposition of CrO_3 [18]. Oxygen pressure (e.g., ~100 MPa at 300 °C) is also available by the vaporization of liquid O_2 accumulated in a high-pressure bomb by cooling of O_2 gas using liquid N_2. In addition, using the oxygen gas compressor, in which water was chosen as a pressure transmitting medium, synthesis under a pressure of 200 MPa at 600 °C was conducted [19]. In fact, oxygen gas pressure in this range enables the control of oxygen stoichiometry in Pv-related oxides such as $BaFeO_{3-\delta}$ [20], $SrFeO_{3-\delta}$ [21], and the synthesis of ReNiO$_3$, (Re = Pr, Nd, Sm) [22,23] and ReCoO$_3$, (Re = Ho, Er) [24] with $GdFeO_3$-type perovskite structure.

Furthermore, hydrothermal systems have been used for synthesis and crystal growth of inorganic materials under the condition of several hundred degree Celsius and several hundred megapascals [15]. In the hydrothermal synthesis, water is usually used as a solvent and pressure medium. The pressure is then controlled by the degree of water filling in a vessel and temperature, which is estimated from P–V–T data of water. For example, ReO_3-type oxyfluoride

Table 3.1 Apparatus and pressure medium used for synthesis of high-pressure perovskites.

Used apparatus	Sample volume	Pressure range (GPa)	Pressure medium	Heater	Temperature range
Piston cylinder (PC)	10^1–10^2 mm^3	2–3.5	PP	Carbon	800–1100 °C
Belt	10^1–10^2 mm^3	1–9	PP	Carbon	800–1700 °C
Multiple anvil					
Tetrahedral (TA)	10^1–10^2 mm^3	3–7	PP	Carbon	500–1500 °C
Cubic (CA)	10^1–10^2 mm^3	3–10	PP	Carbon	600–1700 °C
Octahedral (Kawai-type) (KOA)	1–10^1 mm^3	12–28	PP, MgO	Pt, Ta, Re, $LaCrO_3$	1100–1600 °C
Diamond-anvil cell (DAC)	10^{-3}–1 mm^3	20–120		Laser heating	RT–2100 °C

PP: pyrophyllite.

$MO_{3-x}F_x$ (M = Mo, W) has been prepared under hydrothermal condition with 700 °C and 300 MPa [25].

3.2.3
Synthesis under Pressure in the GPa Range

The change of coordination number of anion around metal cation, that is, phase transformation, usually demands a pressure over 1 GPa, as mentioned in Section 3.2.4. Almost all gases would be solidified under GPa pressure range at room temperature, which means that liquid or solid compression in order to generate pressure in the gigapascal range is necessary. In order to generate pressure on the order of gigapascal, several high-pressure apparatuses have been developed: (a) piston cylinder type, (b) belt type, (c) girdle type, (d) multianvil type high-pressure apparatus in which there are tetrahedral, cubic, and octahedral anvil apparatus, (e)diamond anvil cell (DAC), (f) shock apparatus, and so on [13–15]. The characteristics of high-pressure apparatus and pressure medium used for synthesis of high-pressure perovskites mentioned below are listed in Table 3.1, which refers to Table 4.4 in the book by Navrotsky [14]. An example of cell assembly for cubic multianvil-type high-pressure apparatus is shown in Figure 3.2.

In order to obtain high-pressure phase, that is, to quench metastable high-pressure phase, the sample should be rapidly cooled at the same pressure because slow cooling may bring in transformation from high-pressure phase to low-pressure phase. Samples are usually heated in electrical resistance furnace

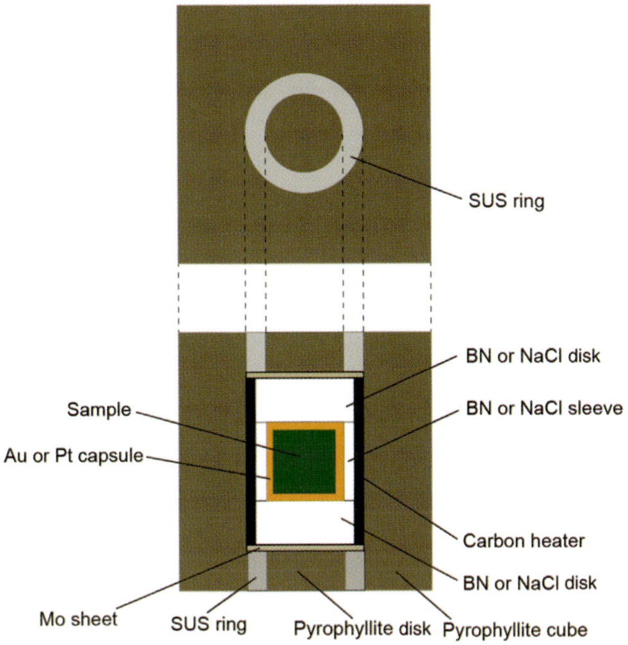

Figure 3.2 An example of cell assembly for cubic multi-anvil-type high-pressure apparatus.

composed of carbon, metal, or alloy such as Pt, Ta, or Re with high melting point and relatively high electronic resistivity(but very conductive), and cooling rate can be easily controlled by the electric current.

The monitoring of phase formation under high pressure and high temperature is very useful to clarify the reaction process, to understand the phase relation at high temperature and pressure, and to develop new materials by high-pressure synthesis. In order to monitor the evolution of phases, *in situ* X-ray diffraction experiments have been performed. At first, the experiment under pressure has been conducted using laboratory X-ray diffractometers, but it is a time-consuming task because of the weak intensity of X-ray radiation. The progress in synchrotron facilities in these 30 years has enabled us to do in a much shorter time owing to intense synchrotron X-ray radiation [13].

3.2.4 Why Can Perovskites Be Stabilized by High-Pressure Synthesis?

The answer for above question can be primarily demonstrated by thermodynamics.

The derivative of Gibbs free energy is given by the following equation:

$$dG = VdP - SdT. \tag{3.2}$$

When temperature is fixed, $dG = VdP$, that is, the derivative of G with respect to P is volume. With increasing pressure, the phase with smaller volume would be more stable than the phase with greater volume. Therefore, high-pressure synthesis can be usually employed to stabilize a high-density phase. As the Pv-type compounds are relatively dense, high pressure is effective to stabilize Pv-type phases.

Furthermore, Pv-type phase generally possesses high entropy. For instance, the change in coordination number of X for A from 6 to 8 or 12 accompanied by the phase transition from ilmenite to Pv increases the entropy. At a fixed pressure, Eq. (3.2) is described as $dG = -SdT$, which indicates that a phase with greater entropy would be stabilized with an increase in temperature. Similar phase relation to that shown in Figure 3.3 on $CdTiO_3$ [26] is often observed in ABX_3 compounds, and as a result Pv phase would be observed under high-pressure and high-temperature conditions. For example, standard enthalpy change ΔH°, standard entropy change ΔS°, and volume change ΔV° of $CdTiO_3$ from ilmenite to perovskite are +15.0 kJ/mol, +14.2 J/(molK), and −4.09 cm^3/mol (−6.9%), respectively [14,27–29]. Supposing that the temperature dependence of entropy and the pressure dependence of volume is negligible, the transition pressure at 500 °C can be estimated from these data to be 1.7 GPa, which is consistent with the experimental data. The coordination number of Cd varies from six to eight with the transformation from ilmenite to Pv in $CdTiO_3$. Considering that the values of thermochemical parameters for other compounds are in the same order, the transformation accompanied by the change of coordination number requires pressure greater than 1 GPa.

Another merit of high-pressure synthesis is that the synthesis in a closed space can be performed. Adding oxidizing agents such as $KClO_3$, $KClO_4$, CrO_3, and

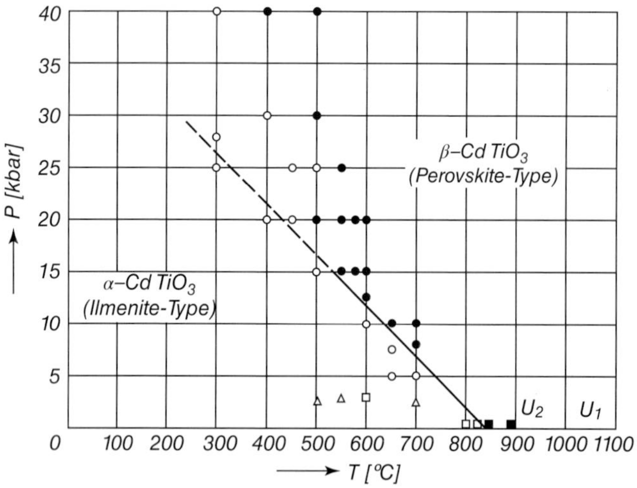

Figure 3.3 Pressure–temperature ($P - T$) phase diagram of $CdTiO_3$ (after Ref. [26]). α- and β-forms correspond to ilmenite- and perovskite-type phases, respectively.

CaO_2, which decompose into oxygen and salt or oxide according to the following equation at high temperatures, to starting materials, followed by placing and sealing in metal capsule, synthesis under oxidized condition can be realized, resulting in the phases containing cations with high valence state such as Fe^{5+}, Pb^{4+}, Cu^{3+}, and Pd^{3+} in perovskites, for example, $CaFeO_3$, $LaLi_{1/2}Fe_{1/2}O_3$, $BiNiO_3$, $LaCuO_3$, and $LaPdO_3$ (see Table 3.2).

$$\begin{aligned} &2\,KClO_3 \rightarrow KCl + KClO_4 + O_2 \\ &KClO_4 \rightarrow KCl + 2\,O_2 \\ &4\,CrO_3 \rightarrow 2Cr_2O_3 + 3\,O_2 \\ &2CaO_2 \rightarrow 2CaO + O_2 \end{aligned} \tag{3.3}$$

Furthermore, synthesis under inert condition can be performed by encapsulating the starting material in glove box filled with inert gas.

The drawback is that only a small amount of sample, for example, at most sub-gram of sample, is obtained, and the calibration of pressure and temperature is needed.

3.3 High-Pressure Perovskite ABX_3

3.3.1 Phase Stability of Perovskite

The stability of Pv-type phases in a wide pressure range allows us to synthesize a lot of high-pressure Pvs. Table 3.2 lists ABX_3-type perovskites synthesized with the aid of pressure. In order to predict the stability of Pv-type phase, the Goldschmidt tolerance factor [157] defined by Eq. (3.4) is useful, although it is based on the simple model in which atoms are represented by hard spheres and Pvs tolerate the deviation from perfect contact of atoms as hard spheres:

$$t = \frac{r_A + r_X}{\sqrt{2}(r_B + r_X)}. \tag{3.4}$$

Here, r_A, r_B, and r_X are ionic radii of A, B, and X, respectively.

Increasing deviation of t from unity indicates increasing strain on the Pv structure. When the tolerance factor ranges from $0.85 < t < 1.01$ (here, ionic radii of eight coordination for A ion and six coordination for B ion are employed), the ABX_3 compounds usually adopt Pv-type structure under ambient pressure. Pressure extends the stability of Pv and realizes the Pv-type phase even beyond the this range of t. Figure 3.4 shows the Goldschmidt diagram for various ABO_3 compounds. Closed and open circles indicate Pvs with $GdFeO_3$-type structure and Pvs with other type structures synthesized at ambient pressure, respectively. Closed and open squares indicate Pvs with $GdFeO_3$-type structure and Pvs with other type structures synthesized with the aid of pressure, respectively. Open and closed triangles denote the LN-type oxides synthesized at ambient pressure and with the aid of pressure, respectively. Half-closed triangles denote LN-type

Table 3.2 ABX_3 type compounds obtained by high-pressure synthesis.

Compound	*t*	Phase	Apparatus	Capsule	Synthetic condition	Oxidizing agent	Remarks	Reference
$BaIrO_3$	0.985	Pv tetra $I4/mcm$	KOA		25 GPa, 1150 °C	—		[30]
$BaOsO_3$	0.982	Pv cubic $Pm\bar{3}m$	KOA		17 GPa, 1600 °C 30 m	—		[31]
$BaRuO_3$	0.987	Pv cubic $Pm\bar{3}m$	KOA		18 GPa, 1100 °C	—		[32,33]
$BiAlO_3$	0.939	Pv rhomb $R3c$	Belt	Au	6 GPa, 1000 °C 40 m	—	Polar	[34]
$BiCoO_3$	0.934	Pv tetragonal $P4mm$	Belt	Au	6 GPa, 970 °C 100 m	—	Polar	[35]
$BiCrO_3$	0.902	Pv triclinic	TA	Graphite	4 GPa, 700 °C			[36]
$BiGaO_3$	0.900	Pv monoclinic Cm	DAC	—	6.3 GPa, RT	—	Pyroxene → Pv	[37]
$BiGa_{1-x}M_xO_3$ (M = Cr, Mn, Fe)	—	Pv $R3c$, $C2/c$, Cm Pv $C2/c$, Cm	Belt	Pt (M = Cr, Fe) Au (M = Mn)	6 GPa, 1700 K, 2 h 6 GPa, 1300 K, 2 h	— —		[38]
$BiInO_3$	0.826	Pv $Pna2_1$	Belt	Au	6 GPa, 1000 °C, 80 m	—	Polar $GdFeO_3$type	[39]
$BiMnO_3$	0.889	Pv triclinic Pv monoclinic $C2$	TA CA	Graphite Au	4 GPa, 700 °C 6 GPa, 700 °C, 3 h	— —		[36] [40]
$BiNiO_3$	0.800*	Pv triclinic $P\bar{1}$	CA	Au	6 GPa, 1000 °C, 30 m	$KClO_4$	*$(Bi^{3+}{}_{1/2}Bi^{5+}{}_{1/2})NiO_3$ columnar-type order, see Section 3.4.2.2	[41]
$BiRhO_3$	0.880	Pv ortho. $Pnma$	Belt	Au	6 GPa, 1600 K, 2 h	—	$GdFeO_3$-type	[42]
$BiScO_3$	0.847	Pv monocl. $C2/c$	Belt	—	6 GPa, 1140 °C, 40 m	—	Noncentrosymmetric	[43]

$BiYO_3$	0.790	Pv			7 GPa, 750 °C			[44]
$Bi_{1/2}Ag_{1/2}TiO_3$	0.926	Pv rhomb. *R3c*	—	Quartz glass	0.7 MPa, 1000 °C, 3 h	Ag_2O	Polar	[16]
$Bi(Co_{1/2}Ti_{1/2})O_3$	0.876	Pv monoclinic	CA	Au	6 GPa, 1050–1300 °C, 10–30 m	—	Superstructure	[17]
$Bi(Cr_{1/2}Ni_{1/2})O_3$	0.905	Pv ortho. *Pnma*	Belt	Au	5.5 GPa, 1100 °C, 30 m	—	—	[45]
$Bi(Mg_{1/2}Ti_{1/2})O_3$	0.881	Pv monoclinic Pv ortho. *Pnnm*	CA Anvil	Au —	6 GPa, 1050–1300 °C, 10–30 m 5–6 GPa, 1270 K	— —	Superstructure Antiferroelectric distortion	[17] [46]
$Bi(Ni_{1/2}Ti_{1/2})O_3$	0.888	Pv monoclinic	CA	Au	6 GPa, 1050–1300 °C, 10–30 m	—	Superstructure	[17,47]
$Bi(Zn_{1/2}Ti_{1/2})O_3$	0.877	Pv tetragonal *P4mm*	WMA	*Pt	6 GPa, 900 °C, 1 h	—	Polar *Pt-lined Al_2O_3	[48]
$Bi(Zn_{1/2}V_{1/2})O_3$	0.882	Pv tetragonal *P4mm*	CA	Au	9 GPa, 1100 °C, 30 m	—	—	[49]
$CaFeO_3$	0.898	Pv tetragonal Pv tetragonal	PC girdle	Au Pt	2 GPa, 1100 °C, 1 h >2 GPa, 1000 °C, 1 h	CrO_3 CrO_3, CaO_2	$CaFe^{4+}O_3$ $Fe^{4+} \rightarrow Fe^{3+} + Fe^{5+}$	[50] [51]
$CaGeO_3$	0.923	Pv cubic Pv Pv ortho. *Pbnm* Pv ortho. *Pbnm*	Bridgman TA TA	Pt Graphite	12 GPa, 900 °C, 3–5 m 6.5 GPa, 1200–1400 °C, 4 h 6.5 GPa, 1200 °C, 15–100 m	— — —	Double cell parameter 6–7 GPa, 700–1300 °C	[52] [53] [54] [55]

(continued)

Table 3.2 (*Continued*)

Compound	*t*	Phase	Apparatus	Capsule	Synthetic condition	Oxidizing agent	Remarks	Reference
$CaPbO_3$	0.819	Pv, ortho. *Pbnm*	CA	Au	5 GPa, 1000–1100 °C, 1 h	—	$GdFeO_3$-type	[56]
$CaRhO_3$	0.891	Pv, ortho. *Pbnm*	Belt	Pt	6 GPa, 1500 °C, 1 h	$KClO_4$	$GdFeO_3$-type	[57]
$CaSiO_3$	0.990	Pv cubic	DAC	—	16 GPa, 1500 °C	—	Unquenchable	[58]
		Pv ortho. *Pbnm* or *Cmcm*	KOA	Amorphous B, epoxy resin	13 GPa, 1400 K	—	*In situ* X-ray diffraction	[59]
$CdGeO_3$	0.916	Pv distorted	Squeezer		13 GPa, 900 °C	—	Single-crystal $GdFeO_3$-type	[52]
		Pv ortho. *Pbnm*	KOA		12 GPa, 1200 °C			[60]
$CdSnO_3$	0.846	Pv ortho. *Pbnm*	TA	Pt	6.5 GPa, 1500 °C, 1 h	—	Single crystal	[61]
$CdTiO_3$	0.882	Pv	Bridgman	—	2 GPa, 600 °C, 4 h	—	Ilmenite (AP, >850 °C)	[26]
		Pv ortho. *Pbnm*	PC		2.5 GPa, 800 °C, 1.5 h	—	*P–T* $GdFeO_3$-type	[62]
$CuTaO_3$	0.821	LN	TA	Pt	6.5 GPa, 1000–1200 °C	—	Single crystal	[63]
$DyNiO_3$	0.876	Pv ortho. *Pbnm*	Belt	Pt	6 GPa, 950 °C, 12 m	$KClO_3$	$GdFeO_3$-type	[64]
			PC	Au	2 GPa, 900 °C, 20 m	$KClO_4$	$GdFeO_3$-type	[65]
$ErCoO_3$	0.874	Pv ortho. *Pbnm*	Belt	Pt	6 GPa, 930 °C, 15 m	$KClO_3$	$GdFeO_3$-type	[66]
					20 MPa, 900 °C, 12 h	O_2		[24]
$ErMnO_3$	0.831	Pv ortho. *Pbnm*	CA	Au	5 GPa, 800 °C, 30 m	—	$GdFeO_3$-type	[67]
$ErNiO_3$	0.867	Pv ortho. *Pbnm*	Belt	Pt	6 GPa, 950 °C, 12 m	$KClO_3$	$GdFeO_3$-type	[64]
$EuNiO_3$	0.890	Pv ortho. *Pbnm*	Belt	Pt	6 GPa, 950 °C, 12 m	$KClO_3$	$GdFeO_3$-type	[64]
$FeGeO_3$	0.839	LN Pv Ortho. (HP)	DAC	—	33 GPa, RT		clinopyroxene → perovskite	[68]
$FeTiO_3$	0.769	LN rhomb. *R3c*	KOA	Pt	15 GPa, 1000 °C	—		[69]

			DAC	—	16 GPa at RT	—	Akaogi and Akimoto (1977), unpublished work	[70]
			KOA	Fe	15–16 GPa, 1200 °C, 30–60 m			[71]
$GaFeO_3$	0.750	Pv LN	DAC	—	42 GPa at room temperature Decompression 24 GPa	—		[72]
$GdGaO_3$	0.859	Pv ortho. *Pnma*	Girdle		7.2 GPa, 1000 °C, 1 h	—	Single-crystal $GdFeO_3$-type	[73]
$GdNiO_3$	0.885	Pv ortho. *Pbnm*	Belt	Pt	6 GPa, 950 °C, 12 m	$KClO_3$	$GdFeO_3$-type	[64]
			—	Au	90 MPa, 850 °C, 24 h	O_2		[65]
$HgPbO_3$	0.826	Pv rhomb.	TA	Au	6.5 GPa, 600–1000 °C	—		[74]
$HgSnO_3$	0.859	Pv rhomb. $R\bar{3}c$	CA	Au	7 GPa, 700–1100 °C, 30 m	—		[75]
$HgTiO_3$	0.896	Pv rhomb. *R3c*	TA	Au	6.5 GPa, 800–1000 °C	—	Polar	[74]
$HoCoO_3$	0.878	Pv ortho. *Pbnm*	Belt	Pt	6 GPa, 930 °C, 15 m	$KClO_3$	$GdFeO_3$-type	[66]
					20 MPa, 900 °C, 12 h	O_2		[24]
$HoMnO_3$	0.835	Pv ortho. *Pbnm*	Belt	Au	4.5 GPa, 950 °C, 2 h	—	$GdFeO_3$-type	[76]
		Pv ortho. *Pbnm*	CA		5 GPa, 800 °C, 30 m	—	$GdFeO_3$-type	[67]
$HoNiO_3$	0.871	Pv ortho. *Pbnm*	Belt	Pt	6 GPa, 950 °C, 12 m	$KClO_3$	$GdFeO_3$-type	[64]
		Pv monocl. $P2_1/n$	PC	Au	2 GPa, 900 °C, 20 m	$KClO_4$	$GdFeO_3$-type charge disproportionation	[77]
$InCrO_3$	0.814	Pv ortho. *Pnma*	TA	Au	6.5 GPa, 1250 °C	—	$GdFeO_3$-type	[78]
		Pv ortho. *Pnma*	Belt		6 GPa, 1500 K, 2 h	—	$GdFeO_3$-type	[79]

(*continued*)

Table 3.2 *(Continued)*

Compound	t	Phase	Apparatus	Capsule	Synthetic condition	Oxidizing agent	Remarks	Reference
$InFeO_3$	0.802	LN rhomb. $R3c$	KOA	Pt	15 GPa, 1450 °C, 30 m	—	HP phase: Pv $Pnma$ or $Pn2_1a$	[80]
$InRhO_3$	0.794	Pv ortho. $Pnma$	TA	Au	6.5 GPa, 1350 °C	—	$GdFeO_3$-type	[78]
		Pv monocl. $P2_1/n$	Belt		6 GPa, 1220 °C, 2 h		Superstructure, $GdFeO_3$-type	[81]
$(In_{1-y}M_y)MnO_3$ $(1/9 \leq y \leq 1/3)$	*0.802	Pv monocl. $P2_1/n$	Belt	Pt	6 GPa, 1500 °C, 40 m	—	$GdFeO_3$-type *estimated as $InMnO_3$	[82]
$(In_{1x}M_x)MO_3$	—	Pv rhomb. $R3c$	Belt	Pt	6 GPa, 1500 °C, 30 m	—	$x = 0.112$–0.176	[83]
$(M = Fe_{0.5}Mn_{0.5})$		LN rhomb. $R3c$	Belt	Pt	6 GPa, 1500 °C, 30 m	—	$x = 0.143$, HP phase: Pv	[84]
$K_{1-x}Bi_{1+x}O_3$	*0.953	Pv cubic $Pm\bar{3}m$	CA	Au	5 GPa, 1000 °C, 1 h	$KClO_4$	$x = 0.0$–0.2 *$KBiO_3$	[85,86]
$LaCuO_3$	0.933	Pv	Belt		6.5 GPa, 900 °C, 10 m	$KClO_3$		[87]
$La(Mn_{1/2}Ir_{1/2})O_3$	0.878	Pv ortho. $Pnma$	Belt		6 GPa, 1000–1100 °C, 15 m		$GdFeO_3$-type	[88]
$LaPdO_3$	0.838	Pv	Belt		5 GPa, 1100–1150 °C, 10–15 m	$KClO_3$		[89]
$LiOsO_3$	0.831	LN	Belt	Pt	6 GPa, 1200 °C, 1 h	$LiClO_4$	1600 °C, 5 h for single crystal	[90]
$LuCoO_3$	0.879	Pv ortho. $Pbnm$	Belt	Pt	6 GPa, 930 °C, 15 m	$KClO_3$	$GdFeO_3$-type	[66]
			PC	Au	2 GPa, 900 °C, 25 m	$KClO_4$		[24]

$LuMnO_3$	0.822	Pv ortho. *Pbnm*	CA	Au	5 GPa, 800 °C, 30 m	—		[67]
$LuNiO_3$	0.858	Pv ortho. *Pbnm*	Belt	Pt	6 GPa, 950 °C, 12 m	$KClO_3$	$GdFeO_3$-type	[64]
$LuRhO_3$	0.814	Pv ortho., *Pnma*	Belt	Au	6 GPa, 1300 K, 2 h	—	$GdFeO_3$-type	[42]
$(Lu_{1/4}Mn_{3/4})VO_3$	0.838	Pv ortho., *Pnma*	CA	Au	9 GPa, 900 °C, 30 m	—	A-site disordered $(Lu^{3+}{}_{1/4}Mn^{2+}{}_{3/4})V^{3.75+}O_3$	[91]
$MgGeO_3$	0.839	LN	KOA	—	25 GPa, 1000 °C	—	HP phase: Pv	[68]
$MgSiO_3$	0.900	Pv ortho. *Pbnm*	DAC	—	30 GPa, 1000 °C	—	$GdFeO_3$-type	[92]
		Pv ortho. *Pbnm*	KOA	Pt	28 GPa, 1000 °C, 60–90 m	—	Starting mater.: $MgSiO_3$ glass	[93]
$MgTiO_3$	0.808	LN, rhomb. *R3c*	KOA	Pt	21 GPa, 1200 °C, 3 h	—	HP phase: Pv	[94]
$MnGeO_3$	0.865	Pv ortho.	DAC	—	25 GPa, 1400–1800 °C	—	—	[95]
$Mn_4Nb_2O_9$	0.793	LN rhomb. *R3c*	Toroid	—	9 GPa, 1700 °C, 10 m	—	$Mn(Mn_{1/3}Nb_{2/3})O_3$, B-site disordered	[96]
$MnSnO_3$	0.798	LN rhomb. *R3c*	TA	C	7 GPa, 1000 °C,	—	UQ HP phase: Pv	[97]
		Pv ortho. *Pbnm*	DAC	—	15–40 m	—	LN → Pv	[70]
		LN rhomb. *R3c*	CA	Au	7 GPa at RT		Polar	[98]
					7 GPa, 800 °C, 30 m			
$MnTiO_3$	0.832	LN rhomb. *R3c*	TA	C	7 GPa, 1100 °C,		$P(GPa) = 11 - 0.0044\ T$	[99]
		LN rhomb. *R3c*	CA	Au	15–40 m		(°C)	[100]
		Pv ortho. *Pbnm*	DAC		6 GPa, 1300 °C, 1 h		Crystal	[101]
		LN rhomb. *R3c*	CA		2–3 GPa at RT		HP phase: Pv $GdFeO_3$-type	[98]
					7 GPa, 700 °C, 2 h		Polar	
$MnVO_3$	0.843	Pv ortho., *Pbnm*	TA	Pt	>4.5 GPa, 900 °C, 30–60 m	—	$GdFeO_3$-type	[102]

(*continued*)

Table 3.2 (*Continued*)

Compound	*t*	Phase	Apparatus	Capsule	Synthetic condition	Oxidizing agent	Remarks	Reference
$NaOsO_3$	0.924	Pv ortho., *Pbnm*	Belt	Pt	6 GPa, 1700 °C, 2.5 h		$GdFeO_3$-type	[103]
$NaSbO_3$	0.912	Pv ortho., *Pbnm*	MA	Au	10.5 GPa, 1150 °C, 4 h	—	$GdFeO_3$-type, Ilmenite	[104]
$NdNiO_3$	0.905	Pv ortho. *Pbnm*	Belt	Pt	15–20 MPa, 1000 °C, several days	O_2	$GdFeO_3$-type	[22]
					6 GPa, 950 °C, 12 m	$KClO_3$		[65]
$NdRhO_3$	0.859	Pv ortho., *Pnma*	Belt	Au	6 GPa, 1300 K, 2 h	—	$GdFeO_3$-type	[42]
$PbCoO_3$	0.839	Pv Cubic $Pn\overline{3}$	KOA	Au	12 GPa 1473 K 30 m	—	1: 3 A-site order, see Table 3.5 $Pb^{2+}Pb^{4+}{}_3Co^{2+}{}_2Co^{3+}{}_2O_{12}$	[105]
$PbCrO_3$	0.883*	Pv cubic	Belt	Au	>5 GPa, 1150 °C	—	$^*Pb^{2+}{}_{0.5}Pb^{4+}{}_{0.5}Cr^{3+}O_3$	[106]
		Pv charge glass state	CA		8 GPa, 800 °C, 20 m	—		[107]
$PbFeO_3$	0.870*	Pv	CA	Au	6–7 GPa, 1000–1300 °C, 30 m	—	$^*Pb^{2+}{}_{0.5}Pb^{4+}{}_{0.5}Fe^{3+}O_3$	[108]
$PbMnO_3$	0.986	Pv	KOA		15 GPa, 1500 °C, 30 m	—	6 H phase (8 GPa)	[109]
$PbNiO_3$	0.792	Pv ortho. *Pnma*	CA	Au	3 GPa, 800 °C, 1 h	—	HP phase: Pv($GdFeO_3$-type)	[110]
		LN rhomb. *R3c*			Annealed	—		
$PbRuO_3$	0.942	Pv ortho. *Pnma*	Belt		9 GPa, 1400 °C, 15 m		$GdFeO_3$-type	[111,112]

$PbSnO_3$	0.910	Pv monoclinic	TA		6–7 GPa, 500 °C, 40 m	—		[113]
$PbVO_3$	0.961	Pv tetragonal *P4mm*	Lens	Au or Pt	4–6 GPa, 700–750 °C	—	—	[114]
		Pv tetragonal *P4mm*	CA		6 GPa, 700–950 °C, 30–120 m	—	—	[115]
$PbZnO_3$	0.773	LN rhomb. *R3c*	CA	Au	6 GPa, 1100 °C, 30 m	$KClO_3$	—	[116]
$Pb(Fe_{1/2}V_{1/2})O_3$	0.955	Pv tetragonal *P4mm*	CA	Au	7 GPa, 950 °C, 30 m	—		[117]
$PrNiO_3$	0.911	Pv ortho. *Pbnm*			15–20 MPa, 1000 °C, several days	O_2		[22]
$Rb_{1-x}Bi_{1+x}O_3$	0.985	Pv cubic $Pm\bar{3}m$	CA	Au	5 GPa, 1000 °C, 1 h	$KClO_4$	$x = 0.0–0.4$	[86]
$ScAlO_3$	0.830	Pv ortho. *Pbnm*	PC	Pt	12 GPa, 1000 C			[118]
					3. 5 GPa			[119]
$ScCrO_3$	0.794	Pv, ortho. *Pnma*	KOA	Au	4.5 GPa, 1200 °C, 3 h	—	$GdFeO_3$-type	[120]
			Belt	Au	6 GPa, 1500 K, 2 h	—		[79]
$ScFeO_3$	0.785	LN rhomb. *R3c*	KOA	Pt	15 GPa, 1450 °C, 30 m	—	HP phase: Pv Bixbyite → Pv	[121]
$ScMnO_3$	0.785	Pv monocl. $P2_1/n$	MOA	Au	12.5 GPa, 1100 °C, 1 h	—	Superstructure, $GdFeO_3$-type	[122]
$ScRhO_3$	0.777	Pv monocl. $P2_1/n$	Belt	Au	6 GPa, 1220 °C, 2 h	—	Superstructure, $GdFeO_3$-type	[81]
$ScVO_3$	0.787	Pv ortho. *Pnma*	Belt	Au	8 GPa, 800 °C, 30 m			[123]

(*continued*)

Table 3.2 (*Continued*)

Compound	*t*	Phase	Apparatus	Capsule	Synthetic condition	Oxidizing agent	Remarks	Reference
$SmNiO_3$	0.894	Pv ortho. *Pbnm*	Belt	Pt	6 GPa, 950 °C, 12 m	$KClO_3$	$GdFeO_3$-type	[64]
					15–20 MPa, 1000 °C, several days	O_2		[22]
$SrBiO_3$	0.802	Pv monocl. $P2_1/n$			10 MPa, 850 °C, 35 h	O_2	Charge-ordered, $GdFeO_3$-type	[124]
$Sr_{1-x}M_xBiO_3$ (M = K, Rb)		Pv tetra. *I4/mcm* (M = K, $x = 0.6$)	Belt	Pt	2 GPa, 700 °C, 1 h	KO_2	Superconductor	[124]
$SrCrO_3$	0.965	Pv cubic	TA		6–6.5 GPa, 800 °C	—		[125]
$SrFeO_3$	0.948	Pv cubic	Autoclave		50 MPa, 300 °C	O_2	$Sr_2Fe_2O_5+O_2 \rightarrow SrFeO_3$	[126]
$SrFe_{1-x}Co_xO_3$	—	Pv cubic	CA	Au	6 GPa, 600 °C, 30 m	$KClO_4$		[127]
$SrGeO_3$	0.975	Pv cubic $Pm\bar{3}m$			6 GPa, 1300 K	—		[128]
$SrIrO_3$	0.929	Pv, ortho. *Pbnm*	Belt	Au or Pt	>4.5 GPa, 1000 °C	—	—	[129,130]
$SrOsO_3$	0.927	Pv			6–6.5 GPa, 1200 °C, 2 h			[131]
$SrRhO_3$	0.940	Pv ortho. *Pnma*	Belt	Pt	6 GPa, 1500 °C, 1 h	$KClO_4$	$GdFeO_3$-type	[132]
$SrRu_{1-x}Cr_xO_3$ (x>0.2)	—	$x = 0.4$ $R\bar{3}c$ $x = 0.5$–1 $Pm\bar{3}m$	PC, WMA	Au BN	3.5–10.5 GPa, 1000 °C, 1 h 1100 °C, 1 h		PC $x = 0.4$, 0.6 WMA $x = 0.5$, 0.8, 1	[133]
$TlCrO_3$	0.835	Pv ortho. *Pnma*	Belt	Au	6 GPa, 1500 K, 2 h	—	$GdFeO_3$-type	[134]
$TlMnO_3$	0.823	Pv triclinic $P\bar{1}$	Belt	Au	6 GPa, 1500 K, 2 h	—	Jahn–Teller distortion	[135]
$Tl_{1/2}Na_{1/2}TiO_3$	0.875	Pv tetra. *P4/mmm*	CA	Au	5.5 GPa, 1000 °C, 30 s	—		[136]

$(Tl_{1-x}Sc_x)ScO_3$	0.785*	LN rhomb. *R3c*	Belt		*6 GPa, 1300 °C, 1–2 h	—	*Condition for $TlScO_3$	[137]
$TmCoO_3$	0.870	Pv ortho. *Pbnm*	Belt	Pt	6 GPa, 930 °C, 15 m	$KClO_3$	$GdFeO_3$-type	[66]
			PC	Au	2 GPa, 900 °C, 25 m	$KClO_4$		[24]
$TmMnO_3$	0.828	Pv ortho. *Pbnm*	CA	Au	5 GPa, 800 °C, 30 m	—	$GdFeO_3$-type	[67]
$TmNiO_3$	0.864	Pv ortho. *Pbnm*	Belt	Pt	6 GPa, 950 °C, 12 m	$KClO_3$	$GdFeO_3$-type	[64]
$YCoO_3$	0.879	Pv ortho. *Pbnm*	Belt	Pt	6 GPa, 930 °C, 15 m	$KClO_3$	$GdFeO_3$-type	[66]
$YGaO_3$	0.847	Pv ortho. *Pnma*	Girdle		7.2 GPa, 1000 °C, 1 h	—	Single-crystal growth	[73]
$YInO_3$	0.777	Pv ortho. *Pnma*	TA		6.5 GPa, 1250 °C	—	$GdFeO_3$-type	[78]
$YMnO_3$	0.836	Pv ortho. *Pbnm*	CA	Au	5 GPa, 800 °C, 30 m	—	$GdFeO_3$-type	[67]
$YNiO_3$	0.873	Pv ortho. *Pbnm*	Belt	Pt	6 GPa, 950 °C, 12 m	$KClO_3$	$GdFeO_3$-type	[64]
		Pv monocl. *P2₁/n*	PC	Au	2 GPa, 900 °C, 20 m	$KClO_4$	$GdFeO_3$-type charge disproportionation	[77]
$(Y_{1/4}Mn_{3/4})VO_3$	0.842	Pv ortho. *Pnma*	CA	Au	9 GPa, 900 °C, 30 m	—	A-site disordered $(Y^{3+}{}_{1/4}Mn^{2+}{}_{3/4})V^{3.75+}O_3$	[91]
$YbCoO_3$	0.867	Pv ortho. *Pbnm*	Belt	Pt	6 GPa, 930 °C, 15 m	$KClO_3$	$GdFeO_3$-type	[66]
			PC	Au	2 GPa, 900 °C, 25 m	$KClO_4$		[24]
$YbGaO_3$	0.835	Pv ortho. *Pnma*	Girdle		7.2 GPa, 1000 °C, 1 h	—	$GdFeO_3$-type single crystal	[73]
$YbMnO_3$	0.825	Pv ortho. *Pbnm*	CA	Au	5 GPa, 1100 °C, 30 m	—	$GdFeO_3$-type	[138]
		Pv ortho. *Pbnm*	CA	Au	5 GPa, 800 °C, 30 m		$GdFeO_3$-type	[67]
$YbNiO_3$	0.860	Pv ortho. *Pbnm*	Belt	Pt	6 GPa, 950 °C, 12 m	$KClO_3$	$GdFeO_3$-type	[64]
$ZnGeO_3$	0.843	LN rhomb. *R3c*	KOA	Re	21-23 GPa, 1200–1600 °C, 1–3 h	—		[139]
		Pv ortho. *Pbnm*	DAC	—	19 GPa at RT	—	LN → Pv IL → Pv	[140]

(*continued*)

Table 3.2 (*Continued*)

Compound	t	Phase	Apparatus	Capsule	Synthetic condition	Oxidizing agent	Remarks	Reference
$ZnPbO_3$	0.748	LN rhomb. $R3c$	CA	Au	8 GPa, 1000 °C, 30 m	—		[141]
$ZnSnO_3$	0.778	LN rhomb. $R3c$	CA	Au	7 GPa, 1000 °C, 30 m	—	Polar	[142]
$ZnTiO_3$	0.811	LN rhomb. $R3c$	KOA	Pt	16 GPa, 1200 °C, 30 m	—	Polar, HP phase: Pv	[143]
$Zn(Fe_{1/2}Ta_{1/2})O_3$	0.796	LN rhomb. $R3c$	WMA	Pt	9 GPa, 1350 °C, 1 h	—	B-site disordered	[144]
$CsFeF_3$	1.029	Pv cubic $Pm\bar{3}m$	Belt	Au	8 GPa, 700 °C, 30 m	—	6 H → 3C (see Table 3.3)	[145]
$CsMnF_3$	1.005	Pv cubic $Pm\bar{3}m$	TA	Graphite tube	<3 GPa	—	6 H → 3C (see Table 3.3)	[146]
		Pv cubic $Pm\bar{3}m$	Belt	Au	3 GPa, 700 °C, 30 m	—	6 H → 3C (see Table 3.3)	[145]
$RbNiF_3$	1.029	Pv cubic $Pm\bar{3}m$	Belt	Au	2.5 GPa, 300 °C	—	6 H → 3C (see Table 3.3)	[149]
		Pv cubic $Pm\bar{3}m$	TA	Graphite tube	<3 GPa	—		[146]
$TlNiF_3$	1.022	Pv cubic $Pm\bar{3}m$	TA	Graphite tube	<3 GPa	—	6 H → 3C (see Table 3.3)	[146,148]
$RbMnCl_3$	0.916	Pv cubic $Pm\bar{3}m$	Belt	Au	0.7 GPa, 700 °C, 30 m	—	6 H → 3C (see Table 3.3)	[149]
$CsMnCl_3$	0.951	Pv cubic $Pm\bar{3}m$	Belt	Au	2.7 GPa, 700 °C, 30 m	—	9R → 6 H → 3C (see Table 3.3)	[149]
$BaFeO_2F$	0.978	Pv cubic $Pm\bar{3}m$		Ni	5 GPa, 1000 °C		$BaO + BaF_2 + Fe_2O_3$	[150,151]
$BaNbO_2N$	—	Pv cubic $Pm\bar{3}m$		Ni	5 GPa, 1000 °C	BaO_2	$BaO + BaO_2 + NbN$	[150]
$KTiO_2F$	1.030	Pv cubic $Pm\bar{3}m$	TA		6–6.5 GPa, 1000 °C		$KF + TiO_2$	[152]

$MnTaO_2N$	—	LN rhomb. *R3c*	CA		6 GPa, 1400 °C, 1 h	—	MnO + TaON, polar	[153]
$PbFeO_2F$	0.933	Pv		Ni Au	5 GPa, 1000 °C		$PbO + PbF_2 + Fe_2O_3$	[150,151]
$PbMnO_2F$	0.933	Pv tetragonal	CA	Au	4–7 GPa, 1000 °C, 30 m	—	$a \times a \times 4a$	[154]
$PbScO_2F$	0.889	Pv cubic $Pm\bar{3}m$	CA	Au	4 GPa, 1000 °C, 30 m	—		[155]
$TlTlO_2F$	0.927	Pv cubic $Pm\bar{3}m$		Au	0.5 GPa, 500 °C	—	$Ti^{+}Tl^{3+}OF_2$	[156]

Apparatus: PC: piston cylinder, belt: belt-type, girdle: girdle–type, TA: tetrahedral anvil, CA: cubic anvil, KOA: Kawai-type octahedral anvil, WMA: multianvil with Walker-type module, DAC: diamond anvil cell, Lens: lens-type.

oxides, which are confirmed to be $GdFeO_3$-type Pv at high pressures. The value of tolerance factor represented by the dotted line was estimated using ionic radii after Shannon [158] of A ion in eightfold coordination, and O^{2-} and X ions in sixfold coordination, respectively. The Pvs with small tolerance factor is accompanied by the tilting of MX_6 octahedra, and most of them possess $GdFeO_3$-type structure with tilting system (Glazer notation [3]: $a^-b^+a^-$, space group: *Pnma*) (alternative representation: $a^-a^-c^+$, *Pbnm*)). As seen in Table 3.2 and Figure 3.4,

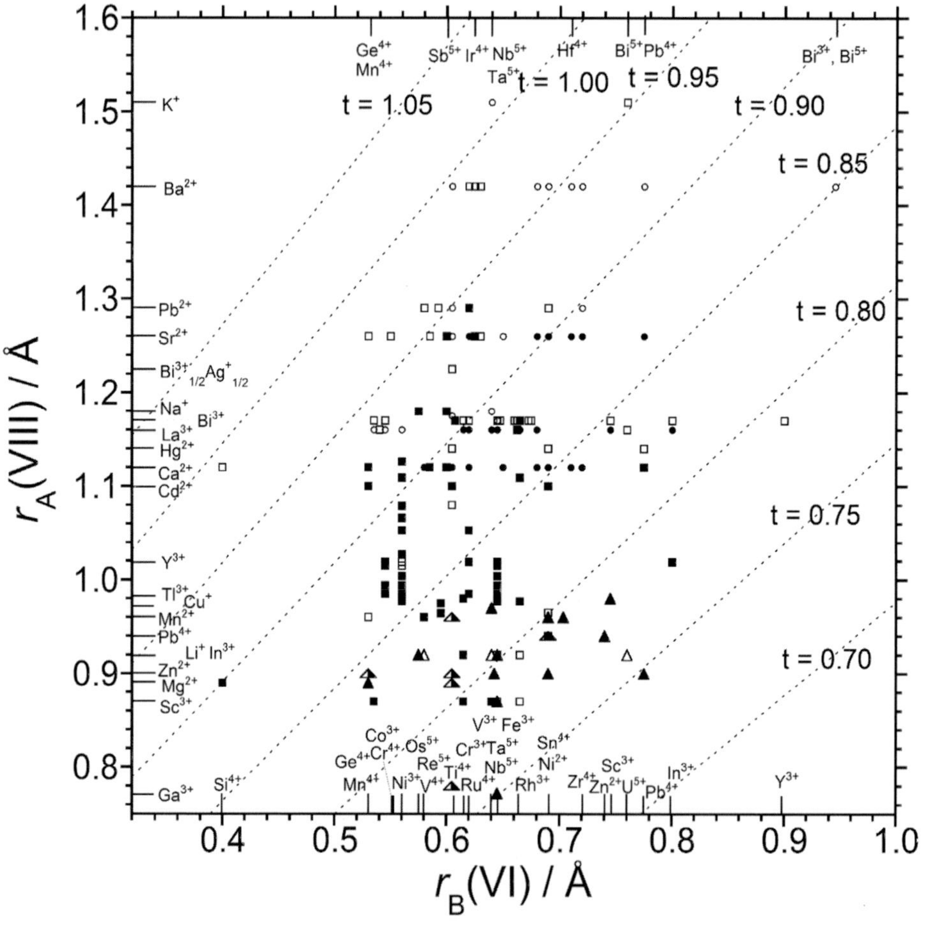

Figure 3.4 Goldschmidt diagram for various ABO_3 compounds. Closed and open circles indicate perovskites (Pvs) with $GdFeO_3$-type structure and Pvs with other type structures synthesized at ambient pressure, respectively. Closed and open squares indicate Pvs with $GdFeO_3$-type structure and Pvs with other type structures synthesized with aid of pressure, respectively. Open and closed triangles denote the LN-type oxides synthesized at ambient pressure and with aid of pressure, respectively. Half-closed triangles denote LN-type oxides that are confirmed to be $GdFeO_3$-type Pv at high pressures. The value of tolerance factor represented by the dotted line was estimated using ionic radii after Shannon of A ion in eightfold coordination, and O^{2-} and X ions in sixfold coordination, respectively.

Pvs possess $GdFeO_3$-type structure over wide range of tolerance factor, which indicates that $GdFeO_3$-type structure is thermodynamically stable. Watching from vantage point, there is a boundary in the vicinity $t = 0.85$ between Pv and LN-type compounds. As shown in Figure 3.4, it has been confirmed that several LN-type oxides synthesized under high pressure are recovered as retrograde from Pv-type oxides on release of pressure, which will be described in Section 3.3.3. As seen in Figure 3.4, pressure surely helps to stabilize Pv-type oxides, including LN-type oxides with smaller tolerance factor down to $t = 0.75$ than that of Pv synthesized at ambient pressure, but there is not an explicit tolerance factor as a boundary which makes clear the pressure effect. This means that the stability cannot be explained by only the simple ion packing model. There are some groups of compounds for which high-pressure synthesis is required to stabilize Pv phase even though their tolerance factors are relatively large. Though silicates ($MgSiO_3$, $CaSiO_3$) and germinates ($CaGeO_3$, $SrGeO_3$) have appropriate tolerance factors for Pv of $0.90 < t < 1.00$, we need high pressure to stabilize Pv phase. Some Pv-type compounds containing Cd^{2+}, Hg^{2+}, Tl^{3+}, Pb^{2+}, and Bi^{3+} ions in A-site are also stabilized by high-pressure synthesis. Navrotsky [159] and Akaogi *et al.* [139] have discussed the stability of $A^{2+}B^{4+}O_3$ perovskite in terms of enthalpy as well as tolerance factor. Navrotsky first pointed out that the standard formation enthalpy ΔH_f° data of Pvs increased almost linearly along a straight line with increasing deviation of tolerance factor from unity. Figure 3.5 shows the relationship between enthalpies of formation and tolerance factors for $A^{2+}B^{4+}O_3$ perovskites given by Akaogi *at al.* [139]. The figure shows that though

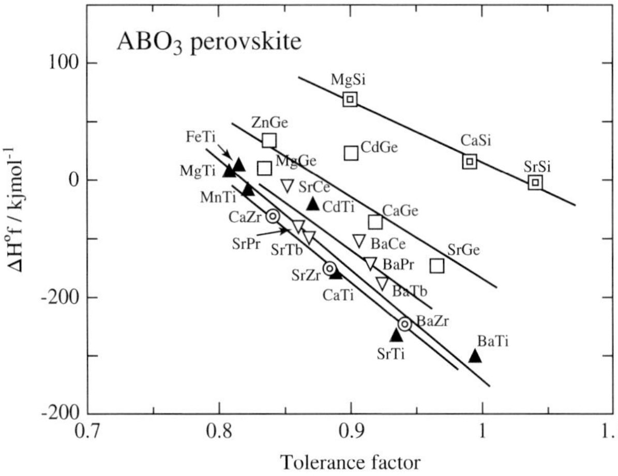

Figure 3.5 A relationship of enthalpy of formation of $A^{2+}B^{4+}O_3$ perovskite with tolerance factor, $t = (R_A + R_O)/\sqrt{2}(R_B + R_O)$. R_O, R_A, and R_B are Shannon–Prewitt ionic radii of O^{2-}, A^{2+} in eightfold coordination and B^{4+} in sixfold coordination, respectively. Symbols with AB indicate enthalpies of formation of $A^{2+}B^{4+}O_3$ perovskites. *Closed triangles* and *double circles* indicate enthalpies of formation of $ATiO_3$ and $AZrO_3$ perovskites, respectively. *Open reverse triangles* represent enthalpies of formation of $ALnO_3$ perovskites. *Open and double squares* show enthalpies of formation of $AGeO_3$ and $ASiO_3$ perovskites, respectively. (Reproduced with permission from Ref. [139]. Copyright 2005, Springer.)

the ΔH_f° generally increases linearly with decreasing tolerance factor, the data are on the different lines dependent on B ion, Si, Ge Ti, Zr, and Ln (Ln = Ce, Pr, and Tb), indicating that the tolerance factor is not the only predominant factor determining the stability of Pv phase. As seen in Figure 3.5, ΔH_f° of $ASiO_3$, $AGeO_3$ are much greater than other series. Furthermore, ΔH_f° of Cd compounds, $CdTiO_3$ and $CdGeO_3$, deviate upwardly from the lines for $ATiO_3$ and $AGeO_3$, respectively, indicating that Pv-phase is more unstable than it is expected. Akaogi *et al.* addressed that the electronic configuration of $(4d)^{10}$ of Cd^{2+} weakens the stability of Pv-type phase. Though the thermochemical data on $Hg^{2+}MO_3$ and $Tl_{1/2}Na_{1/2}TiO_3$ were not available, Hg^{2+} and Tl^{3+}(electronic configuration: $(5d)^{10}$) with the similar d^{10} electronic configuration as Cd^{2+} may give the same effect. Furthermore, though the data are not shown in Figure 3.5, ΔH_f° of lead compounds $PbTiO_3$ (−36.7 ± 2.7 kJ/mol) and $PbZrO_3$ (0.7 ± 4.3 kJ/mol) [160] also deviates upwardly from the lines for $ATiO_3$ and $AZrO_3$. Regarding Pv-type compounds containing Pb^{2+} and Bi^{3+} ions in A-site, Pb^{2+} and Bi^{3+} possess common electronic configuration of $[Xe](4f)^{14}(5d)^{10}(6s)^2$. The lone pair electrons $(6s)^2$ with steric effect destabilizes Pv-type structure under ambient pressure, while under high pressure the steric effect is suppressed. Consequently, high pressure leads to the stabilization of Pv-type compounds containing Pb^{2+} and Bi^{3+} ions in A-site.

3.3.2
Evolution from Layered Structure to Perovskite-Type Structure with Pressure

As seen in Figure 3.4, several compounds such as $BaOsO_3$ and $BaRuO_3$ do not crystallize into Pv at ambient pressure even though the tolerance factor is close to unity, and the stabilization of Pv requires high pressure. When $t > 0.97$, ABX_3 compounds tend to adopt Pv-related hexagonal or rhombohedral layered structure with face-sharing BX_6 octahedra and the structure changes with pressure as listed in Table 3.3. Longo and Kafalas [33] and Syono *et al.* [146] independently demonstrated the structural change of ABX_3 compounds with the hexagonal or rhombohedral layered structure to Pv with pressure. They pointed out that with pressure, the packing sequence of A and X ions varied from hexagonal closed packing (hcp) to cubic closed packing (ccp), through their mixed packing, accompanied by the decrease in volume. The sequence change is as follows: 2 H (*hh*, ab) corresponding to hcp, → 9R (*hhchhchhc*, ababcbcac) → 4 H (*hchc*, abac) → 6 H(*hcchcc*, abcacb) → 3C(*ccc*, abc) corresponding to ccp (perovsite). A detailed explanation and discussion of these structures was given by Katz and Ward [161]. The average number of face-shared BX_6 in an octahedron chain decreases in the order of ∞ (2 H) → 3 (9R) → 2 (4 H) → 1.5 (6 H) → 1 (3C). The face-shared BX_6 can be stabilized when an attractive metal–metal (B-B) interaction, that is, B–B covalent bonding, overcomes the repulsive B–B ionic interaction [11,161]. Large A cation, which brings in large tolerance factor, also helps to expand the B–B interatomic distance and stabilize the face-shared octahedra. Pressure works their competition to lead to the corner-shared octahedra, that is, the Pv-type structure since B–B interatomic distance decreases with increase in

Table 3.3 ABX_3-type compounds with hexagonal- or rhombohedral-layered structure at ambient pressure and the structural variation with pressure.

Compound	t	2 H	9R	4 H	6 H	3C	Remarks	Reference
$BaCoO_{3-x}$	1.033	AP						[162]
$BaCrO_3$	1.023		*	*	*		(*) Mixed phases at 6–6.5 GPa	[163]
$BaIrO_3$	0.985		AP	3 GPa	5 GPa	25 GPa		[30,164–166]
$BaMnO_3$	1.033	AP	3–7.5 GPa	9 GPa				[146,167]
$BaNiO_3$	1.061	O_2 200 MPa						[168]
$BaOsO_3$	0.982				AP	17 GPa		[31]
$BaRuO_3$	0.987		AP	3 GPa	5 GPa	18 GPa		[32,33]
$BaSiO_3$	1.108		27. 9 GPa		48.5 GPa		DAC 9R(1500–1700 K), 6 H (1700–2700 K)	[169]
$BaTiO_3$(HT)	0.995				*AP	AP	(*) Oxygen-deficient	[170]
$SrIrO_3$	0.929				AP	4.5 GPa		[130]
$SrMnO_3$	0.975			AP	5–9.5 GPa			[146]
$SrNiO_3$	1.000	O_2 5 MPa						[171]
$SrSiO_3$	1.045				35 GPa		DAC 1450 ± 150 K, 10 m	[172]
$PbMnO_3$	0.986				8 GPa	15 GPa		[109]
$CsCoF_3$	1.046		AP		7 GPa		700 °C	[145]
$CsFeF_3$	1.029				AP	7 GPa	700 °C	[145]
$CsMgF_3$	1.059	*	3- 4 GPa		4–6.5 GPa		700 °C (*) No stable phase to 3 GPa	[145]
$CsMnF_3$	1.005				AP	2.6 GPa	700 °C	[145,146]
$CsNiF_3$	1.075	AP	0.5 GPa		4.7 GPa		700 °C	[145]
$CsZnF_3$	1.049	*			3–8 GPa		700 °C (*) No stable phase to 3 GPa	[145]
$RbNiF_3$	1.029				AP	2 GPa	700 °C	[146,147]
$TlNiF_3$	1.022				AP	<3 GPa		[146,148]
$CsMnCl_3$	0.951		AP		0.8	2.7 GPa	700 °C	[149]
$RbMnCl_3$	0.916				AP	0.7 GPa	700 °C	[149]

pressure and the Pv phase possesses smaller volume than layered phase. Table 3.3 shows that critical tolerance factor for phase transformation depends on the kind of cations and anions. This also supports the idea that the stabilization of face-shared octahedra is strongly related to the strength of the metal–metal interaction. The metal–metal attractive interaction is dependent on the electronic configuration of B ions and their interatomic distance. Furthermore, the covalency between metal and anion weakens the metal–metal repulsive interaction, and the metal–oxygen bond is more covalent than metal–halide bond, which weakens the metal–metal repulsive interaction. In addition, the repulsive metal–metal interaction is not well screened by monovalent anion.

3.3.3
$LiNbO_3$ as Derivatives of Perovskite

As mentioned in Section 3.3.1, when the tolerance factor of Pv, t, is not more than 0.85, the $LiNbO_3$(LN)-type phase generally appears, although there are some exceptions. The LN-type structure can be described as a derivative of the Pv-type structure [173,174] because both LN- and Pv-type compounds (general formula: ABX_3) possess three-dimensional corner-sharing BX_6 octahedra (Figure 3.6). LN-type compounds exhibit polar structures with a rhombohedral polar space group $R3c$ and the tilting of octahedra with the Glazer notation of $a^-a^-a^-$. The cooperative cation shift along the hexagonal c-direction (corresponding to the 111 direction in the primitive lattice of Pv) against closed-packed anions, for example, the oxygen layer, results in spontaneous polarization (Figure 3.6a). However, studies on functional LN-type oxides relative to Pv-type oxides have been limited because, to the best of our knowledge, except the well-known cases of $LiNbO_3$ and $LiTaO_3$, only a few LN-type compounds, such as $LiUO_3$ [175–177], $LiReO_3$ [178], $Li_{1-x}Cu_xNbO_3$ [179], (Li,Cu)TaO_3 [180] have been synthesized under ambient conditions. In contrast, in the field of earth science and high-pressure science, LN-type compounds are considered to be retrograde products of high-pressure Pv phases with $GdFeO_3$-type structure during decompression, and several LN-type oxides, such as $MnMO_3$ (M = Ti [98–101], Sn [70, 97, 98]), $FeMO_3$ (M = Ti [69–71], Ge [68]), $MgMO_3$ (M = Ti [94], Ge [95]), $ZnGeO_3$ [139,140], the phase close to (Ca, Mg, Fe) Al_2SiO_6 [181], and

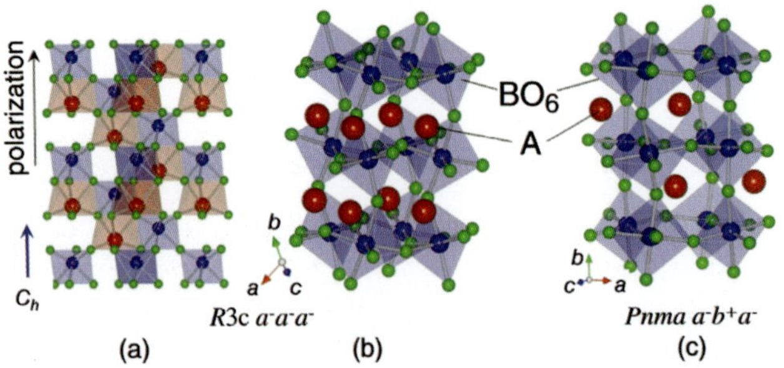

Figure 3.6 Comparison of structures of $LiNbO_3$-type ABO_3 (a and b) and $GdFeO_3$-type perovskite (c). Figures (b) and (c) are drawn from the viewpoint of tilting of corner-shared octahedra BO_6. The crystal structures were drawn by VESTA [182].

$CuTaO_3$ [63], have been reported as metastable quenched phases. Among them, Sleight *et al.* [63] and Syono *et al.* [97, 99] were the first to focus on the high-pressure synthesis of $LiNbO_3$-type oxides as functional materials. Following them more than 30 years later, novel LN-type oxides have been recently synthesized under high pressures and high temperatures. LN-type oxides containing transition metal ions, such as $FeTiO_3$ [183] and $MnMO_3$ (M = Ti, Sn) [98], show the dielectric and magnetic coupling behavior and have been examined as multiferroics. Furthermore, among the LN-type oxides, $ZnSnO_3$ [142] is composed of only d^{10} ions, Zn^{2+} and Sn^{4+}, without second-order Jahn–Teller (SOJT) d^0 ions or stereoactive $(6s)^2$ ions, indicating that the polarity is primarily originated from a polar LN-type structure obtained using high pressure and temperature. In addition, the intensities of second harmonic generation originated from noncentrosymmetry and dielectric permittivity of LN-type polar oxides with the SOJT ions with d^0 electronic configuration (Ti^{4+}, Nb^{5+}, Ta^{5+}), such as $MnTiO_3$, $FeTiO_3$, $LiNbO_3$, $LiTaO_3$, and $ZnTiO_3$, are greater than those of LN-type stannates such as $MnSnO_3$ and $ZnSnO_3$. This is attributable to the high electronic polarization of SOJT active ions [143]. These findings suggest that we might find attractive functional properties by the selection of constituent ions based on their having a naturally occurring polar LN-type structure.

As already mentioned, it has been confirmed that several high-pressure LN-type oxides are recovered as retrograde from Pv-type oxides on release of pressure, and under high pressure Pv phase possesses $GdFeO_3$-type structure. In this case, Pv is more stable phase under high pressure and LN phase is a meta-stable phase. As seen in Figure 3.6, the transformation from Pv- to LN-type phase can be attained by the change of tilting of BO_6 octahedra and the displacement of A ion in the cage, and does not need the long-range diffusion. Therefore, the transformation from Pv- to LN-type phase easily occurs on depression process. However, it is not yet clear if high-pressure LN-type phase always appears via Pv or some LN-type compounds directly appears. In addition, there was a rare case. In $Pb^{4+}Ni^{2+}O_3$[110], Pv phase was obtained by high-pressure synthesis and the obtained Pv-type phase irreversibly transformed to a LN-type phase by heat treatment at ambient pressure.

Meanwhile, $InMO_3$ (M = Cr [78, 79], Rh [78, 81]) and $ScMO_3$ (M = Cr [79, 120], Rh [81]) crystallize into Pv structure by high-pressure synthesis, although their tolerance factors make us expect the appearance of LN-type phase. The high cation repulsion between $In^{3+}(Sc^{3+})$ and M^{3+} ion of the face-shared octahedra in LN phase (see Figure 3.6a) may diminish the phase stability of LN relative to that of Pv.

3.4
Ordered High-Pressure Perovskites

The ordered arrangements of A and B cations in Pv further expand richness of Pv world [1,9,184–187]. In this section, among ordered Pvs, we focus our attention on 1:1 B-site-ordered, 1 : 3 A-site-ordered, and 1 : 1 A-site-ordered perovskites synthesized under high pressure.

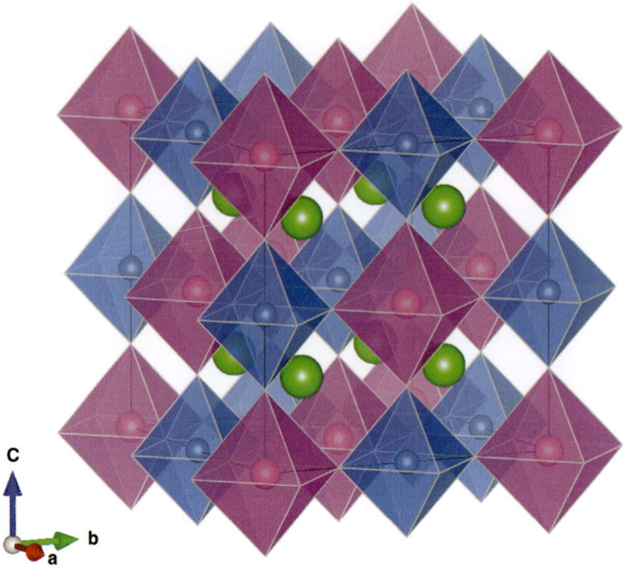

Figure 3.7 The structure of rock salt-type 1 : 1 B-site-ordered perovskite $A_2BB'O_6$. The crystal structure was drawn by VESTA [182].

3.4.1
B-Site-Ordered High-Pressure Perovskite

In Pvs with two different B ions, especially oxide $A_2BB'O_6$, there could be several valence pairs such as $A^{2+}B^{1+}B'^{7+}O_6$, $A^{2+}B^{2+}B'^{6+}O_6$, $A^{2+}B^{3+}B'^{5+}O_6$, and $A^{2+}B^{4+}B'^{4+}O_6$. If the B–B′ ion arrangement is in ordered manner, for example, in rock salt-type manner (Figure 3.7), the gain in Madelung energies increases with increasing difference in charge of the B cations [188], which leads to a lot of B-site-ordered Pvs with two different B ions with different charges. There are also a lot of reports on B-site-ordered Pvs synthesized under pressure, which are listed in Table 3.4. As seen in Table 3.4, the ordered Pvs of $A_2BB'O_6$ usually exhibit rock salt-type ordered arrangement. The ordered arrangement in $A_2BB'X_6$ would bring in the decrease in volume as well as the gain in electrostatic energy. On the other hand, in the disordered arrangement of B ions, high probability that the same B ions with greater charge are adjacent can be expected, which gives rise to the repulsive interaction between B ions with greater charge, resulting in increase in volume. Therefore, pressure would stabilize the ordered arrangement. As seen in Table 3.4, rock salt-type-ordered perovskites, including A ion with relatively small ionic radius such as Ca^{2+}, Mn^{2+}, and Sc^{3+}, exhibit monoclinic structures with space group $P2_1/n$ whose tilting of octahedra (Glazer notation: $a^-a^-c^+$) is of the same manner as $GdFeO_3$-type perovskite. As $GdFeO_3$-type structure is also observed when ionic radius of A ion is relatively small, monoclinic structure with space group $P2_1/n$ can be considered as B-site-ordered version of $GdFeO_3$-type structure. In addition, there is

Table 3.4 B-site-ordered perovskites $A_2BB'O_6$ obtained by high-pressure synthesis.

Compound	Ordering	Structure	Apparatus	Capsule	Synthetic condition	Oxidizing agents	Remarks	Reference
Ba_2CaIrO_6	RS	Cubic $Fm\bar{3}m$	Gas		60 MPa, 880 °C, 48 h	O_2		[189]
Ba_2CrNbO_6	RS	Cubic $Fm\bar{3}m$			8 GPa, 930 °C		8 H → Pv	[190]
Ba_2CrTaO_6	RS	Cubic $Fm\bar{3}m$	Belt		8 GPa, 900 °C, 40 m		8 H → Pv,	[191]
Ba_2CuOsO_6	RS	Tetragonal $I4/m$	Belt	Pt	6 GPa, 1500 °C, 1 h	$KClO_4$		[192]
Ba_2CuTeO_6	RS	Tetragonal $I4/m$	CA	Au	5 GPa, 900 °C, 10 m		4 H → Pv	[193]
Ba_2BiFeO_6	RS	Cubic $Fm\bar{3}m$	Gas, belt		70 MPa, + 4 GPa, 900 °C	O_2, $KClO_3$	$Ba_2Bi^{5+}Fe^{3+}O_6$	[194]
Bi_2NiMnO_6	RS	Monoclinic $C2$	CA	Au	6 GPa, 800 °C, 30 m			[195]
Ca_2CaOsO_6	RS	Monoclinic $P2_1/n$	Belt	Pt	6 GPa, 1400 °C, 1 h	CaO_2	Single crystal	[196]
Ca_2FeOsO_6	RS	Monoclinic $P2_1/n$	Belt	Pt	6 GPa, 1500 °C, 1 h	$KClO_4$		[197]
Ca_2InOsO_6	RS	Monoclinic $P2_1/n$	Belt	Pt	6 GPa, 1400 °C, 1 h	$KClO_4$		[198]
In_2NiMnO_6	RS	Monoclinic $P2_1/n$	Belt	Au	6 GPa, 1600 K, 2 h			[199]
La_2CuZrO_6	L	Monoclinic $P2_1/m$	CA		6 GPa, 1100 °C, 30 m		$La_2O_3 + CuO + ZrO_2$	[200]
Ln_2CuSnO_6 (Ln = Pr, Nd, Sm)	L	Monoclinic $P2_1/m$	CA		Pr: 6 GPa, 1000 °C, 30 m Nd: 6 GPa, 1200 °C, 30 m Sm: 8 GPa, 1200 °C, 30 m	—	Sm: considerable impurity phases	[200]
La_2LiFeO_6	RS	Rhombohedral, $R\bar{3}$	Belt CA	Pt	6 GPa, 900 °C, 15 m 8 GPa, 900 °C, 30 m	$KClO_3$ $KClO_4$		[201] [202]
La_2LiIrO_6	RS	Orthorhombic, $Pmm2$	Gas app. Belt		200 MPa 6 GPa, 800 °C	O_2 $KClO_3$	Pv orthorhombic	[203,204]
La_2LiVO_6	RS	Cubic $Fm\bar{3}m$	Belt		6 GPa, 900 °C, 15 m	$KClO_3$		[205]
Mn_2CrSbO_6	RS	Monoclinic $P2_1/n$	Conac-type	Au	8 GPa, 1200 °C, 15 m			[206]

(continued)

Table 3.4 (*Continued*)

Compound	Ordering	Structure	Apparatus	Capsule	Synthetic condition	Oxidizing agents	Remarks	Reference
Mn_2FeReO_6	RS	Monoclinic $P2_1/n$	MA		5 GPa, 1350 °C, 1 h			[207]
			WMA		11 GPa, 1400 °C, 20 m			[208]
Mn_2FeSbO_6	RS	Monoclinic $P2_1/n$	?	Pt	6 GPa, 900–1000 °C, 15, 30–45 m	—		[209,210]
Mn_2MnReO_6	RS	Monoclinic $P2_1/n$		Pt	5 GPa, 1400 °C, 1 h	—	—	[211]
$MnReFeSbO_6$ (Re = Eu, Gd)	RS	Monoclinic $P2_1/n$	WMA	Pt	10 GPa, 1200 °C, 20 m		A-site disordered	[212]
Sc_2NiMnO_6	RS	Monoclinic $P2_1/n$	Belt	Pt	6 GPa, 1350 °C, 2 h			[213]
Sr_2CaIrO_6	RS	Monoclinic $P2_1/n$	Belt	Pt	6 GPa, 900 °C, 5 m	$KClO_3$		[214]
Sr_2CuMoO_6	RS	Tetragonal $I4/m$	CA	Au	4 GPa, 900 °C, 1 h			[215]
Sr_2FeBiO_6	RS	Cubic $Fm\bar{3}m$	Belt		7.5 GPa, 1000 °C, 10 m	$KClO_3$	$Sr_2Fe^{3+}Bi^{5+}O_6$	[216]
Sr_2FeOsO_6	RS	Cubic $Fm\bar{3}m$	Autoclave	Au	120 MPa, 500 °C, 2 days	O_2		[217,218]
Sr_2MgIrO_6	RS	Cubic $Fm\bar{3}m$	Belt	Pt	6 GPa, 900 °C, 5 m	$KClO_3$		[214]
Sr_2MIrO_6 (M = Ni, Zn)	RS	Monoclinic $P2_1/n$			20 MPa, 900 °C, 48 h	O_2		[219]

Ordering type, RS: rock salt, L: layered.

an exception regarding ordering. $LnCuSnO_6$ (Ln = Pr, Nd, Sm) and $LaCuZrO_6$ exhibit layered ordering configuration along *c*-axis [199] the same as $LaCuSnO_6$ synthesized at ambient pressure [187,220]. The layered ordering along *c*-axis is related to the first-order Jahn-Tellar (FOJT) effect of Cu^{2+}. When the counter B ion is relatively large such as Sn^{4+} and Zr^{4+}, the rock salt-type arrangement may not be coexistent with elongated CuO_6 attributable to FOJT effect.

3.4.2 A-Site-Ordered High-Pressure Perovskite

Among A-site-ordered Pv, 1 : 3-ordered, and 1 : 1 columnar-type-ordered Pvs have been synthesized under high pressures. On the other hand, only $K_{2/3}Th_{1/3}TiO_3$ [221] has been reported as high-pressure A-site-ordered Pv with layered arrangement along *c*-axis. In this part, the former two ordered Pvs will be mentioned.

3.4.2.1 1: 3 A-Site-Ordered High-Pressure Perovskite $AA'_3B_4O_{12}$

The compounds with chemical formula of $AA'_3B_4O_{12}$ have been extensively investigated because they exhibit their rich physical properties attributable to the orbital and magnetic interactions between A or A′ and B ions in addition to those between B ions [222–225]. The structures are characterized by the A-site ordering and the tilting of corner-shared BO_6 octahedra with the Glazer notation of $a^+a^+a^+$, which doubles the Pv primitive cell, $2a_p \times 2a_p \times 2a_p$ (a_p: cell parameter of primitive Pv), as shown in Figure 3.8. When our attention is focused on A-site-ordered arrangement and the species of A ions, A ions that are alkaline metal, alkaline-earth metal, lanthanides, and bismuth ions occupy the corner and the body center of double Pv lattice, while A′ ions that are often

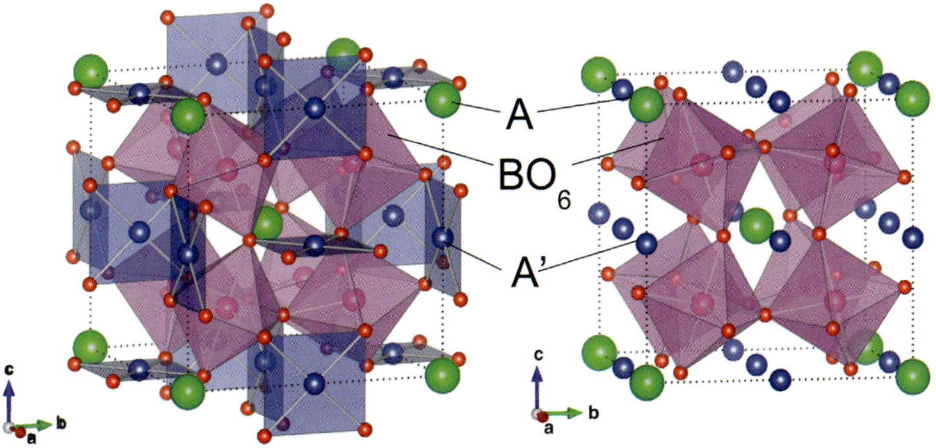

Figure 3.8 The structure of 1: 3 A-site-ordered perovskite $AA'_3B_4O_{12}$. The crystal structures were drawn by VESTA [182].

Jahn-Teller ions, such as Cu^{2+} and Mn^{3+}, occupy the face center and the centers of edge of the lattice, resulting in the ratio of A to A′ of 1 : 3. A ions are 12 coordinated by oxygen ions, while A′ ions are square coordinated. The more details are described in Chapter "Perovskite." Considerable number of compounds of this type as well as rock salt-type B-site-ordered Pvs have been synthesized under high pressure and high temperature. The compounds and synthetic conditions are listed in Table 3.5. Among them, there are the compounds that exhibit the rock salt-type 1 : 1 B-site ordering as well as 1 : 3 A-site ordering such as $CaCu_3Ga_2B_2O_{12}$ (B = Nb, Ru, Sb, Ta) and $CaCu_3Fe_2M_2O_{12}$ (M = Re, Sb, Os) (for references, see Table 3.5). Furthermore, aided by high pressure, the compounds including $Fe^{2+}(3d^6)$ and $Pd^{2+}(4d^8)$, with a preference of square coordination, as A′-site ion such as $CaFe_3Ti_4O_{12}$ [226] and $CaPd_3B_4O_{12}$ (B = Ti, V) [227], have been stabilized.

3.4.2.2 Columnar-Type 1 : 1 A-Site-Ordered High-Pressure Perovskite $AA'B_2O_6$

There are fewer reports on this type of compounds in contrast to $AA'_3B_4O_{12}$-type compounds. $CaFeTi_2O_6$ [226] has been first reported as a compound exhibiting this type of ordering. Following this, $BiNiO_3$ with the columnar-type charge ordering of Bi^{3+} and Bi^{5+} has been found [41]. Recently, $CaMnTi_2O_6$ [228] and $CaZnTi_2O_6$ were synthesized under high pressure and high temperature. More recently, (Mn*Re*)(MnSb)O_6 (*Re* = La, Pr, Nd, Sm) [212], which exhibits 1: 1 rock salt-type B-site order with cations Mn^{2+} and Sb^{5+} as well as A-site order with cations $Mn^{2+}$and Re^{3+}, was reported. All the reported compounds have been synthesized under high pressure at this time, as listed in Table 3.6. The structure of these compounds except for $BiNiO_3$ are characterized by the columnar ordering of A and A′ ions along *c*-axis and the tilting of corner-shared BO_6 octahedra with the Glazer notation of $a^+a^+c^-$, which doubles the Pv primitive cell, $2a_p \times 2a_p \times 2a_p$ (a_p: cell parameter of primitive Pv), as shown in Figure 3.9. The smaller A ion such as Fe^{2+} and Mn^{2+} have alternatively tetrahedral and square planar coordination along *c*-axis, while larger A ion such as Ca^{2+} ions have decahedral coordination. Though this type of compounds are considered to be stabilized by A/A′ ion size mismatch such as Ca^{2+} and Mn^{2+} and $a^+a^+c^-$ octahedral tilting [185], the detail has not been clarified due to few reports on this type of compounds. On the other hand, in $BiNiO_3$, the columnar type charge ordering and the different tilting of NiO_6 (the Glazer notation of $a^-b^+a^-$ ($GdFeO_3$-type)) coexist [41].

3.5 Perovskite-Related Layered Compounds Synthesized under High Pressures

As Pv-related layered compounds, Ruddelesden–Popper(RP)-type, Dion-Jacobson-type, and Aurivius-type compounds have been intensively studied. Among them, only high-pressure synthesis of RP-type compounds have been reported. RP-type compounds $A_{n+1}B_nX_{3n+1}$ composed of *n* Pv blocks interleaved one AX

Table 3.5 $AA'_3B_4O_{12}$-type compounds obtained by high-pressure synthesis.

Compound	Structure (RT)	Apparatus	Capsule	Synthetic condition	Remarks	Reference
$BiCu_3Fe_4O_{12}$	$Im\bar{3}$	CA	—	10 GPa, 1100 °C, 1 h	$KClO_4$, $BiCu^{3+}Fe^{3+}{}_4O_{12} \leftrightarrow BiCu^{2+}Fe^{3.75+}{}_4O_{12}$	[229]
$BiCu_3Mn_4O_{12}$	$Im\bar{3}$	CA	Au	6 GPa, 1000 °C, 30 m		[230]
$Bi_{0.67}Cu_3V_4O_{12}$	$Im\bar{3}$	Toroid	Pt	6–8 GPa, 700–1000 °C		[231]
$BiMn_7O_{12}$	$I2/m$	CA	Au	5 GPa, 1100 °C, 30 min		[232]
$CaCu_3Cr_4O_{12}$	$Im\bar{3}$	TA	Pt	6 GPa, 1100 °C		[233]
$CaCu_3Cr_2Sb_2O_{12}$	$Pn\bar{3}$	KOA	Au	10 GPa, 1000 °C, 1 h	B-site NaCl-type ordered	[234]
$CaCu_3Cr_2Ru_2O_{12}$	$Pn\bar{3}$	KOA	Au	12 GPa, 1150 °C, 1 h	B-site NaCl-type ordered	[235]
$CaCu_3Fe_4O_{12}$	$Im\bar{3}$	KOA	Pt	15 GPa, 1300 K, 30 m	$KClO_4$	[236]
$CaCu_3Fe_2Os_2O_{12}$	$Pn\bar{3}$	CA	Pt	8–10 GPa, 1300 °C, 30 m +800 °C, 6 h	$KClO_4$, B-site NaCl-type ordered	[237]
$CaCu_3Fe_2Re_2O_{12}$	$Pn\bar{3}$	CA		10 GPa, 1500 K, 1 h	B-site NaCl-type ordered	[238]
$CaCu_3Fe_2Sb_2O_{12}$	$Pn\bar{3}$	CA	Au	10 GPa, 1500 K, 1 h, $KClO_4$	B-site NaCl-type ordered	[239]
$CaCu_3Ga_2B_2O_{12}$ (B = Sb, Ta)	$Pn\bar{3}$	KOA	Au	12–12.5 GPa, 1100 °C, 3 h	B-site NaCl-type ordered	[240]
$CaCu_3Ga_2Nb_2O_{12}$	$Im\bar{3}$	KOA	Au	12–12.5 GPa, 1100 °C, 3 h	B-site disordered	[240]
$CaCu_3Ga_2Ru_2O_{12}$	$Im\bar{3}$	KOA	Au	12.5 GPa, 1200 °C	B-site disordered	[241]
$CaCu_3Ge_4O_{12}$	$Im\bar{3}$	CA	Au	6 GPa, 1000 °C, 30 m		[242]
$CaCu_3Ir_4O_{12}$	$Im\bar{3}$	WMA	Au	9 GPa, 900 °C, 30 m		[243]
$CaCu_3Mn_4O_{12}$	$Im\bar{3}$	Belt	Au or Pt	5 GPa, 1250 °C, 1 h		[244]
$CaCu_3Pt_4O_{12}$	$Im\bar{3}$	KOA	Au	12 GPa, 1250 °C, 30 m		[245]
$CaCu_3Sn_4O_{12}$	$Im\bar{3}$	CA	Au	6 GPa, 1000 °C, 30 m		[242]

(continued)

Table 3.5 *(Continued)*

Compound	Structure (RT)	Apparatus	Capsule	Synthetic condition	Remarks	Reference
$CaCu_3V_4O_{12}$	$Im\bar{3}$	Toroid	Pt	8 GPa, 700 °C, 3 min+, 1100 °C, 10 m		[246]
		CA	Au	9 GPa, 1000 °C, 30 m		[247]
$CaCu_2CoV_4O_{12}$	$Im\bar{3}$	Toroid	Pt	7 GPa, 1100 °C, 15 m	$Ca^{2+}(Co^{2+}Cu^{+}{}_2)(V^{4+}{}_2V^{5+}{}_2)O_{12}$	[248]
$CaFe_3Ti_4O_{12}$	$Im\bar{3}$	KOA	Fe	15 GPa, 1150 °C, 3 h		[226]
$CaMn_7O_{12}$	$R\bar{3}$	Belt	Au or Pt	8 GPa, 1000 °C, 1 h	$Ca^{2+}Mn^{3+}{}_3(M^{3+}{}_3Mn^{4+})O_1$	[249]
$Ca(Mn_{3-x}Cu_x)Mn_4O_{12}$	$Im\bar{3}$			5 GPa, 1570 K	$x = 2.2$	[250]
$CaPd_3Ti_4O_{12}$	$Im\bar{3}$	KOA	Au	15 GPa, 1000 °C, 30 m	$CaPd^{2+}{}_3T^{i4+}{}_4O_{12}$	[227]
$CaPd_2V_4O_{12}$	$Im\bar{3}$	KOA	Au	15 GPa, 1000 °C, 30 m	$CaPd^{2+}{}_2V^{4+}{}_4O_{12}$	[227]
$CdMn_7O_{12}$	$R\bar{3}$	Belt	Au or Pt	8 GPa, 1000 °C, 1 h	$Cd^{2+}Mn^{3+}{}_3\ (M^{3+}{}_3Mn^{4+})O_1$	[249]
$DyMn_3Al_4O_{12}$	$Im\bar{3}$	CA	Au	9 GPa, 900 °C, 1 h	$Dy^{3+}Mn^{3+}{}_3Al^{3+}{}_3O_1$	[251]
$LaCu_3Fe_4O_{12}$	$Im\bar{3}$	CA	—	10 GPa, 1400 K	$KClO_4$, $LaCu^{3+}Fe^{3+}{}_4O_{12} \leftrightarrow LaCu^{2+}Fe^{3.75+}{}_4O_{12}$	[252]
$LnCu_3Fe_4O_{12}$ (Ln = La –Lu)	$Im\bar{3}$	KOA	Pt	15 GPa 1000–1200 °C, 30 m, $KClO_4$	$3Cu^{2+} + 4\,Fe^{3.75+} \rightarrow 3Cu^{3+} + 4\,Fe^{3+}$ (Ln = La–Tb) $8\,Fe^{3.75+} \rightarrow 5\,Fe^{3+} + 3\,Fe^{5+}$ (Ln = Dy–Lu)	[253]
$LaCu_3Mn_4O_{12}$	$Im\bar{3}$	PC	Au	2 GPa, 1000 °C, 1 h		[254]
$LaMn_7O_{12}$	$I2/m$	Belt	Au or Pt	4 GPa, 1000 °C, 1 h	$La^{3+}Mn^{3+}{}_3(M^{3+}{}_4)O_{12}$	[249]
$LaMn_3Cr_4O_{12}$	$Im\bar{3}$	CA	—	8–10 GPa, 1000 °C, 30 m	$La^{3+}Mn^{3+}{}_3Cr^{3+}{}_4O_{12}$	[255]
$LaMn_3Ti_4O_{12}$	$Im\bar{3}$	CA	—	8–10 GPa, 1000 °C, 3 0 m	$La^{3+}Mn^{1.67+}{}_3Ti^{4+}{}_4O_{12}$	[255]
$La_xCu_3V_4O_{12}$	$Im\bar{3}$	Toroid	Pt	8 GPa, 1000 °C, 15 m	$x = 0.67, 0.84$	[256]
$LnMn_3V_4O_{12}$ (Ln = La, Nd, Gd)	$Im\bar{3}$	CA	Au	9 GPa, 900 °C, 30 m	$Ln^{3+}Mn_3V_4O_{12}$	[91]
$MnCu_3V_4O_{12}$	$Im\bar{3}$	KOA	Pt	12 GPa, 1100 °C, 20 m	$Mn^{2+}Cu_3V_4O_{12}$	[257]

$NaCu_3V_4O_{12}$	$Im\bar{3}$	Toroidal CA	Pt Au	8 GPa, 700 °C, 3 min + 1100 °C, 10 m 9 GPa, 1000 °C, 30 m		[246] [247]
$NaMn_7O_{12}$	$Im\bar{3}$	Belt	Pt	8 GPa, 1000 °C, 1 h	$Na^+Mn^{3+}{}_3(M^{3+}{}_2Mn^{4+}{}_2)O_{12}$	[258]
$NdMn_7O_{12}$	$I2/m$	Belt	Au or Pt	8 GPa, 1000 °C, 1 h	$Nd^{3+}Mn^{3+}{}_3(M^{3+}{}_4)O_{12}$	[249]
$PbMn_7O_{12}$	$R\bar{3}$			7.5 GPa, 1000 °C		[259]
$PbCoO_3$	0.839	Pv Cubic $Pn\bar{3}$ KOA	Au	12 GPa 1473 K 30 m	1: 3 A-site order, see Table 3.5 $^*Pb^{2+}Pb^{4+}{}_3Co^{2+}{}_2Co^{3+}{}_2O_{12}$	[105]
$SrCu_3Fe_4O_{12}$	$Im\bar{3}$	KOA	—	15 GPa, 1000 °C, 30 m	$KClO_4$, $Sr^{2+}Cu^{2+}{}_3Fe^{4+}{}_4O_{12}$	[260]
$SrCu_3Ti_4O_{12}$	$Im\bar{3}$	CA	Au	7.7 GPa, 1000–1200 °C, 30 m		[261]
$SrMn_7O_{12}$	$R\bar{3}$	Belt	Au or Pt	5 GPa, 1000 °C, 1 h	$Sr^{2+}Mn^{3+}{}_3(M^{3+}{}_3Mn^{4+})O_1$	[249]
$TbCu_3Mn_4O_{12}$	$Im\bar{3}$			5 GPa, 1570 K		[250]
$TmCu_3Mn_4O_{12}$	$Im\bar{3}$			5 GPa, 1570 K		[250]
$YCu_3Fe_4O_{12}$	$Im\bar{3}$	KOA	Pt	15 GPa, 1000 °C, 30 m		[262]
$YCu_3V_4O_{12}$	$Im\bar{3}$	CA	Au	9 GPa, 1000 °C, 30 m	$Y^{3+}Cu^{2+}{}_3Fe^{3.75+}{}_4O_{12} \leftrightarrow Y^{3+}Cu^{2+}{}_3Fe^{3+}{}_2Fe^{3+}{}_{0.5}Fe^{5+}{}_{1.5}O_{12}$	[247]
$YMn_3Al_4O_{12}$	$Im\bar{3}$	CA	Au	9 GPa, 900 °C, 1 h	$Y^{3+}Mn^{3+}{}_3Al^{3+}{}_3O_1$	[251,263]
$YbMn_3Al_4O_{12}$	$Im\bar{3}$	CA	Au	9 GPa, 900 °C, 1 h	$Yb^{3+}Mn^{3+}{}_3Al^{3+}{}_3O_1$	[251]

Table 3.6 Columnar-type and layered A-site-ordered perovskites $A'A''B_2O_6$ obtained by high-pressure synthesis.

	Structure	Apparatus	Capsule	Synthetic condition	Remarks	Reference
Columnar						
$BiNiO_3$	Triclin. $P\bar{1}$	CA	Au	6 GPa, 1000 °C, 30 m	$KClO_4$, $(Bi^{3+}{}_{1/2}Bi^{5+}{}_{1/2})NiO_3$	[41]
$CaFeTi_2O_6$	Tetragonal, $P4_2/nmc$	KOA	Fe	14–15 GPa, 1150–1200 °C	Centrosymmetric	[226,264]
$CaMnTi_2O_6$	Tetragonal, $P4_2mc$	CA	Au, Pt	7 GPa, 1200–1700 °C, 30 m	Polar, ferroelectric	[228]
$CaZnTi_2O_6$	Tetragonal, $P4_2mc$?	CA	Au	7.5 GPa, 1200 °C, 30 m	Noncentrosymmetric	Inaguma *et al.*, private communication
$MnRMnSbO_6$ (R = La, Pr, Nd, Sm).	Tetragonal, $P4_2/n$	WMA	Pt	10 GPa, 1200 °C, 20 m	B-site NaCl-type ordered	[212]
Layered						
$K_{2/3}Th_{1/3}TiO_3$	Tetragonal, $P4/mmm$			6 GPa, 1200 °C	Partially ordered	[221]

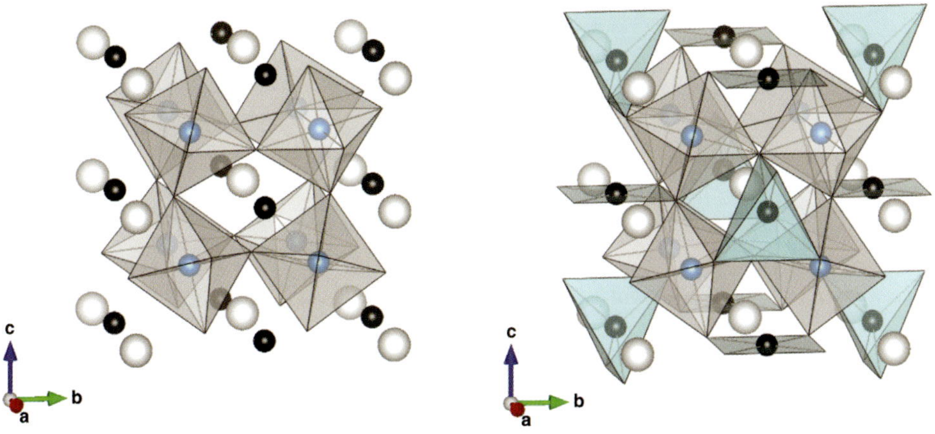

Figure 3.9 The structure of columnar-type 1 : 1 A-site-ordered perovskite $AA'B_2O_6$. The crystal structures were drawn by VESTA [182].

rock-salt layer and the chemical formula can be rewritten as $(AX)(ABX_3)_n$. A_2BX_4 ($n = 1$), which corresponds to K_2NiF_4-type structure, $A_3B_2X_7$ ($n = 2$), $A_4B_3X_{10}$ ($n = 3$) [265,266]. When $n = \infty$, the compounds correspond to Pv. Compared with Pv-type compounds, there have been very fewer reports on this type compounds synthesized under high pressure. Table 3.7 lists RP-type compounds obtained by high-pressure synthesis and the relatives. Among RP phases, the stability of K_2NiF_4-type structure in A_2BO_4 at ambient pressure has been discussed [267–269]. As for Pvs, Poix has defined the tolerance factor as follows:

$$t(\text{Poix}) = \psi_A / \sqrt{2}\beta_B \tag{3.5}$$

Here, ψ_A and β_B are constants associated with A–O and B–O distances in nine and six coordination, respectively. According to Poix, when $0.85 < t$ (Poix) < 1.02, the tetragonal K_2NiF_4-type structure would be predicted. Furthermore, Ganguli has found that the K_2NiF_4 structure appears when the ratio of ionic radius of A ion in nine coordination to that of B ion in six coordination, r_A/r_B is in the range from 1.7 to 2.4. In Table 3.7, the values of t (Poix) and r_A/r_B are also shown. If the value β_B is not given in the paper by Poix, we take the sum of ionic radii of B ion and oxide ion for β_B. For oxyfluorides and oxychlorides, we calculated the value assuming as oxides. As seen in Table 3.7, although the compounds except for Ca_2SiO_4 can be predicted to adopt K_2NiF_4-type structure at ambient pressure, the stabilization of K_2NiF_4-type structure for several compounds requires high pressure. This means that only the tolerance factor and the ratio of ionic radii cannot predict the stability of K_2NiF_4-type structure as it is for Pvs, and the covalent bonding between B and anion is also vital. Furthermore, in both $Sr_{n+1}Ir_nO_{3n+1}$ and $Sr_{n+1}Rh_nO_{3n+1}$, the phases with $n = 1$ (K_2NiF_4-type) can be synthesized at ambient pressure, while high pressure is required to stabilize the compound with n greater than unity. This is related to the

Table 3.7 Ruddelesden–Popper-type compounds obtained by high-pressure synthesis.

Compound	t(Poix)	r_A/r_B	Structure	Apparatus	Capsule	Synthetic condition	Remarks	Reference
Ba_2RuO_4	1.016	2.371	K_2NiF_4, *I4/mmm*	Belt	Au or Pt	6.5 GPa, 1200 °C		[129]
Ca_2GeO_4	0.933	2.226	K_2NiF_4, tetragonal	Bridgeman	—	10–12 GPa, 900 °C, 3–5 m		[270]
Ca_2SiO_4	1.001	2.950	K_2NiF_4, tetragonal	DAC	—	22–26 GPa, 1000 °C	Recovered phase, amorphous	[271]
$Sr_{n+1}Cu_nO_{2n+1+\delta}$	0.886	1.795	Deficient RP, tetra.	CA	Au	6 GPa, 800–900 °C, 30 m	$KClO_4$	[272]
Sr_2CrO_4	1.014	2.382	K_2NiF_4, tetra., *I4/mmm*	Belt	Au or Pt	6.5 GPa, 1000 °C		[129]
$Sr_3Cr_2O_7$	—	—	RP($n=2$), tetra., *I4/mmm*	Belt Belt	Au or Pt Au	6.5 GPa, 1000 °C 6.5–8 GPa, 1050 °C, 10 m		[129] [273]
$SrCrO_3$	—	—	Pv, cubic, $Pm\bar{3}m$	TA	—	6–6.5 GPa, 800 °C		[125]
Sr_2IrO_4	0.932	2.096	K_2NiF_4, *I4/mmm*	—		AP, 800 °C in air		[129]
$Sr_3Ir_2O_7$	—	—	RP($n=2$), tetra., *I4/mmm*	Belt	Au or Pt	>1 GPa, 1000 °C		[129]
$Sr_4Ir_3O_{10}$	—	—	RP($n=3$), tetra., *I4/mmm*	Belt	Au or Pt	>3.5 GPa, 1000 °C		[129]
$SrIrO_3$	—	—	Pv, ortho., *Pbnm*	Belt	Au or Pt	>4.5 GPa, 1000 °C	$GdFeO_3$-type	[129,130]
Sr_2RhO_4	0.943	2.183	K_2NiF_4, tetra., $I4_1/acd$	—	—	AP 1250 °C, 36 h, O_2 flow		[274,275]
$Sr_3Rh_2O_7$	—	—	RP ($n=2$), ortho., *Bbcb*	Belt	Pt	6 GPa, 1500 ° C, 1 h		[276]

$Sr_4Rh_3O_{10}$	—	—	RP($n = 3$), ortho., *Pbam*	Belt	Pt	6 GPa, 1500 °C, 1 h		[277]
$SrRhO_3$	—	—	Pv, ortho., *Pnma*	Belt	Pt	6 GPa, 1500 °C, 1 h	$KClO_4$	[132]
$Ba_2PdO_2Cl_2$	0.882	1.709	K_2NiF_4, tetra., *I4/mmm*	Belt	Pt	3 GPa, 1400 °C, 1 h	PdO_4, square plane	[278]
Sr_2CoO_3F	0.939	2.148	K_2NiF_4, tetra., *I4/mmm*	Belt	Pt	6 GPa, 1700 °C, 1.5 h		[279]
$Sr_2MnO_2Cl_2$	0.846	1.578	K_2NiF_4, tetra., *I4/mmm*	Belt	Pt	6 GPa, 1500 °C, 1 h	MnO_4, square plane	[278]
$Sr_2NiO_2Cl_2$	0.903	1.899	K_2NiF_4, tetra., *I4/mmm*	Belt	Pt	3 GPa, 1400 °C, 1 h	NiO_4, square plane	[278]
Sr_2NiO_3Cl	0.963	2.339	K_2NiF_4, tetra., *P4/mmm*	Belt	Pt	3 GPa, 1500 °C	O/Cl ordered	[280]
Sr_2NiO_3F	0.963	2.339	K_2NiF_4, tetra, *I4/mmm*	Belt	Pt	6 GPa, 1500 °C	O/F disordered at apical site	[280]

instability of Pv structure with corner-shared BO_6 octahedra at ambient pressure for their compounds, as mentioned in Section 3.3.2. As *n* decreases, the degree of corner-sharing decreases, which is associated with the stability of K_2NiF_4-type structure.

In addition, please note that many high T_c superconducting cuprates with Pv-related layered structures have been synthesized under high pressures, although this chapter has not dealt with them.

3.6
Post-Perovskite: More Dense Than Perovskite

In this section, we mention a recent topic on a derivative of high-pressure Pv, the so called "post-perovskite" (hereafter abbreviated as pPv). PPv with a chemical formula of ABX_3 in fact does not possess Pv-type structure but the same structure as $CaIrO_3$ (orthorhombic, space group *Cmcm*) as shown in Figure 3.10, and is known as more dense phase than Pv-type phase. As seen in Figure 3.10, the structure consists of layers perpendicular to *b*-axis, which are formed from corner-shared (along *c*-axis) and edge-shared BX_6 octahedra (along *a*-axis) and separated by A ions. The Earth's lower mantle is believed to be composed mainly of (Mg, Fe)SiO_3 Pv. *Ab initio* simulation and high-pressure

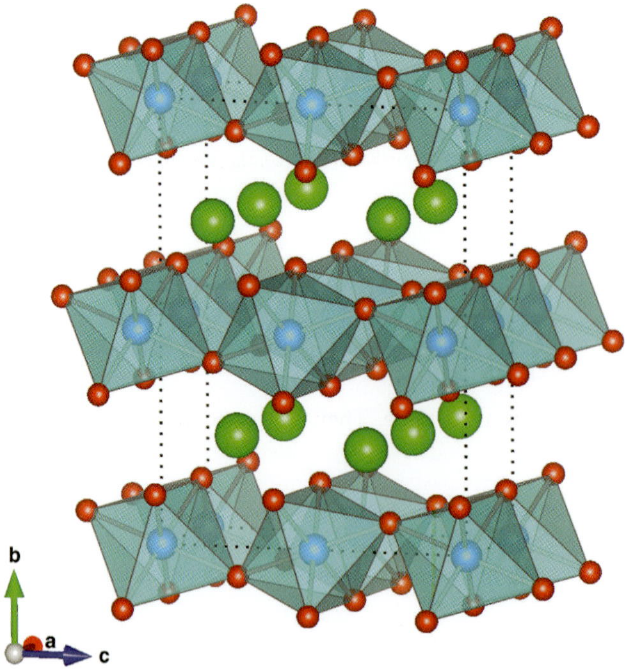

Figure 3.10 The structure of post-perovskite ABX_3. The crystal structures were drawn by VESTA [182].

experiments [281,282] have demonstrated that at pressures and temperatures of the lowermost part of the lower mantle called D″ layer (above 120 GPa and 2000 °C), $MgSiO_3$ transforms from Pv into a layered $CaIrO_3$-type phase. Therefore, the compounds with a $CaIrO_3$-type structure are generally called "pPv" among geoscientists. The discovery of pPv-type $MgSiO_3$ not only shed light on the constitution and dynamics of the D″ layer of the earth's lower mantle but also gave us a new insight on high-pressure transitions in Pv-type compounds. On the other hand, there are difficulties in experimental studies due to high transition pressure of $MgSiO_3$. Therefore, several pPvs as analogue materials with lower transition pressure than $MgSiO_3$ have been investigated. The volume change associated with Pv to pPv transition for all the reported pPv phases is about −1 to −2%. In addition, in the pressure–temperature (P–T) phase diagram, Clapeyron slope of Pv–pPv phase boundary, dP/dT is positive, which originates from a little bit greater entropy of Pv phase than that of pPv-type phase. Furthermore, in the field of materials science, pPv is considered to be a quasi-two-dimensional metal–insulator system as corner- and edge-shared BX_6 octahedra in each layer give rise to a rectangular lattice of B ions with d electrons [283–285]. The known pPv compounds at the present time are listed in Table 3.8.

The Pv to pPv phase transition and the predominant factors for the stabilization of pPv have been discussed in terms of tolerance factor, tilting of BX_6 octahedron, and covalent character of B–X bond [289,292,297,306]. As seen in Table 3.8, the Pvs with value of tolerance factor t from 0.8 to 0.9 transforms to pPvs [289]. However, several Pvs with close values of t to those of pPvs such as $ATiO_3$ (A = Ca, Fe, Mn, Cd, Zn) and $CdGeO_3$ do not transform to pPv, which implies that there are other predominant factors for the stabilization of pPvs. The transition to pPv then involves in tilting of BX_6 octahedron, and when the average tilt angle is greater than 15° at ambient pressure and increases to around 25° with pressure, the transition of Pv to pPv occurs [289,299,306]. Furthermore, Ohgushi *et al.* [292], Bremholm *et al.* [297], and Akaogi *et al.* [306] pointed out that covalent character of B–X bonds plays an important role in the stabilization of pPv structure. They compared the difference in electronegativity between B and X for pPvs and suggested that covalent character of B–X bonds is favorable for stabilization of pPv structure. Among pPvs, $CaRhO_3$, $CaIrO_3$, $CaPtO_3$, $NaOsO_3$, and $NaIrO_3$ with B ions of platinum group elements that transform to pPv at relatively low pressures are unique. They do not adopt Pv structure at ambient pressure and $CaPtO_3$ and $NaIrO_3$ with Pv-type structure have not been reported yet, although the corresponding tolerance factors predict that they crystalize into Pv phases. These compounds prefer bending B–X–B geometry approaching to 90° in pPv to linear B–X–B geometry close to 180° in Pv. This is a common feature appearing in platinum group compounds such as $SrIrO_3$, $BaRuO_3$, and $BaOsO_3$ with face-shared BO_6 octahedra as mentioned in Section 3.3.2. Therefore, the preference is attributable to covalent bonding of B–O associated with the electronic configuration of d orbital, that is, π bonding between t_{2g} orbital of platinum group B ion and 2p orbital of oxygen.

Table 3.8 Post-perovskite with $CaIrO_3$-type structure obtained by high-pressure synthesis.

Compound	*t*	Valence electron of B ion	Stable structure at AP	Transformation	Apparatus	Capsule	Synthetic condition	Remarks	Reference
$MgSiO_3$	0.900	$(2p)^6$	Orthopyroxene	Pv → pPv (113 GPa, 2400 K)	DAC	—	120 GPa, 2400 K	UQ	[286]
$MgGeO_3$	0.839	$(3d)^{10}$	Orthopyroxene	Pv → pPv (63 GPa, 1800 K)	DAC	—	70 GPa, 1800 K	UQ	[287]
$MnGeO_3$	0.865	$(3d)^{10}$	Orthopyroxene	Pv → pPv (60 GPa, 1800 K)	DAC	—	72 GPa, 1800 K	UQ	[288]
$CaSnO_3$	0.853	$(4d)^{10}$	Pv	Pv → pPv (40 GPa, 2000 K)	DAC	—	41 GPa, 1860 K	Q	[289]
$CaRuO_3$	0.882	$(4d)^4$, $(t_{2g})^4$	Pv	Pv → pPv (24 GPa, 1200 °C)	KOA	Pt	26 GPa, 1200 °C, 30 m	Q	[290]
$CaRhO_3$	0.891	$(4d)^5$, $(t_{2g})^5$	—	Pv → (Intermediate phase) → pPv,	KOA	Pt	12 GPa, 1000 °C, 1 h	Q	[291]
$CaPtO_3$	0.880	$(5d)^6$, $(t_{2g})^6$	—	pPv pPv	CA CA	Pt Au	4 GPa, 800 °C, 30 min 7 GPa, 1000 °C, 30 min	QQ	[292] [293]
$CaIrO_3$	0.880	$(5d)^5$, $(t_{2g})^5$	pPv (<1400 °C) Pv (>1400 °C)	pPv Pv-pPv (0.9–3.0 GPa, 1350–1550 °C) Pv-pPv (*P* (GPa) = 0.040, *T* (K) = 67.1)	— PC KOA	— Pt	AP 900–1100 °C 0.9 GPa, 1350 °C, 9 h 1.5 GPa, 900 °C, 1 h	Q QQ	[294] [295] [296]
$NaIrO_3$	0.926	$(5d)^4$, $(t_{2g})^4$	—	$Na_2O_2 + Ir + NaClO_3$ → pPv	CA	Au	4.5 GPa, 800 °C, 30 min	Q	[297]
$NaOsO_3$	0.924	$(5d)^3$, $(t_{2g})^3$	—	Pv → pPv $Na_2OsO_4 + KSbO_3$-type phase → pPv	KOA	Pt	16 GPa, 1135 K 6 GPa, 1100 K	Q	[298]
$NaFeF_3$	0.808	$(3d)^6$, $(t_{2g})^4(e_g)^2$	Pv	Pv → pPv	KOA	h-BN	10 GPa, 700 K	Q	[299]

$NaMgF_3$	0.866	$(2p)^6$	Pv	Pv → pPv	DAC	—	31. 3 GPa at RT	Q	[300]
				Pv → pPv	DAC	—	30 GPa, 1700 °C	PQ	[301]
$NaNiF_3$	0.879	$(3d)^8$, $(t_{2g})^6(e_g)^2$	Pv	Pv → pPv	KOA	Au or Pt	22 GPa, 700 °C, 25 h	QQ	[302]
				(16–18 GPa, 1000–1200 °C)	KOA	Au	17 GPa, 1000 °C, 30 m		[303]
$NaCoF_3$	0.855	$(3d)^7$, $(t_{2g})^5(e_g)^2$	Pv	Pv → pPv	KOA	Au or Pt	18 GPa, 700 °C, 14 h	PQ	[302]
				Pv → pPv (16.5 GPa, 1000 °C)	KOA	Au	18 GPa, 1000 ° C, 1 h	PQ	[304]
$NaZnF_3$	0.857	$(3d)^{10}$	Pv	Pv → pPv	DAC	—	25 GPa at RT	—	[305]
				Pv → pPv (16 GPa, 1000 °C)	KOA	Au	16 GPa, 8000 °C, 2 h	PQ	[306]

Remarks: Quenchability of pPv to ambient conditions, Q: completely quenchable, PS: partly quenchable. UQ: unquenchable

3.7 Final Remarks

High-pressure perovskites have been synthesized since 1960s. During the past decades, a considerable number of high-pressure perovskites have been reported. When checking the number of the references in this chapter, there seems to be three booms of high-pressure perovskite: first, late 1960s to 1975; second, 1995–2000 (when including perovskite-related high-temperature superconductors, second boom may have started from the late 1980s), and, third, from 2005 until now, which includes "post-perovskite" boom.

Their booms must be related to the improvement of high-pressure apparatus and the availability as well as discovery of novel materials such as high-temperature superconductor. Through the collaboration between earth science and materials science, solid-state chemist could have accessed a pressure over 10 GPa using Kawai-type octahedral multianvil, which would bring us high-pressure novel perovskites. One possibility of synthesis aided by pressure is to synthesize mixed anion perovskites, on which there have been few reports. Then, another possibility of high-pressure synthesis is the stabilization of metastable quenched phase such as $LiNbO_3$-type phase. In addition, as mentioned in Section 3.2.3, the monitoring of high-pressure synthesis can be performed using synchrotron X-ray with high intensity. On the other hands, with the help of the recent improvement of the X-ray detector, we again have a chance to try the *in situ* experiment in our laboratory.

Through the studies on synthesis of high-pressure perovskite, we could again realize that the thermodynamic approach and the consideration of chemical bonding are vital to elucidate the stability of perovskite, although the tolerance factor is a good parameter. The theoretical approach to evaluate the stability and predict high-pressure phase will be increasingly required, although this chapter did not deal with it.

References

1 Mitchell, R.H. (2002) *Pervoskites: Modern and Ancient*, Almaz Press.

2 Aleksandrov, K.S. (1976) The sequences of structural phase transitions in perovskites. *Ferroelectrics*, **14**, 801–805.

3 Glazer, A. (1972) The classification of tilted octahedra in perovskites. *Acta. Crystallogr. B*, **28**, 3384–3392.

4 Glazer, A. (1975) Simple ways of determining perovskite structures. *Acta Crystallogr. A*, **31**, 756–762.

5 Howard, C.J. and Stokes, H.T. (1998) Group-theoretical analysis of octahedral tilting in perovskites. *Acta. Crystallogr. B*, **54**, 782–789.

6 Howard, C.J. and Stokes, H.T. (2002) Group-theoretical analysis of octahedral tilting in perovskites. *Acta Crystallogr. B (Erratum).*, **58**, 565.

7 Megaw, H.D. (1973) *Crystal Structures: A Working Approach*, Saunders, Philadelphia.

8 O'Keeffe, M. and Hyde, B.G. (1977) Some structures topologically related to cubic perovskite ($E2_1$), ReO_3 ($D0_9$) and Cu_3Au ($L1_2$). *Acta. Crystallogr. B*, **33**, 3802–3813.

9 Woodward, P. (1997) Octahedral tilting in perovskites. I. Geometrical considerations. *Acta. Crystallogr. B*, **53**, 32–43.
10 Woodward, P. (1997) Octahedral tilting in perovskites. II. Structure stabilizing forces. *Acta. Crystallogr. B*, **53**, 44–66.
11 Goodenough, J.B., Kafalas, J.A., and Longo, J.M. (1972) High-pressure synthesis, in *Preparative Methods in Solid State Chemistry* (ed. P. Hagenmuller), Academic Press, pp. 1–69.
12 Rao, C.N.R. (1994) *Chemical Approaches to the Synthesis of Inorganic Materials*, John Wiley & Sons, Inc., New York.
13 Ito, E. (2007) Theory and practice: multianvil cells and high-pressure experimental methods, in *Treatise on Geophysics* (ed. G. Schubert), Elsevier, Amsterdam, pp. 197–230.
14 Navrotsky, A. (1994) *Physics and Chemistry of Earth Materials*, Cambridge University Press.
15 Rooymans, C.J.M. (1972) High-pressure techniques in preparative chemistry, in *Preparative Methods in Solid State Chemistry* (ed. P. Hagenmuller), Academic Press, pp. 71–132.
16 Inaguma, Y., Katsumata, T., Wang, R., Kobashi, K., Itoh, M., Shan, Y.-J., and Nakamura, T. (2001) Synthesis and dielectric properties of a perovskite $Bi_{1/2}Ag_{1/2}TiO_3$. *Ferroelectrics*, **264**, 127–132.
17 Inaguma, Y., Miyaguchi, A., and Katsumata, T. (2003) Synthesis and lattice distortion of ferroelectric/antiferroelectric Bi (III)-containing perovskites. *Mater. Res. Soc. Symp. Proc.*, **755**, 471–476.
18 Kubota, B. (1961) Decomposition of higher oxides of chromium under various pressures of oxygen. *J. Am. Ceram. Soc.*, **44**, 239–248.
19 Kume, S., Kanamaru, F., Shibasaki, Y., Koizumi, M., Yasunami, K., and Fukuda, T. (1971) Generation of high pressure oxygen and synthesis of $CoCrO_4$ with $CrVO_4$ type structure. *Rev. Sci. Instrum.*, **42**, 1856–1858.
20 MacChesney, J.B., Potter, J.F., Sherwood, R.C., and Williams, H.J. (1965) Oxygen stoichiometry in the barium ferrates: its effect on magnetization and resistivity. *J. Chem. Phys.*, **43**, 3317–3322.
21 MacChesney, J.B., Sherwood, R.C., and Potter, J.F. (1965) Electric and magnetic properties of the strontium ferrates. *J. Chem. Phys.*, **43**, 1907–1913.
22 Lacorre, P., Torrance, J.B., Pannetier, J., Nazzal, A.I., Wang, P.W., and Huang, T.C. (1991) Synthesis, crystal structure, and properties of metallic $PrNiO_3$: comparison with metallic $NdNiO_3$ and semiconducting $SmNiO_3$. *J. Solid State Chem.*, **91**, 225–237.
23 Moriga, T., Usaka, O., Nakabayashi, I., Hirashima, Y., Kohno, T., Kikkawa, S., and Kanamaru, F. (1994) Reduction of the perovskite-type $LnNiO_3$ (Ln=Pr, Nd) to $Ln_3Ni_3O_7$ with monovalent nickel ions. *Solid State Ionics*, **74**, 211–217.
24 Alonso, J.A., Martinez-Lope, M.J., de la Calle, C., and Pomjakushin, V. (2006) Preparation and structural study from neutron diffraction data of $RCoO_3$ (R=Pr, Tb, Dy, Ho, Er, Tm, Yb, Lu) perovskites. *J. Mater. Chem.*, **16**, 1555–1560.
25 Sleight, A.W. (1969) Tungsten and molybdenum oxyfluorides of the type $MO_{3-x}F_x$. *Inorg. Chem.*, **8**, 1764–1767.
26 Liebertz, J. and Rooymans, C.J.M. (1965) Die ilmenit/perowskit-phasenumwandlung von $CdTiO_3$ unter hohem Druck. *Z. Phys. Chem.*, **44**, 242–249.
27 Kennedy, B.J., Zhou, Q., and Avdeev, M. (2011) The ferroelectric phase of $CdTiO_3$: a powder neutron diffraction study. *J. Solid State Chem.*, **184**, 2987–2993.
28 Navrotsky, A. (1987) High pressure transitions in silicates. *Prog. Solid State Chem.*, **17**, 53–86.
29 Neil, J.M., Navrotsky, A., and Kleppa, O.J. (1971) Enthalpy of ilmenite–perovskite transformation in cadmium titanate. *Inorg. Chem.*, **10**, 2076–2077.
30 Cheng, J.G., Ishii, T., Kojitani, H., Matsubayashi, K., Matsuo, A., Li, X., Shirako, Y., Zhou, J.S., Goodenough, J.B., Jin, C.Q. *et al.* (2013) High-pressure synthesis of the $BaIrO_3$ perovskite: a Pauli paramagnetic metal with a Fermi liquid ground state. *Phys. Rev. B*, **88**, 205114.

31 Shi, Y., Guo, Y., Shirako, Y., Yi, W., Wang, X., Belik, A.A., Matsushita, Y., Feng, H.L., Tsujimoto, Y., Arai, M. *et al.* (2013) High-pressure synthesis of 5d cubic perovskite $BaOsO_3$ at 17GPa: ferromagnetic evolution over 3d to 5d series. *J. Am. Chem. Soc.*, **135**, 16507–16516.

32 Jin, C.Q., Zhou, J.S., Goodenough, J.B., Liu, Q.Q., Zhao, J.G., Yang, L.X., Yu, Y., Yu, R.C., Katsura, T., Shatskiy, A. *et al.* (2008) High-pressure synthesis of the cubic perovskite $BaRuO_3$ and evolution of ferromagnetism in $ARuO_3$ (A=Ca, Sr, Ba) ruthenates. *Proc. Natl. Acad. Sci. USA*, **105**, 7115–7119.

33 Longo, J.M. and Kafalas, J.A. (1968) Pressure-induced structural changes in the system $Ba_{1-x}Sr_xRuO_3$. *Mater. Res. Bull.*, **3**, 687–692.

34 Belik, A.A., Wuernisha, T., Kamiyama, T., Mori, K., Male, M., Nagai, T., Matsui, Y., and Takayama-Muromachi, E. (2006) High-pressure synthesis, crystal structures, and properties of perovskite-like $BiAlO_3$ and pyroxene-like $BiGaO_3$. *Chem. Mater.*, **18**, 133–139.

35 Belik, A.A., Iikubo, S., Kodama, K., Igawa, N., Shamoto, S.-i., Niitaka, S., Azuma, M., Shimakawa, Y., Takano, M., Izumi, F. *et al.* (2006) Neutron powder diffraction study on the crystal and magnetic structures of $BiCoO_3$. *Chem. Mater.*, **18**, 798–803.

36 Sugawara, F., Iida, S., Syono, Y., and Akimoto, S.-i. (1965) New magnetic perovskites $BiMnO_3$ and $BiCrO_3$. *J. Phys. Soc. Jpn.*, **20**, 1529–1529.

37 Yusa, H., Belik, A.A., Takayama-Muromachi, E., Hirao, N., and Ohishi, Y. (2009) High-pressure phase transitions in $BiMO_3$ (M=Al, Ga, and In): *in situ* X-ray diffraction and Raman scattering experiments. *Phys. Rev. B*, **80**, 214103.

38 Belik, A.A., Rusakov, D.A., Furubayashi, T., and Takayama-Muromachi, E. (2012) $BiGaO_3$-based perovskites: a large family of polar materials. *Chem. Mater.*, **24**, 3056–3064.

39 Belik, A.A., Stefanovich, S.Y., Lazoryak, B.I., and Takayama-Muromachi, E. (2006) $BiInO_3$: a polar oxide with $GdFeO_3$-type perovskite structure. *Chem. Mater.*, **18**, 1964–1968.

40 Atou, T., Chiba, H., Ohoyama, K., Yamaguchi, Y., and Syono, Y. (1999) Structure determination of ferromagnetic perovskite $BiMnO_3$. *J. Solid State Chem.*, **145**, 639–642.

41 Ishiwata, S., Azuma, M., Takano, M., Nishibori, E., Takata, M., Sakata, M., and Kato, K. (2002) High pressure synthesis, crystal structure and physical properties of a new Ni(II) perovskite $BiNiO_3$. *J. Mater. Chem.*, **12**, 3733–3737.

42 Yi, W., Liang, Q., Matsushita, Y., Tanaka, M., Hu, X., and Belik, A.A. (2013) Crystal structure and properties of high-pressure-synthesized $BiRhO_3$, $LuRhO_3$, and $NdRhO_3$. *J. Solid State Chem.*, **200**, 271–278.

43 Belik, A.A., Iikubo, S., Kodama, K., Igawa, N., Shamoto, S.I., Maie, M., Nagai, T., Matsui, Y., Stefanovich, S.Y., Lazoryak, B.I. *et al.* (2006) $BiScO_3$: centrosymmetric $BiMnO_3$-type oxide. *J. Am. Chem. Soc.*, **128**, 706–707.

44 Tomashpol'skii, Y.Y., Zubova, E.V., Burdina, K.P., and Venevtsev, Y.N. (1968) X-ray diffraction study of new perovskites synthesized under high pressures. *Kristallografiya*, **13**, 987–990.

45 Arévalo-López, Á.M., Dos santos-García, A.J., Levin, J.R., Attfield, J.P., and Alario-Franco, M.A. (2015) Spin-glass behavior and incommensurate modulation in high-pressure perovskite $BiC_{r0.5}Ni_{0.5}O_3$. *Inorg. Chem.*, **54**, 832–836.

46 Khalyavin, D.D., Salak, A.N., Vyshatko, N.P., Lopes, A.B., Olekhnovich, N.M., Pushkarev, A.V., Maroz, I.I., and Radyush, Y.V. (2006) Crystal structure of metastable perovskite $Bi(Mg_{1/2}Ti_{1/2})O_3$: Bi-based structural analogue of antiferroelectric $PbZrO_3$. *Chem. Mater.*, **18**, 5104–5110.

47 Inaguma, Y. and Katsumata, T. (2003) High pressure synthesis, lattice distortion, and dielectric properties of a perovskite $Bi(Ni_{1/2}Ti_{1/2})O_3$. *Ferroelectrics*, **286**, 111–117.

48 Suchomel, M.R., Fogg, A.M., Allix, M., Niu, H., Claridge, J.B., and Rosseinsky, M.J. (2006) Bi_2ZnTiO_6: a lead-free closed-shell polar perovskite with a

calculated ionic polarization of 150 μC cm^{-2}. *Chem. Mater.*, **18**, 4987–4989.

49 Yu, R., Hojo, H., Oka, K., Watanuki, T., Machida, A., Shimizu, K., Nakano, K., and Azuma, M. (2015) New $PbTiO_3$-type giant tetragonal compound Bi_2ZnVO_6 and its stability under pressure. *Chem. Mater.*, **27**, 2012–2017.

50 Kanamaru, F., Miyamoto, H., Mimura, Y., Koizumi, M., Shimada, M., Kume, S., and Shin, S. (1970) Synthesis of a new perovskite $CaFeO_3$. *Mater. Res. Bull.*, **5**, 257–261.

51 Takeda, Y., Naka, S., Takano, M., Shinjo, T., Takada, T., and Shimada, M. (1978) Preparation and characterization of stoichiometric $CaFeO_3$. *Mater. Res. Bull.*, **13**, 61–66.

52 Ringwood, A.E. and Major, A. (1967) Some high-pressure transformations of geophysical significance. *Earth Planet Sci. Lett.*, **2**, 106–110.

53 Prewitt, C.T. and Sleight, A.W. (1969) Garnet-like structures of high-pressure cadmium germanate and calcium germanate. *Science*, **163**, 386.

54 Sasaki, S., Prewitt, C.T., and Liebermann, R.C. (1983) The crystal structure of $CaGeO_3$ perovskite and the crystal chemistry of the $GdFeO_3$-type perovskites. *Am. Mineral.*, **68**, 1189–1198.

55 Ross, N.L., Akaogi, M., Navrotsky, A., Susaki, J.-i., and McMillan, P. (1986) Phase transitions among the $CaGeO_3$ polymorphs (wollastonite, garnet, and perovskite structures): studies by high-pressure synthesis, high-temperature calorimetry, and vibrational spectroscopy and calculation. *J. Geophys. Res. Solid Earth*, **91**, 4685–4696.

56 Yamamoto, A., Khasanova, N.R., Izumi, F., Wu, X.J., Kamiyama, T., Torii, S., and Tajima, S. (1999) Crystal structure and its role in electrical properties of the perovskite $CaPbO_3$ synthesized at high pressure. *Chem. Mater.*, **11**, 747–753.

57 Yamaura, K. and Takayama-Muromachi, E. (2006) High-pressure synthesis of the perovskite rhodate $CaRhO_3$. *Physica C Supercond. Appl.*, **445–448**, 54–56.

58 Liu, L.G. and Ringwood, A.E. (1975) Synthesis of a perovskite-type polymorph of $CaSiO_3$. *Earth Planet Sci. Lett.*, **28**, 209–211.

59 Uchida, T., Wang, Y., Nishiyama, N., Funakoshi, K.i., Kaneko, H., Nozawa, A., Von Dreele, R.B., Rivers, M.L., Sutton, S.R., Yamada, A. *et al.* (2009) Non-cubic crystal symmetry of $CaSiO_3$ perovskite up to 18GPa and 1600K. *Earth Planet Sci. Lett.*, **282**, 268–274.

60 Susaki, J.i. (1989) $CdGeO_3$-phase transformations at high pressure and temperature and structural refinement of the perovskite polymorph. *Phys. Chem. Miner.*, **16**, 634–641.

61 Shannon, R.D., Gillson, J.L., and Bouchard, R.J. (1977) Single crystal synthesis and electrical properties of $CdSnO_3$, Cd_2SnO_4, In_2TeO_6 and $CdIn_2O_4$. *J. Phys. Chem. Solids*, **38**, 877–881.

62 Liebermann, R.C. (1976) Elasticity of the ilmenite–perovskite phase transformation in $CdTiO_3$. *Earth Planet Sci. Lett.*, **29**, 326–332.

63 Sleight, A.W. and Prewitt, C.T. (1970) Preparation of $CuNbO_3$ and $CuTaO_3$ at high pressure. *Mater. Res. Bull.*, **5**, 207–211.

64 Demazeau, G., Marbeuf, A., Pouchard, M., and Hagenmuller, P. (1971) Sur une série de composés oxygènes du nickel trivalent derivés de la perovskite. *J. Solid State Chem.*, **3**, 582–589.

65 Alonso, J.A., Martínez-Lope, M.J., Casais, M.T., Martínez, J.L., Demazeau, G., Largeteau, A., García-Muñoz, J.L., Muñoz, A., and Fernández-Díaz, M.T. (1999) High-pressure preparation, crystal structure, magnetic properties, and phase transitions in $GdNiO_3$ and $DyNiO_3$ perovskites. *Chem. Mater.*, **11**, 2463–2469.

66 Demazeau, G., Pouchard, M., and Hagenmuller, P. (1974) Sur de nouveaux composés oxygénés du cobalt +III dérivés de la perovskite. *J. Solid State Chem.*, **9**, 202–209.

67 Uusi-Esko, K., Malm, J., Imamura, N., Yamauchi, H., and Karppinen, M. (2008) Characterization of $RMnO_3$ (R=Sc, Y, Dy-Lu): high-pressure synthesized metastable perovskites and their hexagonal precursor phases. *Mater. Chem. Phys.*, **112**, 1029–1034.

68 Hattori, T., Matsuda, T., Tsuchiya, T., Nagai, T., and Yamanaka, T. (1999). Clinopyroxene-perovskite phase transition of $FeGeO_3$ under high pressure and room temperature. *Phys Chem Miner* **26**, 212–216.

69 Ito, E. and Matsui, Y. (1979) High-pressure transformations in silicates, germanates, and titanates with ABO_3 stoichiometry. *Phys. Chem. Miner.*, **4**, 265–273.

70 Leinenweber, K., Utsumi, W., Tsuchida, Y., Yagi, T., and Kurita, K. (1991) Unquenchable high-pressure perovskite polymorphs of $MnSnO_3$ and $FeTiO_3$. *Phys. Chem. Miner.*, **18**, 244–250.

71 Mehta, A., Leinenweber, K., Navrotsky, A., and Akaogi, M. (1994) Calorimetric study of high pressure polymorphism in $FeTiO_3$: stability of the perovskite phase. *Phys. Chem. Miner.*, **21**, 207–212.

72 Arielly, R., Xu, W.M., Greenberg, E., Rozenberg, G.K., Pasternak, M.P., Garbarino, G., Clark, S., and Jeanloz, R. (2011) Intriguing sequence of $GaFeO_3$ structures and electronic states to 70GPa. *Phys. Rev. B*, **84**, 094109.

73 Marezio, M., Remeika, J.P., and Dernier, P.D. (1966) High pressure synthesis of $YGaO_3$, $GdGaO_3$ and $YbGaO_3$. *Mater. Res. Bull.*, **1**, 247–255.

74 Sleight, A.W. and Prewitt, C.T. (1973) High-pressure $HgTiO_3$ and $HgPbO_3$: preparation, characterization, and structure. *J. Solid State Chem.*, **6**, 509–512.

75 Yoshida, M., Katsumata, T., and Inaguma, Y. (2008) High-pressure synthesis, crystal and electronic structures, and transport properties of a novel perovskite $HgSnO_3$. *Inorg. Chem.*, **47**, 6296–6302.

76 Waintal, A., Capponi, J.J., Bertaut, E.F., Contré, M., and François, D. (1966) Transformation sous haute pression de la forme ferroelectrique de $MnHoO_3$ en une forme perovskite paraelectrique. *Solid State Commun.*, **4**, 125–127.

77 Alonso, J.A., Martínez-Lope, M.J., Casais, M.T., Aranda, M.A.G., and Fernández-Díaz, M.T. (1999) Metal–insulator transitions, structural and microstructural evolution of $RNiO_3$ (R=Sm, Eu, Gd, Dy, Ho, Y) perovskites: evidence for room-temperature charge disproportionation in monoclinic $HoNiO_3$ and $YNiO_3$. *J. Am. Chem. Soc.*, **121**, 4754–4762.

78 Shannon, R.D. (1967) Synthesis of some new perovskites containing indium and thallium. *Inorg. Chem.*, **6**, 1474–1478.

79 Belik, A.A., Matsushita, Y., Tanaka, M., and Takayama-Muromachi, E. (2012) Crystal structures and properties of perovskites $ScCrO_3$ and $InCrO_3$ with small ions at the A site. *Chem. Mater.*, **24**, 2197–2203.

80 Fujita, K., Kawamoto, T., Yamada, I., Hernandez, O., Hayashi, N., Akamatsu, H., Lafargue-Dit-Hauret, W., Rocquefelte, X., Fukuzumi, M., Manuel, P. *et al.* (2016) $LiNbO_3$-type $InFeO_3$: room-temperature polar magnet without second-order Jahn–Teller active ions. *Chem. Mater*, **28** (18), 6644–6655.

81 Belik, A.A., Matsushita, Y., Tanaka, M., and Takayama-Muromachi, E. (2013) High-pressure synthesis, crystal structures, and properties of $ScRhO_3$ and $InRhO_3$ perovskites. *Inorg. Chem.*, **52**, 12005–12011.

82 Belik, A.A., Matsushita, Y., Tanaka, M., and Takayama-Muromachi, E. (2010) $(In_{1-y}Mn_y)MnO_3$ ($1/9 \leq y \leq 1/3$): unusual perovskites with unusual properties. *Angew. Chem., Int. Ed.*, **49**, 7723–7727.

83 Belik, A.A., Furubayashi, T., Matsushita, Y., Tanaka, M., Hishita, S., and Takayama-Muromachi, E. (2009) Indium-based perovskites: a new class of near-room-temperature multiferroics. *Angew. Chem., Int. Ed.*, **48**, 6117–6120.

84 Belik, A.A., Furubayashi, T., Yusa, H., and Takayama-Muromachi, E. (2011) Perovskite, $LiNbO_3$, corundum, and hexagonal polymorphs of $(In_{1-x}M_x)MO_3$. *J. Am. Chem. Soc.*, **133**, 9405–9412.

85 Khasanova, N.R., Izumi, F., Kamiyama, T., Yoshida, K., Yamamoto, A., and Tajima, S. (1999) Crystal structure of the $(K_{0.87}Bi_{0.13})BiO_3$ superconductor. *J. Solid State Chem.*, **144**, 205–208.

86 Khasanova, N.R., Yamamoto, A., Tajima, S., Wu, X.J., and Tanabe, K. (1998)

Superconductivity at 10.2K in the K–Bi–O system. *Physica C Supercond. Appl.*, **305**, 275–280.

87 Demazeau, G., Parent, C., Pouchard, M., and Hagenmuller, P. (1972) Sur deux nouvelles phases oxygenees du cuivre trivalent: $LaCuO_3$ et $La_2Li_{0.50}Cu_{0.50}O_4$. *Mater. Res. Bull.*, **7**, 913–920.

88 Demazeau, G., Siberchicot, B., Matar, S., Gayet, C., and Largeteau, A. (1994) A new ferromagnetic oxide La_2MnIrO_6: synthesis, characterization, and calculation of its electronic structure. *J. Appl. Phys.*, **75**, 4617–4620.

89 Kim, S., Lemaux, S., Demazeau, G., Kim, J., and Choy, J. (2001) $LaPdO_3$: the first PdIII oxide with the perovskite structure. *J. Am. Chem. Soc.*, **123**, 10413–10414.

90 Shi, Y., Guo, Y., Wang, X., Princep, A.J., Khalyavin, D., Manuel, P., Michiue, Y., Sato, A., Tsuda, K., Yu, S. *et al.* (2013) A ferroelectric-like structural transition in a metal. *Nat. Mater.*, **12**, 1024–1027.

91 Shimakawa, Y., Zhang, S., Saito, T., Lufaso, M.W., and Woodward, P.M. (2014) Order–disorder transition involving the A-site cations in $Ln^{3+}Mn_3V_4O_{12}$ perovskites. *Inorg. Chem.*, **53**, 594–599.

92 Liu, L.-G. (1975) Post-oxide phases of forsterite and enstatite. *Geophys. Res. Lett.*, **2**, 417–419.

93 Ito, E. and Matsui, Y. (1978) Synthesis and crystal-chemical characterization of $MgSiO_3$ perovskite. *Earth Planet Sci. Lett.*, **38**, 443–450.

94 Linton, J.A., Yingwei, F., and Navrotsky, A. (1997) Complete Fe–Mg solid solution in lithium niobate and perovskite structures in titanates at high pressures and temperatures. *Am. Mineral.*, **82**, 639–642.

95 Liu, L.-G. (1977) Post-ilmenite phases of silicates and germanates. *Earth Planet Sci. Lett.*, **35**, 161–168.

96 Tyutyunnik, A.P., Zubkov, V.G., D'Yachkova, T.V., Zainulin, Y.G., Tarakina, N.V., Sayagués, M.J., and Svensson, G. (2002) High-pressure high-temperature synthesis of $Mn_4Nb_2O_9$: a XRD and TEM study. *Solid State Sci.*, **4**, 941–949.

97 Syono, Y., Sawamoto, H., and Akimoto, S. (1969) Disordered ilmenite $MnSnO_3$ and its magnetic property. *Solid State Commun.*, **7**, 713–716.

98 Aimi, A., Katsumata, T., Mori, D., Fu, D., Itoh, M., Kyômen, T., Hiraki, K., Takahashi, T., and Inaguma, Y. (2011) High-pressure synthesis and correlation between structure, magnetic, and dielectric properties in $LiNbO_3$-type $MnMO_3$ (M=Ti, Sn). *Inorg. Chem.*, **50**, 6392–6398.

99 Syono, Y., Akimoto, S.-i., Ishikawa, Y., and Endoh, Y. (1969) A new high pressure phase of $MnTiO_3$ and its magnetic property. *J. Phys. Chem. Solids*, **30**, 1665–1672.

100 Ko, J. and Prewitt, C.T. (1988) High-pressure phase transition in manganese titanate ($MnTiO_3$) from the ilmenite to the lithium niobate ($LiNbO_3$) structure. *Phys. Chem. Miner.*, **15**, 355–362.

101 Ross, N.L., Ko, J., and Prewitt, C.T. (1989) A new phase transition in $MnTiO_3$: $LiNbO_3$-perovskite structure. *Phys. Chem. Miner.*, **16**, 621–629.

102 Syono, Y., Akimoto, S.-i., and Endoh, Y. (1971) High pressure synthesis of ilmenite and perovskite type $MnVO_3$ and their magnetic properties. *J. Phys. Chem. Solids*, **32**, 243–249.

103 Shi, Y.G., Guo, Y.F., Yu, S., Arai, M., Belik, A.A., Sato, A., Yamaura, K., Takayama-Muromachi, E., Tian, H.F., Yang, H.X. *et al.* (2009) Continuous metal-insulator transition of the antiferromagnetic perovskite $NaOsO_3$. *Phys. Rev. B Condens. Matter. Mater. Phys.*, **80**, 169907.

104 Mizoguchi, H., Woodward, P.M., Byeon, S.-H., and Parise, J.B. (2004) Polymorphism in $NaSbO_3$: structure and bonding in metal oxides. *J. Am. Chem. Soc.*, **126**, 3175–3184.

105 Sakai, Y., Yang, J., Yu, R., Hojo, H., Yamada, I., Miao, P., Lee, S., Torii, S., Kamiyama, T., Ležaić, M., Long, Y., Azuma, M. et al. (2017). A-Site and B-Site Charge Orderings in an s–d Level Controlled Perovskite Oxide $PbCoO_3$. *J Am Chem Soc.*, DOI: 10.1021/jacs.7b01851

106 Roth, W.L. and Devries, R.C. (1967) Crystal and magnetic structure of $PbCrO_3$. *J. Appl. Phys.*, **38**, 951–952.

107 Yu, R., Hojo, H., Watanuki, T., Mizumaki, M., Mizokawa, T., Oka, K., Kim, H., Machida, A., Sakaki, K., Nakamura, Y. *et al.* (2015) Melting of Pb charge glass and simultaneous Pb–Cr charge transfer in $PbCrO_3$ as the origin of volume collapse. *J. Am. Chem. Soc.*, **137**, 12719–12728.

108 Tsuchiya, T., Saito, H., Yoshida, M., Katsumata, T., Ohba, T., Inaguma, Y., Tsurui, T., and Shikano, M. (2007) High-pressure synthesis of a novel $PbFeO_3$. *Mater. Res. Soc. Symp. Proc.*, **988E**, 0988-QQ0909-0916.

109 Oka, K., Azuma, M., Hirai, S., Belik, A.A., Kojitani, H., Akaogi, M., Takano, M., and Shimakawa, Y. (2009) Pressure-induced transformation of 6H hexagonal to 3C perovskite structure in $PbMnO_3$. *Inorg. Chem.*, **48**, 2285–2288.

110 Inaguma, Y., Tanaka, K., Tsuchiya, T., Mori, D., Katsumata, T., Ohba, T., Hiraki, K., Takahashi, T., and Saitoh, H. (2011) Synthesis, structural transformation, thermal stability, valence state, and magnetic and electronic properties of $PbNiO_3$ with perovskite- and $LiNbO_3$-type structures. *J. Am. Chem. Soc.*, **133**, 16920–16929.

111 Cheng, J.G., Zhou, J.S., and Goodenough, J.B. (2009) Metal–metal transition in perovskite $PbRuO_3$. *Phys. Rev. B*, **80**, 174426.

112 Kafalas, J.A. and Longo, J.M. (1970) Pressure-induced pyrochlore to perovskite transformations in the $Sr_{1-x}Pb_xRuO_3$ system. *Mater. Res. Bull.*, **5**, 193–198.

113 Sugawara, F., Syono, Y., and Akimoto, S. (1968) High pressure synthesis of a new perovskite $PbSnO_3$. *Mater. Res. Bull.*, **3**, 529–532.

114 Shpanchenko, R.V., Chernaya, V.V., Tsirlin, A.A., Chizhov, P.S., Sklovsky, D.E., Antipov, E.V., Khlybov, E.P., Pomjakushin, V., Balagurov, A.M., Medvedeva, J.E. *et al.* (2004) Synthesis, structure, and properties of new perovskite $PbVO_3$. *Chem. Mater.*, **16**, 3267–3273.

115 Belik, A.A., Azuma, M., Saito, T., Shimakawa, Y., and Takano, M. (2005) Crystallographic features and tetragonal phase stability of $PbVO_3$, a new member of $PbTiO_3$ family. *Chem. Mater.*, **17**, 269–273.

116 Mori, D., Tanaka, K., Saitoh, H., Kikegawa, T., and Inaguma, Y. (2015) Synthesis, direct formation under high pressure, structure, and electronic properties of $LiNbO_3$-type oxide $PbZnO_3$. *Inorg. Chem.*, **54**, 11405–11410.

117 Tsuchiya, T., Katsumata, T., Ohba, T., and Inaguma, Y. (2009) High-pressure synthesis and characterization of a novel perovskite $PbFe_{1/2}V_{1/2}O_3$. *J. Ceram. Soc. Jpn.*, **117**, 102–105.

118 Reid, A.F. and Ringwood, A.E. (1975) High-pressure modification of $ScAlO_3$ and some geophysical implications. *J. Geophys. Res.*, **80**, 3363–3370.

119 Sinclair, W., Eggleton, R., and Ringwood, A. (1979) Crystal synthesis and structure refinement of high-pressure $ScAlO_3$ perovskite. *Z. Kristallogr. Cryst. Mater.*, **149**, 307.

120 Park, J.H. and Parise, J.B. (1997) High pressure synthesis of a new chromite, $ScCrO_3$. *Mater. Res. Bull.*, **32**, 1617–1624.

121 Kawamoto, T., Fujita, K., Yamada, I., Matoba, T., Kim, S.J., Gao, P., Pan, X., Findlay, S.D., Tassel, C., Kageyama, H. *et al.* (2014) Room-temperature polar ferromagnet $ScFeO_3$ transformed from a high-pressure orthorhombic perovskite phase. *J. Am. Chem. Soc.*, **136**, 15291–15299.

122 Chen, H., Yu, T., Gao, P., Bai, J., Tao, J., Tyson, T.A., Wang, L., and Lalancette, R. (2013) Synthesis and structure of perovskite $ScMnO_3$. *Inorg. Chem.*, **52**, 9692–9697.

123 Castillo-Martínez, E., Bieringer, M., Shafi, S.P., Cranswick, L.M.D., and Alario-Franco, M.Á. (2011) Highly stable cooperative distortion in a weak Jahn–Teller d^2 cation: perovskite-type $ScVO_3$ obtained by high-pressure and high-temperature transformation from bixbyite. *J. Am. Chem. Soc.*, **133**, 8552–8563.

124 Kazakov, S.M., Chaillout, C., Bordet, P., Capponi, J.J., Nunez-Regueiro, M., Rysak, A., Tholence, J.L., Radaelli, P.G., Putilin, S.N., and Antipov, E.V. (1997) Discovery of a second family of bismuth-oxide-based superconductors. *Nature*, **390**, 148–150.

125 Chamberland, B.L. (1967) Preparation and properties of $SrCrO_3$. *Solid State Commun.*, **5**, 663–666.

126 Takeda, Y., Kanno, K., Takada, T., Yamamoto, O., Takano, M., Nakayama, N., and Bando, Y. (1986) Phase relation in the oxygen nonstoichiometric system, $SrFeO_x$ ($2.5 \leq x \leq 3.0$). *J. Solid State Chem.*, **63**, 237–249.

127 Kawasaki, S., Takano, M., and Takeda, Y. (1996) Ferromagnetic properties of $SrFe_{1-x}Co_xO_3$ synthesized under high pressure. *J. Solid State Chem.*, **121**, 174–180.

128 Shimizu, Y., Syono, Y., and Akimoto, S. (1970) High-pressure transformations in $SrGeO_3$, $SrSiO_3$, $BaGeO_3$, and $BaSiO_3$. *High Temp. High Press.*, **2**, 113–120.

129 Kafalas, J.A. and Longo, J.M. (1972) High pressure synthesis of (ABX_3) $(AX)_n$ compounds. *J. Solid State Chem.*, **4**, 55–59.

130 Longo, J.M., Kafalas, J.A., and Arnott, R.J. (1971) Structure and properties of the high and low pressure forms of $SrIrO_3$. *J. Solid State Chem.*, **3**, 174–179.

131 Chamberland, B.L. (1978) Solid state preparations and reactions of ternary alkaline-earth osmium oxides. *Mater. Res. Bull.*, **13**, 1273–1280.

132 Yamaura, K. and Takayama-Muromachi, E. (2001) Enhanced paramagnetism of the 4d itinerant electrons in the rhodium oxide perovskite $SrRhO_3$. *Phys. Rev. B*, **64**, 224424.

133 Williams, A.J., Gillies, A., Attfield, J.P., Heymann, G., Huppertz, H., Martínez-Lope, M.J., and Alonso, J.A. (2006) Charge transfer and antiferromagnetic insulator phase in $SrRu_{1-x}Cr_xO_3$ perovskites: solid solutions between two itinerant electron oxides. *Phys. Rev. B*, **73**, 104409.

134 Yi, W., Matsushita, Y., Katsuya, Y., Yamaura, K., Tsujimoto, Y., Presniakov, I.A., Sobolev, A.V., Glazkova, Y.S., Lekina, Y.O., Tsujii, N. *et al.* (2015) High-pressure synthesis, crystal structure and magnetic properties of $TlCrO_3$ perovskite. *Dalton Trans.*, **44**, 10785–10794.

135 Yi, W., Kumagai, Y., Spaldin, N.A., Matsushita, Y., Sato, A., Presniakov, I.A., Sobolev, A.V., Glazkova, Y.S., and Belik, A.A. (2014) Perovskite-structure $TlMnO_3$: a new manganite with new properties. *Inorg. Chem.*, **53**, 9800–9808.

136 Shan, Y.J., Nakamura, T., Miyata, M., Kobashi, K., Inaguma, Y., and Itoh, M. (1999) New perovskite oxide $(Tl_{1/2}Na_{1/2})TiO_3$. *Mater. Res. Soc. Symp. Proc.*, **547**, 105–114.

137 Belik, A.A., Yi, W., Kumagai, Y., Katsuya, Y., Tanaka, M., and Oba, F. (2016) $LiNbO_3$-type oxide $(Tl_{1-x}Sc_x)ScO_3$: high-pressure synthesis, crystal structure, and electronic properties. *Inorg. Chem.*, **55**, 1940–1945.

138 Huang, Y.H., Fjellvåg, H., Karppinen, M., Hauback, B.C., Yamauchi, H., and Goodenough, J.B. (2006) Crystal and magnetic structure of the orthorhombic perovskite $YbMnO_3$. *Chem. Mater.*, **18**, 2130–2134.

139 Akaogi, M., Kojitani, H., Yusa, H., Yamamoto, R., Kido, M., and Koyama, K. (2005) High-pressure transitions and thermochemistry of $MGeO_3$ (M=Mg, Zn and Sr) and Sr-silicates: systematics in enthalpies of formation of $A^{2+}B^{4+}O_3$ perovskites. *Phys. Chem. Miner.*, **32**, 603–613.

140 Yusa, H., Akaogi, M., Sata, N., Kojitani, H., Yamamoto, R., and Ohishi, Y. (2006) High-pressure transformations of ilmenite to perovskite, and lithium niobate to perovskite in zinc germanate. *Phys. Chem. Miner.*, **33**, 217–226.

141 Yu, R., Hojo, H., Mizoguchi, T., and Azuma, M. (2015) A new $LiNbO_3$-type polar oxide with closed-shell cations: $ZnPbO_3$. *J. Appl. Phys.*, **118**, 094103.

142 Inaguma, Y., Yoshida, M., and Katsumata, T. (2008) A polar oxide $ZnSnO_3$ with a $LiNbO_3$-type structure. *J. Am. Chem. Soc.*, **130**, 6704–6705.

143 Inaguma, Y., Aimi, A., Shirako, Y., Sakurai, D., Mori, D., Kojitani, H., Akaogi, M., and Nakayama, M. (2014)

High-pressure synthesis, crystal structure, and phase stability relations of a $LiNbO_3$-type polar titanate $ZnTiO_3$ and its reinforced polarity by the second-order Jahn–Teller effect. *J. Am. Chem. Soc.*, **136**, 2748–2756.

144 Li, M.-R., Stephens, P.W., Retuerto, M., Sarkar, T., Grams, C.P., Hemberger, J., Croft, M.C., Walker, D., and Greenblatt, M. (2014) Designing polar and magnetic oxides: Zn_2FeTaO_6 – in search of multiferroics. *J. Am. Chem. Soc.*, **136**, 8508–8511.

145 Longo, J.M. and Kafalas, J.A. (1969) The effect of pressure and B-cation size on the crystal structure of $CsBF_3$ compounds (B=Mn, Fe, Co, Ni, Zn, Mg). *J. Solid State Chem.*, **1**, 103–108.

146 Syono, Y., Akimoto, S.-i., and Kohn, K. (1969) Structure relations of hexagonal perovskite-like compounds ABX_3 at high pressure. *J. Phys. Soc. Jpn.*, **26**, 993–999.

147 Kafalas, J.A. and Longo, J.M. (1968) Effect of pressure on the structure and magnetic properties of $RbNiF_3$. *Mater. Res. Bull.*, **3**, 501–506.

148 Kohn, K., Fukuda, R., and Iida, S. (1967) A new ferrimagnetic compound $TlNiF_3$. *J. Phys. Soc. Jpn.*, **22**, 333–333.

149 Longo, J.M. and Kafalas, J.A. (1971) Effect of pressure on the crystal structure of $CsMnCl_3$ and $RbMnCl_3$. *J. Solid State Chem.*, **3**, 429–433.

150 Troyanchuk, I.O., Kasper, N.V., Mantytskaya, O.S., and Shapovalova, E.F. (1995) High-pressure synthesis on some perovskite-like compounds with a mixed anion type. *Mater. Res. Bull.*, **30**, 421–425.

151 Troyanchuk, I.O., Kasper, N.V., Mantytskaya, O.S., Shapovalova, E.F., Virchenko, V.A., and Karpei, A.L. (1994) High-pressure synthesis of the oxyfluorides $PbFeO_2F$, $BaFeO_2F$ with the perovskite structure. *Neorg. Mater.*, **30**, 992–994.

152 Chamberland, B.L. (1971) New oxyfluoride perovskite, $KTiO_2F$. *Mater. Res. Bull.*, **6**, 311–315.

153 Tassel, C., Kuno, Y., Goto, Y., Yamamoto, T., Brown, C.M., Hester, J., Fujita, K., Higashi, M., Abe, R., Tanaka, K. *et al.* (2015) $MnTaO_2N$: polar $LiNbO_3$-type oxynitride with a helical spin order. *Angew. Chem., Int. Ed.*, **54**, 516–521.

154 Katsumata, T., Nakashima, M., Inaguma, Y., and Tsurui, T. (2012) Synthesis of new perovskite-type oxyfluoride, $PbMnO_2F$. *Bull. Chem. Soc. Jpn.*, **85**, 397–399.

155 Katsumata, T., Nakashima, M., Umemoto, H., and Inaguma, Y. (2008) Synthesis of the novel perovskite-type oxyfluoride $PbScO_2F$ under high pressure and high temperature. *J. Solid State Chem.*, **181**, 2737–2740.

156 Demazeau, G., Grannec, J., Marbeuf, A., Portier, J., and Hagenmuller, P. (1969) Perovskite-type thallium oxyfluoride. *C. R. Acad. Sci. Ser. C*, **269**, 987–988.

157 Goldschmidt, V.M. (1926) Die Gesetze der Krystallochemie. *Naturwissenschaften*, **14**, 477–485.

158 Shannon, R.D. (1976) Revised effective ionic radii and systematic studies of interatomic distances in halides and chalcogenides. *Acta Crystallogr. A*, **32**, 751–767.

159 Navrotsky, A. (1981) *Energetics of Phase Transitions in AX, ABO_3, and AB_2O_4 Compounds*, Academic Press.

160 Rane, M.V., Navrotsky, A., and Rossetti, G.A. (2001) Enthalpies of formation of lead zirconate titanate (PZT) solid solutions. *J. Solid State Chem.*, **161**, 402–409.

161 Katz, L. and Ward, R. (1964) Structure relations in mixed metal oxides. *Inorg. Chem.*, **3**, 205–211.

162 Gushee, B.E., Katz, L., and Ward, R. (1957) The preparation of a barium cobalt oxide and other phases with similar structures. *J. Am. Chem. Soc.*, **79**, 5601–5603.

163 Chamberland, B.L. (1969) The preparation and crystallographic properties of $BaCrO_3$ polytypes. *Inorg. Chem.*, **8**, 286–290.

164 Chamberland, B.L. (1991) A study on the $BaIrO_3$ system. *J. Less-Common Met.*, **171**, 377–394.

165 Cheng, J.G., Alonso, J.A., Suard, E., Zhou, J.S., and Goodenough, J.B. (2009) A new perovskite polytype in the high-pressure

sequence of $BaIrO_3$. *J. Am. Chem. Soc.*, **131**, 7461–7469.

166 Siegrist, T. and Chamberland, B.L. (1991) The crystal structure of $BaIrO_3$. *J. Less-Common Met.*, **170**, 93–99.

167 Chamberland, B.L., Sleight, A.W., and Weiher, J.F. (1970) Preparation and characterization of $BaMnO_3$ and $SrMnO_3$ polytypes. *J. Solid State Chem.*, **1**, 506–511.

168 Shimada, M., Takeda, Y., Taguchi, H., Kanamaru, F., and Koizumi, M. (1975) Growth of single crystals of $BaFe^{4+}O_3$(12L), $BaNi^{4+}O_3$(2L) and $BaCo^{4+}O_3$(2L) under high oxygen pressure. *J. Cryst. Growth*, **29**, 75–76.

169 Yusa, H., Sata, N., and Ohishi, Y. (2007) Rhombohedral (9R) and hexagonal (6H) perovskites in barium silicates under high pressure. *Am. Mineral.*, **92**, 648–654.

170 Arend, H. and Kihlborg, L. (1969) Phase composition of reduced and reoxidized barium titanate. *J. Am. Ceram. Soc.*, **52**, 63–65.

171 Takeda, Y., Hashino, T., Miyamoto, H., Kanamaru, F., Kume, S., and Koizumi, M. (1972) Synthesis of $SrNiO_3$ and related compound, $Sr_2Ni_2O_5$. *J. Inorg. Nucl. Chem.*, **34**, 1599–1601.

172 Yusa, H., Akaogi, M., Sata, N., Kojitani, H., Kato, Y., and Ohishi, Y. (2005) Unquenchable hexagonal perovskite in high-pressure polymorphs of strontium silicates. *Am. Mineral.*, **90**, 1017–1020.

173 Megaw, H. (1968) A note on the structure of lithium niobate, $LiNbO_3$. *Acta Crystallogr. A*, **24**, 583–588.

174 Navrotsky, A. (1998) Energetics and crystal chemical systematics among ilmenite, lithium niobate, and perovskite structures. *Chem. Mater.*, **10**, 2787–2793.

175 Hinatsu, Y., Fujino, T., and Edelstein, N. (1992). Magnetic susceptibility of $LiUO_3$. *J Solid State Chem* **99**, 182–188.

176 Kemmler, S. (1965). Zur Kristallstruktur von $LiUO_3$. *Z Anorg Allg Chem* **338**, 9–14.

177 Miyake, C., Fuji, K., and Imoto, S. (1979). Electron spin resonance spectra and magnetic susceptibilities of mixed oxides, MUO_3 (M; Li, Na, K, and Rb), of pentavalent uranium. *Chem Phys Lett* **61**, 124–126.

178 Cava, R.J., Santoro, A., Murphy, D.W., Zahurak, S., and Roth, R.S. (1982). The structures of lithium-inserted metal oxides: $LiReO_3$ and Li_2ReO_3. *J Solid State Chem* **42**, 251–262.

179 Kumada, N., and Kinomura, N. (1990). Topochemical preparation of $LiNbO_3$ type $Li_{1-x}Cu_xNbO_3$ from $CuNb_2O_6$. *Mater Res Bull* **25**, 881–889.

180 Kumada, N., Hosoda, S., Muto, F., and Kinomura, N. (1989). A new lithium insertion compound (lithium, copper) tantalum oxide, (Li,Cu)TaO_3, with the $LiNbO_3$-type structure. *Inorg Chem* **28**, 3592–3595.

181 Funamori, N., Yagi, T., Miyajima, N., and Fujino, K. (1997). Transformation in Garnet from Orthorhombic Perovskite to $LiNbO_3$ Phase on Release of Pressure. *Science* **275**, 513–515.

182 Momma, K. and Izumi, F. (2011) *VESTA 3* for three-dimensional visualization of crystal, volumetric and morphology data, *J. Appl. Crystallogr.*, **44**, 1272–1276.

183 Varga, T., Kumar, A., Vlahos, E., Denev, S., Park, M., Hong, S., Sanehira, T., Wang, Y., Fennie, C.J., Streiffer, S.K., et al. (2009). Coexistence of Weak Ferromagnetism and Ferroelectricity in the High Pressure $LiNbO_3$-Type Phase of $FeTiO_3$. *Phys Rev Lett* **103**, 047601.

184 Davies, P.K. (1999) Cation ordering in complex oxides. *Curr. Opin. Solid State Mater. Sci.*, **4**, 467–471.

185 King, G. and Woodward, P.M. (2010) Cation ordering in perovskites. *J. Mater. Chem.*, **20**, 5785–5796.

186 Vasala, S. and Karppinen, M. (2015) $A_2B'B''O_6$ perovskites: a review. *Prog. Solid State Chem.*, **43**, 1–36.

187 Anderson, M.T., Greenwood, K.B., Taylor, G.A., and Poeppelmeier, K.R. (1993) B-cation arrangements in double perovskites. *Prog. Solid State Chem.*, **22**, 197–233.

188 Rosenstein, R.D. and Schor, R. (1963) Superlattice Madelung energy of idealized ordered cubic perovskites. *J. Chem. Phys.*, **38**, 1789–1790.

189 Demazeau, G., Jung, D.Y., Sanchez, J.P., Colineau, E., Blaise, A., and Fournes, L. (1993) Iridium(VI) stabilized in a perovskite-type lattice: Ba_2CaIrO_6. *Solid State Commun.*, **85**, 479–484.

190 Choy, J.-H., Hong, S.-T., and Choi, K.-S. (1996) Crystal structure, magnetism and phase transformation in perovskites A_2CrNbO_6(A=Ca, Sr, Ba). *J. Chem. Soc. Faraday Trans.*, **92**, 1051–1059.

191 Hong, S.-T., Park, J.-H., and Choy, J.-H. (1995) Unusual magnetic property of Ba_2CrTaO_6 and phase transformation under high pressure. *J. Phys. Chem.*, **99**, 6176–6181.

192 Feng, H.L., Arai, M., Matsushita, Y., Tsujimoto, Y., Yuan, Y., Sathish, C.I., He, J., Tanaka, M., and Yamaura, K. (2014) High-pressure synthesis, crystal structure and magnetic properties of double perovskite oxide Ba_2CuOsO_6. *J. Solid State Chem.*, **217**, 9–15.

193 Iwanaga, D., Inaguma, Y., and Itoh, M. (1999) Crystal structure and magnetic properties of B-site ordered perovskite-type oxides $A_2CuB'O_6$ (A=Ba, Sr; B′=W, Te). *J. Solid State Chem.*, **147**, 291–295.

194 Byeon, S.H., Demazeau, G., Jin-Ho, C., and Fournes, L. (1991) High-pressure synthesis and characterization of ordered cubic perovskite $Ba_2Bi(V)Fe(III)O_6$. *Mater. Lett.*, **12**, 163–167.

195 Azuma, M., Takata, K., Saito, T., Ishiwata, S., Shimakawa, Y., and Takano, M. (2005) Designed ferromagnetic, ferroelectric Bi_2NiMnO_6. *J. Am. Chem. Soc.*, **127**, 8889–8892.

196 Feng, H.L., Shi, Y., Guo, Y., Li, J., Sato, A., Sun, Y., Wang, X., Yu, S., Sathish, C.I., and Yamaura, K. (2013) High-pressure crystal growth and electromagnetic properties of 5d double-perovskite Ca_3OsO_6. *J. Solid State Chem.*, **201**, 186–190.

197 Feng, H.L., Arai, M., Matsushita, Y., Tsujimoto, Y., Guo, Y., Sathish, C.I., Wang, X., Yuan, Y.H., Tanaka, M., and Yamaura, K. (2014) High-temperature ferrimagnetism driven by lattice distortion in double perovskite Ca_2FeOsO_6. *J. Am. Chem. Soc.*, **136**, 3326–3329.

198 Feng, H.L., Sathish, C.I., Li, J., Wang, X., and Yamaura, K. (2013) Structure, and magnetic properties of a new double perovskite Ca_2InOsO_6. *Physics Procedia*, **45**, 117–120.

199 Azuma, M., Kaimori, S., and Takano, M. (1998) High-pressure synthesis and magnetic properties of layered double perovskites Ln_2CuMO_6 (Ln=La, Pr, Nd, and Sm; M=Sn and Zr). *Chem. Mater.*, **10**, 3124–3130.

200 Yi, W., Liang, Q., Matsushita, Y., Tanaka, M., and Belik, A.A. (2013) High-pressure synthesis, crystal structure, and properties of In_2NiMnO_6 with antiferromagnetic order and field-induced phase transition. *Inorg. Chem.*, **52**, 14108–14115.

201 Demazeau, G., Buffat, B., Ménil, F., Fournès, L., Pouchard, M., Dance, J.M., Fabritchnyi, P., and Hagenmuller, P. (1981) Characterization of six-coordinated iron(V) in an oxide lattice. *Mater. Res. Bull.*, **16**, 1465–1472.

202 Xiong, P., Seki, H., Guo, H., Hosaka, Y., Saito, T., Mizumaki, M., and Shimakawa, Y. (2016) Geometrical spin frustration of unusually high valence Fe^{5+} in the double perovskite La_2LiFeO_6. *Inorg. Chem.*, **55**, 6218–6222.

203 Darriet, J., Demazeau, G., and Pouchard, M. (1981) Une modelisation du comportement magnetique d'un compose de l'iridium pentavalent: $LaLi_{1/2}Ir_{1/2}O_3$. *Mater. Res. Bull.*, **16**, 1013–1017.

204 Hayashi, K., Demazeau, G., Pouchard, M., and Hagenmuller, P. (1980) Preparation and magnetic study of a new iridium(V) perovskite: $LaLi_{0.5}Ir_{0.5}O_3$. *Mater. Res. Bull.*, **15**, 461–467.

205 Demazeau, G., Oh-Kim, E., Choy, J., and Hagenmuller, P. (1987) A vanadate(V) oxide with perovskite structure: La_2LiVO_6, comparison with homologous La_2LiMO_6 phases (M=Fe, Nb, Mo, Ru, Ta, Re, Os, Ir). *Mater. Res. Bull.*, **22**, 735–740.

206 Dos santos-Garcia, A.J., Solana-Madruga, E., Ritter, C., Avila-Brande, D., Fabelo, O., and Saez-Puche, R. (2015) Synthesis, structures and magnetic properties of the

dimorphic Mn_2CrSbO_6 oxide. *Dalton Trans.*, **44**, 10665–10672.

207 Li, M.-R., Retuerto, M., Deng, Z., Stephens, P.W., Croft, M., Huang, Q., Wu, H., Deng, X., Kotliar, G., Sánchez-Benítez, J. *et al.* (2015) Giant magnetoresistance in the half-metallic double-perovskite ferrimagnet Mn_2FeReO_6. *Angew. Chem., Int. Ed.*, **54**, 12069–12073.

208 Arévalo-López, A.M., McNally, G.M., and Attfield, J.P. (2015) Large magnetization and frustration switching of magnetoresistance in the double-perovskite ferrimagnet Mn_2FeReO_6. *Angew. Chem., Int. Ed.*, **54**, 12074–12077.

209 Bazuev, G.V., Golovkin, B.G., Lukin, N.V., Kadyrova, N.I., and Zainulin, Y.G. (1996) High pressure synthesis and polymorphism of complex oxides Mn_2BSbO_6 (B=Fe, V, Cr, Ga, Al). *J. Solid State Chem*, **124**, 333–337.

210 Tyutyunnik, A.P., Bazuev, G.V., Kuznetsov, M.V., and Zainulin, Y.G. (2011) Crystal structure and magnetic properties of double perovskite Mn_2FeSbO_6. *Mater. Res. Bull.*, **46**, 1247–1251.

211 Li, M.-R., Hodges, J.P., Retuerto, M., Deng, Z., Stephens, P.W., Croft, M.C., Deng, X., Kotliar, G., Sánchez-Benítez, J., Walker, D. *et al.* (2016) Mn_2MnReO_6: synthesis and magnetic structure determination of a new transition-metal-only double perovskite canted antiferromagnet. *Chem. Mater.*, **28**, 3148–3158.

212 Solana-Madruga, E., Arévalo-López, Á. M., Dos Santos-García, A.J., Urones-Garrote, E., Ávila-Brande, D., Sáez-Puche, R., and Attfield, J.P. (2016) Double double cation order in the high-pressure perovskites $MnRMnSbO_6$. *Angew. Chem., Int. Ed.*, **55**, 9340–9344.

213 Yi, W., Princep, A.J., Guo, Y., Johnson, R.D., Khalyavin, D., Manuel, P., Senyshyn, A., Presniakov, I.A., Sobolev, A.V., Matsushita, Y. *et al.* (2015) Sc_2NiMnO_6: a double-perovskite with a magnetodielectric response driven by multiple magnetic orders. *Inorg. Chem.*, **54**, 8012–8021.

214 Jung, D.-Y. and Demazeau, G. (1995) High oxygen pressure and the preparation of new iridium(VI) oxides with perovskite structure: Sr_2MIrO_6 (M=Ca, Mg). *J. Solid State Chem.*, **115**, 447–455.

215 Vasala, S., Cheng, J.G., Yamauchi, H., Goodenough, J.B., and Karppinen, M. (2012) Synthesis and characterization of $Sr_2Cu(W_{1-x}Mo_x)O_6$: a quasi-two-dimensional magnetic system. *Chem. Mater.*, **24**, 2764–2774.

216 Byeon, S.H., Nakamura, T., Itoh, M., and Matsuo, M. (1992) A new Fe(III) perovskite antiferromagnetically ordered via superexchange interaction at fairly high temperature ($T_N \cong 91K$). *Mater. Res. Bull.*, **27**, 1065–1072.

217 Paul, A.K., Jansen, M., Yan, B., Felser, C., Reehuis, M., and Abdala, P.M. (2013) Synthesis, crystal structure, and physical properties of Sr_2FeOsO_6. *Inorg. Chem.*, **52**, 6713–6719.

218 Sleight, A.W., Longo, J., and Ward, R. (1962) Compounds of osmium and rhenium with the ordered perovskite structure. *Inorg. Chem.*, **1**, 245–250.

219 Kayser, P., Martínez-Lope, M.J., Alonso, J.A., Retuerto, M., Croft, M., Ignatov, A., and Fernández-Díaz, M.T. (2013) Crystal structure, phase transitions, and magnetic properties of iridium perovskites Sr_2MIrO_6 (M=Ni, Zn). *Inorg. Chem.*, **52**, 11013–11022.

220 Anderson, M.T. and Poeppelmeier, K.R. (1991) Lanthanum copper tin oxide (La_2CuSnO_6): a new perovskite-related compound with an unusual arrangement of B cations. *Chem. Mater.*, **3**, 476–482.

221 Chakhmouradian, A.R. and Mitchell, R.H. (2001) Crystal structure of novel high-pressure perovskite $K_{2/3}Th_{1/3}TiO_3$, a possible host for Th in the upper mantle. *Am. Mineral.*, **86**, 1076–1080.

222 Shimakawa, Y. (2008) A-site-ordered perovskites with intriguing physical properties. *Inorg. Chem.*, **47**, 8562–8570.

223 Shimakawa, Y. and Mizumaki, M. (2014) Multiple magnetic interactions in A-site-

ordered perovskite-structure oxides. *J. Phys. Condens. Matter*, **26**, 473203.

224 Vasil'ev, A.N. and Volkova, O.S. (2007) New functional materials $AC_3B_4O_{12}$ (review). *Low Temp. Phys.*, **33**, 895–914.

225 Yamada, I. (2014) High-pressure synthesis, electronic states, and structure–property relationships of perovskite oxides, $ACu_3Fe_4O_{12}$ (A: divalent alkaline earth or trivalent rare-earth ion). *J. Ceram. Soc. Jpn.*, **122**, 846–851.

226 Leinenweber, K., Linton, J., Navrotsky, A., Fei, Y., and Parise, J.B. (1995) High-pressure perovskites on the join $CaTiO_3$–$FeTiO_3$. *Phys. Chem. Miner.*, **22**, 251–258.

227 Shiro, K., Yamada, I., Ikeda, N., Ohgushi, K., Mizumaki, M., Takahashi, R., Nishiyama, N., Inoue, T., and Irifune, T. (2013) Pd^{2+}-incorporated perovskite $CaPd_3B_4O_{12}$ (B=Ti, V). *Inorg. Chem.*, **52**, 1604–1609.

228 Aimi, A., Mori, D., Hiraki, K., Takahashi, T., Shan, Y.J., Shirako, Y., Zhou, J., and Inaguma, Y. (2014) High-pressure synthesis of A-site ordered double perovskite $CaMnTi_2O_6$ and ferroelectricity driven by coupling of A-site ordering and the second-order Jahn–Teller effect. *Chem. Mater.*, **26**, 2601–2608.

229 Long, Y., Salto, T., Tohyama, T., Oka, K., Azuma, M., and Shimakawa, Y. (2009) Intermetallic charge transfer in A-site-ordered double perovskite $BiCu_3Fe_4O_{12}$. *Inorg. Chem.*, **48**, 8489–8492.

230 Takata, K., Yamada, I., Azuma, M., Takano, M., and Shimakawa, Y. (2007) Magnetoresistance and electronic structure of the half-metallic ferrimagnet $BiCu_3Mn_4O_{12}$. *Phys. Rev. B*, **76**, 024429.

231 Kadyrova, N.I., Zainulin, Y.G., Volkov, V.L., Zakharova, G.S., Mikhalev, K.N., D'Yachkova, T.V., Zubkov, V.G., and Tyutyunnik, A.P. (2005) High-pressure defect phase $Bi_{0.67}Cu_3V_4O_{12}$. *Russ. J. Inorg. Chem.*, **50**, 655–658.

232 Imamura, N., Karppinen, M., Motohashi, T., Fu, D., Itoh, M., and Yamauchi, H. (2008) Positive and negative magnetodielectric effects in A-site ordered $(BiMn_3)Mn_4O_{12}$ perovskite. *J. Am. Chem. Soc.*, **130**, 14948–14949.

233 Subramanian, M.A., Marshall, W.J., Calvarese, T.G., and Sleight, A.W. (2003) Valence degeneracy in $CaCu_3Cr_4O_{12}$. *J. Phys. Chem. Solids*, **64**, 1569–1571.

234 Byeon, S.-H., Lee, S.-S., Parise, J.B., Woodward, P.M., and Hur, N.H. (2005) New ferrimagnetic oxide $CaCu_3Cr_2Sb_2O_{12}$:high-pressure synthesis, structure, and magnetic properties. *Chem. Mater.*, **17**, 3552–3557.

235 Byeon, S.-H., Lee, S.-S., Parise, J.B., and Woodward, P.M. (2006) New perovskite oxide $CaCu_3Cr_2Ru_2O_{12}$: comparison with structural, magnetic, and transport properties of the $CaCu_3B_2B'_2O_{12}$ perovskite family. *Chem. Mater.*, **18**, 3873–3877.

236 Yamada, I., Takata, K., Hayashi, N., Shinohara, S., Azuma, M., Mori, S., Muranaka, S., Shimakawa, Y., and Takano, M. (2008) A perovskite containing quadrivalent iron as a charge-disproportionated ferrimagnet. *Angew. Chem., Int. Ed.*, **47**, 7032–7035.

237 Deng, H., Liu, M., Dai, J., Hu, Z., Kuo, C., Yin, Y., Yang, J., Wang, X., Zhao, Q., Xu, Y. *et al.* (2016) Strong enhancement of spin ordering by A-site magnetic ions in the ferrimagnet $CaCu_3Fe_2Os_2O_{12}$. *Phys. Rev. B Condens. Matter. Mater. Phys.*, **94**, 024414.

238 Chen, W.-T., Mizumaki, M., Seki, H., Senn, M.S., Saito, T., Kan, D., Attfield, J.P., and Shimakawa, Y. (2014) A half-metallic A- and B-site-ordered quadruple perovskite oxide $CaCu_3Fe_2Re_2O_{12}$ with large magnetization and a high transition temperature. *Nat. Commun.*, **5**, 3909.

239 Chen, W.-T., Mizumaki, M., Saito, T., and Shimakawa, Y. (2013) Frustration relieved ferrimagnetism in novel A- and B-site-ordered quadruple perovskite. *Dalton Trans.*, **42**, 10116–10120.

240 Byeon, S.-H., Lufaso, M.W., Parise, J.B., Woodward, P.M., and Hansen, T. (2003) High-pressure synthesis and characterization of perovskites with simultaneous ordering of both the A- and

B-site cations, $CaCu_3Ga_2M_2O_{12}$ (M=Sb, Ta). *Chem. Mater.*, **15**, 3798–3804.

241 Byeon, S.-H., Lee, S.-S., Parise, J.B., Woodward, P.M., and Hur, N.H. (2004) High-pressure synthesis of metallic perovskite ruthenate $CaCu_3Ga_2Ru_2O_{12}$. *Chem. Mater.*, **16**, 3697–3701.

242 Shiraki, H., Saito, T., Yamada, T., Tsujimoto, M., Azuma, M., Kurata, H., Isoda, S., Takano, M., and Shimakawa, Y. (2007) Ferromagnetic cuprates $CaCu_3Ge_4O_{12}$ and $CaCu_3Sn_4O_{12}$ with A-site ordered perovskite structure. *Phys. Rev. B*, **76**, 140403.

243 Cheng, J.G., Zhou, J.S., Yang, Y.F., Zhou, H.D., Matsubayashi, K., Uwatoko, Y., MacDonald, A., and Goodenough, J.B. (2013) Possible kondo physics near a metal–insulator crossover in the A-site ordered perovskite $CaCu_3Ir_4O_{12}$. *Phys. Rev. Lett.*, **111**, 176403.

244 Chenavas, J., Joubert, J.C., Marezio, M., and Bochu, B. (1975) The synthesis and crystal structure of $CaCu_3Mn_4O_{12}$: a new ferromagnetic-perovskite-like compound. *J. Solid State Chem.*, **14**, 25–32.

245 Yamada, I., Takahashi, Y., Ohgushi, K., Nishiyama, N., Takahashi, R., Wada, K., Kunimoto, T., Ohfuji, H., Kojima, Y., Inoue, T. *et al.* (2010) $CaCu_3Pt_4O_{12}$: the first perovskite with the B site fully occupied by Pt^{4+}. *Inorg. Chem.*, **49**, 6778–6780.

246 Kadyrova, N.I., Zakharova, G.S., Zainulin, Y.G., Volkov, V.L., D'yachkova, T.V., Tyutyunnik, A.P., and Zubkov, V.G. (2003) Synthesis and properties of the new compounds $NaCu_3V_4O_{12}$ and $CaCu_3V_4O_{12}$ obtained under uniform compression. *Dokl. Chem.*, **392**, 251–253.

247 Shiraki, H., Saito, T., Azuma, M., and Shimakawa, Y. (2008) Metallic behavior in A-site-ordered perovskites $ACu_3V_4O_{12}$ with A=Na^+, Ca^{2+}, and Y^{3+}. *J. Phys. Soc. Jpn.*, **77**, 064705.

248 Kadyrova, N.I., Zainulin, Y.G., Zakharova, G.S., Tyutyunnik, A.P., and Mel'nikova, N.V. (2011) Synthesis and properties of the high-pressure phase $CaCu_2CoV_4O_{12}$. *Russ. J. Inorg. Chem.*, **56**, 1717–1722.

249 Bochu, B., Chenavas, J., Joubert, J.C., and Marezio, M. (1974) High pressure synthesis and crystal structure of a new series of perovskite-like compounds CMn_7O_{12} (C=Na, Ca, Cd, Sr, La, Nd). *J. Solid State Chem.*, **11**, 88–93.

250 Troyanchuk, I.O., Lobanovsky, L.S., Kasper, N.V., Hervieu, M., Maignan, A., Michel, C., Szymczak, H., and Szewczyk, A. (1998) Magnetotransport phenomena in $A(Mn_{3-x}Cu_x)Mn_4O_{12}$ (A=Ca, Tb, Tm) perovskites. *Phys. Rev. B*, **58**, 14903–14907.

251 Saito, T., Tohyama, T., Woodward, P.M., and Shimakawa, Y. (2011) Material design and high-pressure synthesis of novel *A*-site-ordered perovskites $AMn_3Al_4O_{12}$ (*A*=Y, Yb, and Dy) with square-planar-coordinated Mn^{3+}. *Bull. Chem. Soc. Jpn.*, **84**, 802–806.

252 Long, Y.W., Hayashi, N., Saito, T., Azuma, M., Muranaka, S., and Shimakawa, Y. (2009) Temperature-induced A–B intersite charge transfer in an A-site-ordered $LaCu_3Fe_4O_{12}$ perovskite. *Nature*, **458**, 60–63.

253 Yamada, I., Etani, H., Tsuchida, K., Marukawa, S., Hayashi, N., Kawakami, T., Mizumaki, M., Ohgushi, K., Kusano, Y., Kim, J. *et al.* (2013) Control of bond-strain-induced electronic phase transitions in iron perovskites. *Inorg. Chem.*, **52**, 13751–13761.

254 Alonso, J.A., Sánchez-Benítez, J., De Andrés, A., Martínez-Lope, M.J., Casais, M.T., and Martínez, J.L. (2003) Enhanced magnetoresistance in the complex perovskite $LaCu_3Mn_4O_{12}$. *Appl. Phys. Lett.*, **83**, 2623–2625.

255 Long, Y., Saito, T., Mizumaki, M., Agui, A., and Shimakawa, Y. (2009) Various valence states of square-coordinated Mn in A-site-ordered perovskites. *J. Am. Chem. Soc.*, **131**, 16244–16247.

256 Kadyrova, N.I., Zainulin, Y.G., Volkov, V.L., Zakharova, G.S., and Korolev, A.V. (2007) High-pressure defect phase $La_xCu_3V_4O_{12}$. *Russ. J. Inorg. Chem.*, **52**, 825–828.

257 Akizuki, Y., Yamada, I., Fujita, K., Nishiyama, N., Irifune, T., Yajima, T., Kageyama, H., and Tanaka, K. (2013) A-site-ordered perovskite $MnCu_3V_4O_{12}$

with a 12-coordinated manganese(II). *Inorg. Chem.*, **52**, 11538–11543.

258 Marezio, M., Dernier, P.D., Chenavas, J., and Joubert, J.C. (1973) High pressure synthesis and crystal structure of $NaMn_7O_{12}$. *J. Solid State Chem.*, **6**, 16–20.

259 Locherer, T., Dinnebier, R., Kremer, R.K., Greenblatt, M., and Jansen, M. (2012) Synthesis and properties of a new quadruple perovskite: A-site ordered $PbMn_3Mn_4O_{12}$. *J. Solid State Chem.*, **190**, 277–284.

260 Yamada, I., Tsuchida, K., Ohgushi, K., Hayashi, N., Kim, J., Tsuji, N., Takahashi, R., Matsushita, M., Nishiyama, N., Inoue, T. *et al.* (2011) Giant negative thermal expansion in the iron perovskite $SrCu_3Fe_4O_{12}$. *Angew. Chem., Int. Ed.*, **50**, 6579–6582.

261 Mori, D., Shimoi, M., Kato, Y., Katsumata, T., Hiraki, K.-I., Takahashi, T., and Inaguma, Y. (2011) High-pressure synthesis, structure, dielectric and magnetic properties for $SrCu_3Ti_4O_{12}$. *Ferroelectrics*, **414**, 180–189.

262 Etani, H., Yamada, I., Ohgushi, K., Hayashi, N., Kusano, Y., Mizumaki, M., Kim, J., Tsuji, N., Takahashi, R., Nishiyama, N. *et al.* (2013) Suppression of intersite charge transfer in charge-disproportionated perovskite $YCu_3Fe_4O_{12}$. *J. Am. Chem. Soc.*, **135**, 6100–6106.

263 Tohyama, T., Saito, T., Mizumaki, M., Agui, A., and Shimakawa, Y. (2010) Antiferromagnetic interaction between A′-site Mn spins in A-site-ordered perovskite $YMn_3Al_4O_{12}$. *Inorg. Chem.*, **49**, 2492–2495.

264 Leinenweber, K. and Parise, J. (1995) High-pressure synthesis and crystal structure of $CaFeTi_2O_6$, a new perovskite structure type. *J. Solid State Chem.*, **114**, 277–281.

265 Ruddlesden, S.N. and Popper, P. (1957) New compounds of the K_2NiF_4-type. *Acta. Crystallogr.*, **10**, 538–539.

266 Ruddlesden, S.N. and Popper, P. (1958) The compound $Sr_3Ti_2O_7$ and its structure. *Acta Crystallogr.*, **11**, 54–55.

267 Ganguli, D. (1979) Cationic radius ratio and formation of K_2NiF_4-type compounds. *J. Solid State Chem.*, **30**, 353–356.

268 Ganguly, P. and Rao, C.N.R. (1984) Crystal chemistry and magnetic properties of layered metal oxides possessing the K_2NiF_4 or related structures. *J. Solid State Chem.*, **53**, 193–216.

269 Poix, P. (1980) Etude de la structure K_2NiF_4 par la méthode des invariants, I. Cas des oxydes A_2BO_4. *J. Solid State Chem.*, **31**, 95–102.

270 Reid, A.F. and Ringwood, A.E. (1970) The crystal chemistry of dense M_3O_4 polymorphs: high pressure Ca_2GeO_4 of K_2NiF_4 structure type. *J. Solid State Chem.*, **1**, 557–565.

271 Liu, L.-G. (1978) High pressure Ca_2SiO_4, the silicate K_2NiF_4-isotype with crystalchemical and geophysical implications. *Phys. Chem. Miner.*, **3**, 291–299.

272 Hiroi, Z., Takano, M., Azuma, M., and Takeda, Y. (1993) A new family of copper oxide superconductors $Sr_{n+1}Cu_nO_{2n+1+\delta}$ stabilized at high pressure. *Nature*, **364**, 315–317.

273 Castillo-Martínez, E. and Alario-Franco, M.A. (2007) Revisiting the Sr-Cr(IV)-O system at high pressure and temperature with special reference to $Sr_3Cr_2O_7$. *Solid State Sci.*, **9**, 564–573.

274 Itoh, M., Shimura, T., Inaguma, Y., and Morii, Y. (1995) Structure of two-dimensional conductor of Sr_2RhO_4. *J. Solid State Chem.*, **118**, 206–209.

275 Shimura, T., Itoh, M., and Nakamura, T. (1992) Novel two-dimensional conductor Sr_2RhO_4. *J. Solid State Chem.*, **98**, 198–200.

276 Yamaura, K., Huang, Q., Young, D.P., Noguchi, Y., and Takayama-Muromachi, E. (2002) Crystal structure and electronic and magnetic properties of the bilayered rhodium oxide $Sr_3Rh_2O_7$. *Phys. Rev. B*, **66**, 134431.

277 Yamaura, K., Huang, Q., Young, D.P., and Takayama-Muromachi, E. (2004) Crystal structure and magnetic properties of the trilayered perovskite

$Sr_4Rh_3O_{10}$: a new member of the strontium rhodate family. *Chem. Mater.*, **16**, 3424–3430.

278 Tsujimoto, Y., Sathish, C.I., Matsushita, Y., Yamaura, K., and Uchikoshi, T. (2014) New members of layered oxychloride perovskites with square planar coordination: $Sr_2MO_2Cl_2$ (M=Mn, Ni) and $Ba_2PdO_2Cl_2$. *Chem. Commun.*, **50**, 5915–5918.

279 Tsujimoto, Y., Li, J.J., Yamaura, K., Matsushita, Y., Katsuya, Y., Tanaka, M., Shirako, Y., Akaogi, M., and Takayama-Muromachi, E. (2011) New layered cobalt oxyfluoride, Sr_2CoO_3F. *Chem. Commun.*, **47**, 3263–3265.

280 Tsujimoto, Y., Yamaura, K., and Uchikoshi, T. (2013) Extended Ni(III) oxyhalide perovskite derivatives: Sr_2NiO_3X (X=F, Cl). *Inorg. Chem.*, **52**, 10211–10216.

281 Murakami, M., Hirose, K., Kawamura, K., Sata, N., and Ohishi, Y. (2004) Post-perovskite phase transition in $MgSiO_3$. *Science*, **304**, 855–858.

282 Oganov, A.R. and Ono, S. (2004) Theoretical and experimental evidence for a post-perovskite phase of $MgSiO_3$ in Earth's D″ layer. *Nature*, **430**, 445–448.

283 Ohgushi, K., Gotou, H., Yagi, T., Kiuchi, Y., Sakai, F., and Ueda, Y. (2006) Metal–insulator transition in $Ca_{1-x}Na_xO_3$ with post-perovskite structure. *Phys. Rev. B*, **74**, 241104.

284 Shirako, Y., Satsukawa, H., Wang, X.X., Li, J.J., Guo, Y.F., Arai, M., Yamaura, K., Yoshida, M., Kojitani, H., Katsumata, T. *et al.* (2011) Integer spin-chain antiferromagnetism of the 4d oxide $CaRuO_3$ with post-perovskite structure. *Phys. Rev. B*, **83**, 174411.

285 Yamaura, K., Shirako, Y., Kojitani, H., Arai, M., Young, D.P., Akaogi, M., Nakashima, M., Katsumata, T., Inaguma, Y., and Takayama-Muromachi, E. (2009) Synthesis and magnetic and charge-transport properties of the correlated 4d post-perovskite $CaRhO_3$. *J. Am. Chem. Soc.*, **131**, 5010–5010.

286 Hirose, K., Sinmyo, R., Sata, N., and Ohishi, Y. (2006) Determination of post-perovskite phase transition boundary in $MgSiO_3$ using Au and MgO pressure standards. *Geophys. Res. Lett.*, **33**, L01310.

287 Hirose, K., Kawamura, K., Ohishi, Y., Tateno, S., and Sata, N. (2005) Stability and equation of state of $MgGeO_3$ post-perovskite phase. *Am. Mineral.*, **90**, 262–265.

288 Tateno, S., Hirose, K., Sata, N., and Ohishi, Y. (2006) High-pressure behavior of $MnGeO_3$ and $CdGeO_3$ perovskites and the post-perovskite phase transition. *Phys. Chem. Miner.*, **32**, 721–725.

289 Tateno, S., Hirose, K., Sata, N., and Ohishi, Y. (2010) Structural distortion of $CaSnO_3$ perovskite under pressure and the quenchable post-perovskite phase as a low-pressure analogue to $MgSiO_3$. *Phys. Earth Planet. Inter.*, **181**, 54–59.

290 Kojitani, H., Shirako, Y., and Akaogi, M. (2007) Post-perovskite phase transition in $CaRuO_3$. *Phys. Earth Planet. Inter.*, **165**, 127–134.

291 Shirako, Y., Kojitani, H., Akaogi, M., Yamaura, K., and Takayama-Muromachi, E. (2009) High-pressure phase transitions of $CaRhO_3$ perovskite. *Phys. Chem. Miner.*, **36**, 455–462.

292 Ohgushi, K., Matsushita, Y., Miyajima, N., Katsuya, Y., Tanaka, M., Izumi, F., Gotou, H., Ueda, Y., and Yagi, T. (2008) $CaPtO_3$ as a novel post-perovskite oxide. *Phys. Chem. Miner.*, **35**, 189–195.

293 Inaguma, Y., Hasumi, K., Yoshida, M., Ohba, T., and Katsumata, T. (2008) High-pressure synthesis, structure, and characterization of a post-perovskite $CaPtO_3$ with $CaIrO_3$-type structure. *Inorg. Chem.*, **47**, 1868–1870.

294 McDaniel, C.L. and Schneider, S.J. (1972) Phase relations in the $CaO–IrO_2–Ir$ system in air. *J. Solid State Chem.*, **4**, 275–280.

295 Hirose, K. and Fujita, Y. (2005) Clapeyron slope of the post-perovskite phase transition in $CaIrO_3$. *Geophys. Res. Lett.*, **32**, L13313.

296 Kojitani, H., Furukawa, A., and Akaogi, M. (2007) Thermochemistry and high-pressure equilibria of the post-perovskite phase transition in $CaIrO_3$. *Am. Mineral.*, **92**, 229–232.

297 Bremholm, M., Dutton, S.E., Stephens, P.W., and Cava, R.J. (2011) $NaIrO_3$: a pentavalent post-perovskite. *J. Solid State Chem.*, **184**, 601–607.

298 Crichton, W.A., Yusenko, K.V., Riva, S., Mazzali, F., and Margadonna, S. (2016) An alternative route to pentavalent postperovskite. *Inorg. Chem.*, **55**, 5738–5740.

299 Bernal, F.L., Yusenko, K.V., Sottmann, J., Drathen, C., Guignard, J., Løvvik, O.M., Crichton, W.A., and Margadonna, S. (2014) Perovskite to postperovskite transition in $NaFeF_3$. *Inorg. Chem.*, **53**, 12205–12214.

300 Liu, H.Z., Chen, J., Hu, J., Martin, C.D., Weidner, D.J., Häusermann, D., and Mao, H.K. (2005) Octahedral tilting evolution and phase transition in orthorhombic $NaMgF_3$ perovskite under pressure. *Geophys. Res. Lett.*, **32**, 1–5.

301 Hustoft, J., Catalli, K., Shim, S.-H., Kubo, A., Prakapenka, V.B., and Kunz, M. (2008) Equation of state of $NaMgF_3$ postperovskite: implication for the seismic velocity changes in the D″ region. *Geophys. Res. Lett.*, **35**, L10309.

302 Dobson, D.P., Hunt, S.A., Lindsay-Scott, A., and Wood, I.G. (2011) Towards better analogues for $MgSiO_3$ post-perovskite: $NaCoF_3$ and $NaNiF_3$, two new recoverable fluoride post-perovskites. *Phys. Earth Planet. Inter.*, **189**, 171–175.

303 Shirako, Y., Shi, Y.G., Aimi, A., Mori, D., Kojitani, H., Yamaura, K., Inaguma, Y., and Akaogi, M. (2012) High-pressure stability relations, crystal structures, and physical properties of perovskite and post-perovskite of $NaNiF_3$. *J. Solid State Chem.*, **191**, 167–174.

304 Yusa, H., Shirako, Y., Akaogi, M., Kojitani, H., Hirao, N., Ohishi, Y., and Kikegawa, T. (2012) Perovskite-to-postperovskite transitions in $NaNiF_3$ and $NaCoF_3$ and disproportionation of $NaCoF_3$ postperovskite under high pressure and high temperature. *Inorg. Chem.*, **51**, 6559–6566.

305 Yakovlev, S., Avdeev, M., Sterer, E., Greenberg, Y., and Mezouar, M. (2009) High-pressure structural behavior and equation of state of $NaZnF_3$. *J. Solid State Chem.*, **182**, 1545–1549.

306 Akaogi, M., Shirako, Y., Kojitani, H., Nagakari, T., Yusa, H., and Yamaura, K. (2014) High-pressure transitions in $NaZnF_3$ and $NaMnF_3$ perovskites, and crystal-chemical characteristics of perovskite–postperovskite transitions in ABX_3 fluorides and oxides. *Phys. Earth Planet. Inter.*, **228**, 160–169.

4
Solvothermal Methods

Nobuhiro Kumada

University of Yamanashi, Center for Crystal Science and Technology, Miyamae-cho 7-32, 400-8511 Kofu Japan

4.1
Solvothermal Methods for Crystal Growth and Synthesis of New Compounds

A solvothermal reaction is defined as a chemical reaction that occurs in an aqueous or nonaqueous solvent at high temperature and pressure in a closed system. The definition includes other reactions at high temperature and pressure such as a hydrothermal reaction using water, an ammonothermal one using ammonia, a glycothermal one using glycol, and an ionothermal one using an ionic liquid as solvent. The term "hydrothermal" originated in geochemistry where hydrothermal reactions were employed for crystal growth under conditions near the critical point of water. Hydrothermal techniques have been successful in the industrial production of large quartz single crystals. Another important application of hydrothermal reactions is the synthesis of zeolitic aluminosilicates. In the current decade, hydrothermal synthesis is applicable to a variety of solvents and hence has also been termed "solvothermal". This expansion has been accompanied by developments in chemical engineering, materials science, geology, mineralogy, and, particularly, nanoscience, where facile control of the size and morphology of crystalline particles provides many opportunities to employ solvothermal reactions as described in other chapter. In addition to nanoscience, solvothermal reactions have offered possibility of emerging of new compounds such as an organic–inorganic complex. This chapter introduces some examples of bulk single-crystal growth and synthesis of new compounds using solvothermal methods.

4.2
α-Quartz

A piezoelectric α-quartz is used as an oscillator in important electric devices. Hydrothermal growth of α-quartz single crystal was described by Spezia at the

Handbook of Solid State Chemistry, First Edition. Edited by Richard Dronskowski, Shinichi Kikkawa, and Andreas Stein.

beginning of the twentieth century. After World War II, tons of α-quartz single crystals were produced on an industrial scale [1]. The history of hydrothermal growth of α-quartz single crystal was reviewed by Iwaski [1]. The hydrothermal growth of an α-quartz single crystal is typically carried out in alkaline solution (NaOH or Na_2CO_3) using two temperature zones, in which seed crystals are placed at the upper portion (∼380 °C) and nutrients are placed at the lower portion (∼400 °C) as shown in Figure 4.1. This autoclave was designed by Walker in 1950s [2] and is a prototype for current industrial scale autoclaves. The upper and lower portions are separated by a baffle that plays an important role for convection control. The inner pressure is ∼100 MPa when the temperatures are 380 °C in the upper portion and 400 °C in the lower portion and the volume fraction of solvent is ∼75%. The growth rate depends strongly on crystal face of the seed and the nutrient, temperatures, solution composition, and volume fraction of solvent. Figure 4.2 shows an α-quartz single crystal produced by hydrothermal reaction in an industrial facility. For oscillator applications, incorporation of OH^- and impurity metal ions such as Fe^{3+} and Al^{3+} ions in a single crystal influences the acoustic Q-value. Attempts to obtain a high quality single crystal with high Q-value have been reported [3]. On the other hand, amethyst, which is a purple colored single crystal of SiO_2, can be synthesized by

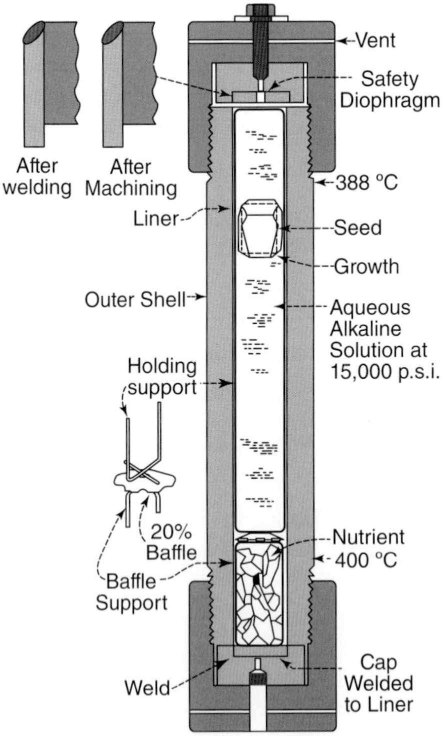

Figure 4.1 Cross-section of an autoclave used to grow α-quartz single crystals. (Figure 8 in Ref. [2].)

Figure 4.2 An α-quartz single crystal produced by hydrothermal reaction in an industrial facility.

adding NH_4F to the alkaline solution used for hydrothermal growth of α-quartz [4]. A single crystal of stishovite-type SiO_2, which is a high-pressure allotrope of silicon dioxide, has also been prepared by hydrothermal reaction under the condition of 9–9.5 GPa at 770–1170 K [5].

4.3 GaN

Gallium nitride, GaN, is an important semiconductor used in LED lamps and high-power, high-frequency, heterojunction field-effect transistors. Crystal growth of GaN single crystals has been attempted by several techniques including high-pressure solution growth, Na flux, and ammonothermal methods [6]. The ammonothermal crystal growth of GaN was first carried out by Dwilinski *et al.* [7] by separating inside of the autoclave for hydrothermal growth of α-quartz as shown in Figure 4.3. Figure 4.4 shows a GaN single crystal synthesized by ammonothermal reaction. The reaction temperature and pressure were 673–773 K and 200–400 MPa, respectively. Three types of mineralizers were used: (i) basic (NH^{2-}) [8,9], (ii) acidic (NH^{4+}) [10–12], and (iii) neutral. The mineralizer influences the solubility for GaN in ammonia. The solubility of GaN in supercritical ammonia with 1.5 mol% $NaNH_2$ [8] showed the maximum value at 550–600 °C, as represented in Figure 4.5. On the other hand, for acidic mineralizers (NH_4F, NH_4I, NH_4Br, and NH_4Cl), the highest growth rate is observed when NH_4F is used [12], as shown in Figure 4.6.

4.4 ZnO

A zinc oxide, ZnO, is an important semiconductor material and is also used in photovoltaic devices. Its single crystal growth by hydrothermal reaction has been

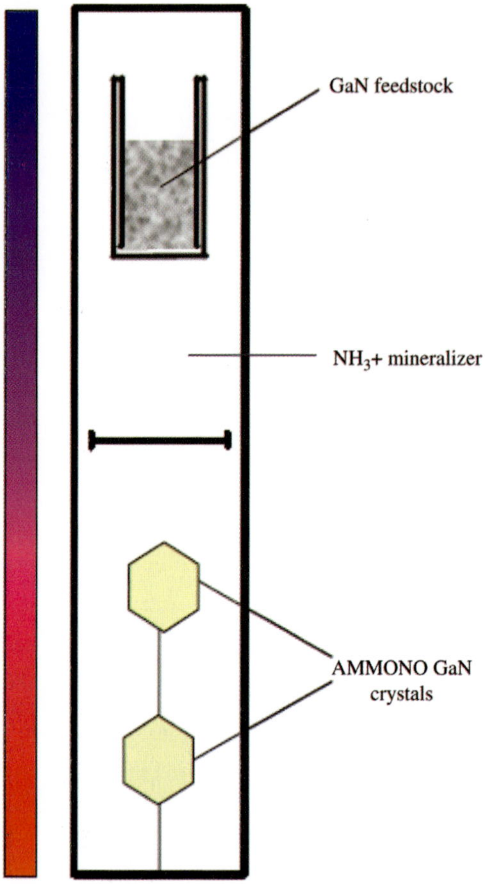

Figure 4.3 Ammonothermal crystal growth of GaN. The bar on the left represents the temperature gradient: red – highest, blue – lowest. (Figure 2 in Ref. [7].)

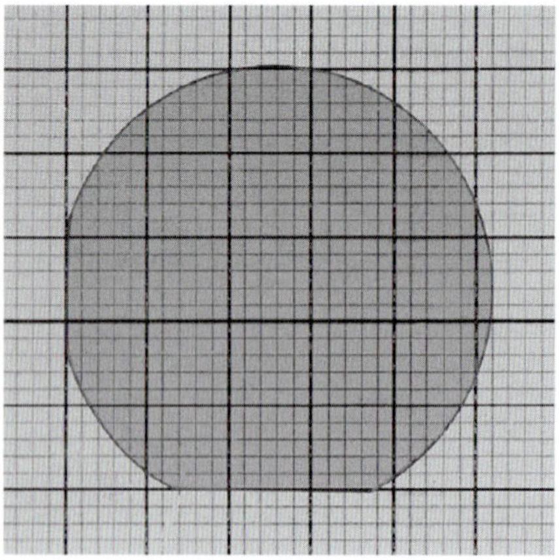

Figure 4.4 A 1 in. GaN substrate synthesized by ammonothermal reaction. (Figure 3 in Ref. [7].)

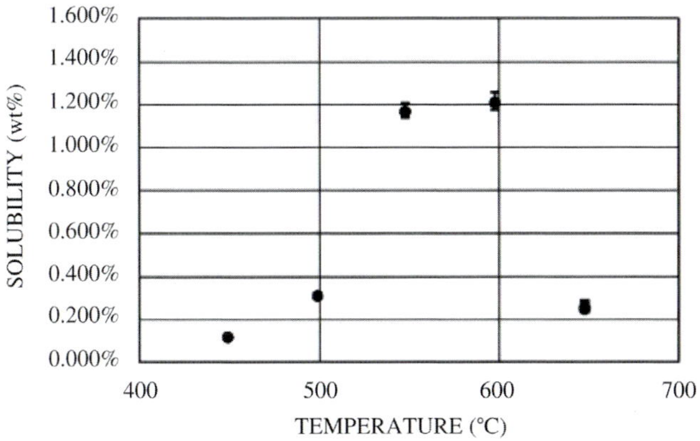

Figure 4.5 Solubility of GaN in supercritical ammonia with 1.5 mol% $NaNH_2$ as a function of temperature. (Figure 3 in Ref. [8].)

investigated for several decades [13–17]. The conditions for hydrothermal ZnO single crystal growth are very similar to those employed for α-quartz. The interior design and temperature distribution of the autoclave are shown in Figure 4.7. The inner volume of this autoclave is 200 ml and the typical experimental conditions are as follows. The growth temperature is 633–653 K and the temperature of the lower portion is 658–678 K, and the solution is 3.0 M KOH and 1.0–2.0 M LiOH is added to enhance the growth rate. The Li^+-ion was incorporated into the substrate and the concentration of Li^+-ion was critical issue. Fukuda and co-workers succeeded in growing high-quality crystals of ZnO by hydrothermal reaction using platinum-lined autoclaves and KOH

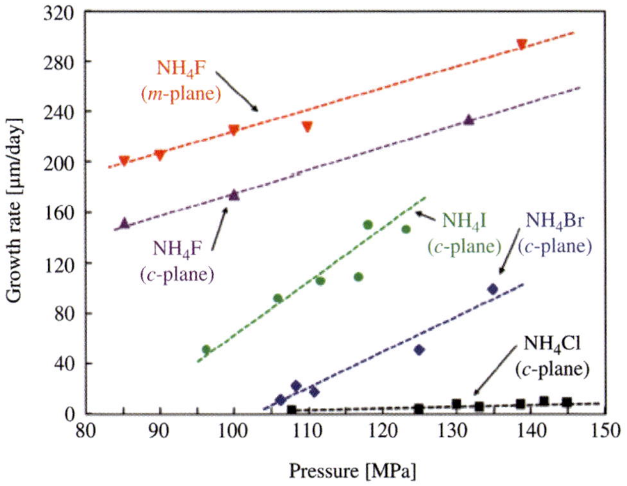

Figure 4.6 Pressure dependence of GaN crystal growth rate using different acidic mineralizers. (Figure 4 in Ref. [12].)

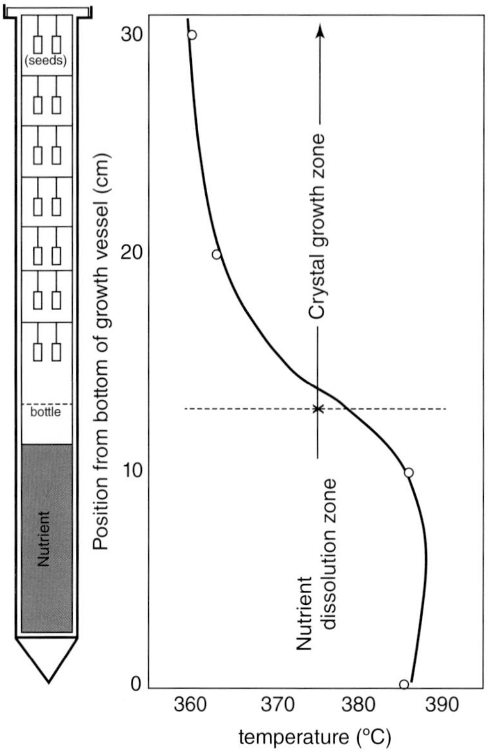

Figure 4.7 Typical temperature distribution in an autoclave for hydrothermal crystal growth of ZnO. (Figure 1 in Ref. [14].)

solutions [17]. Figure 4.8 shows ZnO single crystals prepared at temperature of 603–653 K in the growth zone and 613–643 K for the dissolution zone with the growth rate ranging from 150 to 300 μm per day.

4.5
YAG ($Y_3Al_5O_{12}$)

Yttrium aluminate, $Y_3Al_5O_{12}$ (YAG) with a garnet-type structure, is a well-known laser material. Its single crystals are grown by Czochralski and flux methods. Hydrothermal growth of YAG was demonstrated by Nilsen and co-workers [18], and the hydrothermal growth of $Y_3Fe_5O_{12}$ (YIG) [19] and $Y_3Ga_5O_{12}$ (YGG) [20] was investigated by Laudise *et al.* in the 1960s. There was no subsequent report on hydrothermal growth of garnet-type oxide for a half of century. However, Kolis and co-workers reinvestigated the hydrothermal growth of YAG [21] and, on the basis of the data, succeeded in growing $Lu_3Al_5O_{12}$ (LuAG)

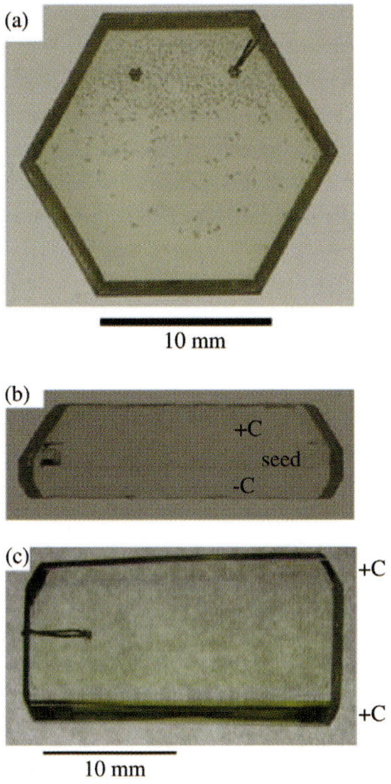

Figure 4.8 A (0 0 0 1) ZnO single crystal grown using KOH mineralizer (a), its cross-sectional view perpendicular to (0 0 0 1) (b), and a (1 0 −1 0) crystal (c). (Figure 1 in Ref. [17].)

hydrothermal [22]. Figure 4.9 shows a high-quality LuAG single crystal prepared at the growth temperatures between 873 and 923 K using 2–8 M K_2CO_3 solution as a mineralizer [22].

4.6 $AlPO_4$ (Berlinite) and $GaPO_4$

Both of $AlPO_4$ (berlinite) and $GaPO_4$ exhibit a modification similar to SiO_2. Their low-temperature forms with the α-quartz crystal structure have been proposed to be a postoscillator material for α-quartz. The hydrothermal growth of $AlPO_4$ [23,24] and $GaPO_4$ [25–27] single crystals has been investigated since the early 1980s. Three types of hydrothermal methods [24,27] have been proposed to overcome the negative temperature coefficient of solubility for $AlPO_4$ and $GaPO_4$ in H_3PO_4 solution: (i) slow heating method (SH), (ii) vertical reverse temperature gradient method (VTG, Figure 4.10), and (iii) horizontal

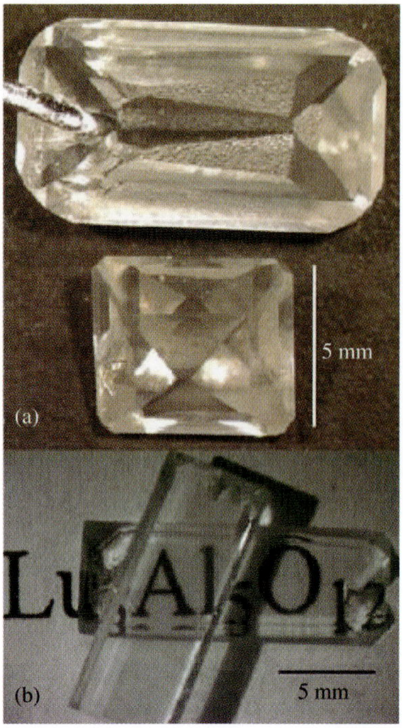

Figure 4.9 (a) Fully faceted LuAG single crystals with a silver mounting wire clearly visible in the upper picture, and (b) LuAG crystals harvested from YAG surrogate seeds. (Figure 5 in Ref. [22].)

temperature gradient method (HTG, Figure 4.11). The VTG and a variant of HTG methods provided good results for $AlPO_4$ and were viewed as being capable of producing large crystals with a high degree of perfection when compared with the slow heating method [24]. For $GaPO_4$, the reflux method was superior to VTG and HTG methods [27].

4.7 $ABe_2BO_3F_2$ (ABBF, A = K, Rb, Cs, Tl)

Alkaline (thallium) fluroberyllium borates, $ABe_2BO_3F_2$ (ABBF, A = K, Rb, Cs, Tl), exhibit an excellent property for nonlinear optical applications in the deep-UV region (below 200 nm). In 2006, Ning and Tang [28] succeeded in the hydrothermal crystal growth of KBBF using KF and H_2BO_3 as a mineralizer. As-grown crystal of KBBF is shown in Figure 4.12. Single crystals for other members [29] of this family were prepared by McMillen and Kolis using fluoride mineralizers. Other attempts at hydrothermal growth of $ABe_2BO_3F_2$ have been performed using alkaline hydroxide or carbonate as a mineralizer [30,31].

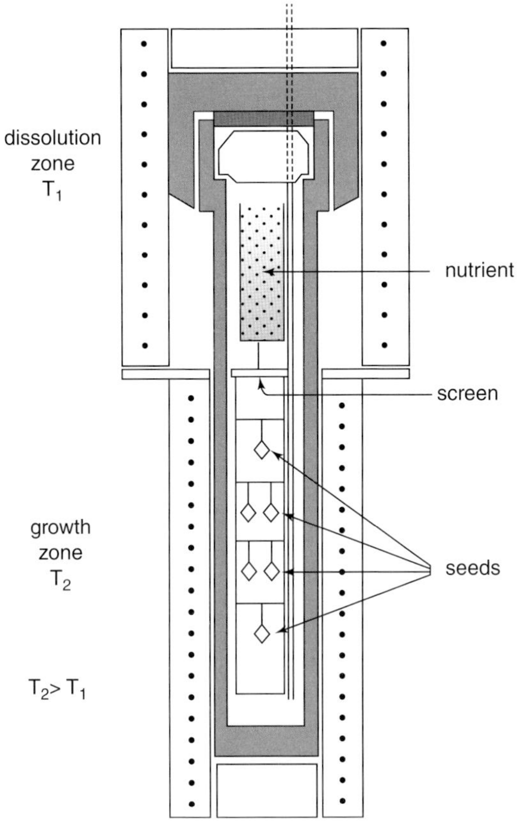

Figure 4.10 Schematic diagram of the autoclave used for the vertical reverse temperature gradient (VTG) method. (Figure 2 in Ref. [24].)

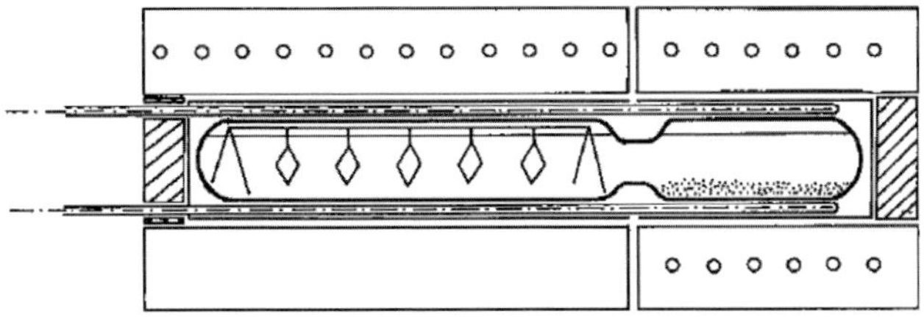

Figure 4.11 Hydrothermal system primarily used for the horizontal temperature gradient (HTG) method. (Figure 4 in Ref. [24].)

4.8
Synthesis of Zirconium Phosphates

Solvothermal reaction is a useful method to crystallize organic–inorganic complexes in phosphates and silicates as new compounds. The diversity of

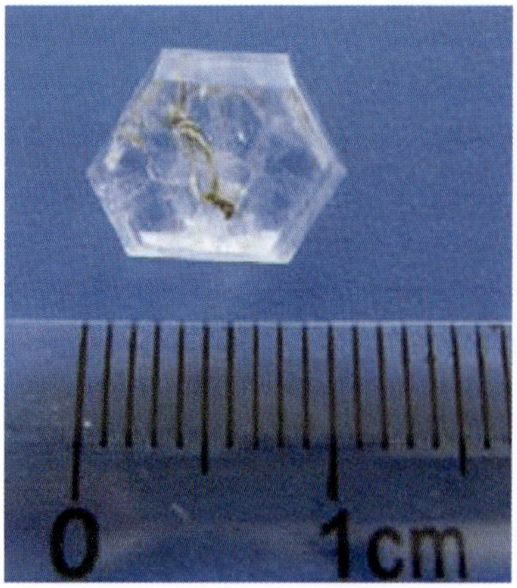

Figure 4.12 Hydrothermally as-grown KBBF crystal. (Figure 4 in Ref. 28)

connection of metal-centered polyhedra and the combination of organic compounds could produce a variety of new compounds based on ZrO_6 octahedra and PO_4 tetrahedra. Recently, Vivani *et al.* reviewed the crystal structures of a variety of compounds based on zirconium phosphate [32]. Among this family of compounds, two types of layered zirconium phosphates, α-$Zr(HPO_4)_2{\cdot}H_2O$ [33] and γ-$Zr(PO_4)(H_2PO_4){\cdot}2H_2O$ [34] are well-known as "basic compounds." Various zirconium phosphates can be easily derived by intercalation, ion-exchange, or direct solvothermal reactions. Seven types of new zirconium phosphates (hereafter, ZrPO-1–ZrPO-7) have been prepared by solvothermal reaction using ethylene or diethylene glycol as solvent [35–39]. The crystal structures of ZrPO-1–ZrPO-7 are shown in Figure 4.13. The structural features of ZrPO-1–ZrPO-7 may be described as 1D, 2D, or 3D, built by the inorganic parts of corner-sharing ZrO_6 octahedra and PO_4 tetrahedra incorporating the organic molecules in them. The crystal structure depends on the employed synthetic conditions. Four types of zirconium phosphates, ZrPO-3, -4, -6, and -7 were synthesized under similar solvothermal conditions using $ZrOCl_2{\cdot}8H_2O$ and H_3PO_4 as starting compounds and diethylene glycol as solvent. By changing the P/Zr ratio in the starting compounds and the reaction temperature, different crystal structures were formed. ZrPO-3 in a 1D structure was prepared with P/Zr = 1.0 at 180 °C. ZrPO-4 in a 2D layered structure was also prepared at the same temperature by increasing the P/Zr ratio to 1.8. Both higher temperatures and P/Zr ratio produced ZrPO-6 with a 2D layered structure different from ZrPO-4. ZrPO-7 in a 3D structure was crystallized at an increasing reaction temperature of 260 °C. Increasing the reaction temperature and the P/Zr ratio in solvothermal reactions could promote condensation of phosphates to enhance the dimensionality of the corresponding crystal structures.

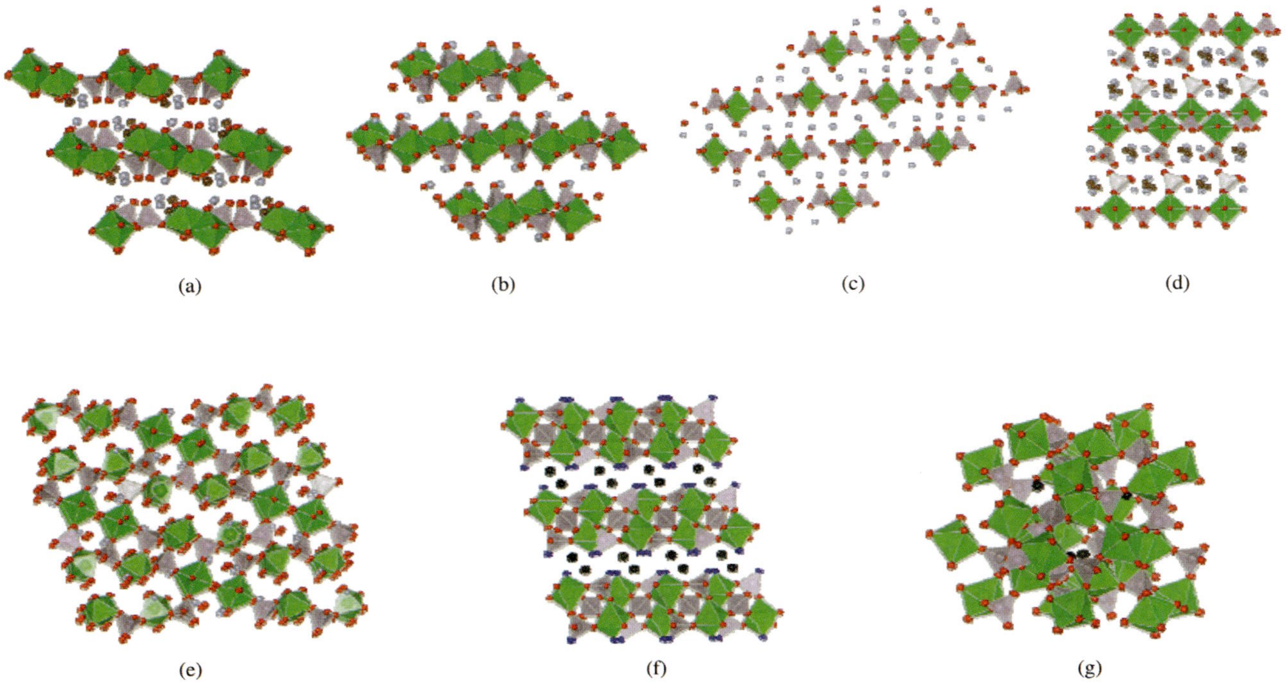

Figure 4.13 Crystal structures of (a) $[NH_4]_2[enH_2]_2Zr_3(OH)_6(PO_4)_4$ (ZrPO-1), (b) $[NH_4]_2Zr(OH)_3(PO_4)$ (ZrPO-2), (c) $[NH_4]_3Zr(OH)_2(PO_4)(HPO_4)$ (ZrPO-3), (d) $[enH_2]Zr(OH)(PO_4)(HPO_4)$ (ZrPO-4), (e) $(NH_4)_5[Zr_3(OH)_9(PO_4)_2(HPO_4)]$ (ZrPO-5), (f) $(NH_4)_2[Zr_2(OH)_2(PO_4)_2(HPO_4)]$ (ZrPO-6), and (g) $(NH_4)Zr_2(PO_4)_3$ (ZrPO-7). en: ethylenediamine.

4.9 Synthesis of Bismuth Oxides

Bismuth can be in trivalent and pentavalent states in oxides. Some mixed valence bismuth oxides are well-known superconductors with a perovskite-type structure such as $Ba(Bi,Pb)O_3$ [40] and $(Ba,K)BiO_3$ [41]. However, it is difficult to prepare pentavalent bismuth oxides through high-temperature reactions, except for the system containing barium or alkaline metal oxides because most bismuth oxides possess trivalent bismuth atoms. Other superconductive perovskite-type bismuth oxides, such as $(Sr,A)BiO_3$ (A = K, Rb) [42], $K_{1-x}Bi_{1+x}O_3$ ($x = 0.0$–0.1) [43], and $(K,Bi)BiO_3$ [44], have been prepared only under high-pressure conditions. Hydrothermal reactions using a hydrated sodium bismuth oxide $NaBiO_3{\cdot}nH_2O$ could allow the formation of pentavalent bismuth oxides [45]. This method permits to crystallize new superconductive bismuth oxides with the double perovskite-type structure [46–48] together with a variety of pentavalent bismuth oxides such as Bi_2O_4 [49], $LiBiO_3$ [50], $AgBiO_3$ [51], and ABi_2O_6 (A = Mg, Zn, Sr, Ba, Cd) [52–55]. Figure 4.14 shows the crystal structures of a superconductive bismuth oxide with a double perovskite-type structure, Bi_2O_4, $LiBiO_3$, $MgBi_2O_6$, and $SrBi_2O_6$. Bi_2O_4 possesses the crystal structure of β-Sb_2O_4 and distinct crystallographic sites; Bi^{3+} has eightfold coordination while Bi^{5+} has octahedral

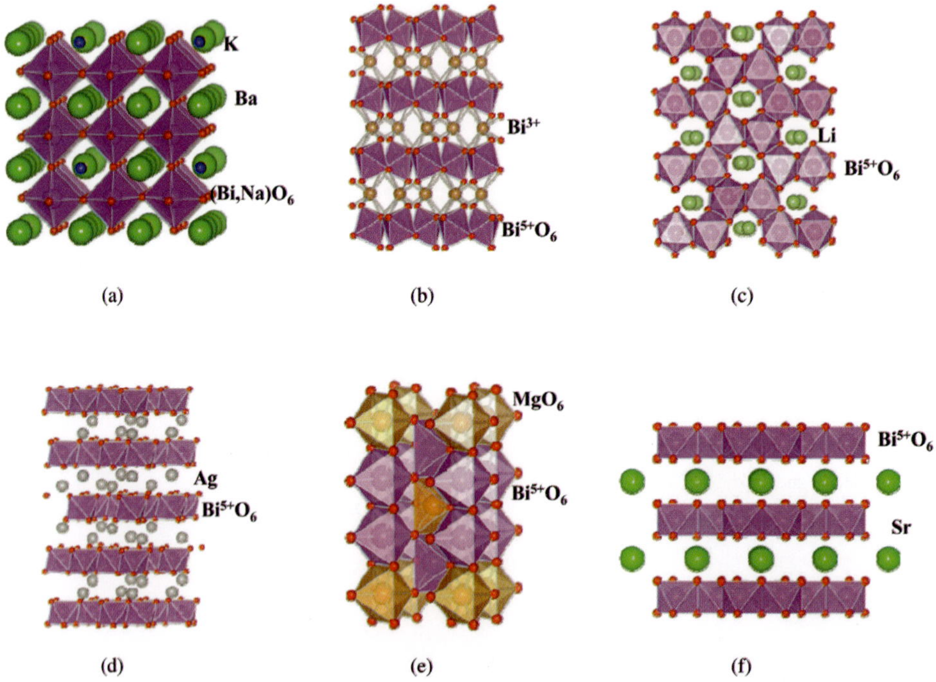

Figure 4.14 Crystal structures of (a) $(K_{1.00})(Ba_{1.00})_3(Bi_{0.89}Na_{0.11})_4O_{12}$, (b) Bi_2O_4, (c) $LiBiO_3$, (d) $AgBiO_3$, (e) $MgBi_2O_6$, and (f) $SrBi_2O_6$.

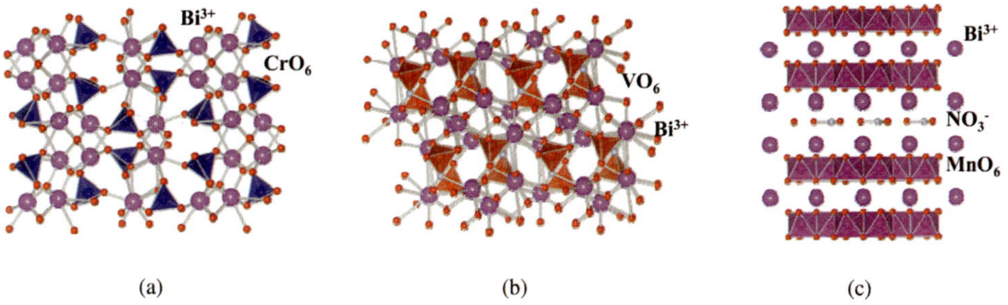

Figure 4.15 Crystal structures of (a) $Bi_8(CrO_4)O_{11}$, (b) $Bi_{3.33}(VO_4)_2O_2$, and (c) $Bi_3Mn_4O_{12}(NO_3)$.

coordination. This is the first example of crystal structure analyzed on any binary bismuth oxide in mixed valence. The crystal structure of $LiBiO_3$ is closely related to that of $LiSbO_3$ [56] and $AgBiO_3$ has an ilmenite-type structure [51]. In the case of $MgBi_2O_6$ and $ZnBi_2O_6$, trirutile-type compounds were obtained while $SrBi_2O_6$, $BaBi_2O_6$, and $CdBi_2O_6$ adopted a $PbSb_2O_6$-type structure. Similar to pentavalent bismuth oxides, hydrothermal reactions using $NaBiO_3{\cdot}nH_2O$ produced various trivalent bismuth oxides such as $HBi_3(CrO_4)_2O_3$ [57], $Bi_8(CrO_4)O_{11}$ [58], $Bi_{3.33}(VO_4)_2O_2$ [59], $Bi_3Mn_4O_{12}(NO_3)$ [60], and $(Ln,Bi)_3O_4NO_3$ (Ln = Y, Sm ~ Dy) [61]. Figure 4.15 shows the crystal structures of $Bi_8(CrO_4)O_{11}$ [58], $Bi_{3.33}(VO_4)_2O_2$ [59], and $Bi_3Mn_4O_{12}(NO_3)$ [60]. While Bi^{5+} has octahedral coordination, Bi^{3+} prefers irregular shape coordination with high coordination number (>6) owing to the stereoactive lone pair. The synthesis of these trivalent bismuth oxides by conventional high-temperature reactions is also not trivial. Therefore, hydrothermal synthesis using $NaBiO_3{\cdot}nH_2O$ is useful to obtain mixed valency of Bi^{5+} and/or Bi^{3+} in a variety of bismuth oxides.

References

1 Iwaski, F. and Iwasaki, H. (2002) Historical review of quartz crystal growth. *J. Cryst. Growth*, **237–239**, 820–827.
2 Walker, A.C. (1953) Hydrothermal synthesis of quartz crystals. *J. Am. Ceram. Soc.*, **36**, 250–256.
3 Lias, N.C., Grudenski, Ms.E.E., Kolb, E.D., and Laudise, R.A. (1973) The growth of high acoustic Q quartz at high growth rates. *J. Cryst. Growth*, **18**, 1–6.
4 Balitsky, V.S. (1977) Growth of large amethyst crystals from hydrothermal fluoride solutions. *J. Cryst. Growth*, **41**, 100–102.
5 Lityagina, L.M., Dyuzheva, T.I., Nikolaev, N.A., and Bendeliani, N.A. (2001) Hydrothermal crystal growth of stishovite (SiO_2). *J. Cryst. Growth*, **222**, 627–629.
6 Amano, H. (2013) Progress and prospect of the growth of wide-band-gap group III nitrides: development of the growth method for single-crystal bulk GaN. *Jpn. J. Appl. Phys*, **52**, 050001-1–050001-10.
7 Dwiliński, R., Doradziński, R., Garczyński, J., Sierzputowski, L.P., Puchalski, A., Kanbara, Y., Yagi S K., Minakuchi, H., and Hayashi, H. (2008) Excellent crystallinity of truly bulk ammonothermal GaN. *J. Cryst. Growth*, **310**, 3911–3916.
8 Hashimoto, T., Saito, M., Fujito, K., Wu, F., Speeck, S.J., and Nakamura, S. (2007) Seeded growth of GaN by the basic

ammonothermal method. *J. Cryst. Growth*, **305**, 311–316.
9 Hashimoto, T., Wu, F., Saito, M., Fujito, K., Speeck, S.J., and Nakamura, S. (2010) Status and perspectives of the ammonothermal growth of GaN substrates. *J. Cryst. Growth*, **310**, 876–880.
10 Yoshida, K., Aoki, K., and Fukuda, T. (2014) High-temperature acidic ammonothermal method for GaN crystal growth. *J. Cryst. Growth*, **393**, 93–97.
11 Tomida, D., Kuroda, K., Hoshino, N., Suzuki, K., Kagamitani, Y., Ishiguro, Y., Fukuda, T., and Yokoyama, C. (2010) Solubility of GaN in supercritical ammonia with ammonium chloride as a mineralizer. *J. Cryst. Growth*, **312**, 3161–3164.
12 Bao, Q., Saito, M., Hazu, K., Furusawa, K., Kagamitani, Y., Kayano, R., Tomida, D., Qiao, K., Ishiguro, T., Yokoyama, C., and Chichibu, S.F. (2013) Ammonothermal crystal growth of GaN using an NH_4F mineralizer. *Cryst. Growth Des.*, **13**, 4158–4161.
13 Laudise, R.A. and Ballman, A.A. (1960) Hydrothermal synthesis of zinc oxide and zinc sulfide. *J. Phys. Chem.*, **64**, 688–691.
14 Sakagami, N. and Wada, M. (1974) Growth kinetics and morphology of ZnO single crystal grown under hydrothermal conditions. *Yogyo-Kyokai-Shi*, **82**, 405–413.
15 Sakagami, N. and Shibayama, K. (1981) Hydrothermal growth and characterization of ZnO single crystals. *Jpn. J. Appl. Phys*, **S-20-4**, 201–205.
16 Croxall, D.F., Ward, R.C.C., Wallace, C.A., and Kell, R.C. (1974) Hydrothermal growth and investigation of Li-doped zinc oxide crystals of high purity and perfection. *J. Cryst. Growth*, **22**, 117–124.
17 Ehrentraut, D., Maeda, K., Kano, M., Fujii, K., and Fukuda, T. (2011) Next-generation hydrothermal ZnO crystals. *J. Cryst. Growth*, **320**, 18–22.
18 Puttbach, R.C., Monchamp, R.R., and Nilson, J.W. (1967) *Hydrothermal Growth of $Y_3Al_5O_{12}$ in Crystal Growth*, Pergamon, Oxford, pp. 569–571.
19 Laudise, R.A. and Kolb, E.D. (1962) Hydrothermal crystallization of yttrium-iron garnet on a seed. *J. Am. Ceram. Soc.*, **45**, 51–53.
20 Laudise, R.A., Crockett, J.H., and Ballman, A.A. (1961) The hydrothermal crystallization of yttrium iron garnet and yttrium gallium garnet and a part of the crystallization diagram Y_2O_3-Fe_2O_3-H_2O-Na_2CO_3. *J. Phys. Chem.*, **65**, 359–361.
21 McMillen, C.D., Mann, M., Fan, J., Zhu, L., and Kolis, J.W. (2012) Revisiting the hydrothermal growth of YAG. *J. Cryst. Growth*, **356**, 58–64.
22 Moore, C.A., McMillen, C.D., and Kolis, J.W. (2013) Hydrothermal growth of single crystals of $Lu_3Al_5O_{12}$ (LuAG) and its doped analogues. *Cryst. Growth Des.*, **13**, 2298–2306.
23 Kolb, E.D., Grenier, J.C., and Laudise, R.A. (1981) Solubility and growth of $AlPO_4$ in a hydrothermal solvent: HCl. *J. Cryst. Growth*, **51**, 178–182.
24 Jumas, J.C., Goiffon, A., Capelle, B., Zarka, A., Doukhan, J.C., and Philippot, E. (1987) Crystal growth of berlinite, $AlPO_4$: physical characterization and comparison with quartz. *J. Cryst. Growth*, **80**, 133–148.
25 Hirano, S., Miwa, K., and Naka, S. (1986) Hydrothermal synthesis of gallium orthophosphate crystals. *J. Cryst. Growth*, **79**, 215–218.
26 Hirano, S. and Kim, P.C. (1991) Growth of gallium orthophosphate single crystals in acidic hydrothermal solutions. *J. Mater. Sci*, **26**, 2805–2808.
27 Balitsky, D.V., Philippot, E., Papet, Ph., Balitsky, V.S., and Pey, F. (2005) Comparative crystal growth of $GaPO_4$ crystals in the retrograde and direct solubility range by hydrothermal methods of temperature gradient. *J. Cryst. Growth*, **275**, e887–e894.
28 Ning, Y. and Tang, D. (2006) Hydrothermal growth of $KBe_2BO_3F_2$ crystals. *J. Cryst. Growth*, **293**, 233–235.
29 McMillen, C.D. and Kolis, J.W. (2008) Hydrothermal crystal growth of $ABe_2BO_3F_2$ (A = K, Rb, Cs, Tl) NLO crystals. *J. Cryst. Growth*, **310**, 2033–2038.
30 Zhou, H.T., He, X.L., Zhou, W.N., Hu, Z.G., Zhang, C.L., Huo, H.D., Wang, J.L., Qin, S.J., Zuo, Y.B., Lu, F.H., Liu, L.J., Wang, X.Y., Liu, Y.C., Li, D.P., Zhang, H.X., and Chen, X.Y. (2011)

Hydrothermal growth of KBBF crystals from KOH solution. *J. Cryst. Growth*, **318**, 613–617.

31 Liu, L., Zhou, H., He, X., Zhang, X., Wang, X., Lu, F., Zhang, C., Zhou, W., and Chen, C. (2012) Hydrothermal growth and optical properties of $RbBe_2BO_3F_2$ crystals. *J. Cryst. Growth*, **348**, 60–64.

32 Vivani, R., Alberti, G., Costantino, F., and Nocchetti, M. (2008) New advances in zirconium phosphate and phosphoonate chemistry: structural archetypes. *Microporous Mesoporous Mater.*, **107**, 58–70.

33 Clearfield, A. and Smith, G.D. (1969) Crystallography and structure of .alpha.-zirconium bis(monohydrogen orthophosphate) monohydrate. *Inorg. Chem.*, **8**, 431–436.

34 Poojary, M., Shpeizer, B., and Clearfield, A. (1995) X-ray powder structure and Rietveld refinement of γ-zirconium phosphate, $Zr(PO_4)(H_2PO_4){\cdot}2H_2O$. *J. Chem. Soc. Dalton Trans.*, **1995**, 111–113.

35 Wang, D., Yu, R., Kumada, N., and Kinomura, N. (2000) Nonaqueous synthesis and characterization of a novel layered zirconium phosphate templated with mixed organic and inorganic cations. *Chem. Mater.*, **12**, 956–960.

36 Kumada, N., Nakatani, T., Yonesaki, Y., Takei, T., and Kinomura, N. (2008) Preparation of zew zirconium phosphates by solvothermal reaction. *J. Mater. Sci.*, **43**, 2206–2212.

37 Wang, D., Yu, R., Kumada, N., Kinomura, N., Yanagisawa, K., Matsumura, Y., and Yashima, T. (2002) A novel layered zirconium phosphate $[NH_4]_2[Zr(OH)_3(PO_4)]$ synthesized through non-aqueous route. *Chem. Lett.*, **2002**, 804–805.

38 Wang, D., Yu, R., Takei, T., Kumada, N., Kinomura, N., Onda, A., Kajiyoshi, K., and Yanagisawa, K. (2002) Non-aqueous synthesis and structure of a novel monodimensional zirconium phosphate: $[NH_4]_3[Zr(OH)_2(PO_4)(HPO_4)]$. *Chem. Lett.*, **2002**, 398–399.

39 Kumada, N., Hinata, J., Dong, Q., Yonesaki, Y., Takei, T., and Kinomura, N. (2011) Preparation and crystal structure of two types of zirconium phosphates by hydrothermal reaction. *J. Ceram. Soc. Jpn.*, **119**, 412–416.

40 Sleight, A.W., Gillson, J.L., and Bierstedt, P.E. (1975) High-temperature superconductivity in the $BaPb_{1-x}Bi_xO_3$ systems. *Solid State Commun.*, **17**, 27–28.

41 Cava, R.J., Batlogg, B., Krajewski, J.J., Farrow, R., Rupp, L.W., Jr, White, A.E., Short, K., Peck, W.F., and Kometani, T. (1998) Superconductivity near 30 K without copper: the $Ba_{0.6}K_{0.4}BiO_3$ perovskite. *Nature*, **332**, 814–816.

42 Kazakov, S.M., Chaillout, C., Bordet, P., Capponi, J.J., Nunez-Regueiro, M., Rysak, A., Tholence, J.L., Radaelli, P.G., Putilin, P.G., and Antipov, E.V. (1997) Discovery of a second family of bismuth-oxide-based superconductors. *Nature*, **390**, 148–150.

43 Khasanova, N.R., Yamamoto, A., Tajima, S., Wu, X.J., and Tanabe, K. (1998) Superconductivity at 10.2 K in the K–Bi–O system. *Physica C*, **305**, 275–280.

44 Khasanova, N.R., Izumi, F., Kamiyama, K., Yoshida, K., Yamamoto, A., and Tajima, S. (1999) Crystal structure of the $(K_{0.87}Bi_{0.13})BiO_3$ superconductor. *J. Solid State Chem.*, **144**, 205–208.

45 Kumada, N. (2013) Preparation and crystal structure of new inorganic compounds by hydrothermal reaction. *J. Ceram. Soc. Jpn*, **121**, 135–141.

46 Rubel, M.H.K., Miura, A., Takei, T., Kumada, N., Ali, M.M., Nagao, M., Watauchi, S., Tanaka, I., Oka, K., Azuma, M., Magome, E., Moriyoshi, C., Kuroiwa, Y., and Islam, A.K.M.A. (2014) Superconducting double pervoskite bismuth oxide $(Na_{0.25}K_{0.45})(Ba_{1.00})_3(Bi_{1.00})_4O_{12}$ prepared by low temperature hydrothermal reaction. *Angew. Chem., Int. Ed.*, **147**, 3599–3603.

47 Rubel, M.H.K., Miura, A., Takei, T., Kumada, N., Ali, M.M., Oka, K., Azuma, M., Magome, E., Moriyoshi, C., and Kuroiwa, Y. (2015) Low-temperature hydrothermal synthesis of a new Bi based $(Ba_{0.83}K_{0.17})(Bi_{0.54}Pb_{0.46})O_3$ superconductor. *J. Alloys. Compd.*, **634**, 208–214.

48 Rubel, M.H.K., Takei, T., Kumada, N., Ali, M.M., Miura, A., Tadanaga, K., Oka, K., Azuma, M., Yashima, M., Fujii, K., Magome, E., Moriyoshi, C., Kuroiwa, Y., Hester, J.R., and Avdeev, M. (2016)

Hydrothermal synthesis, crystal structure, and superconductivity of a double-perovskite Bi oxide. *Chem. Mater.*, **28**, 459–465.

49 Kumada, N., Kinomura, N., Woodward, P.M., and Sleight, A.W. (1995) Preparation of new mixed valent bismuth oxides from sodium bismuth oxide. *J. Solid State Chem.*, **116**, 281–285.

50 Kumada, N., Takahashi, N., Kinomura, N., and Sleight, A.W. (1996) Preparation and crystal structure of a new lithium bismuth oxide: $LiBiO_3$. *J Solid State Chem.*, **126**, 121–126.

51 Kumada, N., Kinomura, N., and Sleight, A.W. (2000) Neutron powder diffraction refinement of ilmenite-type bismuth oxides: $ABiO_3$ (A = Na, Ag). *Mater. Res. Bull.*, **35**, 2397–2402.

52 Kumada, N., Takahashi, N., Kinomura, N., and Sleight, A.W. (1997) Preparation and crystal structure of ABi_2O_6 (A = Mg, Zn) with the trirutile-type structure. *Mater. Res. Bull.*, **32**, 1003–1008.

53 Kumada, N., Kinomura, N., and Sleight, A.W. (1999) Ion-exchange reaction of Na^+ in $NaBiO_3{\cdot}nH_2O$ with Sr^{2+} and Ba^{2+}. *Solid State Ion.*, **122**, 183–189.

54 Kumada, N., Miura, A., Takei, T., and Yashima, M. (2014) Crystal structure of a pentavalent bismuthate, $SrBi_2O_6$ and a fluorite-type $(Pb_{1/3}Bi_{2/3})O_{1.6}$. *J. Asian Ceram. Soc.*, **2**, 150–153.

55 Kumada, N., Miura, A., Takei, T., Nishimoto, S., Kameshima, Y., Miyake, M., Kuroiwa, Y., and Moriyoshi, C. (2015) Hydrothermal synthesis and crystal structure of two new cadmium bismuthates, $CdBi_2O_6$ and $Cd_{0.37}Bi_{0.63}O_{1.79}$. *J. Asian Ceram. Soc.*, **3**, 251–254.

56 Edstrand, M. and Ingri, N. (1954) The crystal structure of the double lithium antimony(V) oxide $LiSbO_3$. *Acta Chem. Scand.*, **8**, 1021–1031.

57 Kodialam, S., Kumada, N., Mackey, R., and Sleight, A.W. (1994) Crystal structure of a new hydrogen bismuth chromate $HBi_3(CrO_4)_2O_3$. *Euro. J. Solid State Inorg. Chem.*, **31**, 739–746.

58 Kumada, N., Takei, T., Kinomura, N., and Walles, G. (2006) Preparation and crystal structure of $Bi_8(CrO_4)O_{11}$. *J. Solid State Chem.*, **179**, 821–827.

59 Kumada, N., Takei, T., Haramoto, R., Yonesaki, Y., Dong, Q., Kinomura, N., Nishimoto, S., Kameshima, Y., and Michihiro, M. (2011) Preparation and crystal structure of a new bismuth vanadate, $Bi_{3.33}(VO_4)_2O_2$. *Mater. Res. Bull.*, **46**, 962–965.

60 Smirnova, O.A., Azuma, M., Kumada, N., Kusano, Y., Matsuda, M., Shimakawa, Y., Takei, T., Yonesaki, Y., and Kinomura, N. (2009) Synthesis, crystal strcuture and magnetic properties of $Bi_3Mn_4O_{12}(NO_3)$ oxynitrate comprising S = 3/2. *J. Am. Chem. Soc.*, **131**, 8313–8317.

61 Kumada, N., Takahashi, N., Kinomura, N., and Sleight, A.W. (1998) Preparation and crystal structure of a new ytrrium bismuth oxynitrate: $(Y,Bi)_3O_4NO_3$. *J. Solid State Chem.*, **139**, 321–325.

5
High-Throughput Synthesis Under Hydrothermal Conditions

Nobuaki Aoki, Gimyeong Seong, Tsutomu Aida, Daisuke Hojo, Seiichi Takami, and Tadafumi Adschiri

Tohoku University, World Premier International Research Center-Advanced Institute for Materials Research (WPI-AIMR), 2-1-1 Katahira, 980-8577 Sendai, Japan

5.1
Introduction

Hydrothermal synthesis has been carried out since the end of nineteenth century, mainly in growing crystals based on the natural geothermal processes (ore/mine formation). Precursor is usually metal oxides, hydroxides, or sometimes salts. Dissolution and recrystallization of precursor by changing temperature in a reactor is a classical method. It takes more than a few days for the crystal growth.

When a metal salt is used as a precursor, particle formation reactions expressed by the following equations are expected just by increasing temperature.

$$M(NO_3)_x + xH_2O = M(OH)_x + xHNO_3 \tag{5.1}$$

$$M(OH)_x = x/2\,H_2O + MO_{x/2} \tag{5.2}$$

The particle size and shape can be controlled by the changing temperature, pressure, or coexisting species. This method can be used for variety of metal oxides, sulphide, and phosphates. Also for the powder synthesis, batch type autoclave reactors are used. The objective of this method is mainly for obtaining homogeneous size particles or controlling shape through Ostwald ripening, requiring several hours. For this reason, continuous flow-type reactors where sufficient ripening time cannot be achieved have been rarely used. However, the continuous system is effective for industrial scale production.

In this context, Adschiri group has proposed a flow-type reaction system with rapid heating, as shown in Figure 5.1, in view of nanoparticle (NP) synthesis where the required ripening time is drastically reduced [1].

Handbook of Solid State Chemistry, First Edition. Edited by Richard Dronskowski, Shinichi Kikkawa, and Andreas Stein.

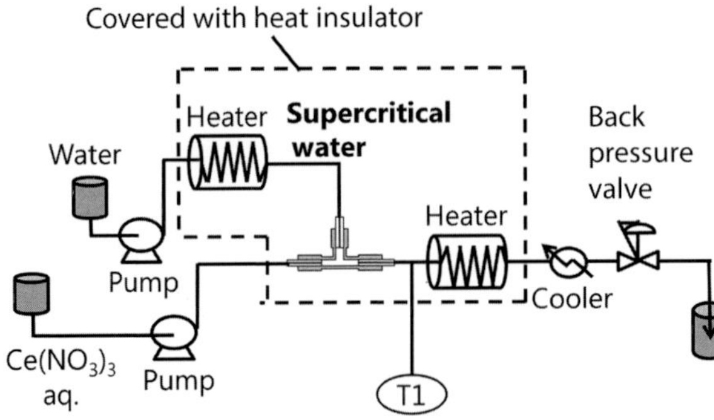

Figure 5.1 Schematic representation of a flow-type reactor for the hydrothermal synthesis of CeO_2 nanoparticles (NPs). The precursor solution (e.g., $Ce(NO_3)_3$ for CeO_2 particle synthesis) and distilled water are fed into the reactor using high-pressure pumps. The water was heated and mixed with the precursor solution in a T-mixer. Cartridge heaters were used to preheat the water and maintain a constant reaction temperature. The symbol T1 denotes a thermocouple to set reaction temperatures with a temperature controller. To cool the solution following the reaction, a shell-type cooler was connected to the end of the reactor. The reaction pressure is maintained above the critical point using a back-pressure valve.

This chapter summarizes the specific features of continuous flow-type reaction system, which enables high-throughput synthesis of nanoparticles. First, the mechanism of the hydrothermal synthesis method is summarized with reviewing the water properties, solubility, and reaction rate and equilibrium. Nanoparticles synthesis, metallic nanoparticles formation, and organic modification are explained. Next, phenomena in the reactor are described, including mixing of fluids and kinetics of reaction, which are essential to design flow-type reactors. Then, applications of the methods are illustrated, which covers catalyst, metamaterials, medical applications, semiconductor applications, and hybrid materials.

5.2 Mechanism of Hydrothermal Nanoparticle Synthesis

5.2.1 Properties of Water

The density of water around the critical point (374 °C, 22.1 MPa) is show in Figure 5.2 [2]. Around the critical point, the density changes suddenly with a little change of temperature and/or pressure. With this great change of density, most of the water properties change sharply around the critical point. Figure 5.3 shows the dielectric constant, which is one of the major controlling factors of solubility, reaction equilibrium, and reaction rate, changes significantly [3].

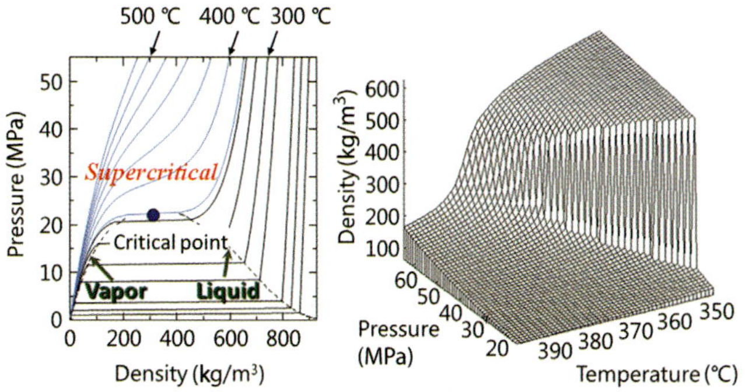

Figure 5.2 Density of water around the critical point.

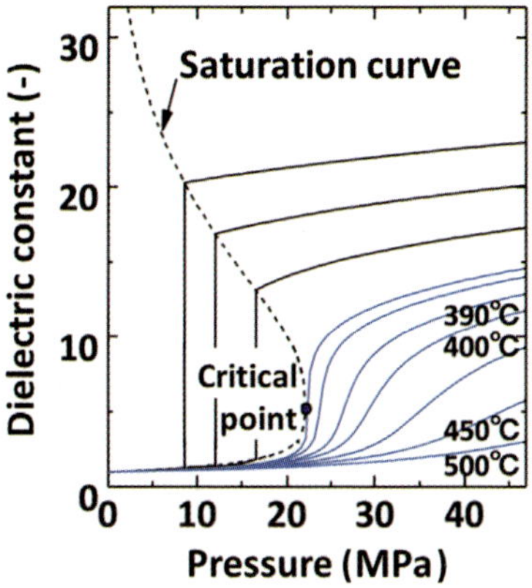

Figure 5.3 Dielectric constant of water around the critical point.

5.2.2
Reaction and Solubility in Water

Solubility of silica in high temperature/pressure water is reported in a literature [4]. At a constant temperature, the solubility rises with increasing pressure. At a high pressure, the solubility increases with temperature. At a low pressure, it first increases and then decreases with elevating temperature. This trend can be explained by considering the density change of water. At a higher density, the solubility increases with temperature, but at a constant pressure, the density of water decreases with increasing temperature, which leads to a decline in the solubility.

The solubility of metal oxides can be estimated by using the Helgeson–Kirkham–Flowers (HKF) model [5]. The following equations are examples of the chemical equations related to the metal oxide solubilization in water [1].

$$CuO(s) + 2H^+ = Cu^{2+} + H_2O \quad (5.3)$$

$$CuO(s) + H^+ = CuOH^+ \tag{5.4}$$

$$CuO(s) + H_2O = Cu(OH)_2 + H^+ \tag{5.5}$$

$$CuO(s) + 2H_2O = Cu(OH)_3{}^- + H^+ \tag{5.6}$$

By the combination of charge balance, mass balance, and reaction equilibrium, the concentration of each species can be determined. For solving the equations, the reaction equilibrium constant K at high temperature and pressure is estimated. The HKF model is a way to predict the reaction equilibrium constant over a wide range of temperature and pressure (25–1000 °C, 0.1–500 MPa). A simplified one that reduces the necessary initial information and thus can be used more widely is shown in Eq. (5.7) [6,7].

$$\begin{aligned}\ln K_{T,r} = {} & \ln K_{T_r,r_r} - \frac{\Delta H^0_{T,r_r}}{R}\left(\frac{1}{T} - \frac{1}{T_r}\right) \\ & + \frac{\beta(1.0-\rho^*)^{2/3} + \alpha\Delta\omega_{T_r,r_r} T_r}{R}\left(\frac{1}{T} - \frac{1}{T_r}\right), \\ & - \frac{\Delta\omega_{T,r}}{RT}\left(\frac{1}{\varepsilon} - 1\right) + \frac{\Delta\omega_{T_r,r_r}}{RT}\left(\frac{1}{\varepsilon_{T_r,r_r}} - 1\right)\end{aligned} \tag{5.7}$$

where T is the absolute temperature, ρ is the density, subscript r denotes the values at the critical point, R is the gas constant, ΔH is the heat of reaction, ρ^* is the relative water density (density divided by that at a reference condition (25 °C and 0.1 MPa)), α is a constant, $6.385 \times 10^{-5}\,K^{-1}$, $\Delta\omega$ is a parameter determined by the reaction system, and ε is the dielectric constant. The β is a reaction dependent constant and is given by

$$\beta = \lambda_1\left(\Delta C^0_{P,T_r,\rho_r} + \lambda_2\Delta\omega_{T_r,\rho_r}\right) + \lambda_3, \tag{5.8}$$

where $\Delta C^0_{P,T_r,\rho_r}$ is the heat capacity, $\lambda_1 = 9.766 \times 10^1$ K, $\lambda_2 = 2.0 \times 10^{-4}\,K^{-1}$, and $\lambda_3 = -3.317 \times 10^2$ J/mol.

Figure 5.4 shows the solubilities of CuO and PbO estimated by the simplified HKF model without any fitting parameters [8]. The plots shown in those figures are the experimental data. As explained above, the solubility increases first and then decreases near the critical temperature of water.

From Eq. (5.7), which is the function of density and dielectric constant, the change of reaction equilibrium and thus of solubility can be understood. The effect of dielectric constant on the reaction rate is expressed by the Born type equation [11].

$$\ln k = \ln k_0 - \frac{E}{RT} + \frac{\Psi}{RT}\left(\frac{1}{\varepsilon} - \frac{1}{\varepsilon_0}\right), \tag{5.9}$$

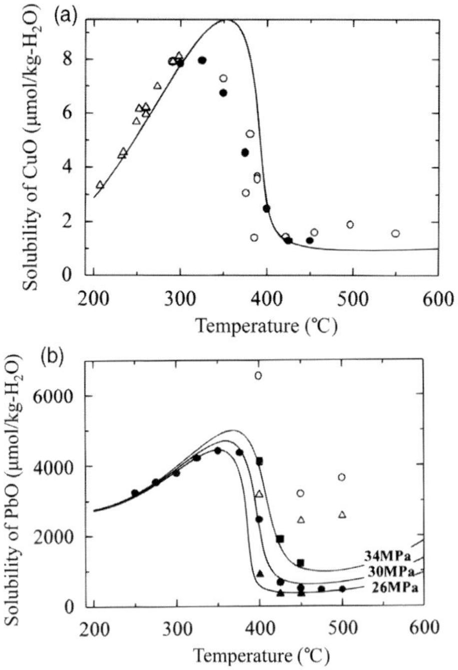

Figure 5.4 Solubilities of metal oxides from experiments and the estimation using the HKF model. (a) Solubility of CuO. ●: 28 MPa [8], ○: 28 MPa [8], Δ: saturation [9], –: HKF model. (b) Solubility of PbO. ■: 34 MPa (this work), ●: 30 MPa [8], ▲: 26 MPa [8], ○: 30 MPa [10], Δ: 26 MPa [10], –: HKF model.

where R is the gas constant, Ψ is the constant depending on the reaction system, ε is the dielectric constant, and ε_0 is the dielectric constant at a reference temperature. For hydrothermal reaction, the polarity of the activated state is lower than that of ionic reactants. Thus, the equation suggests that the reaction rate increases with decreasing dielectric constant. The reaction rates drastically increase above the critical point. The rate above the critical point will be estimated in Section 5.4.

5.2.3 Nucleation and Particle Formation

For explaining the formation of nanoparticle, the nucleation theory is useful. Figure 5.5 illustrates the LaMer mechanism [12]. With proceeding the hydrothermal synthesis reaction, the concentration of monomer increases and goes beyond the solubility limit estimated using the HKF model and then the critical supersaturation [13–15] to start the nucleation. Once the nucleation occurs, the concentration of metal oxide in water drastically decreases.

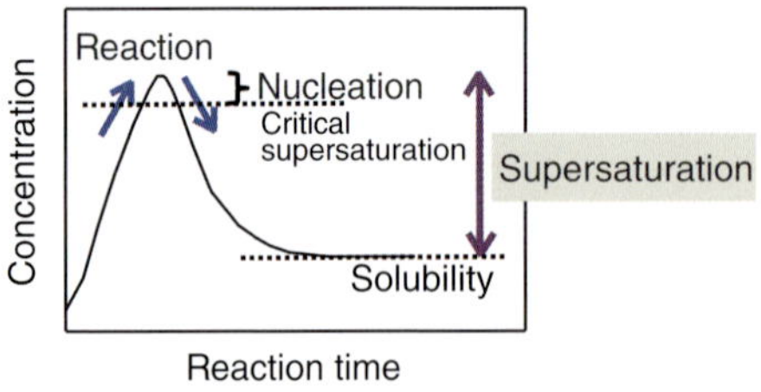

Figure 5.5 LaMer mechanism of nucleation.

The nucleation rate has been theoretically established [15,16]. The nucleation rate I is expressed as a function of critical free energy ΔG_{crit}, which is the highest energy to keep a product dissolved in a solution.

$$I = A \exp\left(-\frac{\Delta G_{\text{crit}}}{k_B T}\right), \tag{5.10}$$

$$\Delta G_{\text{crit}} = \frac{4}{3}\pi\gamma r_{\text{crit}}^2, \tag{5.11}$$

where A is the proportional constant, k_B is the Boltzmann's constant, T is the temperature, γ is the surface energy, and r_{crit} is the critical radius. The critical radius is the size of particle at the beginning of nucleation and is given by

$$r_{\text{crit}} = \frac{2\gamma v}{k_B T \ln S}, \tag{5.12}$$

$$S = \frac{C}{C_{\text{sol}}}, \tag{5.13}$$

where v is the molar volume, S is the supersaturation ratio, C is the concentration of the product, and C_{sol} is the solubility of product. The substitution of ΔG_{crit} in Eq. (5.10) by Eqs. (5.11) and (5.12) gives

$$I = A \exp\left(\frac{16\pi\gamma v}{3k_B^3 T^3 (\ln S)^2}\right). \tag{5.14}$$

For the nanoparticles synthesis, to obtain high supersaturation ratio is essential. To achieve high value of S, high reaction rate is required.

Since the solubility is low and the reaction rate is elevated in supercritical conditions, extremely high supersaturation degree and rapid nucleation reaction are expected to be obtained. For achieving a reaction-controlled condition to form small and monodisperse NPs, rapid heating of aqueous solution to the supercritical state is critical. However, a batch type reaction system, or even a flow reactor system with heat exchanger cannot make rapid heating. In contrast, the flow-type reaction system including the direct mixing of preheated supercritical water and precursor solution shown in Figure 5.1 achieved rapid heating up to the supercritical state. This is the key to synthesize nanoparticles continuously.

Table 5.1 Nanoparticles prepared through hydrothermal synthesis (the values in () are for particles with organic surface modification).

Precursors	Products	Particle size (nm)	References
$Al(NO_3)_3$	AlOOH	20–1000 (5–20)	[17–20]
$Ce(NO_3)_3$, $Ce(OH)_4$	CeO_2	20–300 (2–10)	[21–25]
$Co(NO_3)_2$	Co_3O_4	~100 (5–20)	[18,26]
$Cu(OH)_2$, $Cu(Gly)_2$	CuO, Cu_2O	50~100 (5)	[27]
$Fe(NO_3)_3$, $Fe_2(SO_4)_3$, $FeCl_2$	α-Fe_2O_3	~50 (5–20)	[18,28]
$Gd(NO_3)_2$	$Gd(OH)_3$	~20	[18]
$Ni(NO_3)_2$	NiO	~200 (5–20)	[29]
Ni(phetanthororin) + H_2	Ni	~500	[29]
$Ti(SO_4)_2$	TiO_2	~20 (5–20)	[18,30]
$Zn(NO_3)_2$	ZnO	~20	[31–33]
$ZrOCl_2$	ZrO_2	~20	[18]
$Al(NO_3)_3$, $Co(NO_3)_2$	$CoAl_2O_4$	5 (5)	[34]
$Ba(OH)_2{\cdot}8\,H_2O$, TiO_2	$BaTiO_3$	~20	[35]
$Mg(OH)_2$, $Fe(OH)_3$	$MgFe_2O_4$	~20	[36]
LiOH, $Co(NO_3)_2$	$LiCoO_2$	20–500	[37,38]
$Al(NO_3)_3$, $Y(NO_3)_3$	YAG ($Al_5Y_3O_{12}$)	~100 (~300)	[39,40]

5.2.4 Mechanism of Nanoparticle Formation

5.2.4.1 Hydrothermal Synthesis

Hydrothermal synthesis is a method to produce metal oxide and to grow crystals. Examples of particles obtained are shown in Table 5.1. Various nanoparticles can be synthesized, including single-, double-, and triple-component nanoparticles. Moreover, this reaction can be used for sulphate, phosphate, and metallic nanoparticles. The products cover magnetic, dielectric, high refractive index, Li ion battery electrode, and fluorescent materials.

5.2.4.2 Metal Nanoparticle Formation

Metal nanoparticles can be obtained as well as a wide range of metal oxides using the supercritical hydrothermal synthesis method with formic acid, glycerol, or hydrogen as a reducing agent. However, the direct use of hydrogen has a drawback in a safety issue because of the high risk of explosion [41], and glycerol has difficulty in producing cobalt and iron nanoparticles because of weaker reducing potential than that of hydrogen [42]. Adschiri group has succeeded in synthesizing a variety of metal nanoparticles by using a sub-/supercritical condition from early 1990s to the present [43–47], and has conducted in-depth study of the reaction system through thermodynamic [48] and kinetic analyses [49].

In general, the formation of metal nanoparticles in an aqueous solution has been thought as an impractical method, since water promotes the oxidation of metal and water itself can oxidize metal. However, metal is formed when the concentration of formic acid is sufficiently high. The decomposition of formic acid provides hydrogen and the hydrogen can reduce metal ions/metal oxides as well as suppress the oxidation of metal from air or water [50]. Moreover, under supercritical conditions, the reduction potential of hydrogen greatly increases because of the homogenous phase between hydrogen and water and the elevation of fugacity of H_2 (hydrogen is nearly insoluble in water below the critical point of water) [51]. Principle of the reductive supercritical hydrothermal synthesis is slightly different from a general hydrothermal synthesis. This reaction system is that the hydrogen formation from the decomposition of formic acid is followed by the reduction of metal oxides with hydrogen. The chemical equations are expressed as follows:

$$\mathrm{HCOOH(g)} \rightleftarrows \mathrm{H_2(g)} + \mathrm{CO_2(g)} \tag{5.15}$$

$$\left(\mathrm{CO(g)} + \mathrm{H_2O(g)}\text{À}\ \mathrm{HCOOH(g)}\right) \tag{5.16}$$

$$\mathrm{MO}_x\mathrm{(s)} + x\mathrm{H_2(g)}\text{À}\ \mathrm{M(s)} + x\mathrm{H_2O(g)} \tag{5.17}$$

Equation (5.15), which is a part of water–gas shift reaction, is very fast, and the equilibrium is shifted to right hand side under subcritical/supercritical water condition [52,53]. Thus, the required amount of hydrogen to form metal can be estimated from Eq. (5.17).

Thermodynamically important information in redox reaction can be obtained from the electrochemical series. The electrochemical series shows that a reducing component of the redox couple (hydrogen) has a tendency to reduce the oxidizing component of any redox couple (Ni^{2+}, Fe^{2+}, . . .) placed on in sequence (minus direction). This analysis only shows whether the reaction proceeds spontaneous or nonspontaneous in the specific temperature conditions, and the related thermodynamic equations can be obtained from the Gibbs free energy of the reaction as follows:

$$E^0 = -\Delta_r G^0, \tag{5.18}$$

$$E = -\Delta_r G/\nu F, \tag{5.19}$$

$$E = E^0 - \frac{RT}{\nu F} \ln Q, \tag{5.20}$$

where ν is the stoichiometric coefficient of electron, F is the Faraday constant, 96.48 kC/mol and Q is the reaction coefficient. Equation (5.19) is the well-known Nernst equation.

For a reaction system with $E > 0$ and $\Delta_r G < 0$, the reaction takes place spontaneously. Since $E = 0$ and $Q = K$ at the equilibrium state, the relation between the standard potential and the equilibrium constant at the specific temperature T is expressed as follows:

$$\ln K = \frac{\nu F E^0}{RT} \left(K = K_\mathrm{p}^{-1} = \frac{P_{\mathrm{H_2O}}}{P_{\mathrm{H_2}}} \right). \tag{5.21}$$

Thus, the partial pressure of hydrogen is calculated. The standard reduction potential for calculating Eq. (5.21) is well summarized in some inorganic chemistry books (for instance [54]).

To obtain a partial pressure of hydrogen in water, the Gibbs free energy of reaction should be applied to Eq. (5.17):

$$\Delta_\mathrm{r} G^0 = RT \ln K_\mathrm{p} \quad (K_\mathrm{p} = f_{\mathrm{H_2}}/f_{\mathrm{H_2O}}) \tag{5.22}$$

where K_p is the equilibrium constant of the reduction, $f_{\mathrm{H_2}}$ is the fugacity of hydrogen and $f_{\mathrm{H_2O}}$ is the fugacity of water. The values of the standard Gibbs free energy of formation of metal oxides at various temperatures are obtained from the Ellingham diagram [55].

The required amount of formic acid drastically increases between cobalt and iron in the list of redox potential of metal. This means copper, nickel, and cobalt have a possibility to be synthesized, but iron, zinc, and silicon are very difficult. Figure 5.6 shows the similar results in consistent with this estimation. In the case of iron, iron is rapidly oxidized by air or water, even if iron is synthesized with this method. To prevent reoxidation of iron by air or water, appropriate modifiers should be needed (coating agent), which will be described later.

When the revised HKF model is used, more precise values for those metal syntheses can be obtained. In addition, the Predictive Soave-Redlich-Kwong Equation of State (PSRK EOS) gives more precise values of fugacity of hydrogen, since the PSRK EOS more accurately expresses hydrogen and water system than other equations of state including ideal gas law [56]. Figure 5.7 shows the mole fraction of formic acid required for obtaining metallic phase of Co over water obtained from experimental results, and the results are fitted using different EOSs (assumed as $\mathrm{HCOOH:H_2 = 1:1}$).

In the reductive supercritical hydrothermal synthesis method, the reaction pathway including two main routes is expected [52]. The first main route is the formation of metal oxide from the variety of precursors such as metal complexes or metal hydroxides, which is followed by the reduction of the metal oxides with hydrogen (hereafter called reaction path A) to form metal nanoparticles. The second reaction path is the direct reduction of metal ions from the aqueous solution (hereafter called reaction path B). The reaction mechanism of the metal nanoparticle formation can be understood better from the kinetic study of the cobalt nanoparticle formation [49]. For an autoclave system, it takes several minutes to several hours to reach the desired reaction temperature depending on the size and material of the reactors (up to 380 °C, about 5 min for 5 cc hastelloy reactor). When the formic acid is used as a source of H_2, it is necessary to heat the reactor up to the minimum temperature for the decomposition of formic

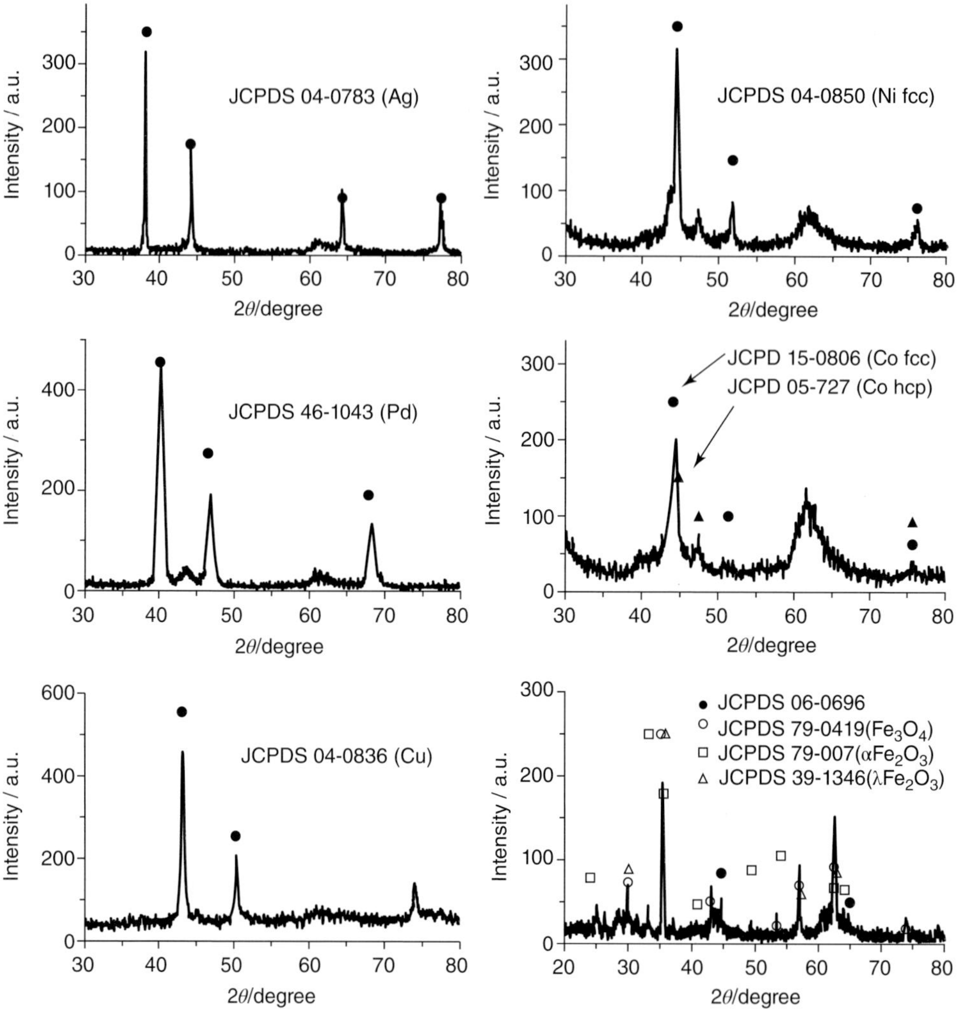

Figure 5.6 Powder XRD diffraction patterns of metal nanoparticles synthesized by the reductive supercritical hydrothermal method. The diffraction patterns of JCPDS files are also depicted for comparison. The primary particle diameters of each nanoparticle was calculated using the Scherrer's equation, and the diameters were 72.4, 26.7, 42.3, 36.2, 19.2, and 71.7 nm for Ag, Pd, Cu, Ni, Co, and Fe_3O_4, respectively [46].

acid. Under the conditions, hydrogen deficiency occurs in the initial reaction stage, and it leads the formation of metal oxide instead of metal nanoparticles. Later, cobalt oxide is reduced by hydrogen to form metallic cobalt. For enhancing the contact of cobalt oxide and hydrogen, the size of the intermediate cobalt oxide becomes an important key to efficiently proceed the reduction to metal cobalt. When the different sizes of the cobalt oxides were loaded instead of original precursor (cobalt (II) acetate·4 H_2O), the cobalt oxide NP of 93 nm size was

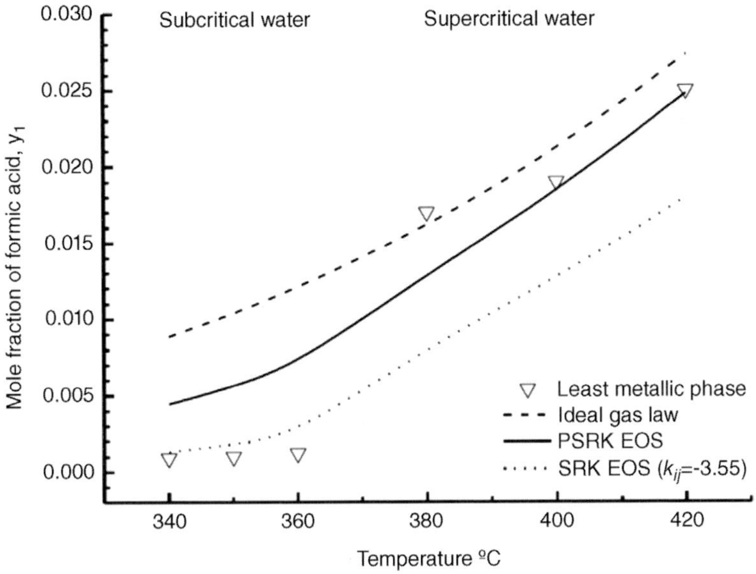

Figure 5.7 Plot of mole fraction of formic acid required for obtaining metallic phase of Co versus temperature with calculated results. (∇) experimental data; (– – –) estimation using ideal gas law; (–) PSRK EOS; and (. . .) SRK EOS with $k_{ij} = -3.55$ [48].

reduced to be in 40 mol% of cobalt, but that of 407 nm size was not effectively reduced (1.37 M formic acid injected).

Summarizing the results of the kinetic analysis and the thermodynamic analysis shown above, we can simply describe the total reaction mechanism of the metal nanoparticle using reductive hydrothermal synthesis method in the batch type reactor as shown in Figure 5.8. Here, the reaction mechanism of cobalt nanoparticles is a typical case. As mentioned earlier, reaction path A is the main reaction pathway because of the initial concentration and the rate of decomposition of formic acid. This reaction path requires a higher concentration of formic acid than the minimum required concentration of formic acid to reduce the metal oxide formed during the heating process. Moreover, it has the disadvantage of decreasing in the yield of the nanoparticles because of increased total reaction time.

In order to overcome this advantage, the flow type reactor is again more appropriate for the metal nanoparticle synthesis rather than the batch type reactor. A flow-type reactor system has the advantage in preventing the formation of metal oxides by rapidly raising the temperature for the decomposition of formic acid. This method can solve the hydrogen shortage during the temperature rising interval. Sue *et al.* [44] has synthesized nickel fine particles using a flow-type reactor. Homogeneous nucleation of Ni from the precursor solution is difficult. They used 1,10-phenanthroline to prevent Ni ions from being hydrated and synthesized magnetite simultaneously for the seed material of Ni. In the case

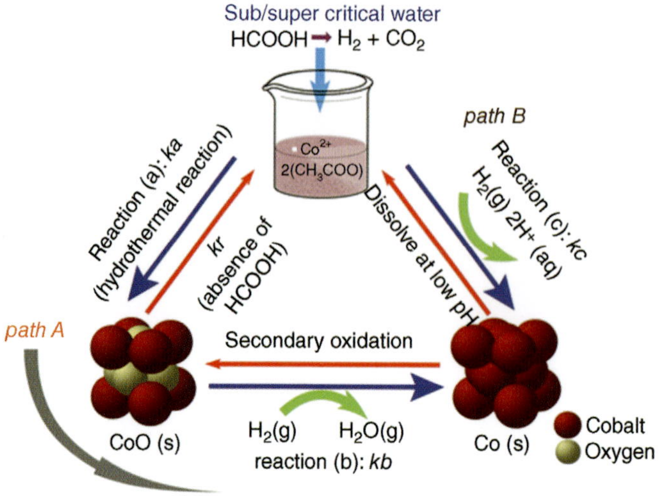

Figure 5.8 The proposed reaction mechanism for cobalt nanoparticle formation in the reductive supercritical hydrothermal process.

of cobalt, additional seed material is not necessary since the nucleation of cobalt is easier than Ni.

Copper, nickel, and cobalt are synthesized using the reductive hydrothermal process with/without surface modifiers, since these metals are relatively stable in the air and water. However, iron and zinc are very difficult to synthesize by this method. Unsubstantial hydrogen partial pressure would be required for synthesizing these metals. Moreover, keeping metal phase is very difficult because phase separation between hydrogen and water occurs in the cooling step. Therefore, the oxidation of metal using air and water should be effectively prevented using *in situ* surface capping method with a silane coupling agent, polymers, and organic substances.

5.2.4.3 **Organic Modified Nanoparticles**

For the nanoparticle synthesis, especially for metallic nanoparticles synthesis or sulfide/selenide nanoparticle synthesis by reverse micelle method and solvothermal method including hot-soap method, capping agents are used to stabilize the nanoparticles and suppress the particle growth. Also for metal oxide nanoparticles, there are some reports to use capping agents [57–59].

For hydrothermal synthesis, organic modifier needs to dissolve into water, and therefore the applicable organic modifier is limited to short-chain hydrophilic molecules. The concentration of modifier cannot be high because of the solubility in water, which limit the precursor concentration critical for determining the productivity.

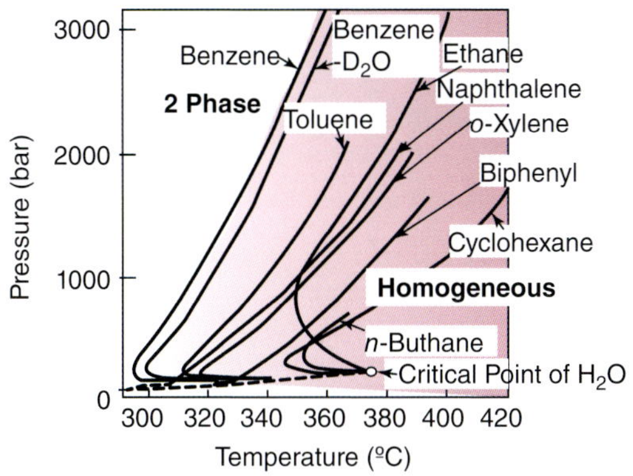

Figure 5.9 Critical loci for organic compounds–water binary systems.

Capping method except the silane capping agent method is through the physical adsorption. Adsorption occurs on the basis of the adsorption equilibrium as follows:

$$k_a\, CN(1-\theta) = k_d N\theta, \tag{5.23}$$

where C is the concentration of modifier, k_a and k_d are the rate constants of adsorption and desorption, respectively, N is the total amount of adsorption site, and θ is the fraction of adsorption. From Eq. (5.23), the amount of adsorption $N\theta$ is expressed as

$$N\theta = Nk_aC/(k_aC + k_d) = NKC/(1 + KC), \tag{5.24}$$

where $K = k_a/k_d$ is the adsorption constant. Generally speaking, since the concentration of organic modifier in the aqueous solution C is limited except hydrophilic organic modifier, high surface coverage ratio cannot be expected for the organic modifiers.

Figure 5.9 shows the phase behavior of water-organic species in two component systems [60]. Curves are the critical loci. In the right hand side of these curves, homogeneous phase is formed. Under the high temperature/pressure condition, oil and water are miscible. If this condition is selected, organic modifiers can be introduced into the reaction atmosphere in high concentration to achieve high surface coverage. The size of the nanoparticles should be determined to minimize the surface energy of nanoparticles through the adsorption of modifier.

Another important issue for the surface modification is the stability of the modifier on the surface. Above discussion suggests that after recovering the organic capped nanoparticles in the other pure solvent where no organic modifiers are contained, namely, the concentration is zero, most of the organic modifiers should desorb from the surface.

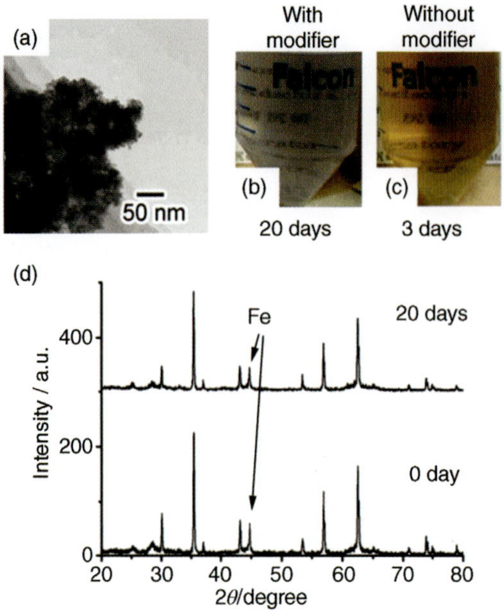

Figure 5.10 (a) TEM image of Fe nanoparticles synthesized with 0.1 M of hexanoic acid. (b, c) Image of the dispersed Fe NPs in water synthesized with and without hexanoic acid, respectively, and (d) XRD peak patterns of the surface-modified Fe NPs [46].

Under the high temperature/pressure condition, chemical bond formation takes place. Through the highly concentrated surface modification, the local concentration of water on the surface of nanoparticles can be extremely reduced. Under the condition, water acts as an acid catalyst to promote dehydration.

$$M - OH + RCOOH = RCOO - M + H_2O \tag{5.25}$$

For the surface modification in hydrothermal conditions, the pH of precursor solution and the dissociation of capping agents are also significant [20]. Generally speaking, the zeta potential of nanoparticles shifts from positive to negative by swinging the pH from low to high. As an example of the dissociation, carboxylic acid dissociates to form carbanion COO^- above the pK_a. Matching the surface charge and dissociation is the key to achieve high concentrations of reactants shown in the left-hand side of the following reaction:

$$M - OH^+ + RCOO^- = RCOO - M + H_2O \tag{5.26}$$

As mentioned above, surface modification is important for the metal nanoparticle synthesis. Figure 5.10a shows the iron nanoparticles synthesized using 0.1 M of hexanoic acid. Complete iron phase was not obtained in this condition. However, the yield of iron increased up to 7.6 wt% when hexanoic acid was injected to the reaction system. Also, hexanoic acid modified iron nanoparticles were compared with nonmodified sample under Ar atmosphere for several days

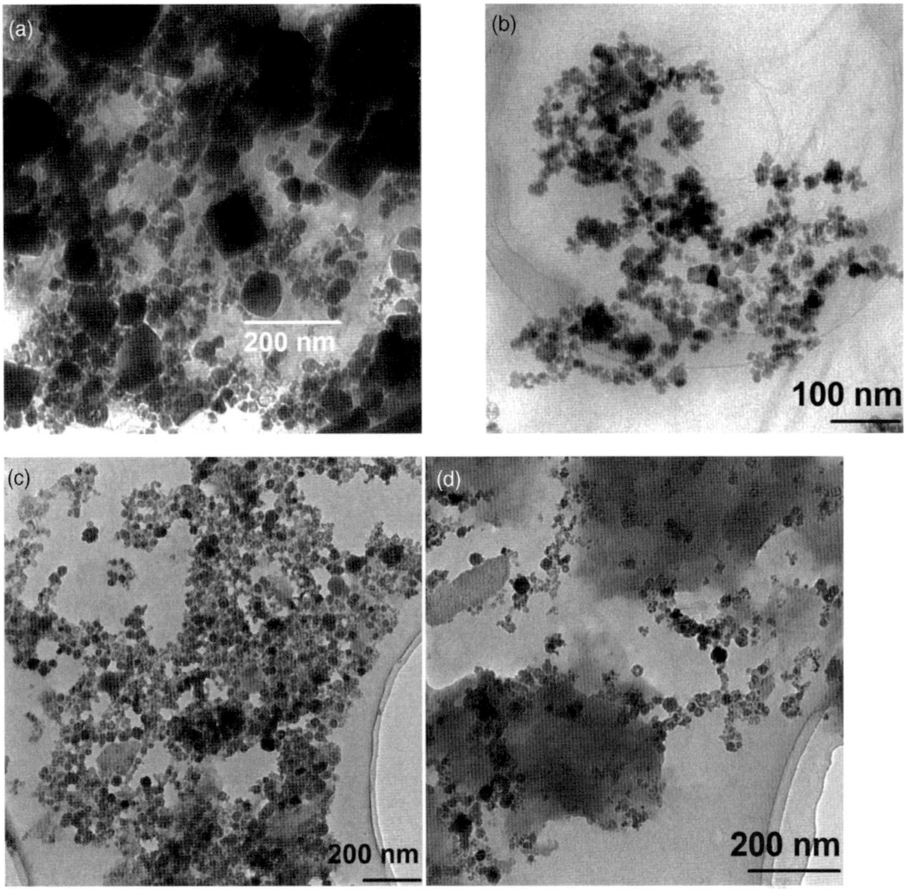

Figure 5.11 TEM images of the TOP modified cobalt nanoparticles synthesized at 380 °C and 25 MPa using a flow-type reactor. The molar concentration of TOP was (a) 0.08 M; (b) 0.20 M; (c) 0.40 M; and (d) 0.80 M.

as shown in Figure 10b and c. Both samples were dispersed in water. For the nonmodified sample, the color of solution was changed from transparent to dark yellow in 3 days. In contrast, the hexanoic acid, modified one, maintained its color for 20 days. From XRD peak patterns, iron phase (44.5°) was also confirmed as shown in Figure 5.10d. Thus, the oxidation of iron nanoparticles in water was effectively prevented using this *in situ* surface capping method.

Cobalt nanoparticles were also modified in a flow reactor system [61]. Similar to the iron nanoparticle system, choosing a modifier having high affinity with cobalt particles is an important issue for this reaction system to form nanoparticles less than 10 nm. Trioctyl-phosphine (TOP) has strong interaction with metal. Therefore, the reaction of cobalt nanoparticle formation using TOP is rapid, and such reaction is suitable to apply in the flow-type reactor. Figure 5.11 shows the synthesized cobalt nanoparticles with changing molar concentration of TOP. When TOP was added in the reaction system, the particle size of cobalt drastically decreased. With increasing concentration of TOP, the size of cobalt particles gradually decreased.

5.3
Flow Visualization Inside Supercritical Hydrothermal Process

As we discussed above, the high-throughput synthesis of metal oxide nanoparticles was realized by using a flow-type supercritical hydrothermal reactor. In this reactor, the aqueous solution of metal salt was mixed with supercritical water to produce metal oxide and metallic nanoparticles. We have found that the operational conditions including the flow rates of supercritical water and precursor solution largely affect the size and its distribution of produced metal oxide nanoparticles. One of the main reasons is that these conditions affected the mixing behavior of the two streams, resulting in the faster or slower rate of mixing as compared with that of reaction. Nowadays, we can perform flow-dynamics simulation to estimate the mixing time. However, the specific heat of water has a cusp at the critical point, which makes reliable simulation difficult. Therefore, the experimental studies that verify the validity of the simulation should be performed. To identify the mixing behavior of supercritical water in a stainless-steel reactor, we proposed a visualization method using neutron radiography. Neutron is largely scattered by hydrogen atoms and can transmit in most metals. Therefore, neutron radiography can visualize the density of water in a reactor made from metal alloys. Because the density of water is a function of temperature at a constant pressure, we would obtain the distribution of temperature of water stream in the stainless tube reactor. On the basis of this idea, we first performed neutron radiography experiments on the mixing between supercritical water and room-temperature water [62,63]. In this study, we used conventional flow-type reactor for supercritical hydrothermal synthesis. To obtain neutron radiography images, neutron beam from a nuclear reactor was irradiated to the junction of two streams flowing into the flow-type reactor. The transmitted neutron beam was converted to fluorescence light by a $^{6}LiF/ZnS$ converter plate and the fluorescence light was monitored by a charge-coupled device camera. The results clearly showed that the neutron beam visualizes the density of water, that is, the distribution of temperature in the mixing piece and connecting tubes made of SUS 316 (Figure 5.12).

This result demonstrated that the formation of a density-stratified layer when the flow rate from the side was small. Therefore, a small side flow rate should be avoided when producing nanoparticles with a small particle-size distribution, because it might prohibit the rapid increase in the temperature of precursor solutions. We also confirmed the room-temperature water from side went downward on the wall of reactor tube after mixing. This information provides us valuable information on the reaction process that leads to experimental conditions with enhanced mixing rate. However, the obtained

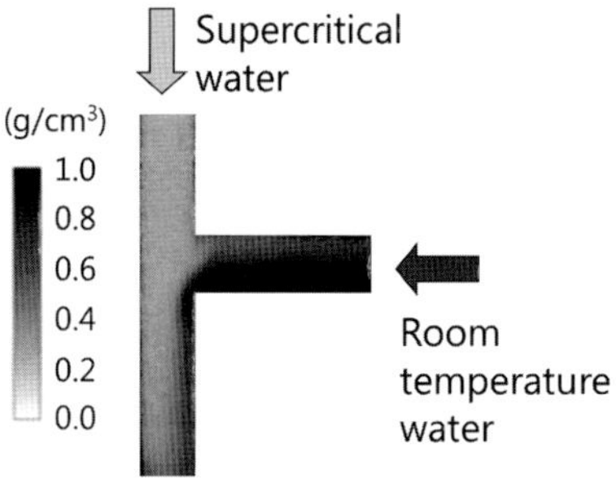

Figure 5.12 Neutron radiography image of the mixing of supercritical water and room temperature water.

images were two-dimensional images and showed the averaged density of water toward the direction of neutron. Thus, we then performed computed-tomography (CT) neutron radiography measurements.

In this CT measurement, we used an experimental setup similar to our previous studies except for using a rotational stage for tomography measurement. In this study, supercritical water was supplied from either top or side of the junction. Imaging of the mixing piece was repeated for 200 times while rotating the mixing piece, feeding and mixing supercritical water and room temperature water. Figure 5.13 shows a representative cross sectional image at the center of the mixer with supplying supercritical water from top [64]. Figure 5.13 clearly shows the change in the density of water streams in the mixer and tubes. The supercritical water fed from top was mixed with the room temperature water from side at the mixing point. After the mixing, the room temperature water with high density flowed along the side-wall of the vertical tube.

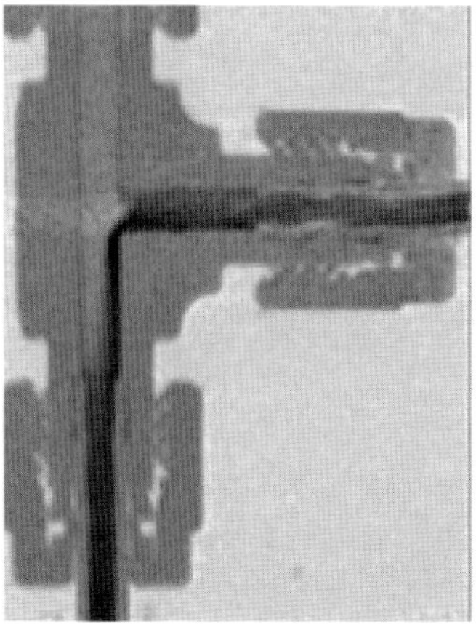

Figure 5.13 Neutron CT image of the center of mixing piece for supercritical hydrothermal synthesis.

The obtained images showed the three dimensional distribution of the density of water in stainless-steel tubes and the tee junction. This result demonstrated that neutron tomography clearly visualized the mixing behavior of supercritical water and room temperature water. We have also discussed the relationship between the mixing behavior and the size and its distribution of produced nanoparticles [65]. We believe that these results can be used to verify the validity of fluid dynamics simulation.

5.4 Process Design under a Reaction-controlled Condition Using Reynolds and Damköhler Numbers

5.4.1 Reaction Rate of Hydrothermal Reaction

In Section 5.4.3, the precise flow phenomena inside supercritical hydrothermal process were described. Process design methodology with considering flow phenomena, especially mixing, and reaction is then established. Reactions for the supercritical hydrothermal synthesis of metal oxide nanoparticles are usually rapid because of the low density and dielectric constant of supercritical water [7,8]. For example, the rate constants of consumption of $Ce(NO_3)_3$ as a precursor in CeO_2 nanoparticle synthesis were measured and estimated as shown in Figure 5.14 [66]. The rate constant represented by the dash line is estimated using Eq. (5.9). Here, we chose ε_0 as the dielectric constant of water at 200 °C [3]. Using the obtained rate constants, we fitted the parameters k_0, E, and Ψ using the least square method, to give $k_0 = 2.8 \times 10^7\ s^{-1}$, $E = 84$ kJ/mol, and $\Psi = 4.5 \times 10^2$ kJ/mol. The rate constant sharply increases in the sub- and supercritical temperature range ($1000/T < 1.7\ K^{-1}$). The estimated rate constant under supercritical conditions is on the order of $10^3\ s^{-1}$. The order of reciprocal number of the rate constant, that is, the time constant, is 1 ms. Thus, to achieve a

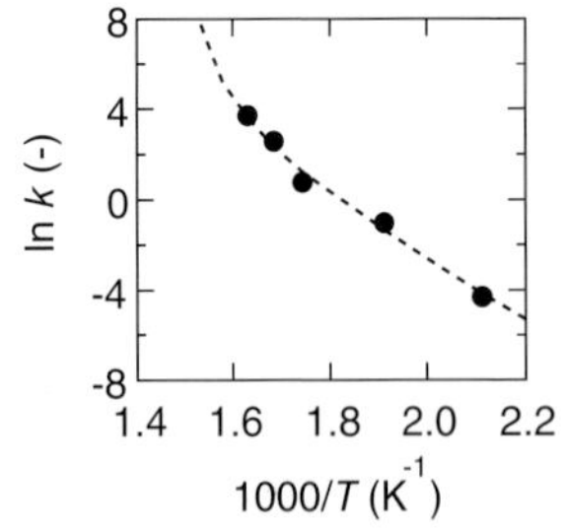

Figure 5.14 Estimation of rate constant. The rate constants determined from experiments were denoted by solid circle (•). The reaction temperature ranges from 200 to 340 °C. The constant shown by the dashed line is estimated from Eq. (5.9) to include the effect of decrease in dielectric constant.

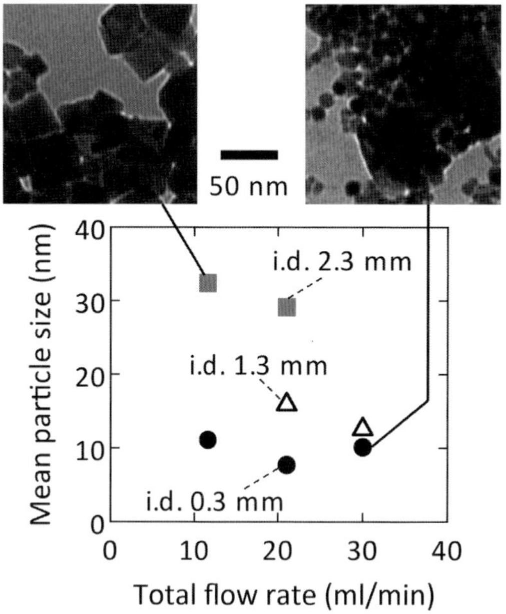

Figure 5.15 Effect of mixing conditions on mean nanoparticle size (380 °C, conversion of precursor ~1.0). The mean particle size was determined using the TEM images.

reaction-controlled condition, instantaneous mixing whose time constant is shorter than 1 ms is required for supercritical hydrothermal syntheses.

5.4.2
Effect of Mixing Rate of Particle Size

Figure 5.15 shows an example of effect of mixing rate on produced CeO_2 particle size in the flow-type reactor depicted in Figure 5.1. This figure also shows typical TEM images of the obtained nanoparticles. A precursor solution and preheated water are fed into a T-mixer (e.g., Swagelok union tee). The channel inner diameters (i.d.) of mixers are shown in Figure 5.15. The particle size was found to decrease with increasing total flow rate and decreasing channel diameter. As the conversion of the precursor was ~1.0 in these experiments, the amount of precursor consumed was considered equal in each experiment. Furthermore, as smaller particles contain less precursors, the total amount of precursor consumed is calculated by multiplying the amount of precursor included in each particle by the number of particles formed. This therefore indicates that the number of formed nuclei increases under enhanced mixing conditions, with smaller particles.

On the basis of the above results, the formation mechanism of ceria nanoparticles related to mixing rate is illustrated in Figure 5.16. When the flow rate is low or the channel size is large, the effect of convection to split fluids is weak

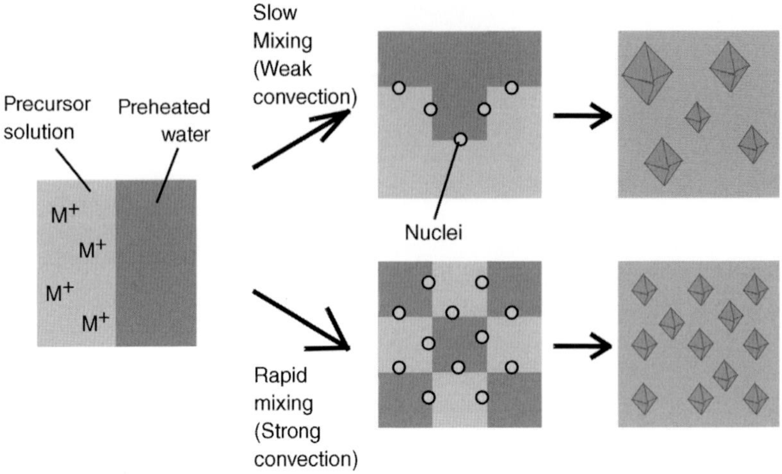

Figure 5.16 Effects of mixing rate on particle formation. When the effect of convection to split fluids is strong (e.g., high flow rate or small mixer channel size) and the fluid segment is small. This results in a large contacting volume of precursor solution and preheated water and rapid mixing. The rapid mixing promotes nucleation and decreases the produced nanoparticle size.

and the fluid segment is large. This results in a small contacting volume of precursor solution and preheated water and slow mixing. Under such conditions, primary nucleation occurs sparingly. The precursor was mainly consumed in particle growth, thus resulting in the formation of larger particles. In contrast, when mixing is rapid, the precursor was mainly consumed in the primary nucleation step. This resulted in the formation of smaller nanoparticles. Thus, the mixing rate is significant on determining the particle size.

In addition, to achieve a large degree of supersaturation, a sudden temperature rise is essential. As shown in Figure 5.17, the particle size decreases with increasing temperature. This can be attributed to the increase in the nucleation rate

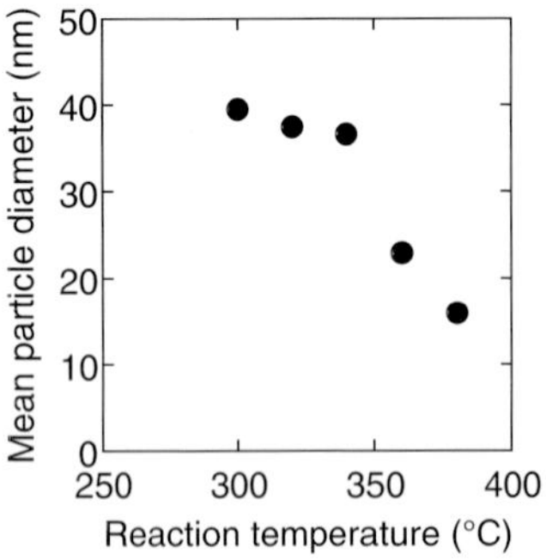

Figure 5.17 Effect of reaction temperature on particle size (mixer channel i.d. 1.3 mm, total flow rate 21.0 ml/min, conversion of precursor ~1.0).

with temperature, particularly in the supercritical region, due to the sudden decrease in metal oxide solubility [6,8,67]. Therefore, rapid heat transfer is also important for producing small and homogeneous nanoparticles.

5.4.3
Determination of Reaction-Controlled Condition Using Dimensionless Numbers

Considering the above issues, a continuous flow-type system is suitable for nanoparticle synthesis with a high throughput. Furthermore, for large-scale production, nanoparticles were synthesized using flow-type reactors under subcritical and supercritical conditions [18,68–70]. Use of a flow-type reactor containing submillimeter- to millimeter-scale channels (called microreactors) has been reported to enhance both mass and heat transfers and mixing rate [71–76]. The enhanced performance is due to increased surface and interface area in a miniaturized channel. Several microreactors for supercritical hydrothermal syntheses have been developed [75–77].

Systematical development of a high-throughput process for industrial production is one of the main goals of basic research of chemical engineering. In this context, to implement flow-type reactors in industrial applications, the establishment of design methodology is essential. However, to date, flow-type reactors have been developed by trial and error based on each reaction system. Therefore, to enhance the effectiveness and accelerate the development of continuous processes, establishing a design method to determine the optimal operating conditions and sizes of flow-type reactors is required.

To establish a design guideline for the flow-type reactor for nanoparticle synthesis, we examined the effects of operating parameters on produced nanoparticle size. From the thorough optimization of operating parameters for the flow-type reactor, in particular, parameters related to mixing rate (i.e., flow rate and mixer channel size), a design guideline using dimensionless numbers (i.e., Reynolds number and Damköhler number) was established. For processes with identical dimensionless numbers and different scales, there exists an analogy of phenomena in these processes. Therefore, a design methodology using dimensionless numbers is useful to scale up a process from a lab scale.

Figure 5.18 shows the relationship between the mean particle size and the Reynolds number, which represents the ratio of inertial forces to viscous forces, immediately after the mixing point. The physical properties provided in the NIST Chemistry WebBook [78] were used for the calculation of Reynolds number. The Reynolds number increased with increasing flow rate and temperature, resulting in an improvement in turbulence intensity and mixing rate. With respect to temperature, this is likely due to a sharp decrease in viscosity with increasing temperature around the critical point. In addition, the Reynolds number can be related to the effect of mixing on particle formation. The particle size is stable when the Reynolds number is higher than a specific value (e.g., 2.5×10^4 for 380 °C), which depends on the reaction temperature.

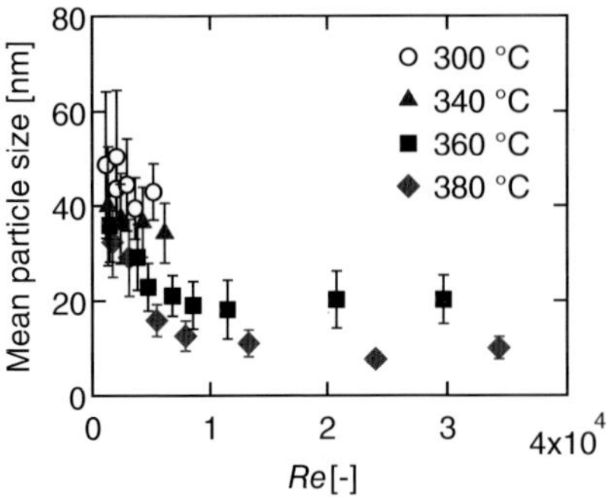

Figure 5.18 CeO_2 particle size as a function of Reynolds number immediately after the mixing point. The data of each reaction temperature was obtained by varying the flow rate (11.6–37.5 ml/min) and the mixer channel diameter (0.3, 1.3, and 2.3 mm). The temperature ranges from 300 to 380 °C.

However, as the Reynolds number is independent of the reaction rate, we also correlated the results using the Damköhler number, Da, which represents the ratio of reaction rate to mixing rate. When the mixing rate is represented as u/d [79], the Damköhler number can be expressed by the following equation:

$$\mathrm{Da} = kd/u, \tag{5.27}$$

where k is the rate constant, d is the diameter of the T-mixer, and u is the mean flow velocity immediately after the mixing point. The value of the Damköhler number increases with rising temperature at a fixed feed rate of reactant fluids. This is because the reaction rate increases sharply with temperature, although the mixing rate (u/d) increases more slowly than the reaction rate, as it is reflected by the decrease in fluid density with temperature.

Using the obtained rate constants, we calculated the value of *Da* for each experimental condition, as summarized in Figure 5.19. The particle size reduced with decreasing Da, with the size becoming stable at $\mathrm{Da} < 10^{-1}$. Thus, $\mathrm{Da} \sim 10^{-1}$ was used as the threshold to determine whether the examined condition was

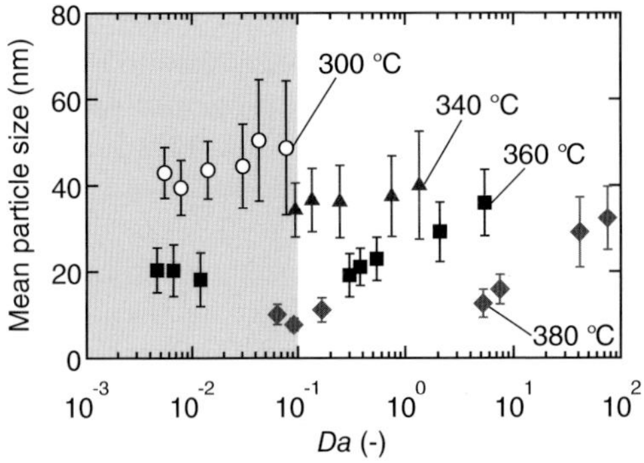

Figure 5.19 CeO_2 particle size as a function of Da to determine the threshold of reaction-controlled conditions.

reaction-controlled for synthesis of CeO_2 nanoparticles. While the threshold value of the Reynolds number for a reaction-controlled condition depends on the reaction temperature (see Figure 5.18), the threshold value of Da is effective for a fixed reaction system regardless of the reaction temperature. However, the threshold value is dependent on the ratio of reaction rates of multiple reactions [80]. In addition, the threshold value tends to be low for systems with high rates of subsequent reactions. Nanoparticle formation can thus be considered a multiple reaction system, consisting of reaction, nucleation, and particle growth stages. Therefore, in this system, the rate of the subsequent reaction, namely, the particle growth reaction, is inferred to be relatively high.

Thus, from the examination of effects of operating parameters in the flow-type reactor, a design guideline is established using the dimensionless number Da with respect to mixing operation. The design guideline will contribute to establishing the engineering of nanoparticles and development of continuous systems for the industrial scale production of nanoparticles.

The mixer channel geometry also affects the mixing rate. However, for an actual process with supercritical hydrothermal synthesis, the geometry would have little impact on a process design that enables a reaction-controlled condition. To validate this, we evaluate the Reynolds number in an industrial scale process using supercritical water. The Reynolds number can be rewritten as

$$Re = \frac{\rho u d}{\mu} = \frac{1}{\mu/\rho} u d, \tag{5.28}$$

where ρ is the density of reactant fluid, u is the flow velocity, d is the channel diameter, and μ is the viscosity of reactant fluid. In the actual process, for instance, the production scale is assumed to be 1000 t/year, d is on the order of 0.1 m, and u ranges from 1 to 10 m/s. Compared with gas phase (both μ and ρ are low) and liquid phase (both μ and ρ are high), the kinematic viscosity μ/ρ of supercritical water is extremely low because μ of supercritical water is as low as that of gas phase, while ρ is as high as that of liquid phase. This leads to a high value of Re on the order of 10^5, which this value is well above the threshold of reaction-controlled condition shown in Figure 5.18. This tendency suggests that the supercritical hydrothermal reaction in an actual process takes place under reaction-controlled conditions, and the process thus provides nanoparticles regardless of the mixer channel geometry.

5.5 Applications of Nanoparticles

5.5.1 Surface-Modified Nanoparticles by Organic Molecules

The organic modification method is used for not only producing particles dispersed in organic solvent but also controlling the structure of the particles. The structure of cerium oxide particles made under a supercritical condition is

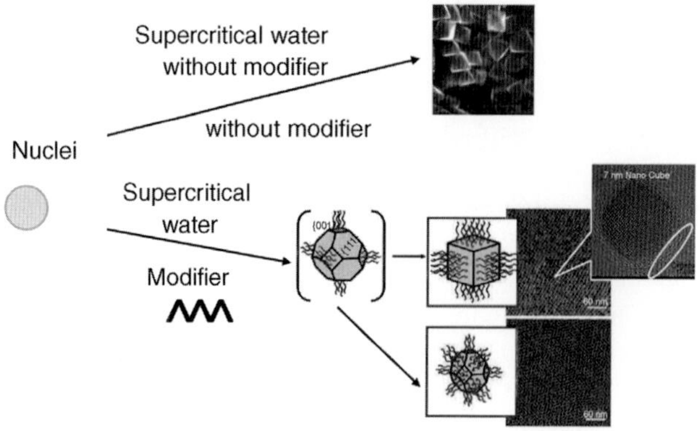

Figure 5.20 Size and morphology change of CeO_2 NPs with and without organic ligands.

octahedral without surface modifier and cubic with surface modifier (e.g., decanoic acid for cerium oxide). With increasing the amounts of surface modifier, the shape of the particles is changed from cubic to truncated octahedral. This is because the most reactive (001) surface is modified with organic modifier, which suppresses the growth of this surface and forms the cubic shape nanoparticles. By introducing more amount of organic modifier, less reactive (111) surface is also reacted with the organic modifier, which leads to the formation of polyhedron nanoparticles as shown in Figure 5.20 [81].

5.5.2 Superhybrid Materials

Recently, the size of the devices has reduced while the power supply has increased. Thus, how to release the heat from power devices is the bottle neck in developing devices with further power supply. For constructing such devices, thermal conductive materials that have high thermal conductivity, electrical insulation, adhesiveness, and fabricability are required. Polymer-ceramic (BN, AlN, Si_3N_4, or Al_2O_3) composites are now being developed. To enhance the thermal conductivity of the hybrid materials, the filler content should be increased. However, when the ceramic filler content is increased above 50 wt%, the viscosity of the hybrid materials rises drastically. Such a high viscosity reduces the fabricability. Furthermore, void formation occurs and decreases the electrical insulation as well as thermal conductivity. Thus, because of these incompatible functions, creating high heat-transfer hybrid materials is a difficult task. The surface modification of BN particles by the supercritical hydrothermal method improves the affinity for the polymers, resulting in high BN content of the hybrid materials (more than 90 wt%) without any void formation. The thermal conductivity is as high as 40 W/(m K), which is one order of magnitude higher than the currently used materials (4 W/(m K)). The devices made from these BN particles have high fabricability, electrical insulation, and flexibility as

Figure 5.21 Flexible heat-transfer thin film using a polymer-ceramic hybrid material.

shown in Figure 5.21. Controlling the refractive index with maintaining the fabricability of the polymer matrix is needed in optical materials. Homogeneous dispersion of high refractive index NPs, such as TiO_2 and ZrO_2, in the polymers increases the refractive index of the polymers since the transparency is kept with this dispersion. Also in this case, surface control is the key to disperse NPs. The organic surface modification of NPs increases the affinity for the polymer matrix and affords the fabrication of flexible polymer films that have high refractive-index transparency [82].

5.5.3 Catalysts

For cerium oxide, the most active surface is known to be the (100). By adding surface modifier (e.g., decanoic acid) into the supercritical hydrothermal synthesis, (100)-exposed cerium oxide, in other words cubic CeO_2 nanoparticles were successfully synthesized. The particles showed an extremely high oxygen storage capacity (OSC) even at a lower temperature (Figure 5.22), which suggests a high catalytic activity for environmental clean-up [25].

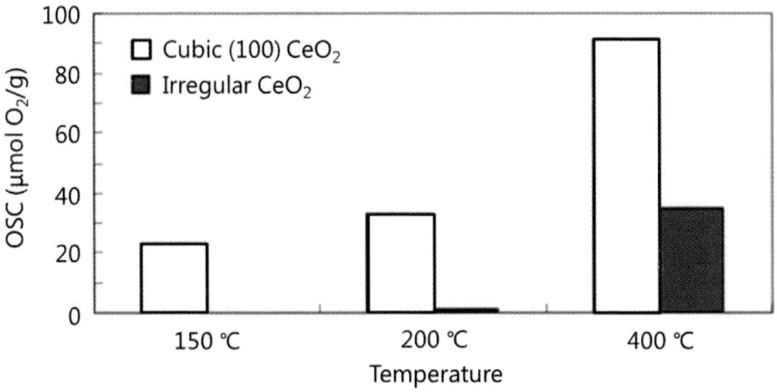

Figure 5.22 Oxygen storage capacity of CeO_2 nanoparticles.

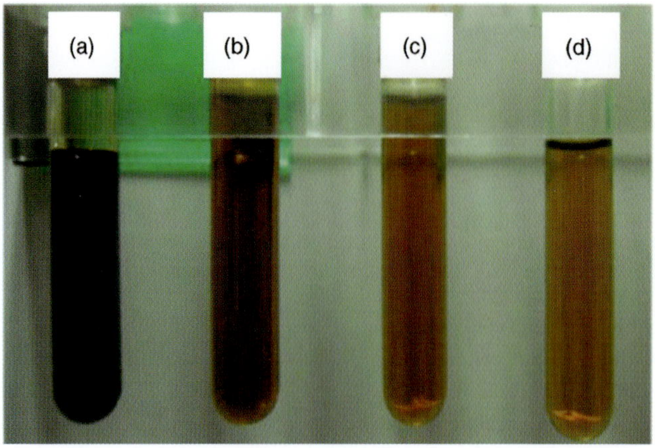

Figure 5.23 Bitumen solution and recovered oil mixed with 1-methyl naphthalene. The upgrading of bitumen was carried out under 450 °C for 1 h. (a) Unreacted bitumen, (b) recovered oil without using a catalyst, (c) recovered oil using the octahedral CeO_2, and (d) recovered oil using the cubic CeO_2.

As catalysts, nanoparticles are useful because of their higher surface-to-volume ratio. The supercritical method is a promising method to synthesize the nanocatalysts with controlled exposed surface. The waste treatment of unused heavy oil such as bitumen from the petroleum industry and black liquor from the pulp and paper industry or other biomass conversion processes are of great concern. (100)-exposed cerium oxide catalysts were used for upgrading bitumen under 450 °C for 1 h (Figure 5.23). The bitumen upgrading reaction was occurred with low coke formation. This is also one example of a green material synthesized using supercritical water, an environmentally benign solvent [68].

We also created silicon-iron oxide and zirconium-iron oxide nanoparticle catalyst. For the silicon-iron oxide catalyst, the value of OSC at 300 °C is the same as that of a conventional cerium oxide catalyst. However, the silicon-iron oxide is unstable at high temperature because of the instability of silicon under such a condition. In contrast, the OSC of the zirconium-iron oxide catalyst is almost 10 times higher than that of previous automotive catalysts and stable over 500 °C.

5.5.4 Nanoink for 3D Printing

Surface modified nanoparticles with organic compounds can be dispersed in organic solvents. We have succeeded in the preparation of decanoic acid-capped ceria NPs dispersed in cyclohexane up to 60 wt% [50]. This solution also shows low viscosity (8 cP) even at 60 wt% concentration. Recently, we are trying to make nanoink for 3D printer using the prepared surface modified nanoparticles. The nanoink needs low viscosity, high concentration of materials, and stability. However, nanoparticle solutions usually give extremely high viscosities at high concentrations, because inorganic nanoparticles have low affinity to organic solvents and form clusters. The nanoparticle solution including surface modified nanoparticles offers an opportunity to overcome these difficulties. If nanoink for 3D printer is created, we can make many kinds of object by inorganic materials controlled with the precision of nanometer order.

References

1 Adschiri, T., Hakuta, Y., and Arai, K. (2000) Hydrothermal synthesis of metal oxide fine particles at supercritical conditions. *Ind. Eng. Chem. Res.*, **39** (12), 4901–4907.

2 Uematsu, M. (2002) Thermophysical properties of supercritical fluids, in *Supercritical Fluids – Molecular Interaction, Physical Properties, and New Applications* (eds Y. Arai, T. Sako, and Y. Takebayashi), Springer, Berlin, pp. 71–77.

3 Uematsu, M. and Franck, E.U. (1980) Static dielectric constant of water and steam. *J. Phys. Chem. Ref. Data.*, **9** (4), 1291–1306.

4 Karásek, P., Šťavíková, L., Planeta, J., Hohnová, B., and Roth, M. (2013) Solubility of fused silica in sub- and supercritical water: estimation from a thermodynamic model. *J. Supercrit. Fluids*, **83**, 72–77.

5 Helgeson, H.C., Kirkham, D.H., and Flowers, G.C. (1981) Theoretical prediction of the thermodynamic behavior of aqueous electrolytes at high pressures and temperatures: IV. Calculation of activity coefficients, osmotic coefficients, and apparent molal and standard and relative partial molal properties to 600 °C and 5 KB. *Am. J. Sci.*, **281**, 1249–1516.

6 Adschiri, T., Hakuta, Y., Sue, K., and Arai, K. (2001) Hydrothermal synthesis of metal oxide nanoparticles at supercritical conditions. *J. Nanopart. Res.*, **3** (2), 227–235.

7 Sue, K., Adschiri, T., and Arai, K. (2002) Predictive model for equilibrium constants of aqueous inorganic species at subcritical and supercritical conditions. *Ind. Eng. Chem. Res.*, **41** (13), 3298–3306.

8 Sue, K., Hakuta, Y., Smith J Jr., R.L., Adschiri, T., and Arai, K. (1999) Solubility of lead(II) oxide and copper(II) oxide in subcritical and supercritical water. *J. Chem. Eng. Data*, **44** (6), 1422–1426.

9 Hearn, B., Hunt, M., and Hayward, A. (1969) Solubility of cupric oxide in pure subcritical and supercritical water. *J. Chem. Eng. Data*, **14** (4), 442–447.

10 Yokoyama, C., Iwabuchi, A., and Takahashi, S. (1993) Solubility of PbO in supercritical water. *Fluid Phase Equilib.*, **82**, 323–331.

11 Amis, E.S. and Hinton, J.F. (1973) *Solvent Effects on Chemical Phenomena*, Academic Press, New York.

12 LaMer, V.K. and Dinegar, R.H. (1950) Theory, production and mechanism of formation of monodispersed hydrosols. *J. Am. Chem. Soc.*, **72** (11), 4847–4854.

13 Peters, B. (2011) Supersaturation rates and schedules: nucleation kinetics from isothermal metastable zone widths. *J. Cryst. Growth*, **317** (1), 79–83.

14 Sugimoto, T., Shiba, F., Sekiguchi, T., and Itoh, H. (2000) Spontaneous nucleation of monodisperse silver halide particles from homogeneous gelatin solution I: silver chloride. *Colloids Surf. A*, **164** (2–3), 183–203.

15 Kashchiev, D., Verdoes, D., and van Rosmalen, G.M. (1991) Induction time and metastability limit in new phase formation. *J. Cryst. Growth*, **110** (3), 373–380.

16 Thanh, N.T.K., Maclean, N., and Mahiddine, S. (2011) Mechanisms of nucleation and growth of nanoparticles in solution. *Chem. Rev.*, **114** (15), 7610–7630.

17 Adschiri, T., Kanazawa, K., and Arai, K. (1992) Rapid and continuous hydrothermal synthesis of boehmite particles in subcritical and supercritical

water. *J. Am. Ceram. Soc.*, **75** (9), 2615–2618.

18 Adschiri, T., Kanazawa, K., and Arai, K. (1992) Rapid and continuous hydrothermal crystallization of metal oxide particles in supercritical water. *J. Am. Ceram. Soc.*, **75** (4), 1019–1022.

19 Hakuta, Y., Adschiri, T., Hirakoso, H., and Arai, K. (1999) Chemical equilibria and particle morphology of boehmite (AlOOH) in sub and supercritical water. *Fluid Phase Equilib.*, **158–160**, 733–742.

20 Mousavand, T., Ohara, S., Umetsu, M., Zhang, J., Takami, S., Naka, T., and Adschiri, T. (2007) Hydrothermal synthesis and *in situ* surface modification of boehmite nanoparticles in supercritical water. *J. Supercrit. Fluids*, **40** (3), 397–401.

21 Hakuta, Y., Onai, S., Terayama, H., Adschiri, T., and Arai, K. (1998) Production of ultra-fine ceria particles by hydrothermal synthesis under supercritical conditions. *J. Mater. Sci. Lett.*, **17** (14), 1211–1213.

22 Zhang, J., Ohara, S., Umetsu, M., Naka, T., Hatakeyama, Y., and Adschiri, T. (2007) Colloidal ceria nanocrystals: tailor-made crystal shape in supercritical water. *Adv. Mater.*, **19** (2), 203–206.

23 Takami, S., Ohara, S., Adschiri, T., Wakayama, Y., and Chikyow, T. (2008) Continuous synthesis of organic–inorganic hybridized cubic nanoassemblies of octahedral cerium oxide nanocrystals and hexanedioic acid. *Dalton Trans*, (40), 5442–5446.

24 Taguchi, M., Takami, S., Naka, T., and Adschiri, T. (2009) Growth mechanism and surface chemical characteristics of dicarboxylic acid-modified CeO_2 nanocrystals produced in supercritical water: tailor-made water-soluble CeO_2 nanocrystals. *Cryst. Growth Des.*, **9** (12), 5297–5303.

25 Zhang, J., Kumagai, H., Yamamura, K., Ohara, S., Takami, S., Morikawa, A., Shinjoh, H., Kaneko, K., Adschiri, T., and Suda, A. (2011) Extra-low-temperature oxygen storage capacity of CeO_2 nanocrystals with cubic facets. *Nano Lett.*, **11** (2), 361–364.

26 Mousavand, T., Naka, T., Sato, K., Ohara, S., Umetsu, M., Takami, S., Nakane, T., Matsushita, A., and Adschiri, T. (2009) Crystal size and magnetic field effects in Co_3O_4 antiferromagnetic nanocrystals. *Phys. Rev. B: Condens. Matter*, **79** (14), 144411.

27 Togashi, T., Hitaka, H., Ohara, S., Naka, T., Takami, S., and Adschiri, T. (2010) Controlled reduction of Cu^{2+} to Cu+ with an N,O-type chelate under hydrothermal conditions to produce Cu_2O nanoparticles. *Mater. Lett.*, **64** (9), 1049–1051.

28 Takami, S., Sato, T., Mousavand, T., Ohara, S., Umetsu, M., and Adschiri, T. (2007) Hydrothermal synthesis of surface-modified iron oxide nanoparticles. *Mater. Lett.*, **61** (26), 4769–4772.

29 le Cleroq, M., Adschiri, T., and Arai, K. (2001) Hydrothermal processing of nickel containing biomining or bioremediation biomass. *Biomass Bioenergy*, **21** (1), 73–80.

30 Mousavand, T., Zhang, J., Ohara, S., Umetsu, M., Naka, T., and Adschiri, T. (2007) Organic-ligand-assisted supercritical hydrothermal synthesis of titanium oxide nanocrystals leading to perfectly dispersed titanium oxide nanoparticle in organic phase. *J. Nanopart. Res.*, **9** (6), 1067–1071.

31 Ohara, S., Mousavand, T., Sasaki, T., Umetsu, M., Naka, T., and Adschiri, T. (2008) Continuous production of fine zinc oxide nanorods by hydrothermal synthesis in supercritical water. *J. Mater. Sci.*, **43** (7), 2393–2396.

32 Ohara, S., Mousavand, T., Umetsu, M., Takami, S., Adschiri, T., Kuroki, Y., and Tanaka, M. (2004) Hydrothermal synthesis of fine zinc oxide particles under supercritical conditions. *Solid State Ionics*, **172** (1–4), 261–264.

33 Mousavand, T., Ohara, S., Naka, T., Umetsu, M., Takami, S., and Adschiri, T. (2010) Organic-ligand-assisted hydrothermal synthesis of ultrafine and hydrophobic ZnO nanoparticles. *J. Mater. Res.*, **25** (2), 219–223.

34 Rangappa, D., Naka, T., Kondo, A., Ishii, M., Kobayashi, T., and Adschiri, T. (2007) Transparent $CoAl_2O_4$ hybrid nano pigment by organic ligand-assisted supercritical water. *J. Am. Chem. Soc.*, **129** (36), 11061–11066.

35 Atashfaraz, M., Shariaty-Niassar, M., Ohara, S., Minami, K., Umetsu, M., Naka, T., and Adschiri, T. (2007) Effect of titanium dioxide solubility on the formation of $BaTiO_3$ nanoparticles in supercritical water. *Fluid Phase Equilib.*, **257** (2), 233–237.
36 Sasaki, T., Ohara, S., Naka, T., Vejpravova, J., Sechovsky, V., Umetsu, M., Takami, S., Jeyadevan, B., and Adschiri, T. (2010) Continuous synthesis of fine $MgFe_2O_4$ nanoparticles by supercritical hydrothermal reaction. *J. Supercrit. Fluids*, **53** (1–3), 92–94.
37 Kanamura, K., Goto, A., Ho, R.Y., Umegaki, T., Toyoshima, K., Okada, K., Hakuta, Y., Adschiri, T., and Arai, K. (2000) Preparation and electrochemical characterization of $LiCoO_2$ particles prepared by supercritical water synthesis. *Electrochem. Solid State Lett.*, **3** (6), 256–258.
38 Shin, Y.H., Koo, S.-M., Kim, D.S., Lee, Y.-H., Veriansyah, B., Kim, J., and Lee, Y.-W. (2009) Continuous hydrothermal synthesis of HT-$LiCoO_2$ in supercritical water. *J. Supercrit. Fluids*, **50** (3), 250–256.
39 Hakuta, Y., Seino, K., Ura, H., Adschiri, T., Takizawa, H., and Arai, K. (1999) Production of phosphor (YAG:Tb) fine particles by hydrothermal synthesis in supercritical water. *J. Mater. Chem.*, **9** (10), 2671–2674.
40 Sahraneshin, A., Takami, S., Minami, K., Hojo, D., Arita, T., and Adschiri, T. (2012) Synthesis and morphology control of surface functionalized nanoscale yttrium aluminum garnet particles via supercritical hydrothermal method. *Prog. Cryst. Growth Charact. Mater.*, **58** (1), 43–50.
41 Carcassi, M.N. and Fineschi, F. (2005) Deflagrations of H_2-air and CH_4-air lean mixtures in a vented multi-compartment environment. *Energy*, **30** (8), 1439–1451.
42 Kim, M., Son, W., Ahn, K.H., Kim, D.S., Lee, H., and Lee, Y. (2014) Hydrothermal synthesis of metal nanoparticles using glycerol as a reducing agent. *J. Supercrit. Fluids*, **90**, 53–59.
43 le Clercq, M., Adschiri, T., and Arai, K. (2001) Hydrothermal processing of nickel containing biomining or bioremediation biomass. *Biomass Bioenergy*, **21** (1), 73–80.
44 Sue, K., Kakinuma, N., Adschiri, T., and Arai, K. (2004) Continuous production of nickel fine particles by hydrogen reduction in near-critical water. *Ind. Eng. Chem. Res.*, **43** (9), 2073–2078.
45 Ohara, S., Hitaka, H., Zhang, J., Umetsu, M., Naka, T., and Adschiri, T. (2007) Hydrothermal synthesis of cobalt nanoparticles in supercritical water. *J. Jpn. Soc. Powder Powder Metall.*, **54** (9), 635–638.
46 Arita, T., Hitaka, H., Minami, K., Naka, T., and Adschiri, T. (2011) Synthesis of iron nanoparticle: challenge to determine the limit of hydrogen reduction in supercritical water. *J. Supercrit. Fluids*, **57** (2), 183–189.
47 Arita, T., Hitaka, H., Minami, K., Naka, T., and Adschiri, T. (2011) Synthesis and characterization of surface-modified FePt nanocrystals by supercritical hydrothermal method. *Chem. Lett.*, **40** (6), 588–590.
48 Seong, G., Takami, S., Arita, T., Minami, K., Hojo, D., Yavari, A.R., and Adschiri, T. (2011) Supercritical hydrothermal synthesis of metallic cobalt nanoparticles and its thermodynamic analysis. *J. Supercrit. Fluids*, **60**, 113–120.
49 Seong, G. and Adschiri, T. (2014) The reductive supercritical hydrothermal process, a novel synthesis method for cobalt nanoparticles: synthesis and investigation on the reaction mechanism. *Dalton Trans.*, **43** (28), 10778–10786.
50 Adschiri, T., Takami, S., Arita, T., Hojo, D., Minami, K., Aoki, N., and Togashi, T. (2013) Supercritical hydrothermal synthesis, in *Handbook of Advanced Ceramics: Materials, Applications, Processing, and Properties*, 2nd edn (ed. S. Somiya), Elsevier Inc., Tokyo, pp. 949–978.
51 Frank, E.U. (1981) Special aspects of fluid solutions at high pressures and sub- and supercritical temperatures. *Pure Appl. Chem.*, **53** (7), 1401–1416.
52 Yu, J. and Savage, P.E. (1998) Decomposition of formic acid under hydrothermal conditions. *Ind. Eng. Chem. Res.*, **37** (1), 2–10.
53 Akiya, N. and Savage, P.E. (1998) Role of water in formic acid decomposition. *AIChE J.*, **44** (2), 405–415.

54 Shriver, D.F., Atkins, P.W., Overton, T.L., Rourke, J.P., Weller, M.T., and Armstrong, F.A. (2006) Oxidation and reduction, in *Inorganic Chemistry*, 4th edn, Oxford University Press, Oxford, pp. 141–168.

55 Ellingham, H.J.T. (1944) Reducibility of oxides and sulphides in metallurgical processes. *J. Soc. Chem. Ind.*, **63**, 125.

56 Horstmann, S., Jabłoniec, A., Krafczyk, J., Fischer, K., and Gmehling, J. (2005) PSRK group contribution equation of state: comprehensive revision and extension IV, including critical constants and (-function parameters for 1000 components. *Fluid Phase Equilib.*, **227** (2), 157–164.

57 Nilsing, M., Lunell, S., Persson, P., and Ojamäe, L. (2005) Phosphonic acid adsorption at the TiO_2 anatase (1 0 1) surface investigated by periodic hybrid HF-DFT computations. *Surf. Sci.*, **582** (1–3), 49–60.

58 Zhao, N., Pan, D., Nie, W., and Ji, X. (2006) Two-phase synthesis of shape-controlled colloidal zirconia nanocrystals and their characterization. *J. Am. Chem. Soc.*, **128** (31), 10118–10124.

59 Sato, K., Abe, H., and Ohara, S. (2010) Selective growth of monoclinic and tetragonal zirconia nanocrystals. *J. Am. Chem. Soc.*, **132** (8), 2538–2539.

60 Schneider, G.M. (1972) Phase behavior and critical phenomena in fluid mixtures under pressure. *Ber. Bunsenges. Phys. Chem.*, **76** (3–4), 325–331.

61 Seong, G. (2012) Reductive supercritical hydrothermal synthesis of metal nanoparticles. Dissertation. Tohoku University.

62 Takami, S., Sugioka, K., Tsukada, T., Adschiri, T., Sugimoto, K., Takenaka, N., and Saito, Y. (2012) Neutron radiography on tubular flow reactor for hydrothermal synthesis: *in situ* monitoring of mixing behavior of supercritical water and room-temperature water. *J. Supercrit. Fluids*, **63**, 46–51.

63 Sugioka, K., Ozawa, K., Tsukada, T., Takami, S., Adschiri, T., Sugimoto, K., Takenaka, N., and Saito, Y. (2014) Neutron radiography and numerical simulation of mixing behavior in a reactor for supercritical hydrothermal synthesis. *AIChE J.*, **60** (3), 1168–1175.

64 Takami, S., Sugioka, K., Ozawa, K., Tsukada, T., Adschiri, T., Sugimoto, K., Takenaka, N., and Saito, Y. (2015) *In situ* neutron tomography on mixing behavior of supercritical water and room temperature water in a tubular flow reactor. *Phys. Procedia*, **69**, 564–569.

65 Sugioka, K., Ozawa, K., Kubo, M., Tsukada, T., Takami, S., Adschiri, T., Sugimoto, K., Takenaka, N., and Saito, Y. (2016) Relationship between size distribution of synthesized nanoparticles and flow and thermal fields in a flow-type reactor for supercritical hydrothermal synthesis. *J. Supercrit. Fluids*, **109**, 43–50.

66 Aoki, N., Sato, A., Sasaki, H., Litwinowicz, A.-A., Seong, G., Aida, T., Hojo, D., Takami, S., and Adschiri, T. (2016) Kinetics study to identify reaction-controlled conditions for supercritical hydrothermal nanoparticle synthesis with flow-type reactors. *J. Supercrit. Fluids*, **110**, 161–166.

67 Sue, K., Suzuki, M., Arai, K., Ohashi, T., Ura, H., Matsui, K., Hakuta, Y., Hayashi, H., Watanabe, M., and Hiaki, T. (2006) Size-controlled synthesis of metal oxide nanoparticles with a flow-through supercritical water method. *Green Chem.*, **8** (7), 634–638.

68 Dejhosseini, M., Aida, T., Watanabe, M., Takami, S., Hojo, D., Aoki, N., Arita, T., Kishita, A., and Adschiri, T. (2013) Catalytic cracking reaction of heavy oil in the presence of cerium oxide nanoparticles in supercritical water. *Energy Fuel*, **27** (8), 4624–4631.

69 de Tercero, M.D., Martínez, I.G., Herrmann, M., Bruns, M., Kübel, C., Jennewein, S., Fehrenbacher, U., Barner, L., and Türk, M. (2013) Synthesis of *in situ* functionalized iron oxide nanoparticles presenting alkyne groups via a continuous process using near-critical and supercritical water. *J. Supercrit. Fluids*, **82**, 83–95.

70 Veriansyah, B., Park, H., Kim, J., Min, B.K., Shin, Y.H., Lee, Y., and Kim, J. (2009) Characterization of surface-modified ceria oxide nanoparticles synthesized continuously in supercritical methanol. *J. Supercrit. Fluids*, **50** (3), 283–291.

71 Weng, X., Zhang, J., Wu, Z., Liu, Y., Wang, H., and Darr, J.A. (2011) Continuous syntheses of highly dispersed composite nanocatalysts via simultaneous co-precipitation in supercritical water. *Appl. Catal.*, **103** (3–4), 453–461.

72 Aoki, N., Fukuda, T., Maeda, N., and Mae, K. (2013) Design of confluence and bend geometry for rapid mixing in microchannels. *Chem. Eng. J.*, **227**, 198–202.

73 Aoki, N., Kitajima, R., Itoh, C., and Mae, K. (2008) Microreactor for synthesis via intermediates with assembled units enabling rapid operations. *Chem. Eng. Technol.*, **31** (8), 1140–1145.

74 Aoki, N., Tanigawa, S., and Mae, K. (2011) Design and operation of gas–liquid slug flow in miniaturized channels for rapid mass transfer. *Chem. Eng. Sci.*, **66** (24), 6536–6543.

75 Mae, K., Suzuki, A., Maki, T., Hakuta, Y., Sato, H., and Arai, K. (2007) A new micromixer with needle adjustment for instant mixing and heating under high pressure and high temperature. *J. Chem. Eng. Jpn*, **40** (12), 1101–1107.

76 Wakashima, Y. and Suzuki, A. (2008) Numerical investigation of Joule-heating microtube heater for supercritical state water generation: heat transfer characteristics of single microtube. *J. Chem. Eng. Jpn*, **41** (4), 227–237.

77 Wakashima, Y., Suzuki, A., Kawasaki, S., Matsui, K., and Hakuta, Y. (2007) Development of a new swirling micro mixer for continuous hydrothermal synthesis of nano-size particles. *J. Chem. Eng. Jpn*, **40** (8), 622–629.

78 The National Institute of Standards and Technology, (2015) NIST Chemistry WebBook, NIST Standard Reference Database Number 69, Thermophysical Properties of Fluid Systems. Available at http://webbook.nist.gov/chemistry/fluid/ (accessed December 17, 2015).

79 Mao, K.W. and Toor, H.L. (1971) Second-order chemical reactions with turbulent mixing. *Ind. Eng. Chem. Fundam.*, **10** (2), 192–197.

80 Aoki, N., Hasebe, S., and Mae, K. (2006) Geometric design of fluid segments in microreactors using dimensionless numbers. *AIChE J.*, **52** (4), 1502–1515.

81 Adschiri, T. (2007) Supercritical hydrothermal synthesis of organic–inorganic hybrid nanoparticles. *Chem. Lett.*, **36** (10), 1188–1193.

82 Adschiri, T., Lee, Y.-W., Goto, M., and Takami, S. (2011) Green materials synthesis with supercritical water. *Green Chem.*, **13** (6), 1380–1390.

6
Particle-Mediated Crystal Growth

R. Lee Penn

University of Minnesota – Twin Cities, 207 Pleasant Street SE, Minneapolis, MN 55455, USA

6.1
Introduction

Crystal growth is a core area of research in solid-state chemistry and other fields such as mineralogy, geochemistry, biomineralization, materials science, and more. Historically, developments in solid-state chemistry have been driven by industrial applications and technology, and the diversity of crystalline materials in engineered settings is remarkable. Properties, such as conductivity, reactivity, catalytic activity, and mechanical strength, depend strongly on crystal shape and size, phase purity, composition, and structure as well as the presence of defects. The physicochemical environment in which crystals form can strongly impact these properties. Indeed, controlling these properties through synthetic variables such as temperature and precursor chemistry has lead to the availability of a vast library of synthetic crystals for a wide range of applications, including electronics, catalysis, magnetic recording, medical devices, pigments, and more.

The ultimate goal of crystal growth is to rationally and reproducibly produce a crystal or crystals with desired properties. A deep understanding of crystal growth mechanisms can improve our ability to rationally control crystal formation and growth, so we can purposefully produce new types of structures with tunable properties. The ability to produce a large number of nanocrystals, each with the same number of atoms, identical composition, shape, surface chemistry, and defect content, is highly desirable. However, we are far from achieving that goal.

At the most basic level, crystal growth is the process of arranging atomic and molecular scale objects (i.e., atoms, monatomic ions, polyatomic ions, clusters, and molecules) into ordered, three-dimensional structures. A perfect crystal has both short-range and long-range order and lacks defects. The growth of single crystals without significant defects requires specialized methods that favor slow growth without nucleation of new crystallites.

Handbook of Solid State Chemistry, First Edition. Edited by Richard Dronskowski, Shinichi Kikkawa, and Andreas Stein.

Figure 6.1 Photographs of a synthetic quartz crystal resting on a slab of granite. The quartz crystal is $19.5 \times 8.5 \times 1.2\,cm^3$ in size. In the lower image, the crystal is photographed from above, and the cloudy zone that runs through the center is the result of the seed crystal used during synthesis.

Examples of methods for growth of large crystals with few defects span both solution phase (e.g., hydrothermal crystal growth) and gas phase (e.g., atomic layer deposition) methods. The idea is to prevent nucleation while favoring slow, monomer-by-monomer growth on an existing crystal. An example is the seeded growth of quartz crystals (Figure 6.1) under hydrothermal conditions, which was developed at the commercial scale by Walker and Buehler in 1950 at Bell Laboratories [1]. Such crystals are high quality in the sense that they are optically clear and do not contain significant concentrations of defects such as stacking faults and dislocations.

Real crystals, however, are never perfect. They contain a range of defects, including point defects and extended line and plane defects, and these defects can disrupt both short-range and long-range order. The presence of defects can dramatically change the physical and chemical properties of crystals, and defects (e.g., stacking faults) can serve as nucleation sites for phase transformations. The mechanisms by which defects are introduced into initially defect-free (except point defects) vary and are generally poorly understood. However, particle-mediated crystal growth mechanisms are known to facilitate the incorporation of defects into growing crystals.

This chapter's primary focus is on crystal growth mechanisms that are particularly relevant for nanocrystals and provides a brief discussion of classical crystal growth and more in-depth discussion solid-phase particle-mediated crystal growth. A critical step in the science and engineering of crystal growth is characterization. Thus, the chapter ends with a discussion of characterization methods that are essential for elucidating and quantifying crystal growth. Finally, a particularly challenging aspect of elucidating crystal growth mechanisms is that multiple mechanisms often operate simultaneously. Thus, time-resolved methods that enable characterization of crystals as they exist in their growth media across growth stages are essential.

6.2 Classical Crystal Growth

Classical crystal growth can be well described as monomer-by-monomer addition of molecular-scale species to existing crystals. These species range from single atoms and ions to polyatomic ions, clusters, and molecules. The thermodynamic description of classical crystal nucleation is based on Gibbs theory. The balance between the decrease in free energy upon nucleation and increase in free energy upon formation of the new interface means that nuclei will only be stable once a critical size has been reached, with particles smaller than this critical size dissolving and particles of or larger than this size persisting. As supersaturation approaches unity, continued growth is governed by the distribution of size and shape because solubility is both size and shape dependent. As the degree of supersaturation decreases, the critical size increases. This growth of large crystals at the expense of smaller crystals is often referred to as Ostwald ripening or coarsening and is well described by the Gibbs–Thomson relation. Crystals formed by classical nucleation and growth are typically expected to be largely free of defects (except for point defects) and to have morphologies consistent with the symmetry of the crystal structure, and plots of r^3 versus time are fit well with linear fits and facilitate quantitative comparisons as a function of crystal growth conditions [2–4].

There is a rich literature describing classical nucleation and crystal growth (e.g., Ref. [5] and references contained therein), and such work has been tremendously successful. In fact, classical crystal growth theory describes reasonably well the growth of crystals in dilute solutions as well as crystals that have relatively high solubilities in the growth media of interest.

Promoting rapid nucleation and inhibiting growth has been a productive method for producing nanocrystals with fairly monodisperse particle size distributions in high yield. An example is the hot injection method, which has become a widely employed approach for producing high quality nanocrystals of, in particular, semiconducting materials (e.g., Ref. [6]). Precursor solutions are injected into a hot solvent that can actively coordinate the nanocrystal surface and provide nutrients for crystal nucleation and growth. Solutes that can cap nanocrystals are sometimes added. A key feature is that precursor molecules pyrolyze upon injection into the hot solvent/solution. The goal is to initiate fast and homogeneous nucleation. Nanocrystals can be harvested immediately after nucleation or after subsequent growth, often at a lower temperature. It is the separation of nucleation and growth steps that lead to controlled size and size distribution.

A recent example is the hot injection synthesis of the mixed transition metal sulfide kesterite (Cu_2ZnSnS_4). In this example, the variable that leads to control over nanocrystal size is temperature, with average size increasing from just a few nanometers to tens of nanometers by increasing the temperature from 150 to 340 °C [7]. Indeed, temperature is one of the most easily controlled and most important variables that can lead to control over nanocrystal size. Other ways to modify products of a hot injection synthesis include varying the concentration of precursors, additives such as capping agents, ionic strength, solvent choice, precursor choice, rate of injection, and more.

6.3 Nonclassical Crystal Growth

Unexpected structures, microstructures, morphologies, and textures that cannot be explained by classical nucleation and crystal growth are commonly seen in the literature. These observations have lead to a reexamination of how crystals form and grow. A few of the myriad examples of crystals with features that cannot be explained by classical crystal growth include the symmetry-defying anatase crystals of Penn and Banfield (Figure 6.2) [8], lead selenide crystals of Cho *et al.* [9], the "winding polycrystalline nanoparticle chains" of Pt_3Fe of Liao *et al.* [10], and the "quasi-continuous sheets" of PbS of Schliehe *et al.* [11].

In fact, there is no one universal description of how crystals grow. Similar lines of evidence (e.g., observed morphologies, textures, and microstructures) have been interpreted differently [12]. The preparation of pseudocubic hematite (α-Fe_2O_3) has been described by several groups, and two examples highlight the difficulty in interpreting electron micrographs of the crystals. Kandori *et al.* [13] described their about 1.2 μm pseudocubic hematite crystals, which were formed by aging an acidic solution of ferric chloride, as polycrystals comprised of oriented subcrystals, and their X-ray diffraction data clearly demonstrated a difference in the average crystal domain size, which was smaller as quantified by line broadening analysis, and the size of the pseudocubic hematite crystals, which was larger as measured from electron micrographs. They concluded that the initial precipitation and aggregation of hydrous ferric oxide crystallites was

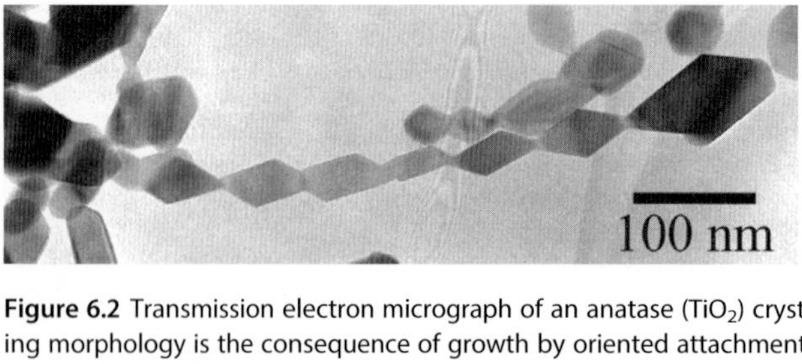

Figure 6.2 Transmission electron micrograph of an anatase (TiO_2) crystal. This symmetry defying morphology is the consequence of growth by oriented attachment. (Reprinted with permission from Ref. [8]. Copyright 1999, Elsevier.

followed by recrystallization and dehydration. An open question not addressed in their work was whether the initial aggregation produced an aggregate of oriented or randomly oriented subcrystals.

In contrast, Sugimoto *et al.* [14] described their 1.6 μm pseudocubic hematite crystals, which were produced by aging a partially neutralized solution of ferric chloride, as grown by monomer-by-monomer addition, concluding that the observed texture arose because two-dimensional growth on existing surfaces was blocked from forming continuous layers by adsorbed species from solution (e.g., chloride anions and chloroferric complexes). Interestingly, Sugimoto *et al.* even described the similarity of textures observed in their pseudocubic hematite crystals to those of Kandori *et al.* However, they stated that the crystallographic registry between the subcrystals composing the pseudocubic hematite crystals precluded the possibility of a particle-mediated growth mechanism [14].

Nonclassical crystal growth mechanisms have been proposed as pathways by which unique structures, microstructures, morphologies, and textures come about. Particle-based crystallization in particular has featured prominently in the recent crystal growth literature (Refs [15–22] and references contained within those critical reviews). It has become clear that both classical crystal growth and particle-based crystal growth are important in solid-state chemistry, and numerous reports of both types of crystal growth occurring simultaneously exist in the literature. Figure 6.3 is a simplified schematic representation highlighting particle-mediated crystal growth mechanisms, with solid-phase primary particles ranging from crystallites with well-defined facets (a, b) to amorphous particles (e).

6.4
Oriented Attachment

Oriented attachment (Figure 6.3a), which has also been referred to as oriented aggregation and epitaxial assembly, has been recognized as an important crystal growth mechanism since at least the late nineteenth century [17], although it has become widely recognized only in the last two decades. First, primary crystallites reversibly attach to one another to form structural intermediates that are analogous to outersphere complexes. There is no direct contact between primary crystallites in this complexed state, and solvent molecules and other molecular-scale species reside in the spaces separating them. The primary crystallites can rearrange and reorient through Brownian motion, and a structural intermediate composed of oriented crystallites can result. The schematic representation shown in Figure 6.4a highlights the difference between the fully oriented structural intermediate and the final secondary crystal. In both cases, building block crystallites are oriented with respect to one another, but, in the intermediate state, species like solvent and solutes reside in the spaces separating the building block crystals. Removal of those solvent molecules and other species can result in the formation of a new single crystal. This single crystal is a secondary object in the sense that it is composed of primary building block crystallites, but it has

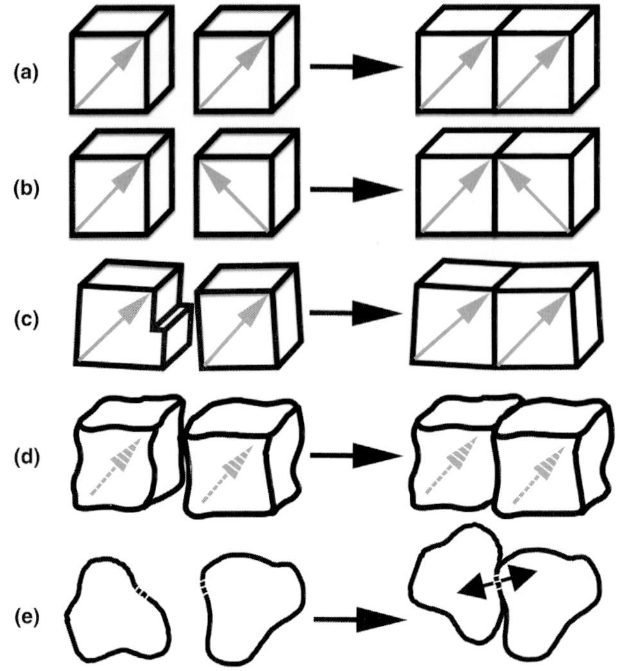

Figure 6.3 Schematic highlighting the many solid-particle-mediated growth mechanisms. Primary particles can range from well-ordered and smoothly faceted crystallites (a, b) to crystallites with steps or other surface features (c) to poorly crystalline particles (d) to amorphous particles (e). Gray arrows indicate the crystallographic orientation of crystals, and the hashed gray arrows represent poorly crystalline particles. In the case of (e), the hash marks represent structural units that match, and such structural units could serve to nucleate crystallization.

crystallographic registry throughout its entirety. These new single crystals can have morphologies that vary markedly from expectations based on symmetry.

The frequency of oriented attachment depends strongly on solution conditions. For example, Penn and Banfield qualitatively described the relative frequency of oriented attachment at moderately acidic pH and under hydrothermal conditions as most frequent on the anatase {112}, less frequent on the anatase (001), and rare on the anatase {101}. At higher and lower pHs, the frequency of oriented attachment was substantially reduced. This led the authors to conclude that electrostatics play an important role in determining the surfaces across which oriented attachment might occur [8]. In that study, the difference in the acid concentration was about an order of magnitude between experiments. Indeed, it seems likely that very small perturbations in solution conditions (e.g., pH) could result in large differences in the relative rates of oriented attachment along specific crystallographic directions.

A second example involves the growth of goethite nanocrystals by oriented attachment. In the work by Penn and coworkers, they demonstrated that the

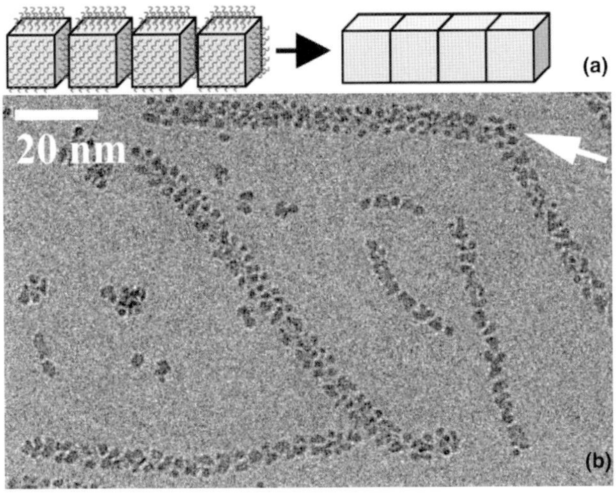

Figure 6.4 (a) is a schematic highlighting the arrangement of crystallites in the secondary intermediate structure, in which crystallites are separated by a layer of solvent molecules or other dissolved species (represented with red squiggles), and the final product crystal resulting from crystal growth by oriented attachment. (b) is a Cryo-TEM image of goethite crystals in vitrified water. Higher resolution images demonstrate that the crystallites residing in the same intermediate structure share the same crystallographic alignment. The white arrow serves to highlight an interface across which the crystallites crystallographic alignments match the observed alignment in twinned goethite crystals.

rate of crystal growth by oriented attachment increases as the difference between the isoelectric point of the nanocrystals and the pH of the solution decreases [23,24]. They further showed that the rate of crystal growth by oriented attachment increases with decreasing particle size [24–26]. These two results are consistent with trends for rates of aggregation as predicted by DLVO (Derjaguin, Landau, Verwey, and Overbeek) theory, which describes aggregation using a balance between repulsive electrostatic and attractive van der Waals forces. However, DLVO theory cannot predict oriented attachment.

The size dependence of oriented attachment explains why oriented attachment is often described as most important during the earliest stages of crystal growth. In fact, the use of classical nucleation and growth theory to fit experimental data often fails to produce acceptable fits to data tracking particle size over time, and this is often most pronounced at the early stages of growth (e.g., ZnO nanoparticle growth in alcohol-based solutions) [27,28]. At the early stages of growth, when the average crystallite size is small, crystal growth by oriented attachment will be faster than at later stages of crystal growth, when crystallite size is larger. Similarly, once oriented attachment has occurred, continued crystal growth can occur by oriented attachment by addition of primary crystallites or by oriented attachment between larger oriented aggregates. However, oriented attachment between product crystals would be expected to occur far less often than the oriented attachment between primary particles.

It is important to remember that oriented attachment and coarsening commonly operate simultaneously, and evidence can consist of fits to crystal growth data using kinetic models [28–30] and/or obtaining electron micrographs of individual crystals [8,9,31–33]. Recently, De Yoreo and coworkers were able to image this process by using liquid cell transmission electron microscopy. Their samples consisted of ferrihydrite nanoparticles in water. They observed the transition from isolated primary crystallites to the intermediate structure, with the particles maintaining close but not direct contact while undergoing rearrangements until a favorable orientation was achieved. At that instant, they observed a jump to contact event, which involved the formation of a direct interface between the two primary crystallites. Immediately upon the formation of the new interface, rapid growth by coarsening was observed at regions with negative radii of curvature that formed at the new contact interface [34].

6.4.1 Role of Attachment in the Development of Microstructure

Oriented attachment provides a route by which microstructural features such as stacking faults, defects, and twin boundaries can be produced. This is because oriented attachment requires structural accord only at the interface between primary crystallites. It is even possible to envision cases in which oriented attachment between heterogeneous materials could produce intricate nanostructures of two or more crystalline phases. In the next sections, different types of oriented attachment are described.

6.4.1.1 Twinned Attachment

In the case of two surfaces oriented such that two-dimensional structural accord is achieved with the introduction of a new symmetry element (e.g., a mirror plane), a twin or stacking fault can form (Figure 6.3b). This has been experimentally observed for oriented attachment across the anatase {112}. When complexed crystallites achieve coalignment in all three directions, attachment results in production of a new single crystal. However, a fraction of the attachments result in the formation of twins, with a mirror plane introduced at the {112} contact interface between the two crystallites [35,36].

One can estimate the free energy difference between the single crystal versus the twinned crystal by way of statistical mechanics. In the goethite crystals described by Yuwono *et al.*, approximately 10^3 single crystal oriented attachments for each twinned attachment were observed (semiquantitative estimate from electron micrographs published [37] and images from the same sample not heretofore published, Figure 6.4b). This three order of magnitude difference in the frequency of twinned attachment versus oriented attachment (using the aging temperature of 80 °C) yields an estimated free-energy difference between the two processes of just 20 kJ per mole of particle–particle contacts (5 kcal/mol). Just 6 kJ per mole decrease in the energy difference between these two conditions increases the expected frequency of twinning to 1% of particle–

particle attachments. This semiquantitative analysis produces an estimate of the energy difference between the single crystal and twinned cases of oriented attachment, although no known systematic and rigorously quantitative studies have been published to date.

6.4.1.2 Nearly Oriented Attachment

Nearly oriented attachment is depicted in Figure 6.3 c and has also been referred to as imperfect oriented attachment [31]. Features that can cause nearly oriented attachment, as opposed to oriented attachment, include steps on one or both of the surfaces involved (e.g., the tilt boundary resulting from attachment involving one stepped surface and one "flat" surface, as depicted in Figures 6.3b and 6.5). Further attachment events across such a tilt boundary can result in the production of screw dislocations as well as dislocations of mixed character [31]. Indeed, nearly oriented attachment is a mechanism by which defects can be incorporated into initially defect-free (except point defects) nanocrystals.

The incorporation of defects like dislocations and stacking faults has important implications from the perspectives of reactivity and mechanical properties. The properties of a crystalline material are governed by the nature of bonding between atoms and ions, crystal structure, composition (including the identity and concentration of trace components), and defects. Crystal growth involving rough surfaces and conditions favoring oriented attachment could result in high defect concentrations in the resulting product crystals, especially if growth by other mechanisms like coarsening is minimal. In contrast, if oriented attachment involves very smoothly faceted crystallites, one could produce product materials with fewer defects. Achieving control over the surface across which attachment occurs coupled with the use of smoothly faceted or roughly faceted precursor

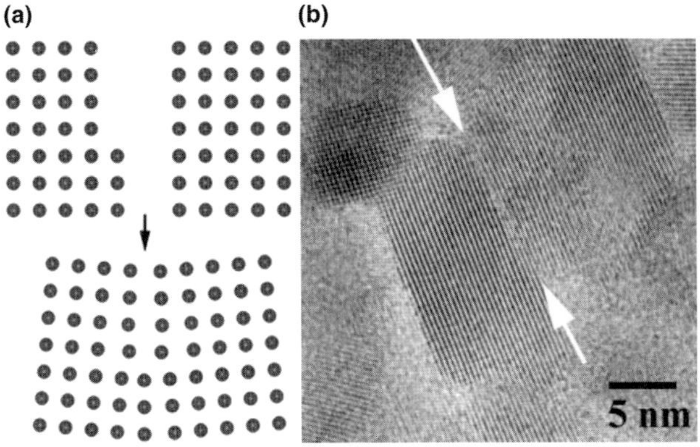

Figure 6.5 (a) A schematic highlighting how a tilt boundary can result upon oriented attachment involving a stepped surface. The result is the incorporation of an edge dislocation. (b) A TEM image of anatase, and the white arrows serve to highlight a tilt boundary between two crystallites of anatase. (Right-hand image is reprinted with permission from Ref. [31]. Copyright 1998, The American Association for the Advancement of Science.)

nanocrystals could yield symmetry-defying morphologies with controlled defect concentrations.

6.4.1.3 Oriented Attachment between Poorly Crystalline Particles

Oriented attachment between poorly crystalline particles could occur when suitable structural accord at the new interface has been achieved (Figure 6.3d). All four of these examples: oriented attachment, twinned attachment, nearly oriented attachment, and attachment of poorly crystalline particles have in common the formation of a complexed state in which particles can rearrange and reorient. When a favorable arrangement of primary particles has been achieved, the complex can irreversibly convert by removal of species from the spaces separating the primary particles and formation of direct bonds between the surfaces. The nature of the product depends on the nature of the primary particles and their relative orientations. Oriented attachment of poorly crystalline materials results in the formation of a new and larger poorly crystalline particle. One could envision the possibility of twinning and the incorporation of other features, as observed with oriented attachment between crystalline particles.

Structural Recognition

A more general perspective of oriented attachment, perhaps, is structural recognition – with an emphasis on the nature of structural units produced at the interface. Regardless, the generation of an intermediate structure within which primary crystals can rearrange and reorient seems an essential feature of oriented attachment. Once structural accord is achieved at the interface, irreversible conversion to the attached state can proceed via elimination of solvent and other dissolved species from the spaces separating primary particles and formation of direct chemical bonds between the two surfaces. If one considers oriented attachment from the perspective of structural units formed across the interface of attachment, one could extend oriented attachment to include amorphous particles. Thus, oriented attachment can be described as particle–particle attachment via structural accord at the interface to produce new secondary objects composed of primary particles that range from well crystallized to amorphous. Finally, one could consider structural accord between heterogeneous surfaces, which could provide a route by which complicated, nanostructured materials could be produced.

Major Control Parameters

From the perspective of coarsening, the rate of crystal growth increases by 10 orders of magnitude when increasing molecular-scale nutrient concentration from submicromolar to molar. Thus, solubility is a major control parameter. The result of oriented attachment of two primary particles is essentially a doubling of size. Thus, one could examine data tracking particle size versus time and estimate the frequency of oriented attachment required to keep pace with classical crystal growth. In the case of titanium dioxide hydrothermally coarsened at 200 °C, the doubling time is on the order of an hour at an acidic pH of about 1

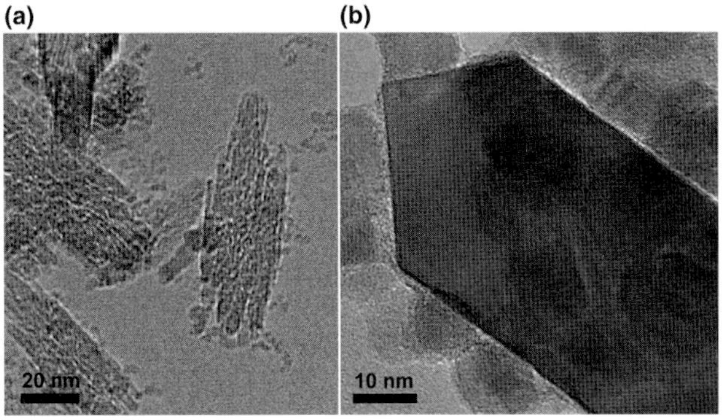

Figure 6.6 TEM images of rutile crystals. (a) A cryo-TEM image of a rutile crystal produced in low solubility conditions, and (b) a dry TEM image of a rutile crystal produced in high solubility conditions. In both cases, the small about 5–10 particles are anatase, and the larger crystals are rutile. (The left-hand image is reprinted with permission from Ref. [39]. Copyright 2015, Cambridge University Press.)

and several hours at the more mild pH of about 3. In the case of ZnO in low-molecular-weight alcohols, volume doubling occurs in a matter of minutes [38]. Figure 6.6 shows electron micrographs of rutile crystals grown under low (a) and high solubility (b) conditions. In Figure 6.6a, the product rutile crystals are highly porous and have many features consistent with a particle-mediated crystal growth mechanism [39]. In Figure 6.6b, the product rutile crystal has a morphology that is consistent with that expected from symmetry.

The initial size of the primary particles plays an important role in the rates of growth by oriented attachment. In the case of goethite growth by oriented attachment, the rate of crystal growth by oriented attachment increases dramatically with decreasing primary particle size. Decreasing the primary particle size from 7 to 5 nm in length resulted in a two-order of magnitude increase in the rate of crystal growth by oriented attachment [26].

Molecular dynamics simulations (using the ReaxFF reactive force field) by the group of Fichthorn and coworkers provide insight into oriented attachment at the molecular scale. Specifically, they compared results from simulations of attachment in the presence and absence (i.e., in vacuum) of water. Their results demonstrate that the presence of the water molecules is essential for the rearrangement of the primary particles to occur. In vacuum, irreversible attachment occurred regardless of the relative orientation of the primary crystallites, and a polycrystal resulted. With water present, crystallites did not immediately undergo irreversible attachment. Rather, the particles maintained close but not direct contact while undergoing rearrangements until oriented attachment or twinned attachment occurred. The authors concluded that a dynamic network of hydrogen bonds form between surface hydroxyls and surface oxygens of the attaching crystallites. They further concluded that the propensity for oriented or twinned attachment is directly related to the ability of a particular surface to dissociate water. That is to say, oriented or twinned attachment is most strongly favored on crystal surfaces that can dissociate water readily and most weakly favored on those that cannot [40].

Experimental work by Burrows *et al.* [41] demonstrated marked differences the nature of nanocrystal aggregates formed upon transfer into a variety of solvents. Solvent molecules that could coordinate with crystal surfaces produced the most open, least dense aggregates. Based on the cryo-TEM results of Yuwono *et al.* [37,42], it is possible to hypothesize that intraaggregate rearrangements are most likely within the most open aggregates. Thus, one can conclude that nature of the solvent–surface interaction is an important control parameter for oriented versus random attachment.

Oriented attachment provides a route by which symmetry-defying morphologies can be realized, but many open questions remain that prevent realization of tight control over morphology by exploiting it. Through the judicious selection of solvent, additives, and temperature, one can enhance or inhibit crystal growth by oriented attachment. Further, rates are strongly impacted by the initial size of primary particles, increasing rates with decreasing primary particle size. However, we are far from able to predict what conditions will favor oriented attachment.

6.4.2
Role of Attachment in Phase Transformation

Phase transformation involves two critical steps: nucleation and growth of the new phase at the expense of the old. Nucleation can occur homogeneously in the growth medium, at the medium-particle interface, within the bulk of the particle, and at the interface between two particles. Furthermore, many materials exhibit size-dependent phase stability. Thus, particle-mediated crystal growth can play two important roles in the kinetics of phase transformations. First, particle-mediated crystal growth can indirectly impact the kinetics of phase transformation by simply contributing to crystal growth, and, second, the attachment event can produce an interface from which nucleation of the product phase is facilitated.

The alumina, titania, zirconia, and iron oxides systems are examples in which particle size-dependent phase stability has been observed [43–46]. Once small nanoparticles grow to some threshold size, phase transformation to a phase with lower bulk-free energy can occur. Size-dependent phase stability is a consequence of a difference in both the surface and bulk energies of the phases in question. The phase with the lower surface energy will be favored at small size while the phase with the lower bulk energy will be favored at larger size. The balance between these two energy terms determines the threshold size at which transformation to the more thermodynamically stable phase occurs. Mechanisms that impact the kinetics of crystal growth can thus impact not only the initial phase transformation but also the growth of the new phase at the expense of the old. Further, microstructure (e.g., defect concentration) and morphology play important roles in phase transformations, and particle-mediated crystal growth is a route by which complicated microstructures and morphologies can arise.

6.5 The Anatase to Rutile Phase Transformation

In the case of anatase and rutile, two industrially important polymorphs of TiO_2, anatase is the most stable at small size due to its comparatively low surface energy. At larger particle size, rutile that is the most stable due to its comparatively has low bulk energy [47,48].

Zhang and Banfield demonstrated the importance of nucleation at anatase–anatase particle contacts in their work studying the anatase to rutile phase transformation by aging dry samples at high temperature in air. They used kinetic models to fit their data tracking crystallite size and phase composition as a function of time, temperature, and an effective concentration of anatase–anatase interfaces. They demonstrated that the activation energy for interface nucleation, which refers to nucleation at anatase–anatase contacts, was lower than for both surface nucleation and bulk nucleation. This means that the phase transformation kinetics were dominated by interface nucleation at the lower range of the temperatures employed in their study. They hypothesized that the rate of phase transformation was related to the number of anatase–anatase contacts and tested this hypothesis by "diluting" their solid samples with alumina particles. This dilution effectively reduced the number of anatase–anatase contacts as compared to the sample composed of 100% anatase, and their data demonstrated decreased rates of phase transformation for samples diluted with alumina. Their data further demonstrated that nucleation of new rutile was the rate-limiting step, with growth on existing rutile fast by comparison [49].

In related work, Penn and Banfield examined the rate of the anatase to rutile-phase transformation under acidic hydrothermal conditions. Under these conditions, extensive evidence for growth by oriented attachment and coarsening were observed by way of electron microscopy. Evidence for growth by oriented attachment included observations of twinned crystallites as well as crystallites that had complicated microstructures and morphologies. They proposed that twin boundaries across the anatase {112} could serve as nuclei for the anatase to rutile-phase transformation and demonstrated that a two-unit cell wide slab of material at the twin boundary contained structural elements common to the rutile crystal structure. In addition, they imaged a twinned anatase nanocrystal that contained a structurally unique, about 2.5 nm by 7 nm slab of material at the twin boundary. This slab had a quite different appearance than the rest of the twinned nanocrystal, and the observation lead to the hypothesis that the slab of material was a nucleus of rutile. They concluded that oriented attachment played a critical role in the mechanism of the phase transformation under hydrothermal conditions [36].

One important difference between the two studies just described is the experimental conditions. In the Zhang and Banfield study [49], the experiments were performed in air while the experiments of Penn and Banfield [36] were performed under hydrothermal conditions. In a liquid medium, Brownian motion means that particles are in constant motion and can reorient and rearrange far more readily than one would expect in vacuum or gas environments. One might hypothesize that the experimental conditions that favor the formation of twins, in the case of the anatase and rutile system, could yield increased rates of phase transformation as compared to experimental conditions that inhibit the formation of twins. Furthermore, one could expect the rate of phase transformation to be highly dependent on the mechanism by which the particles grow. All other things held constant (e.g., titania solubility), one would expect faster rates of phase transformation under conditions that favor particle-mediated crystal growth and slower rates of phase transformation under conditions that inhibit particle-mediated crystal growth.

6.6
Two Examples of Phase Transformations in the Iron Oxides

The iron oxides comprise a broad diversity of crystal structures and compositions and include the iron oxides, oxyhydroxides, and hydroxides. The iron oxides are important planetary (e.g., products of rock and mineral weathering) and industrial materials (e.g., medical imaging, pigments, and catalysts). Oriented attachment plays important roles in the crystal growth, generation of complicated morphologies and microstructures, and phase transformations in the iron oxides.

Previous work has demonstrated particle size-dependent phase stability in the iron oxides [46]. The kinetics of crystal growth and phase transformation are directly linked, with the mechanisms of crystal growth strongly impacting the kinetics of phase transformation. When considering crystal growth dominated by oriented attachment, one could consider whether phase transformation precedes or follows oriented attachment (Figure 6.7), and evidence for both has been observed.

The first example is the formation of hematite (α-Fe_2O_3) from akaganeite (β-$FeO(OH)$) in aqueous suspensions. The authors concluded that phase transformation followed akaganeite crystal growth by oriented attachment based on results obtained using cryogenic transmission electron microscopy (cryo-TEM) and X-ray diffraction (XRD) to characterize the size, microstructure, and morphology of particles as a function of the extent of phase transformation. Before any hematite had formed, the authors observed that akaganeite crystallites had grown predominantly by oriented attachment. They tracked the kinetics of phase transformation and examined crystals as a function of time, and they concluded that the initial hematite formed via phase transformation of larger akaganeite crystals. They reported two key observations. First, the hematite crystals

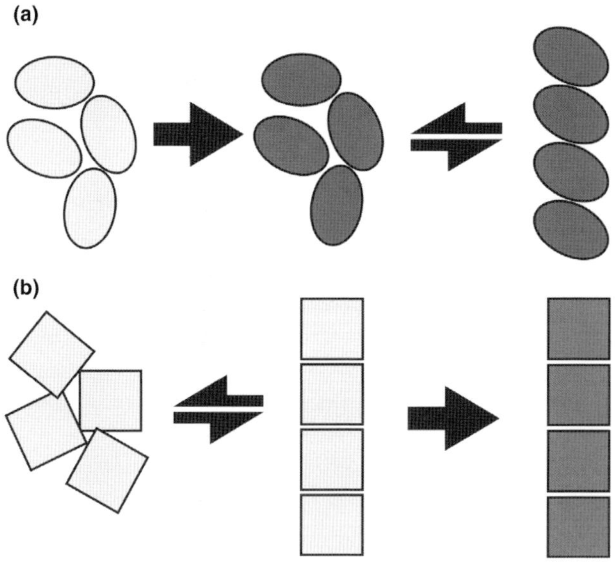

Figure 6.7 Schematic showing phase transformation occurring before (a) or after (b) crystal growth by oriented attachment. In the scheme (a), the elliptical shapes were selected based on the observation that phase transformation from ferrihydrite to goethite most likely precedes goethite crystal growth by oriented attachment. In the scheme (b), the square shapes were selected based on the observation that phase transformation from akaganeite to hematite occurs after akaganeite crystal growth by oriented attachment.

appeared to be pseudomorphs, with size, morphology, and microstructure similar to the larger akaganeite crystals, which had grown by oriented attachment. Second, no evidence for hematite with crystal domain size consistent with the primary crystallite size was detected. Thus, they concluded that crystal growth by oriented attachment precedes the phase transformation [32]. Purposeful growth by oriented attachment of a precursor phase could be exploited to produce a pseudomorph composed of a new phase by inducing phase transformation after oriented attachment.

In contrast, results tracking the ferrihydrite, a hydrous iron oxyhydroxide that occurs only as nanoparticles, to goethite (α-FeOOH) phase transformation in aqueous suspension are most consistent with phase transformation preceding crystal growth by oriented attachment. In experiments characterizing samples as a function of aging time, key observations were, first, that the X-ray diffraction pattern peaks for the first-detected goethite exhibited line broadening consistent with the initial ferrihydrite size, and, second, the TEM images of isolated about 5 nm particles exhibited lattice fringes consistent with the goethite crystal structure. Furthermore, in elongated structures that had substantial texture consistent with particle-mediated growth, lattice fringes uniquely consistent with ferrihydrite d-spacings were not observed [23]. In later work by Yuwono *et al.* [37], cryo-TEM images of vitrified samples of the aqueous suspensions provided additional evidence for phase transformation followed by crystal growth by oriented attachment. At the earliest stages, aggregates were characterized by open, fractal morphologies. At this stage, no evidence for goethite was detected by way of X-ray diffraction, which lead to the conclusion that the fractal aggregates were composed of ferrihydrite primary particles. With time, aggregates with lengths, widths, and overall morphologies matching the final goethite crystals – including morphologies similar to twins common to goethite – appeared (Figure 6.4b). These intermediate structures exhibited lattice fringe spacings, which spanned the entire length and width of the structures, consistent with the goethite crystal structure in high-resolution cryo-TEM images. The authors hypothesized that these intermediate structures, which appeared to have fairly linear arrangements, formed by intraaggregate rearrangement of the primary particles residing within the fractal aggregates after transformation from ferrihydrite to goethite [37].

The ferrihydrite to goethite phase transformation is an interesting one because it is defies the expectation that the initial material must grow to a threshold size prior to transformation to the more stable phase, which is goethite in this case. In the case of akaganeite growth by oriented attachment and subsequent phase transformation to hematite, particle size-dependent phase stability is consistent with the results of Navrotsky and coworkers, in which they demonstrated that akaganeite would be more stable at smaller size [46]. Ferrihydrite, however, is not as well understood from a crystallographic standpoint nor even compositional or thermodynamic standpoints. Indeed, it is likely that even the hydration state might vary as a function of particle size and shape. However, the observation of intermediates with size and shape similar to the goethite product single crystals and twinned in the cryo-TEM images strongly supports the possibility that ferrihydrite is not necessarily the most stable phase at smaller size but that it may represent a kinetically trapped state.

6.7 Growth of Silicalite-1 from Amorphous Primary Particles

At first glance, it might seem that amorphous nanoparticles cannot grow by oriented attachment. However, by reframing oriented attachment as structural recognition, as discussed earlier, one could envision growth proceeding by a mechanism analogous to oriented attachment. In parallel with the above discussion regarding phase transformations, structural elements produced at interfaces between amorphous nanoparticles (Figure 6.3e) could serve as nuclei for crystallization.

In the case of TPA-silicalite-1, which is hereafter referred to simply as silicalite-1, results tracking growth from suspensions containing silica, water, and the structure-directing agent (SDA) tetrapropylammonium (TPA) hydroxide are consistent with particle-mediated growth [50–54]. In the typical synthesis, amorphous silica nanoparticles in the few nm size range form spontaneously upon mixing precursor solutions. These initial particles have a core-shell structure, with cores rich in silica and shells rich in the SDA [55]. With aging at temperatures as low as room temperature, the particles structurally evolve, become progressively less colloidally stable, and aggregates of the particles form. At some point, crystalline silicalite-1 appears.

Key observations include that aggregation of the precursor, amorphous nanoparticles precedes crystallization of silicalite-1 [53] and that the silicalite-1 product crystals often have textures and morphologies consistent with particle-mediated growth (e.g., low angle tilt boundaries and rounded protrusions and pores with sizes similar to the primary particle size, as shown in Ref. [50] and later reports). Experiments that selectively dissolved amorphous silica but not crystalline silicalite-1 resulted in silicalite-1 crystals that were porous with textures dominated by rounded protrusions with sizes similar to the primary particle size. The presence of such features is commonly interpreted as consistent with particle-mediated crystal growth. Based on the observation that the aggregates form prior to the first detection of crystalline silicalite-1 combined with the observed textures and microstructures of silicalite-1 crystals before and after selective dissolution, the authors concluded that crystallization nucleated within the amorphous aggregates [53]. It is certainly conceivable that crystalline silicalite-1 nucleated from structural elements produced at a contact interface between amorphous particles.

Just as with oriented attachment with crystalline and poorly crystalline materials, there is evidence that, in addition to the particle-mediated growth mechanism, crystal growth by coarsening also occurs. At low temperature, at which growth by monomer addition is slow, crystal textures are dominated by features consistent with particle-mediated growth. When suspensions are aged at higher temperatures, growth by monomer-by-monomer addition is substantially faster. Under those conditions, the early-stage crystals had features consistent with particle-mediated growth but later stage crystals typically had few such features and morphologies consistent with the silicalite-1 crystal symmetry [54]. This is consistent with recent work, in which *in situ* atomic force microscopy to detect crystal growth by particle attachment and by addition of molecular scale species was used [56]. These examples also serve to highlight how textures arising from particle-mediated growth may not be retained in the final product crystals.

6.8 Quantifying Crystal Growth

One can pose several questions regarding how one can elucidate crystal growth mechanisms.

1) What must we characterize in order to elucidate and quantify nonclassical crystal growth?
2) What are reliable markers of crystal growth by a particle-mediated mechanism?
3) What constitutes convincing evidence for nonclassical crystal growth?

The critical role of holistic materials characterization cannot be understated. Useful characterization methods enable the determination of purity, composition, size, shape, and microstructure of the product crystal(s) as well as the spatial distribution of elements within crystal(s) and their surface chemistry. Additional characteristics include identifying facets across which attachment occurs, quantifying the frequency of different types of attachments, characterizing the degree to which particles are crystalline, and quantifying the kinetics of phase transformation. In the case of preparing a large number of crystals, as is typically the goal with nanocrystals, the distributions of purity, size, shape, and microstructure must also be understood. A final difficulty is characterization of the species that reside in the spaces separating primary particles in intermediate structures.

Contributions to crystal growth by multiple mechanisms can be difficult to quantify and even detect because features produced are often transient. For example, when crystals grow by oriented attachment, simultaneous growth by diffusion of molecular scale species from areas of high radius of curvature to areas of low radius of curvature (i.e., coarsening) can essentially erase physical evidence for oriented attachment. Overgrowth by coarsening or atomic-scale rearrangement via recrystallization or phase transformation could result in loss of features like dimpled boundaries and defects that serve as relics of a particle-mediated crystal growth mechanism. Thus, the presence or absence of particular features does not necessarily demonstrate that a particular growth mechanism was key in a crystal's growth history [15]. In solvothermal crystal growth, for example, oriented attachment can dominate at the early stages and coarsening at later stages of crystal growth [30]. Furthermore, recrystallization and phase transformation could further complicate matters. Thus, the characterization of crystals at different stages of growth is essential to elucidating the mechanisms at play.

Such quantitative data are difficult and time consuming to obtain. Direct imaging using methods such as transmission electron microscopy (TEM) have long been primary methods for obtaining evidence for particle-mediated crystal growth. TEM in particular continues to be a core tool in the toolbox for characterizing crystal growth. However, high-resolution imaging methods are less than ideal for a few reasons. First, the methods sample vanishingly small amounts of material, which means one must take care that data are representative of the whole sample. Second, identifying the facet(s) across which attachment has occurred can be a time-consuming process. If orientations in which twin boundaries and so on are discernable are rare, datasets might not accurately reflect the frequency with which such twin boundaries form. In addition, many observations are required to obtain statistically significant datasets. Third, in the case of TEM, images are two-dimensional representations of three-dimensional objects that mean that conclusive identification of the interface between two crystalline domains is often challenging and sometimes impossible. Even with the three-dimensional images possible through electron tomography, the best achievable resolution is not yet adequate for detailed examination of subnanometer features that can be present in crystals grown by an aggregative mechanism.

Additional drawbacks of using TEM include the potential for beam damage, artifacts introduced upon sample preparation, and the necessity of placing the sample into a high vacuum. The latter means that TEM is inherently an ex situ technique, and one must beware of artifacts potentially introduced as a result of insertion into the vacuum environment as well as with sample preparation. However, no other reasonably available technique can provide images and diffraction patterns that enable quantitative characterization of crystallographic orientations, sizes, morphologies, and microstructure of crystals at the few nanometer lengthscale.

A key element to holistic characterization is in situ characterization. Advancements in materials characterization methods, such as cryogenic and fluid cell TEM, have enabled the direct observation of crystals in their growth media and, in the case of fluid cell TEM, even direct observation of crystals growing by oriented attachment and coarsening.

Cryo-TEM sample preparation prevents introduction of artifacts upon drying and/or insertion of the sample in to the TEM vacuum. Sample preparation involves vitrification of thin films of liquid suspensions using cryogenic liquids (e.g., liquid ethane or liquid nitrogen) followed by imaging under cryogenic conditions [57,58]. Cryo-TEM has enabled direct imaging of the structures formed during crystal growth by oriented attachment [32,37] as well as characterization of aggregation state in a variety of solvents [57]. Cryogenic TEM (cryo-TEM) images provide snapshots of how particles are arranged in the liquid state, and rapid vitrification is essential to preserving the arrangement of the suspended objects.

Sample preparation for liquid cell TEM [34,59,60] simply involves trapping liquid suspension between two electron-transparent (e.g., graphene sheets or thin films of GaN) surfaces so that the sample can be inserted into the high vacuum environment of the TEM. This approach was used to directly image crystal growth by oriented attachment and coarsening of ferrihydrite in water [34] as well as platinum in a solution of oleylamine in *o*-dichlorobenzene [60]. Liquid cell TEM enables quantitative characterization of the motions of particles, but how the electron beam perturbs liquid suspensions of nanoparticles is not yet well understood. In one example, exposure of goethite nanocrystals in water to the electron beam resulted in rapid dissolution, most likely due to reduction of Fe (III) to Fe (II), which is far more soluble than Fe (III) [12].

To obtain three-dimensional information about the size, texture, and shape of crystals, one can turn to electron tomography, which involves collection of images at numerous tilt conditions and subsequent image reconstruction to generate a three-dimensional image of the object. This technique is difficult, especially when working with crystalline materials due to changing contrast as a function of tilt, and does not have sufficient spatial resolution in order to work with extremely small nanocrystals. However, it has been very effectively used to characterize akaganeite crystal growth by oriented attachment with subsequent phase transformation to hematite [32].

A significant drawback of the above direct methods is that they sample vanishingly small amounts of material. This begs the question as to whether results are representative of the overall system; thus, correlative methods that enable characterization of comparatively large amounts of material are essential.

X-ray diffraction [61–63] can be used to identify crystalline phases, quantify average domain sizes along specific crystallographic directions, and even assess particle size distribution. Scattering techniques such as dynamic light scattering and small angle X-ray scattering enable characterization of size and size distribution in the suspended state. A challenge with these methods is that appropriate models of particle size, size distribution, and even aggregation state must be developed in order to fit scattering data. Thus, methods like DLS and SAXS are best used in combination with a direct imaging method such as TEM.

Methods that enable the determination of crystal size as a function of time can yield data that can be fit using a variety of crystal growth models. Kinetic models can provide a method by which particle-mediated crystal growth can be detected, and they can often enable quantification of the relative contributions to growth by more than one mechanism [19,22,28,30,64–72]. Unfortunately, these models cannot typically identify the nature of the particle-mediated crystal growth (e.g., oriented attachment versus random aggregation followed by recrystallization). Detailed description of these models is beyond the scope of this chapter; thus, the reader is referred to the aforementioned references for detailed descriptions of models that can be used to fit crystal growth data. The use of semiquantitative and quantitative models to fit kinetic crystal growth data improves drastically the ability to detect crystal growth by particle-based mechanisms, even when features that could serve as indicators of such growth have been erased by subsequent processes.

6.9 Concluding Remarks

Particle-mediated crystal growth has become widely recognized as an important and common crystal growth mechanism. It provides a route by which unique morphologies and complicated microstructure can be produced. This chapter specifically addresses crystal growth by attachment of solid-phase particles. Crystal growth often involves more than one mechanism, with classical growth often occurring simultaneously with particle-mediated mechanisms. Solvent and adsorbed species can have dramatic and unexpected impacts on crystal growth, with some species arresting growth altogether, some inhibiting coarsening while promoting particle-mediated growth and vice versa, and still others enhancing both coarsening and particle-mediated growth. Systematic studies quantifying the contributions to growth by both classical and nonclassical mechanisms will lead to predictive models that can be used to tailor synthesis and growth methods to produce materials with desired size, shape, and microstructure, as well as control the distribution of those properties. Holistic materials characterization is essential, and direct images of final crystals have yielded deep insights into the mechanisms by which crystals grow. However, reliance upon such data cannot produce a complete picture of how a crystal grew since features produced during the earliest stages of growth could be erased by processes dominating the later stages of growth. Thus, the best materials characterization approach employs both time-resolved and *in situ* methods so as to examine how crystals evolve from their initial nucleation to their final state.

Acknowledgments

Parts of this work were carried out in the Institute of Technology Characterization Facility, University of Minnesota, a member of the NSF-funded Materials Research Facilities Network (www.mrfn.org). We acknowledge that support was provided by the University of Minnesota, the National Science Foundation (#0957696), and the Nanostructural Materials and Processes Program at the University of Minnesota – Twin Cities.

References

1 McWhan, D. (2012) *Sand and Silicon: Science That Changed the World*, Oxford University Press, Inc., New York.
2 Oskam, G., Hu, Z.S., Penn, R.L., Pesika, N., and Searson, P.C. (2002) *Phys. Rev. E*, **66**, 011403.
3 Wagner, C. (1961) *Z. Elektrochem.*, **65**, 581–591.
4 Lifshitz, I.M. and Slyozov, V.V. (1961) *J. Phys. Chem. Solids*, **19**, 35–50.
5 Iggland, M. and Mazzotti, M. (2012) *Cryst. Growth Des.*, **12**, 1489–1500.
6 Murray, C.B., Norris, D.J., and Bawendi, M.G. (1993) *J. Am. Chem. Soc.*, **115**, 8706–8715.
7 Chernomordik, B.D., Beland, A.E., Trejo, N.D., Gunawan, A.A., Deng, D.D., Mkhoyan, K.A., and Aydil, E.S. (2014) *J. Mater. Chem. A*, **2**, 10389–10395.
8 Penn, R.L. and Banfield, J.F. (1999) *Geochim. Cosmochim. Acta*, **63**, 1549–1557.
9 Cho, K.-S., Talapin, D.V., Gaschler, W., and Murray, C.B. (2005) *J. Am. Chem. Soc.*, **127**, 7140–7147.
10 Liao, H.-G., Cui, L., Whitelam, S., and Zheng, H. (2012) *Science*, **336**, 1011–1014.
11 Schliehe, C., Juarez, B.H., Pelletier, M., Jander, S., Greshnykh, D., Nagel, M., Meyer, A., Foerster, S., Kornowski, A., Klinke, C., and Weller, H. (2010) *Science*, **329**, 550–553.
12 Penn, R.L., Li, D., and Soltis, J. *New Perspectives on Mineral Nucleation and Growth* (ed. L.G. Benning), Springer.
13 Kandori, K., Kawashima, Y., and Ishikawa, T. (1991) *J. Chem. Soc. Faraday Trans.*, **87**, 2241–2246.
14 Sugimoto, T., Muramatsu, A., Sakata, K., and Shindo, D. (1993) *J. Colloid Interface Sci.*, **158**, 420–428.
15 J. J., DeYoreo., Gilbert, P.U.P.A., Sommerkijk, N.A.J.M., Penn, R.L., Whitelam, S., Joester, D., Zhang, H., Rimer, J.D., Navrotsky, A., Banfield, J.F., Wallace, A.G., Michel, F.M., Meldrum, F.C., Coelfen, H., and Dove, P.M. (2015) *Science*, **349**. doi: 10.1126/science.aaa6760
16 Dalmaschio, C.J., Ribeiro, C., and Leite, E.R. (2010) *Nanoscale*, **2**, 2336–2345.
17 Ivanov, V.K., Fedorov, P.P., Baranchikov, A.Y., and Osiko, V.V. (2014) *Russ. Chem. Rev.*, **83**, 1204–1222.
18 Niederberger, M. and Colfen, H. (2006) *Phys. Chem. Chem. Phys.*, **8**, 3271–3287.

19 Penn, R.L. (2004) *J. Phys. Chem. B*, **108**, 12707–12712.

20 Xiong, Y. and Tang, Z. (2012) *Science China Chem.*, **55**, 2272–2282.

21 Zhang, H. and Banfield, J.F. (2012) *J. Phys. Chem. Lett.*, **3**, 2882–2886.

22 Zhang, J., Huang, F., and Lin, Z. (2010) *Nanoscale*, **2**, 18–34.

23 Burleson, D. and Penn, R.L. (2006) *Langmuir*, **22**, 402–409.

24 Burrows, N.D., Hale, C.R.H., and Penn, R.L. (2013) *Cryst. Growth Des.*, **13**, 3396–3403.

25 Burrows, N.D., Hale, C.R.H., and Penn, R.L. (2012) *Cryst. Growth Des.*, **12**, 4787–4797.

26 Penn, R.L., Tanaka, K., and Erbs, J. (2007) *J. Cryst. Growth*, **309**, 97–102.

27 Hu, Z.S., Ramirez, D.J.E., Cervera, B.E.H., Oskam, G., and Searson, P.C. (2005) *J. Phys. Chem. B*, **109**, 11209–11214.

28 Ratkovich, A.S. and Penn, R.L. (2008) *J. Solid State Chem.*, **181**, 1600–1608.

29 Huang, F., Zhang, H., and Banfield, J.F. (2003) *J. Phys. Chem. B*, **107**, 10470–10475.

30 Huang, F., Zhang, H.Z., and Banfield, J.F. (2003) *Nano. Lett.*, **3**, 373–378.

31 Penn, R.L. and Banfield, J.F. (1998) *Science*, **281**, 969–971.

32 Frandsen, C., Legg, B.A., Comolli, L.R., Zhang, H., Gilbert, B., Johnson, E., and Banfield, J.F. (2014) *CrystEngComm*, **16**, 1451–1458.

33 Banfield, J.F., Welch, S.A., Zhang, H., Ebert, T.T., and Penn, R.L. (2000) *Science*, **289**, 751–754.

34 Li, D., Nielsen, M.H., Lee, J.R.I., Frandsen, C., Banfield, J.F., and Yoreo, J.J.D. (2012) *Science*, **336**, 1014–1018.

35 Penn, R.L. and Banfield, J.F. (1998) *Am. Mineral.*, **83**, 1077–1082.

36 Penn, R.L. and Banfield, J.F. (1999) *Am. Mineral.*, **84**, 871–876.

37 Yuwono, V.M., Burrows, N.D., Soltis, J.A., and Penn, R.L. (2010) *J. Am. Chem. Soc.*, **132**, 2163.

38 Ratkovich, A.S. and Penn, R.L. (2007) *J. Phys. Chem. C*, **111**, 14098–14104.

39 Sabyrov, K., Yuwono, V., and Penn, R.L. (2015) *MRS Proceedings*, vol. **1721**, Cambridge University Press, pp. mrsf14–1721.

40 Raju, M., Duin, A.C.T.v., and Fichthorn, K.A. (2014) *Nano. Lett.*, **14**, 1836–1842.

41 Burrows, N.D., Kesselman, E., Sabyrov, K., Stemig, A., Talmon, Y., and Penn, R.L. (2014) *CrystEngComm*, **16**, 1472–1481.

42 Yuwono, V.M., Burrows, N.D., Soltis, J.A., Do, T.A., and Penn, R.L. (2012) *Faraday Discuss.*, **159**, 235–245.

43 Zhang, H.Z. and Banfield, J.F. (2000) *J. Phys. Chem. B*, **104**, 3481–3487.

44 Chraska, T., King, A.H., and Berndt, C.C. (2000) *Mater. Sci. Eng. A Struct. Mater.*, **286**, 169–178.

45 Navrotsky, A. (2003) *Geochem. Trans.*, **4**, 34–37.

46 Navrotsky, A., Mazeina, L., and Majzlan, J. (2008) *Science*, **319**, 1635–1638.

47 Gribb, A.A. and Banfield, J.F. (1997) *Am. Mineral.*, **82**, 717–728.

48 Ranade, M.R., Navrotsky, A., Zhang, H.Z., Banfield, J.F., Elder, S.H., Zaban, A., Borse, P.H., Kulkarni, S.K., Doran, G.S., and Whitfield, H.J. (2002) *Proc. Natl. Acad. Sci. USA*, **99**, 6476–6481.

49 Zhang, H.Z. and Banfield, J.F. (2000) *J. Mater. Res.*, **15**, 437–448.

50 Davis, T.M., Drews, T.O., Ramanan, H., He, C., Schnablegger, H., Katsoulakis, M.A., Kokkoli, E., Penn, R.L., and Tsapatsis, M. (2006) *Nat. Mater.*, **5**, 400–408.

51 Kumar, S., Davis, T.M., Ramanan, H., Penn, R.L., and Tsapatsis, M. (2007) *J. Phys. Chem. B*, **111**, 3398–3403.

52 Kumar, S., Penn, R.L., and Tsapatsis, M. (2011) *Microporous Mesoporous Mat.*, **144**, 74–81.

53 Kumar, S., Wang, Z., Penn, R.L., and Tsapatsis, M. (2008) *J. Am. Chem. Soc.*, **130**, 17284.

54 Kumar, S., Wang, Z., Penn, R.L., and Tsapatsis, M. (2008) *J. Am. Chem. Soc.*, **130**, 17284–17286.

55 Davis, T.M., Drews, T.O., Ramanan, H., He, C., Dong, J.S., Schnablegger, H., Katsoulakis, M.A., Kokkoli, E., McCormick, A.V., Penn, R.L., and Tsapatsis, M. (2006) *Nat. Mater.*, **5**, 400–408.

56 Lupulescu, A.I. and Rimer, J.D. (2014) *Science*, **344**, 729–732.

57 Burrows, N.D., Kesselman, E., Sabyrov, K., Stemig, A., Talmon, Y., and Penn, R.L. (2014) *CrystEngComm*, **16**, 1472–1481.

58 Burrows, N.D. and Penn, R.L. (2013) *Microsc. Microanal.*, **19**, 1542–1553.

59 Nielsen, M.H., Aloni, S., and De Yoreo, J.J. (2014) *Science*, **345**, 1158–1162.

60 Yuk, J.M., Park, J., Ercius, P., Kim, K., Hellebusch, D.J., Crommie, M.F., Lee, J.Y., Zettl, A., and Alivisatos, A.P. (2012) *Science*, **336**, 61–64.

61 Chiche, D., Digne, M., Revel, R., Chaneac, C., and Jolivet, J.P. (2008) *J. Phys. Chem. C*, **112**, 8524–8533.

62 Hapiuk, D., Masenelli, B., Masenelli-Varlot, K., Tainoff, D., Boisron, O., Albin, C., and Mélinon, P. (2013) *The J. Phys. Chem. C*, **117**, 10220–10227.

63 Huang, F., Zhang, H.Z., and Banfield, J. (2003) *Nano. Lett.*, **3**, 373–378.

64 Gilbert, B., Zhang, H., Huang, F., Finnegan, M.P., Waychunas, G.A., and Banfield, J.F. (2003) *Geochem. Trans.*, **4**, 20–27.

65 Huang, F., Zhang, H.Z., and Banfield, J.F. (2003) *J. Phys. Chem. B*, **107**, 10470–10475.

66 Zhang, H.Z. and Banfield, J.F. (2004) *Nano. Lett.*, **4**, 713–718.

67 Sabyrov, K., Burrows, N.D., and Penn, R.L. (2013) *Chem. Mater.*, **25**, 1408–1415.

68 Penn, R.L. and Soltis, J.A. (2014) *Crystengcomm*, **16**, 1409–1418.

69 Sabyrov, K., Adamson, V., and Penn, R.L. (2014) *CrystEngComm*, **16**, 1488–1495.

70 Zhang, H., Penn, R.L., Lin, Z., and Coelfen, H. (2014) *CrystEngComm*, **16**, 1407–1408.

71 Xue, X., Penn, R.L., Leite, E.R., Huang, F., and Lin, Z. (2014) *CrystEngComm*, **16**, 1419–1429.

72 Burrows, N.D., Yuwono, V.M., and Penn, R.L. (2010) *MRS Bull.*, **35**, 133–137.

7
Sol–Gel Synthesis of Solid-State Materials

Guido Kickelbick[1,2] *and Patrick Wenderoth*[1]

[1]*Saarland University, Inorganic Solid Sate Chemistry, Campus Building C4 1, 66123 Saarbrücken, Germany*
[2]*INM – Leibniz Institute for New Materials, Campus D2 2, 66123 Saarbrücken, Germany*

7.1
Introduction

The sol–gel process is a major toolbox for the mild preparation of inorganic oxide materials during the last decades [1]. Contrary to traditional solid-state methods which often start with preformed powders applying high temperatures for the production of the targeted compounds, the sol–gel approach starts from molecular precursors in solution which will be crosslinked to three-dimensional networks. Because the process is based on the irreversible formation of bonds between the molecular species, the final products are in the first stage of the kinetic products. In many cases, this leads to amorphous structures in a first place, particularly if the process is carried out under mild conditions ($T < 100\,°C$). Kinetically controlled reactions are sensitive to changes of the reaction parameters; therefore, variations in temperature, solvents, and the composition of the precursors have a strong influence on the final products in a sol–gel process.

The name of the process represents the different morphologies that are passed from the molecular educts to the final materials. In a solution, the precursors start to react with each other and a sol is formed, which is defined as a stable suspension of colloidal particles in a liquid. The particles are still reactive on their surface and have the tendency to react with each other in forming a gel, which consists in the majority of the cases of the primary sol particles that reacted with each other forming covalent bonds. The obtained three-dimensional porous solid structure incorporates the continuous liquid phase. This so-called wet gel can be heated and transformed to a dry product by thermal treatment or other processing steps, which will

Handbook of Solid State Chemistry, First Edition. Edited by Richard Dronskowski, Shinichi Kikkawa, and Andreas Stein.

be explained in a further paragraph. Applying particular reaction conditions, for example, pH value or presence of electrolytes, the sol can be prevented from gelation, which makes it possible to isolate the solid particles in the form of nanoparticles.

The plethora of molecular precursors that can be used in combination with different additives and structure directing agents allows various compositions and morphologies [2–5]. Therefore, the sol–gel technique became an important technology in academic as well as industrial research and applications. In this chapter, we will give an overview of the basic principles of the sol–gel process, its mechanisms, precursors and the final products. Particularly, the emerging field of nanotechnology brought a new boost to this process.

7.1.1
Requirement for Low-Temperature Processes

As mentioned in Section 7.1, traditional solid-state chemistry typically applies high temperature processes in the synthesis of compounds. In most of the cases the preparative procedure starts from solid precursors, a high temperature treatment is applied, and the targeted materials are again solids-in many cases in their thermodynamic equilibrium. Several aspects make it important to switch to milder reaction processes, so called soft chemistry approaches or *chimie douce* [6]. Important parameters are the decrease in energy consumption, which means lower temperatures, and the need for nonequilibrium phases. Other aspects that became more and more important in the last decades are the availability of tailored morphologies of the final materials such as thin films and coatings, which are in many cases not available applying classical high-temperature approaches. Soft templating of materials, for example, its porosity, is another feature, which is only easily realizable if the process consists of the opportunity to work in a liquid phase. Furthermore, in many modern applications the solid-state materials need to be attached to thermal sensitive supports such as polymers or structures in microelectronics that are influenced by the high temperatures. Sol–gel chemistry as a solvent-based method can fill here gaps. Solid-state materials preparation under mild conditions and in solvents also opens new routes for the formation of templated structures and materials at the borderline to biological/living materials. In addition, the sol–gel method allows the bottom-up production of nanoparticles and is therefore an important methodology to promote this technology [7,8]. As a real bottom-up approach starting from a molecular precursor and ending up in macroscopic structures controlling the morphology over several length scales, makes this technique an ideal approach for hierarchically ordered materials [9,10]. Finally, also new patterning and processing technologies are feasible applying sol–gel chemistry, for example, inkjet printing, soft lithography, or rapid prototyping [11–19].

7.1.2 Inorganic–Organic Hybrid Materials

Hybrid materials became one of the most developed new materials classes in the last decades [20,21]. Driving forces of this development are the promising combinations of different functions in one material and the synergistic properties that can be achieved in the final material. Through a high control of composition and reaction parameters, it is possible to merge inorganic and organic or even biological components in one solid material. Due to the mild reaction conditions and the possibility to incorporate also thermally labile organic functions during the formation of the three-dimensional structure, the sol–gel approach became the enabling technology for high-tech materials in fields such as microelectronics, photonics and optical materials, (nano)composites, sensors, and biomedical applications [22–27]. Following two methods are available in the formation of hybrid materials:

1) The organic functions are embedded in the sol–gel matrix without formation of strong chemical interactions. Practically, this type of materials is formed if the organic or biological materials are mixed with a solvent and the sol–gel precursors and afterwards the sol–gel process is started. The inorganic network is thus formed around the organic moieties and the molecules are entrapped. This type of materials is often called *class 1* hybrids. Important for the embedment of the molecules is that their polarity fits the sol–gel process. If the polarity difference is too large, phase separation can occur and the efficiency of embedment is decreased.
2) Covalent linkage between the organic moieties and the inorganic sol–gel network can be obtained by modifying the molecules with groups that can interact with the inorganic network, for example, $Si(OR)_3$ groups. The resulting materials are called *class II* hybrid materials.

7.2 Mechanisms

7.2.1 Hydrolytic Sol–Gel Process

The by far best understood sol–gel mechanism is the hydrolytic condensation of silica-based materials. Starting compounds in this process are alkoxysilanes that contain Si-OR groups. In a first step, these groups are hydrolyzed to form Si-OH groups. These so-called silanols are in the majority of cases not stable and undergo condensation reactions, either with other alkoxides or with further silanol groups in the solution (Eq. (7.1)). The most common precursors are tetraalkoxysilanes with the general formula $Si(OR)_4$ such as tetramethoxysilane (TMOS) and tetraethoxysilane (TEOS). Generally, also other species with the tendency to form silanol groups can be used, for example, water glass solutions under specific pH conditions.

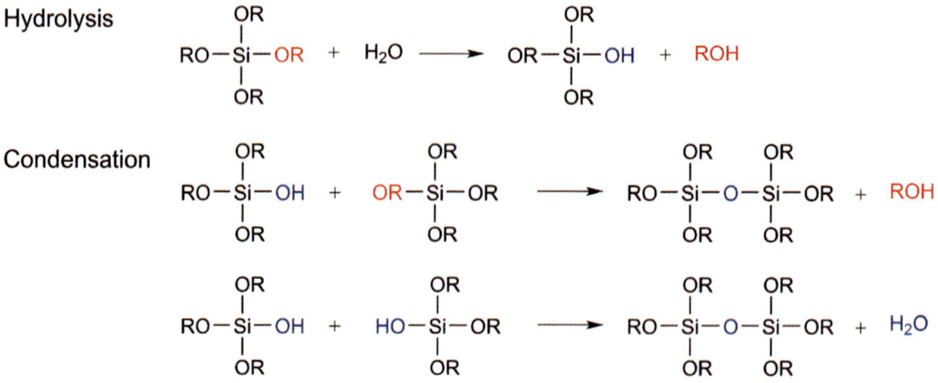

(7.1)

Alkoxysilanes react quite slow with water; therefore, the hydrolysis and condensation reaction is catalyzed either by an acid or a base [28]. Under acidic catalysis, the Si—OH or Si—OR groups are protonated in a first step and the alcohol or water is eliminated from the intermediate, after a water or a silanol group has attacked the intermediate (Eq. (7.2)). Under base catalysis, first a OH^- or $Si{-}O^-$ ion reacts under nucleophilic attack with the silicon atom forming a five-membered transition state like in S_N2 reactions and a RO^- ion is released.

In both mechanisms the electron density at the silicon atom is crucial. It changes in the following order: $\equiv Si{-}R' > \equiv Si{-}OR > \equiv Si{-}OH > \equiv Si{-}O{-}Si\equiv$. For acid catalysis, the electron density at the silicon atom should be high because the protonated species is then stabilized and hydrolysis and condensation is accelerated. Under basic conditions it is *vice versa*, less electron density at the silicon center is preferred for a nucleophilic attack.

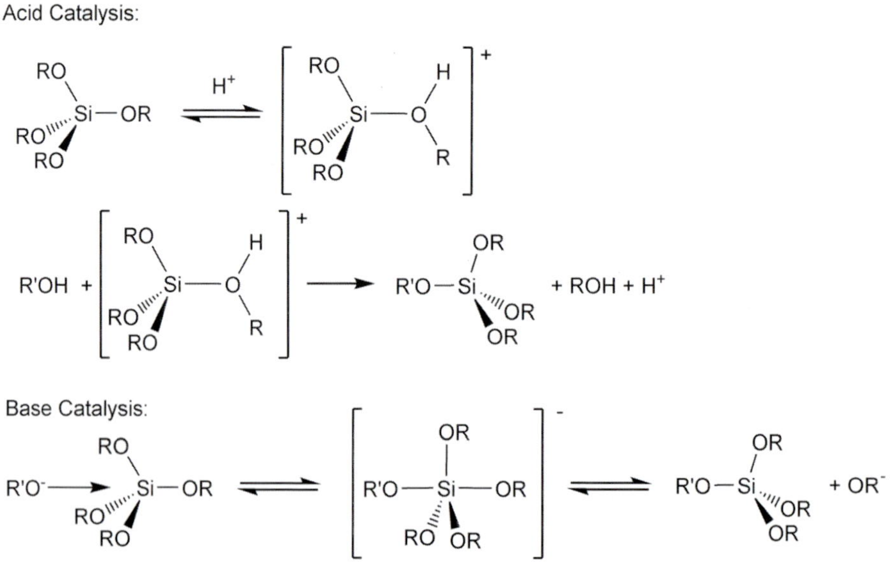

(7.2)

R' = H or $(RO)_{3-x}R''_xSi$

The mechanistic differences of the catalytic processes lead to consequences in the hydrolysis and condensation depending on the catalytic conditions and the kind of substitution at the silicon atom. Many of the principal differences of the processes were studied by ^{29}Si NMR, which is a powerful tool to determine the initial steps as well as the final materials with respect to their composition and crosslinking density [29–32]. The two most profound differences are the morphology of the network and the reactivity differences in the case when mixed precursors are used. Under basic conditions, more branched networks are obtained and therefore particle like structures are preferred. Contrary, under acid conditions more chain-like polymeric structures are formed. Organically functionalized precursors $R'Si(OR)_3$ react faster under acid conditions and slower under basic conditions compared to the tetraalkoxysilanes $Si(OR)_4$. In the beginning of the condensation process many intermediates are formed, such as oligomers, rings, and chains, which are transformed in the three-dimensional structures. A rough scheme of the morphology changes in the silicon-based sol–gel process depending on the pH is presented in Figure 7.1.

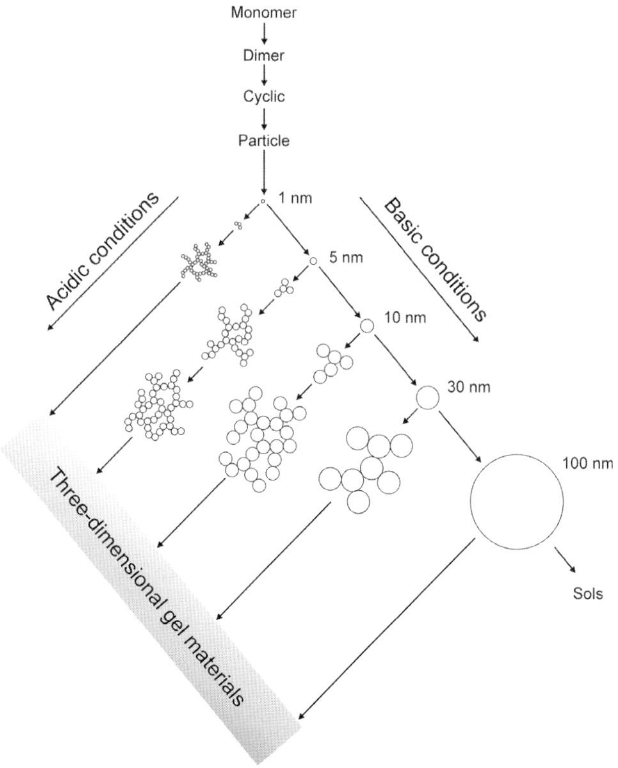

Figure 7.1 Evolution of the materials morphology in silica-based sol–gel processes depending on the pH value. (Adapted from Ref. [33].)

The Si—C bond in organically substituted alkoxysilanes, for example, R′-Si(OR)$_3$, is hydrolytically stable. Therefore, if such compounds are used as precursors in the sol–gel process they allow the incorporation of organic groups in the network. Depending on the number of organic substituents in the precursors $R'_xSi(OR)_{4-x}$ ($x = 1$–3), which are directly related to the number of alkoxide groups, different materials are obtained by hydrolysis and condensation reactions of such precursors alone. Monoalkoxysilanes $R'_3Si(OR)$ form dimers, dialkoxysilanes $R'_2Si(OR)_2$ form linear polymers, and trialkoxysilanes $R'Si(OR)_3$ result in the formation of silsesquioxanes with the general formula $R'SiO_{1.5}$. Silsesquioxanes are a unique class of materials, they exist as amorphous three-dimensional structures, ladder like molecules or discrete molecules, known as polyhedral oligomeric silsesquioxanes (POSS) [34]. Tetraalkoxysilanes form three-dimensional networks and represent the most general case of a silicon-based sol–gel process where a purely inorganic network is formed. If organically functionalized alkoxysilanes are used in mixtures with tetraalkoxysilanes, the Si—O—Si network is tailored as well as the organic content of the final hybrid material. The incorporation of organic moieties in the sol–gel derived material leads to the following alterations:

- the degree of crosslinking of the inorganic network is reduced,
- polarity changes are induced,
- the reactivity of the remaining alkoxide groups can be changed.

Overall, many parameters influence the hydrolysis and condensation reactions of the precursors and thus the formation of the final product. The major parameters are as follows:

- the type of precursor,
- the catalytic conditions (e.g., pH),
- the ratio between water and alkoxide groups,
- the temperature,
- the solvent,
- the presence of ions,
- the concentrations of the different compounds in the reacting solution.

Applying metal alkoxides instead of alkoxysilanes leads to metal oxides as final materials [3]. Metal alkoxides are much more reactive than alkoxysilanes in the hydrolytic sol–gel process thus usually no catalyst is required and the reaction often needs more control, which will be explained in the precursor chapter.

7.2.2 Nonhydrolytic Sol–Gel Process

Contrary to the conventional hydrolytic sol–gel processes, in nonhydrolytic (sometimes also called nonaqueous) sol–gel processes the source of the oxygen required to form the three-dimensional network is not water but another reagent, which can be a metal precursor itself or an organic molecule [8,35–37]. Two different mechanisms can be distinguished: (i) nonhydrolytic hydroxylation reactions or (ii) aprotic condensation reactions. In the first reaction type, the hydroxyl groups are formed by internal elimination reactions in the metal alkoxides or by reaction of an alcohol with a metal halide. Usually, nonhydrolytic sol–gel processes require higher temperatures (20 °C–120 °C) for the condensation reaction to occur and sometimes catalysts such as Lewis acids are added (Eq. (7.3)). The kinetics of this process is often slower than the conventional aqueous process. In case of the aprotic condensation reactions the process has to be carried out in aprotic solvents and the ratio between the different metal species has to be controlled very precisely to obtain fully condensed products. The latter method can be particularly used for the preparation of nanoparticles [8].

$$\equiv M{-}OR + X{-}M\equiv \longrightarrow \equiv M{-}O{-}M\equiv + R{-}X$$

$$X = Cl, Br, I, \text{ Kat.: Lewis Acid: } FeCl_3, AlCl_3, ZrCl_4, \text{ etc.}$$

$$\equiv M{-}OR + RO{-}M\equiv \longrightarrow \equiv M{-}O{-}M\equiv + R{-}O{-}R \qquad (7.3)$$

$$\equiv M{-}O{-}C({=}O)R' + RO{-}M\equiv \longrightarrow \equiv M{-}O{-}M\equiv + R{-}O{-}C({=}O)R'$$

Aprotic condensation reactions are usually carried out in aprotic media and the oxygen atom for the condensation reaction is provided by the metal precursors themselves eliminating a small molecule such as an organohalide, an ether, or an ester. Generally, compared to the hydrolytic sol–gel process in the nonhydrolytic process reactivity differences between the various precursors are less important. A plethora of different metal oxides, mixed oxides, as well as hybrid materials can be produced with this method [36–38]. The latter class of materials is produced if organoalkoxysilanes are mixed with another precursor for the nonhydrolytic sol–gel process.

Although the sol–gel method is mainly used for the synthesis of oxidic materials, in principle the underlying chemical principles can also be applied to other materials. Particularly, the nonhydrolytic sol–gel approach was also used for metal sulfides or nitrides [39–46]. In these approaches sulfur- or nitrogen-containing species were used as nucleophiles. In recent years, nanoscale metal fluorides were also synthesized with a so-called fluoridic sol–gel process that is based on the reaction between metal alkoxides and anhydrous HF [47].

7.2.3 Aging and Drying

Although it may seem that after gelation of a bulk sol–gel material the reaction stops, this is actually not the case. At the sol–gel transition a rapid increase of the viscosity by several orders of magnitude can be observed [48–51]. But this does not mean that reactions stop in the gel state. Unreacted precursors that are still in the liquid and M-OH and M-OR groups on the surface of the sol particles can react with each other. Therefore, the gel ages with time, which is accompanied by a shrinking process. This has to be considered particularly in the case when bulk materials are prepared. Aging of the gel cannot be avoided but it might be accelerated if higher temperatures are applied during the sol–gel process. Nonhydrolytic sol–gel routes therefore often produce much higher condensed materials, because the increased temperatures which are usually required speed up the condensation of unreacted species [38,52–54].

Conventional drying of a gel can be achieved by increasing the temperature or decreasing the pressure. During this process usually the gel shrinks, cracks are formed, and finally the network collapses and a so-called xerogel powder is formed. This happens particularly with bulk samples. The drying process can be much better controlled in the case of thin films. Therefore, coating technology is one of the most promising areas where sol–gel materials are applied.

There are possibilities to avoid the collapse of the gel during the drying process, for example, by exchanging the solvent in the pores of the gel against a supercritical medium. The final materials consist only of the solid content of the former gel and the liquid, which was originally in the pores is removed. These so-called aerogels are lightweight structures with very high specific surfaces and extremely low thermal conductivities [55–60].

7.3 Precursors

The sol–gel process represents a transformation of a molecular precursor to a three-dimensional solid. In the majority of the cases, metal or semimetal alkoxides are used as precursors. The advantage of the molecular precursor-based approach is that beside the reaction parameters (temperature, solvent, pH, etc.) also the defined molecular structure of the precursor has an influence on the reactivity, for example, the organic substituents or ligands. In addition, the molecular compounds can be obtained in high purity, which makes the sol–gel reaction interesting for technologies like opto- and microelectronics. The kind of metal precursor or mixture of different precursors decides the composition of the final metal oxide. In other molecular precursor routes, such as thermal decomposition processes, the structure of ligands stabilizing the precursors do not play an important role because they decompose during the process. This is different in the sol–gel process where the substituents show a major influence on the final products. They can regulate the homogeneity of the final material through compatibilization within the reaction medium, they can control the kinetics of the whole process, and they can introduce additional functions in the final material.

7.3.1
Silicon-Based Precursors

Tetraalkoxysilanes are broadly used as precursors for sol–gel reactions. As mentioned in Section 7.2, the number of alkoxides substituents as well as the type of alkoxides has a general influence on the reaction rate in the sol–gel process. Generally, the reaction rate decreases in the following row: $Si(OMe)_4 > Si(OEt)_4 > Si(O^nPr)_4 > Si(O^iPr)_4$. This means that sterically more hindered alkoxides react slower in the sol–gel process than less bulky rests. Substituents that are bound via the hydrolytically stable Si—C bond have also an influence on the hydrolysis and condensation rate depending on the catalytic conditions. The most prominent precursors for the silica network formation are tetramethoxysilane (TMOS) and tetraethoxysilane (TEOS). Organically functionalized precursors can be used to modify the silica network, for example, to turn it more hydrophobic. In most cases the organic groups incorporate additional functions in the final sol–gel material, such as polymerizable groups or reactivity centers for further modification reactions. Such organic moieties are particularly required in the formation of inorganic–organic hybrid materials [21]. A random selection of different organically functionalized precursors is shown in Figure 7.2. These precursors are usually mixed with TMOS or TEOS and the sol–gel process is started. After gelation and aging a hybrid material is obtained which contains the organic functions. If polymerizable groups are incorporated, they can be additionally crosslinked by thermal- or UV-curing [61–65].

The functionalization of sol–gel materials can be also achieved by post processing methods. Here, mainly organically functionalized molecules of the type $R''SiR'_2X$ are used, where X is either an alkoxide or a chlorine. These molecules cannot form networks because they have only one group that can be hydrolyzed but they can modify the surface of the silica network. Typical agents for the surface hydrophobization of hydrophilic silica materials are trimethylsilyl chloride (Me_3SiCl) or hexamethyldisilazene ($Me_3SiNHSiMe_3$) [66–68].

Many of the trialkoxysilane precursors are commercially available, but they also can be easily synthesized by hydrosilylation reactions. For this approach, the organic molecule requires an unsaturated bond that should be ideally electron rich; terminal alkenes are ideal for this type of reaction. The trimethoxy- or

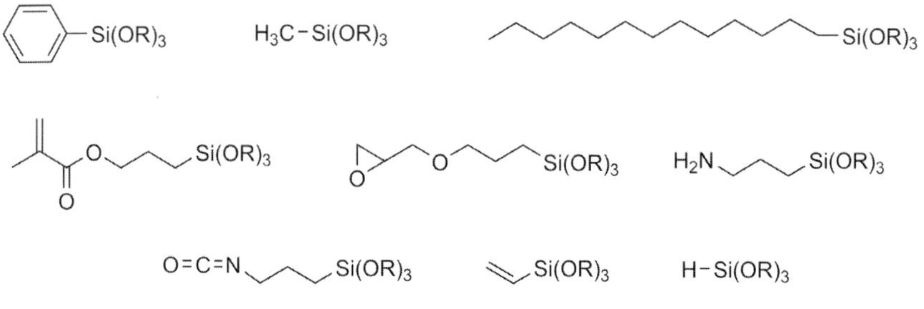

Figure 7.2 A selection of organically functionalized trialkoxysilanes.

triethoxysilane are reacted with the unsaturated bond by platinum-catalyzed hydrosilylation (Eq. (7.4a)). Alternatively, 3-isocyanatopropyltrialkoxysilane can be used for a trialkoxysilane functionalization if it is added to a nucleophilic group of an organic molecule such as an alcohol or an amine (Eq. (7.4b)).

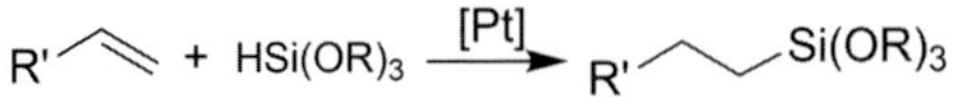

(7.4a)

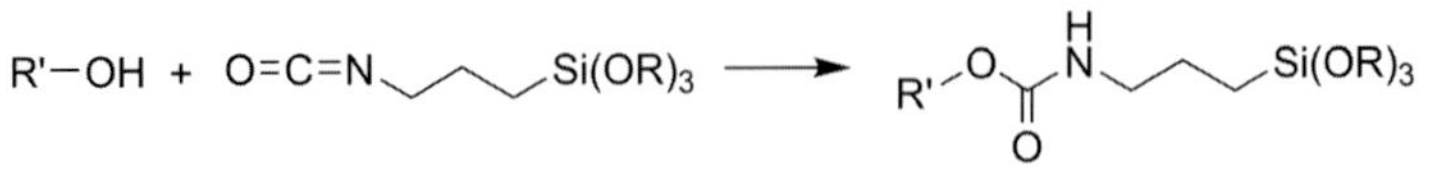

(7.4b)

Beside monofunctionalized precursors, also molecules with more than one trialkoxysilane functionality can be applied in the sol–gel process [69]. In the resulting materials the organic functionality becomes part of the network by an integration in the network structure. Examples of such molecules are shown in Figure 7.3. Due to the fact that two trialkoxysilanes are attached to an organic backbone and thus each of these molecules contains six potential crosslinking positions, the kinetic of the sol–gel reaction is usually significantly increased with these precursors, which is expressed in much faster gelation times compared to monofunctional precursors. In combination with templates periodic mesoporous organosilicas (PMOs) are obtained, which have recently attracted much interest as a class of porous hybrid materials with the opportunity to incorporate organic functions in the regular pores [70–74].

7.3.2
Metal Precursors

Nearly every metal oxide material can be prepared by the sol–gel process if the metal precursors are available and the reaction conditions can be tailored in such a way that the reaction kinetics can be controlled. In most cases again metal alkoxides are used as precursors, but there are also other methodologies how the active M-OH species can be formed to start the condensation reaction. Compared to silicon as precursors, metals show some inherent differences in their reactivity. First, they are more electropositive (Lewis acid character) than silicon; second, their coordination number is usually higher than their valence. Therefore, their reactivity with regard to the attack of nucleophiles, such as

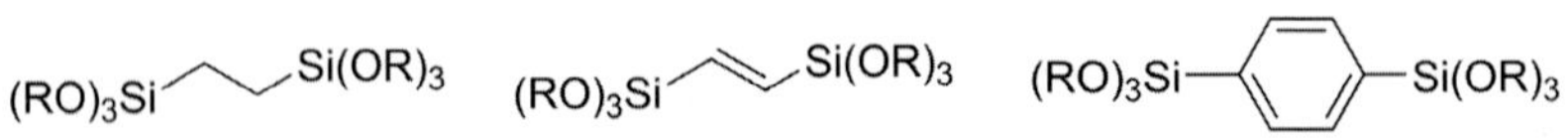

Figure 7.3 Difunctionalized organic molecules with two trialkoxysilane groups that can be used for the formation of hybrid materials.

water, is dramatically increased compared to their alkoxysilane counterparts. A comparison of reaction rates in the hydrolytic sol–gel process of silicon tetra-isopropoxide compared to metal iso-propoxides reveals the accelerated reactivity differences: $Si(O^iPr)_4$ <<< $Sn(O^iPr)_4$, $Ti(O^iPr)_4 < Zr(O^iPr)_4 < Ce(O^iPr)_4$. Therefore, reaction rate control is a major point in the sol–gel reaction of metal alkoxide-based sol–gel process, particularly if different precursors with various reactivity rates should form a homogeneous material in the end. If the reaction rates do not fit each other, a loss of microstructural control over the final oxide material is often the case, which means that homogeneity is lost due to phase separation. One solution to this problem is the use of organic additives that act as chelating ligands (e.g., carboxylic acids, β-diketones, etc.) and allow to modify the reactivity of the precursors [75–77]. Typically, these bidentate ligands substitute partially the alkoxide groups (Eq. (7.5)) and thus limit the reactivity of the precursors mainly by blocking reaction sites at the metal center. These additives can be compared to organic substituents in the case of silicon-based materials, they also reduce the degree of condensation in the material and they allow the incorporation of organic functions in the formed inorganic network.

$$M(OR)_n + BLH \longrightarrow M(OR)_{n-1}BL + HOR \qquad (7.5)$$

BLH =

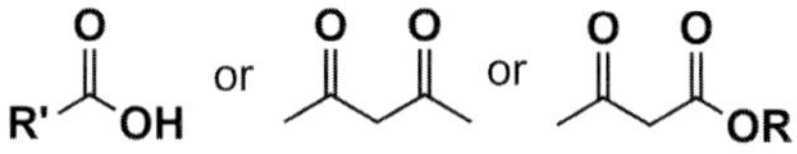

Another method, if mixed metal oxides are required and the reaction rates of the components do not fit is the use of single source precursors [78–81]. These compounds consist of a mixture of the metals in the targeted stoichiometry of the metals in the metal oxides. Usually the ligands are chosen in such a way that a decomposition during the sol–gel process or a subsequent thermal treatment results in a low contamination of the final solids by other than the targeted materials.

Beside the typically used metal alkoxides also metal salts can form the reactive M-OH species, which is necessary for the condensation reaction. Metal ions usually form aqua complexes in solution. Under accurate control of the pH conditions the formation of M-OH groups occurs depending on the oxidation state of the metal ions and thus condensation can occur. Sometimes this method is called in literature the forced hydrolysis approach [82]. This technique is particularly useful for the preparation of nanophase materials.

7.4 Materials Based on Sol–Gel Chemistry

The sol–gel process grants access to a large number of different oxide materials. Recently, other materials like sulfides, fluorides, or nitrides are also targeted but still the research and applications are mainly focused on oxides. Due to the broadly controllable process parameters (precursors, temperature, solvent, additives), many morphologies and chemical compositions are feasible. Particularly, the possibility in processing the derived materials in form of bulk materials, thin films, porous materials, hybrids, and so on leads to a plethora of applications which are partially already commercialized [83,84].

7.4.1
Bulk Materials

The sol–gel process can be seen as an inorganic polycondensation reaction which ends in a solid material. As such it should in principle be possible to form a bulk material that can be used directly in an application. Looking closer the difficulties in forming bulk materials become obvious. The most apparent challenge is the shrinkage of the material after gelation during aging and drying, which is accompanied by crack formation and complete collapse of the material. Realizing these limits there are essentially three possibilities to obtain bulk samples from sol–gel materials:

1) The liquid in the pores is extracted either in a supercritical medium or by surface functionalization of the three-dimensional network with hydrophobic species and a subsequent release of the hydrophilic liquid from the material. With this process, highly porous bulk materials, so called aerogels, are obtained that consist of the solid content of the former gel [58].
2) The sol–gel material is usually aged and dried, which in many cases leads to a powder, the so-called xerogel [85]. Applying heat treatment like in the ceramic process the powder can be transferred to a solid. Of course, this technique would abolish many of the advantages of the sol–gel process, because a high temperature step is required. Despite this disadvantage, the sol–gel route is the method of choice if specific compositions of the final material and/or high purity materials are required.
3) In hybrid materials, the shrinkage and cracking can be controlled much better, because the organic moieties provide some elasticity. Another approach to obtain low shrinkage materials is the use of the released alcohols as organic moiety in the hybrid material. For this purpose, the alkoxides should contain polymerizable groups that are polymerized during the inorganic sol–gel process. The resulting bulk materials are homogenous and shrink to a comparable low extend if the reaction conditions and the precursors are chosen well (Figure 7.4) [86]. During the sol–gel process, the alkoxysilane derivatives release four equivalents of polymerizable alcohol that are transformed to an organic polymer in presence of the appropriate initiator or catalyst. Usually free monomer is present as a cosolvent in such systems. Since both the cosolvent and the released alcohol polymerize, no evaporation is necessary and large-scale shrinkages are avoided. The concept proved useful for free radical polymerizations and ring opening metathesis polymerization. Recently, this concept was extended to polycondensation reactions, the so-called twin polymerizations (Figure 7.4c) [87].

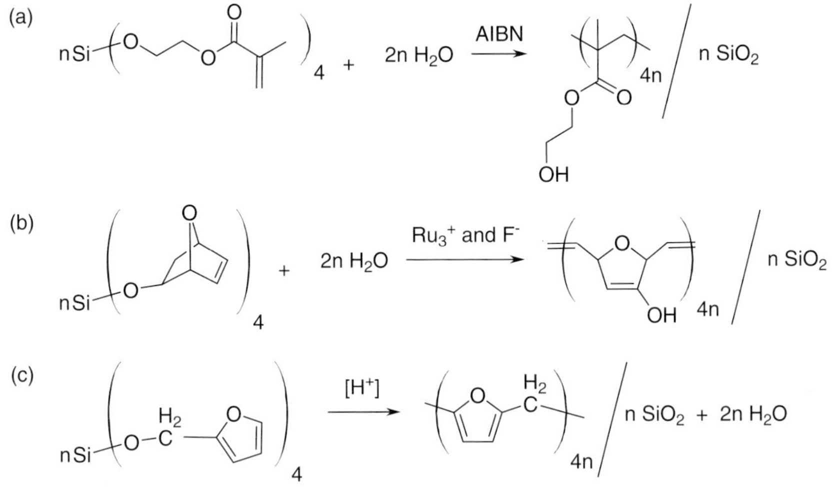

Figure 7.4 Silicon alkoxides precursors for the formation of low shrinkage bulk sol–gel materials. The precursors are divided by the type of polymerization (a) free radical, (b) ring opening metathesis, and (c) polycondensation.

7.4.2
Thin Films and Coatings

Many of the disadvantages of bulk materials formation can be overcome if thin films or coatings are produced. Depending on the thickness, processing, and adhesion of the film to the support, shrinkage and cracking can be much better controlled. Therefore, the formation of sol–gel films is probably the most widely used processing method for metal oxide sol–gel materials. It is important to mention that particularly the high optical transparency of the films produced by the sol–gel process make them very interesting for various high-tech applications in areas like scratch resistant, antireflective or electrical conductive coatings. Thin film preparation via the sol–gel process involves preparation of the sol, deposition of the sol on a support, formation of a gel, and drying of the gel. Depending on the sol parameters and deposition conditions, the structure of the gel and the drying conditions may vary. All three stages are interconnected with each other and determine the final structure of the film. The coating can be applied via dip-, spin-, or spray coating technology. Many coating applications of sol–gel materials are based on hybrid mixtures of silica and/or metal oxides and organically modified alkoxysilanes as shown in Section 7.3.1.

7.4.3
Nanoparticles

With the constant growing possibilities for applications of nanoparticles, their controlled synthesis gets more and more important. One of the major synthetic pathways for the production of oxidic nanoparticles is the sol–gel process. A milestone in this field was the preparation of silica nanoparticles from tetraalkoxysilanes in a base-catalyzed reaction. The method was named after the scientist who discovered it, Stöber. It produces quite easily silica spheres with a narrow particle size distribution from the size of a couple of nanometers up to microns, mainly controlled by the availability of the precursors in the solution [88,89].

The size of the prepared particles depends on the type of silicon alkoxides, the ratio between the various components in the sol–gel synthesis, and the alcohol used as solvent. Since the reactions are carried out at high $pH > 7$, the formed particles are stable toward gelation because of the repulsion effects from the negatively charged particle surface. The process was used and modified in the last decades for the production of many types of silica and mixed oxide particles. The obtained colloids are also interesting because they can be easily surface-functionalized with organic functions and polymers. Because of the high control over their morphology and composition silica colloids acted as model systems for many other nanoparticles. More recent approaches use the concept of preparation of silica particles by a sol–gel approach for the production of mesoporous SiO_2 nanoparticles which can be used as adsorbents, in catalysis, as polymer fillers, and in biomedical applications such as bioimaging and drug delivery [90]. Additionally, the methodology can also be used for the preparation of core–shell structures either with a silica core and metal shells or *vice versa* [91].

Besides silica, many other metal oxide particles can be produced via hydrolysis and condensation either starting from their halogen compounds, such as $TiCl_4$, $ZrOCl_2$, or using metal alkoxides of the type $M(OR)_n$. While the silicon-based sol–gel process usually results in amorphous silica as a product, the metal-based sol–gel process often shows crystalline products even at lower temperatures. The metal oxide particles are often peptized, for example, with HNO_3 in the case of titania production, to electrostatically stabilize their surface and get a more homogeneous and more stable dispersion.

A large variety of metal oxide nanoparticles were produced in the last couple of years applying nonhydrolytic sol–gel reactions [8]. One of the most used methods is the so-called "benzyl alcohol route," in which various metal oxide precursors are reacted with benzyl alcohol as the oxygen donor for the network formation. Generally, this process involves a precursor, which is either an inorganic metal salt (e.g., metal halides) or a metal compound such as metal alkoxides, acetates, or acetylacetonates. The reactions require increased temperatures in a range from 50 to 250 °C and work either in autoclaves or under microwave-assisted conditions. A broad variety of mainly crystalline nanoparticles with a binary or ternary composition were obtained. Examples of metal oxides produced with the benzyl alcohol route are SnO_2, $InNbO_4$, TiO_2, CeO_2, and Al:ZnO to mention a selection. The formed metal oxide nanoparticles show often very small diameters of a few nanometers. They form different shapes including spheres, platelets, and rods. In most cases the particles are crystalline.

Beside the formation of metal oxides there were also efforts for the production of nonoxidic nanoparticles. Only a few examples were reported in literature, for example, the formation of sulfides and nitrides was achieved using a so-called thio sol–gel (TSG) and a hydrazide sol–gel (HSG) route, respectively [92–95]. In addition, recently also metal fluorides were produced by a route, which is similar to a sol–gel reaction.

The formation and crystallization mechanisms of the obtained particles seem to be complex and are still under investigation. These complex mechanisms might be the reason for the large variety of sizes and shapes of the obtained nanoparticles.

7.4.4 Porous Materials

Porosity is a major property of materials, which is highly important for applications in fields where interface phenomena play an important role, such as catalysis or absorbance. The sol–gel approach is the enabling technology for the preparation of so-called mesoporous materials consisting of a solid network structure and highly regular porosities in the size of a few nanometers in diameter [96,97]. For the preparation of such materials a template is necessary, which can be removed after the inorganic network is formed [98]. In the case of mesoporous materials, the template for the pores can be soft organic matter in form of amphiphilic surfactants that organize into micelles and more complicated three-dimensional arrangements such as hexagonal arranged rod-like structure, cubic interpenetrating networks, or lamellar structures (Figure 7.5). The latter are usually not used for the structural engineering, because the inorganic layers formed by this templating process usually collapse after removal of the template. At the interface between the hydrophobic and the hydrophilic regions of the formed structures, the sol–gel process is carried out and thus the structure of the obtained phases is imprinted into the solid sol–gel derived material (Figure 7.6). As long as the organic surfactants are still incorporated in the formed inorganic matrix, the materials can be considered as inorganic–organic hybrid materials. After removal of the template, which commonly occurs via calcinations, the materials are purely inorganic in nature. The pure silicate or metal oxide walls of the mesoporous materials can also be substituted by hybrid

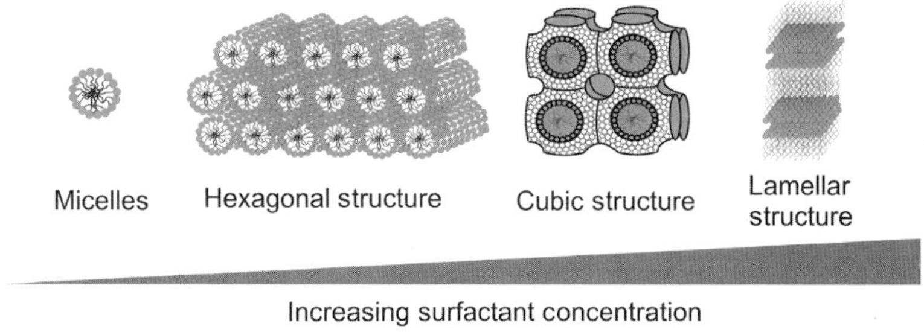

Figure 7.5 Examples of ordered 3D structures obtained by supramolecular self-assembly of surfactant molecules. (Adapted from Ref. [21].)

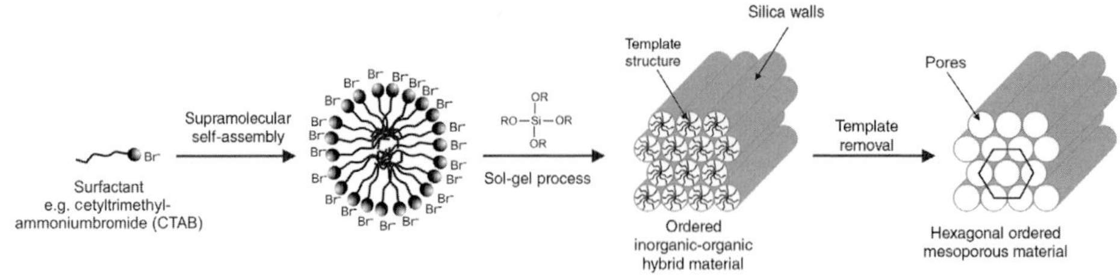

Figure 7.6 Formation of well-ordered mesoporous materials by a templating approach. (Adapted from Ref. [21].)

materials formed by bridged silsesquioxanes and so-called periodically mesoporous organosilicas (PMOs) are built [73].

Another templating process of sol–gel materials is the use of organic colloidal particles with narrow size distributions, which are able to order in three-dimensional objects so-called colloidal crystals [99–101]. If organic colloids such as latex spheres are used, the voids in the colloidal crystal can be infiltrated by the metal or semimetal alkoxides. After formation of the inorganic material by the sol–gel reaction, a hybrid material is obtained that incorporates the organic spheres. The organic template is usually removed after this process and a porous material is formed as a replica of the colloidal crystal (Figure 7.7).

7.4.5
Hybrid Materials

Materials that consist of both an inorganic and an organic or biological part are referred to as hybrid materials [21]. As mentioned in Section 7.1, there are two possibilities, the organic or biological species is incorporated in the material only by physical entrapment that generates class I type hybrids or the organic species is connected to the inorganic network by covalent bonds and forms a class II type material. Both have found a widespread use in modern applications. By the synergetic combination of two different compounds, a new material is formed that benefits from both sources. With this approach, organic materials that have a good processability can be combined with inorganic materials with a high thermal stability, refractive index, conductivity, and so on. The final compounds are

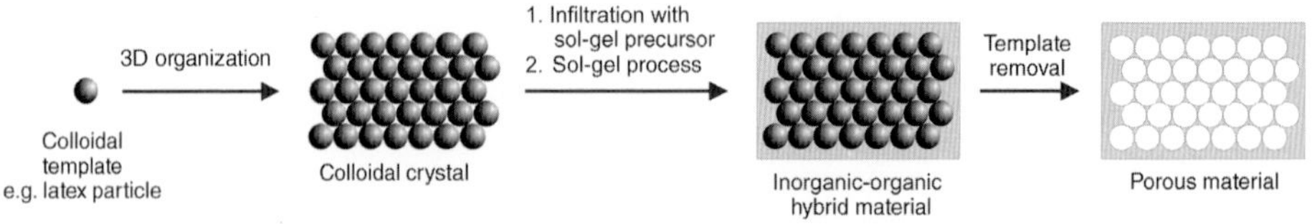

Figure 7.7 Colloidal crystal templating. (Adapted from Ref. [21].)

used in a plethora of different applications. Particularly, hybrids with polymerizable bonds included are often combined with polymer matrices or polymer supports [27].

Inorganic-biological hybrid materials are realized by the entrapment of enzymes and proteins in porous silica matrices obtained from the sol–gel process [102–104]. This approach allows a protection of the bioactive molecules from environmental influences and thus the incorporation of these sensitive biological materials in various devices. In the preparation of such materials, it is often necessary to control carefully the reaction condition such as the pH because of the sensitivity of biomolecules. It was shown that the biological active molecules are often astonishing stable in the inorganic matrix. The porous nature of the sol–gel matrix still has the ability to transport the substrates to the active center and thus the enzymes still have a high activity. The derived sol–gel systems are, for example, used as catalysts in sensors or screening devices [105–107].

Another topic in the field of hybrid materials which is based on the combination of inorganic and organic species as well as the porous nature of the, particularly, silica sol–gel derived materials is microencapsulation [108]. The technology allows the incorporation of molecules in capsules and therefore their stabilization. The method is particularly useful for the purpose of protection and controlled release in biomedical applications. The process is based on the enclosing of micrometer-sized particles of solids or droplets of liquids or gases in an inert shell. The latter in turn isolates and protects the ingredients from the external environment. The sol–gel approach allows the templated microparticle formation where the template is a lyotopic mesophase, a foam, or an emulsion. The method allows for the incorporation of hydrophobic molecules such as essential oils, flavors, vitamins, proteins (including enzymes), and many other biomolecules in a silica matrix.

7.5 Applications

The variety of accessible substances and material classes leads to a vast diversity of possible applications. We report here only a few examples and not every field of application will be covered.

The modification of surfaces affects directly their properties and is therefore of great importance in the production process. Commonly used methods for deposition of thin films are, for example, sputtering or vapor deposition that needs relatively high cost infrastructure and complex procedures. Compared to that, the sol–gel method is rather simple and cheap. First, a sol has to be prepared which can be applied on the desired surface, where it forms a gel that is dried afterwards. Applying this method, various applications can be mentioned. Pure silica films are mainly interesting for low-κ materials for microelectronic applications, particularly if a porous structure can be achieved [109]. Other examples for thin film applications are scratch and abrasion resistant coatings. These systems are mainly based on hybrid materials mixtures and are often used to protect polymeric materials such as polycarbonates, polyacrylates, polyethylene, and many others [110–113]. Particularly, the optical transparency is important for many of these applications. If it comes to optical transparent materials, another beneficial property is antireflectivity. This property is mainly influenced by three parameters: the refractive index, the thickness of the used (multi)layers, and the surface structure of the film. All three parameters can be controlled in the sol–gel process, for example, alternating layers of low (SiO_2) and high refractive index (transition metal oxides) materials can be deposited on a support that leads to antireflective properties of a coating. If, in addition, sol–gel particles are included in the film, a surface structure is achieved [114–117]. Such coatings are very important in solar panel production to increase the efficiency of these devices. Another possibility is the use of sol–gel produced titania in its anatase modification as photocatalytic active coatings in self-cleaning applications [118,119]. Due to the modular character of the sol–gel process also multifunctional coatings can be prepared which merge several before mentioned functionalities in one material [120,121].

Other typical examples for the application of sol–gel coatings are in the field of corrosion protection or ferroelectric films [122–127].

The fabrication of light-emitting diodes (LEDs) is one of fields where sol–gel materials also became quite popular, particularly in the electronic transparent packaging. All these devices need an encapsulation to protect the diode from influences of the environment. State of the art is the use of polysiloxanes as encapsulants. An improvement in thermal stability and refractive index of the nowadays used simple polysiloxanes is achieved if sol–gel-based hybrids are used [128,129].

Many other applications can be mentioned where the sol–gel methods open new pathways for materials use and the interested reader may refer to the literature and patents in this field.

7.6 Conclusion

The sol–gel process as a bottom-up solution based methodology to prepare three-dimensional networks from molecular precursors represents an important reaction with a broad applicability for many different materials. The exceptional number of parameters that can be changed and the rather mild processing conditions allow particularly the formation of different morphologies that can be obtained, starting from nanoparticles, over thin films to bulk materials. Porosity can be generated over several length scales and the elemental composition can be varied like in no other process. This makes the sol–gel technique to a unique tool for the preparation of many high-tech materials.

References

1 Brinker, C. and Scherer, G. (1990) *Sol-Gel Science: The Physics and Chemistry of Sol-Gel Processing*, Academic Press.

2 Hench, L.L. and West, J.K. (1990) *Chem. Rev.*, **90**, 33–72.

3 Livage, J., Henry, M., and Sanchez, C. (1988) *Prog. Solid State Chem.*, **18**, 259–341.

4 Sanchez, C. and Ribot, F. (1994) *New J. Chem.*, **18**, 1007–1047.

5 Wen, J. and Wilkes, G.L. (1996) *Chem. Mater.*, **8**, 1667–1681.

6 Sanchez, C., Rozes, L., Ribot, F., Laberty-Robert, C., Grosso, D., Sassoye, C., Boissiere, C., and Nicole, L. (2010) *C. R. Chim.*, **13**, 3–39.

7 Cushing, B.L., Kolesnichenko, V.L., and O'Connor, C.J. (2004) *Chem. Rev.*, **104**, 3893–3946.

8 Niederberger, M. (2007) *Acc. Chem. Res.*, **40**, 793–800.

9 Imai, H., Oaki, Y., and Kotachi, A. (2006) *Bull. Chem. Soc. Jpn.*, **79**, 1834–1851.

10 Nakanishi, K., Takahashi, R., Nagakane, T., Kitayama, K., Koheiya, N., Shikata, H., and Soga, N. (2000) *J. Sol-Gel Sci. Technol.*, **17**, 191–210.

11 Seraji, S., Wu, Y., Jewell-Larson, N.E., Forbess, M.J., Limmer, S.J., Chou, T.P., and Cao, G. (2000) *Adv. Mater.*, **12**, 1421–1424.

12 Yu, M., Lin, J., Wang, Z., Fu, J., Wang, S., Zhang, H.J., and Han, Y.C. (2002) *Chem. Mater.*, **14**, 2224–2231.

13 Orsi, G., De Maria, C., Montemurro, F., Chauhan, V.M., Aylott, J.W., and Vozzi, G. (2015) *Curr. Top. Med. Chem.*, **15**, 271–278.

14 Pan, Z., Wang, Y., Huang, H., Ling, Z., Dai, Y., and Ke, S. (2015) *Ceram. Int.*, **41**, 12515–12528.

15 Costa, C., Pinheiro, C., Henriques, I., and Laia, C.A.T. (2012) *ACS Appl. Mater. Interfaces*, **4**, 5266–5275.

16 Hossain, S.M.Z., Luckham, R.E., Smith, A.M., Lebert, J.M., Davies, L.M., Pelton, R.H., Filipe, C.D.M., and Brennan, J.D. (2009) *Anal. Chem.*, **81**, 5474–5483.

17 Kim, A., Song, K., Kim, Y., and Moon, J. (2011) *ACS Appl. Mater. Interfaces*, **3**, 4525–4530.

18 Wang, J., Bowie, D., Zhang, X., Filipe, C., Pelton, R., and Brennan, J.D. (2014) *Chem. Mater.*, **26**, 1941–1947.

19 Yun, H.-s., Kim, S.-e., and Hyeon, Y.-t. (2007) *Chem. Commun.*, 2139–2141.

20 Comez-Romero, P and Sanchez, C. (eds) (2004) *Functional Hybrid Materials*, Wiley-VCH Verlag GmbH.

21 Kickelbick, G. (ed.) (2007) *Hybrid Materials: Synthesis, Characterization, and Applications*, Wiley-VCH Verlag GmbH.

22 Belleville, P. (2010) *C. R. Chim.*, **13**, 97–105.

23 Gvishi, R., Narang, U., Ruland, G., Kumar, D.N., and Prasad, P.N. (1997) *Appl. Organomet. Chem.*, **11**, 107–127.

24 Lebeau, B. and Innocenzi, P. (2011) *Chem. Soc. Rev.*, **40**, 886–906.

25 Ovsianikov, A., Viertl, J., Chichkov, B., Oubaha, M., MacCraith, B., Sakellari, I., Giakoumaki, A., Gray, D., Vamvakaki, M., Farsari, M., and Fotakis, C. (2008) *ACS Nano*, **2**, 2257–2262.

26 Tran-Thi, T.-H., Dagnelie, R., Crunaire, S., and Nicole, L. (2011) *Chem. Soc. Rev.*, **40**, 621–639.

27 Kickelbick, G. (2002) *Prog. Polym. Sci.*, **28**, 83–114.

28 Brinker, C.J. (1988) *J. Non-Cryst. Solids*, **100**, 31–50.

29 Assink, R.A. and Kay, B.D. (1991) *Annu. Rev. Mater. Sci.*, **21**, 491–513.

30 Komori, Y., Nakashima, H., Hayashi, S., and Sugahara, Y. (2005) *J. Non-Cryst. Solids*, **351**, 97–103.

31 Dirken, P.J., Smith, M.E., and Whitfield, H.J. (1995) *J. Phys. Chem.*, **99**, 395–401.

32 Glaser, R.H., Wilkes, G.L., and Bronnimann, C.E. (1989) *J. Non-Cryst. Solids*, **113**, 73–87.

33 Iler, R.K. (1979) *The Chemistry of Silica: Solubility, Polymerization, Colloid and Surface Properties and Biochemistry*, John Wiley and Son, Ltd.

34 Kickelbick, G. (2014) *Struct. Bond.*, **155**, 1–28.

35 Vioux, P.A. (1997) *Chem. Mater.*, **9**, 2292–2299.

36 Hay, J.N. and Raval, H.M. (2001) *Chem. Mater.*, **13**, 3396–3403.

37 Debecker, D.P. and Mutin, P.H. (2012) *Chem. Soc. Rev.*, **41**, 3624–3650.
38 Andrianainarivelo, M., Corriu, R., Leclercq, D., Mutin, P.H., and Vioux, A. (1996) *J. Mater. Chem.*, **6**, 1665–1671.
39 Mutin, P.H. and Vioux, A. (2013) *J. Mater. Chem. A*, **1**, 11504–11512.
40 Zhou, X., Heinrich, C.P., Kluenker, M., Dolique, S., Mull, D.L., and Lind, C. (2014) *J. Sol-Gel Sci. Technol.*, **69**, 596–604.
41 Zhou, X., Soldat, A.C., and Lind, C. (2014) *RSC Adv.*, **4**, 717–726.
42 Cheng, F., Archibald, S.J., Clark, S., Toury, B., Kelly, S.M., Bradley, J.S., and Lefebvre, F. (2003) *Chem. Mater.*, **15**, 4651–4657.
43 Cheng, F., Clark, S., Kelly, S.M., Bradley, J.S., and Lefebvre, F. (2004) *J. Am. Ceram. Soc.*, **87**, 1413–1417.
44 Cheng, F., Kelly, S.M., Clark, S., Bradley, J.S., Baumbach, M., and Schuetze, A. (2006) *J. Membr. Sci.*, **280**, 530–535.
45 Cheng, F., Kelly, S.M., Lefebvre, F., Clark, S., Supplit, R., and Bradley, J.S. (2005) *J. Mater. Chem.*, **15**, 772–777.
46 Cheng, F., Toury, B., Lefebvre, F., and Bradley, J.S. (2003) *Chem. Commun.*, 242–243.
47 Ruediger, S. and Kemnitz, E. (2008) *Dalton Trans.*, 1117–1127.
48 Breyer, T., Breitbarth, F.W., and Vogelsberger, W. (2000) *Bull. Pol. Acad. Sci. Chem.*, **48**, 325–336.
49 Winter, R., Hua, D.W., Song, X., Mantulin, W., and Jonas, J. (1990) *J. Phys. Chem.*, **94**, 2706–2713.
50 Yamane, M., Inoue, S., and Yasumori, A. (1984) *J. Non-Cryst. Solids*, **63**, 13–21.
51 Ponton, A., Warlus, S., and Griesmar, P. (2002) *J. Colloid Interface Sci.*, **249**, 209–216.
52 Popa, A.F., Mutin, P.H., Vioux, A., Delahay, G., and Coq, B. (2004) *Chem. Commun.*, 2214–2215.
53 Leong, K.H., Monash, P., Ibrahim, S., and Saravanan, P. (2014) *Sol. Energy*, **101**, 321–332.
54 Andrianainarivelo, M., Corriu, R.J.P., Leclercq, D., Mutin, P.H., and Vioux, A. (1997) *J. Mater. Chem.*, **7**, 279–284.
55 Rolison, D.R. and Dunn, B. (2001) *J. Mater. Chem.*, **11**, 963–980.
56 Randall, J.P., Meador, M.A.B., and Jana, S.C. (2011) *ACS Appl. Mater. Interfaces*, **3**, 613–626.
57 Pierre, A.C. and Pajonk, G.M. (2002) *Chem. Rev.*, **102**, 4243–4265.
58 Husing, N. and Schubert, U. (1998) *Angew. Chem., Int. Ed.*, **37**, 22–45.
59 Hrubesh, L.W. (1998) *J. Non-Cryst. Solids*, **225**, 335–342.
60 Gesser, H.D. and Goswami, P.C. (1989) *Chem. Rev.*, **89**, 765–788.
61 Wu, C., Wu, Y., Xu, T., and Fu, Y. (2008) *J. Appl. Polym. Sci.*, **107**, 1865–1871.
62 Sangermano, M., Amerio, E., Epicoco, P., Priola, A., Rizza, G., and Malucelli, G. (2007) *Macromol. Mater. Eng.*, **292**, 634–640.
63 Lee, B.K., Hong, L.-Y., Lee, H.Y., Kim, D.-P., and Kawai, T. (2009) *Langmuir*, **25**, 11768–11776.
64 Han, Y.-H., Taylor, A., Mantle, M.D., and Knowles, K.M. (2007) *J. Sol-Gel Sci. Technol.*, **43**, 111–123.
65 Chang, C.-J. and Tzeng, H.-Y. (2006) *Polymer*, **47**, 8536–8547.
66 Mahadik, S.A., Kavale, M.S., Mukherjee, S.K., and Rao, A.V. (2010) *Appl. Surf. Sci.*, **257**, 333–339.
67 Shang, H.M., Wang, Y., Limmer, S.J., Chou, T.P., Takahashi, K., and Cao, G.Z. (2005) *Thin Solid Films*, **472**, 37–43.
68 Venkateswara Rao, A., Latthe, S.S., Nadargi, D.Y., Hirashima, H., and Ganesan, V. (2009) *J. Colloid Interface Sci.*, **332**, 484–490.
69 Loy, D.A. and Shea, K.J. (1995) *Chem. Rev.*, **95**, 1431–1442.
70 Yang, Q., Liu, J., Zhang, L., and Li, C. (2009) *J. Mater. Chem.*, **19**, 1945–1955.
71 Van Der Voort, P., Esquivel, D., De Canck, E., Goethals, F., Van Driessche, I., and Romero-Salguero, F.J. (2013) *Chem. Soc. Rev.*, **42**, 3913–3955.
72 Hunks, W.J. and Ozin, G.A. (2005) *J. Mater. Chem.*, **15**, 3716–3724.
73 Hoffmann, F., Cornelius, M., Morell, J., and Froeba, M. (2006) *Angew. Chem., Int. Ed.*, **45**, 3216–3251.
74 Fujita, S. and Inagaki, S. (2008) *Chem. Mater.*, **20**, 891–908.
75 Ribot, F., Toledano, P., and Sanchez, C. (1991) *Chem. Mater.*, **3**, 759–764.

76 Chatry, M., Henry, M., In, M., Sanchez, C., and Livage, J. (1994) *J. Sol-Gel Sci. Technol.*, **1**, 233–240.
77 Leaustic, A., Babonneau, F., and Livage, J. (1989) *Chem. Mater.*, **1**, 248–252.
78 Veith, M. (2002) *J. Chem. Soc., Dalton Trans.*, 2405–2412.
79 Rupp, W., Huesing, N., and Schubert, U. (2002) *J. Mater. Chem.*, **12**, 2594–2596.
80 Meyer, F., Hempelmann, R., Mathur, S., and Veith, M. (1999) *J. Mater. Chem.*, **9**, 1755–1763.
81 Mathur, S., Veith, M., Haas, M., Shen, H., Lecerf, N., Huch, V., Hufner, S., Haberkorn, R., Beck, H.P., and Jilavi, M. (2001) *J. Am. Ceram. Soc.*, **84**, 1921–1928.
82 Matijevic, E. and Sapieszko, R.S. (2000) *Surfactant Sci. Ser.*, **92**, 2–34.
83 Frenzer, G. and Maier, W.F. (2006) *Annu. Rev. Mater. Res.*, **36**, 281–331.
84 Arkles, B. (2001) *MRS Bull.*, **26**, 402–403, 405–408.
85 Roy, R. (1987) *Science*, **238**, 1664–1669.
86 Novak, B.M. (1993) *Adv. Mater.*, **5**, 422–433.
87 Ebert, T., Seifert, A., and Spange, S. (2015) *Macromol. Rapid Commun.*, **36**, 1623–1639.
88 Stöber, W., Fink, A., and Bohn, E. (1968) *J. Colloid Interface Sci.*, **26**, 62–69.
89 Nozawa, K., Gailhanou, H., Raison, L., Panizza, P., Ushiki, H., Sellier, E., Delville, J.P., and Delville, M.H. (2005) *Langmuir*, **21**, 1516–1523.
90 Wu, S.-H., Mou, C.-Y., and Lin, H.-P. (2013) *Chem. Soc. Rev.*, **42**, 3862–3875.
91 Jankiewicz, B.J., Jamiola, D., Choma, J., and Jaroniec, M. (2012) *Adv. Colloid Interface Sci.*, **170**, 28–47.
92 Ludi, B., Olliges-Stadler, I., Rossell, M.D., and Niederberger, M. (2011) *Chem. Commun.*, **47**, 5280–5282.
93 Sriram, M.A. and Kumta, P.N. (1994) *J. Am. Ceram. Soc.*, **77**, 1381–1384.
94 Kim, I.-s. and Kumta, P.N. (2003) *J. Mater. Chem.*, **13**, 2028–2035.
95 Kim, I.-s. and Kumta, P.N. (2003) *Mater. Sci. Eng., B*, **B98**, 123–134.
96 Taguchi, A. and Schueth, F. (2004) *Microporous Mesoporous Mater.*, **77**, 1–45.
97 Ying, J.Y., Mehnert, C.P., and Wong, M.S. (1999) *Angew. Chem., Int. Ed.*, **38**, 56–77.
98 Kresge, C.T., Leonowicz, M.E., Roth, W.J., Vartuli, J.C., and Beck, J.S. (1992) *Nature*, **359**, 710–712.
99 Schroden, R.C., Al-Daous, M., and Stein, A. (2001) *Chem. Mater.*, **13**, 2945–2950.
100 Warren, S.C., Perkins, M.R., Adams, A.M., Kamperman, M., Burns, A.A., Arora, H., Herz, E., Suteewong, T., Sai, H., Li, Z., Werner, J., Song, J., Werner-Zwanziger, U., Zwanziger, J.W., Graetzel, M., Di Salvo, F.J., and Wiesner, U. (2012) *Nat. Mater.*, **11**, 460–467.
101 Waterhouse, G.I.N., Metson, J.B., Idriss, H., and Sun-Waterhouse, D. (2008) *Chem. Mater.*, **20**, 1183–1190.
102 Avnir, D., Braun, S., Lev, O., and Ottolenghi, M. (1994) *Chem. Mater.*, **6**, 1605–1614.
103 Gupta, R. and Chaudhury, N.K. (2007) *Biosens. Bioelectron.*, **22**, 2387–2399.
104 Jin, W. and Brennan, J.D. (2002) *Anal. Chim. Acta*, **461**, 1–36.
105 Aylott, J.W., Richardson, D.J., and Russell, D.A. (1997) *Analyst*, **122**, 77–80.
106 Park, C.B. and Clark, D.S. (2002) *Biotechnol. Bioeng.*, **78**, 229–235.
107 Reetz, M.T., Tielmann, P., Wiesenhoefer, W., Koenen, W., and Zonta, A. (2003) *Adv. Synth. Catal.*, **345**, 717–728.
108 Ciriminna, R., Sciortino, M., Alonzo, G.; de., Schrijver, A., and Pagliaro, M. (2011) *Chem. Rev.*, **111**, 765–789.
109 Guleryuz, H., Kaus, I., Filiatre, C., Grande, T., and Einarsrud, M.-A. (2010) *J. Sol-Gel Sci. Technol.*, **54**, 249–257.
110 Amerio, E., Fabbri, P., Malucelli, G., Messori, M., Sangermano, M., and Taurino, R. (2008) *Prog. Org. Coat.*, **62**, 129–133.
111 Hauk, R., Frischat, G.H., and Ruppert, K. (1999) *Glass Sci. Technol.*, **72**, 386–392.
112 Toselli, M., Marini, M., Fabbri, P., Messori, M., and Pilati, F. (2007) *J. Sol-Gel Sci. Technol.*, **43**, 73–83.
113 Wu, L.Y.L., Chwa, E., Chen, Z., and Zeng, X.T. (2008) *Thin Solid Films*, **516**, 1056–1062.
114 Bautista, M.C. and Morales, A. (2003) *Sol. Energy Mater. Sol. Cells*, **80**, 217–225.
115 Chen, D. (2001) *Sol. Energy Mater. Sol. Cells*, **68**, 313–336.

116 Lien, S.-Y., Wuu, D.-S., Yeh, W.-C., and Liu, J.-C. (2006) *Sol. Energy Mater. Sol. Cells*, **90**, 2710–2719.

117 Manca, M., Cannavale, A., De Marco, L., Arico, A.S., Cingolani, R., and Gigli, G. (2009) *Langmuir*, **25**, 6357–6362.

118 Paz, Y. and Heller, A. (1997) *J. Mater. Res.*, **12**, 2759–2766.

119 Paz, Y., Luo, Z., Rabenberg, L., and Heller, A. (1995) *J. Mater. Res.*, **10**, 2842–2848.

120 Faustini, M., Nicole, L., Boissiere, C., Innocenzi, P., Sanchez, C., and Grosso, D. (2010) *Chem. Mater.*, **22**, 4406–4413.

121 Liu, Z., Zhang, X., Murakami, T., and Fujishima, A. (2008) *Sol. Energy Mater. Sol. Cells*, **92**, 1434–1438.

122 Lev, O., Wu, Z., Bharathi, S., Glezer, V., Modestov, A., Gun, J., Rabinovich, L., and Sampath, S. (1997) *Chem. Mater.*, **9**, 2354–2375.

123 Wang, D. and Bierwagen, G.P. (2009) *Prog. Org. Coat.*, **64**, 327–338.

124 Zheludkevich, M.L., Salvado, I.M., and Ferreira, M.G.S. (2005) *J. Mater. Chem.*, **15**, 5099–5111.

125 Zheludkevich, M.L., Shchukin, D.G., Yasakau, K.A., Moehwald, H., and Ferreira, M.G.S. (2007) *Chem. Mater.*, **19**, 402–411.

126 Kato, K., Zheng, C., Finder, J.M., Dey, S.K., and Torii, Y. (1998) *J. Am. Ceram. Soc.*, **81**, 1869–1875.

127 Yi, G., Wu, Z., and Sayer, M. (1988) *J. Appl. Phys.*, **64**, 2717–2724.

128 Kim, J.-S., Yang, S., and Bae, B.-S. (2010) *Chem. Mater.*, **22**, 3549–3555.

129 Kim, J.-S., Yang, S.C., Kwak, S.-Y., Choi, Y., Paik, K.-W., and Bae, B.-S. (2012) *J. Mater. Chem.*, **22**, 7954–7960.

8
Templated Synthesis for Nanostructured Materials

Yoshiyuki Kuroda[1] *and Kazuyuki Kuroda*[2,3]

[1]*Waseda University, Waseda Institute for Advanced Study, 1-6-1 Nishiwaseda, Shinjuku-ku, Tokyo, 169-8050, Japan*
[2]*Waseda University, Faculty of Science and Engineering, 3-4-1 Ohkubo, Shinjuku-ku, Tokyo, 169-8555, Japan*
[3]*Waseda University, Kagami Memorial Research Institute for Materials Science and Technology, 2-8-26 Nishiwaseda, Shinjuku-ku, Tokyo, 169-0051, Japan*

8.1
Introduction

Nanostructural design of solids is a quite important issue in the solid-state chemistry, because of the potential applications in wide areas of nanotechnology, including catalysis, sorption, separation, optics, energy storage, drug delivery, and so on. In order to design nanostructures in solid-state materials, the use of molecules, molecular assemblies, and porous solids for guided solidification of targeted materials (*templated synthesis*) attracts much attention. Templated synthesis is a common terminology in various fields of chemistry, including organic chemistry [1,2], inorganic chemistry [3], solid-state chemistry [4–8], supramolecular chemistry [9], and biochemistry [10]. In this chapter, the templated synthesis in solid-state chemistry, relating to the creation of nanospace, periodic structures, molecular recognition sites, and morphologies, which is quite important in materials science, is reviewed.

Templated synthesis in solid-state chemistry is replication of a template with a targeted wall component; the product is therefore a negative replica of the template. Because the nanostructures of negative replica are predictable according to those of templates, templated synthesis is a powerful tool to design morphology, nanostructures, and properties of solid-state materials.

General concept of templated synthesis is as follows (Figure 8.1). A wall component is infiltrated within the void space of a template and solidified. The template is selectively removed from the composite. The terminologies of exo- and endoskeletons have also been found in templated syntheses [11]. Figure 8.1a represents the endotemplating, in which a template consists of assembly of

Handbook of Solid State Chemistry, First Edition. Edited by Richard Dronskowski, Shinichi Kikkawa, and Andreas Stein.

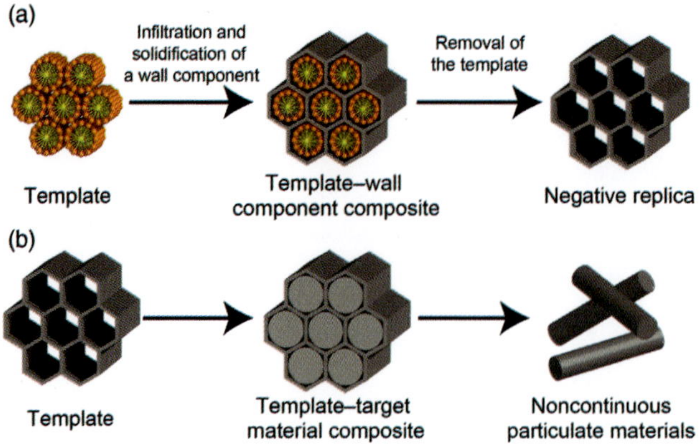

Figure 8.1 General concept of the templated syntheses. (a) Endotemplating. (b) Exotemplating.

nanoscale objects with void spaces among them and a wall component forms an exoskeletal framework. On the other hand, a material having regular nanospace in its interior works as exotemplate, in which a particulate object is formed as a negative replica (Figure 8.1b).

Templated synthesis can intuitively be understood as a miniaturized technique of casting (the so-called nanocasting) in which molten metal is infiltrated in a die, cooled to solidification, and removed from the die. However, better understanding will be obtained by regarding templated synthesis as a wider concept for synthetic chemistry than that of mechanical casting. For example, zeolites, microporous aluminosilicates, can be synthesized by hydrothermal treatment of aluminosilicate gel containing alkali or quaternary ammonium cations [12,13]. These cations are surrounded by aluminosilicate frameworks and determine their crystallized structures. The cations are called as structure-directing agents (SDA) rather than templates because the process does not involve direct replication (casting) of the three-dimensional structures of the atoms and molecules, although this well-established procedure is closely related to the templated synthesis.

In order to understand the mechanisms of templated syntheses, the following factors (i.e., physical confinement, chemical interactions, and self-organization) are important. Physical confinement is the most fundamental concept of templated synthesis. Because wall components cannot penetrate into the main framework of templates, wall components are *physically confined* within the void spaces of templates. Functional group-selective interactions (e.g., electrostatic interactions, hydrogen bonds, and π–π interactions) and bonds (e.g., ester bond, amide bond, and coordination bond) can be used to fix other molecules on templates, in which geometrical locations are controlled. Such *chemical interactions* are essential to the replication of molecules and molecular assemblies, such as DNA, and they can also be used for the control of interfaces between templates and wall components, in the case of solid-state materials. Templates are often self-assembled materials or their derivatives. *Self-assembly* is quite useful to form nanoscale structures whose periodicity is much larger than those of molecular templates with atomic-scale structures. In this chapter, templates, preparative methods, and mechanisms of templated synthesis are summarized.

8.2
Variety of Templates

Many of inorganic and organic substances in a wide range of length scales have been used as templates of solid-state materials with three-dimensional nanostructures (Figure 8.2). In simple cases, single-atomic ions are regarded as templates. Organic molecules, their assemblies, and porous solids are also frequently used as templates. Porous solids are mostly prepared by using such templates; therefore, they are classified as secondary templates. In this context, the length scales of the structures introduced in this chapter are defined according to those of porous materials, where micro-, meso-, and macroscales mean the length scales below 2, 2–50, and above 50 nm, respectively [14].

8.2.1
Atomic-Scale Templates

Metal cations, such as alkali and alkaline earth metals, are one of the smallest templates used for the structural control of open-framework metal oxides [15–21]. Inorganic ion exchange materials that selectively exchange specific cations have been prepared by extraction of an alkali cation (template) from a mixed metal

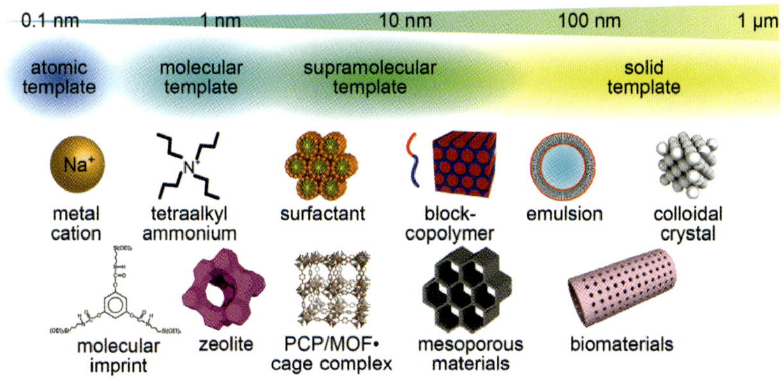

Figure 8.2 Typical examples of templates summarized according to the length scales of their characteristic structures.

oxide. For instance, lithium cation within a lithium–manganese spinel oxide is extracted by acid treatment to form λ-MnO_2. The A-site of the spinel oxide becomes vacant due to the extraction, and the site acts as an ion-exchange site highly selective for Li^+ [15].

Furthermore, layered manganese oxides intercalating various alkali and alkaline earth cations can be converted into a spinel and various tunnel oxides by hydrothermal treatment (Figure 8.3) [19,20]. The spinel and tunnel oxides have three-dimensional and one-dimensional void spaces, respectively, in their frameworks. The widths of the void spaces are similar to those of the hydrated/non-hydrated metal cations intercalated in the precursors, in which size matching probably determines the nanostructures of the products. Metal cations are also used as structure-directing agent of zeolites (explained below).

8.2.2
Microscale Templates

Both molecules and microporous solids have been used as templates. In this length scale as small as atoms and molecules, the chemical interactions between templates and wall components are as important as physical confinement effects. And the self-assembly processes are often accompanied with crystallization of the template–wall component composites.

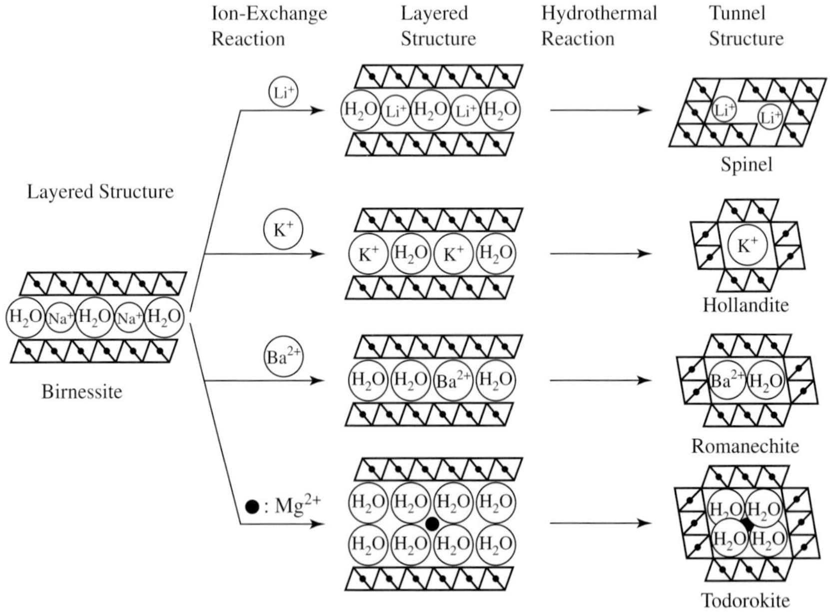

Figure 8.3 Transformation of layered manganese oxides into tunnel structures by using atomic templates. (Adapted with permission from Ref. [20]. Copyright 1998, Springer.)

8.2.2.1 Molecular Templates

A zeolite is a microporous aluminosilicate, which is prepared by hydrothermal treatment of aluminosilicate gel containing mineralizers (e.g., alkali cations and fluoride anion) and/or organic molecules [4]. The number of known framework types of zeolites is over 200 [22,23]. The structures of zeolites depend on the organic molecules that often work as templates; that is, organic molecules are included within micropores of the resultant frameworks. At the initial formation process of tetra-*n*-propylammonium (TPA)–Si-ZSM-5, preorganized inorganic–organic composite structure is formed between a TPA cation and oligomeric silicate anions, in which the TPA molecules adopt a conformation similar to that they have in the zeolite product [24,25]. The composites further aggregate into a crystal (Figure 8.4), which implies the templated formation of the zeolitic structure. On the other hand, the templated formation is not always observed; for example, the VPI-5 aluminophosphate does not contain the organic molecule within micropores [26]. Additionally, an organic molecule can direct the crystallization of more than one zeolitic structures (structure-directing agent (SDA)) [4]. Recently, several zeolites have been synthesized without using SDAs by the seeded growth method [27–32] or hydrothermal conversion [33]. In these cases, it is suggested that common composite-building units formed by dissolution of preformed zeolites also determine the resultant structures [34].

8.2.2.2 Molecular imprinting

The molecular imprinting is the mimicking of highly specific binding sites of enzymes on solid-state materials. Such binding sites with specific shapes and surface functional groups are designed on organic polymers, using molecular templates [35]. A monomer is polymerized under the presence of the template that interacts strongly with the monomer, leading to definite orientation of the monomer around the template. After the removal of the template, there is a molecular-imprinted void space that possesses specifically oriented functional groups. In order to accelerate a targeted reaction, a molecular template whose structure is an analogue of the transition state of the reactants is used [36,37], which is also applicable for enantioselective synthesis [38,39]. Noncovalent interactions (electrostatic interactions and hydrogen bonds) [36,37], covalent bonds [40,41], and coordination bonds [42] are used as interactions between monomers and templates.

Molecular imprinting on surfaces of silica is also useful for highly selective catalysis. The so-called footprint silica gel catalysts are prepared by rearrangement of framework in the presence of template molecules (Figure 8.5) [43–45]. Silica gel is doped with Al^{3+}, forming surface Lewis acid sites. A template molecule with Lewis base sites is adsorbed on the surfaces, associating with the Lewis acid sites. The aluminosilicate framework is rearranged by aging, which gives specifically oriented surface acid sites and silanol groups on the surfaces. Such structures can accelerate specific reactions.

The bifunctional template having abilities for templating and for sol–gel condensation or silylation is used. Such a bifunctional template consists of a

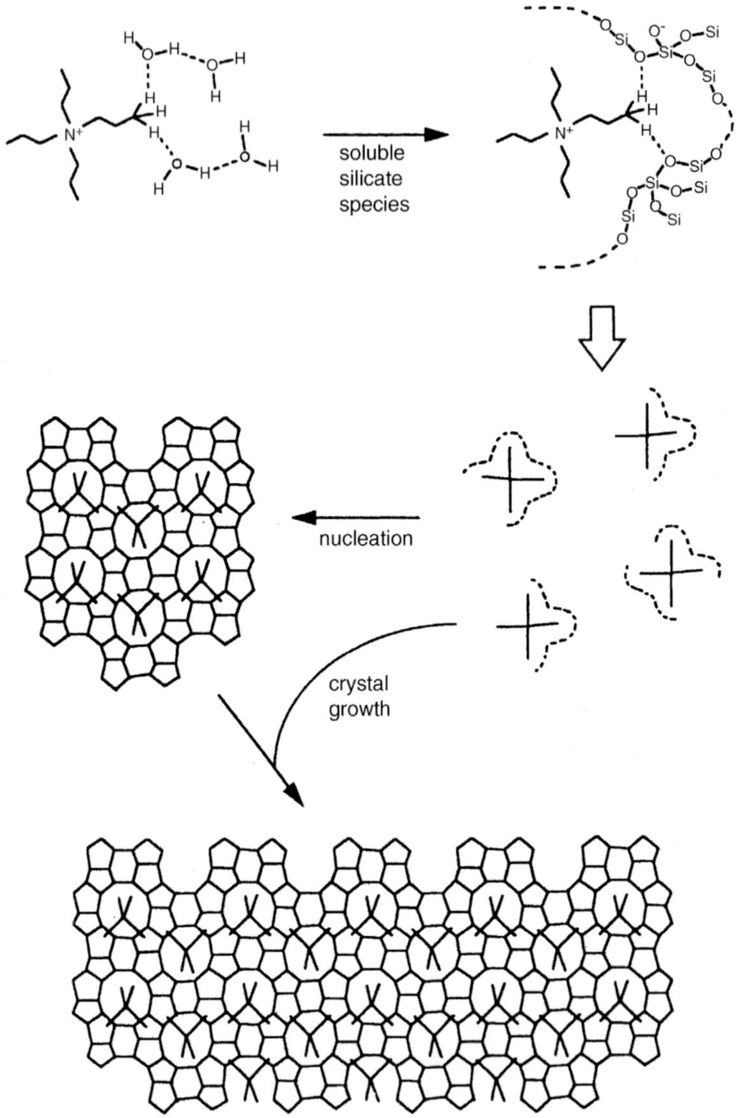

Figure 8.4 Mechanism of the structure-directed synthesis of TPA–Si-ZSM-5. (Adapted with permission from Ref. [25]. Copyright 1999, American Chemical Society.)

template moiety and condensable silyl groups. These parts are connected through cleavable groups, such as imine [46] and carbamate [47] linkages.

8.2.2.3 **Microporous Solid Templates**

Zeolite is also used as a template with atomic- or molecular-scale void spaces. The carbonization of zeolite/poly(acrylonitrile) [48] or zeolite/poly(furfuryl

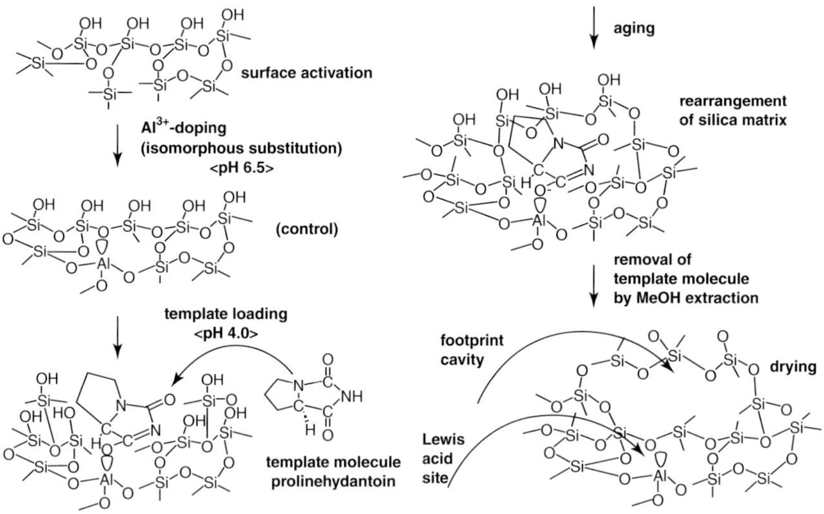

Figure 8.5 Schematic illustration of the preparation of "footprint" silica gel catalysts. (Adapted with permission from Ref. [45]. Copyright 1998, American Chemical Society.)

alcohol) [48] composite and chemical vapor deposition of propylene [49] in zeolite led to the filling of micropores in zeolites with graphitic carbon framework. After the removal of templates, the obtained zeolite-templated carbon showed diffraction peaks typical of zeolitic structures, suggesting the negative replication of the structures with graphitic carbon. The zeolite-templated carbon showed extremely high surface areas and were useful as electric double-layer capacitors.

Other organic materials with microscale void spaces are also used as templates. Compared with zeolites, porous organic materials are suitable for functionalization of the pore surfaces, and selective incorporation of metallic species is possible. A porous cage complex with nanometer scale interior has been used as a template of inorganic nanoparticles, such as silica [50], and core–shell silica/titania and silica zirconia [51]. The inner space of the cage complex was modified with sugar residue that is used for reversible condensation reaction with inorganic species. The obtained inorganic nanoparticles have quite narrow size distribution (i.e., $M_w/M_n < 1.01$).

8.2.3 Mesoscale Templates

Mesoscale templates include mostly molecular assembly and related porous materials. Molecular assembly provides not only larger nanospaces in products than those of microscale templates, but also diverse geometries of nanospaces because the geometries of templates depend simply on the arrangements of constituting molecules. Therefore, *self-assembly* of templates is an important factor to discuss on the templated syntheses in this length scale.

8.2.3.1 Liquid Crystalline Templates

Amphiphilic molecules, such as surfactants, are self-assembled into micelles and liquid crystals. Such liquid crystalline phases of surfactants can also be used as templates of mesoporous materials (Figure 8.6). The liquid crystalline templating of mesoporous silica has been discovered in the early 1990s [52–57], and this concept was applied for quite a wide range of wall components, including metal oxides [5,58], metals [59], polymers [60] (and can be converted into carbons), and inorganic–organic hybrids [61].

The formation of mesoporous silicas by sol–gel processes involves cooperative assembly of a surfactant template and silicate species to form mesostructured composites. At the initial stage of the synthesis of MCM-41, hexadecyltrimethylammonium cation and silicate anion are cooperatively assembled into the lamellar phase, and gradually transformed into the 2D hexagonal phase along with the condensation of silicates [62]. The mesostructure (the phase of the assembly) is dependent on various parameters. The packing parameter ($g = V/a_0 l$, where V is the total volume of the surfactant chains plus any cosolvent organic molecules, a_0 is the effective head group area at the micelle surface, and l is the kinetic surfactant tail length or the curvature elastic energy) [63] is useful

Figure 8.6 TEM image of mesoporous silica FSM-16. (Adapted with permission from Ref. [54]. Copyright 1993, Royal Society of Chemistry.)

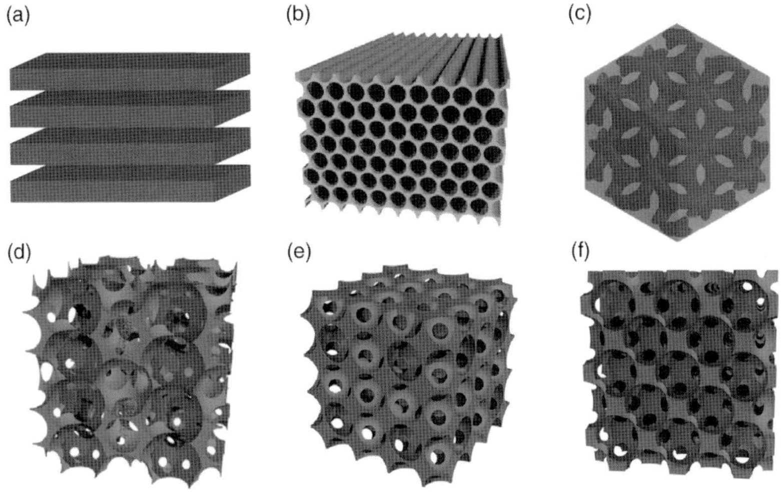

Figure 8.7 Models of surfactant-templated structures. (a) Lamellar. (b) 2D hexagonal (*p6mm*). (c) Double gyroid ($Ia\bar{3}d$). (d) Cubic ($Pm\bar{3}n$). (e) cubic ($Im\bar{3}m$). (f) Cubic ($Fm\bar{3}m$).

to understand the phase behavior, As the packing parameter increases, the shapes of micelles change from cages to rods, and sheets, leading the formation of cubic ($Pm\bar{3}n$) to hexagonal (*p6mm*), cubic ($Ia\bar{3}d$), and lamellar (Figure 8.7) [64]. Not only surfactant itself, but also associated silicates, cosolvents, and additives affect the porous geometry.

The pore sizes can be controlled by the molecular weight of a surfactant, that is, surfactants with longer alkyl chains tend to provide larger mesopores, and block copolymer surfactants give much larger mesopores than molecular surfactants [65–67]. The addition of oils as a micelle expander is useful to enlarge the resultant mesopore sizes [57].

Because mesoporous metal oxides are synthesized on the basis of sol–gel reactions, their morphologies can readily be controlled. Films [68–71], fibers [72], monoliths [73], and particles with various shapes (e.g., sphere [74–78] and rod [79]) have been reported.

The direct templating process is an alternative way of synthesizing mesoporous materials. A lyotropic liquid crystalline phase of a surfactant template, formed under highly concentrated condition, is replicated by silica [80] or metals [81,82]. This process is useful to prepare mesoporous metals because metallic frameworks are able to be manually deposited after the formation of lyotropic liquid crystalline templates [81–91].

Mesoporous silica has also been prepared by the transformation of layered silicate, kanemite ($NaHSi_2O_5 \cdot 3H_2O$), intercalating alkyltrimethylammonium cations by heat treatment [52–55]. Upon heat treatment, the silicate layers are fragmented and reconstructed while surfactant molecules are assembled into highly ordered cylindrical micelles. Mild acid treatment of the kanemite–alkyltrimethylammonium composite also provided a mesoporous silica with square channels [92,93]. In this case, kanemite sheet was folded together with the formation of highly ordered cylindrical micelles in the interlayer space upon acid treatment, which shows the uniqueness of the layered silicate precursor.

Similar to the surfactant-templated syntheses, mesostructured siloxane and/or mesoporous silica are synthesized by self-direction of surfactant-like alkoxy silanes. An alkoxy(alkyl)silane with a long alkyl chain ($C_nH_{2n+1}Si(OEt)_3$, $n = 12$, 14, 16, or 18) is hydrolyzed to be an amphiphilic molecule, where hydrolyzed silyl group ($—Si(OH)_3$) and long-chain alkyl groups are hydrophilic and hydrophobic, respectively. The molecule is self-assembled into mesostructured hybrid silica with a lamellar structure [94] or hybrid silica films by cocondensation with tetraethoxysilane [95]. The self-assembled structures can be varied into 2D hexagonal [96–98], 2D monoclinic [97], 3D cage-type [99], and inverse micelles [100], by introducing oligomeric alkoxysiloxane moieties as larger head groups. The inverse micelles, in which alkyl chains cover the outer surface of each micelle, can be disassembled into particles in nonpolar solvents.

8.2.3.2 Microphase-Separated Block Copolymer Templates

Microphase-separated nanostructures of block copolymers have also been used as mesoscale templates. Thin films of block copolymers possessing immiscible blocks are annealed to undergo microphase separation, forming subdomains of the immiscible blocks [101]. The microphase-separated structures can be varied as lamellar, cylindrical, spherical, and so on. The periodicity of the structure is usually larger than those of surfactant liquid crystals because of their much larger molecular weight. Precursors of wall components, such as metal alkoxides [102], or metal nanoparticles immiscible with one block of the template [103,104] are introduced within the microphase-separated film for replication.

8.2.3.3 Mesoporous Solid Templates

Mesoporous silica is used as templates of various solid-state materials. Such templates are called as hard templates in contrast to soft templates consisting of surfactants and block copolymers [105]. Because hard templates are nanostructured before the replication process, it is not necessary to consider self-assembly of templates and wall components. Negative replicas are obtained simply by infiltration of a wall precursor, solidification, and subsequent removal of templates. Therefore, hard templates are frequently used as templates of mesoporous materials with various compositions (e.g., metal oxides [106–108], carbons [109–112], non-oxide ceramics [113], and metals [114–117]) even though the preparative process is of at least two steps. On the other hand, the morphology of the products prepared by using hard templates is usually limited in powders, in contrast to that those of mesoporous materials prepared by using soft templates are variable as powders, films, fibers, monoliths, and so on.

Because most mesoporous solid templates consist of continuous pore walls, the negative replicas have discontinuous pore walls; for example, mesoporous silica SBA-15 with hexagonally arranged straight channels is used as templates, and the resultant negative replica consists of hexagonally arranged nanowires or nanotubes [110,117]. The interstices among the nanowires are regarded as mesopores; thus, the pore size of the negative replicas can be controlled by wall thickness of mesoporous solid templates [118].

8.2.4 Macroscale Templates

In macroscale, the nanostructures of products are intuitively understandable according to *physical confinement* by templates, whereas chemical interactions and self-assembly processes are also important if we look into the mechanisms in detail.

8.2.4.1 Emulsion/Reverse Micelle Templates

Emulsions, formed in water–oil–surfactant systems, have been used as macroscale templates. Assemblies of emulsions are used as porogens like liquid crystals, whereas the produced pores are quite large (typically larger than 100 nm) [119]. Various metal oxides, such as SiO_2, TiO_2, Al_2O_3, and ZrO_2, have been used as wall components [119,120]. Emulsions or reverse micelles are also used as nanoreactors in which nanoparticles are synthesized [6,121–126]. The nanospaces of emulsions and reverse micelles provide physical confinement to suppress excess growth of targeted components and/or suppress feeds of precursors; therefore, the size and size distributions of the resultant nanoparticles can precisely be controlled.

8.2.4.2 Water Droplets and Ice Templates

Even self-assembled water droplets are used as templates of macropores, which is called the Breath Figure process [127–130]. The periodic array of water droplet is formed on a substrate during the evaporation of a volatile solvent containing polymer in the presence of moist air flowing across the surface of the substrate. The polymer deposited around the water droplets spontaneously forms a macroporous film. Various polymers and polyion complexes are used as polymer sources, and inorganic nanoparticles can also be incorporated within polymer matrices.

Ice also works as templates of solid-state materials [131]. Silica fibers with polygonal cross sections were prepared by cooling aqueous polysilicic acid [132]. The polysilicic acid is excluded from the domain of ice, and proceeded concentration-dependent condensation to form silica fiber. Macroporous silica has also been prepared by cooling precursor silica hydrogels under a condition where pseudo-steady-state growth of ice rod continues [133]. Crystalline ice rods formed and grew along the temperature gradient. They expelled silica framework around them to form honeycomb structure. After thawing ice templates, macroporous silica monoliths with packed polygonal channels were obtained. As wall components, metal oxides, such as alumina [134,135], iron oxide [135], and titania [136], carbon [137], and polymers [138–140] have been used. Other hierarchical structures, such as lamellar and brick-and-mortar structure, were also prepared by ice templating [141].

8.2.4.3 Colloidal Templates

Colloidal crystals are periodically assembled monodispersed colloidal particles, such as latex particles and colloidal silica whose sizes are typically in submicrometer scale (Figure 8.8a) [142]. They are formed by self-assembly of the colloidal particles by solvent evaporation [143], filtration [144], centrifugation [145], and vertical deposition [146,147] methods. The colloidal particles are usually assembled into face-centered cubic (fcc) and/or hexagonal close-packed (hcp) structures. Wall precursors are infiltrated in the interstices of the particles. The templates are removed by calcination, solvent extraction, or chemical etching to provide highly ordered macroporous replicas (Figure 8.8b). Because colloidal templates have characteristics of hard templates, various wall components such

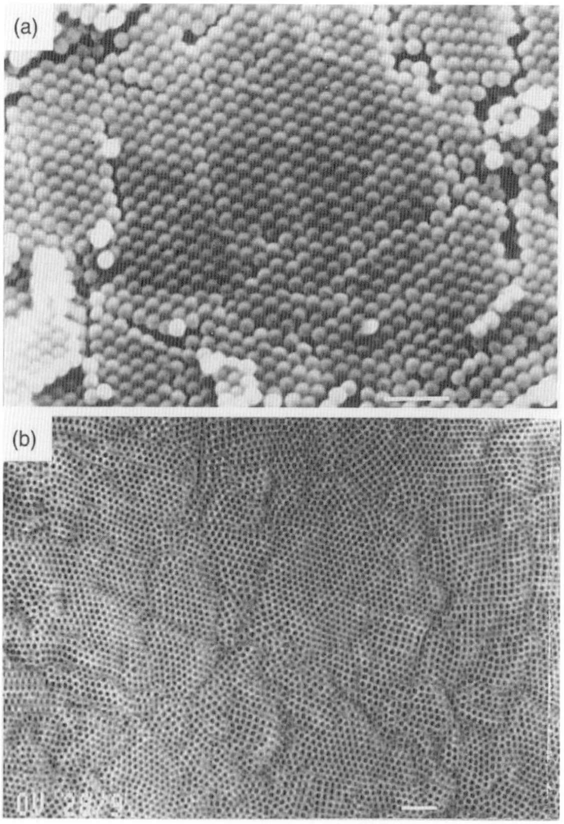

Figure 8.8 SEM images of (a) the colloidal crystal template and (b) macroporous silica replica. Scale bars: 1 μm. (Adapted with permission from Ref. [144]. Copyright 1997, Nature Publishing Group.)

as metal oxides [145,148–150], hydroxides [151], metal salts [152], carbon [153], diamond [153], metals [154,155], semiconductors [156,157], polymers [158], and nonoxide ceramics [159] can be used. The templating method using colloidal crystals are especially called as colloidal templating, and the negative replicas are called as three-dimensionally ordered macroporous (3DOM) materials, possessing interconnected spherical macropores. Recently, silica nanoparticles about 12 nm in size have been reported [160–162], and they are also used as colloidal templates of mesoporous materials [160–168].

Individual colloidal particles are also used as single-particle templates of hollow spheres. The surface of colloidal particles is coated with various wall components by layer-by-layer technique [169–173] or sol–gel reaction of silica/organosilica [174–176], retaining their dispersibilities.

Metal nanocrystals are also used as templates of metal nanocages. When silver nanocrystals with specific morphologies, such as sphere, cube, rod, plate, and prism, are reacted with $HAuCl_4$, gold is deposited on the surface of the silver nanocrystals to replicate their morphologies. The silver nanocrystals are oxidized for dissolution through the deposition of gold; thus, gold nanocrystals with hollow interior are obtained [177,178]. This procedure is also applicable for platinum and palladium as noble metal shells [179].

8.2.4.4 Biotemplates

Biominerals, such as a biosilica like diatom, possess unique three-dimensional structures in submicrometer to meter scales [180–182]. Such materials are used as templates of macroporous materials with very low cost. For example, electroless deposition of gold on a diatom followed by the etching of silica framework with HF solution provided gold replica of diatom [181].

8.3 Synthetic Methods

Replication of templates is conducted by impregnation of precursors, solidification of wall components, and removal of template. In this section, the techniques applicable for above processes are described.

8.3.1 Impregnation of Precursors

Precursors of wall components are small substances, such as molecules, oligomers [183], and nanoparticles [103,104,184]. Thus, they should be impregnated within the void spaces of templates. The impregnation of precursors of wall components within preformed templates is usually conducted by incipient wetness method [185] that has been used for preparation of supported metal catalysts. A precursor is dissolved in an appropriate solvent and added to a template. The volume of the solution is usually the same as the pore volume of the template. The solution is impregnated within the template by capillary action. The volume of a precursor deposited within the void space of a template is usually smaller than the total pore volume; thus, the impregnation is frequently repeated for several times to achieve sufficient impregnation of a precursor.

Another important impregnation method is the vapor-phase transport or chemical vapor deposition, which is especially useful for templates in thin film form. In this method, a template is exposed to vapor or stream of a volatile precursor of carbons (e.g., styrene and acrylonitrile) [48,49,186], metal oxides (metal alkoxides) [187], and semiconductor (disilane) [157]. Accordingly, precursor, template, and catalyst if necessary are packed in a closed vessel, or template is placed in a tube furnace and precursor is flown with inert gas. A precursor is diffused within templates to fill the void spaces. Various templates, such as zeolites [48,49], mesoporous silicas [186], and liquid crystalline surfactants [187], can be used in this method.

8.3.2
Solidification of Wall Components

The methods for solidification are dependent on the types of wall components. As typical cases, (i) metal oxides, (ii) polymers, (iii) carbon, and (iv) metals are described.

8.3.2.1 Metal Oxides

The sol–gel method is one of the most frequently used methods for the synthesis of oxide frameworks. The sol–gel method is suitable for controlled solidification without deterioration of fragile templates, and also is applicable for various compositions [184,188,189]. Zeolites are synthesized by hydrothermal method, in which dissolution and recrystallization of aluminosilicate gels proceed [25]. For stable hard templates, thermal decomposition (calcination) of amorphous oxides and/or metal salts is also frequently used [105].

8.3.2.2 Polymers

Polymer frameworks are formed by typical radical polymerization [190]. For the soft-templating synthesis of mesoporous polymers, sols of phenol/formaldehyde (PF) resin or resorcinol/formaldehyde (RF) resin are frequently used as wall components because their condensation reactions are highly controllable like sol–gel reactions of metal oxides, enabling even coassembly of RF sols and silica sols with block copolymer templates [191,192].

8.3.2.3 Carbon

Carbon is formed by carbonization of sucrose [193] and polymer frameworks such as propylene [48], poly(acrylonitrile) [194], poly(furfuryl alcohol) [195], and PF and RF resins [191,192,196]. Carbonization is carried out under inert atmosphere at higher than 800 °C. At such a temperature, the carbonized product is amorphous. Graphitized framework was achieved by using a perylene derivative as carbon source [197]. The carbon framework is doped with nitrogen when a carbon source containing nitrogen is used.

8.3.2.4 Metals

Metal is deposited by electronic [85,86,198] or electroless reduction [81,87,114,199,200] of metal cations or complexes. Electronic reduction is advantageous to control the degree of reaction by potential, current, and reaction time. For example, three-dimensionally ordered macroporous gold is prepared by electrodeposition of gold among colloidal silica constituting a colloidal crystal. The thickness of the 3DOM gold is controllable by the parameters [198]. However, electronic reduction is applicable only for films to ensure sufficient conduction paths. Electroless reduction is carried out with hydrogen gas [114] or chemical reductants (e.g., ascorbic acid [200], sodium borohydride [87], and dimethylamineborane [87,199]), which is applicable for templates in various forms.

8.3.3 Removal of Templates

Templates are usually removed by (i) calcination, (ii) solvent extraction, and (iii) chemical etching. Appropriate methods that do not deteriorate the nanostructures of targeted wall components should be chosen.

8.3.3.1 Calcination

Calcination is normally used when wall components are metal oxides. Because the frameworks of metal oxides are further condensed by thermal treatment, calcination is useful to form robust porous materials. Calcination is applicable for organic templates, such as quaternary ammoniums, surfactants, block copolymers, and latex particles. On the other hand, calcination also causes crystallization of amorphous metal oxide frameworks; however, volume change of the frameworks due to crystallization often destroys their nanostructures. Temperature and/or ramp rate should be optimized. The reinforcements of nanostructures by using a stable block copolymer as a template [201], and incorporation of siloxane materials [202] or carbon [203] within pores are also quite effective.

8.3.3.2 Solvent Extraction

Solvent extraction is used if thermal treatment deteriorates targeted wall components. For instance, wall components containing organic moieties (i.e., organosiloxane [150,204,205]) are decomposed and most metals are oxidized or melted upon thermal treatments. Solvent extraction is applicable for organic templates, such as surfactants, block copolymers, and latex particles, whereas quaternary ammoniums are usually not extracted because their alkyl chains are confined in micropores of zeolitic products. Organic solvents, such as toluene, tetrahydrofuran (THF), and acetone, are used to remove latex particles [150]. To remove surfactant templates incorporated in as-synthesized mesoporous materials, water, alcohol, or THF-containing HCl [204,205], or supercritical CO_2 [206] are used.

Recently, solvent extraction combined with dialysis is used in order to remove surfactant templates from colloidal mesoporous silica nanoparticles, retaining their dispersibility [78]. If usual solvent extraction processes are applied for colloidal mesoporous silica nanoparticles dispersed in a solvent, the colloidal mesoporous silica nanoparticles must be separated from the solvent by centrifugation or filtration, causing their aggregation. Such aggregated nanoparticles are difficult to be redispersed. An aqueous dispersion of colloidal mesoporous silica nanoparticles are placed in a dialysis tube and soaked in an ethanolic solution of acetic acid.

8.3.3.3 Chemical Etching

If a template is decomposed by chemical treatments, such processes are used for selective chemical etching of templates. Representative examples are hydrofluoric acid [114] and sodium hydroxide [109] for silica, acids for melamine formaldehyde [170], and calcium carbonate [207]. Chemical etching is useful when metals, carbons, and polymers are used as the wall components.

8.3.4 Interfacial Control of Templated Synthesis

To prepare homogeneous template–wall component composites, the formation of stable interfaces between templates and wall components, which is caused by strong interactions between them, is important. Molecular templates tend to cause segregation if appropriate interface is not formed. Infiltration of wall components into solid templates is also affected by the interactions between template and wall component. In this section, approaches to form stable interfaces are introduced from the viewpoint of the interactions between templates and wall components.

Ionic species, such as quaternary ammonium cations and cationic and anionic surfactants, are used as templates of zeolites and mesoporous materials, respectively. These templates form highly stable interfaces with oppositely charged wall components [4,19,20,24,25]. Zeolites and mesoporous silica are prepared from soluble silicate anions that strongly interact with quaternary ammonium (for zeolites) [4,24,25] or cationic surfactants (for mesoporous silica) [208], forming self-assembled composites. In the latter case, electrostatic interactions are described as S^+I^- manner (S^+: cationic surfactant, and I^-: anionic inorganic species). Contrarily, cationic metal oxide species, consisting of Mg, Al, Mn, Fe, Co, Ni, and Zn, for example, likely interact with anionic surfactants (e.g., long-chain alkylphosphate esters) to provide self-assembled composites in the S^-I^+ manner (S^-: anionic surfactant and I^+: cationic inorganic species). If templates and wall components have the same electric charges, they can interact with each other through oppositely charged ionic species. For example, silicate species are positively charged when pH is below their isoelectric point. Both the positively charged silicate and cationic surfactant interact with the counteranion of the surfactant in the $S^+X^-I^+$ manner (X^-: halogen anion). Anionic surfactant also interacts with negatively charged zinc oxide species at pH higher than 12.5 in the $S^-M^+I^-$ manner (M^+: metal cation). The electrostatic interactions are also important for the above-mentioned layer-by-layer method for the preparation of core–shell and hollow particles by using colloidal templates [169,170].

Such a control of electrostatic interactions between templates and wall components contributes to further nanostructural control of products. Tatsumi and Che *et al.* have used 3-aminopropyltriethoxysilane as a costructure directing agent (CSDA) for anionic surfactant-templated synthesis of mesoporous silica [209]. Silicate anions do not interact with anionic surfactant, although silicate anions connect with the CSDA via sol–gel process. The $—NH_2$ group of CSDA is protonated and interacts with anionic surfactant to form stable interface. Applying this concept for anionic surfactants with amino acid moieties, chiral mesoporous silicas with helical mesochanels have also been reported [210].

Hydrogen bonds are usually concerned between neutral template and wall components [208,211,212]. For example, nonionic surfactants possessing a poly (ethylene oxide) chain as a hydrophilic group interact with silanol groups under acidic conditions, which provides stable interface to form self-assembled composites.

8.4 Morphological and Hierarchical Control

Morphologies, outer shapes, of porous materials are quite important factors for their practical use. The control of hierarchical internal structures is also important for formation of fractal surfaces, improvement of guest diffusivity, and integration of functions. In this section, the synthetic approaches for morphological and hierarchical control of porous materials are described.

8.4.1 Nanoparticles

Nanoparticles of porous materials are expected to be useful as fillers [213], drug carriers [214], building blocks of hierarchically structured materials [215], and so on [216]. The dispersibility of nanoparticulate porous materials is also important issues for their practical use. Crystalline porous materials, such as zeolites, are synthesized through crystal growth. The growth of zeolite is controlled by synthetic conditions, solvent, and additives, providing nanoparticles by suppressed crystal growth [216]. Further precise control of their morphologies is achieved by using templates. The deposition of zeolite in 3DOM carbon provides assembly of uniform and spherical zeolite nanoparticles several tens of nanometers in size [163]. The primary particles can be dispersed by dialysis against basic solutions and/or sonication [217].

Nanoparticles of mesoporous silica are also intensively investigated. The two major methods for the synthesis of mesoporous silica nanoparticles are spray drying [74] and Stöber method [76,77]. In the spray drying method, a precursor solution of mesoporous silica is sprayed into a preheated chamber to form microdroplets. Surfactants and silicate species are self-assembled into mesostructures upon drying of the solvent. The Stöber method has first been reported for the synthesis of monodispersed colloidal silica [218], in which TEOS is hydrolyzed and condensed in aqueous and/or ethanolic basic solutions. Mesoporous silica nanoparticles prepared by the method is highly monodispersed. Recently, mesoporous silica nanoparticles about 10 nm in size [78] and monodispersed mesoporous silica nanoparticles smaller than 100 nm, which can be assembled into colloidal crystals, have been reported on the basis of the Stöber method (Figure 8.9) [219].

8.4.2
Nanorods and Nanofibers

Nanorods and nanofibers with one-dimensional morphology are provided by templating [220–222] and electrospinning [72,223], respectively. Anodic aluminum oxide is a useful template for the preparation of nanorods, applying straight channels passing through the membrane. The channel width is controllable during the preparation process [224]. Moreover, liquid crystalline templates are aligned in unique structures within the channels, depending on the channel widths [220].

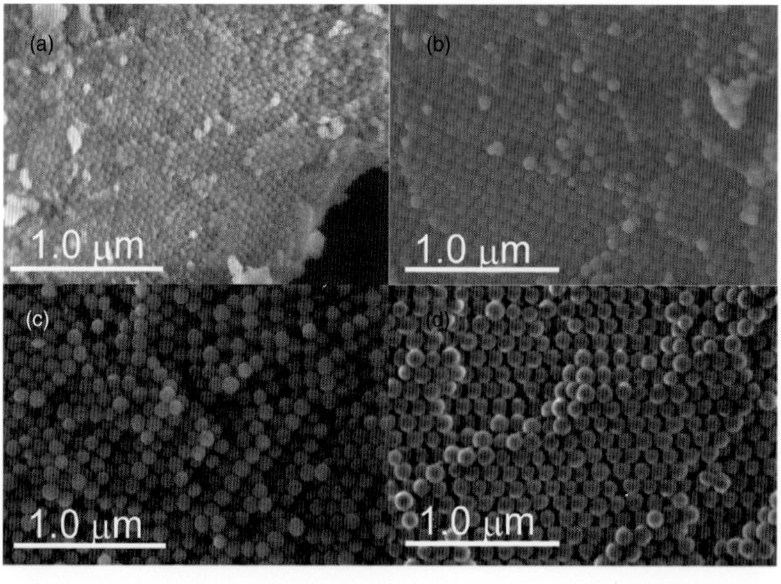

Figure 8.9 SEM images of colloidal crystals consisting of monodispersed mesoporous silica nanoparticles whose diameters are (a) 73, (b) 82, (c) 92, and (d) 108 nm. (Adapted with permission from Ref. [219]. Copyright 2014, American Chemical Society.)

8.4.3
Films and Membranes

Films and membranes are quite important targets in the morphological control of porous materials for their applications in coatings, devices, separation media, and so on. Crystalline porous materials, such as zeolites, are prepared as films by hydrothermal method with heterogeneous nucleation on substrates, although such a film consists of aggregates of microcrystals with intercrystalline void spaces. In order to use zeolite films for separation, membranes such a void must be excluded. Recently, voidless zeolite membranes have been developed by secondary growth [225], rapid thermal treatment [226], rubbing method [227], and use of nanoparticulate seeds [216,228].

Mesoporous/mesostructured films with various compositions have been prepared via sol–gel processes [68,70,189]. Such films are deposited on substrates by spin-coating, dip-coating, and hydrothermal methods. During the spin-coating or dip-coating, the surfactants and wall components are self-assembled into mesostructures together with the solvent evaporation, which is called evaporation-induced self-assembly (EISA) process [115].

The alignment of mesostructures with interfaces (i.e., template–substrate and template–air interfaces) is an important factor of the mesoporous/mesostructured films. The rod-like micelles of the 2D hexagonal (*p6mm*) mesostructure are aligned parallel to the interfaces. The in-plane alignment (direction of rod-like micelles) is usually not strictly controlled. The in-plane alignment is controlled by the assistance of substrates, such as mica [229–231], graphite [230], Si substrate [232], Langmuir–Blodgett film [233], and rubbing-treated polyimide-coated substrates (Figure 8.10) [234–237], which are known to have ability to

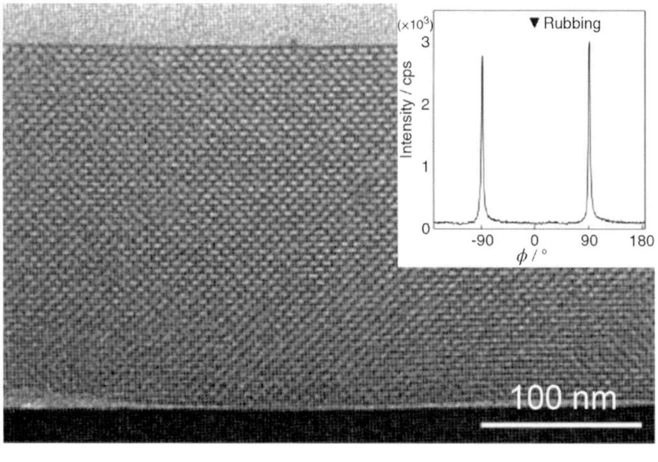

Figure 8.10 Cross-sectional TEM image of the mesoporous silica film with uniaxially aligned mesochannels, prepared on a rubbing-treated polyimide-coated substrate. Inset shows its ϕ scanning in-plane XRD pattern $2\theta_\chi = 2.48°$. (Adapted with permission from Ref. [236]. Copyright 2005, American Chemical Society.)

align mesostructures, which is due to interactions-ordered surface structures of substrates and surfactant molecules. Single-crystal-like arrangement of cage-type mesopores in the 3D hexagonal structure has been achieved by using a rubbing-treated polyimide-coated substrate [238].

Vertical alignment of mesochannels (e.g., 2D hexagonal structure) is also intensively studied because such mesochannels should have excellent accessibility for guest species or can be used for nanopatterned media with very small repeating distances. In order to achieve the vertical alignments, magnetic field [239–243], electrodeposition [244], epitaxial growth [245,246], and Stöber solution growth [247,248] have been reported.

Larger templates, such as colloidal templates, are also used as templates of macroporous films. Colloidal crystal films are prepared by the vertical deposition method [146,147], and they are used as templates of 3DOM films [198,249].

8.4.4 Hierarchical Structures by Combined Use of Templates

The hierarchical integration of nanospaces with different length scales is useful for synergistic functions. For example, introduction of mesopores and/or macropores into microporous zeolite is quite useful to enhance catalytic lifetime and diffusivity of substrates [250,251]. Also, cocontinuous macroporous monolith gels possessing hierarchical mesopores in the macropore walls show both low pressure loss and high separation efficiency as chromatography columns [252]. Such hierarchical structures can be controlled by combining multiple templates. Although there are a lot of combinations of templates, examples important for the understanding of concepts are introduced here.

8.4.4.1 Combination of Hard and Soft Templates

The most simple way to control hierarchical structures is the combination of hard and soft templates. When combining different templates, collapse of one template or mixing of the templates into homogeneous phase often occurs [253]. Because hard templates are stable and cannot be mixed with other molecules, they are ideal for the control of hierarchical structures.

As a hard template, colloidal crystals are frequently used in order to form three-dimensionally ordered macropores that facilitate diffusion of guest molecules to the macropore walls. In the macropore walls, micropores and/or mesopores are formed by using molecular soft templates, such as quaternary ammoniums [215,254], and surfactants [164,253,255]. Use of porous building blocks as components of macropore walls has also been reported [256,257].

8.4.4.2 Combination of Different Soft Templates

In limited cases, different soft templates are combined to form hierarchical structures. The use of ionic liquid-type surfactant (1-hexadecyl-3-methylimidazolium chloride, C_{16}mimCl) with block copolymers, such as hydrogenated poly(butadiene-*b*-ethylene oxide) (KLE) and poly(styrene-*b*-ethylene oxide)

(PS-PEO), provides hierarchical assembly of the templates without phase separation or formation of mixed micelles [67,253].

8.4.4.3 Use of Spontaneous Phase Separation

The sol–gel reaction of alkoxysilanes forms monolith gels with macroporous cocontinuous structures by spinodal decomposition [258,259]. Such monoliths are also formed in the presence of a block copolymer template to form hierarchical meso- and macroporous structures, and they are useful as chromatography columns.

8.4.4.4 Use of Bifunctional Templates

Molecular design of organic templates is critical to the control of hierarchical structures, including zeolitic frameworks. Hexadecyltrimethylammonium chloride ($C_{16}TAC$) is a typical surfactant template of mesoporous silica. A $C_{16}TAC$ derivative that has a (trimethoxysilyl)alkyl group on the head group has been used as a bifunctional template to form mesopores within a MFI-type ("MFI" is a framework type of zeolites [22,23]) zeolite [260]. The bifunctional template was added in a synthetic gel of MFI-type zeolite, and hydrothermally treated. The product possessed MFI-type zeolitic structure and mesoscale void spaces within the crystallites because the bifunctional template is connected with the zeolitic framework through Si–O–Si linkage and forms micellar assembly within the crystallites.

Diquaternary ammonium-type surfactant ($C_{22}H_{45}$-$N^+(CH_3)_2$-C_6H_{12}-$N^+(CH_3)_2$-C_6H_{13}, $C_{22\text{-}6\text{-}6}$) is also a bifunctional template that has abilities for molecular assembly and structure-directing agents (SDAs) of zeolites [251]. When hydrothermal synthesis of zeolite using $C_{22\text{-}6\text{-}6}$ as a SDA is conducted, the head group is incorporated within the micropores of MFI zeolite, forming stable interface (Figure 8.11a); therefore, the growth of MFI zeolite is suppressed along the *b*-axis. Accordingly, the MFI zeolite formed sheet-like morphology (Figure 8.11b). The zeolitic sheets were stacked along *b*-axis. MFI zeolites with channel-type

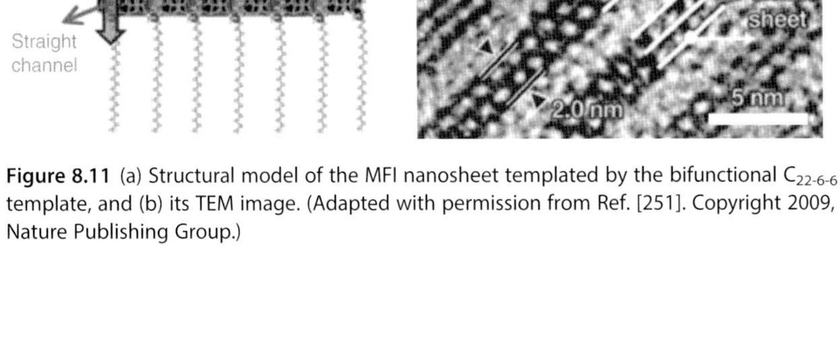

Figure 8.11 (a) Structural model of the MFI nanosheet templated by the bifunctional $C_{22\text{-}6\text{-}6}$ template, and (b) its TEM image. (Adapted with permission from Ref. [251]. Copyright 2009, Nature Publishing Group.)

mesopores have also been prepared by using similar bifunctional template, such as $C_{18}H_{37}$–$N^+(CH_3)_2$–C_6H_{12}–$N^+(CH_3)_2$–C_6H_{12}–$N^+(CH_3)_2$–$C_{18}H_{37}(Br^-)_3$ [261].

8.5 Applications of Nanostructured Materials Synthesized by Templating Methods

Porous materials synthesized by templating methods are applied in various fields. For example, molecular-imprinted polymers can be regarded as mimicking of enzymes [36–39]. Zeolites are used as catalysts and catalyst supports [262]. Because they possess molecular-sieving micropores, zeolites show size or shape selectivity. Such molecular sieving effects are also important for gas separation membranes. Zeolites are used as separation membranes for gaseous molecules, for example, H_2/CO_2, H_2/N_2, H_2/CH_4, and O_2/N_2, and catalytic membrane reactors [263]. Recently, very thin zeolite membranes have been prepared, using zeolite nanosheets as seed crystals [264].

Mesoporous silica and mesoporous materials with various compositions are also promising as catalysts and catalyst supports [262,265]. Due to their relatively large pore size, they are used as catalyst supports for metal cations, organic molecules, complexes, metal nanoparticles, and enzymes. Framework composition is also important for catalytic activities and photocatalysis [266]. Light-harvesting mesoporous organosilica has been demonstrated by incorporation of organic chromophores within frameworks [267]. Photocatalytic activity of Re complex is enhanced by energy transfer from organic chromophore in the framework (Figure 8.12) [268]. Mesoporous materials are used as chromatography media [259] and electrode-active materials of secondary batteries [269]. Mesoporous carbon is applied for supercapacitors [270].

Morphological control of mesoporous silica is quite important for their applications. Mesoporous thin films are promising as optical materials. For example, mesoporous thin films with uniaxially aligned mesochannels are used for polarized emission [271]. Spherical mesoporous silica nanoparticles are investigated as nanomedicine including drug delivery carriers [272].

Three-dimensionally ordered macroporous materials are used as catalysts, solar cells, lithium-ion batteries, fuel cells, and supercapacitors with high diffusivity [273]. Because 3DOM materials show structural colors depending on the periodicities, they are investigated for colorimetric sensors [273]. Hierarchically ordered materials consisting of macropores and micro/mesopores are reported to be useful as catalysts and separation media with high diffusivity [250–252].

8.6 Summary and Outlook

Templated synthesis is a versatile concept for nanostructural control of solid-state materials with various compositions in a wide range of length scales.

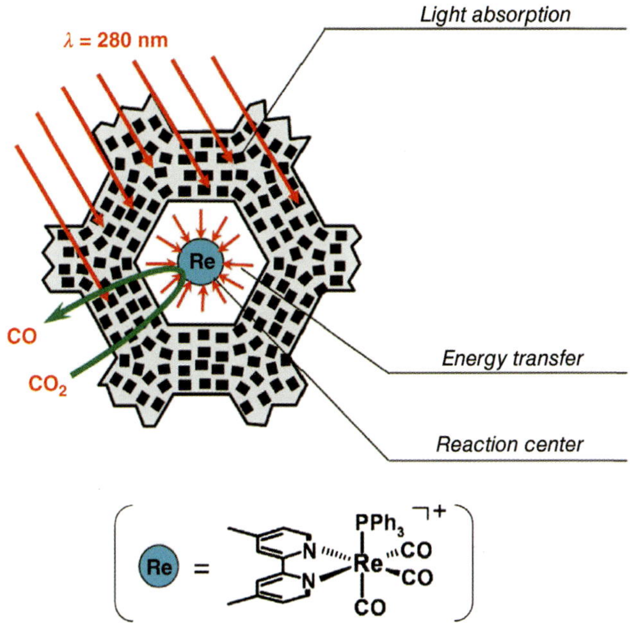

Figure 8.12 Schematic illustration of light-harvesting by periodic mesoporous organosilica and enhancing of photocatalysis of Re complex. (Adapted with permission from Ref. [268]. Copyright 2010, American Chemical Society.)

Materials provided by the templated synthesis are not only porous materials but also nanomaterials with unique morphologies and hierarchical structures. Nanostructured materials with rigid frameworks are suitable as templates, whereas ions, organic molecules, molecular assemblies, water drops, and so on, which temporary form ordered structures, can also be used as templates. It extensively enhances the diversity of nanostructures in solid-state materials. Templated synthesis has potentials to control desired nanostructures and morphologies in any materials by understanding the processes of impregnation of precursors into templates, solidification of the precursors, and selective removal of templates.

Although many of the templated syntheses have been developed to date, there are still lots of issues. Biological organisms in nature provide diverse materials from very limited resources. In fact, there have been many researches of templated synthesis, mimicking natural and biological systems; however, most of them have mimicked only a part of the natural and biological systems, such as enzymes, photosynthesis, separation membranes, and so on. One of our goals would be a perfect control of nonequilibrium and dynamic processes in materials synthesis and create highly functionalized materials that rise above the natural and biological systems.

References

1 Anderson, S., Anderson, H.L., and Sanders, J.K.M. (1993) *Acc. Chem. Res.*, **26**, 469–475.

2 Li, X. and Liu, D.R. (2004) *Angew. Chem., Int. Ed.*, **43**, 4848–4870.

3 Vilar, R. (2003) *Angew. Chem., Int. Ed.*, **42**, 1460–1477.

4 Lobo, R., Zones, S., and Davis, M. (1995) *J. Incl. Phenom. Macrocycl. Chem.*, **21**, 47–78.

5 Wan, Y. and Zhao, D. (2007) *Chem. Rev.*, **107**, 2821–2860.

6 Landfester, K. (2001) *Adv. Mater.*, **13**, 765–768.

7 Wulff, G. (2002) *Chem. Rev.*, **102**, 1–27.

8 Stein, A., Li, F., and Denny, N.R. (2008) *Chem. Mater.*, **20**, 649–666.

9 Beves, J.E., Blight, B.A., Campbell, C.J., Leigh, D.A., and McBurney, R.T. (2011) *Angew. Chem., Int. Ed.*, **50**, 9260–9327.

10 Erlich, H.A., Gelfand, D., and Sninsky, J.J. (1991) *Science*, **252**, 1643–1651.

11 Schüth, F. (2003) *Angew. Chem., Int. Ed.*, **42**, 3604–3622.

12 Cundy, C.S. and Cox, P.A. (2003) *Chem. Rev.*, **103**, 663–701.

13 Cundy, C.S. and Cox, P.A. (2005) *Microporous Mesoporous Mater.*, **82**, 1–78.

14 IUPAC (1997) *Compendium of Chemical Terminology*, 2nd edn (the *Gold Book*, compiled by A.D. McNaught and A. Wilkinson), Blackwell Scientific Publications, Oxford.

15 Feng, Q., Miyai, Y., Kanoh, H., and Ooi, K. (1992) *Langmuir*, **8**, 1861–1867.

16 Brock, S.L., Duan, N., Tian, Z.R., Giraldo, O., Zhou, H., and Suib, S.L. (1998) *Chem. Mater.*, **10**, 2619–2628.

17 Feng, Q., Kanoh, H., and Ooi, K. (1999) *J. Mater. Chem.*, **9**, 319–333.

18 Shen, Y.F., Zerger, R.P., DeGuzman, R.N., Suib, S.L., McCurdy, L., Potter, D.I., and O'Young, C.L. (1993) *Science*, **260**, 511–515.

19 Shen, Y.F., Suib, S.L., and O'Young, C.L. (1994) *J. Am. Chem. Soc.*, **116**, 11020–11029.

20 Feng, Q., Yanagisawa, K., and Yamasaki, N. (1998) *J. Porous Mater.*, **5**, 153–161.

21 Ueda, W., Vitry, D., and Katou, T. (2005) *Catal. Today*, **99**, 43–49.

22 Baerlocher, Ch., Meier, W.M., and Olson, D.H. (2001) *Atlas of Zeolite Framework Types*, 5th edn, Elsevier, Amsterdam.

23 Robson, H. and Lillerud, K.P. (2001) *Verified Syntheses of Zeolitic Materials*, 2nd edn, Elsevier, Amsterdam.

24 Chang, C.D. and Bell, A.T. (1991) *Catal. Lett.*, **8**, 305–316.

25 Burkett, S.L. and Davis, M.E. (1994) *J. Phys. Chem.*, **98**, 4647–4653.

26 Davis, M.E., Montes, C., Hathaway, P.E., Arhancet, J.P., Hasha, D.L., and Garces, J.M. (1989) *J. Am. Chem. Soc.*, **111**, 3919–3924.

27 Xie, B., Song, J., Ren, L., Ji, Y., Li, J., and Xiao, F.-S. (2008) *Chem. Mater.*, **20**, 4533–4535.

28 Kamimura, Y., Chaikittisilp, W., Itabashi, K., Shimojima, A., and Okubo, T. (2010) *Chem. Asian J.*, **5**, 2182–2191.

29 Majano, G., Darwiche, A., Mintova, S., and Valtchev, V. (2009) *Ind. Eng. Chem. Res.*, **48**, 7084–7091.

30 Yokoi, T., Yoshioka, M., Imai, H., and Tatsumi, T. (2009) *Angew. Chem., Int. Ed.*, **48**, 9884–9887.

31 Iyoki, K., Kamimura, Y., Itabashi, K., Shimojima, A., and Okubo, T. (2010) *Chem. Lett.*, **39**, 730–731.

32 Zhang, H., Guo, Q., Ren, L., Yang, C., Zhu, L., Meng, X., Li, C., and Xiao, F.-S. (2011) *J. Mater. Chem.*, **21**, 9494–9497.

33 Jon, H., Nakahata, K., Lu, B., Oumi, Y., and Sano, T. (2006) *Microporous Mesoporous Mater.*, **96**, 72–78.

34 Itabashi, K., Kamimura, Y., Iyoki, K., Shimojima, A., and Okubo, T. (2012) *J. Am. Chem. Soc.*, **134**, 11542–11549.

35 Wulff, G. (1995) *Angew. Chem., Int. Ed.*, **34**, 1812–1832.

36 Wulff, G., Gross, T., and Schönfeld, R. (1997) *Angew. Chem., Int. Ed.*, **36**, 1962–1964.

37 Sellergren, B., Lepistö, M., and Mosbach, K. (1988) *J. Am. Chem. Soc.*, **110**, 5853–5860.

38 Wulff, G. and Vietmeier, J. (1989) *Makromol. Chem.*, **190**, 1717–1726.

39 Wulff, G. and Vietmeier, J. (1989) *Makromol. Chem.*, **190**, 1727–1735.

40 Wulff, G., Sarhan, A., and Zabrocki, K. (1973) *Tetrahedron Lett.*, **44**, 4329–4332.

41 Alexander, C., Smith, C.R., Whitcombe, M.J., and Vulfson, E.N. (1999) *J. Am. Chem. Soc.*, **121**, 6640–6651.

42 Fujii, Y., Kikuchi, K., Matsutani, K., Ota, K., Adachi, M., Syohi, M., Haneishi, I., and Kuwana, Y. (1984) *Chem. Lett.*, 1487–1490.

43 Morihara, K., Kurihara, S., and Suzuki, J. (1988) *Bull. Chem. Soc. Jpn.*, **61**, 3991–3998.

44 Shimada, T., Nakanishi, K., and Morihara, K. (1992) *Bull. Chem. Soc. Jpn.*, **65**, 954–958.

45 Morihara, K. (1998) *Molecular and Ionic Recognition with Imprinted Polymers*, ACS Symposium Series, vol. **703** (eds R.A. Bartsch and M. Maeda), American Chemical Society, Washington, DC, pp. 300–313.

46 Wulff, G., Heide, B., and Helfmeier, G. (1986) *J. Am. Chem. Soc.*, **108**, 1089–1091.

47 Katz, A. and Davis, M.E. (2000) *Nature*, **403**, 286–289.

48 Kyotani, T., Nagai, T., Inoue, S., and Tomita, A. (1997) *Chem. Mater.*, **9**, 609–615.

49 Itoi, H., Nishihara, H., Kogure, T., and Kyotani, T. (2011) *J. Am. Chem. Soc.*, **133**, 1165–1167.

50 Suzuki, K., Sato, S., and Fujita, M. (2010) *Nat. Chem.*, **2**, 25–29.

51 Suzuki, K., Takao, K., Sato, S., and Fujita, M. (2011) *Angew. Chem., Int. Ed.*, **50**, 4858–4861.

52 Yanagisawa, T., Shimizu, T., Kuroda, K., and Kato, C. (1990) *Bull. Chem. Soc. Jpn.*, **63**, 988–992.

53 Yanagisawa, T., Shimizu, T., Kuroda, K., and Kato, C. (1990) *Bull. Chem. Soc. Jpn.*, **63**, 1535–1537.

54 Inagaki, S., Fukushima, Y., and Kuroda, K. (1993) *J. Chem. Soc., Chem. Commun.*, 680–682.

55 Inagaki, S., Koiwai, A., Suzuki, N., Fukushima, Y., and Kuroda, K. (1996) *Bull. Chem. Soc. Jpn.*, **69**, 1449–1457.

56 Kresge, C.T., Leonowicz, M.E., Roth, W.J., Vartuli, J.C., and Beck, J.S. (1992) *Nature*, **359**, 710–712.

57 Beck, J.S., Vartuli, J.C., Roth, W.J., Leonowicz, M.E., Kresge, C.T., Schmitt, K.D., Chu, C.T.-W., Olson, D.H., Sheppard, E.W., McCullen, S.B., Higgins, J.B., and Schlenker, J.L. (1992) *J. Am. Chem. Soc.*, **114**, 10834–10843.

58 Gu, D. and Schüth, F. (2014) *Chem. Soc. Rev.*, **43**, 313–344.

59 Yamauchi, Y. and Kuroda, K. (2008) *Chem. Asian J.*, **3**, 664–676.

60 Wan, Y., Shi, Y., and Zhao, D. (2008) *Chem. Mater.*, **20**, 932–945.

61 Hoffman, F., Cornelius, M., Morell, J., and Fröba, M. (2006) *Angew. Chem., Int. Ed.*, **45**, 3216–3251.

62 Monnier, A., Schüth, F., Huo, Q., Kumar, D., Margolese, D., Maxwell, R.S., Stucky, G.D., Krishnamurty, M., Petroff, P., Firouzi, A., Janicke, M., and Chmelka, B.F. (1993) *Science*, **261**, 1299–1303.

63 Israelachvili, J.N., Mitchell, D.J., and Ninham, B.W. (1977) *Biochim. Biophys. Acta*, **470**, 185–201.

64 Huo, Q., Margolese, D.I., and Stucky, G.D. (1996) *Chem. Mater.*, **8**, 1147–1160.

65 Zhao, D., Feng, J., Huo, Q., Melosh, N., Fredrickson, G.H., Chmelka, B.F., and Stucky, G.D. (1998) *Science*, **279**, 548–552.

66 Yu, K., Hurd, A.J., Eisenberg, A., and Brinker, C.J. (2001) *Langmuir*, **17**, 7961–7965.

67 Kuang, D., Brezesinski, T., and Smarsly, B. (2004) *J. Am. Chem. Soc.*, **126**, 10534–10535.

68 Ogawa, M. (1994) *J. Am. Chem. Soc.*, **116**, 7941–7942.

69 Ogawa, M. (1996) *Chem. Commun.*, 1149–1150.

70 Lu, Y., Ganguli, R., Drewien, C.A., Anderson, M.T., Brinker, C.J., Gong, W., Guo, Y., Soyez, H., Dunn, B., Huang, M.H., and Zink, J.I. (1997) *Nature*, **389**, 364–368.

71 Yang, H., Coombs, N., Sokolov, I., and Ozin, G.A. (1996) *Nature*, **381**, 589–592.

72 Madhugiri, S., Zhou, W., Ferraris, J.P., and Balkus, K.J., Jr. (2003) *Microporous Mesoporous Mater.*, **63**, 75–84.

73 Feng, P., Bu, X., Stucky, G.D., and Pine, D.J. (2000) *J. Am. Chem. Soc.*, **122**, 994–995.

74 Lu, Y., Fan, H., Stump, A., Ward, T.L., Rieker, T., and Brinker, C.J. (1999) *Nature*, **398**, 223–226.
75 Büchel, G., Grün, M., Unger, K.K., Matsumoto, A., and Tsutsumi, K. (1998) *Supramol. Sci.*, **5**, 253–259.
76 Yano, K., Suzuki, N., Akimoto, Y., and Fukushima, Y. (2002) *Bull. Chem. Soc. Jpn.*, **75**, 1977–1982.
77 Yano, K. and Fukushima, Y. (2004) *J. Mater. Chem.*, **14**, 1479–1584.
78 Urata, C., Aoyama, Y., Tonegawa, A., Yamauchi, Y., and Kuroda, K. (2009) *Chem. Commun.*, **34**, 5094–5096.
79 Wu, Y., Cheng, G., Katsov, K., Sides, S.W., Wang, J., Tang, J., Fredrickson, G.H., Moskovits, M., and Stucky, G.D. (2004) *Nat. Mater.*, **3**, 816–822.
80 Attard, G.S., Glyde, J.C., and Göltner, C.G. (1995) *Nature*, **378**, 366–368.
81 Attard, G.S., Göltner, C.G., Corker, J.M., Henke, S., and Templer, R.H. (1997) *Angew. Chem., Int. Ed.*, **36**, 1315–1317.
82 Attard, G.S., Bartlett, P.N., Coleman, N.R.B., Elliott, J.M., Owen, J.R., and Wang, J.H. (1997) *Science*, **278**, 838–840.
83 Nelson, P.A., Elliott, J.M., Attard, G.S., and Owen, J.R. (2002) *Chem. Mater.*, **14**, 524–529.
84 Kijima, T., Yoshimura, T., Uota, M., Ikeda, T., Fujikawa, D., Mouri, S., and Uoyama, S. (2004) *Angew. Chem., Int. Ed.*, **43**, 228–232.
85 Thepkaew, J., Therdthianwong, S., Kucernak, A., and Therdthianwong, A. (2012) *J. Electroanal. Chem.*, **685**, 41–46.
86 Yamauchi, Y., Momma, T., Fuziwara, M., Nair, S.S., Ohsuna, T., Terasaki, O., Osaka, T., and Kuroda, K. (2005) *Chem. Mater.*, **17**, 6342–6348.
87 Yamauchi, Y., Momma, T., Yokoshima, T., Kuroda, K., and Osaka, T. (2005) *J. Mater. Chem.*, **15**, 1987–1994.
88 Yamauchi, Y., Nair, S.S., Momma, T., Ohsuna, T., Osaka, T., and Kuroda, K. (2006) *J. Mater. Chem.*, **16**, 2229–2234.
89 Yamauchi, Y., Ohsuna, T., and Kuroda, K. (2007) *Chem. Mater.*, **19**, 1335–1342.
90 Yamauchi, Y., Komatsu, M., Fuziwara, M., Nemoto, Y., Sato, K., Yokoshima, T., Sukegawa, H., Inomata, K., and Kuroda, K. (2009) *Angew. Chem., Int. Ed.*, **48**, 7792–7797.
91 Yamauchi, Y., Tonegawa, A., Komatsu, M., Wang, H., Wang, L., Nemoto, Y., Suzuki, N., and Kuroda, K. (2012) *J. Am. Chem. Soc.*, **134**, 5100–5109.
92 Kimura, T., Kamata, T., Fuziwara, M., Takano, Y., Kaneda, M., Sakamoto, Y., Terasaki, O., Sugahara, Y., and Kuroda, K. (2000) *Angew. Chem., Int. Ed.*, **39**, 3855–3859.
93 Kimura, T., Tamura, H., Tezuka, M., Mochizuki, D., Shigeno, T., Ohsuna, T., and Kuroda, K. (2008) *J. Am. Chem. Soc.*, **130**, 201–209.
94 Shimojima, A., Sugahara, Y., and Kuroda, K. (1997) *Bull. Chem. Soc. Jpn.*, **70**, 2847–2853.
95 Shimojima, A., Sugahara, Y., and Kuroda, K. (1998) *J. Am. Chem. Soc.*, **120**, 4528–4529.
96 Shimojima, A. and Kuroda, K. (2003) *Angew. Chem., Int. Ed.*, **42**, 4057–4060.
97 Shimojima, A., Liu, Z., Ohsuna, T., Terasaki, O., and Kuroda, K. (2005) *J. Am. Chem. Soc.*, **127**, 14108–14116.
98 Shimojima, A., Goto, R., Atsumi, N., and Kuroda, K. (2008) *Chem. Eur. J.*, **14**, 8500–8506.
99 Sakamoto, S., Shimojima, A., Miyasaka, K., Ruan, J., Terasaki, O., and Kuroda, K. (2009) *J. Am. Chem. Soc.*, **131**, 9634–9635.
100 Sakamoto, S., Tamura, Y., Hata, H., Sakamoto, Y., Shimojima, A., and Kuroda, K. (2014) *Angew. Chem., Int. Ed.*, **53**, 9173–9177.
101 Bates, F.S. (1991) *Science*, **251**, 898–905.
102 Pai, R.A., Humayun, R., Schulberg, M.T., Sengupta, A., Sun, J.-N., and Watkins, J.J. (2004) *Science*, **303**, 507–510.
103 Warren, S.C., DiSalvo, F.J., and Wiesner, U. (2007) *Nat. Mater.*, **6**, 156–161.
104 Warren, S.C., Messina, L.C., Slaughter, L.S., Kamperman, M., Zhou, Q., Gruner, S.M., DiSalvo, F.J., and Wiesner, U. (2008) *Science*, **320**, 1748–1752.
105 Lu, A.-H. and Schüth, F. (2006) *Adv. Mater.*, **18**, 1793–1805.
106 Yang, H., Shi, Q., Tian, B., Lu, Q., Gao, F., Xie, S., Fan, J., Yu, C., Tu, B., and Zhao, D. (2003) *J. Am. Chem. Soc.*, **125**, 4724–4725.
107 Tian, B., Liu, X., Solovyov, L.A., Liu, Z., Liu, Z., Yang, H., Zhang, Z., Xie, S.,

Zhang, F., Tu, B., Yu, C., Terasaki, O., and Zhao, D. (2004) *J. Am. Chem. Soc.*, **126**, 865–875.

108 Li, W.-C., Lu, A.-H., Weidenthaler, C., and Schüth, F. (2004) *Chem. Mater.*, **16**, 5676–5681.

109 Ryoo, R., Joo, S.H., and Jun, S. (1999) *J. Phys. Chem. B*, **103**, 7743–7746.

110 Jun, S., Joo, S.H., Ryoo, R., Kruk, M., Jaroniec, M., Liu, Z., Ohsuna, T., and Terasaki, O. (2000) *J. Am. Chem. Soc.*, **122**, 10712–10713.

111 Kawashima, D., Aihara, T., Kobayashi, Y., Kyotani, T., and Tomita, A. (2000) *Chem. Mater.*, **12**, 3397–3401.

112 Lu, A.-H., Schmidt, W., Spliethoff, B., and Schüth, F. (2003) *Adv. Mater.*, **15**, 1602–1606.

113 Shi, Y., Wan, Y., and Zhao, D. (2011) *Chem. Soc. Rev.*, **40**, 3854–3878.

114 Shin, H.J., Ryoo, R., Liu, Z., and Terasaki, O. (2001) *J. Am. Chem. Soc.*, **123**, 1246–1247.

115 Wang, D., Zhou, W.L., McCaughy, B.F., Hampsey, J.E., Ji, X., Jiang, Y.-B., Xu, H., Tang, J., Schmehl, R.H., O'Connor, C., Brinker, C.J., and Lu, Y. (2003) *Adv. Mater.*, **15**, 130–133.

116 Kumai, Y., Tsukada, H., Akimoto, Y., Sugimoto, N., Seno, Y., Fukuoka, A., Ichikawa, M., and Inagaki, S. (2006) *Adv. Mater.*, **18**, 760–762.

117 Takai, A., Doi, Y., Yamauchi, Y., and Kuroda, K. (2010) *J. Phys. Chem. C*, **114**, 7586–7593.

118 Jiao, F., Hill, A.H., Harrison, A., Berko, A., Chadwick, A.V., and Bruce, P.G. (2008) *J. Am. Chem. Soc.*, **130**, 5262–5266.

119 Imhof, A. and Pine, D.J. (1997) *Nature*, **389**, 948–951.

120 Zhang, H., Hardy, G.C., Khimyak, Y.J., and Rosseinsky, M.J., Cooper, A.I. (2004) *Chem. Mater.*, **16**, 4245–4256.

121 Osseo-Asare, K. and Arriagada, F.J. (1990) *Colloids Surf.*, **50**, 321–339.

122 Pileni, M.P., Lisiecki, I., Motte, L., and Petit, C. (1992) *Res. Chem. Intermed.*, **17**, 101–113.

123 Petit, C., Lixon, P., and Pileni, M.-P. (1993) *J. Phys. Chem.*, **97**, 12974–12983.

124 Hopwood, J.D. and Mann, S. (1997) *Chem. Mater.*, **9**, 1819–1828.

125 Lee, Y., Lee, J., Bae, C.J., Park, J.-G., Noh, H.-J., Park, J.-H., and Hyeon, T. (2005) *Adv. Funct. Mater.*, **15**, 503–509.

126 Wang, C.J., Wu, Y.A., Jacobs, R.M.J., Warner, J.H., Williams, G.R., and O'Hare, D. (2011) *Chem. Mater.*, **23**, 171–180.

127 Bunz, U.H.F. (2006) *Adv. Mater.*, **18**, 973–989.

128 Karthaus, O., Maruyama, N., Cieren, X., Shimomura, M., Hasegawa, H., and Hashimoto, T. (2000) *Langmuir*, **16**, 6071–6076.

129 Srinivasarao, M., Collings, D., Philips, A., and Patel, S. (2001) *Science*, **292**, 79–83.

130 Widawski, G., Rawiso, M., and François, B. (1994) *Nature*, **369**, 387–389.

131 Gutiérez, M.C., Ferrer, M.L., and del Monte, F. (2008) *Chem. Mater.*, **20**, 634–648.

132 Mahler, W. and Bechtold, M.F. (1980) *Nature*, **285**, 27–28.

133 Nishihara, H., Mukai, S.R., Yamashita, D., and Tamon, H. (2005) *Chem. Mater.*, **17**, 683–689.

134 Sofie, S.W. and Dogan, F. (2001) *J. Am. Ceram. Soc.*, **84**, 1459–1464.

135 Johnson, D.W. and Schnettler, F.J. (1970) *J. Am. Ceram. Soc.*, **53**, 440–444.

136 Mukai, S.R., Nishihara, H., Shichi, S., and Tamon, H. (2004) *Chem. Mater.*, **16**, 4987–4991.

137 Mukai, S.R., Nishihara, H., Yoshida, T., Taniguchi, K., and Tamon, H. (2005) *Carbon*, **43**, 1557.

138 Chen, G., Uchida, T., and Tateishi, T. (2001) *Biomaterials*, **22**, 2563–2567.

139 Kang, H.-W., Tabata, Y., and Ikada, Y. (1999) *Biomaterials*, **20**, 1339–1344.

140 Daamen, W.F., Van Moerkerk, H.Th.B., Hafmans, T., Buttafoco, L., Poot, A.A., Veerkamp, J.H., and Van Kuppevelt, T.H. (2003) *Biomaterials*, **24**, 4001–4009.

141 Munch, E., Launey, M.E., Alsem, D.H., Saiz, E., Tomsia, A.P., and Ritchie, R.O. (2008) *Science*, **322**, 1516–1520.

142 Xia, Y., Gates, B., Yin, Y., and Lu, Y. (2000) *Adv. Mater.*, **12**, 693–713.

143 Yan, F. and Goedel, W.A. (2004) *Adv. Mater.*, **16**, 911–915.

144 Velev, O.D., Jede, T.A., Lobo, R.F., and Lenhoff, A.M. (1997) *Nature*, **389**, 447–448.

145 Holland, B.T., Blanford, C.F., and Stein, A. (1998) *Science*, **281**, 538–540.
146 Ye, Y.-H., LeBlanc, F., Haché, A., and Truong, V.-V. (2001) *Appl. Phys. Lett.*, **78**, 52–54.
147 Gu, Z.-Z., Fujishima, A., and Sato, O. (2002) *Angew. Chem., Int. Ed.*, **41**, 2068–2070.
148 Holland, B.T., Blanford, C.F., Do, T., and Stein, A. (1999) *Chem. Mater.*, **11**, 795–805.
149 Yan, H., Blanford, C.F., Holland, B.T., Smyrl, W.H., and Stein, A. (2000) *Chem. Mater.*, **12**, 1134–1141.
150 Li, H., Zhang, L., Dai, H., and He, H. (2009) *Inorg. Chem.*, **48**, 4421–4434.
151 Géraud, E., Prévot, V., Ghanbaja, J., and Leroux, F. (2006) *Chem. Mater.*, **18**, 238–240.
152 Li, C. and Qi, L. (2008) *Angew. Chem., Int. Ed.*, **47**, 2388–2393.
153 Zakhidov, A.A., Baughman, R.H., Iqbal, Z., Cui, C., Khayrullin, I., Dantas, S.O., Marti, J., and Ralchenko, V.G. (1998) *Science*, **282**, 897–901.
154 Velev, O.D., Tessier, P.M., Lenhoff, A.M., and Kaler, E.W. (1999) *Nature*, **401**, 548.
155 Jiang, P., Cizeron, J., Bertone, J.F., and Colvin, V.L. (1999) *J. Am. Chem. Soc.*, **121**, 7957–7958.
156 Braun, P.V. and Wiltzius, P. (1999) *Nature*, **402**, 603–604.
157 Blanco, A., Chomski, E., Grabtchak, S., Ibisate, M., John, S., Leonard, S.W., Lopez, C., Meseguer, F., Miguez, H., Mondia, J.P., Ozin, G.A., Toader, O., and van Driel, H.M. (2000) *Nature*, **405**, 437–440.
158 Jiang, P., Hwang, K.S., Mittleman, D.M., Bertone, J.F., and Colvin, V.L. (1999) *J. Am. Chem. Soc.*, **121**, 11630–11637.
159 Sung, I.-K., Christian, MitchellM., Kim, D.-P., and Kenis, P.J.A. (2005) *Adv. Funct. Mater.*, **15**, 1336–1342.
160 Yokoi, T., Sakamoto, Y., Terasaki, O., Kubota, Y., Okubo, T., and Tatsumi, T. (2006) *J. Am. Chem. Soc.*, **128**, 13664–13665.
161 Hartlen, K.D., Athanasopoulos, A.P.T., and Kitaev, V. (2008) *Langmuir*, **24**, 1714–1720.
162 Snyder, M.A., Lee, J.A., Tracy, M., Scriven, L.E., and Tsapatsis, M. (2007) *Langmuir*, **23**, 9924–9928.
163 Fan, W., Snyder, M.A., Kumar, S., Lee, P.-S., Yoo, W.C., McCormick, A.V., Penn, R.L., Stein, A., and Tsapatsis, M. (2008) *Nat. Mater.*, **7**, 984–991.
164 Kuroda, Y., Yamauchi, Y., and Kuroda, K. (2010) *Chem. Commun.*, **46**, 1827–1829.
165 Kuroda, Y. and Kuroda, K. (2010) *Angew. Chem., Int. Ed.*, **49**, 6993–6997.
166 Kuroda, Y., Sakamoto, Y., and Kuroda, K. (2012) *J. Am. Chem. Soc.*, **134**, 8684–8692.
167 Fukasawa, Y., Takanabe, K., Shimojima, A., Antonietti, M., Domen, K., and Okubo, T. (2011) *Chem. Asian J.*, **6**, 103–109.
168 Vu, A., Li, X., Phillips, J., Han, A., Smyrl, W.H., Buhlmann, P., and Stein, A. (2013) *Chem. Mater.*, **25**, 4137–4148.
169 Caruso, F., Caruso, R.A., and Möhwald, H. (1998) *Science*, **282**, 1111–1114.
170 Donath, E., Sukhorukov, G.B., Caruso, F., Davis, S.A., and Möhwald, H. (1998) *Angew. Chem., Int. Ed.*, **37**, 2201–2205.
171 Li, L., Ma, R., Ebina, Y., Iyi, N., and Sasaki, T. (2005) *Chem. Mater.*, **17**, 4386–4391.
172 Correa-Duarte, M.A., Kosiorek, A., Kandulski, W., Giersig, M., and Liz-Marzán, L.M. (2005) *Chem. Mater.*, **17**, 3268–3272.
173 Kuroda, Y. and Kuroda, K. (2008) *Sci. Technol. Adv. Mater.*, **9**, 025018.
174 Imhof, A. (2001) *Langmuir*, **17**, 3579–3585.
175 Ding, S., Chen, J.S., Qi, G., Duan, X., Wang, Z., Giannelis, E.P., Archer, L.A., and Lou, X.W. (2011) *J. Am. Chem. Soc.*, **133**, 21–23.
176 Qi, G., Wang, Y., Estevez, L., Switzer, A.K., Duan, X., Yang, X., and Giannelis, E.P. (2010) *Chem. Mater.*, **22**, 2693–2695.
177 Sun, Y., Mayers, B.T., and Xia, Y. (2002) *Nano Lett.*, **2**, 481–485.
178 Sun, Y. and Xia, Y. (2004) *J. Am. Chem. Soc.*, **126**, 3892–3901.
179 Sun, Y., Mayers, B., and Xia, Y. (2003) *Adv. Mater.*, **15**, 641–646.
180 Nassif, N. and Livage, J. (2011) *Chem. Soc. Rev.*, **40**, 849–859.

181 Yu, Y., Addai-Mensah, J., and Losic, D. (2010) *Langmuir*, **26**, 14068–14072.

182 Cai, Y., Allan, S.M., Sandhage, K.H., and Zalar, F.M. (2005) *J. Am. Ceram. Soc.*, **88**, 2005–2010.

183 Hagiwara, Y., Shimojima, A., and Kuroda, K. (2008) *Chem. Mater.*, **20**, 1147–1153.

184 Fan, J., Boettcher, S.W., and Stucky, G.D. (2006) *Chem. Mater.*, **18**, 6391–6396.

185 Marceau, E., Carrier, X., and Che, M. (2009) *Impregnation and Drying, Synthesis of Solid Catalysts* (ed. K.P. de Jong), Wiley-VCH Verlag GmbH, Weinheim, pp. 59–82.

186 Xia, Y. and Mokaya, R. (2004) *Adv. Mater.*, **16**, 1553–1558.

187 Nishiyama, N., Tanaka, S., Egashira, Y., Oku, Y., and Ueyama, K. (2003) *Chem. Mater.*, **15**, 1006–1011.

188 Tsung, C.-K., Fan, J., Zheng, N., Shi, Q., Forman, A.J., Wang, J., and Stucky, G.D. (2008) *Angew. Chem., Int. Ed.*, **47**, 8682–8686.

189 Sanchez, C., Boissière, C., Grosso, D., Laberty, C., and Nicole, L. (2008) *Chem. Mater.*, **20**, 682–737.

190 Johnson, S.A., Ollivier, P.J., and Mallouk, T.E. (1999) *Science*, **283**, 963–965.

191 Meng, Y., Gu, D., Zhang, F., Shi, Y., Yang, H., Li, Z., Yu, C., Tu, B., and Zhao, D. (2005) *Angew. Chem., Int. Ed.*, **44**, 7053–7059.

192 Liu, R., Shi, Y., Wan, Y., Meng, Y., Zhang, F., Gu, D., Chen, Z., Tu, B., and Zhao, D. (2006) *J. Am. Chem. Soc.*, **128**, 11652–11662.

193 Ryoo, R., Joo, S.H., and Jun, S. (1999) *J. Phys. Chem. B*, **103**, 7743–7746.

194 Kruk, M., Dufour, B., Celer, E.B., Kowalewski, T., Jaroniec, M., and Matyjazewski, K. (2005) *J. Phys. Chem. B*, **109**, 9216–9225.

195 Joo, S.H., Choi, S.J., Oh, I., Kwak, J., Liu, Z., Terasaki, O., and Ryoo, R. (2001) *Nature*, **412**, 169–172.

196 Liang, C., Hong, K., Guiochon, G.A., Mays, J.W., and Dai, S. (2004) *Angew. Chem., Int. Ed.*, **43**, 5785–5789.

197 Liu, R., Wu, D., Feng, X., and Müllen, K. (2010) *Angew. Chem., Int. Ed.*, **49**, 2565–2569.

198 Szamocki, R., Reculusa, S., Ravaine, S., Bartlett, P.N., Kuhn, A., and Hempelmann, R. (2006) *Angew. Chem., Int. Ed.*, **45**, 1317–1321.

199 Yamauchi, Y., Takai, A., Komatsu, M., Sawada, M., Ohsuna, T., and Kuroda, K. (2008) *Chem. Mater.*, **20**, 1004–1011.

200 Wang, H., Jeong, H.Y., Imura, M., Wang, L., Radhakrishnan, L., Fujita, N., Castle, T., Terasaki, O., and Yamauchi, Y. (2011) *J. Am. Chem. Soc.*, **133**, 14526–14529.

201 Grosso, D., Boissière, C., Smarsly, B., Brezesinski, T., Pinna, N., Albouy, P.A., Amenitsch, H., Antonietti, M., and Sanchez, C. (2004) *Nat. Mater.*, **3**, 787–792.

202 Shirokura, N., Nakajima, K., Nakabayashi, A., Lu, D., Hara, M., Domen, K., Tatsumi, T., and Kondo, J.N. (2006) *Chem. Commun.*, 2188–2190.

203 Lee, J., Orilall, M.C., Warren, S.C., Kamperman, M., DiSalvo, F.J., and Wiesner, U. (2008) *Nat. Mater.*, **7**, 222–228.

204 Asefa, T., MacLachlan, M.J., Coombs, N., and Ozin, G.A. (1999) *Nature*, **402**, 867–871.

205 Inagaki, S., Guan, S., Fukushima, Y., Ohsuna, T., and Terasaki, O. (1999) *J. Am. Chem. Soc.*, **121**, 9611–9614.

206 Kawi, S. and Lai, M.W. (2002) *AIChE J.*, **48**, 1572–1580.

207 Huang, X., Yang, J., Mao, S., Chang, J., Hallac, P.B., Fell, C.R., Metz, B., Jiang, J., Hurley, P.T., and Chen, J. (2014) *Adv. Mater.*, **26**, 4326–4332.

208 Huo, Q., Margolese, D.I., Ciesla, U., Feng, P., Gier, T.E., Sieger, P., Leon, R., Petroff, P.M., Schüth, F., and Stucky, G.D. (1994) *Nature*, **368**, 317–321.

209 Che, S., Garcia-Bennett, A.E., Yokoi, T., Sakamoto, K., Kunieda, H., Terasaki, O., and Tatsumi, T. (2003) *Nat. Mater.*, **2**, 801–805.

210 Che, S., Liu, Z., Ohsuna, T., Sakamoto, K., Terasaki, O., and Tatsumi, T. (2004) *Nature*, **429**, 281–284.

211 Tanev, P.T. and Pinnavaia, T.J. (1995) *Science*, **267**, 865–867.

212 Bagshaw, S.A., Prouzet, E., and Pinnavaia, T.J. (1995) *Science*, **269**, 1242–1244.

213 Leo, C.P., Ahmad Kamil, N.H., Junaidi, M.U.M., Kamal, S.N.M., and Ahmad, A.L. (2013) *Sep. Purif. Technol.*, **103**, 84–91.

214 Slowing, I.I., Vivero-Escoto, J.L., Wu, C.-W., and Lin, V.S.-Y. (2008) *Adv. Drug Deliv. Rev.*, **60**, 1278–1288.

215 Rhodes, K.H., Davis, S.A., Caruso, F., Zhang, B., and Mann, S. (2000) *Chem. Mater.*, **12**, 2832–2834.

216 Valtchev, V. and Tosheva, L. (2013) *Chem. Rev.*, **113**, 6734–6760.

217 Lee, P.-S., Zhang, X., Stoeger, J.A., Malek, A., Fan, W., Kumar, S., Yoo, W.C., Hashimi, S.A., Penn, R.L., Stein, A., and Tsapatsis, M. (2011) *J. Am. Chem. Soc.*, **133**, 493–502.

218 Stöber, W., Fink, A., and Bohn, E. (1968) *J. Colloid Interface Sci.*, **26**, 62–69.

219 Yamamoto, E., Kitahara, M., Tsumura, T., and Kuroda, K. (2014) *Chem. Mater.*, **26**, 2927–2933.

220 Wu, Y., Cheng, G., Katsov, K., Sides, S.W., Wang, J., Tang, J., Fredrickson, G.H., Moskovits, M., and Stucky, G.D. (2004) *Nat. Mater.*, **3**, 816–822.

221 Yamauchi, Y., Takai, A., Nagaura, T., Inoue, S., and Kuroda, K. (2008) *J. Am. Chem. Soc.*, **130**, 5426–5427.

222 Sakurai, M., Shimojima, A., Yamauchi, Y., and Kuroda, K. (2008) *Langmuir*, **24**, 13121–13126.

223 Madhugiri, S., Dalton, A., Gutierrez, J., Ferraris, J.P., and Balkus, K.J., Jr. (2003) *J. Am. Chem. Soc.*, **125**, 14531–14538.

224 Thompson, G.E. (1997) *Thin Solid Films*, **297**, 192–201.

225 Lovallo, M.C. and Tsapatsis, M. (1996) *AIChE J.*, **42**, 3020–3029.

226 Choi, J., Jeong, H.-K., Snyder, M.A., Stoeger, J.A., Masel, R.I., and Tsapatsis, M. (2009) *Science*, **325**, 590–593.

227 Lee, J.S., Kim, J.H., Lee, Y.J., Jeong, N.C., and Yoon, K.B. (2007) *Angew. Chem., Int. Ed.*, **46**, 3087–3090.

228 Yoo, W.C., Stoeger, J.A., Lee, P.-S., Tsapatsis, M., and Stein, A. (2010) *Angew. Chem., Int. Ed.*, **49**, 8699–8703.

229 Yang, H., Kuperman, A., Coombs, N., Mamiche-Afara, S., and Ozin, G.A. (1996) *Nature*, **379**, 703–705.

230 Aksay, I.A., Trau, M., Manne, S., Honma, I., Yao, N., Zhou, L., Fenter, P., Eisenberger, P.M., and Gruner, S.M. (1996) *Science*, **273**, 892–898.

231 Suzuki, T., Kanno, Y., Morioka, Y., and Kuroda, K. (2008) *Chem. Commun.*, **28**, 3284–3286.

232 Miyata, H. and Kuroda, K. (1999) *J. Am. Chem. Soc.*, **121**, 7618–7624.

233 Miyata, H. and Kuroda, K. (1999) *Adv. Mater.*, **11**, 1448–1452.

234 Miyata, H. and Kuroda, K. (2000) *Chem. Mater.*, **12**, 49–54.

235 Miyata, H., Noma, T., Watanabe, M., and Kuroda, K. (2000) *Chem. Mater.*, **14**, 766–772.

236 Miyata, H., Kawashima, Y., Itoh, M., and Watanabe, M. (2005) *Chem. Mater.*, **17**, 5323–5327.

237 Miyata, H., Fukushima, Y., Okamoto, K., Takahashi, M., Watanabe, M., Kubo, W., Komoto, A., Kitamura, S., Kanno, Y., and Kuroda, K. (2011) *J. Am. Chem. Soc.*, **133**, 13539–13544.

238 Miyata, H., Suzuki, T., Fukuoka, A., Sawada, T., Watanabe, M., Noma, T., Takada, K., Mukaide, T., and Kuroda, K. (2004) *Nat. Mater.*, **3**, 651–656.

239 Firouzi, A., Schaefer, D.J., Tolbert, S.H., Stucky, G.D., and Chmelka, B.F. (1997) *J. Am. Chem. Soc.*, **119**, 9466–9477.

240 Tolbert, S.H., Firouzi, A., Stucky, G.D., and Chmelka, B.F. (1997) *Science*, **278**, 264–268.

241 Yamauchi, Y., Sawada, M., Noma, T., Ito, H., Furumi, S., Sakka, Y., and Kuroda, K. (2005) *J. Mater. Chem.*, **15**, 1137–1140.

242 Yamauchi, Y., Sawada, M., Sugiyama, A., Osaka, T., Sakka, Y., and Kuroda, K. (2006) *J. Mater. Chem.*, **16**, 3693–3700.

243 Yamauchi, Y., Sawada, M., Komatsu, M., Sugiyama, A., Osaka, T., Hirota, N., Sakka, Y., and Kuroda, K. (2007) *Chem. Asian J.*, **2**, 1505–1512.

244 Walcarius, A., Sibottier, E., Etienne, M., and Ghanbaja, J. (2007) *Nat. Mater.*, **6**, 602–608.

245 Richman, E.K., Brezesinski, T., and Tolbert, S.H. (2008) *Nat. Mater.*, **7**, 712–717.

246 Yamauchi, Y., Nagaura, T., Ishikawa, A., Chikyow, T., and Inoue, S. (2008) *J. Am. Chem. Soc.*, **130**, 10165–10170.

247 Teng, Z., Zheng, G., Dou, Y., Li, W., Mou, C.-Y., Zhang, X., Asiri, A.M., and Zhao, D. (2012) *Angew. Chem., Int. Ed.*, **51**, 2173–2177.

248 Kao, K.-C., Lin, C.-H., Chen, T.-Y., Liu, Y.-H., and Mou, C.-Y. (2015) *J. Am. Chem. Soc.*, **137**, 3779–3782.

249 Gu, Z.-Z., Kubo, S., Fujishima, A., and Sato, O. (2002) *Appl. Phys. A*, **74**, 127–129.

250 Ogura, M., Shinomiya, S., Tateno, J., Nara, Y., Nomura, M., Kikuchi, E., and Matsukata, M. (2001) *Appl. Catal. A*, **219**, 33–43.

251 Choi, M., Na, K., Kim, J., Sakamoto, Y., Terasaki, O., and Ryoo, R. (2009) *Nature*, **461**, 246–250.

252 Minakuchi, H., Nakanishi, K., Soga, N., Ishizuka, N., and Tanaka, N. (1996) *Anal. Chem.*, **68**, 3498–3501.

253 Sel, O., Kuang, D., Thommes, M., and Smarsly, B. (2006) *Langmuir*, **22**, 2311–2322.

254 Valtchev, V. (2002) *Chem. Mater.*, **14**, 4371–4377.

255 Sen, T., Tiddy, G.J.T., Casci, J.L., and Anderson, M.W. (2003) *Angew. Chem., Int. Ed.*, **42**, 4649–4653.

256 Wang, L., Ebina, Y., Takada, K., and Sasaki, T. (2004) *J. Phys. Chem. B*, **108**, 4283–4288.

257 Kuroda, Y., Tamakoshi, M., Murakami, J., and Kuroda, K. (2007) *J. Ceram. Soc. Jpn.*, **115**, 233–236.

258 Nakanishi, K. and Kanamori, K. (2005) *J. Mater. Chem.*, **15**, 3776–3786.

259 Nakanishi, K. and Tanaka, N. (2007) *Acc. Chem. Res.*, **40**, 863–873.

260 Choi, M., Cho, H.S., Srivastava, R., Venkatesan, C., Choi, D.-H., and Ryoo, R. (2006) *Nat. Mater.*, **5**, 718–723.

261 Na, K., Jo, C., Kim, J., Cho, K., Jung, J., Seo, Y., Messinger, R.J., Chmelka, B.F., and Ryoo, R. (2011) *Science*, **333**, 328–332.

262 Corma, A. (1997) *Chem. Rev.*, **97**, 2373–2419.

263 Caro, J., Noack, M., Kölsch, P., and Schäfer, R. (2000) *Microporous Mesoporous Mater.*, **38**, 3–24.

264 Varoon, K., Zhang, X., Elyassi, B., Brewer, D.D., Gettel, M., Kumar, S., Lee, J.A., Maheshwari, S., Mittal, A., Sung, C.-Y., Cococcioni, M., Francis, L.F., McCormick, A.V., Mkhoyan, K.A., and Tsapatsis, M. (2011) *Science*, **334**, 72–75.

265 Taguchi, A. and Schüth, F. (2005) *Microporous Mesoporous Mater.*, **77**, 1–45.

266 Ismail, A.A. and Bahnemann, D.W. (2011) *J. Mater. Chem.*, **21**, 11686–11707.

267 Inagaki, S., Ohtani, O., Goto, Y., Okamoto, K., Ikai, M., Yamanaka, K., Tani, T., and Okada, T. (2009) *Angew. Chem., Int. Ed.*, **48**, 4042–4046.

268 Takeda, H., Ohashi, M., Tani, T., Ishitani, O., and Inagaki, S. (2010) *Inorg. Chem.*, **49**, 4554–4559.

269 Jiao, F. and Bruce, P.G. (2007) *Adv. Mater.*, **19**, 657–660.

270 Zhai, Y., Dou, Y., Zhao, D., Fulvio, P.F., Mayes, R.T., and Dai, S. (2011) *Adv. Mater.*, **23**, 4828–4850.

271 Molenkamp, W.C., Watanabe, M., Miyata, H., and Tolbert, S.H. (2004) *J. Am. Chem. Soc.*, **126**, 4476–4477.

272 Vivero-Escoto, J.L., Slowing, I.I., Trewyn, B.G., and Lin, V.S.-Y. (2010) *Small*, **6**, 1952–1967.

273 Stein, A., Wilson, B.E., and Rudisill, S.G. (2013) *Chem. Soc. Rev.*, **42**, 2763–2803.

9
Bio-Inspired Synthesis and Application of Functional Inorganic Materials by Polymer-Controlled Crystallization

Lei Liu and Shu-Hong Yu

University of Science and Technology of China, Department of Chemistry, Division of Nanomaterials and Chemistry, Hefei National Laboratory for Physical Sciences at Microscale, Jinzhai Road 96, Hefei 230026, P. R. China

9.1
Introduction

Living organisms in nature are able to impose ingenious control over the crystallization and organization of a wide range of inorganic crystals (calcium phosphate (CaP), calcium carbonate ($CaCO_3$), silica (SiO_2), iron oxide (FeO_x), etc.) via biomineralization, creating a diverse of so-called biominerals such as skeleton, tooth, shell, siliceous sponge, and magnetite in magnetotactic bacteria [1,2]. These biominerals are typically inorganic–organic composite materials showing hierarchically ordered superstructures with several levels, from molecular to nanoscale, then to microscale, and finally to macroscale, which endow them with superior properties surpassing their geological and synthetic counterparts to meet specific or multiple functions such as mechanical support and protection [3], predation [4,5], navigation [6], optical sensing [7], and graviperception [8]. Biomineralization is considered to be an optimized process for materials preparation for a given set of requirements and restrictions through billions of years evolution on life and a great reservoir of inspiration for the design and preparation of next-generation advanced functional materials [9–11].

Compared with conventional *in vitro* chemical synthesis methods, biomineralization demonstrates several exceeding advantages: (i) Biomineralization generally occurs under mild conditions, that is, ambient temperature and pressure, aqueous media, and physiological pH near neutral value. (ii) Most of the elements employed by biomineralization for producing biominerals are earth abundant elements encompassing calcium, iron, carbon, oxygen, silicon, phosphorus, and so on. (iii) Precise controls ranging from morphological to spatial, from chemical to crystallographic, from microstructural to superstructural over the growth and assembly of inorganic crystals are involved and perfectly performed during whole biomineralization process. (iv) All the above controls

Handbook of Solid State Chemistry, First Edition. Edited by Richard Dronskowski, Shinichi Kikkawa, and Andreas Stein.

consequently lead to the last superiority, that is, the complex structure with ordered hierarchy within biominerals and the resulting unique and remarkable properties of these biogenic inorganic–organic hybrid materials. One of the most famous examples is nacre, exhibiting a multilayered "brick-and-mortar" structure where aragonite $CaCO_3$ tablets and biopolymers act as "brick" and "mortar," respectively. This well-organized combination of inorganic and organic materials accounts for high damage tolerance derived from improvements in both strength and toughness by a twofold and three orders higher than those of pure aragonite crystals, respectively [12].

Due to the mild synthesis conditions and eco-friendly process of biomineralization and the fascinating architecture and outstanding properties of biominerals, a rich family of research projects focusing on controlled synthesis of inorganic and inorganic–organic composite materials through biomimetic and bio-inspired strategies have been motivated and gained considerable interest and achievements during the past decades. These studies tried to transform the optimized designs and underlying mechanisms in biomineralization to artificial synthesis of inorganic materials, and finally to their practical applications. There exist a large number of excellent review articles summarizing this field from different viewpoints in a broad manner. For example, in 2008, the *Chemical Reviews* journal has published a special issue that was devoted to biomineralization and summed up the progresses in biomimetic and bio-inspired mineralization (Volume 108, Issue 11, November 12, 2008).

As is well known, two main kinds of biopolymers, insoluble organic matrices and soluble biomacromolecules, have been involved into the formation of biominerals. While the former generally makes up ordered frameworks serving as hard templates to induce the nucleation of inorganics and restrict their subsequent growth within confined compartments, the latter acting as soft templates can change the crystallization behaviors of inorganics from their theoretically expected ones and direct the size, shape, polymorph, and organization of primary nanounits by absorbing onto certain crystal faces and lattice matching. All of the roles played by biopolymers in biominerals have been widely simulated by using natural and artificial templates. Particularly, using soluble polymer additives as crystal growth modifiers for controlled crystallization attracts much more attention and has been proved to be a quite versatile method in producing various inorganic materials, including noble metals, metal oxides, metal sulfides, and especially biological minerals [13,14]. For example, a great amount of soluble polymers, such as proteins extracted from biominerals [15], polypeptides [16], double hydrophilic block copolymers (DHBCs) [17,18], polyelectrolytes [19,20], dendrimers [21], and low-molar-mass additives [22,23], have been widely employed as crystal modifiers to control the crystallization of carbonates. A large variety of carbonate particles and aggregates with novel morphologies and/or complex assembly superstructures have been demonstrated, including monodispersed microspheres [24] and microrings [25], pancakes [26], fibers [27], helices [17,28], mesocrystals [29], multilayered structures [30], complex single crystals [31,32], and uniform patterns [33].

The investigations in the area of polymer-controlled crystallization for biological minerals have been extensively and comprehensively summarized by uncountable outstanding review and feature articles [34,35], research news [36], and book chapters [37,38]. However, in recent years, increasing studies have paid more and more attention to the extended application of bio-inspired polymer-controlled crystallization routes in the design and preparation of functional inorganic materials. Therefore, in order to avoid duplication with previous reviews, we devote our efforts in this chapter mainly to recent advances in preparation of functional inorganic materials via biologically inspired synthesis methods. Some basic principles of crystallization will be first introduced briefly (Section 9.2). Emphasis will then be laid on latest developments and important progresses of bio-inspired synthesis of various kinds of functional inorganic materials via polymer-controlled crystallization (Section 9.3). Their resulting improvements in different properties and potential applications in diverse fields will also be discussed and highlighted subsequently (Section 9.4). The challenges of current studies and prospects for future research about this topic will be finally suggested (Section 9.5).

9.2 Basic Principles of Crystal Growth and Aggregation

According to the classical crystallographic theory, crystals have been thought to be formed by gradual addition or attachment of their smallest construction units (atoms, molecules, or ions) to existing primary nanoparticles, as shown in Figure 9.1, path *a*. However, in the past few years, more and more studies have revealed some nonclassical routes for crystal growth, such as oriented attachment [39,40], mesocrystal [41], and solid transformation from amorphous phases [42], which can proceed through particle-based ordered aggregation or programmed organization (Figure 9.1, paths *b–d*) [43,44]. Within nonclassical crystallization processes, nanoparticles with larger size than the monomers of classical crystallization are considered as the primary building blocks to generate ordered superstructures via mesoscopic self-assembly and subsequently fuse to a single crystal. The role of these bigger nanoparticle building blocks can be played by pure crystallites, polymer-stabilized crystallites, or metastable amorphous intermediates.

9.2.1 Morphological Control in Classical Crystallization

Under ideal conditions, the crystal shape is defined by its intrinsic crystallization habit and is the outside embodiment of the unit cell replication and amplification. Dating back to the early twentieth century, Wulff first presented a purely thermodynamic treatment over the crystal shape selection, which is based on the minimization of total surface free energy of a growing crystal. In general,

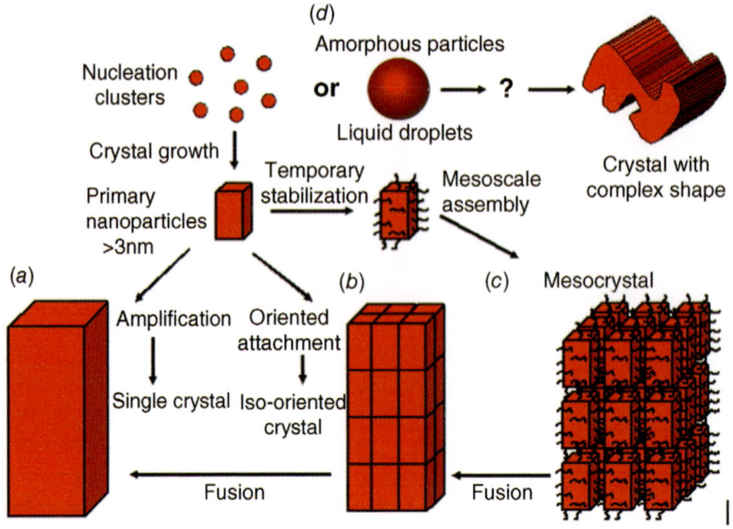

Figure 9.1 Schematic representation of classical (a) and nonclassical crystallization (b–d) pathways. (Reproduced with permission from Ref. [43]. Copyright 2007, Springer.)

different crystallographic planes of a certain crystal exhibit different surface energies that are further related to the growth rates of relevant crystal faces. The higher the surface energy of a certain crystal face, the faster the growth rate along the direction perpendicular to this face. Therefore, with the proceeding of crystal growth, in order to decrease the total surface energy, the exposure of the faces with higher surface energies will be reduced and finally eliminated, while instead lower surface energy faces will become more exposed in area as suggested by Wulff's rule. In other words, since higher energy surfaces grow faster and will vanish in the final crystal, the presented ultima shape of a crystal is bounded by lower energy surfaces (Figure 9.2a).

However, the real case is that inorganic material with the same chemical composition and phase could demonstrate diverse morphologies that are far from that thermodynamically predicted. The shape variations can be induced by various factors, that is, precursor concentration, temperature, solvent media, and especially polymer additives, because of their alteration over the chemical potential of crystallization in liquids. Under the control of polymer additives acting as crystal growth modifiers, the deviation of crystal shape can be attributed to the selective or preferential absorption of the polymer additives onto specific crystal faces, which can accordingly lower their surface energy and consequently reduce the growth rate of these faces. Thus, in this case, the polymer-stabilized crystal faces are generally exposed and determine the final crystal shape, as shown in Figure 9.2b. Increasing studies have been published to confirm the above statement. For example, polyelectrolyte poly(sodium 4-styrenesulfonate) (PSS) can selectively absorb onto calcite {001} faces, leading to the morphological change

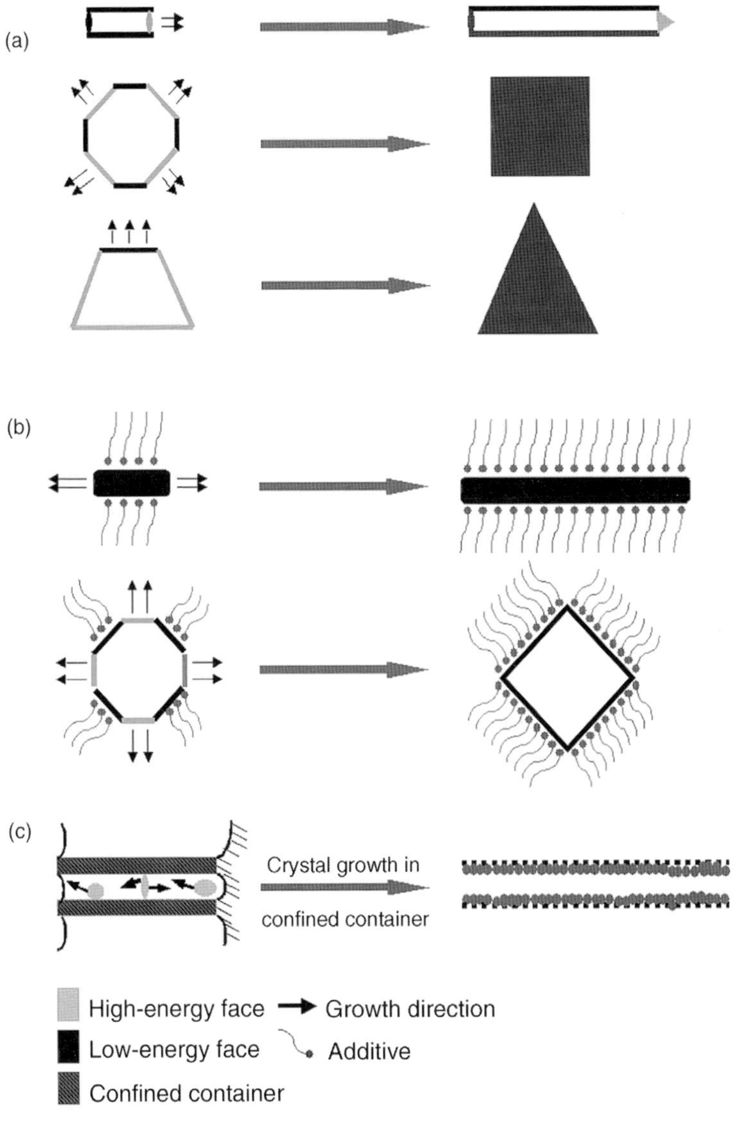

Figure 9.2 Illustrations of crystal growth process. (a) Thermodynamically predicted crystal growth. (b) Polymer additive-controlled crystal growth. (c) Crystal growth within a confined environment. (Reproduced with permission from Ref. [44]. Copyright 2007, Springer.)

from theoretical rhombohedron bounded by six {104} faces to 2D platelet mainly by two {001} faces, which has been certified by both experimental data [19,29] and molecular dynamic simulation results [45]. In addition, the crystal shape can also be modified by using hard templates, such as confined apartment with certain shapes (Figure 9.2c). Meldrum and coworkers have shown an excellent example by using polymer replica of a sea urchin skeletal plate to prepare extraordinary single crystals with complex and macroporous structures and curved and smooth surfaces for a series of inorganic materials, including Cu [46], $CaCO_3$, $PbSO_4$, $CuSO_4$, NaCl [47,48], silica, and titania [49].

9.2.2 Nonclassical Crystallization

The practical crystallization process is the synergistic result of both thermodynamic control and kinetic control by different parameters. Compared with the thermodynamic control-dominated classical crystallization, kinetic control becomes increasingly prominent within nonclassical crystallization that usually covers more actual cases and is more versatile in constructing more complex and emergent superstructures. This particle-based nonclassical crystallization is a newly developed concept with a short history of about 20 years and playing increasingly important role in materials science, as reviewed in several articles [34,39,40]. As schematically illustrated in Figure 9.1, nonclassical crystallization contains three main pathways: oriented attachment (path *b*), mesocrystal (path *c*), and amorphous precursor (path *d*).

9.2.2.1 Oriented Attachment

Oriented attachment refers to the ordered arrangement of primary nanoparticles along the same crystallographic direction into an iso-oriented crystal that will transform to a single crystal via further irreversible coalesce. During the late twentieth century, this mechanism was first proposed by Penn and Banfield in the case of anatase TiO_2 generation under hydrothermal conditions, where adjacent TiO_2 nanoparticles spontaneously approach closer to and assemble with each other, followed by coalescing along the same crystallographic direction in a highly oriented and irreversible fashion [50]. In addition, the eliminated crystallographic faces during fusion are identified to be the high-energy surfaces, which is thermodynamically favorable because that will substantially reduce the total surface free energy and increase the system entropy. The mutual orientation between approaching primary nanoparticles is generally achieved by their collision and rotation. Recently, these supposed evolution stages have been finely captured by direct observation over the crystal growth of iron oxide in real time using *in situ* transmission electron microscopy (TEM) technology [51]. Immediately after the initial studies, this novel crystallization mechanism has been applied to a large number of other material systems, such as noble metals [52,53], metal halides [54,55], metal oxides [56,57] and multimetal oxides [58,59], metal chalcogenides [60], metal–organic frameworks (MOFs) [61], $BiVO_4$ [62], $NaNbO_3$ [63], Zn_2TiO_4 [64], and others [65,66]. For more detailed information on specific materials, the readers are encouraged to refer to some recent reviews [39,67].

Due to the fusion of these exposed high-energy surfaces, oriented attachment offers special advantages to induce the anisotropic growth of nanocrystals, which imposes a profound influence over the final morphologies and structural evolution of inorganic materials. Yang and Zeng summarized the formation of one- (1D), two- (2D), and three-dimensional (3D) structures from accordingly lower dimensional primary nanoparticles via oriented attachment (Figure 9.3) [68]. Oriented attachment process is the most frequently employed or speculated growth mechanism to explain the fabrication of anisotropic (especially 1D) nanomaterials. Classical crystallization generally creates defects in the form of branches during the polymer-controlled crystallization of 1D nanomaterials. However, through oriented attachment, nearly defect-free and highly homogeneous 1D nanostructures, such as nanorods [69], nanowires [52], and nanotubes [70], have been widely and extensively reported for diverse inorganic materials. Worm-like nanowires of Pd [71] and Pt-M (M=Cu, Co, Ni, Fe)[72] alloys have also been achieved through oriented attachment of nanodots under solvothermal conditions in the presence of organic additives. If the primary nanounits exhibit an unusual shape, some very interesting 1D nanowires with intricate and unique substructures can be realized, such as hexapod building blocks for radically branched nanowires [73]. Furthermore, it was also reported that oriented attachment can be applied to the case of 1D heterostructures, when the components in a mixture possess similar crystallographic features, for example, TiO_2-SnO_2[74] and CdS(CdSe)-CdTe [75].

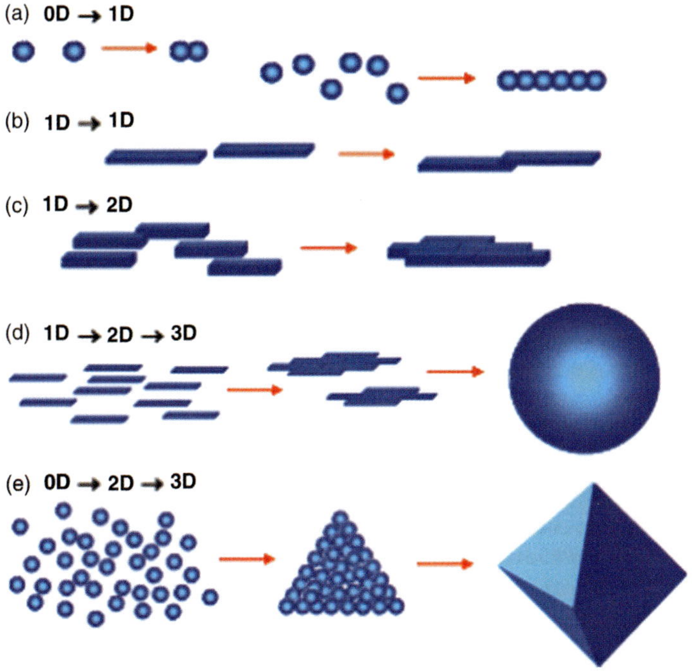

Figure 9.3 Various organizing schemes for self-construction of nanostructures by oriented attachment. (Reproduced with permission from Ref. [68]. Copyright 2004, Wiley-VCH Verlag GmbH.)

Oriented attachment is also advantageous for constructing 2D single crystals, where primary nanoparticles only assemble along lateral directions without thickness increase. Square-shaped PbS nanosheets bounded by {100} facets were produced from PbS nanoparticles via this mechanism with the aid of oleic acid and chlorine-containing cosolvents, even though it tends to form 3D networks rather than 1D or 2D structures due to its cubic crystal symmetry [76]. Oriented attachment can also occur subsequent to a misoriented attachment driven by the relaxation of stress at the incoherently attached interfaces [77]. Very recently, Vanmaekelbergh and coworkers reported a study of oriented attachment of PbSe nanocrystals, leading to the formation of ultrathin PbSe single-crystalline nanosheets with periodicities of planar square or honeycomb geometry [78].

9.2.2.2 **Mesocrystal**

Mesocrystal, short for "mesoscopically structured crystal," which was first introduced by Cölfen and coworkers for minerals generated by bio-inspired crystallization process in the early years of this century, is composed of individual nanocrystals aligned in a common crystallographic register so that a mesocrystal diffracts electrons or X-rays similar to a corresponding single crystal [41,43,79]. According to Figure 9.1, path *c*, mesocrystals are typically hybrid materials of crystalline nanoparticles and interspacing amorphous organic layers. As the primary building blocks are already aligned along the same crystallographic direction, the as-prepared mesocrystal will easily evolve into a single crystal when the interspacing amorphous stabilizer is insufficient. Because the crystalline nanoparticles provide compressive strength and stiffness while the amorphous layers impart tensile strength and toughness, above hybrid design of mesocrystal can solve the conflict between the expected improvements of both stiffness and toughness in one material. Therefore, the mechanism of mesocrystal has been adopted by biomineralization of sea urchin spine [80] for mechanical optimization.

Even though more and more mesocrystals have been reported by researchers for different material systems, most of them are still mainly descriptive focusing on structural characterization and the understanding of underlying ordered assembly mechanisms of primary nanoparticles remains quite poor. There are four main possible mechanisms applied to explain mesocrystal formation, of which different forces driving the 3D mutual alignment of nanoparticles to a mesocrystal have been proposed (Figure 9.4) [79]. The intuitionally simplest mesocrystal formation mechanism is to expropriate the templating effects of an organic matrix with nanosized compartments, where separated crystalline nanoparticles are aligned and oriented by the prestructured organic matrix (Figure 9.4a). This matrix-induced mesocrystal formation was suggested for coral growth [81]. However, there is no report of artificial systems to confirm this mechanism so far. The second mechanism requires external driving forces derived from physical fields such as electric, magnetic, or dipole fields, or possible polarization forces (Figure 9.4b). Kniep and coworkers demonstrated the intrinsic electric field lines around a fluorapatite–gelatin mesocrystal by electron

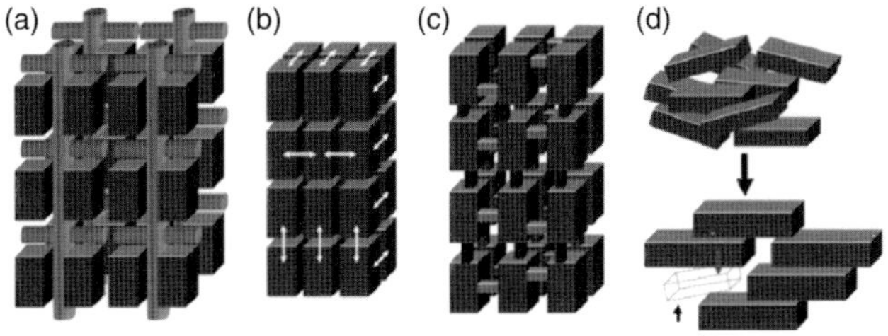

Figure 9.4 Illustrative diagrams of four principal possibilities to explain the 3D ordered assembly of nanoparticles into a mesocrystal. (a) Oriented organic matrices. (b) Physical fields or mutual alignment of identical crystal faces. (c) Epitaxial growth of a nanoparticle employing a mineral bridge connecting the two nanoparticles. (d) Nanoparticle alignment by spatial constraints. (Reproduced with permission from Ref. [79]. Copyright 2010, Wiley-VCH Verlag GmbH.)

holography, which responds to the aligned dipoles between neighboring nanoparticle units [82]. Another possibility for the 3D nanoparticle alignment in crystallographic register is the crystallographic connection between the nanoparticles by the so-called mineral bridges (Figure 9.4c), which was first proposed in explaining the mutual *c*-axis orientation of aragonite tablets in abalone nacre [83]. The last possible mechanism for mesocrystal construction is a simple geometric argument, as shown in Figure 9.4d. When the nanoparticles with anisotropic shape grow bigger, it is favorable from the viewpoint of systematic entropy to align them in order in a constrained reaction environment. Despite some of the above possibilities are still conjectural, they may actually be combined together to implement the growth of mesocrystals in a specific system.

Compared with single crystals, mesocrystals possess some superior properties such as higher porosity and surface area, pore accessibility, better mechanical performance, size-dependent magnetic and optical properties, and unique electrical conductivity [84,85], which could be beneficial to improve their properties and extend their application in related fields. Therefore, exploring mesocrystals of various materials, particularly the functional materials, has received considerable attention. In the past decades, mesocrystals with an ever-increasing list, including organic compounds [86,87], noble metals [88,89], metal oxides [90–92], metal chalcogenides [93,94], and others [95,96], have been successfully synthesized and discovered. Recently, several impressive reviews have been published from diverse viewpoints focusing respectively on different classes of materials [84,85,97,98]. Here, it should be pointed out that mesocrystals are probably much more common than we considered presently, because some of them may be recognized or misinterpreted as single crystal for their well-faceted appearance and single-crystal-like electron diffraction patterns.

9.2.2.3 **Amorphous Precursor**

The third nonclassical crystallization mechanism is the mesoscale solid-phase transformation of disordered amorphous precursors, which is extremely capable of creating complicated and off-equilibrium crystal morphologies because amorphous precursors can be quite easily molded and form any shape for its isotropy and good plasticity. This growth process is first reported by Addadi and coworkers from their research about sea urchin spine regeneration [99]. They found that the regenerated spine was started from the initial deposition of a transient amorphous $CaCO_3$ (ACC) onto the existing scaffold followed by quick transformation into crystalline calcite to keep the smooth convoluted fenestrated morphology. Subsequently, this nonclassical crystallization mechanism was also observed in many other biomineral systems, such as nacre [100], chiton teeth [101], and magnetite in magnetotactic bacteria [102]. More interestingly, under specific conditions (i.e., in the presence of inducer or inhibitor), the amorphous and crystalline phases can mutual transform into the other to implement specific life activities such as shell molt of crustacean [103]. In bio-inspired mineralization systems, amorphous precursor particles are omnipresent, so they are often the early nanoparticle stage in a nonclassical crystallization pathway [14,43]. However, this solid-phase transformation is seldom reported for other functional inorganic materials.

An alternative of amorphous precursor for generating complex morphologies is the so-called polymer-induced liquid precursor (PILP) proposed by Gower *et al.* [104,105]. Such dense liquid precursor in the state of little droplet is formed through liquid–liquid separation in a supersaturated crystallization system, which is generally induced by acidic polyelectrolytes like poly(acrylic acid) (PAA) or poly(aspartic acid) (PAsp). The role played by the polymers appears to be sequestering and concentrating isolated ions, and thus preventing crystallization process in solution. However, the exact reason for the formation of PILP is still a mystery. The main shortcoming of this concept is that it is merely applied to the bio-inspired mineralization of $CaCO_3$ and few organic systems with low molecular weight [106]. It seems powerless in explaining the crystallization of vast other inorganic and organic materials. At least, the report with direct proofs for the formation of PILP of other materials is still missing so far.

Underpinning the work on amorphous precursor or PILP are the increasing studies on the prenucleation behavior (the formation of stable clusters prior to nucleation) of some well-studied systems [107]. One of the prerequisite assumptions for classical crystallization is the separation and homodisperse of monomeric species of a crystal. However, recent work selecting $CaCO_3$ as a model system argues against this notion [108,109]. Using analytical ultracentrifugation (AUC) together with potentiometric titration techniques, it was found that stable ion clusters (the so-called prenucleation clusters (PNCs)) composed of up to several tens of $CaCO_3$ units represented a popular state rather than just free ions or simple ion pairs [108]. Molecular dynamics simulations provided further confirmation about the thermodynamic stability of PNCs and additional information on their structures that may have linear, ring-like, or branched configurations [110]. The occurrence of this kind of prenucleation clusters in homogeneous solutions has also been evidenced for other materials, including CaP [111], silica [112], and amino acid [113].

9.3
Bio-Inspired Polymer-Controlled Crystallization

As emphasized in Section 9.1, biominerals with complex crystalline superstructures are realized with the assistance of organic templates, especially the molecular recognition properties of water-soluble biomacromolecules. Therefore, bio-inspired approaches simulating biomineralization through polymer-controlled crystallization have spurred a great deal of research not only on the important biominerals ($CaCO_3$, CaP, and silica), but also on other functional inorganic materials (Table 9.1). In the following sections, recent developments in the field

Table 9.1 Various functional inorganic materials obtained via bio-inspired polymer-controlled crystallization.

Material	Polymer	Morphology	Application	References
Au	Lactoferrin	Quantum cluster	—	[114]
Au	BSA	Quantum cluster	—	[114,115]
Au	C_{12}-PEP_{Au} with HEPES	Nanoparticle double helices	—	[116]
Au	Protein	Nanoplate	—	[117]
Au	C_6-AA- PEP_{Au}	Hollow nanoparticle superstructure	—	[118]
Ag	Denatured BSA	Quantum cluster	Hg^{2+} sensing	[119]
Ag	Ge8	Curled nanowire	—	[120]
Ag	PAA	Nanobelt, hierarchical nanocolumn	—	[121]
Ag	AG4	Nanoplatelet	—	[122]
Pt	T7	Nanocube	—	[123]
Pt	S7	Nanotetrahedron	—	[123]
Pt	BP7A	Nanopod, nanosphere	—	[124]
Pt	Insulin amyloid	Nanowire	Electrochemical catalysis	[125]
Pt	Ac-IIIK-$CONH_2$	Nanowire	Electrochemical catalysis	[126]
Pd	Pd4-GG- p53 Tet	Coral-like nanostructure	Chemical catalysis	[127]
Pd	Pd-R5	Nanodot, nanopod, nanoworm	Chemical catalysis	[128]
Cu	Lysozyme	Quantum cluster	Cell labeling	[129]
Fe	Hemoglobin	Quantum cluster	—	[130]
AuAg	BSA	Quantum cluster	—	[131]

(*continued*)

Table 9.1 (*Continued*)

Material	Polymer	Morphology	Application	References
CoPt	Dual-affinity peptide	Nanoparticle	Data storage	[132]
CoPt	BP- $C_{12}H_9CO$-PEP_{Co}	Hollow nano-particle superstructure	—	[133]
TiO_2	Silicatein	Nanoparticle	—	[134]
TiO_2	Lysozyme	Nanoparticle network	—	[135]
TiO_2	Silicatein	Nanoparticle	—	[136]
TiO_2	Protamine	Nanoparticle coating	Lithium-ion battery	[137]
TiO_2	Ac-KFFAAK-Am	Nanotube	—	[138]
TiO_2	β-Lactoglobulin	Nanowire	Photovoltaic device	[139]
TiO_2	(Leucine-lysine)$_8$-PEG_{70}	Nanofiber framework	Photocatalysis	[140]
ZnO	ZnO-1-GGGSC	Microflower	—	[141]
ZnO	PAH	Spindle aggregate	Photoluminescence	[142]
ZnO	ZP-1	Nanoparticle	—	[143]
Fe_3O_4	Mms6	Cubo-octahedron	—	[144]
Fe_3O_4	Mms6	Microparticle array	—	[145]
Fe_3O_4	Alginate	Clustered single crystal	MRI	[146]
Fe_3O_4	M6A	Nanoparticle chain	—	[147]
Fe_3O_4	(Spartic acid)$_{50}$-(serine)$_{50}$	Nanoparticle chain	—	[148]
Fe_3O_4	(Glutamic acid)$_{10}$	Nanoparticle chain	—	[148]
α-Fe_2O_3	Silk fibroin	Olive-like mesocrystal	Photocatalysis	[149]
FeOOH	PAA	Array film	Water treatment	[150]
$CoFe_2O_4$	Mms6	Uniform nanocrystal	—	[151]
Co_3O_4	PVP	Porous aggregate	Gas sensing	[152]
α-$Co(OH)_2$	PAA	Film	—	[153]
γ-MnO_2	Citric acid	Lotus-like aggregate	Chemical catalysis	[154]
MnO-CoO	Agar	Mesoporous solid solution	Lithium-ion battery	[155]
$LiMn_2O_4$	Agar	Mesoporous crystal	Lithium-ion battery	[156,157]
MoO_3	PEO	Nanofiber, microsphere	Chemical catalysis	[158]

(*continued*)

Table 9.1 (*Continued*)

Material	Polymer	Morphology	Application	References
SnO	Gelatin	Nanosheet aggregate	—	[159]
SnO_2	Silicatein	Nanoparticle	—	[160]
SnO_2	Spermine	Nanoparticle film	Self-cleaning and antimicrobial surface	[161]
γ-Ga_2O_3	Silicatein	Nanoparticle	—	[162]
β-Ga_2O_3	Bolaamphiphile peptide	Spherical single crystal	—	[163]
GaOOH	Silk fibroin	Rod-shaped particle	—	[164]
ZnS	BB-TrxA::CT43	Quantum dot	—	[165]
ZnS	P22-CNNPMHQNC	Nanodot hollow sphere	—	[166]
ZnS:Mn	BB-TrxA::CT43	Quantum dot	Biosensing, bioimaging	[167]
ZnS:Mn	BSA	Quantum dot	Biosensing	[168]
CdS	P22-SLTPLTTSHLRS	Nanodot hollow sphere	—	[166]
CdS	tRNA	Quantum dot	—	[169]
CdS	DNA plasmid	Quantum dot	Gene delivery	[170]
CdS	Dendron rodcoil molecule	Nanohelice	—	[171]
CdS	Peptide-amphiphile	Nanofiber, nanosphere	—	[172]
CdS	DNA strand	Quantum dot chain	—	[173]
CdSe	Denatured BSA	Quantum dot	Cu^{2+} sensing	[174]
PbS	Nucleotide	Quantum dot		[175]
Zn/PbS	RNA	Quantum dot, nanofiber, nanosheet	—	[176]
Zn/PbSe	RNA	Honeycomb	—	[177]
Ag_2S	BSA	Nanorod	—	[178]
Ag_2S	BSA	Quantum dot	Cancer imaging	[179]
Ag_2S	Cucurbit[7]uril	Aligned quantum dot	—	[180]
CdSe@ZnS	DNA	Core–shell quantum dot	—	[181]
CdSe@ZnS	Bifunctional peptide	Core–shell quantum dot	—	[182]
$BaTiO_3$	Bolaamphiphile peptide	Nanoparticle	—	[183]

(*continued*)

Table 9.1 (*Continued*)

Material	Polymer	Morphology	Application	References
$BaTiO_3$	BT1, BT2	Nanoparticle	—	[184]
$BaTiO_3$	PAA-*b*-PS	Nanoparticle	—	[185]
$BaTiO_3$	PAA-*b*-PVDF	Nanoparticle	—	[186]
$BaTiOF_4$	Silicatein	Microfloret	—	[187]
$FePO_4$	Fmoc-FF	Nanotube	Lithium-ion battery	[188]
$FePO_4$	SWNT-specific peptide	Nanoparticle network	Lithium-ion battery	[189]
$FePO_4$	DNA	Nanoparticle network	Lithium-ion battery	[190]
$FePO_4$	ATP	Nanoparticle	Lithium-ion battery	[191]
$Na_3V_2(PO_4)_3$	ATP	Bundle-like structure	Sodium-ion battery	[192]
$CdWO_4$	PEG-*b*-PMAA	Nanorod, nanofiber, nanoplatelet	—	[193]
$BaWO_4$	PEO-*b*-PMAA	Penniform architecture	—	[194]
$PbWO_4$	Dextran	Belt-, ear-, feather-like architecture	—	[195]
$MnWO_4$	DNA	Nanobundle, nanoflake	Chemical catalysis, supercapacitor	[196]
$ZnWO_4$	DNA	Nanoparticle chain	Chemical catalysis, supercapacitor	[197]
$NiWO_4$	DNA	Nanoparticle chain	Chemical catalysis, supercapacitor	[198]
ZIF-8	Cyt c with PVP	Rhombic dodecahedron	—	[199]
ZIF-8	BSA	Truncated cube	—	[200]

of morphosynthesis of various functional inorganic materials via bio-inspired polymer-controlled crystallization by using biomacromolecules and synthetic polymers will be discussed in detail.

9.3.1 Noble Metal

Due to their unique optical, electrical, electrochemical, and catalytic properties, colloidal noble metal nanoparticles have wide applications in catalysis, nanodevices, antibiosis, and bio-related areas (biosensing, biomedicine, bioimaging, etc.) [201,202]. Bio-inspired polymer-directed routes can produce metal nanostructures with well-controlled size and morphology, high stability, good solubility and biocompatibility. Various organic additives, particularly the biomolecules, have been extensively requisitioned to mediate the morphogenesis of noble metal materials [203,204]. For example, peptide-based biomimetic mineralization is much more promising compared to other polymers. A great deal of noble metal and alloy nanoparticles and complex assemblies with diverse shapes ranging from 0D to 3D have been successfully realized. Very small nanoclusters with strong fluorescence made up of several or tens of atoms can be induced and stabilized by native lactoferrin [114], bovine serum albumin (BSA) [115,131], lysozyme [129], and hemoglobin [130]. In addition to the tiny nanoclusters, noble metal nanodots with bigger size have also been achieved by polymer-controlled crystallization process [124,128]. More interestingly, polymers with special structure-direction capabilities are able to create noble metal nanocrystals with well-defined shapes bounded by particular crystal faces. The generation of these nanocrystals follows the general principles of morphogenesis via classical crystallization, as introduced in Section 9.2.1. Recently, Huang and coworkers reported the use of facet-specific peptides as regulating agents for the controlled synthesis of Pt nanocrystals with selectively exposed crystal surfaces and predictable shapes (Figure 9.5) [123]. As shown in Figure 9.5a, the peptide sequences of T7 (TLTTLTN) and S7 (SSFPQPN) that recognize Pt-{100} and Pt-{111} planes, respectively, have rationally been identified by phage display. Once the face-selective peptides are used in bio-inspired synthesis of Pt nanomaterials, they can effectively limit the growth rates along [100] and [111] directions and thus induce the formation of nanocubes enclosed by six {100} faces and nanotetrahedrons enclosed by four {111} faces as expected (Figure 9.5b). Subsequently, by combining the control over both nucleation and growth processes, the same group prepared two kinds of single-twinned bipyramids for Pt nanocrystals in high yields [205]. In this work, the peptide BP7A (TLHVSSY) was used to nucleate Pt precursors into Pt single-twinned seeds, which was followed by controlled growth in the presence of T7 or S7. Based on the corresponding facet selectivity of the involved peptide, the overgrowth of {100} and {111} faces would occur with respect to the presence of T7 and S7, respectively. Thus, right bipyramid and {111}-bipyramid Pt nanocrystals can be selectively realized with a yield of more than 80% through this bio-inspired route under mild conditions. These

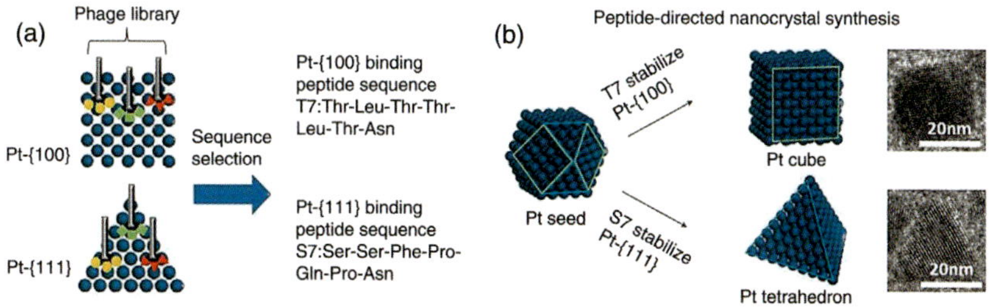

Figure 9.5 Schematic illustrations of facet-specific peptide sequence selection (a) and shape-controlled synthesis of Pt nanocrystals (b). (Reproduced with permission from Ref. [123]. Copyright 2011, Nature Publishing Group.)

presented results strongly imply that facet-specific polymers provide significant advantages in the effective biomimetic synthesis to produce more complex and unconventional nanostructures.

Compared to their 0D counterparts, 1D and 2D noble metal nanostructures exhibit uncommon advantages derived from their anisotropic growth. Therefore, they represent other important research directions, where bio-inspired polymer-controlled growth was also evidenced to be successful. Bai *et al.* demonstrated that by reducing $AgNO_3$ with ascorbic acid in PAA aqueous solution at 4 °C, single-crystalline ultralong Ag nanobelts grown along [110] direction were readily prepared [121]. Further introducing acetic acid into above system, unusual hierarchical Ag nanocolumns consisting of stacked nanoplates can be produced. The single-crystal-like electron diffraction (ED) pattern indicates that the nanocolumn grow along the [111] direction, while HR-TEM image of a single nanoplate confirms that all the face-to-face stacked nanoplates are [111]-oriented, which is a good example to reveal the oriented attachment in nonclassical crystallization. Naik *et al.* applied the peptide-based biomimetic mineralization method to the synthesis of Ag nanomaterials and obtained hexagonal and truncated triangle Ag nanoplatelets from $AgNO_3$ aqueous solution under the control of AG4 peptide sequence (NPSSLFRYLPSD) [122]. However, spherical Ag nanoparticles as by-product in large quantities precipitated as well. Similarly, by adding the extract of a unicellular green alga *Chlorella vulgaris* into $HAuCl_4$ solution, single-crystalline gold nanoplates oriented with their {111} facets as the basal planes were produced [117].

The combination of polymer self-assembly process and their mediation effect over inorganic crystallization has been proved to be powerful in creating complicated noble metal nanostructures. Chen *et al.* demonstrated the generation of Au nanoparticle double helices by combining peptide assembly and peptide-based biomimetic mineralization into one synthesis process, as illustrated in Figure 9.6a [116]. They first functionalized the peptide of PEP_{Au} (AYSS-GAPPMPPF) with succinimide-activated dodecanoic acid at its N-terminus to obtain an amphiphile peptide C_{12}-PEP_{Au}, which could self-assembly into a unique left-handed twisted nanoribbon in HEPES (4-(2-hydroxyethyl)-piperazineethanesulfonic acid) buffer. Then certain amount of $HAuCl_4$ solution was added to the HEPES buffer containing C_{12}-PEP_{Au}. After 30 min of Au mineralization, double-helical assemblies of discrete spherical Au nanoparticles were obtained (Figure 9.6b–d). The distance between Au nanoparticles on the opposite sides of the helices (6.0 ± 0.8 nm) matched very well with the width of the self-assembled peptide double helix (6.1 ± 0.6 nm). Subsequently, using the same protocol but with different modifications, well-defined hollow spherical superstructures made up of Au [118] and magnetic CoPt [133] nanoparticles, respectively, were reported by the same group.

9.3.2
Metal Oxides

Metal oxides represent a highly important class of materials due to their wide ranging optical, electronic, magnetic, chemical, and mechanical properties. The

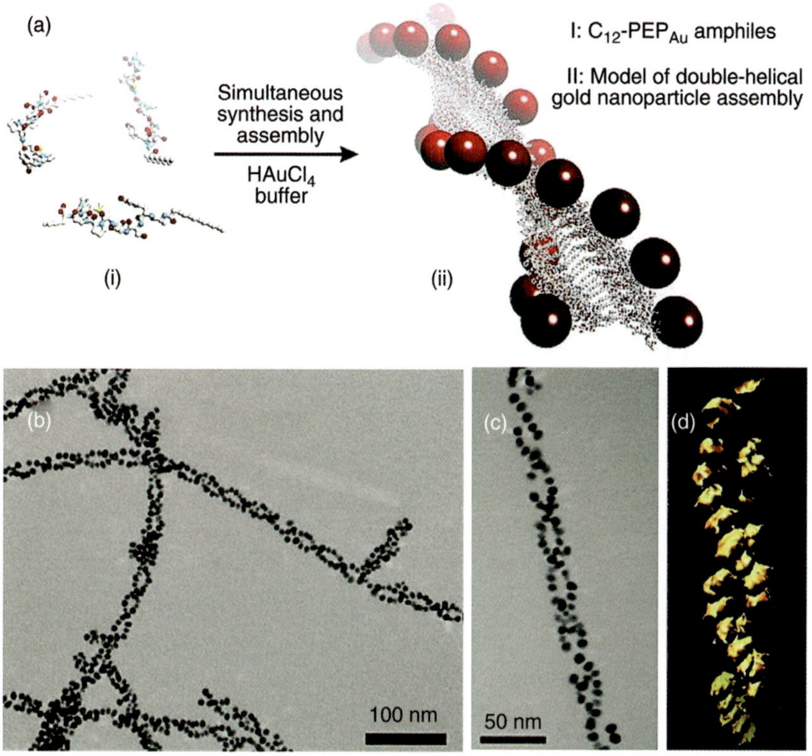

Figure 9.6 Schematic formation process (a), TEM images (b and c), and tomographic 3D reconstruction image (d) of Au nanoparticle double helices. (Reproduced with permission from Ref. [116]. Copyright 2008, American Chemical Society.)

term of metal oxide we used here is a generalized concept, including oxides, oxyhydroxides, and hydroxides. The approaches for the bio-inspired synthesis of metal oxides can be broadly divided into two categories: direct and indirect pathways. The former produces metal oxides from metal ions through only one step, while the latter involves an intermediate that is subsequently treated by a second step to generate the final metal oxides, such as metal oxides from metal carbonates by calcination.

The direct pathways are often used to synthesize metal oxides formed from condensation of their corresponding precursors, where the polymer additives, acting as both catalysts and soft templates, play key roles in accelerating the mineralization process and directing the structures of targeted materials. Since the first report on the synthesis of partially nanocrystalline titania using bio-inspired crystallization by Morse and coworkers [134], a large number of studies have been carried out to prepare various titania and titania-related materials under the control of numerous polymer additives (proteins, peptides, and polyamine) [206,207]. Mms6 (magnetosome membrane-specific protein with a molecular weight of 6 kDa) has exhibited powerful capability in controlling the sizes and morphologies of magnetite nanocrystals and uniform magnetite nanocrystals [151,208] or their array patterns [145] were produced using either coprecipitation or partial oxidation [144] method.

Yu and coworkers have extended the classical slow gas–liquid diffusion method to the preparation of iron oxyhydroxide materials [150]. By diffusing ammonia gas volatilized from the diluted ammonium hydroxide solution into pure $FeSO_4$ aqueous solutions, a lepidocrocite (γ-FeOOH) array film composed of nanosheet building blocks can be produced at the air–water interface that acted as a nucleation template. However, the presence of PAA (0.6 g/l) led to dramatical changes over the shape and size of the building blocks, as shown in Figure 9.7. Two main kinds of nanosized building blocks were observed in this novel array film: nanolaths located at the bottom of the film and nanowires stood nearly perpendicular to film plane. Both of these nanounits were single crystals with high crystallinity and elongated along the [011] direction as confirmed by HR-TEM images (Figure 9.7e and g) and SAED patterns (insets in Figure 9.7d and f), respectively. After being dispersed in water by ultrasonic treatment, these 1D nanostructures exhibited excellent water treatment performance with a high adsorption capacity of 77.2 mg/g toward Cd^{2+} ions. Similarly, by ammonia diffusion into a mixed solution of Fe^{2+} and Fe^{3+} with a molar ratio of 1 : 2, Lenders *et al.* synthesized three kinds of unique chain-like structures with slight variations assembled from uniform magnetite nanoparticles under the assistance of M6A peptide [147], (spartic acid)$_{50}$-(serine)$_{50}$, and (glutamic acid)$_{10}$ [148] respectively, which are compositionally and morphologically similar to natural magnetite assemblies in magnetotactic bacteria.

Similar to the bio-inspired synthesis of noble metal materials with intricate and unconventional nanostructures, polymer self-assemblies have also been

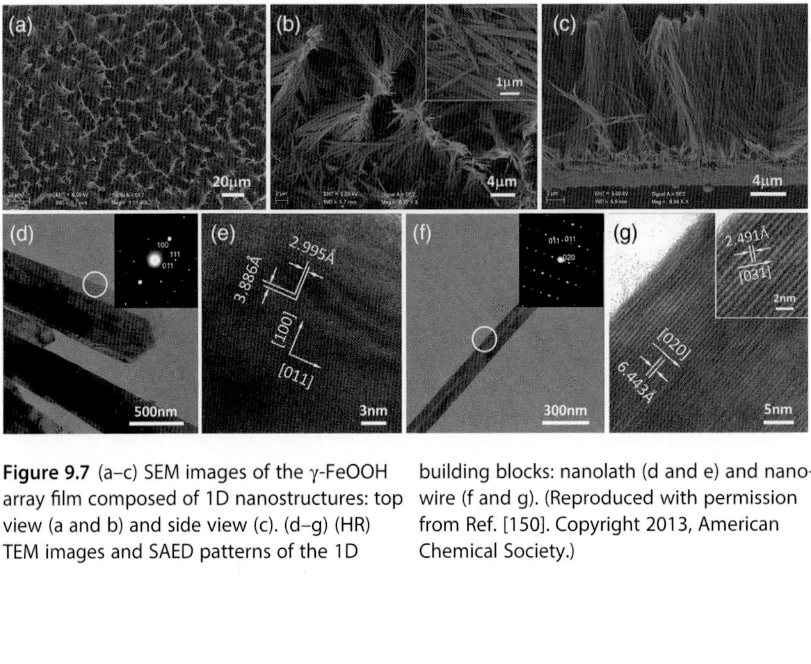

Figure 9.7 (a–c) SEM images of the γ-FeOOH array film composed of 1D nanostructures: top view (a and b) and side view (c). (d–g) (HR) TEM images and SAED patterns of the 1D building blocks: nanolath (d and e) and nanowire (f and g). (Reproduced with permission from Ref. [150]. Copyright 2013, American Chemical Society.)

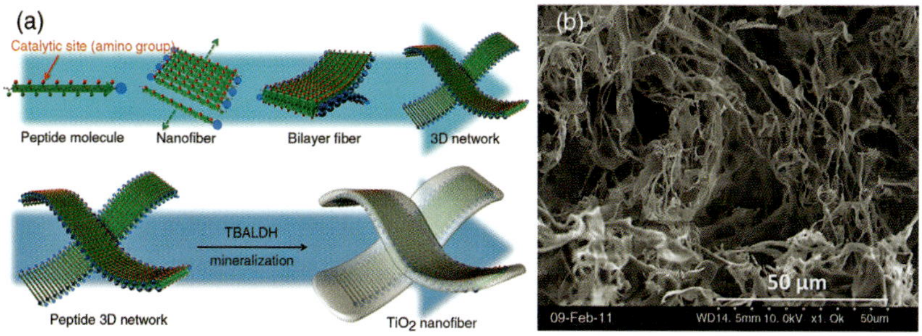

Figure 9.8 (a) Schematic presentation of peptide hierarchical self-assembly and subsequent titania mineralization. (b) SEM images of titania materials induced by $(LK)_8$-PEG_{70}. (Reproduced with permission from Ref. [140]. Copyright 2012, American Chemical Society.)

utilized as both hard templates and growth modifiers for the direct precipitation of more complicated metal oxide nanostructures. In a recent work by Nonoyama *et al.*, it was found that template effects of self-assembled peptides guided the formation of titania nanofibers and nanosheets, which were further intertwined with each other constructing a 3D porous framework (Figure 9.8) [140]. The authors first designed a polyethylene glycol (degree of polymerization = 70, PEG_{70})-modified β-sheet peptide (leucine-lysine)$_8$-PEG_{70} ($(LK)_8$-PEG_{70}) that can spontaneously self-assemble into a 3D nanofibers network under basic conditions, as illustrated in Figure 9.8a. A thin layer of titania coating on the nanofiber surface precipitated (Figure 9.8b) when the $(LK)_8$-PEG_{70} nanofibers network was exposed to titanium bis(ammonium lactato) dihydroxide (Ti-BALDH) solution at 70 °C for 2 days. In this case, the interactions between the amino groups on lysine side chains and cationic titanium complex were presumably considered to be responsible for nucleation of titania on fiber surface. More interestingly, after crystallization by sintering, a mixture of brookite and anatase was detected by XRD for the polylysine-induced titania, while brookite and rutile in the case of $(LK)_8$-PEG_{70}. This phenomenon was attributed to the specific amino sequence in the $(LK)_8$-PEG_{70} β-sheet that might favor the crystal structure of rutile phase. Moreover, during calcinations at 700 °C, elemental (carbon and nitrogen) doping for titania probably occurred, which, together with the high surface area and mixed polymorphs, endowed this composite material with advanced catalytic activity for photocatalytic degradation of rhodamine B.

The indirect pathway usually refers to the preparation of metal oxides generated via an intermediate that can be easily synthesized by bio-inspired polymer-controlled crystallization process under mild conditions. The indirect synthesis of spinel Co_3O_4 using cobalt carbonate as a precursor can be easily achieved through thermal transformation, of which the morphogenesis of cobalt carbonate can be well controlled by polymer additives through gas–liquid diffusion method [209] or under hydrothermal conditions [152,210]. Manganese oxide-based materials have also been synthesized by indirect pathways involving two steps. Imai and coworkers developed a modified gas–liquid diffusion method, where the precursor metal ions were dispersed in agar gel matrix instead of dissolving in water [156,157]. Adopting this method, the authors synthesized highly porous structure of spinel-type $LiMn_2O_4$ from the ordered $MnCO_3$ architecture through nanocrystalline Mn_2O_3 as an intermediate phase [156]. Following this work, they further introduced Co^{3+} into the gel matrix, and finally obtained Co-doped $LiMn_2O_4$ ($LiMn_{2-x}Co_xO_4$, $0 \leq x \leq 0.367$) with hierarchical sponge structures [157]. Recently, the authors changed the calcination conditions to a reduction atmosphere and achieved more exciting results [155]. As already described, when the calcinations of $MnCO_3$ and $CoCO_3$ were performed in air, Mn_2O_3 and Co_3O_4 were obtained through the carbonates decomposition and subsequent partial oxidation. However, calcination of $MnCO_3$ in a reduction atmosphere resulted in MnO by inhibiting the step of partial oxidation. In the case of $MnCO_3$–$CoCO_3$ composite precursors, besides the reduction atmosphere, the calcination temperature also played an important role in the transformation process of carbonate precursors. Lower temperature (e.g., 400 °C) led to the formation of MnO–CoO solid solutions with a few small Co nanocrystals (Figure 9.9a and b), whereas higher temperature (e.g., 600 °C) was responsible for MnO–Co hybrid materials (Figure 9.9c and d). All the materials maintained the macroscopic morphologies of carbonate precursors, but with increased porosity derived from release of CO_2 gases (Figure 9.9), which is the main reason for their enhanced charge–discharge cycle stability and the rate performance as anode materials of lithium-ion batteries (LIBs).

In addition to the above examples, an increasing number of examples for other inorganic compounds, such as ZnO nanocrystals [143], SnO_2 nanodots [161,211],

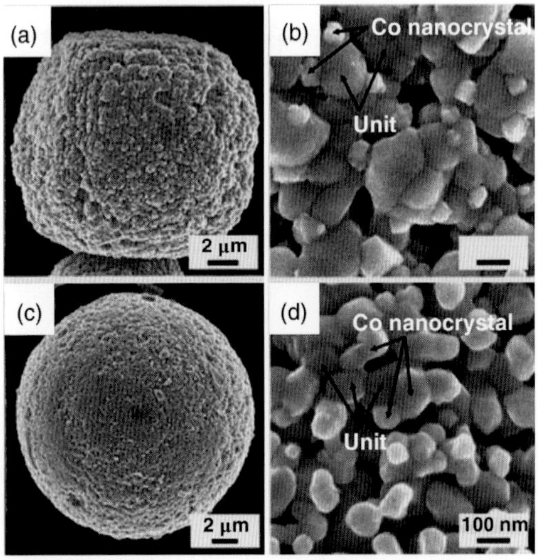

Figure 9.9 SEM images of MnO–CoO solid solutions (a and b) and MnO–Co hybrids (c and d). (Reproduced with permission from Ref. [155]. Copyright 2011, Wiley-VCH Verlag GmbH.)

In_2O_3 nanocubes [212], MoO_3 nanospheres and nanofibers [158], Ga_2O_3 nanoparticles [162] and nanorods [164], have been rationally produced by adopting direct or indirect routes under ambient conditions, indicating the versatility of bio-inspired approaches involving polymer-controlled crystallization. In other words, these approaches could offer a facile but quite useful tool to design functional inorganic materials with anisotropic material properties and could be used for the synthesis of more complex crystalline structures. Moreover, nonclassical crystallization pathways as the growth mechanism have been identified in above examples [143].

9.3.3 Metal Chalcogenides

Metal chalcogenides are another large family of semiconductor materials of great importance for both scientific research and practical applications. Until now, quite a few strategies for the synthesis of nanostructured metal chalcogenides have been reported [213]. However, compared with noble metal and metal oxide materials, the studies reporting the synthesis of metal chalcogenides using bio-inspired crystallization approaches are rather limited and primarily focused on metals such as Zn, Cd, Ag, and Pd.

Both dispersed polymer molecules and their ordered assemblies were exploited to prepare metal chalcogenide nanomaterials. Metal chalcogenide quantum dot (QD), representing one of the most important functional inorganic materials, have been successfully crystallized from corresponding precursors under ambient conditions by using diverse biomolecules as crystal growth modifiers. The folded and structured wild-type transfer RNA (WT-tRNA) yielded CdS QDs with a mean diameter of 4.4 ± 0.4 nm, whereas its unstructured mutant (MT-tRNA) generated CdS QDs with a larger average size (5.5 ± 1.0 nm) and a greater size distribution [169]. Hinds *et al.* have investigated the influences of four different nucleotide triphosphates (adenosine triphosphate (ATP), cytidine triphosphate (CTP), guanosine triphosphate (GTP), and uridine triphosphate (UTP)) over the biomimetic mineralization of PbS nanocrystals and found only GTP can produce IR luminescent PbS QDs with an average diameter of about 4 nm, while the others resulted in larger precipitates and nonemissive solutions [175]. Monodispersed blue-emission Ag_2S nanorods with sizes of about 40 nm in diameter and 220 nm in length precipitated from the solution of $AgNO_3$, BSA, and thioacetamide (TAA) with a higher Ag/BSA ratio [178], while the system of $AgNO_3$, BSA, and Na_2S with a lower Ag/BSA ratio provided ultra-small NIR fluorescent Ag_2S QDs [179].

On the other hand, using organic molecule self-assemblies as both the directing agents and hard templates for metal chalcogenide mineralization, more appealing metal chalcogenide nanomaterials with innovative structures translated from the biomolecule self-assemblies can be rationally realized. Sone *et al.* studied the formation of CdS nanohelices controlled by molecules with a triblock architecture termed “dendron rodcoils” [171]. The semiconductor helices composed of polycrystalline CdS domains of 4–8 nm displayed a pitch of 40–60 nm, which was approximately twice as the period of the twisted organic ribbons assembled from dendron rodcoil molecules. That means the CdS nanocrystals grew along only one face of the twisted ribbon template. Shen *et al.* inserted a peptide sequence CNNPMHQNC with strong affinity for ZnS into the coat protein P22 at the middle of a flexible loop region on the protein, which would assemble to form a procapsid with the engineered peptides located at the center of each of the pentamers and hexamers, as presented in Figure 9.10a [166]. These centered peptides acted as the nucleation sites of ZnS when the P22 procapsids dispersed into precursor solution of ZnS (Figure 9.10b). As the mineralization of ZnS QDs proceeded, ZnS QDs grew bigger and covered the outer surface of the P22 procapsids forming hollow structures composed of wurtzite ZnS QDs (Figure 9.10c and d). Substituting the sequence of CNNPMHQNC with SLTPLTTSHLRS (a CdS binding peptide) resulted in the similar hollow structures composed of CdS nanocrystals [166]. Therefore, the demonstrated strategy can be expanded to serve as a general bio-inspired approach for the design of nanoparticle assemblies with desired components and architectures.

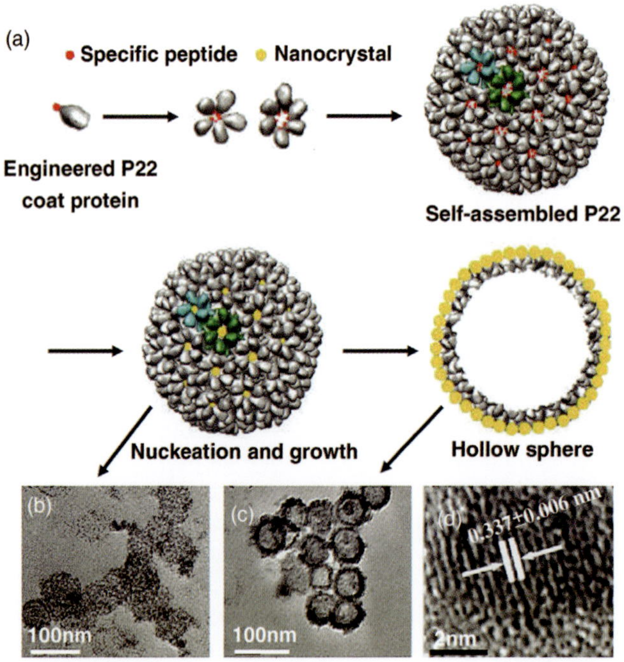

Figure 9.10 Schematic illustration (a) and (HR)TEM images (b–d) of ZnS QD assemblies over self-assembled P22 procapsids: (b) short and (c) long mineralization time and (d) a single ZnS QD. (Reproduced with permission from Ref. [166]. Copyright 2010, American Chemical Society.)

From the above descriptions, it can be easily concluded that by tuning the affinity, specificity, and stereochemical structure of polymer molecules according to the targeted material, bio-inspired polymer-controlled crystallization approaches were evidenced to be quite versatile in tuning the growth of single-phase metal chalcogenides with innovative (super)structures. More than that, these approaches have also been successfully extended to the synthesis of heterogeneous structures composed of two different metal chalcogenides. Wang *et al.* demonstrated that, under the mediation of DNA strands, a thin layer of ZnS could be grown on the surface of CdSe QDs forming CdSe@ZnS core–shell QDs [181]. By choosing different DNA strands, they were able to control the thickness of ZnS shell and hence the diameter of the final core–shell QDs, such as green QD ($\lambda_{Em} = 548$ nm) with a diameter of 4.9 ± 0.2 nm and red QD ($\lambda_{Em} = 617$ nm) with a diameter of 6.9 ± 0.3 nm obtained in the presence of two different DNA strands. The major weakness of this work is that the CdSe QDs were synthesized using traditional method at relatively higher temperatures. Latterly, Singh *et al.* further developed this approach by using a bifunctional peptide to synthesize such core–shell QDs [182]. This bifunctional peptide contained two different domains: sequence CTYSRKHKC and sequence KRRSSEAHNSIV bridged by a proline linker, of which the first domain is CdSe specific, while the second one is ZnS specific. As illustrated in Figure 9.11a, bio-inspired synthesis of CdSe QDs was first performed by peptide-based mineralization, where the CdSe domain can catalytically promote the nucleation of CdSe QDs with precise control over their size, resulting in peptide–CdSe hybrid QDs with a diameter of 4–5 nm. Then the purified peptide–CdSe hybrid QDs were redispersed in the solution of Zn^{2+} and S^{2-}, and ZnS domain of the peptide would induce the mineralization of the ZnS shell around the CdSe core, forming CdSe@ZnS core–shell QDs (Figure 9.11b). Both the core diameter and the shell thickness can be further tuned by using other specific sequences. More importantly, this benign and flexible strategy can be

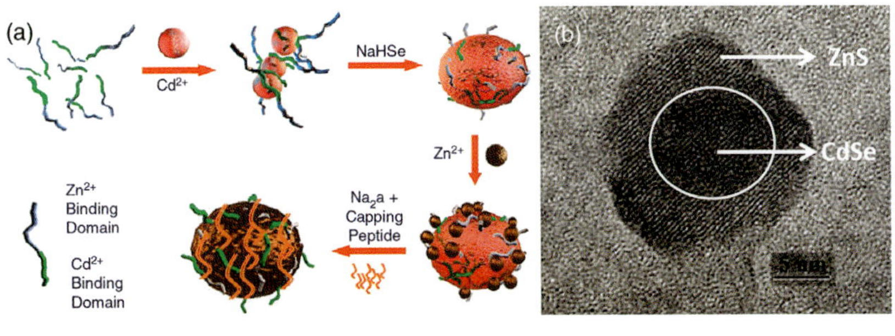

Figure 9.11 Schematic illustration of the synthesis process (a) and HRTEM image (b) of CdSe@ZnS QDs by using a bifunctional peptide. (Reproduced with permission from Ref. [182]. Copyright 2010, Royal Society of Chemistry.)

generalized for other hybrid inorganic nanostructures since the required peptide sequences can be selectively fine-tuned to control the size and composition of different inorganic materials.

9.3.4
Other Functional Inorganic Materials

So far we have mainly discussed the application of bio-inspired polymer-controlled crystallization approaches to the morphosynthesis of functional inorganic materials with simple chemical compositions. In addition to these examples, an ever-increasing number of investigations have been devoted to exploring how to generate inorganic compounds with more complex compositions by using bio-inspired approaches under mild conditions. These materials include perovskites [183,184], metal phosphates [188,214], metal tungstates [193,194], metal–organic frameworks [200], and so on.

Nuraje *et al.* have demonstrated the bio-inspired synthesis of ferroelectric $BaTiO_3$ nanoparticles by using a synthetic peptide-like bolaamphiphile as the nucleation inducer and growth mediator and $BaTi(O_2CC_7H_{15})[OCH(CH_3)_2]_5$ as the precursor of $BaTiO_3$ [183]. Recently, this approach has been further expanded by utilizing synthetic diblock copolymers, such as poly(acrylic acid)-block-polystyrene (PAA-*b*-PS) [185]. Three examples with different diameters of 6.3 ± 1.3 nm, 11.2 ± 1.9 nm, and 27.1 ± 3.1 nm were selectively obtained in the presence of PAA-*b*-PS with molecular weights of PAA blocks of 4500, 8400, and 28 100, respectively. The same group further designed another amphiphilic multiarm star-like block copolymer containing a functional block, poly(acrylic acid)-block-poly(vinylidene fluoride) (PAA-*b*-PVDF), and used it to control the crystallization of monodisperse $BaTiO_3$ nanoparticles, which exhibited the ferroelectric tetragonal structure and displayed high dielectric constant and low dielectric loss [186].

Metal–organic frameworks are crystalline porous solids with supramolecular structures comprised of (clusters of) metal ions and organic molecules linked together by coordination bonds, representing a family of promising materials for a variety of applications due to their amenability to design and fine-tunable and uniform pore structures [215,216]. For the first time, Lyu *et al.* found that cytochrome c (Cyt c), assisted with poly(4-vinylpyridine) (PVP), could effectively catalyze the nucleation of zeolitic imidazolate framework (ZIF-8) and direct its subsequent growth, forming a Cyt c-ZIF-8 hybrid, as illustrated in Figure 9.12a [199]. Here, PVP was used to interact with Cyt c molecules and thus to improve the dispersity of this protein in methanol. After mineralization for 24 h, uniform rhombic dodecahedron hybrid crystals with an average size of approximately 620 nm (Figure 9.12c) were obtained, via some rod-shaped assemblies (Figure 9.12b) as intermediate. Very recently, Liang *et al.* greatly expanded this approach by using various biomacromolecules (proteins, enzymes, and DNA) to control the crystallization of several MOF materials (ZIF-8, $Cu_3(BTC)_2$ (HKUST-1), $Eu_2(1,4\text{-BDC})_3(H_2O)_4$, $Tb_2(1,4\text{-BDC})_3(H_2O)_4$, and Fe(III)

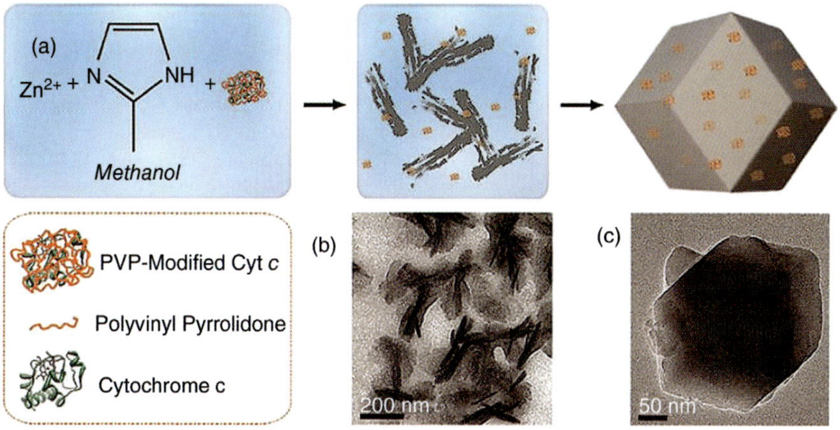

Figure 9.12 (a) Schematic illustration of the synthesis of Cyt c/ZIF-8 hybrid. (b and c) TEM images of the Cyt c/ZIF-8 samples after different mineralization time: (b) 5 min and (c) 24 h. (Reproduced with permission from Ref. [199]. Copyright 2014, American Chemical Society.)

dicarboxylate MOF (MIL-88A)) with homogeneous structures [200]. Still taking ZIF-8 as a typical targeted system, both the morphology and size of the resultant biomacromolecule-ZIF-8 hybrid crystals could be elaborately mediated by finely choosing a proper biomacromolecule under physiological conditions. For example, hybrid crystals with truncated cubic, leaf-like, rhombic dodecahedral, flower-like, and star-like shapes can be selectively synthesized by using BSA, ovalbumin, human serum albumin (HSA), horseradish peroxidase (HRP), and trypsin as crystallization modifiers, respectively. Another important aspect worth to note in both studies is that, after MOF mineralization, the bioactivities of these adopted biomacromolecules still retained or even promoted when being treated under harsh conditions that would normally decompose many biomacromolecules, which implies the efficient protection derived from the mineralized MOF materials [199,200]. A very recent study further confirmed that over a range of temperatures, biomimetic mineralization offers superior protection than other methods such as coprecipitation [217]. These latest studies further highlighted the versatility of this bio-inspired approach involving polymer-controlled crystallization.

9.4 Applications

From the above discussions, we can clearly find that, inspired by the biogenic formation of biominerals, numerous functional inorganic materials with controlled size, specific polymorph, and novel structures have been successfully realized by bio-inspired approaches under polymer mediation. However, the ultimate goal of materials design and synthesis is to gain their practical applications. Therefore, their applications in diverse areas, such as energy transformation and storage, catalysis, environmental management, sensing, (bio)medical science, and so on (as listed in Table 9.1), have also been widely explored and investigated.

As we introduced in Section 9.3.2, Imai and coworkers have prepared several MnO-based porous structures employing their corresponding carbonates biomimetically mineralized within agar gel matrix as the precursors, and further used them as lithium-ion batteries cathode materials [155–157]. The carbon-coated $FePO_4$ nanotubes synthesized by templating peptide assembly nanofibers and subsequent annealing treatment at 350 °C were found to be a promising LIBs cathode material with a high reversible capacity and good capacity retention during cycling [188]. Moreover, a composite of TiO_2–WO_3 with greatly improved photocatalytic activity toward photodegradation of rhodamine B (RhB) was synthesized by coincubation process via lysozyme-mediated bio-inspired mineralization [218]. Hematite mesocrystals obtained by silk fibroin-based bio-inspired mineralization were found to be an efficient photocatalyst for O_2 generation from water oxidation due to their high porosity and surface area derived from their mesocrystal nature [149]. Monodispersed SnO_2 nanospheres with hierarchical structures were synthesized by a step-wise process, including bio-inspired mineralization, vulcanization, and oxidation, and further used as gas sensor materials.

Since the above applications are not so much characteristic of the bio-inspired synthesis, we mainly concentrate on their bio-related applications here. Zhou *et al.* utilized the superior binding affinity of DNA for both Ln^{3+} doped-porous nanoparticles (Ln^{3+}-pNPs) and CaP to synthesize the hybrids of Ln^{3+}-pNP coated with a thin layer of CaP, which can be simultaneously used for magnetic resonance (MR) imaging and stimuli-responsive drug delivery, as illustrated in Figure 9.13 [219]. Mesoporous Ce and Tb codoped $NaGdF_4$ ($NaGdF_4$:Ce/Tb) nanoparticles were first prepared using bio-inspired crystallization method under the control of DNA as superior ligands. The residues of DNA molecules attached on $NaGdF_4$:Ce/Tb nanoparticles can effectively bind Ca^{2+} via electrostatic interaction, and thus acted as nucleation sites of CaP after the introduction of phosphate ions into above system. Thereafter, by precisely controlling subsequent growth process of CaP, a thin layer of CaP shell with pH-dependent biodegradability could be formed on the surface of $NaGdF_4$:Ce/Tb nanoparticles, resulting in the core–shell hybrid of $NaGdF_4$:Ce/Tb@CaP. Furthermore, an additional layer of aptamer (Apt) molecules was attached onto the surface of CaP via chelation interactions, which could produce a stable colloidal nanocomposite (termed as $NaGdF_4$:Ce/Tb@CaP-Apt). Compared to pristine $NaGdF_4$:Ce/Tb, the fluorescence and magnetic properties of $NaGdF_4$:Ce/Tb@CaP-Apt nanocomposites did not change much after CaP coating and Apt attachment, indicating their good potential to serve as efficient luminescent labels in biological imaging and label material for MR imaging contrast. On the other hand, this nanocomposite could be used as nanocarrier for drug delivery and release as

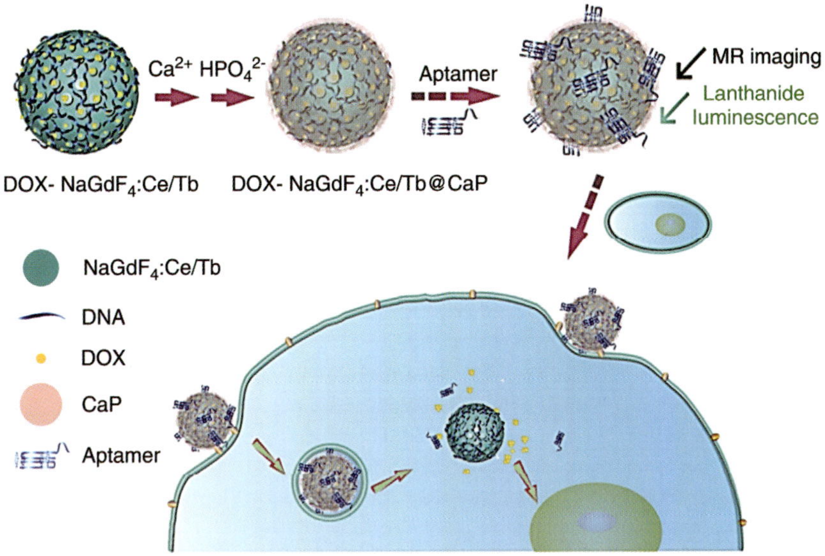

Figure 9.13 Schematic illustration of bio-inspired synthesis of DOX-loaded $NaGdF_4$:Ce/Tb@CaP-Apt hybrid. DOX-loaded $NaGdF_4$:Ce/Tb@CaP-Apt was internalized via receptor-mediated endocytosis. The DOX was released through the dissolving of CaP within intracellular endolysosomal compartment. (Reproduced with permission from Ref. [219]. Copyright 2014, Elsevier.)

well. The involvement of Apt can improve the cellular uptake property of the resultant nanocomposite into specific tumor cells via receptor-mediated endocytosis. Meanwhile, once they were taken by tumor cells, CaP mineral coatings could be readily dissolved into nontoxic ions in the acid environment of tumor cells, which can initiate and facilitate the diffusion-based drug release from these nanocarriers and subsequent cancer cells killing. For example, doxorubicin (DOX)-loaded $NaGdF_4$:Ce/Tb@CaP-Apt can be prepared by soaking pristine $NaGdF_4$:Ce/Tb nanoparticles into DOX solution followed by substantial washing, CaP coating, and Apt attachment consecutively. Their standard methyl thiazolyl tetrazolium (MTT) assay clearly demonstrated that the nanocomposites displayed a high efficiency for killing cancer cells while sparing normal cells. Moreover, Gd_2O_3-based nanocomposites obtained via such polymer-controlled crystallization and hybridation have recently been reported and used for tumor detection and phototherapy as well [220,221].

Another fascinating application of bio-inspired synthesis of inorganic materials is to form a thin shell layer coating the surface of living cells, which can make them more robust and even endow them with new functions [222]. The first example was reported by Tang and coworkers, in which an artificial CaP shell was crystallized on yeast cell under the assistance of PAA [223]. Such encapsulation can protect the cell under harsh conditions and thus make them inert and give them extended lifetime. Later on, the same group prepared CaP-coated

Japanese encephalitis vaccine (JEV) by similar process [224]. It was found that the thermal stability of these CaP-coated JEVs is always more than three times as that of the bare one, and they can be stored at room temperature for at least 1 week. Following these work, a great deal of other mineral cell systems have been successfully performed, such as silica – yeast [225], silica – cyanobacteria [226], TiO_2 – chlorella cell [227], and $LnPO_4$ – zebrafish embryo [228]. This novel application of bio-inspired synthesis is an interesting but quite new research topic and it remains largely unexplored, especially by using other functional materials. Progress in material-based biological engineering will lead to more and more applications throughout the field of biotechnology.

9.5
Summary and Outlook

In summary, recent progress on bio-inspired synthesis and application of diverse functional inorganic materials have been overviewed. Various functional inorganic materials, including noble metals, metal oxides and chalcogenides, metal oxysalts, MOFs, and so on, with precisely controlled size, shape, phase, and organization have been realized under mild conditions. These latest developments have clearly demonstrated that bio-inspired mineralization approaches involving polymer-controlled crystallization are quite promising for the design and preparation of not only biominerals-related inorganic materials but also for the functional inorganic materials.

Since the underlying mechanism in biomineralization is the use of biomacromolecules to mediate the nucleation and growth of biominerals, the main idea in bio-inspired approaches is to design and prepare functional polymers with selectivity over specific functional inorganic materials and further utilize them to control the crystal structure via classical and/or nonclassical crystallization pathways. However, different from biominerals-related inorganic materials, the polymer additives used for bio-inspired mineralization of functional inorganic materials are still mainly focused on biomacromolecules, such as peptides, DNA, RNA, proteins, and their self-assemblies. More importantly, although the potential applications of such functional inorganic materials obtained by bio-inspired mineralization approaches are continuously expanding to diverse fields of science and technology, their large-scale synthesis in a cost-effective fashion is still limited so far. Therefore, exploiting low-cost synthetic polymers mimicking the functionality of biomacromolecules is highly desired from both the scientific viewpoint and the actual requirements for practical applications.

Further exploration in these areas should open new avenues for rationally designing various kinds of functional inorganic materials and their hybrids with organic materials possessing ideal hierarchy at controllable length scales and desirable dimensionality by bio-inspired approaches. In addition, the study of the relationship between the structural specialty/complexity (shape, size, phase, dimensionality, hierarchy, etc.) of the synthetic functional inorganic materials by bio-inspired approaches and their properties will shed new light on the potential but important applications of these materials in various fields in the future.

References

1 Lowenstam, H.A. and Weiner, S. (1989) *On Biomineralization*, Oxford University Press, New York.

2 Chen, P.-Y., McKittrick, J., and Meyers, M.A. (2012) Biological materials: functional adaptations and bioinspired designs. *Prog. Mater. Sci.*, **57** (8), 1492–1704.

3 Wendler, J.E. and Bown, P. (2013) Exceptionally well-preserved cretaceous microfossils reveal new biomineralization styles. *Nat. Commun.*, **4**, 2052.

4 Weaver, J.C., Milliron, G.W., Miserez, A., Evans-Lutterodt, K., Herrera, S., Gallana, I., Mershon, W.J., Swanson, B., Zavattieri, P., DiMasi, E., and Kisailus, D. (2012) The stomatopod dactyl club: a formidable damage-tolerant biological hammer. *Science*, **336** (6086), 1275–1280.

5 Amini, S., Masic, A., Bertinetti, L., Teguh, J.S., Herrin, J.S., Zhu, X., Su, H., and Miserez, A. (2014) Textured fluorapatite bonded to calcium sulphate strengthen stomatopod raptorial appendages. *Nat. Commun.*, **5**, 3187.

6 Chen, L., Bazylinski, D.A., and Lower, B.H. (2010) Bacteria that synthesize nano-sized compasses to navigate using earth's geomagnetic field. *Nat. Educ. Knowl.*, **3** (10), 30.

7 Vukusic, P. and Sambles, J.R. (2003) Photonic structures in biology. *Nature*, **424** (6950), 852–855.

8 Young, L., Oman, C., Watt, D., Money, K., and Lichtenberg, B. (1984) Spatial orientation in weightlessness and readaptation to earth's gravity. *Science*, **225** (4658), 205–208.

9 Wegst, U.G.K., Bai, H., Saiz, E., Tomsia, A.P., and Ritchie, R.O. (2015) Bioinspired structural materials. *Nat. Mater.*, **14** (1), 23–36.

10 Lu, Y., Dong, L., Zhang, L.-C., Su, Y.-D., and Yu, S.-H. (2012) Biogenic and biomimetic magnetic nanosized assemblies. *Nano Today*, **7** (4), 297–315.

11 Yao, H.-B., Fang, H.-Y., Wang, X.-H., and Yu, S.-H. (2011) Hierarchical assembly of micro-/nano-building blocks: bio-inspired rigid structural functional materials. *Chem. Soc. Rev.*, **40** (7), 3764–3785.

12 Jackson, A.P., Vincent, J.F.V., and Turner, R.M. (1988) The mechanical design of nacre. *Proc. R. Soc. Lond. B*, **234** (1277), 415–440.

13 Yao, H.-B., Ge, J., Mao, L.-B., Yan, Y.-X., and Yu, S.-H. (2014) 25th anniversary article: Artificial carbonate nanocrystals and layered structural nanocomposites inspired by nacre: synthesis, fabrication and applications. *Adv. Mater.*, **26** (1), 163–188.

14 Yu, S.H. and Cölfen, H. (2004) Bio-inspired crystal morphogenesis by hydrophilic polymers. *J. Mater. Chem.*, **14** (14), 2124–2147.

15 Natalio, F., Corrales, T.P., Panthöfer, M., Schollmeyer, D., Lieberwirth, I., Müller, W.E.G., Kappl, M., Butt, H.-J., and Tremel, W. (2013) Flexible minerals: self-assembled calcite spicules with extreme bending strength. *Science*, **339** (6125), 1298–1302.

16 Zhang, Z., Gao, D., Zhao, H., Xie, C., Guan, G., Wang, D., and Yu, S.-H. (2006) Biomimetic assembly of polypeptide-stabilized $CaCO_3$ nanoparticles. *J. Phys. Chem. B*, **110** (17), 8613–8618.

17 Yu, S.H., Colfen, H., Tauer, K., and Antonietti, M. (2005) Tectonic arrangement of $BaCO_3$ nanocrystals into helices induced by a racemic block copolymer. *Nat. Mater.*, **4** (1), 51–U5.

18 Yu, S.H., Cölfen, H., Hartmann, J., and Antonietti, M. (2002) Biomimetic crystallization of calcium carbonate spherules with controlled surface structures and sizes by double-hydrophilic block copolymers. *Adv. Funct. Mater.*, **12** (8), 541–545.

19 Geng, X., Liu, L., Jiang, J., and Yu, S.-H. (2010) Crystallization of $CaCO_3$ mesocrystals and complex aggregates in a mixed solvent media using polystyrene sulfonate as a crystal growth modifier. *Cryst. Growth Des.*, **10** (8), 3448–3453.

20 Liu, L., Jiang, J., and Yu, S.-H. (2014) Polymorph selection and structure evolution of $CaCO_3$ mesocrystals under control of poly(sodium 4-styrenesulfonate): synergetic effect of temperature and mixed solvent. *Cryst. Growth Des.*, **14** (11), 6048–6056.

21 Naka, K., Tanaka, Y., and Chujo, Y. D (2002) Effect of anionic starburst dendrimers on the crystallization of $CaCO_3$ in aqueous solution: size control of spherical vaterite particles. *Langmuir*, **18** (9), 3655–3658.

22 Noorduin, W.L., Grinthal, A., Mahadevan, L., and Aizenberg, J. D (2013) Rationally designed complex, hierarchical microarchitectures. *Science*, **340** (6134), 832–837.

23 Kababya, S., Gal, A., Kahil, K., Weiner, S., Addadi, L., and Schmidt, A. D (2015) Phosphate–water interplay tunes amorphous calcium carbonate metastability: spontaneous phase separation and crystallization vs. stabilization viewed by solid state NMR. *J. Am. Chem. Soc.*, **137** (2), 990–998.

24 Guo, X.-H., Yu, S.-H., and Cai, G.-B. D (2006) Crystallization in a mixture of solvents by using a crystal modifier: morphology control in the synthesis of highly monodisperse $CaCO_3$ microspheres. *Angew. Chem., Int. Ed.*, **45** (24), 3977–3981.

25 Gao, Y.-X., Yu, S.-H., Cong, H., Jiang, J., Xu, A.-W., Dong, W.F., and Cölfen, H. (2006) Block-copolymer-controlled growth of $CaCO_3$ microrings. *J. Phys. Chem. B*, **110** (13), 6432–6436.

26 Chen, S.F., Yu, S.H., Wang, T.X., Jiang, J., Cölfen, H., Hu, B., and Yu, B. (2005) Polymer-directed formation of unusual $CaCO_3$ pancakes with controlled surface structures. *Adv. Mater.*, **17** (12), 1461–1465.

27 Olszta, M.J., Gajjeraman, S., Kaufman, M., and Gower, L.B. (2004) Nanofibrous calcite synthesized via a solution–precursor–solid mechanism. *Chem. Mater.*, **16** (12), 2355–2362.

28 Zhu, J.-H., Yu, S.-H., Xu, A.-W., and Colfen, H. (2009) The biomimetic mineralization of double-stranded and cylindrical helical $BaCO_3$ nanofibres. *Chem. Commun.*, (9), 1106–1108.

29 Wang, T., Cölfen, H., and Antonietti, M. (2005) Nonclassical crystallization: mesocrystals and morphology change of $CaCO_3$ crystals in the presence of a polyelectrolyte additive. *J. Am. Chem. Soc.*, **127** (10), 3246–3247.

30 Finnemore, A., Cunha, P., Shean, T., Vignolini, S., Guldin, S., Oyen, M., and Steiner, U. (2012) Biomimetic layer-by-layer assembly of artificial nacre. *Nat. Commun.*, **3**, 966.

31 Finnemore, A.S., Scherer, M.R.J., Langford, R., Mahajan, S., Ludwigs, S., Meldrum, F.C., and Steiner, U. (2009) Nanostructured calcite single crystals with gyroid morphologies. *Adv. Mater.*, **21** (38–39), 3928–3932.

32 Li, C. and Qi, L. (2008) Bioinspired fabrication of 3D ordered macroporous single crystals of calcite from a transient amorphous phase. *Angew. Chem., Int. Ed.*, **47** (13), 2388–2393.

33 Lee, K., Wagermaier, W., Masic, A., Kommareddy, K.P., Bennet, M., Manjubala, I., Lee, S.-W., Park, S.B., Cölfen, H., and Fratzl, P. (2012) Self-assembly of amorphous calcium carbonate microlens arrays. *Nat. Commun.*, **3**, 725.

34 Meldrum, F.C. and Cölfen, H. (2008) Controlling mineral morphologies and structures in biological and synthetic systems. *Chem. Rev.*, **108** (11), 4332–4432.

35 Sommerdijk, N.A.J.M. and de With, G. (2008) Biomimetic $CaCO_3$ mineralization using designer molecules and interfaces. *Chem. Rev.*, **108** (11), 4499–4550.

36 Chen, S.-F., Zhu, J.-H., Jiang, J., Cai, G.-B., and Yu, S.-H. (2010) Polymer-controlled crystallization of unique mineral superstructures. *Adv. Mater.*, **22** (4), 540–545.

37 Yu, S.H. and Chen, S.F. (2008) Biomineralization: self-assembly

processes, in *Nanomaterials: Inorganic and Bioinorganic Perspectives* (eds C.M. Lukehart and R.A. Scott), John Wiley & Sons, Ltd, Chichester, pp. 75–98.

38 Jiang, J., Gao, M.R., Xu, Y.F., and Yu, S.H. (2012) Amorphous calcium carbonate: synthesis and transformation, in *Bioinspiration: From Nano to Micro Scales*, (ed. X.Y. Liu), Springer, pp. 189–220.

39 De Yoreo, J.J., Gilbert, P.U.P.A., Sommerdijk, N.A.J.M., Penn, R.L., Whitelam, S., Joester, D., Zhang, H., Rimer, J.D., Navrotsky, A., Banfield, J.F., Wallace, A.F., Michel, F.M., Meldrum, F.C., Cölfen, H., and Dove, P.M. (2015) Crystallization by particle attachment in synthetic, biogenic, and geologic environments. *Science*, **349** (6247), aaa6760.

40 Zhang, Q., Liu, S.-J., and Yu, S.-H. (2009) Recent advances in oriented attachment growth and synthesis of functional materials: concept, evidence, mechanism, and future. *J. Mater. Chem.*, **19** (2), 191–207.

41 Cölfen, H. and Antonietti, M. (2005) Mesocrystals: inorganic superstructures made by highly parallel crystallization and controlled alignment. *Angew. Chem., Int. Ed.*, **44** (35), 5576–5591.

42 Xiao, J. and Yang, S. (2012) Bio-inspired synthesis: understanding and exploitation of the crystallization process from amorphous precursors. *Nanoscale*, **4** (1), 54–65.

43 Cölfen, H. (2007) Bio-inspired mineralization using hydrophilic polymers. *Top. Curr. Chem.*, **271**, 1–77.

44 Yu, S.H. (2007) Bio-inspired crystal growth by synthetic templates. *Top. Curr. Chem.*, **271**, 79–118.

45 Shen, J.-W., Li, C., van der Vegt, N.F.A., and Peter, C. (2013) Understanding the control of mineralization by polyelectrolyte additives: simulation of preferential binding to calcite surfaces. *J. Phys. Chem. C*, **117** (13), 6904–6913.

46 Lai, M., Kulak, A.N., Law, D., Zhang, Z.B., Meldrum, F.C., and Riley, D.J. (2007) Profiting from nature: macroporous copper with superior mechanical properties. *Chem. Commun.*, (34), 3547–3549.

47 Park, R.J. and Meldrum, F.C. (2002) Synthesis of single crystals of calcite with complex morphologies. *Adv. Mater.*, **14** (16), 1167–1169.

48 Yue, W.B., Kulak, A.N., and Meldrum, F.C. (2006) Growth of single crystals in structured templates. *J. Mater. Chem.*, **16** (4), 408–416.

49 Yue, W.B., Park, R.J., Kulak, A.N., and Meldrum, F.C. (2006) Macroporous inorganic solids from a biomineral template. *J. Cryst. Growth*, **294** (1), 69–77.

50 Penn, R.L. and Banfield, J.F. (1998) Imperfect oriented attachment: dislocation generation in defect-free nanocrystals. *Science*, **281** (5379), 969–971.

51 Li, D., Nielsen, M.H., Lee, J.R.I., Frandsen, C., Banfield, J.F., and De Yoreo, J.J. (2012) Direction-specific interactions control crystal growth by oriented attachment. *Science*, **336** (6084), 1014–1018.

52 Xia, B.Y., Wu, H.B., Yan, Y., Lou, X.W., and Wang, X. (2013) Ultrathin and ultralong single-crystal platinum nanowire assemblies with highly stable electrocatalytic activity. *J. Am. Chem. Soc.*, **135** (25), 9480–9485.

53 Wang, Y., Choi, S.-I., Zhao, X., Xie, S., Peng, H.-C., Chi, M., Huang, C.Z., and Xia, Y. (2014) Polyol synthesis of ultrathin Pd nanowires via attachment-based growth and their enhanced activity towards formic acid oxidation. *Adv. Funct. Mater.*, **24** (1), 131–139.

54 Kozhummal, R., Yang, Y., Gueder, F., Kuecuekbayrak, U.M., and Zacharias, M. (2013) Antisolvent crystallization approach to construction of CuI superstructures with defined geometries. *ACS Nano*, **7** (3), 2820–2828.

55 Tian, Y., Yang, H.-Y., Li, K., and Jin, X. (2012) Monodispersed ultrathin GdF_3 nanowires: oriented attachment, luminescence, and relaxivity for MRI contrast agents. *J. Mater. Chem.*, **22** (42), 22510–22516.

56 Yang, S., Song, X., Zhang, P., Sun, J., and Gao, L. (2014) Self-assembled α-Fe_2O_3

mesocrystals/graphene nanohybrid for enhanced electrochemical capacitors. *Small*, **10** (11), 2270–2279.

57 Zhao, B., Liu, P., Zhuang, H., Jiao, Z., Fang, T., Xu, W., Lu, B., and Jiang, Y. (2013) Hierarchical self-assembly of microscale leaf-like CuO on graphene sheets for high-performance electrochemical capacitors. *J. Mater. Chem. A*, **1** (2), 367–373.

58 Liu, Y., Bai, J., Ma, X., Li, J., and Xiong, S. (2014) Formation of quasi-mesocrystal $ZnMn_2O_4$ twin microspheres via an oriented attachment for lithium-ion batteries. *J. Mater. Chem. A*, **2** (34), 14236–14244.

59 Yu, M., Draskovic, T.I., and Wu, Y. (2014) Understanding the crystallization mechanism of delafossite $CuGaO_2$ for controlled hydrothermal synthesis of nanoparticles and nanoplates. *Inorg. Chem.*, **53** (11), 5845–5851.

60 Ramasamy, K., Sims, H., Butler, W.H., and Gupta, A. (2014) Mono-, few-, and multiple layers of copper antimony sulfide ($CuSbS_2$): a ternary layered sulfide. *J. Am. Chem. Soc.*, **136** (4), 1587–1598.

61 Tsuruoka, T., Furukawa, S., Takashima, Y., Yoshida, K., Isoda, S., and Kitagawa, S. (2009) Nanoporous nanorods fabricated by coordination modulation and oriented attachment growth. *Angew. Chem., Int. Ed.*, **48** (26), 4739–4743.

62 Ren, L., Jin, L., Wang, J.-B., Yang, F., Qiu, M.-Q., and Yu, Y. (2009) Template-free synthesis of $BiVO_4$ nanostructures: I. Nanotubes with hexagonal cross sections by oriented attachment and their photocatalytic property for water splitting under visible light. *Nanotechnology*, **20** (11), 115603.

63 Liu, L., Li, B., Yu, D., Cui, Y., Zhou, X., and Ding, W. (2010) Temperature-induced solid-phase oriented rearrangement route to the fabrication of $NaNbO_3$ nanowires. *Chem. Commun.*, **46** (3), 427–429.

64 Yang, Y., Scholz, R., Fan, H.J., Hesse, D., Goesele, U., and Zacharias, M. (2009) Multitwinned spinel nanowires by assembly of nanobricks via oriented attachment: a case study of Zn_2TiO_4. *ACS Nano*, **3** (3), 555–562.

65 Zhang, X., Zheng, Y., McCulloch, D.G., Yeo, L.Y., Friend, J.R., and MacFarlane, D.R. (2014) Controlled morphogenesis and self-assembly of bismutite nanocrystals into three-dimensional nanostructures and their applications. *J. Mater. Chem. A*, **2** (7), 2275–2282.

66 Viedma, C., McBride, J.M., Kahr, B., and Cintas, P. (2013) Enantiomer-specific oriented attachment: formation of macroscopic homochiral crystal aggregates from a racemic system. *Angew. Chem., Int. Ed.*, **52** (40), 10545–10548.

67 Zhang, H., De Yoreo, J.J., and Banfield, J.F. (2014) A unified description of attachment-based crystal growth. *ACS Nano*, **8** (7), 6526–6530.

68 Yang, H.G. and Zeng, H.C. (2004) Self-construction of hollow SnO_2 octahedra based on two-dimensional aggregation of nanocrystallites. *Angew. Chem., Int. Ed.*, **43** (44), 5930–5933.

69 Hou, L., Zhang, Q., Ling, L., Li, C.-X., Chen, L., and Chen, S. (2013) Interfacial fabrication of single-crystalline ZnTe nanorods with high blue fluorescence. *J. Am. Chem. Soc.*, **135** (29), 10618–10621.

70 Kim, Y.H., Lee, J.H., Shin, D.-W., Park, S.M., Moon, J.S., Nam, J.G., and Yoo, B. (2010) Synthesis of shape-controlled β-In_2S_3 nanotubes through oriented attachment of nanoparticles. *Chem. Commun.*, **46** (13), 2292–2294.

71 Wang, Y., Choi, S.-I., Zhao, X., Xie, S., Peng, H.-C., Chi, M., Huang, C.Z., and Xia, Y. (2014) Polyol synthesis of ultrathin Pd nanowires via attachment-based growth and their enhanced activity towards formic acid oxidation. *Adv. Funct. Mater.*, **24** (1), 131–139.

72 Yu, X., Wang, D., Peng, Q., and Li, Y. (2013) Pt-M (M = Cu, Co, Ni, Fe) nanocrystals: from small nanoparticles to wormlike nanowires by oriented attachment. *Chem. Eur. J.*, **19** (1), 233–239.

73 Fan, X., Xin, M., Gerlein, L.F., and Cloutier, S.G. (2011) Designing and building nanowires: directed nanocrystal self-assembly into radically branched and

zig-zag PbS nanowires. *Nanotechnology*, **22** (26), 265604.

74 Ribeiro, C., Longo, E., and Leite, E.R. (2007) Tailoring of heterostructures in a SnO_2/TiO_2 system by the oriented attachment mechanism. *Appl. Phys. Lett.*, **91** (10), 103105.

75 Nonoguchi, Y., Nakashima, T., Tanaka, A., Miyabayashi, K., Miyake, M., and Kawai, T. (2011) Oligomerization of cadmium chalcogenide nanocrystals into CdTe-containing superlattice chains. *Chem. Commun.*, **47** (40), 11270–11272.

76 Schliehe, C., Juarez, B.H., Pelletier, M., Jander, S., Greshnykh, D., Nagel, M., Meyer, A., Foerster, S., Kornowski, A., Klinke, C., and Weller, H. (2010) Ultrathin PbS sheets by two-dimensional oriented attachment. *Science*, **329** (5991), 550–553.

77 Yu, T., Lim, B., and Xia, Y. (2010) Aqueous-phase synthesis of single-crystal ceria nanosheets. *Angew. Chem., Int. Ed.*, **49** (26), 4484–4487.

78 Boneschanscher, M.P., Evers, W.H., Geuchies, J.J., Altantzis, T., Goris, B., Rabouw, F.T., van Rossum, S.A.P., van der Zant, H.S.J., Siebbeles, L.D.A., Van Tendeloo, G., Swart, I., Hilhorst, J., Petukhov, A.V., Bals, S., and Vanmaekelbergh, D. (2014) Long-range orientation and atomic attachment of nanocrystals in 2D honeycomb superlattices. *Science*, **344** (6190), 1377–1380.

79 Song, R.-Q. and Cölfen, H. (2010) Mesocrystals-ordered nanoparticle superstructures. *Adv. Mater.*, **22** (12), 1301–1330.

80 Seto, J., Ma, Y., Davis, S.A., Meldrum, F., Gourrier, A., Kim, Y.-Y., Schilde, U., Sztucki, M., Burghammer, M., Maltsev, S., Jäger, C., and Cölfen, H. (2012) Structure–property relationships of a biological mesocrystal in the adult sea urchin spine. *Proc. Natl. Acad. Sci. USA*, **109** (10), 3699–3704.

81 Cuif, J.P. and Dauphin, Y. (2005) The environment recording unit in coral skeletons – a synthesis of structural and chemical evidences for a biochemically driven, stepping-growth process in fibres. *Biogeosciences*, **2** (1), 61–73.

82 Simon, P., Zahn, D., Lichte, H., and Kniep, R. (2006) Intrinsic electric dipole fields and the induction of hierarchical form developments in fluorapatite–gelatine nanocomposites: a general principle for morphogenesis of biominerals? *Angew. Chem., Int. Ed.*, **45** (12), 1911–1915.

83 Schäffer, T.E., Ionescu-Zanetti, C., Proksch, R., Fritz, M., Walters, D.A., Almqvist, N., Zaremba, C.M., Belcher, A.M., Smith, B.L., Stucky, G.D., Morse, D.E., and Hansma, P.K. (1997) Does abalone nacre form by heteroepitaxial nucleation or by growth through mineral bridges? *Chem. Mater.*, **9** (8), 1731–1740.

84 Uchaker, E. and Cao, G. (2014) Mesocrystals as electrode materials for lithium-ion batteries. *Nano Today*, **9** (4), 499–524.

85 Tachikawa, T. and Majima, T. (2014) Metal oxide mesocrystals with tailored structures and properties for energy conversion and storage applications. *NPG Asia Mater.*, **6**, e100.

86 Su, Y., Yan, X., Wang, A., Fei, J., Cui, Y., He, Q., and Li, J. (2010) A peony-flower-like hierarchical mesocrystal formed by diphenylalanine. *J. Mater. Chem.*, **20** (32), 6734–6740.

87 Jiang, Y., Gong, H., Volkmer, D., Gower, L., and Cölfen, H. (2011) Preparation of hierarchical mesocrystalline DL-lysine·HCl–poly (acrylic acid) hybrid thin films. *Adv. Mater.*, **23** (31), 3548–3552.

88 Li, T., You, H., Xu, M., Song, X., and Fang, J. (2012) Electrocatalytic properties of hollow coral-like platinum mesocrystals. *ACS Appl. Mater. Interfaces*, **4** (12), 6941–6947.

89 Huang, X., Tang, S., Yang, J., Tan, Y., and Zheng, N. (2011) Etching growth under surface confinement: an effective strategy to prepare mesocrystalline Pd nanocorolla. *J. Am. Chem. Soc.*, **133** (40), 15946–15949.

90 Wu, X.L., Xiong, S.J., Liu, Z., Chen, J., Shen, J.C., Li, T.H., Wu, P.H., and Chu, P.K. (2011) Green light stimulates terahertz emission from mesocrystal microspheres. *Nat. Nanotechnol.*, **6** (2), 102–105.

91 Su, D., Dou, S., and Wang, G. (2014) Mesocrystal Co_3O_4 nanoplatelets as high capacity anode materials for Li-ion batteries. *Nano Res.*, **7** (5), 794–803.

92 Lausser, C., Coelfen, H., and Antonietti, M. (2011) Mesocrystals of vanadium pentoxide: a comparative evaluation of three different pathways of mesocrystal synthesis from tactosol precursors. *ACS Nano*, **5** (1), 107–114.

93 Simon, P., Rosseeva, E., Baburin, I.A., Liebscher, L., Hickey, S.G., Cardoso-Gil, R., Eychmüller, A., Kniep, R., and Carrillo-Cabrera, W. (2012) PbS–organic mesocrystals: the relationship between nanocrystal orientation and superlattice array. *Angew. Chem., Int. Ed.*, **51** (43), 10776–10781.

94 Jin, R., Zhou, J., Guan, Y., Liu, H., and Chen, G. (2014) Mesocrystal Co_9S_8 hollow sphere anodes for high performance lithium ion batteries. *J. Mater. Chem. A*, **2** (33), 13241–13244.

95 Popovic, J., Demir-Cakan, R., Tornow, J., Morcrette, M., Su, D.S., Schloegl, R., Antonietti, M., and Titirici, M.-M. (2011) $LiFePO_4$ mesocrystals for lithium-ion batteries. *Small*, **7** (8), 1127–1135.

96 Hu, B., Wu, L.-H., Liu, S.-J., Yao, H.-B., Shi, H.-Y., Li, G.-P., and Yu, S.-H. (2010) Microwave-assisted synthesis of silver indium tungsten oxide mesocrystals and their selective photocatalytic properties. *Chem. Commun.*, **46** (13), 2277–2279.

97 Bergström, L., Sturm, E.V., Salazar-Alvarez, G., and Cölfen, H. (2015) Mesocrystals in biominerals and colloidal arrays. *Acc. Chem. Res.*, **48** (5), 1391–1402.

98 Fang, J., Ding, B., and Gleiter, H. (2011) Mesocrystals: syntheses in metals and applications. *Chem. Soc. Rev.*, **40** (11), 5347–5360.

99 Politi, Y., Arad, T., Klein, E., Weiner, S., and Addadi, L. (2004) Sea urchin spine calcite forms via a transient amorphous calcium carbonate phase. *Science*, **306** (5699), 1161–1164.

100 Nassif, N., Pinna, N., Gehrke, N., Antonietti, M., Jäger, C., and Cölfen, H. (2005) Amorphous layer around aragonite platelets in nacre. *Proc. Natl. Acad. Sci. USA*, **102** (36), 12653–12655.

101 Wang, Q., Nemoto, M., Li, D., Weaver, J.C., Weden, B., Stegemeier, J., Bozhilov, K.N., Wood, L.R., Milliron, G.W., Kim, C.S., DiMasi, E., and Kisailus, D. (2013) Phase transformations and structural developments in the radular teeth of cryptochiton stelleri. *Adv. Funct. Mater.*, **23** (23), 2908–2917.

102 Baumgartner, J., Morin, G., Menguy, N., Perez Gonzalez, T., Widdrat, M., Cosmidis, J., and Faivre, D. (2013) Magnetotactic bacteria form magnetite from a phosphate-rich ferric hydroxide via nanometric ferric (oxyhydr)oxide intermediates. *Proc. Natl. Acad. Sci. USA*, **110** (37), 14883–14888.

103 Tao, J., Zhou, D., Zhang, Z., Xu, X., and Tang, R. (2009) Magnesium-aspartate-based crystallization switch inspired from shell molt of crustacean. *Proc. Natl. Acad. Sci. USA*, **106** (52), 22096–22101.

104 Gower, L.A. and Tirrell, D.A. (1998) Calcium carbonate films and helices grown in solutions of poly(aspartate). *J. Cryst. Growth*, **191** (1–2), 153–160.

105 Gower, L.B. and Odom, D.J. (2000) Deposition of calcium carbonate films by a polymer-induced liquid-precursor (PILP) process. *J. Cryst. Growth*, **210** (4), 719–734.

106 Wohlrab, S., Cölfen, H., and Antonietti, M. (2005) Crystalline, porous microspheres made from amino acids by using polymer-induced liquid precursor phases. *Angew. Chem., Int. Ed.*, **44** (26), 4087–4092.

107 Gebauer, D., Kellermeier, M., Gale, J.D., Bergstrom, L., and Colfen, H. (2014) Pre-nucleation clusters as solute precursors in crystallisation. *Chem. Soc. Rev.*, **43** (7), 2348–2371.

108 Gebauer, D., Völkel, A., and Cölfen, H. (2008) Stable prenucleation calcium carbonate clusters. *Science*, **322** (5909), 1819–1822.

109 Kellermeier, M., Gebauer, D., Melero-García, E., Drechsler, M., Talmon, Y., Kienle, L., Cölfen, H., García-Ruiz, J.M., and Kunz, W. (2012) Colloidal stabilization of calcium carbonate prenucleation clusters with silica. *Adv. Funct. Mater.*, **22** (20), 4301–4311.

110 Demichelis, R., Raiteri, P., Gale, J.D., Quigley, D., and Gebauer, D. (2011) Stable prenucleation mineral clusters are liquid-like ionic polymers. *Nat. Commun.*, **2**, 590.

111 Habraken, W.J.E.M., Tao, J., Brylka, L.J., Friedrich, H., Bertinetti, L., Schenk, A.S., Verch, A., Dmitrovic, V., Bomans, P.H.H., Frederik, P.M., Laven, J., van der Schoot, P., Aichmayer, B., de With, G., DeYoreo, J.J., and Sommerdijk, N.A.J.M. (2013) Ion-association complexes unite classical and non-classical theories for the biomimetic nucleation of calcium phosphate. *Nat. Commun.*, **4**, 1507.

112 Belton, D.J., Deschaume, O., and Perry, C.C. (2012) An overview of the fundamentals of the chemistry of silica with relevance to biosilicification and technological advances. *FEBS J.*, **279** (10), 1710–1720.

113 Kellermeier, M., Rosenberg, R., Moise, A., Anders, U., Przybylski, M., and Colfen, H. (2012) Amino acids form prenucleation clusters: ESI-MS as a fast detection method in comparison to analytical ultracentrifugation. *Faraday Discuss.*, **159**, 23–45.

114 Chaudhari, K., Xavier, P.L., and Pradeep, T. (2011) Understanding the evolution of luminescent gold quantum clusters in protein templates. *ACS Nano*, **5** (11), 8816–8827.

115 Yu, Y., Luo, Z., Teo, C.S., Tan, Y.N., and Xie, J. (2013) Tailoring the protein conformation to synthesize different-sized gold nanoclusters. *Chem. Commun.*, **49** (84), 9740–9742.

116 Chen, C.-L., Zhang, P., and Rosi, N.L. (2008) A new peptide-based method for the design and synthesis of nanoparticle superstructures: construction of highly ordered gold nanoparticle double helices. *J. Am. Chem. Soc.*, **130** (41), 13555–13557.

117 Xie, J., Lee, J.Y., Wang, D.I.C., and Ting, Y.P. (2007) Identification of active biomolecules in the high-yield synthesis of single-crystalline gold nanoplates in algal solutions. *Small*, **3** (4), 672–682.

118 Song, C., Zhao, G., Zhang, P., and Rosi, N.L. (2010) Expeditious synthesis and assembly of sub-100 nm hollow spherical gold nanoparticle superstructures. *J. Am. Chem. Soc.*, **132** (40), 14033–14035.

119 Guo, C. and Irudayaraj, J. (2011) Fluorescent Ag clusters via a protein-directed approach as a Hg(II) ion sensor. *Anal. Chem.*, **83** (8), 2883–2889.

120 Carter, C.J., Ackerson, C.J., and Feldheim, D.L. (2010) Unusual reactivity of a silver mineralizing peptide. *ACS Nano*, **4** (7), 3883–3888.

121 Bai, J., Qin, Y., Jiang, C., and Qi, L. (2007) Polymer-controlled synthesis of silver nanobelts and hierarchical nanocolumns. *Chem. Mater.*, **19** (14), 3367–3369.

122 Naik, R.R., Stringer, S.J., Agarwal, G., Jones, S.E., and Stone, M.O. (2002) Biomimetic synthesis and patterning of silver nanoparticles. *Nat. Mater.*, **1** (3), 169–172.

123 Chiu, C.-Y., Li, Y., Ruan, L., Ye, X., Murray, C.B., and Huang, Y. (2011) Platinum nanocrystals selectively shaped using facet-specific peptide sequences. *Nat. Chem.*, **3** (5), 393–399.

124 Li, Y. and Huang, Y. (2010) Morphology-controlled synthesis of platinum nanocrystals with specific peptides. *Adv. Mater.*, **22** (17), 1921–1925.

125 Zhang, L., Li, N., Gao, F., Hou, L., and Xu, Z. (2012) Insulin amyloid fibrils: an excellent platform for controlled synthesis of ultrathin superlong platinum nanowires with high electrocatalytic activity. *J. Am. Chem. Soc.*, **134** (28), 11326–11329.

126 Tao, K., Wang, J., Li, Y., Xia, D., Shan, H., Xu, H., and Lu, J.R. (2013) Short peptide-directed synthesis of one-dimensional platinum nanostructures with controllable morphologies. *Sci. Rep.*, **3**, 2565.

127 Janairo, J.I.B., Sakaguchi, T., Hara, K., Fukuoka, A., and Sakaguchi, K. (2014) Effects of biomineralization peptide topology on the structure and catalytic activity of Pd nanomaterials. *Chem. Commun.*, **50** (66), 9259–9262.

128 Bhandari, R. and Knecht, M.R. (2011) Effects of the material structure on the catalytic activity of peptide-templated Pd nanomaterials. *ACS Catal.*, **1** (2), 89–98.

129 Ghosh, R., Sahoo, A.K., Ghosh, S.S., Paul, A., and Chattopadhyay, A. (2014) Blue-

emitting copper nanoclusters synthesized in the presence of lysozyme as candidates for cell labeling. *ACS Appl. Mater. Interfaces*, **6** (6), 3822–3828.

130 Goswami, N., Baksi, A., Giri, A., Xavier, P.L., Basu, G., Pradeep, T., and Pal, S.K. (2014) Luminescent iron clusters in solution. *Nanoscale*, **6** (3), 1848–1854.

131 Mohanty, J.S., Xavier, P.L., Chaudhari, K., Bootharaju, M.S., Goswami, N., Pal, S.K., and Pradeep, T. (2012) Luminescent, bimetallic AuAg alloy quantum clusters in protein templates. *Nanoscale*, **4** (14), 4255–4262.

132 Galloway, J.M., Talbot, J.E., Critchley, K., Miles, J.J., and Bramble, J.P. (2015) Developing biotemplated data storage: room temperature biomineralization of L10 CoPt magnetic nanoparticles. *Adv. Funct. Mater.*, **25** (29), 4590–4600.

133 Song, C., Wang, Y., and Rosi, N.L. (2013) Peptide-directed synthesis and assembly of hollow spherical CoPt nanoparticle superstructures. *Angew. Chem., Int. Ed.*, **52** (14), 3993–3995.

134 Sumerel, J.L., Yang, W., Kisailus, D., Weaver, J.C., Choi, J.H., and Morse, D.E. (2003) Biocatalytically templated synthesis of titanium dioxide. *Chem. Mater.*, **15** (25), 4804–4809.

135 Luckarift, H.R., Dickerson, M.B., Sandhage, K.H., and Spain, J.C. (2006) Rapid, room-temperature synthesis of antibacterial bionanocomposites of lysozyme with amorphous silica or titania. *Small*, **2** (5), 640–643.

136 Tahir, M.N., Theato, P., Muller, W.E.G., Schroder, H.C., Borejko, A., Faib, S., Janshoff, A., Huth, J., and Tremel, W. (2005) Formation of layered titania and zirconia catalysed by surface-bound silicatein. *Chem. Commun.*, (44), 5533–5535.

137 Wang, X., Yan, Y., Hao, B., and Chen, G. (2013) Protein-mediated layer-by-layer synthesis of TiO_2(B)/anatase/carbon coating on nickel foam as negative electrode material for lithium-ion battery. *ACS Appl. Mater. Interfaces*, **5** (9), 3631–3637.

138 Acar, H., Garifullin, R., and Guler, M.O. (2011) Self-assembled template-directed synthesis of one-dimensional silica and titania nanostructures. *Langmuir*, **27** (3), 1079–1084.

139 Bolisetty, S., Adamcik, J., Heier, J., and Mezzenga, R. (2012) Amyloid directed synthesis of titanium dioxide nanowires and their applications in hybrid photovoltaic devices. *Adv. Funct. Mater.*, **22** (16), 3424–3428.

140 Nonoyama, T., Kinoshita, T., Higuchi, M., Nagata, K., Tanaka, M., Sato, K., and Kato, K. (2012) TiO_2 synthesis inspired by biomineralization: control of morphology, crystal phase, and light-use efficiency in a single process. *J. Am. Chem. Soc.*, **134** (21), 8841–8847.

141 Umetsu, M., Mizuta, M., Tsumoto, K., Ohara, S., Takami, S., Watanabe, H., Kumagai, I., and Adschiri, T. (2005) Bioassisted room-temperature immobilization and mineralization of zinc oxide: the structural ordering of ZnO nanoparticles into a flower-type morphology. *Adv. Mater.*, **17** (21), 2571–2575.

142 Begum, G., Manorama, S.V., Singh, S., and Rana, R.K. (2008) Morphology-controlled assembly of ZnO nanostructures: a bioinspired method and visible luminescence. *Chem. Eur. J.*, **14** (21), 6421–6427.

143 Wei, Z., Maeda, Y., and Matsui, H. (2011) Discovery of catalytic peptides for inorganic nanocrystal synthesis by a combinatorial phage display approach. *Angew. Chem., Int. Ed.*, **50** (45), 10585–10588.

144 Amemiya, Y., Arakaki, A., Staniland, S.S., Tanaka, T., and Matsunaga, T. (2007) Controlled formation of magnetite crystal by partial oxidation of ferrous hydroxide in the presence of recombinant magnetotactic bacterial protein Mms6. *Biomaterials*, **28** (35), 5381–5389.

145 Galloway, J.M., Bramble, J.P., Rawlings, A.E., Burnell, G., Evans, S.D., and Staniland, S.S. (2012) Biotemplated magnetic nanoparticle arrays. *Small*, **8** (2), 204–208.

146 Zoppellaro, G., Kolokithas-Ntoukas, A., Polakova, K., Tucek, J., Zboril, R., Loudos, G., Fragogeorgi, E., Diwoky, C., Tomankova, K., Avgoustakis, K., Kouzoudis, D., and Bakandritsos, A.

(2014) Theranostics of epitaxially condensed colloidal nanocrystal clusters, through a soft biomineralization route. *Chem. Mater.*, **26** (6), 2062–2074.

147 Lenders, J.J.M., Altan, C.L., Bomans, P.H.H., Arakaki, A., Bucak, S., de With, G., and Sommerdijk, N.A.J.M. (2014) A bioinspired coprecipitation method for the controlled synthesis of magnetite nanoparticles. *Cryst. Growth Des.*, **14** (11), 5561–5568.

148 Lenders, J.J.M., Zope, H.R., Yamagishi, A., Bomans, P.H.H., Arakaki, A., Kros, A., de With, G., and Sommerdijk, N.A.J.M. (2015) Bioinspired magnetite crystallization directed by random copolypeptides. *Adv. Funct. Mater.*, **25** (5), 711–719.

149 Fei, X., Li, W., Shao, Z., Seeger, S., Zhao, D., and Chen, X. (2014) Protein biomineralized nanoporous inorganic mesocrystals with tunable hierarchical nanostructures. *J. Am. Chem. Soc.*, **136** (44), 15781–15786.

150 Liu, L., Yang, L.-Q., Liang, H.-W., Cong, H.-P., Jiang, J., and Yu, S.-H. (2013) Bio-inspired fabrication of hierarchical FeOOH nanostructure array films at the air–water interface, their hydrophobicity and application for water treatment. *ACS Nano*, **7** (2), 1368–1378.

151 Prozorov, T., Palo, P., Wang, L., Nilsen-Hamilton, M., Jones, D., Orr, D., Mallapragada, S.K., Narasimhan, B., Canfield, P.C., and Prozorov, R. (2007) Cobalt ferrite nanocrystals: out-performing magnetotactic bacteria. *ACS Nano*, **1** (3), 228–233.

152 Li, C.C., Yin, X.M., Wang, T.H., and Zeng, H.C. (2009) Morphogenesis of highly uniform $CoCO_3$ submicrometer crystals and their conversion to mesoporous Co_3O_4 for gas-sensing applications. *Chem. Mater.*, **21** (20), 4984–4992.

153 Oaki, Y., Kajiyama, S., Nishimura, T., and Kato, T. (2008) Selective synthesis and thin-film formation of α-cobalt hydroxide through an approach inspired by biomineralization. *J. Mater. Chem.*, **18** (35), 4140–4142.

154 Pal, P., Pahari, S.K., Giri, A.K., Pal, S., Bajaj, H.C., and Panda, A.B. (2013) Hierarchically ordered porous lotus-shaped nano-structured MnO_2 through $MnCO_3$: chelate mediated growth and shape dependent improved catalytic activity. *J. Mater. Chem. A*, **1** (35), 10251–10258.

155 Kokubu, T., Oaki, Y., Hosono, E., Zhou, H., and Imai, H. (2011) Biomimetic solid-solution precursors of metal carbonate for nanostructured metal oxides: MnO/Co and MnO–CoO nanostructures and their electrochemical properties. *Adv. Funct. Mater.*, **21** (19), 3673–3680.

156 Uchiyama, H., Hosono, E., Zhou, H., and Imai, H. (2009) Three-dimensional architectures of spinel-type $LiMn_2O_4$ prepared from biomimetic porous carbonates and their application to a cathode for lithium-ion batteries. *J. Mater. Chem.*, **19** (23), 4012–4016.

157 Kokubu, T., Oaki, Y., Uchiyama, H., Hosono, E., Zhou, H., and Imai, H. (2010) Biomimetic synthesis of metal ion-doped hierarchical crystals using a gel matrix: formation of cobalt-doped $LiMn_2O_4$ with improved electrochemical properties through a cobalt-doped $MnCO_3$ precursor. *Chem. Asian J.*, **5** (4), 792–798.

158 Muñoz-Espí, R., Burger, C., Krishnan, C.V., and Chu, B. (2008) Polymer-controlled crystallization of molybdenum oxides from peroxomolybdates: structural diversity and application to catalytic epoxidation. *Chem. Mater.*, **20** (23), 7301–7311.

159 Uchiyama, H., Nakanishi, S., and Kozuka, H. (2015) Biomimetic synthesis of nanostructured SnO particles from $Sn_6O_4(OH)_4$ in aqueous solution of gelatin. *CrystEngComm*, **17** (3), 628–632.

160 Andre, R., Tahir, M.N., Schroeder, H.C.C., Mueller, W.E.G., and Tremel, W. (2011) Enzymatic synthesis and surface deposition of tin dioxide using silicatein-α. *Chem. Mater.*, **23** (24), 5358–5365.

161 Andre, R., Natalio, F., Tahir, M.N., Berger, R., and Tremel, W. (2013) Self-cleaning antimicrobial surfaces by bio-enabled growth of SnO_2 coatings on glass. *Nanoscale*, **5** (8), 3447–3456.

162 Kisailus, D., Choi, J.H., Weaver, J.C., Yang, W., and Morse, D.E. (2005) Enzymatic synthesis and nanostructural

control of gallium oxide at low temperature. *Adv. Mater.*, **17** (3), 314–318.

163 Lee, S.-Y., Gao, X., and Matsui, H. (2007) Biomimetic and aggregation-driven crystallization route for room-temperature material synthesis: growth of β-Ga_2O_3 nanoparticles on peptide assemblies as nanoreactors. *J. Am. Chem. Soc.*, **129** (10), 2954–2958.

164 Yan, D., Yin, G., Huang, Z., Liao, X., Kang, Y., Yao, Y., Hao, B., Gu, J., and Han, D. (2009) Biomineralization of uniform gallium oxide rods with cellular compatibility. *Inorg. Chem.*, **48** (14), 6471–6479.

165 Zhou, W., Schwartz, D.T., and Baneyx, F. (2010) Single-pot biofabrication of zinc sulfide immuno-quantum dots. *J. Am. Chem. Soc.*, **132** (13), 4731–4738.

166 Shen, L., Bao, N., Prevelige, P.E., and Gupta, A. (2010) Fabrication of ordered nanostructures of sulfide nanocrystal assemblies over self-assembled genetically engineered P22 coat protein. *J. Am. Chem. Soc.*, **132** (49), 17354–17357.

167 Zhou, W. and Baneyx, F. (2011) Aqueous, protein-driven synthesis of transition metal-doped ZnS immuno-quantum dots. *ACS Nano*, **5** (10), 8013–8018.

168 Wu, P., Zhao, T., Tian, Y., Wu, L., and Hou, X. (2013) Protein-directed synthesis of Mn-doped ZnS quantum dots: a dual-channel biosensor for two proteins. *Chem. Eur. J.*, **19** (23), 7473–7479.

169 Ma, N., Dooley, C.J., and Kelley, S.O. (2006) RNA-templated semiconductor nanocrystals. *J. Am. Chem. Soc.*, **128** (39), 12598–12599.

170 Gao, L. and Ma, N. (2012) DNA-templated semiconductor nanocrystal growth for controlled DNA packing and gene delivery. *ACS Nano*, **6** (1), 689–695.

171 Sone, E.D., Zubarev, E.R., and Stupp, S.I. (2002) Semiconductor nanohelices templated by supramolecular ribbons. *Angew. Chem., Int. Ed.*, **41** (10), 1705–1709.

172 Sone, E.D. and Stupp, S.I. (2004) Semiconductor-encapsulated peptide-amphiphile nanofibers. *J. Am. Chem. Soc.*, **126** (40), 12756–12757.

173 Dong, L., Hollis, T., Connolly, B.A., Wright, N.G., Horrocks, B.R., and Houlton, A. (2007) DNA-templated semiconductor nanoparticle chains and wires. *Adv. Mater.*, **19** (13), 1748–1751.

174 Bu, X., Zhou, Y., He, M., Chen, Z., and Zhang, T. (2013) Bioinspired, direct synthesis of aqueous CdSe quantum dots for high-sensitive copper(II) ion detection. *Dalton Trans.*, **42** (43), 15411–15420.

175 Hinds, S., Taft, B.J., Levina, L., Sukhovatkin, V., Dooley, C.J., Roy, M.D., MacNeil, D.D., Sargent, E.H., and Kelley, S.O. (2006) Nucleotide-directed growth of semiconductor nanocrystals. *J. Am. Chem. Soc.*, **128** (1), 64–65.

176 Kumar, A. and Jakhmola, A. (2009) RNA-templated fluorescent Zn/PbS ($PbS + Zn^{2+}$) supernanostructures. *J. Phys. Chem. C*, **113** (22), 9553–9559.

177 Kumar, A. and Singh, B. (2013) Zn^{2+}-induced folding of RNA to produce honeycomb-like RNA-mediated fluorescing Zn^{2+}/PbSe nanostructures. *J. Phys. Chem. C*, **117** (10), 5386–5396.

178 Yang, L., Xing, R., Shen, Q., Jiang, K., Ye, F., Wang, J., and Ren, Q. (2006) Fabrication of protein-conjugated silver sulfide nanorods in the bovine serum albumin solution. *J. Phys. Chem. B*, **110** (21), 10534–10539.

179 Wang, Y. and Yan, X.-P. (2013) Fabrication of vascular endothelial growth factor antibody bioconjugated ultrasmall near-infrared fluorescent Ag_2S quantum dots for targeted cancer imaging *in vivo*. *Chem. Commun.*, **49** (32), 3324–3326.

180 de la Rica, R. and Velders, A.H. (2011) Biomimetic crystallization of Ag_2S nanoclusters in nanopore assemblies. *J. Am. Chem. Soc.*, **133** (9), 2875–2877.

181 Wang, Q., Liu, Y., Ke, Y., and Yan, H. (2008) Quantum dot bioconjugation during core–shell synthesis. *Angew. Chem., Int. Ed.*, **47** (2), 316–319.

182 Singh, S., Bozhilov, K., Mulchandani, A., Myung, N., and Chen, W. (2010) Biologically programmed synthesis of core–shell CdSe/ZnS nanocrystals. *Chem. Commun.*, **46** (9), 1473–1475.

183 Nuraje, N., Su, K., Haboosheh, A., Samson, J., Manning, E.P., Yang, N.L., and Matsui, H. (2006) Room temperature synthesis of ferroelectric barium titanate nanoparticles using peptide nanorings as templates. *Adv. Mater.*, **18** (6), 807–811.
184 Ahmad, G., Dickerson, M.B., Cai, Y., Jones, S.E., Ernst, E.M., Vernon, J.P., Haluska, M.S., Fang, Y., Wang, J., Subramanyam, G., Naik, R.R., and Sandhage, K.H. (2008) Rapid bioenabled formation of ferroelectric $BaTiO_3$ at room temperature from an aqueous salt solution at near neutral pH. *J. Am. Chem. Soc.*, **130** (1), 4–5.
185 Pang, X., He, Y., Jiang, B., Iocozzia, J., Zhao, L., Guo, H., Liu, J., Akinc, M., Bowler, N., Tan, X., and Lin, Z. (2013) Block copolymer/ferroelectric nanoparticle nanocomposites. *Nanoscale*, **5** (18), 8695–8702.
186 Jiang, B., Pang, X., Li, B., and Lin, Z. (2015) Organic–inorganic nanocomposites via placing monodisperse ferroelectric nanocrystals in direct and permanent contact with ferroelectric polymers. *J. Am. Chem. Soc.*, **137** (36), 11760–11767.
187 Brutchey, R.L., Yoo, E.S., and Morse, D.E. (2006) Biocatalytic synthesis of a nanostructured and crystalline bimetallic perovskite-like barium oxofluorotitanate at low temperature. *J. Am. Chem. Soc.*, **128** (31), 10288–10294.
188 Ryu, J., Kim, S.-W., Kang, K., and Park, C.B. (2010) Mineralization of self-assembled peptide nanofibers for rechargeable lithium ion batteries. *Adv. Mater.*, **22** (48), 5537–5541.
189 Lee, Y.J., Yi, H., Kim, W.-J., Kang, K., Yun, D.S., Strano, M.S., Ceder, G., and Belcher, A.M. (2009) Fabricating genetically engineered high-power lithium-ion batteries using multiple virus genes. *Science*, **324** (5930), 1051–1055.
190 Guo, C.X., Shen, Y.Q., Dong, Z.L., Chen, X.D., Lou, X.W., and Li, C.M. (2012) DNA-directed growth of $FePO_4$ nanostructures on carbon nanotubes to achieve nearly 100% theoretical capacity for lithium-ion batteries. *Energy Environ. Sci.*, **5** (5), 6919–6922.
191 Zhang, X., Bi, Z., He, W., Yang, G., Liu, H., and Yue, Y. (2014) Fabricating high-energy quantum dots in ultra-thin $LiFePO_4$ nanosheets using a multifunctional high-energy biomolecule-ATP. *Energy Environ. Sci.*, **7** (7), 2285–2294.
192 Nie, P., Zhu, Y., Shen, L., Pang, G., Xu, G., Dong, S., Dou, H., and Zhang, X. (2014) From biomolecule to $Na_3V_2(PO_4)_3$/nitrogen-decorated carbon hybrids: highly reversible cathodes for sodium-ion batteries. *J. Mater. Chem. A*, **2** (43), 18606–18612.
193 Yu, S.-H., Antonietti, M., Cölfen, H., and Giersig, M. (2002) Synthesis of very thin 1D and 2D $CdWO_4$ nanoparticles with improved fluorescence behavior by polymer-controlled crystallization. *Angew. Chem., Int. Ed.*, **41** (13), 2356–2360.
194 Shi, H., Qi, L., Ma, J., and Cheng, H. (2003) Polymer-directed synthesis of penniform $BaWO_4$ nanostructures in reverse micelles. *J. Am. Chem. Soc.*, **125** (12), 3450–3451.
195 Yang, J., Lu, C., Su, H., Ma, J., Cheng, H., and Qi, L. (2008) Morphological and structural modulation of $PbWO_4$ crystals directed by dextrans. *Nanotechnology*, **19** (3), 035608.
196 Nithiyanantham, U., Ede, S.R., Kesavan, T., Ragupathy, P., Mukadam, M.D., Yusuf, S.M., and Kundu, S. (2014) Shape-selective formation of $MnWO_4$ nanomaterials on a DNA scaffold: magnetic, catalytic and supercapacitor studies. *RSC Adv.*, **4** (72), 38169–38181.
197 Ede, S.R., Ramadoss, A., Nithiyanantham, U., Anantharaj, S., and Kundu, S. (2015) Biomolecule assisted aggregation of $ZnWO_4$ nanoparticles (NPs) into chain-like assemblies: material for high performance supercapacitor and as catalyst for benzyl alcohol oxidation. *Inorg. Chem.*, **54** (8), 3851–3863.
198 Nithiyanantham, U., Ede, S.R., Anantharaj, S., and Kundu, S. (2015) Self-assembled $NiWO_4$ nanoparticles into chain-like aggregates on DNA scaffold with pronounced catalytic and supercapacitor activities. *Cryst. Growth Des.*, **15** (2), 673–686.

199 Lyu, F., Zhang, Y., Zare, R.N., Ge, J., and Liu, Z. (2014) One-pot synthesis of protein-embedded metal–organic frameworks with enhanced biological activities. *Nano Lett.*, **14** (10), 5761–5765.

200 Liang, K., Ricco, R., Doherty, C.M., Styles, M.J., Bell, S., Kirby, N., Mudie, S., Haylock, D., Hill, A.J., Doonan, C.J., and Falcaro, P. (2015) Biomimetic mineralization of metal–organic frameworks as protective coatings for biomacromolecules. *Nat. Commun.*, **6**, 7240.

201 Zhang, L. and Wang, E. (2014) Metal nanoclusters: new fluorescent probes for sensors and bioimaging. *Nano Today*, **9** (1), 132–157.

202 Miyamura, H. and Kobayashi, S. (2014) Tandem oxidative processes catalyzed by polymer-incarcerated multimetallic nanoclusters with molecular oxygen. *Acc. Chem. Res.*, **47** (4), 1054–1066.

203 Kumar, A. and Kumar, V. (2014) Biotemplated inorganic nanostructures: supramolecular directed nanosystems of semiconductor(s)/metal(s) mediated by nucleic acids and their properties. *Chem. Rev.*, **114** (14), 7044–7078.

204 Ding, Y., Shi, L., and Wei, H. (2014) Protein-directed approaches to functional nanomaterials: a case study of lysozyme. *J. Mater. Chem. B*, **2** (47), 8268–8291.

205 Ruan, L., Chiu, C.-Y., Li, Y., and Huang, Y. (2011) Synthesis of platinum single-twinned right bipyramid and {111}-bipyramid through targeted control over both nucleation and growth using specific peptides. *Nano Lett.*, **11** (7), 3040–3046.

206 Tong, Z., Jiang, Y., Yang, D., Shi, J., Zhang, S., Liu, C., and Jiang, Z. (2014) Biomimetic and bioinspired synthesis of titania and titania-based materials. *RSC Adv.*, **4** (24), 12388–12403.

207 Andre, R., Tahir, M.N., Natalio, F., and Tremel, W. (2012) Bioinspired synthesis of multifunctional inorganic and bio-organic hybrid materials. *FEBS J.*, **279** (10), 1737–1749.

208 Prozorov, T., Mallapragada, S.K., Narasimhan, B., Wang, L., Palo, P., Nilsen-Hamilton, M., Williams, T.J., Bazylinski, D.A., Prozorov, R., and Canfield, P.C. (2007) Protein-mediated synthesis of uniform superparamagnetic magnetite nanocrystals. *Adv. Funct. Mater.*, **17** (6), 951–957.

209 Jiao, Q., Fu, M., You, C., Zhao, Y., and Li, H. (2012) Preparation of hollow Co_3O_4 microspheres and their ethanol sensing properties. *Inorg. Chem.*, **51** (21), 11513–11520.

210 Cong, H.-P. and Yu, S.-H. (2009) Shape control of cobalt carbonate particles by a hydrothermal process in a mixed solvent: an efficient precursor to nanoporous cobalt oxide architectures and their sensing property. *Cryst. Growth Des.*, **9** (1), 210–217.

211 Wang, L., Chen, Y., Ma, J., Chen, L., Xu, Z., and Wang, T. (2013) Hierarchical SnO_2 nanospheres: bio-inspired mineralization, vulcanization, oxidation techniques, and the application for NO sensors. *Sci. Rep.*, **3**, 3500.

212 Shanmugasundaram, A., Ramireddy, B., Basak, P., Manorama, S.V., and Srinath, S. (2014) Hierarchical $In(OH)_3$ as a precursor to mesoporous In_2O_3 nanocubes. *J. Phys. Chem. C*, **118** (13), 6909–6921.

213 Gao, M.-R., Jiang, J., and Yu, S.-H. (2012) Solution-based synthesis and design of late transition metal chalcogenide materials for oxygen reduction reaction (ORR). *Small*, **8** (1), 13–27.

214 Wang, J. and Sun, X. (2015) Olivine $LiFePO_4$: the remaining challenges for future energy storage. *Energy Environ. Sci.*, **8** (4), 1110–1138.

215 Zhang, W., Wu, Z.-Y., Jiang, H.-L., and Yu, S.-H. (2014) Nanowire-directed templating synthesis of metal–organic framework nanofibers and their derived porous doped carbon nanofibers for enhanced electrocatalysis. *J. Am. Chem. Soc.*, **136** (41), 14385–14388.

216 Zhu, Q.-L. and Xu, Q. (2014) Metal–organic framework composites. *Chem. Soc. Rev.*, **43** (16), 5468–5512.

217 Liang, K., Coghlan, C.J., Bell, S.G., Doonan, C., and Falcaro, P. (2016) Enzyme encapsulation in zeolitic imidazolate frameworks: a comparison between controlled co-precipitation and

biomimetic mineralization. *Chem. Commun.*, **52** (3), 473–476.

218 Kim, J.K., Jang, J.-R., Choi, N., Hong, D., Nam, C.-H., Yoo, P.J., Park, J.H., and Choe, W.-S. (2014) Lysozyme-mediated biomineralization of titanium–tungsten oxide hybrid nanoparticles with high photocatalytic activity. *Chem. Commun.*, **50** (82), 12392–12395.

219 Zhou, L., Chen, Z., Dong, K., Yin, M., Ren, J., and Qu, X. (2014) DNA-mediated biomineralization of rare-earth nanoparticles for simultaneous imaging and stimuli-responsive drug delivery. *Biomaterials*, **35** (30), 8694–8702.

220 Wang, Y., Yang, T., Ke, H.T., Zhu, A.J., Wang, Y.Y., Wang, J.X., Shen, J.K., Liu, G., Chen, C.Y., Zhao, Y.L., and Chen, H.B. (2015) Smart albumin-biomineralized nanocomposites for multimodal imaging and photothermal tumor ablation. *Adv. Mater.*, **27** (26), 3874–3882.

221 Wen, Y., Dong, H.Q., Li, Y., Shen, A.J., and Li, Y.Y. (2016) Nano-assembly of bovine serum albumin driven by rare-earth-ion (Gd) biomineralization for highly efficient photodynamic therapy and tumor imaging. *J. Mater. Chem. B*, **4** (4), 743–751.

222 Liu, Z., Xu, X., and Tang, R. (2016) Improvement of biological organisms using functional material shells. *Adv. Funct. Mater.*, **26** (12), 1862–1880.

223 Wang, B., Liu, P., Jiang, W., Pan, H., Xu, X., and Tang, R. (2008) Yeast cells with an artificial mineral shell: protection and modification of living cells by biomimetic mineralization. *Angew. Chem., Int. Ed.*, **47** (19), 3560–3564.

224 Wang, G., Li, X., Mo, L., Song, Z., Chen, W., Deng, Y., Zhao, H., Qin, E., Qin, C., and Tang, R. (2012) Eggshell-inspired biomineralization generates vaccines that do not require refrigeration. *Angew. Chem., Int. Ed.*, **51** (42), 10576–10579.

225 Hong, D., Lee, H., Ko, E.H., Lee, J., Cho, H., Park, M., Yang, S.H., and Choi, I.S. (2015) Organic/inorganic double-layered shells for multiple cytoprotection of individual living cells. *Chem. Sci.*, **6** (1), 203–208.

226 Xiong, W., Yang, Z., Zhai, H., Wang, G., Xu, X., Ma, W., and Tang, R. (2013) Alleviation of high light-induced photoinhibition in cyanobacteria by artificially conferred biosilica shells. *Chem. Commun.*, **49** (68), 7525–7527.

227 Yang, S.H., Ko, E.H., and Choi, I.S. (2012) Cytocompatible encapsulation of individual chlorella cells within titanium dioxide shells by a designed catalytic peptide. *Langmuir*, **28** (4), 2151–2155.

228 Wang, B., Liu, P., Tang, Y., Pan, H., Xu, X., and Tang, R. (2010) Guarding embryo development of zebrafish by shell engineering: a strategy to shield life from ozone depletion. *PLoS One*, **5** (4), e9963.

10
Reactive Fluxes

Oliver Janka and Rainer Pöttgen

Universität Münster, Institut für Anorganische und Analytische Chemie, Corrensstrasse 30, 48149 Münster, Germany

10.1
Introduction

The synthesis of ionic and metallic materials in salt or metal flux media is briefly reviewed in chapter 3 along with high-temperature methods. For these applications, the flux serves solely as a solvent for crystal growth, but it is not consumed during reaction. The flux just provides the solid–liquid interface along with shorter diffusion times compared to standard solid–solid reactions.

This chapter exclusively deals with so-called reactive fluxes that deliver part of the reaction educts and the liquid phase at the same time. In the case of metal fluxes, this can be only one element, for example, a tin flux, while binary and ternary compounds can be used for a decomposition reaction in reactive salt fluxes, often leading to kinetically controlled products. The following sections regroup the different types of reactive fluxes.

General prerequisites for flux growth experiments concern the correct choice of an inert crucible material, the optimal temperature regime for crystal growth, sufficient solubility of the components, a good wetting ability of the flux, and an effective separation of the flux from the product crystals. A huge number of phases have been synthesized with various reactive flux techniques and most results have been summarized in excellent review articles, focusing on the challenges in preparative solid-state chemistry and historical remarks as well. Thus, this chapter emphasizes basically the experimental details with only few examples and gives a broader literature overview for a more detailed reading.

10.2
Reactive Metal Fluxes: Self-Flux Technique

Flux-assisted crystal growth experiments are feasible with most low-melting metals and this research field has been reviewed in detail [1–5]. Herein we

Handbook of Solid State Chemistry, First Edition. Edited by Richard Dronskowski, Shinichi Kikkawa, and Andreas Stein.

discuss only the self-flux technique where the flux medium is consumed during crystal growth. Liquid aluminum, gallium, indium, tin, lead, and zinc are the commonly used flux media. They all have comparatively low melting points and a broader liquidus range. This is an important prerequisite, since a low boiling temperature would lead to evaporation of the flux. It is not surprising that many of the prominent self-flux elements are also components of usual brazing alloys. Both applications deserve a good wetting capability. Self-flux synthesis was also performed with the alkali metals, magnesium, antimony, and bismuth.

Silica tubes and alumina crucibles (vacuum sealed in silica ampoules) are the habitually used container materials. Furthermore, MgO and ZrO_2 crucibles or containers of high-melting inert metals (niobium, tantalum, molybdenum) were used. An overview of usual container materials for diverse solid-state reactions is given in Ref. [6]. The chemical inertness of the container materials toward the flux is a key criterion. One should always keep in mind that the common oxidic containers mostly contain small amounts of sintering additives that might transfer to the flux. In the case of silica ampoules, a silicon uptake from the container wall often occurs and one sometimes observes tiny silicon impurities in the product crystals.

The redox conditions between the crucible material and the flux are mostly not known and they are distinctly different from the standard electrode potentials known from aqueous solutions. In some cases, the oxide ceramics are not that stable. An interesting example is the formation of large well-shaped $U_4O_4Te_3$ crystals during an annealing process of U_3Te_4 in alumina crucibles [7]. The starting temperature of 1830 K was apparently sufficient to release oxygen from the ceramic.

The temperature profile for the annealing sequence is important for optimal results. Many of these parameters have been optimized for certain classes of substances just by serendipity. For details, we refer to Ref. [3] with longer lists of thermal preparation conditions as well as to the original literature. For understanding the fundamental principles of crystal growth, detailed knowledge of the underlying phase diagram is indispensable. A tutorial thermodynamic approach to provide a so-called roadmap for crystal growth has been published recently [8], including binary and multicomponent phase diagrams.

Since the flux medium for the self-flux experiments is always supplied in large excess, often only the phases with the highest content of the flux element can be obtained. This is the case for binary as well as for ternary ones. Typical examples are the stannides $RhSn_4$ [9] and Os_3Sn_7 [10]. For ternaries, the indium self-flux was used for growth of the indium-rich heavy fermion materials $CeTIn_5$ and Ce_2TIn_8 (T = Rh, Ir) [11–13]. Another interesting example concerns the thermoelectric phases $Ba_8Ga_{16}Ge_{30}$ (gallium self-flux) and $Ba_8Ga_{16}Sn_{30}$ (gallium or tin self-flux) [14].

The self-flux technique has also gained access in plutonium chemistry [15]. Intermetallic plutonium compounds exhibit very complex magnetic and transport properties and thus single crystals are requested for direction-dependent measurements and thus for a better understanding of the structure–property relations. All manipulations of the samples can be performed under specific safety conditions. Interesting examples are $PuIn_3$, $PuSn_3$, $PuGa_3$, $PuSb_2$, $PuCoGa_5$, $PuPt_2In_7$ [15], or $PuCoIn_5$ [16], again the p-element-rich phases.

The final problem concerns the separation of flux and crystals. One can distinguish between separation in the liquid and the solid state [17]. This means mechanical or chemical separation. In the mechanical case, one can use the melt centrifugation technique [18,19], mostly applied for aluminum, gallium, and tin fluxes, using silica wool as a kind of sieve. Remaining flux films must additionally be removed by etching. *In situ* decanting [19] is an interesting extension of the centrifugation technique, allowing the flux decanting at high temperatures directly in the furnace.

If the flux is solidified, it must be removed by distillation (in the case of a low boiling point, for example, Mg, Zn, Hg) or by dissolution. If the flux is removed by a chemical reaction, the product crystals need to have higher stability against the solution. Tin, aluminum, and zinc can be dissolved in 1 M hydrochloric acid. For less stable products, one can switch to acetic acid. This, however, deserves longer periods. Other fluxes can be dissolved with mixtures of acetic acid and hydrogen peroxide (lead, bismuth). A mild route is oxidative dissolution with iodine in DMF. The exact procedure needs to be adapted separately to each class of materials.

10.3 Reactive Salt Fluxes

Besides metals, salts can also be utilized as fluxing agents for the synthesis of, for example, novel oxide materials. Here a large diversity of fluxes is readily available; therefore, tuning the reaction conditions regarding synthesis temperature, solubility, volatility of the flux, and even pH control is possible. Basic aspects of salt fluxes such as the physical properties and the structure of molten salts, the solubility of gases, metals, and other salts in molten salts along with a large number of examples are given in Ref. [20]. For a substantial summary of the physical properties of molten salts, we refer to Ref. [21]. While a number of fluxes are inert and therefore only serve as a high-temperature solvent, others can be used to introduce parts of the salt into the final material. Like for metal fluxes, a number of salt flux reactions can be carried out in oxide crucibles (e.g., Al_2O_3) or silica tubes. Where these materials are not applicable, noble metal containers such as silver or gold tubes or platinum crucibles are used. A large number of reactions can be carried out in open containers; however, stabilization of low oxidation states or preventing evaporation sometimes requires fused ampoules or glass jackets as secondary container material.

The fluxing agents can be roughly grouped into (i) the more covalently bonded oxides such as PbO, Bi_2O_3, and transition metal oxides and (ii) classical ionic compounds such as alkali metal and alkaline earth-metal halides, carbonates, hydroxides, borates, and even alkali metal oxides, peroxides, superoxides, or boric acid. For rare earth-containing compounds, the trihalides REX_3 (X = Cl, Br, I) are low melting options. The use of the aforementioned compounds as fluxing agent however is somehow influenced by the individual peculiarities of the respective compound.

While PbO is one of the oldest and most versatile fluxes with a conveniently low melting point (888 °C), its significant volatility and therefore enhanced toxicity hampers the broad use. Bi_2O_3, on the other hand, tends to transfer Bi^{3+} cations, which is a nice feature for reactive flux synthesis. In contrast to metal fluxes, the salt fluxes are usually quite easy to remove. Water, sometimes boiling, can be used in a large number of cases. For moisture-sensitive products, methanol is a suitable substitute. Fluxes with low water solubility such as PbO, Bi_2O_3, or some transition metal oxides can be dissolved in warm acetic acid, hot HNO_3, or warm HCl/ethanol mixtures. Besides solvation, some fluxes, for example, PbO, can also be removed by sublimation.

The following paragraph exemplarily lists some examples for the different reactive fluxes mentioned before. An extensive summary of examples can be found in Ref. [22] and the respective literature listed below. The group of Müller-Buschbaum reported a series of complex lead-containing oxides, for example, $Sr_{1.33}Pb_{0.67}Al_6O_{11}$ [23], $BaPb_{1.5}Mn_6Al_2O_{16}$ [24], and $Pb_3MnAl_{10}O_{20}$ [25], which were synthesized with only a slight excess of PbO, while $Pb_2HoAl_3O_8$ and the isostructural Lu compounds [26] were obtained from a PbO–PbF_2 flux, where the flux was removed by sublimation. A platinum-containing compound, $Sr_4PbPt_4O_{11}$ [27], was synthesized from elemental Pt in Au crucibles, showing that under certain conditions, even Pt is not truly "inert." $Bi_3ScMo_2O_{12}$ [28], $Bi_2Fe_2Mn_2O_{10}$ [29], $Bi_2Fe_2Ga_2O_9$ [30], and the vanadate $BiMg_2VO_6$ [31] should be mentioned as Bi_2O_3 flux products. B_2O_3 by itself has a very high viscosity and therefore is rarely used; however, still a number of borate fluxes, generated *in situ*, were used to transfer alkali or alkaline earth cations. $SrLiCrTi_4O_{11}$ and $SrLiFeTi_4O_{11}$ [32] were prepared from $LiBO_2$, generated by Li_2CO_3 and H_3BO_3. But also binary transition metal oxides with comparably low melting points (e.g., V_2O_5, m.p. 690 °C; MoO_3, m.p. 795 °C; WO_3, m.p. 1473 °C) can be utilized. Vanadates such as $Pb(UO_2)(V_2O_7)$ [33] could be obtained from a V_2O_5 flux, molybdenum compounds (e.g., $Ln(Mo_4Sb_2O_{18})$ [34]) from MoO_3 melts, or tungstates such as $ANbW_2O_9$ (A = K, Rb) from the reaction of Nb_2O_5 in A_2O–WO_3 fluxes [35]. Rare earth halide molybdates $REX[MoO_4]$ (X = Cl [36], Br [37]) and the corresponding tungstates $REX[WO_4]$ (X = Cl [38], Br [39]) could also be obtained by the use of MoO_3 and WO_3, the reactive flux in these cases however was the respective rare earth trihalide REX_3. The trifluorides REF_3 in contrast have extremely high melting points; therefore, the corresponding fluoride molybdates $REF[MoO_4]$ were not accessible by this route but by solution-based chemistry [40]. The low melting trichlorides $RECl_3$ also enable the synthesis of a halide silicates, for example, $RE_6Cl_{10}[Si_4O_{12}]$ [41,42] or $Y_3Cl[SiO_4]_2$ [41]. But not only oxide materials can be obtained from reactive flux reactions. The aforementioned rare earth halides are suitable fluxing agents for other compounds such as sulfide chlorides $RESCl$ [43], $RE_7S_6Cl_9$ [44] or $RE_3S_2Cl_5$ [45].

The following examples were obtained from alkali and alkaline earth halides, carbonate, and hydroxide fluxes. Halide melts can dissolve a wide range of elements, including transition metals, main-group metals, rare earths, and also actinides. For the carbonates, one has to remember that the alkali metal carbonates only dissociate in the melt while the alkaline earth carbonates decompose at elevated temperatures. A LiCl flux was utilized to synthesize the divalent europium silicates $LiEu_3[SiO_4]Cl_3$ and $Li_7Eu_8[SiO_4]_4Cl_7$ [46], while the nitride iodides $NaRE_4N_2I_7$ were grown from NaI melts [47]. Perovskites such as $Ba_6Al_2Rh_2Ho_2O_{15}$ [48] or Ba_3CaRuO_9 [49] were grown from $BaCl_2{\cdot}2\,H_2O$ and $BaCl_2{\cdot}2\,H_2O/CaCO_3$ melts, respectively. $Li_7La_3Zr_2O_{12}$ has been prepared from a Li_2CO_3 melt [50], $Pr_{2.45}Na_{1.55}RhO_6$, and the respective Nd compound from Na_2CO_3 fluxes [51]. Alkali metal hydroxides finally are an in many ways more complex fluxing agent than other simple inorganic salts, since their water content significantly influences the oxoacidity of the melt. They are capable of dissolving nearly every element of the periodic table while stabilizing high oxidation states. Typically, these reactions are hence carried out in silver or alumina crucibles. Sealed ampoules versus open crucibles is an elegant way to control the water content of the system to influence the acidity of the flux. The preparation of rare earth-containing platinum metal oxides by hydroxide melts has been extensively discussed in Ref. [52]. Compounds such as Sr_3NaTO_6 (T = Nb, Ta) can be easily obtained from the dissolution of T_2O_5 in $Sr(OH)_2$-NaOH melts [53], $NaKLaNbO_5$ and $Na_2K_2Gd_4Nb_2O_{13}$ [54] were obtained by adding the corresponding RE_2O_3 to the Nb_2O_5-NaOH-KOH mixture.

Finally, the use of boric acid should be mentioned. This flux has been utilized to synthesize a number of compounds as a reactive flux, for example, actinide compounds are a large field, stabilizing also oxidation states not available via hydrothermal synthesis. As examples, $NpO_2[B_3O_4(OH)_2]$ [55] or $K[(NpO_2)B_{10}O_{14}(OH)_4]$ [56] should be mentioned as well as a report of the recent progress highlighted [57].

10.4 Polychalcogenide Fluxes

As mentioned in the previous sections, a variety of review articles have been published over the last decades for the use of reactive polychalcogenide fluxes. Therefore, we will briefly explain the chemistry and other peculiarities of polychalcogenide fluxes and afterwards list some examples. The properties and advantages of alkali metal polychalcogenides both as reagents and as solvents, which can be considered as a reactive flux, were summarized in Refs [58,59]. One important aspect is that polychalcogenides cannot be prepared during high-temperature syntheses, even in large excess of the chalcogenide Q, due to the thermal instability of the Q_n^{2-} anions (Q = S, Se, Te). Therefore, it becomes clear that low to intermediate temperatures ($T < 500\,°C$) have to be used. The flux is generally generated *in situ*, starting from an alkali chalcogenide A_2Q and the elemental chalcogen Q according to $A_2Q + (n-1)\ Q \rightarrow A_2Q_n$. The A_2Q/Q phase diagrams are well studied and show all similar appearances. The melting point decreases drastically with increasing Q content, with melting points in the case of sulfides even below 100 °C [60]. For the selenides, minima at around 250 °C are found [61–63], the A_2Te_n systems finally show melting points of about 300 °C [61–63]. Besides the low melting point, the polychalcogenides exhibit a large liquidus regime in which they remain liquid without significant chalcogen loss. The melts exhibit highly oxidizing properties and are fairly basic at the same time, enabling the use of elemental metals. The reactions can be carried out in Pyrex or silica ampoules or, if reactive starting materials are involved in alumina crucibles sealed in secondary silica containers. The products are often stable toward hydrolysis and air oxidation; therefore, the flux can be dissolved in water. Investigations by *in situ* Raman and *in situ* X-ray total scattering were used to gain an insight into the molten state between K_2S_3 and K_2S_5 and found that the corresponding S_n^{2-} species of the solid state are mirrored in the melt [64].

Before their use as reactive flux, HgS was recrystallized from sulfide–sulfur melts [65]. Scheel, who also used the polychalcogenides to recrystallize binary chalcogenides such as CdS, CuS, ZnS, and others, commented in one of his papers in 1974: "The polysulfides of larger alkali ions have not been investigated in detail; they are also potential solvents although the tendency of compound formation increases with ionic radius" [66].

In the following paragraphs, we want to list examples of multinary chalcogenides involving reactive fluxes in their syntheses and point out two other reviews listing several of these compounds [55,67,68]. The first reported example for a new compound synthesized via a polychalcogenides flux is $K_4Ti_3S_{14}$, reported by Ibers and coworkers in 1987 [69]. About 35 years earlier, Rüdorff *et al.* already reported on the synthesis of $RbCu_4S_3$ from stoichiometric amounts of Rb_2S, Cu, and S [70]. They however mention that an increase of Rb_2S and S, now acting as a polychalcogenide flux, leads to the formation of other products. Soon after the synthesis of $K_4Ti_3S_{11}$, the compounds $K_4Ta_2S_{11}$ and $K_3Nb_2Se_{11}$ were reported, prepared in a similar way [71]. Since then, the number of reported compounds has increased drastically. As examples, ternary chalcogenide compounds of the early transition metals titanium ($Rb_4Ti_3S_{14}$, $Cs_4Ti_3S_{14}$, $Cs_4Ti_3Se_{13}$ [72]), zirconium ($Rb_4Zr_3Te_{16}$ [73]), hafnium ($Rb_4Hf_3S_{14}$, $Cs_4Hf_3Se_{14}$ [72]), niobium ($K_4Nb_2S_{11}$ [74] and $Rb_6Nb_4S_{22}$ and $Cs_6Nb_4S_{22}$ [75]), and tantalum ($Rb_6Ta_4S_{22}$ [76]) as well as the late transition metals copper (α- and β-$KCuS_4$ [77]), gold (LiAuS and NaAuS [78] and $KAuS_5$, $KAuSe_5$, $NaAuSe_2$, and $KAuSe_2$ [79]), and mercury compounds ($K_2Hg_6S_7$ or $K_2Hg_3S_4$ [80]) should be mentioned. Furthermore, ternary compounds consisting of an alkali metal (*A*), a rare earth (*RE*) or actinide metal (*AM*), and a chalcogen (*Q*) can be prepared from these polychalcogenide melts. For lanthanides, the composition ARE_3Q_8 [81,82], for the actinides KTh_2Q_6 (Q = Se, Te) [83], $CsTh_2Se_6$ [84], and AU_2Se_6 (A = K, Cs) [85] were found.

Another intensively studied group of compounds are rare earth transition metal chalcogenides. Besides the large number of quaternary compounds, only a few examples of ternary compounds have been reported. $EuCu_{0.66}Te_2$, for example, has been prepared from Eu and Cu in a Rb_2Te/Te flux [86], while other members of the $RECu_xTe_2$ series have been prepared from $RECl_3$ melts or by chemical vapor transport [87,88]. The quaternary phases show a large manifold regarding composition, crystal structure, and properties. Here, usually the alkali metals from the flux also get incorporated forming compounds such as the ARE_2CuQ_4, ARE_2CuQ_6, $A_3RE_4Cu_5Q_{10}$, or $ARE_2T_3Q_5$ series. Due to the huge amount of compounds known in this field, we want to refer to Ref. [89] for the primary citations. Besides the rare earth compounds, also actinide representatives have been synthesized. Here numerous transition metal-containing compounds have also been reported, for example, $CsTiUTe_5$ [90] or $CsCuUTe_3$ [91], but also Pd and Pt ($A_2T_3UQ_6$ [92])-containing compounds have been discovered.

The last large group of compounds that should be mentioned here contains complex chalcogen anions $[M_yQ_z]^{n-}$ with M being an elements of group 13–16. The reaction can be carried out in A_2Q/Q fluxes along with the elemental main group components or binary main group chalcogens such as P_2Q_5. The chemistry of chalcophosphates has been summarized in Ref. [93]; a more extensive review including also chalcoarsenates and antimonates is found in Ref. [94]. These compounds have astonishing physical properties making them interesting materials, for example, thermoelectric and solar energy conversion, data storage, or nonlinear optics; therefore, the literature is extensive. Again, the following examples were chosen to give the reader an impression about the current status of this field. For indium as main group element, $CsCdInQ_3$ (Q = Se and Te) have been prepared from Cs_2Q_3/Q melts [95]. For elements of group 14, compounds such as $KEuMS_4$ and K_2EuMSe_5 (M = Si, Ge) [96], Li_2CdMS_4 (M = Ge, Sn) [97], or $Eu_8(Sn_4Se_{14})(Se_3)_2$ [98] were reported. For the group 15 elements, chalcophosphates such as $NaSmP_2S_6$, $KSmP_2S_7$ [99], UP_2S_7, UP_2S_9 [100] or the more complex examples $Cs_4Th_2P_6S_{18}$ and $Rb_7Th_2P_6Se_{21}$ [101] should be listed. The simple $LiAsSe_2$ and $NaAsSe_2$ [102] or the more complex $Cs_2SnAs_2Q_9$ (Q = S, Se) [103] are examples for arsenic compounds. RbU_2SbS_8 and KU_2SbSe_8 [104] or $K_2RE_{2-x}Sb_{4+x}Se_{12}$ (RE = La, Ce, Pr, and Gd) [105] should be noted for antimony. Finally, even interchalcogen compounds between S and Te are known, for example, in the series $ATTeS_3$ (A = K, Rb, Cs; T = Cu, Ag) synthesized from mixed polychalcogenide fluxes [106].

One final aspect concerns the addition of oxygen. The reaction mixtures are usually prepared in inert atmosphere glove boxes and sealed under vacuum in gas-tight ampoules as described above. Oxygen impurities however caused the accidental formation of $Ba_6Ti_5S_{15}O$ [107] and the uranyl compounds $Na_4(UO_2)(S_2)_3{\cdot}Na_2S_3$ and $Cs_4(UO_2)(S_2)_3$ [108]. The compounds were reproduced using either BaO or SeO_2 as oxygen sources. The latter also enabled the synthesis of $Na_4(UO_2)Cu_2S_4$ [109]. $K_4Nb_2Se_{11}O$ and the analogous cesium compound [110] and later $K_4Nb_2S_{10}O$ and $Rb_4Nb_2S_{10}O$ [111] were also prepared from polychalcogenide fluxes. And also oxide containing polychalcogenides of titanium can be prepared by addition of TiO_2 to a reaction mixture similar to what was used to synthesize

$K_4Ti_3S_{14}$, yielding $K_4Ba[Ti_6OS_{20}]$ that contains isolated $[Ti_6OS_8(S_2)_6]^{6-}$ clusters [112].

10.5
Nitride Synthesis via Li_3N and NaN_3

Solid-state nitride materials find broad technical application. Typical fields are hard materials, III–V semiconductors, or luminescence materials in the field of light-emitting diodes. Crystal growth is an important task in explorative synthesis for the discovery of new materials (micrometer-size crystals for structure determination) [113,114] or for semiconductor crystals in centimeter size, for example, GaN [115]. Lithium nitride and sodium azide are the important flux media in this field. Li_3N has a low melting point and serves as lithium and nitrogen source in the reactive self-flux synthesis [116]. In contrast, controlled slow decomposition of sodium azide delivers high-purity sodium and high-purity nitrogen gas at the same time. It is an extreme advantage that the handling of solid NaN_3 is very easy. The purity of the produced nitrogen is higher than that of purified commercial nitrogen [117].

The first examples concern explorative synthesis. Many ternary and quaternary nitridometallates containing lithium were prepared with lithium nitride as a reactive flux [114,116]. Typical examples are the synthesis of Li_6MoN_4 and Li_6CrN_4 [118] or the more complex nitrides $Li_3Ca_2V_{0.79}Nb_{0.21}N_4$, $Li_2Ca_{2.67}Nb_{0.33}N_3$, and $Li_{12}Ca_9W_5N_{20}$ [119]. An overview of the structural chemistry of such nitridometallates is given in [116]. Such lithium nitride fluxes were also used for the synthesis of more complex nitridosilicates. Recent examples are the phases $Li_4Ca_3Si_2N_6$, $LiCa_3Si_2N_5$, and $Li_2Sr_4Si_4N_8O$ [114]. The use of microwave heating (microwave-generated nitrogen plasmas) is also an effective extension in the field of lithium nitride self-flux synthesis [120].

Now we turn to examples for sodium azide self-flux synthesis. The following examples either use the released sodium or nitrogen or even both. Interesting examples are $Ba_3Ga_2N_4$ [121], $Ba_5Si_2N_6$ [122], and the complex nitrides $Sr(Mg_3Ge)N_4$ and $Sr(Mg_2Ga_2)N_4$ [123]. The excess sodium flux can be removed with liquid ammonia or by evaporation under a dynamic vacuum.

High-quality prismatic or needle-shaped GaN crystals can be grown with the assistance of a NaN_3 addition. Here, decomposition of the azide delivers an extremely pure sodium flux (an excellent prerequisite for high-purity semiconductor crystals) and additionally nitrogen as protective gas [115].

The azide route was also used for the synthesis of intermetallic compounds [124,125]. The azide decomposition delivers high-purity sodium and additionally nitrogen as protective gas. A typical product is $Na_8Au_{11}In_6$ [125]. For all synthesis attempts, sodium azide was repeatedly recrystallized from water. Sodium azide along with various amounts of Na-K alloys were used as nitrogen source and flux medium for the preparation and crystal growth of various subnitrides with the cluster unit [$Ba_{14}CaN_6$] embedded in different sodium matrices [126,127].

Finally, we turn to nitrogenation reactions of permanent magnetic materials [128]. The nitrogen content in materials such as $Y_2Fe_{17}N_x$ influences the magnetic properties. Since the nitrogen delivered from sodium azide is of high purity, targeted nitrogenation is possible, designing well-defined magnetic materials. The use of Y_2Fe_{17} powders along with NaN_3 powder leads to an excellent diffusion of the nitrogen into the intermetallic matrix.

10.6 Other Systems

Besides the aforementioned reactive fluxes, reports about the use of $AgNO_3$ for the synthesis of $Ag_9(SiO_4)_2NO_3$ [129], thiourea as reactive flux medium for $Na_4Cu_{32}Sn_{12}S_{48}\cdot 4\ H_2O$ and $K_{11}Cu_{32}Sn_{12}S_{48}\cdot 4\ H_2O$ [130] or NaP for $NaGe_3P_7$ [131] have been published. An interesting approach concerns the use of $KClO_3$ as oxidizing agent under high-pressure high-temperature conditions [132–134]. The reaction product of the decomposition reaction is KCl that forms an eutectic with the nonconsumed $KClO_3$. Thus, we have reactive flux conditions that promote crystal growth.

References

1 Fisk, Z. and Remeika, J.P. (1989) Growth of single crystals from molten metal fluxes, in *Handbook on the Physics and Chemistry of Rare Earths*, vol. **12** (eds K.A. Gschneidner, Jr., and L. Eyring), Elsevier Publishers B.V., Amsterdam, p. 53.

2 Canfield, P.C. and Fisk, Z. (1992) *Philos. Mag. B*, **65**, 1117.

3 Kanatzidis, M.G., Pöttgen, R., and Jeitschko, W. (2005) *Angew. Chem., Int. Ed.*, **44**, 6997.

4 Canfield, P.C. (2012) *Philos. Mag. B*, **92**, 2398.

5 Morrison, G.W., Menard, M.C., Treadwell, L.J., Haldolaarachchige, N., Kendrick, K.C., Young, D.P., and Chan, J.Y. (2012) *Philos. Mag. B*, **92**, 2524.

6 Pöttgen, R. and Johrendt, D. (2014) *Intermetallics: Synthesis, Structure, Function*, De Gruyter, Berlin.

7 Noël, H., Potel, M., Shlyk, L., Kaczorowski, D., and Troć, R. (1995) *J. Alloys Compd.*, **217**, 94.

8 Fisher, I.R., Shapiro, M.C., and Analytis, J.G. (2012) *Philos. Mag. B*, **92**, 2401.

9 Lang, A. and Jeitschko, W. (1996) *J. Mater. Chem.*, **6**, 1897.

10 Künnen, B., Niepmann, D., and Jeitschko, W. (2000) *J. Alloys Compd.*, **309**, 1.

11 Hegger, H., Petrovic, C., Moshopoulou, E.G., Hundley, M.F., Sarrao, J.L., Fisk, Z., and Thompson, J.D. (2000) *Phys. Rev. Lett.*, **84**, 4986.

12 Moshopoulou, E.G., Fisk, Z., Sarrao, J.L., and Thompson, J.D. (2001) *J. Solid State Chem.*, **158**, 25.

13 Macaluso, R.T., Sarrao, J.L., Moreno, N.O., Pagliuso, P.G., Thompson, J.D., Fronczek, F.R., Hundley, M.F., Malinowski, A., and Chan, J.Y. (2003) *Chem. Mater.*, **15**, 1394.

14 Ribeiro, R.A. and Avila, M.A. (2012) *Philos. Mag. B*, **92**, 2492.

15 Bauer, E.D., Tobash, P.H., Mitchell, J.N., and Sarrao, J.L. (2012) *Philos. Mag. B*, **92**, 2466.

16 Bauer, E.D., Altarawneh, M.M., Tobash, P.H., Gofryk, K., Ayala-Valenzuela, O.E., Mitchell, J.N., McDonald, R.D., Mielke, C.H., Ronning, F., Griveau, J.C., Colineau, E., Eloirdi, R., Caciuffo, R., Scott, B.L., Janka, O., Kauzlarich, S.M., and Thompson, J.D. (2012) *J. Phys. Condens. Matter*, **24** 052206.
17 Wolf, T. (2012) *Philos. Mag. B*, **92**, 2458.
18 Boström, M. and Hovmöller, S. (2000) *J. Solid State Chem.*, **153**, 398.
19 Petrovic, C., Canfield, P.C., and Mellen, J.Y. (2012) *Philos. Mag. B*, **92**, 2448.
20 Sundermeyer, W. (1965) *Angew. Chem.*, **77**, 241.
21 Janz, G.J. (1967) *Molten Salts Handbook*, Academic Press, New York.
22 Bugaris, D.E. and zur Loye, H.-C. (2012) *Angew. Chem., Int. Ed.*, **51**, 3780.
23 Pløutz, K.B. and Müller-Buschbaum, Hk. (1982) *Z. Anorg. Allg. Chem.*, **491**, 253.
24 Teichert, A. and Müller-Buschbaum, Hk. (1993) *J. Alloys Compd.*, **202**, 37.
25 Teichert, A. and Müller-Buschbaum, Hk. (1991) *J. Less Common Met.*, **170**, 315.
26 Scheikowski, M. and Müller-Buschbaum, Hk. (1993) *Z. Anorg. Allg. Chem.*, **619**, 1755.
27 Renard, C., Roussel, P., Rubbens, A., Daviero-Minaud, S., and Abraham, F. (2006) *J. Solid State Chem.*, **179**, 2101.
28 Kolitsch, U. and Tillmanns, E. (2003) *Acta Crystallogr. E*, **59**, i43.
29 Giaquinta, D.M. and zur Loye, H.C. (1992) *J. Alloys Compd.*, **184**, 151.
30 Giaquinta, D.M., Papaefthymiou, G.C., Davis, W.M., and Zur Loye, H.C. (1992) *J. Solid State Chem.*, **99**, 120.
31 Huang, J. and Sleight, A.W. (1992) *J. Solid State Chem.*, **100**, 170.
32 Imaz, I., Péchev, S., Koseva, I., Bourée, F., Gravereau, P., Peshev, P., and Chaminade, J.P. (2007) *Acta Crystallogr. B*, **63**, 26.
33 Obbade, S., Dion, C., Saadi, M., Yagoubi, S., and Abraham, F. (2004) *J. Solid State Chem.*, **177**, 3909.
34 Kalpana, G. and Vidyasagar, K. (2007) *J. Solid State Chem.*, **180**, 1708.
35 Yanovsky, V.K. and Voronkova, V.I. (1981) *J. Cryst. Growth*, **52**, 654.
36 Hartenbach, I., Schleid, Th., Strobel, S., and Dorhout, P.K. (2010) *Z. Anorg. Allg. Chem.*, **636**, 1183.
37 Schustereit, T., Henning, H., Schleid, Th., and Hartenbach, I. (2013) *Z. Naturforsch.*, **68b**, 616.
38 Schustereit, T., Schleid, Th., Höppe, H.A., Kazmierczak, K., and Hartenbach, I. (2015) *J. Solid State Chem.*, **226**, 299.
39 Schustereit, T., Schleid, Th., and Hartenbach, I. (2015) *Solid State Sci.*, **48**, 218.
40 Schustereit, T., Schleid, Th., and Hartenbach, I. (2014) *Eur. J. Inorg. Chem.*, **2014**, 5145.
41 Hartenbach, I. and Schleid, Th. (2001) *Z. Anorg. Allg. Chem.*, **627**, 2493.
42 Hartenbach, I., Jagiella, S., and Schleid, Th. (2006) *J. Solid State Chem.*, **179**, 2258.
43 Lissner, F. and Schleid, Th. (1998) *Z. Anorg. Allg. Chem.*, **624**, 452.
44 Lissner, F. and Schleid, Th. (1998) *Z. Anorg. Allg. Chem.*, **624**, 1903.
45 Lissner, F. and Schleid, Th. (2001) *Z. Anorg. Allg. Chem.*, **627**, 507.
46 Jacobsen, H. and Meyer, G. (1994) *Z. Anorg. Allg. Chem.*, **620**, 1351.
47 Schurz, C.M. and Schleid, Th. (2010) *J. Solid State Chem.*, **183**, 2253.
48 Schlüter, D. and Müller-Buschbaum, Hk. (1993) *J. Alloys Compd.*, **191**, 305.
49 Wilkens, J. and Müller-Buschbaum, Hk. (1991) *J. Alloys Compd.*, **177**, L31.
50 Awaka, J., Kijima, N., Hayakawa, H., and Akimoto, J. (2009) *J. Solid State Chem.*, **182**, 2046.
51 Macquart, R.B., Gemmill, W.R., Davis, M.J., Smith, M.D., and zur Loye, H.-C. (2006) *Inorg. Chem.*, **45**, 4391.
52 Mugavero, S.J., III, Gemmill, W.R., Roof, I.P., and zur Loye, H.-C. (2009) *J. Solid State Chem.*, **182**, 1950.
53 Bharathy, M., Rassolov, V.A., and zur Loye, H.C. (2008) *Chem. Mater.*, **20**, 2268.
54 Liao, J.-H. and Tsai, M.-C. (2002) *Crystal Growth Des.*, **2**, 83.
55 Wang, S., Alekseev, E.V., Miller, H.M., Depmeier, W., and Albrecht-Schmitt, T.E. (2010) *Inorg. Chem.*, **49**, 9755.

56 Wang, S., Alekseev, E.V., Depmeier, W., and Albrecht-Schmitt, T.E. (2012) *Inorg. Chem.*, **51**, 7.

57 Wang, S., Alekseev, E.V., Depmeier, W., and Albrecht-Schmitt, T.E. (2011) *Chem. Commun.*, **47**, 10874.

58 Kanatzidis, M.G. (1990) *Chem. Mater.*, **2**, 353.

59 Stein, A., Keller, S.W., and Mallouk, T.E. (1993) *Science*, **259**, 1558.

60 Pearson, T.G. and Robinson, P.L. (1931) *J. Chem. Soc.*, 1304.

61 Mathewson, C.H. (1907) *J. Am. Chem. Soc.*, **29**, 867.

62 Klemm, W., Sodomann, H., and Langmesser, P. (1939) *Z. Anorg. Allg. Chem.*, **241**, 281.

63 Anonymous (1966) Sodium, in *Gmelin's Handbuch der anorganischen Chemie*, Wiley-VCH Verlag GmbH, Weinheim, Suppl. Part 3, pp. 1202–1205 and references therein.

64 Shoemaker, D.P., Chung, D.Y., Mitchell, J.F., Bray, T.H., Soderholm, L., Chupas, P.J., and Kanatzidis, M.G. (2012) *J. Am. Chem. Soc.*, **134**, 9456.

65 Garner, R.W. and White, W.B. (1970) *J. Crystal Growth*, **7**, 343.

66 Scheel, H.J. (1974) *J. Crystal Growth*, **24/25**, 669.

67 Kanatzidis, M.G. and Sutorik, A.C. (1995) *Prog. Inorg. Chem.*, **43**, 151.

68 Pell, M.A. and Ibers, J.A. (1997) *Chem. Ber.*, **130**, 1.

69 Sunshine, S.A., Kang, D., and Ibers, J.A. (1987) *J. Am. Chem. Soc.*, **109**, 6202.

70 Rüdorff, W., Schwarz, H.G., and Walter, M. (1952) *Z. Anorg. Allg. Chem.*, **269**, 141.

71 Schreiner, S., Aleandri, L.E., Kang, D., and Ibers, J.A. (1989) *Inorg. Chem.*, **28**, 392.

72 Huang, F.Q. and Ibers, J.A. (2001) *Inorg. Chem.*, **40**, 2346.

73 Anderson, A.B., Wang, R.-J., and Li, J. (2000) *Acta Crystallogr. C*, **56**, 2.

74 Bensch, W. and Dürichen, P. (1996) *Eur. J. Solid State Inorg. Chem.*, **33**, 309.

75 Bensch, W. and Dürichen, P. (1996) *Z. Anorg. Allg. Chem.*, **622**, 1963.

76 Stoll, P., Näther, C., Jeß, I., and Bensch, W. (2000) *Z. Anorg. Allg. Chem.*, **626**, 959.

77 Kanatzidis, M.G. and Park, Y. (1989) *J. Am. Chem. Soc.*, **111**, 3767.

78 Axtell J III, E.A., Liao, J.-H., and Kanatzidis, M.G. (1998) *Inorg. Chem.*, **37**, 5583.

79 Park, Y. and Kanatzidis, M.G. (1997) *J. Alloys Compd.*, **257**, 137.

80 Kanatzidis, M.G. and Park, Y. (1990) *Chem. Mater.*, **2**, 99.

81 Foran, B., Lee, S., and Aronson, M.C. (1993) *Chem. Mater.*, **5**, 974.

82 Patschke, R., Heising, J., Schindler, J., Kannewurf, C.R., and Kanatzidis, M. (1998) *J. Solid State Chem.*, **135**, 111.

83 Wu, E.J., Pell, M.A., and Ibers, J.A. (1997) *J. Alloys Compd.*, **255**, 106.

84 Cody, J.A. and Ibers, J.A. (1996) *Inorg. Chem.*, **35**, 3836.

85 Chan, B.C., Hulvey, Z., Abney, K.D., and Dorhout, P.K. (2004) *Inorg. Chem.*, **43**, 2453.

86 Patschke, R., Brazis, P., Kannewurf, C.R., and Kanatzidis, M.G. (1999) *J. Mater. Chem.*, **9**, 2293.

87 Dung, N.-H., Pardo, M.-P., and Boy, P. (1983) *Acta Crystallogr. C*, **39**, 668.

88 Huang, F.Q., Brazis, P., Kannewurf, C.R., and Ibers, J.A. (2000) *J. Am. Chem. Soc.*, **122**, 80.

89 Mitchell, K. and Ibers, J.A. (2002) *Chem. Rev.*, **102**, 1929.

90 Cody, J.A., Mansuetto, M.F., Chien, S., and Ibers, J.A. (1994) *Mater. Sci. Forum*, **152–153**, 35.

91 Cody, J.A. and Ibers, J.A. (1995) *Inorg. Chem.*, **34**, 3165.

92 Oh, G.N., Choi, E.S., and Ibers, J.A. (2012) *Inorg. Chem.*, **51**, 4224.

93 Kanatzidis, M.G. (1997) *Curr. Opin. Solid State Mater. Sci.*, **2**, 139.

94 Drake, G.W. and Kolis, J.W. (1994) *Coord. Chem. Rev.*, **137**, 131.

95 Li, H., Malliakas, C.D., Peters, J.A., Liu, Z., Im, J., Jin, H., Morris, C.D., Zhao, L.-D., Wessels, B.W., Freeman, A.J., and Kanatzidis, M.G. (2013) *Chem. Mater.*, **25**, 2089.

96 Evenson, C.R., IV, and Dorhout, P.K. (2001) *Inorg. Chem.*, **40**, 2409.

97 Lekse, J.W., Moreau, M.A., McNerny, K.L., Yeon, J., Halasyamani, P.S., and Aitken, J.A. (2009) *Inorg. Chem.*, **48**, 7516.

98 Evenson, C.R., IV, and Dorhout, P.K. (2001) *Z. Anorg. Allg. Chem.*, **627**, 2178.
99 Goh, E.-Y., Kim, E.-J., and Kim, S.-J. (2001) *J. Solid State Chem.*, **160**, 195.
100 Babo, J.-M., Jouffret, L., Lin, J., Villa, E.M., and Albrecht-Schmitt, T.E. (2013) *Inorg. Chem.*, **52**, 7747.
101 Chan, B.C., Hess, R.F., Feng, P.L., Abney, K.D., and Dorhout, P.K. (2005) *Inorg. Chem.*, **44**, 2106.
102 Bera, T.K., Jang, J.I., Song, J.-H., Malliakas, C.D., Freeman, A.J., Ketterson, J.B., and Kanatzidis, M.G. (2010) *J. Am. Chem. Soc.*, **132**, 3484.
103 Iyer, R.G., Do, J., and Kanatzidis, M.G. (2003) *Inorg. Chem.*, **42**, 1475.
104 Choi, K.-S. and Kanatzidis, M.G. (1999) *Chem. Mater.*, **11**, 2613.
105 Chen, J.H. and Dorhout, P.K. (1997) *J. Alloys Compd.*, **249**, 199.
106 Zhang, X. and Kanatzidis, M.G. (1994) *J. Am. Chem. Soc.*, **116**, 1890.
107 Sutorik, A.C. and Kanatzidis, M.G. (1994) *Chem. Mater.*, **6**, 1700.
108 Sutorik, A.C. and Kanatzidis, M.G. (1997) *Polyhedron*, **16**, 3921.
109 Sutorik, A.C. and Kanatzidis, M.G. (1997) *J. Am. Chem. Soc.*, **119**, 7901.
110 Dürichen, P., Krause, O., and Bensch, W. (1998) *Chem. Mater.*, **10**, 2127.
111 Krause, O., Dürichen, P., Näther, C., and Bensch, W. (1999) *Eur. J. Inorg. Chem.*, **1999**, 1295.
112 Huang, F.Q. and Ibers, J.A. (2001) *Angew. Chem.*, **113**, 2583.
113 DiSalvo, F.J. (1990) *Science*, **247**, 649.
114 Pagano, S., Lupart, S., Zeuner, M., and Schnick, W. (2009) *Angew. Chem.*, **121**, 6453.
115 Ehrentraut, D. and Meissner, E. (2010) Chapter 11, in *Technology of Gallium Nitride Crystal Growth* (eds D. Ehrentraut, E. Meissner, and M. Bockowski), Springer, Berlin, pp. 235–244.
116 Niewa, R. and Jacobs, H. (1996) *Chem. Rev.*, **96**, 2053.
117 Aubry, J. and Streiff, R. (1971) *J. Electrochem. Soc.*, **118**, 650.
118 Gudat, A., Haag, S., Kniep, R., and Rabenau, A. (1990) *Z. Naturforsch.*, **45b**, 111.
119 Hunting, J.L., Szymanski, M.M., Lowalsick, A.L., Downie, C.M., and DiSalvo, F.J. (2013) *J. Solid State Chem.*, **197**, 288.
120 Houmes, J.D. and zur Loye, H.-C. (1997) *J. Solid State Chem.*, **130**, 266.
121 Yamane, H. and DiSalvo, F.J. (1996) *Acta Crystallogr. C*, **52**, 760.
122 Yamane, H. and DiSalvo, F.J. (1996) *J. Alloys Compd.*, **240**, 33.
123 Park, D.G., Dong, Y., and DiSalvo, F.J. (2008) *Solid State Sci.*, **10**, 1846.
124 Zachwieja, U. (1993) *J. Alloys Compd.*, **196**, 171.
125 Zachwieja, U. (1996) *Z. Anorg. Allg. Chem.*, **622**, 1581.
126 Steinbrenner, U. and Simon, A. (1996) *Angew. Chem.*, **108**, 595.
127 Simon, A. and Steinbrenner, U. (1996) *J. Chem. Soc., Faraday Trans.*, **92**, 2117.
128 Ezekwenna, P., Febri, M., l'Héritier, P., and Joubert, J.-C. (1996) *J. Mater. Chem.*, **6**, 1165.
129 Zhu, X., Wang, Z., Huang, B., Wie, W., Dai, Y., Zhang, X., and Quin, X. (2015) *APL Mater.*, **3**, 104413.
130 Zhang, X., Wang, Q., Ma, Z., He, J., Wang, Z., Zheng, C., Lin, J., and Huang, F. (2015) *Inorg. Chem.*, **54**, 5301.
131 Feng, K., Yin, W., He, R., Lin, Z., Jin, S., Yao, J., Fu, P., and Wu, Y. (2012) *Dalton Trans.*, **41**, 484.
132 Mac-Chesney, J.B., Williams, H.J., Sherwood, R.C., and Potter, J.F. (1966) *J. Chem. Phys.*, **44**, 596.
133 Alonso, J.A., Martinez-Lope, M.J., Casais, M.T., Munoz, A., Largeteau, A., and Demazeau, G. (2005) Symposium Y: Solvothermal Synthesis of Materials Proceedings 872E, Y3.4-1, MRS Spring Meeting, San Francisco, March 28–30.
134 Demazeau, G., Baranov, A., Presniakov, I., and Sobolev, A. (2006) *Z. Naturforsch.*, **61b**, 1527.

11
Glass Formation and Crystallization

T. Komatsu

Nagaoka University of Technology, Department of Materials Science and Technology, 1603-1 Kamitomioka-cho, Nagaoka 940-2188, Japan

11.1
Introduction

Oxide glasses based on SiO_2, B_2O_3, and P_2O_5 have various excellent features such as shape forming ability (e.g., fibers and thin plates), homogeneous microstructure and optical transparency, wide design and control in chemical/physical properties, and high chemical durability, and therefore, have been used widely in advanced technologies as key and innovative materials. Glasses, which are thermodynamically metastable (nonequilibrium) solids with random atomic arrangements, also include various scientific important points such as random structure, structural relaxation, phase separation, and crystallization. New glasses are basically developed through the design and control of glass systems and chemical compositions. Even nowadays, numerous glasses with different compositions have been synthesized from the basic scientific interests and practical applications. Besides oxide glasses, other kinds of glasses such as chalcogenide glasses (e.g., As–S–Se), fluoride glasses (e.g., GdF_3–BaF_2–ZrF_4), organic and polymer glasses (e.g., polystyrene), and metallic glasses (e.g., Fe–Co–Si–B) have been synthesized. Usually, glasses are fabricated by melt-quenching method, but other techniques such as sol–gel and vapor deposition methods also have been widely used. In this chapter, oxide glasses synthesized by conventional melt-quenching method are focused.

Glass formation is realized when liquid is cooled continuously into the solid state without causing any detectable crystallization, and thus, liquid must be cooled or quenched with the cooling rate being necessary to avoid crystal nucleation and growth. In this sense, the structure and viscosity of liquids, which depend largely on the chemical composition of liquids and are related to the kinetics of crystal nucleation and growth, are key points for the understanding of glass formation and for the development of new glasses. Compared with crystals, glasses have advantage and flexibility in the design of chemical composition, and thus different multicomponent glasses with continuous composition

Handbook of Solid State Chemistry, First Edition. Edited by Richard Dronskowski, Shinichi Kikkawa, and Andreas Stein.

changes are synthesized. In this chapter, the basic scenario for glass formation, the fragility concept for glass formation, and the nanoscale structure in oxide glasses are described, and then some interesting glass-forming oxide systems from the viewpoint of the relation between glass formation and crystallization are focused [1,2].

11.2
Basic Scenario for Glass Formation

Glass is a solid with the thermodynamically nonequilibrium (metastable) state, in which the structure is random, that is, rack of the symmetry and periodicity in atomic arrangements, and the glass transition phenomenon is observed. Let us consider the glass formation from the quenching (cooling) of melt. Figure 11.1 shows the schematic illustration for the free energies of glass, stable crystal, super-cooled liquid, and stable liquid as a function of temperature. T_g and T_m are the glass transition temperature of glass and the melting temperature of stable crystal, respectively. If a stable liquid is cooled slowly or is held for a long time at around T_m, it must transform into a crystal thermodynamically. The most well known method to synthesize single crystal is to use this slow cooling process at near T_m. On the other hand, if a stable liquid is cooled rapidly and pass the temperature region of around T_m quickly, the stable liquid enters into the super-cooled liquid (the under-cooled liquid) state. The viscosity η of super-cooled liquids decreases with decreasing temperature. If super-cooled liquid has no chance to transform into crystal during cooling process, a super-cooled liquid becomes a solid with the viscosity of $\eta = 10^{12}$ Pa s. Such a solid has atomic arrangements similar to the precursor super-cooled liquid in its structure. That

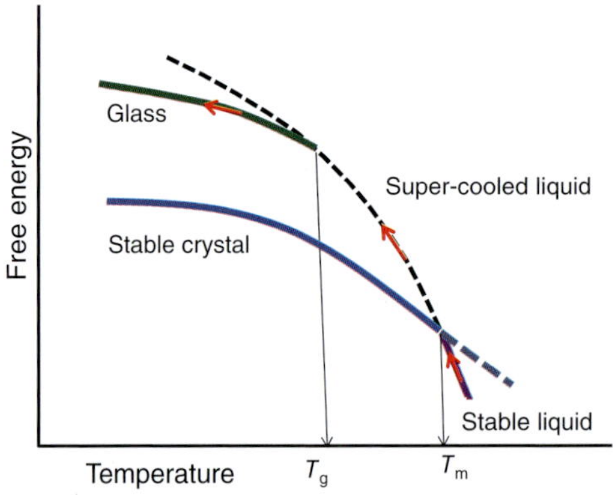

Figure 11.1 Schematic illustration for the free energies of glass, stable crystal, super-cooled liquid, and stable liquid as a function of temperature. T_g and T_m are the glass transition temperature and the melting temperature of stable crystal, respectively.

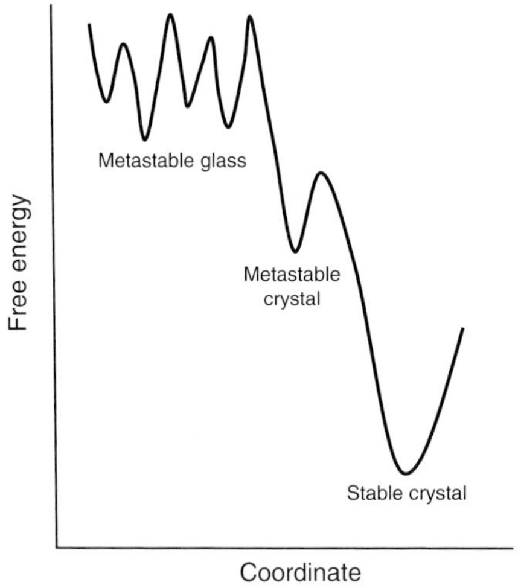

Figure 11.2 Schematic illustration for the free energies of the thermodynamically metastable glassy state, metastable crystalline state, and the stable crystalline state.

is, roughly speaking the structure of glass is random. For glass formation, therefore, crystallization in super-cooled liquids must be avoided kinetically. In this sense, the glass formation and crystallization behavior are closely related to each other, and to understand glass formation is to understand crystallization behavior. In other words, the glassy state is regarded as the metastable (nonequilibrium) state thermodynamically, as shown in Figure 11.2. In the crystallization of super-cooled liquids, metastable crystalline phase is frequently formed, and such a situation is described in Figure 11.2.

In a super-cooled liquid, its viscosity decreases with increasing temperature and atoms and structural units can move or diffuse. The value of viscosity in the super-cooled liquid state depends largely on the glass system and chemical composition. If sufficient energy (i.e., temperature) and time for breaking of bonds and diffusion of atoms are supplied, a crystalline phase with the thermodynamically equilibrium state emerges from a super-cooled liquid. Crystallization proceeds through two steps of nucleation and crystal growth (Figure 11.3). As well-known, in the classical nucleation theory, there are two key factors for the tendency of nucleation and crystal growth in glasses, that is, the liquid-crystal interfacial energy (tension) per unit area γ and the free-energy change per unit volume that is produced by the formation of nuclei (the difference in the free energy between the glassy phase and crystalline phase) ΔG_v. There are two approaches for the origin of γ. One is based on enthalpy, in which the atomic bonding state at the interface is considered to be different from that in the inside of crystals and γ is scaled by the melting enthalpy ΔH_m [3]. The other one is based on entropy, in which some amount of entropy is lost due to the ordered

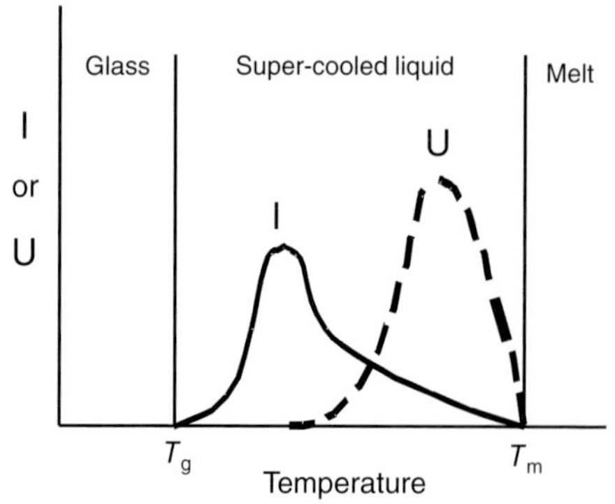

Figure 11.3 Schematic illustration for the crystal nucleation rate *I* and crystal growth rate *U* in a super-cooled liquid state as a function of temperature.

atomic arrangement at the interface and γ is scaled by the melting entropy ΔS_m [4]. The model of the entropy origin for interfacial energy is mainly applied to the crystal nucleation in metallic melts [4,5]. In the entropy origin, the interfacial energy is expressed by Eq. (11.1),

$$\gamma = \alpha \frac{\Delta S_m T}{(N_A V_m^2)^{1/3}}, \tag{11.1}$$

where α is a scaling factor related to the structure of crystal, N_A is the Avogadro's number, V_m is the molar volume, and T is temperature. Equation (11.1) proposes that the structural similarity at the liquid–crystal interface lowers the height of the free-energy barrier (ΔG^*) for crystal nucleation, and consequently, in super-cooled liquids with small ΔS_m values, easy nucleation and crystal growth would be expected.

The normal growth model based on the classical nucleation theory provides the following expressions for the kinetics of homogeneous (steady-state) nucleation rate I and crystal growth rate U [6,7],

$$I \propto \left(D' \text{ or } \frac{1}{\eta'}\right) \exp\left(-\frac{\Delta G^*}{kT}\right), \tag{11.2}$$

$$U \propto \left(D'' \text{ or } \frac{1}{\eta''}\right) \left[1 - \exp\left(-\frac{\Delta G_v}{kT}\right)\right], \tag{11.3}$$

where D' and D'' are diffusion coefficients of atoms in super-cooled liquid and at the interface between liquid and crystal, respectively, and η' and η'' are viscosities of super-cooled liquid and at the interface between liquid and crystal, respectively. k is the Boltzmann constant. In Eqs. (11.2) and (11.3), jump frequency and diffusion length of atoms are transformed to diffusion coefficient or viscosity. In Eqs. (11.2) and (11.3), the first term is for atomic diffusions (transport) and the second term is for thermodynamic barrier and driving force for atomic rearrangements. An excellent review article on the kinetics of the glass crystallization has been published by Fokin *et al.* [6]. The structure and also viscosity of super-cooled liquids change largely with their composition. In order to form glasses and to avoid crystallization during cooling, the viscosity of super-cooled liquids would be required to be high and the structure of super-cooled liquids would be also required to be deviated largely from the structure of crystalline phases (which are expected in a given system and composition) in an atomic nanoscale level.

11.3 Fragility Concept for Glass Formation

As described in Section 11.2, the temperature dependence of viscosity of super-cooled liquids is one of the key factors for glass formation. The concept of strong and fragile classifications for super-cooled liquids proposed by Angell [8] is very important for the understanding of glass formation (Figure 11.4). For example, SiO_2 glass, which is classified into a typical strong glass, exhibits a very weak temperature dependence of viscosity at temperatures above T_g, that is, the Arrhenius-type temperature dependence, and the super-cooled liquid of SiO_2 is regarded as a viscous liquid. The structure of SiO_2 glass consists of three-dimensional network combined with SiO_4 tetrahedral units having strong Si—O—Si

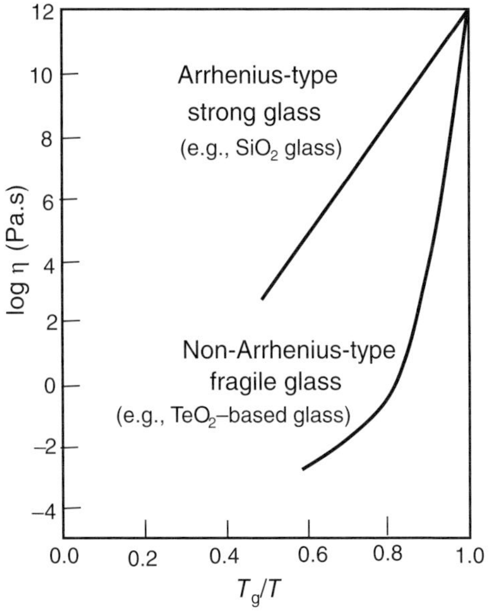

Figure 11.4 Schematic illustration for the temperature dependence of viscosity for strong and fragile glasses in the fragility concept.

bonds. In case of SiO_2, therefore, it is extremely difficult to induce arrangements of atoms or structural units for crystallization, and even in very slow cooling rates of super-cooled liquid, the glass formation of SiO_2 is established very easily. On the other hand, for example, TeO_2-based glasses are classified into the category of fragile glasses. Although the network structure consisting of TeO_3 and TeO_4 structural units is created in TeO_2-based glasses, the strength of Te—O—Te bonds is small, and therefore, the temperature dependence of the viscosity of super-cooled liquids is becoming large, consequently providing a non-Arrhenius-type temperature dependence as shown in Figure 11.4. Basically, rapid quenching rates are required to form glasses in TeO_2-based systems [9].

Sun [10] suggested a bond energy criterion for the glass formation and reported the comprehensive data on the single bond strength $B_{M\text{-}O}$ for various simple oxides based on their dissociation energy. Oxides are divided into three groups, namely, glass-formers ($B_{M\text{-}O} = 498$–339 kJ/mol), intermediates ($B_{M\text{-}O} = 305$–251 kJ/mol), and modifiers ($B_{M\text{-}O} = 251$–42 kJ/mol) in accordance with the value of their single bond strength. The criterion is simple. That is, high value of single bond strength increases glass-forming tendency. Glass-forming oxides such as SiO_2 ($B_{M\text{-}O} = 443$ kJ/mol), B_2O_3 ($B_{M\text{-}O} = 498$ kJ/mol), and P_2O_5 ($B_{M\text{-}O} = 464$ kJ/mol), have high $B_{M\text{-}O}$ values, and their bonding character is mainly covalent. On the other hand, modifiers such as Na_2O ($B_{M\text{-}O} = 84$ kJ/mol) and K_2O ($B_{M\text{-}O} = 54$ kJ/mol) have low $B_{M\text{-}O}$ values, and their bonding character is ionic. Dimitrov and Komatsu [11–13] proposed an approach for the calculation of average single bond strength $B_{M\text{-}O}$ of oxide glasses based on Sun's data for simple oxides. It has been established that the electronic polarizability of glasses increases with decreasing single bond strength. Usually, oxide glasses are synthesized through the combination of glass-forming oxides, intermediates, and modifiers, for example, Li_2O–Al_2O_3–SiO_2 glasses. In multicomponent systems, for example, binary, ternary, and quaternary systems, glasses are regarded as solids with different bonding strengths and different electronic polarizability states locally. The addition of intermediates and modifiers into glass-formers decreases the degree of three-dimensional network and such super-cooled liquids (glasses) obtained are becoming more fragile. In practical glass fabrication processing, optimal or proper viscosities are required for the handing of super-cooled liquids to get glasses with various shapes such as thick plate, thin plate, bottle, fiber, bead. The chemical composition of glasses is determined from the balance of the glass formation and viscosity of super-cooled liquids.

11.4
Nanoscale Structure in Oxide Glasses

In order to understand the glass formation more deeply, information on the structure of glasses is inevitable. Numerous studies and attempts using various experimental techniques such as nuclear magnetic resonance (NMR), Raman scattering spectroscopy, X-ray photoelectron spectroscopy (XPS), and also computer simulations have been carried out to clarify the structure and chemical bonding state in glasses. However, still the detailed information on short- and medium-range orders of glasses is insufficient, and very tough works are required in each glass with a given composition. Glasses have random and homogeneous structures in a microscale level, providing excellent optical transparencies. Nowadays, however, it has been well recognized that in a nanoscale level, glasses are not homogeneous, that is, the presence of nanoscale heterogeneous structure in glasses.

Here, some supports for the nanoscale heterogeneous structure in glasses are described. The well-known relationship between the absorption loss and wavelength (λ) of light in silica glass optical fibers clearly proposes the following features of the structure of glasses. The Rayleigh scattering loss indicates the presence of the density fluctuation in atomic arrangements in glasses, providing the decrease in light scattering loss as a function of λ^{-4}. Looking from the visible light with short wavelengths, glass is a solid with a large (strong) density fluctuation. On the other hand, looking from the infrared light with long wavelengths, glass is a solid with a small (weak) density fluctuation. In other words, glass has a heterogeneous structure in a nanoscale level, and contrary glass is regarded as a solid with a homogeneous structure in a microscale level. The well-known heterogeneous structure model of glasses in a nanoscale level has been proposed by Greaves [14], that is, the modified random network model. The presence of nanoscale heterogeneous structure in glasses has been discussed from Boson peaks appearing at the low frequency (energy) side in Raman scattering spectra and from the first sharp diffraction peak (FSDP) appearing at the structure factor in X-ray or neutron diffraction patterns [15,16]. For example, it has been proposed that Boson peaks are connected with some correlation radius of glass structure in the scale of ~1 nm [16].

It should be pointed out that the degree of nanoscale heterogeneity would vary with glass composition [17]. The arrangement of structural units (short-range order) in glasses is not completely random, but has some correlations on a scale of 1–1.5 nm [17]. Darguad *et al.* [18] examined nanoscale structural heterogeneities in SiO_2–Al_2O_3–MgO–ZnO–ZrO_2 glasses from scanning transmission electron microscope observations in a high angle annular dark-field imaging mode (STEM-HAADF) and proposed a model explaining how glass heterogeneities are affecting devitrification. Their model is that crystal nucleation must not be viewed as processing from a homogeneous media but that static and chemical fluctuations must be taken into account to describe nucleation processes quantitatively. Tokuda *et al.* [19] investigated Na^+ environments in sodium silicate and mixed alkali silicate glasses using ^{23}Na multiple-quantum magic-angle spinning (MQMAS) NMR spectroscopy, *ab initio* molecular orbital (MO) calculation, and Na^+ elution analysis and proposed an inhomogeneous distribution of local structures around Na^+, that is, aggregated and isolated Na^+ sites. Vargheese *et al.* [20] investigated the heterogeneous dynamics of CaO–Al_2O_3–SiO_2 glasses by using the isoconfigurational ensemble method (computer simulation) and found that the heterogeneous dynamics is dominated by concentration fluctuations with regions of high propensity (i.e., mobility) having a larger concentration of both Ca^{2+} and Al^{3+} and regions of low propensity being SiO_2 rich. Zhang *et al.* [21] approached the structural heterogeneity in a hyperquenched $CaO–MgO–SiO_2$ glass by using the hyperquenched-annealing differential scanning calorimeter (DSC) method (i.e., the sub-T_g enthalpy relaxation) and found that the glass consists of two distinct structural domains of higher and lower potential energies. In particular, they proposed that the higher energy nanoscale domains are so unstable that they become ordered during hyperquenching. Recently, Takahashi *et al.* [22] performed *in situ* observations of Boson peaks in crystallizing glasses while elevating temperature in $15K_2O–15Nb_2O_5–70TeO_2$ and $25K_2O–25Nb_2O_5–50GeO_2$ glasses and found that a drastic decrease in elasticity and damping of Boson peaks takes place during the crystallization process at around the glass transition temperature. They proposed that denser regions in the heterogeneous structure of the glasses transform into crystal nuclei.

The structure model proposed by Phillips [23] and Thorpe [24] in chalcogenide glasses is regarded as another expression for nanoscale heterogeneous structure, that is, they use the terms of the over-constrained rigid region and unconstrained floppy region. Bocker *et al.* [25] tried to link the crystal nucleation in glass-forming melts based on $BaO–TiO_2–SiO_2$ and the concept of rigid and floppy regions. They proposed that the size of a floppy region changes with the composition and the nucleation of fresnoite $Ba_2TiSi_2O_8$ crystals takes place predominantly in some small floppy regions [25]. It is known that that the initial crystallization of BaF_2 and CaF_2 nanocrystals in $Al_2O_3–SiO_2$-based oxyfluoride glasses starts rapidly in a floppy region and then further crystallization propagates inside the rigid region slowly (e.g., the formation of $Al_2O_3–SiO_2$-based oxide crystals) [26]. The size of the floppy region would be around 4.5 nm and is becoming larger if the concentration of network modifiers such as Na_2O in oxyfluoride glasses is increased. These studies [23–26] also strongly support the nanoscale heterogeneous structural model in glasses. The well-known concept of the fragility for the glass proposed by Angell [8] is for the whole part (average) of the glass itself. Here, it is proposed to extend the concept of the fragility of the glass itself to nanoscale heterogeneous structure and bonding, that is, the introduction of "nano-scale fragility." In other words, it is proposed that glass consists of both of nanoscale strong and fragile structures. It is considered that the nanoscale strong part is closely related to the nanoscale structure consisting of mainly glass formers such as SiO_2 and B_2O_3 with strong chemical bonds. On the other hand, the nanoscale fragile part would be related to the nanoscale structure consisting of mainly glass modifiers such as Na_2O and BaO with weak chemical bonds. The so-called intermediates such as Al_2O_3 and TiO_2 would take part in a nanoscale strong part or a nanoscale fragile part, depending on the glass system and composition. The existence of topological crystalline-like ordering in melt-quenched glasses is still a subject of debate in the glass structure model [27]. The topological crystalline-like ordering is named as "frozen-in cybotactic groupings," and recently, Wright [27] proposes that there is strong circumstantial evidence for its existence, although it is extremely difficult to obtain unambiguous direct experimental proof of the existence of crystalline-like ordering in glasses.

11.5 Glass Formation and Crystallization in Oxide Glasses

Materials, being irrespective of amorphous and crystal, are synthesized and used in practical applications from the viewpoint of functions. Numerous oxide glasses (glass formation) with different systems and compositions, that is, the design and control of chemical/physical/thermal/mechanical properties, have been reported so far. Furthermore, new oxide glasses have been developed and investigated from the scientific point of view such as random structure, structural relaxation, phase separation, and crystallization, which are typical features and phenomena in glassy materials. For example, many glasses have been developed for the design and control of glass-ceramics (crystallized glasses) with excellent functions, that is, glass-crystal hybrid materials. In Tables 11.1 and

Table 11.1 Glass compositions and crystalline phases formed in the crystallization of stoichiometric and near-stoichiometric glasses. The data are mainly reported by the present authors' group [1]. RE are rare earth elements.

Glass composition (mol%) Stoichiometric or near-stoichiometric	Crystalline phase
33.3 Li_2O–66.7 SiO_2	$Li_2Si_2O_5$
5 $Na_2O.Fe_2O_3.8$ SiO_2	$Na_5FeSi_4O_{12}$
40BaO–20 TiO_2–40 SiO_2	$Ba_2TiSi_2O_8$
33.3BaO–33.3 TiO_2–33.3B_2O_3	$BaTi(BO_3)_2$
25 La_2O_3–50 WO_3–25B_2O_3	$LaWBO_6$
25 La_2O_3–50 MoO_3–25B_2O_3	$LaMoBO_6$
8 Sm_2O_3–42BaO–50B_2O_3	β-BaB_2O_4
50BaF_2–25Al_2O_3–25B_2O_3	$BaAlBO_3F_2$
40BaO–20 TiO_2–40GeO_2	$Ba_2TiGe_2O_8$
25$(La,RE)_2O_3$–25B_2O_3–50GeO_2	$(La,RE)BGeO_5$
25 Li_2O–25B_2O_3–50GeO_2	$LiBGeO_4$
16.7BaO–33.3Ga_2O_3–16.7 La_2O_3–33.3GeO_2	$BaGa_4La_2Ge_2O_{14}$
20 K_2O–20 TiO_2–60GeO_2	$K_2TiGe_3O_9$
25 K_2O–25 Nb_2O_5–50GeO_2	$K_{3.8}Nb_5Ge_3O_{20.4}$
25 Na_2O–25 Nb_2O_5–50GeO_2	$NaNbGeO_5$
26 Li_2O–43 Fe_2O_3–5 Nb_2O_5–26P_2O_5	$LiFePO_4$
33.3 Li_2O–33.3 V_2O_5–33.3P_2O_5	$LiVOPO_4$
37.5 Li_2O–25 V_2O_5–37.5P_2O_5	$Li_3V_2(PO_4)_3$
33.3 Na_2O–33.3 Fe_2O_3–33.3P_2O_5	$Na_2FeP_2O_7$
16.7Bi_2O_3–33.3 SrO–16.7CaO–33.3CuO	$Bi_2Sr_2CaCu_2O_x$
$Bi_{0.8}Pb_{0.2}SrCaCu_{1.5}O_{4.9}$	$(Bi,Pb)_2Sr_2Ca_2Cu_3O_y$

Table 11.2 Glass compositions and crystalline phases formed in the crystallization of off-stoichiometric glasses. The data are mainly reported by the present authors' group [1]. RE are rare earth elements. The chemical composition shown in this table is an example among various compositions.

Glass system Off-stoichiometric composition	Crystalline phase
$10\,MO{-}10\,Fe_2O_3{-}30\,Na_2O{-}50\,SiO_2$ (M=Ni, Co, Fe)	MFe_2O_4
$40\,Li_2O{-}32\,Nb_2O_5{-}28\,SiO_2$	$LiNbO_3$
$30\,Na_2O{-}25\,Nb_2O_5{-}45\,SiO_2$	$NaNbO_3$
$40\,K_2O{-}25\,Nb_2O_5{-}25B_2O_3{-}10Al_2O_3$	$KNbO_3$
$40BaO{-}40\,TiO_2{-}15\,SiO_2{-}5Al_2O_3$	$BaTiO_3$
$15BaO{-}15B_2O_3{-}70\,TeO_2$	BaB_4O_7
$40BaO{-}40\,TiO_2{-}20B_2O_3$	$Ba_3Ti_3O_6(BO_3)_2$
$16\,SrO{-}16BaO{-}32\,Nb_2O_5{-}36B_2O_3$	$Sr_xBa_{1-x}Nb_2O_6$
$2.3RE_2O_3{-}27.4BaO{-}34.3\,Nb_2O_5{-}36B_2O_3$	$Ba_{1-x}RE_{2x/3}Nb_2O_6$
$10BaO{-}10RE_2O_3{-}80\,TeO_2$	$RE_2Te_6O_{15}$, $RE_2Te_5O_{13}$
$21.25RE_2O_3{-}63.75\,MoO_3{-}15B_2O_3$	β'-$RE_2(MoO_4)_3$
$20\,SrBi_2Ta_2O_9{-}80\,Li_2B_4O_7$	$SrBi_2Ta_2O_9$
$16.66\,SrO{-}16.66Bi_2O_3{-}16.66\,Nb_2O_5{-}50\,Li_2B_4O_7$	$SrBi_2Nb_2O_9$
$30BaO{-}15\,TiO_2{-}55GeO_2$	$Ba_2TiGe_2O_8$
$23\,Na_2O{-}12Y_2O_3{-}25\,Fe_2O_3{-}20\,SiO_2{-}20GeO_2$	$Y_3Fe_5O_{12}$
$22.5RE_2O_3{-}47.5\,WO_3{-}30B_2O_3$	$RE_2(WO_4)_3$
$30\,MoO_3{-}50ZnO{-}20B_2O_3$	α-$ZnMoO_4$
$10\,La_2O_3{-}35\,SrO{-}25\,MnO_2{-}30B_2O_3$	$La_{0.7}Sr_{0.3}MnO_3$
$15\,K_2O{-}15\,Sm_2O_3{-}70P_2O_5$	$KSm(PO_3)_4$
$25CaF_2{-}5CaO{-}20Al_2O_3{-}50\,SiO_2$	CaF_2
$0.5CuO{-}40\,Li_2O{-}32\,Nb_2O_5{-}28\,SiO_2$	Cu metal
$72\,SnO{-}28P_2O_5$	Sn metal
$15\,K_2O{-}15\,Nb_2O_5{-}70\,TeO_2$	$K[Nb_{1/3}Te_{2/3}]_2O_{4.8}$
$12.5RE_2O_3{-}30Bi_2O_3{-}57.5B_2O_3$	$RE_xBi_{1-x}BO_3$
$40\,K_2O{-}20\,Nb_2O_5{-}40\,SiO_2$	$KNbSi_2O_7$, $K_3Nb_3Si_2O_{13}$

11.2, various glasses, which have been synthesized for the study of crystallization and for the development of new functional glass-ceramics mainly reported by the present authors' group [2], are summarized. In this section, some typical glasses that are concerned and attractive from the viewpoint of glass formation and crystallization behavior are described.

11.5.1 $BaO{-}TiO_2{-}SiO_2/GeO_2$ Glasses

Most important commercially available oxide glasses such as optical fiber, thin plate, and window glasses are based on SiO_2, and thus numerous SiO_2-based glasses with desired properties have been developed so far. The structure of SiO_2-based glasses consists of three-dimensional network combined with SiO_4 tetrahedral units having strong Si—O—Si bonds and with intermediate oxides such as Al_2O_3 and modifiers such as CaO and Na_2O. The addition of other oxides such as Na_2O, CaO, and Al_2O_3 into SiO_2 enables to control and design various properties of SiO_2-based glasses such as the glass-forming ability, temperature dependence of viscosity, mechanical strength, and thermal expansion coefficient. Germanate (GeO_2) glasses are of practical and scientific interest because of their unique physical properties such as comparatively high refractive index exceeding $n = 1.7$ and the Abbe number of about 24–28. Various GeO_2-based glasses have been, therefore, synthesized. In this section, some features of $BaO–TiO_2–SiO_2$ and $BaO–TiO_2–GeO_2$ glasses are described, because these glasses show the crystallization of $Ba_2TiSi_2O_8$ and $Ba_2TiGe_2O_8$ crystals with excellent second-order optical nonlinearities.

It is known that $BaO–TiO_2–SiO_2$ glasses, in particular, a glass with the composition corresponding to the stoichiometric $Ba_2TiSi_2O_8$ crystalline phase (BTS), that is, $40BaO–20TiO_2–40SiO_2$ glass (BTS glass: $T_g = 730\,°C$, crystallization peak temperature $T_p = 832\,°C$), shows a prominent bulk crystallization, forming nonlinear optical BTS nanocrystals [28,29]. Photoluminescence (PL) of rare earth (RE) ions in glassy or crystalline solids is one of the fundamental light-matter interactions. Dy^{3+} ions in solids show yellow (~570 nm) emissions resulting from the f–f transition of $^4F_{9/2} \rightarrow {}^6H_{13/2}$ and 1.3 μm emissions due to the $^6F_{11/2}, {}^6H_{9/2} \rightarrow {}^6H_{15/2}$ transition, and these emissions have been considered to be utilized as visible solid-state lasers and optical amplifiers in broad band telecommunication systems. Since PL properties of RE ions depend on the structural and chemical bonding states of RE ions in a given host material, it is of interest and importance to search materials giving intense yellow or 1.3 μm emissions for photonic device applications of Dy^{3+} ions. The optical absorption spectra at room temperature for the precursor BTS glass and crystallized (780 °C, 30 min) glasses containing 0.5 mol% Dy_2O_3 are shown in Figure 11.5, in which the optical photographs for these samples are included. The typical transitions between the ground state of $^6H_{15/2}$ and the excited states of ^6F and $^6H_{11/2}$ in Dy^{3+} ions are observed in both samples, and the peak assignments are given in the figure. As can be seen in Figure 11.5, the crystallized sample keeps a good optical transparency, but the absorption edge shifts toward a longer wavelength due to the crystallization. Similar optical absorption spectra were obtained in other crystallized (760–800 °C) glasses. Furthermore, it is seen that the position (~1280 nm) of the peak assigned to the $^6H_{15/2} \rightarrow {}^6F_{11/2}$ transition shifts to a longer wavelength. The average particle sizes of $Ba_2TiSi_2O_8$ crystals in Dy_2O_3-doped crystallized glasses were estimated from the peak width of X-ray diffraction (XRD) patterns by using the Scherrer's equation and were found to be 70–150 nm. Schneider *et al.* [30] examined the coordination of Ti^{4+} in BTS-based glasses from XPS and found that fivefold coordinated TiO_5 structural units predominate (~60%). As reported by Cabral *et al.* [31], BTS glass has an extremely high nucleation rate of $\sim 10^{17}/m^3\,s$. One of the reasons for such a high

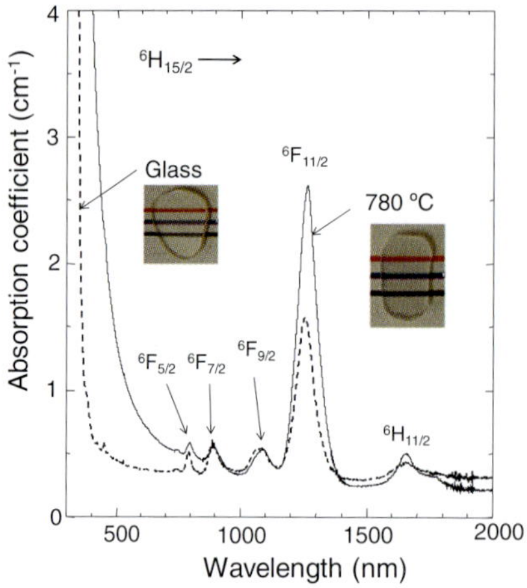

Figure 11.5 Optical absorption spectra at room temperature for the precursor and crystallized (780 °C, 30 min) glasses containing Dy^{3+} ions in $40BaO–20TiO_2–40SiO_2$. The optical photographs for these samples are also shown [28].

nucleation rate would be a nanoscale structural similarity between BTS glass and BTS crystals, in particular, for fivefold coordinated TiO_5 structural units. The PL spectra in the range of 450–700 nm obtained in the experiments of quantum yield measurements for the precursor BTS and crystallized (at 770 and 790 °C, for 30 min) glasses with $0.5Dy_2O_3$ are shown in Figure 11.6. Three peaks assigned to the f–f transitions of $^4F_{9/2} \rightarrow {}^6H_{15/2}$ (blue: 484 nm), $^4F_{9/2} \rightarrow {}^6H_{13/2}$ (yellow: 575 nm), and $^4F_{9/2} \rightarrow {}^6H_{11/2}$ (red: 669 nm) are observed. It is seen that the intensity of these peaks increases due to the crystallization. In particular, the crystallized samples show the strong intensity for the peak corresponding to the

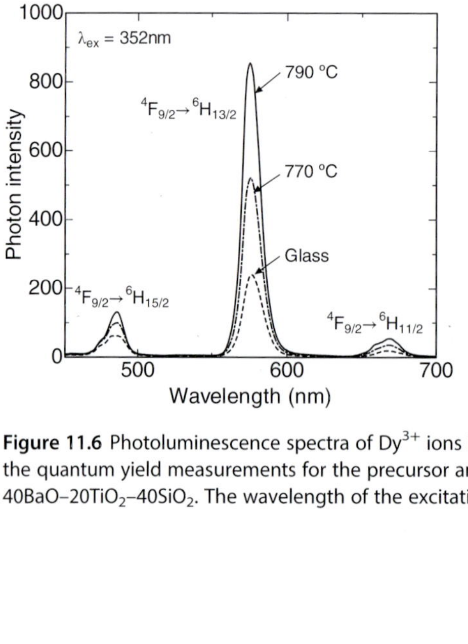

Figure 11.6 Photoluminescence spectra of Dy^{3+} ions in the range of 450–700 nm obtained in the quantum yield measurements for the precursor and crystallized (790 °C, 30 min) glasses in $40BaO–20TiO_2–40SiO_2$. The wavelength of the excitation light was 352 nm [28].

$^4F_{9/2} \rightarrow {}^6H_{13/2}$ transition compared with the precursor BTS glass. It was reported that $40BaO–20TiO_2–40SiO_2$ glass have the values of density $d = 4.294\,g/cm^3$, Young's modulus (estimated from the cube resonance method) $E = 79.9\,GPa$, the shear modulus of $G = 31.1\,GPa$, the bulk modulus of $K = 61.3\,GPa$, the Poisson's ratio $\nu = 0.28$, and the Debye temperature of $\theta_D = 522\,K$ and the crystallized glasses with BTS nanocrystals exhibit the increases in these values, for example, in the crystallized (780 °C, 1 h) glass with the volume fraction of 43.7% BTS crystals, $d = 4.362\,g/cm^3$, $E = 99.6\,GPa$, $G = 38.7\,GPa$, $K = 78.1\,GPa$, $\nu = 0.29$, and $\theta_D = 580\,K$, that is, the elastic properties of $40BaO–20TiO_2–40SiO_2$ glass are largely enhanced due to the nanocrystallization.

According to the glass-forming region in the $BaO–TiO_2–GeO_2$ system reported by Imaoka and Yamazaki [32], the composition corresponding to the stoichiometric $Ba_2TiGe_2O_8$, that is, $40BaO–20TiO_2–40GeO_2$, locates at just out of the glass-forming region. Indeed, it is difficult to prepare bulk glasses for this stoichiometric composition by using the conventional melt-quenching method. Some glasses with modified compositions such as $30BaO–15TiO_2–55GeO_2$ (denoted BTG55) glasses have been synthesized. The BTG55 glass has the values of $T_g = 678$ °C and $T_p = 823$ °C and shows the surface crystallization. The sample obtained by a heat treatment at 720 °C for 3 h is optically transparent and fresnoite-type $Ba_2TiGe_2O_8$ crystals formed at the surface are highly oriented, that is, c-axis orientation. The Maker fringe pattern of second harmonic generations (SHGs) measured for the transparent BTG55 surface-crystallized glass is shown in Figure 11.7, and the value of the second-order optical nonlinearity $d_{33} = 22$ pm/V was obtained [33]. It should be pointed out that this value is comparable to that (~30 pm/V) of $LiNbO_3$ single crystal. In the crystal structure of $Ba_2TiGe_2O_8$, corner-linked TiO_5 pentahedra and pyrogermanate groups Ge_2O_7 comprise flat sheets perpendicular to the [001] direction, and these sheets are interconnected by 10-fold coordinated barium ions. It has been proposed that the origin of second-order optical nonlinearities in the fresnoite-type crystals is the presence of interconnected TiO_5 pyramidal units.

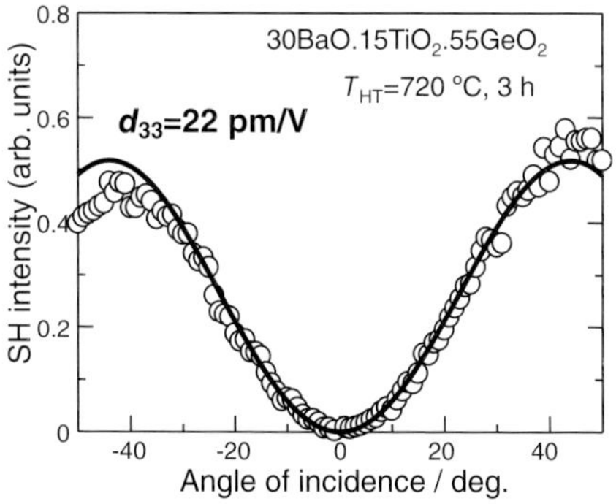

Figure 11.7 Maker fringe pattern of second harmonic intensity for transparent glass-ceramics with the composition of $30BaO–15TiO_2–55GeO_2$. Open circles are experimental data and the solid line is the theoretical fitting pattern [33].

The glasses of $25Li_2O–25B_2O_3–50GeO_2$ and $25La_2O_3–25B_2O_3–50GeO_2$ are also synthesized by using the conventional melt-quenching method, and transparent surface-crystallized glasses consisting of nonlinear optical $LiBGeO_4$ and $LaBGeO_5$ crystals have been developed [34]. It is known that the $K_2O–TiO_2–GeO_2$ system has a wide glass-forming region [32], meaning the possibility of the fabrication of glasses with a large amount of TiO_2. Since all oxides of K_2O, TiO_2, and GeO_2 have large oxide ion electronic polarizabilities, it is expected that $K_2O–TiO_2–GeO_2$ glasses would exhibit high refractive indices and might have a possibility of the formation of optical nonlinear crystals through crystallization. Fukushima *et al.* [35] prepared the glasses of $xK_2O–yTiO_2–zGeO_2$, $x = 15–25$, $y = 15–25$, and $z = 50–60$, by using the conventional melt-quenching method. For example, $20K_2O–20TiO_2–60GeO_2$ glass has the values of $T_g = 551\,°C$ and $T_p = 631\,°C$, $d = 3.501\,g/cm^3$, and refractive index at $\lambda = 632.8$ nm $n = 1.711$. The glasses consist of the network of TiO_6 and GeO_4 polyhedra and give the formation of c-axis oriented $K_2TiGe_3O_9$ crystals with SHGs due to the crystallization. The glass of $25K_2O–25Nb_2O_5–50GeO_2$ was synthesized, and the values of $T_g = 622\,°C$, $d = 3.811\,g/cm^3$, $n = 1.814$, $E = 64.1$ GPa, $G = 26.3$ GPa, $K = 38.2$ GPa, $\nu = 0.215$, and $\theta_D = 500$ K [36]. This glass shows the formation of $K_{3.8}Nb_5Ge_3O_{20.4}$ nanocrystals (diameter: ~30 nm) due to the crystallization. It should be pointed out that the initial crystalline phase is the stable crystalline phase of $K_{3.8}Nb_5Ge_3O_{20.4}$ consisting of all components of K_2O, Nb_2O_5, and GeO_2, and the chemical compositions of the glass and crystal are close to each other.

11.5.2
$ZnO–Bi_2O_3–B_2O_3$-Based Glasses

Since the first report by Dumbaugh [37] on Bi_2O_3-based glasses, it is well established that the addition of Bi_2O_3 has a strong effect on lowering of melting temperatures in glasses and the glasses containing Bi_2O_3 exhibit large third order nonlinear optical susceptibilities $\chi^{(3)}$ of the order 10^{-11} esu [38,39]. So far, numerous glasses containing Bi_2O_3 such as $ZnO–Bi_2O_3–B_2O_3$ and $RE_2O_3–Bi_2O_3–B_2O_3$ (RE: rare-earth such as Sm and Er) have been developed, and the glass formation, structure, and physical/thermal/mechanical properties have been reported. The crystallization of Bi_2O_3-based glasses has also received much attention, because some functional crystals containing Bi_2O_3 are formed through crystallization. For instance, high-T_c Bi_2O_3-based copper superconducting glass-ceramics have been synthesized through the crystallization of glasses with Bi_2O_3. Nonlinear optical $RE_xBi_{1-x}BO_3$ crystals showing SHGs have been also synthesized from $RE_2O_3–Bi_2O_3–B_2O_3$ glasses. The design and control of new Bi_2O_3-based glasses and their crystallization are at the frontiers of the glass science and technology.

First, in this section, the glass formation and some thermal/optical properties of $ZnO–Bi_2O_3–B_2O_3$ glasses are described as a typical example of Bi_2O_3-based glasses. The glasses with the compositions of $xZnO–yBi_2O_3–zB_2O_3$ glasses

Table 11.3 Compositions, glass transition temperature T_g, density d, refractive index at 642.8 nm n, molar polarizability α, oxide ion polarizability α_{O2-}, and optical basicity Λ for ZnO–Bi_2O_3–B_2O_3 glasses [40]. The experimental uncertainties of T_g, d, and n are ±2 °C, ±0.003 g/cm³, and ±0.001, respectively.

Composition (mol%)			T_g (°C)	d (g/cm³)	n	α (Å³)	α_{O2-} (Å³)	Λ
ZnO	Bi_2O_3	B_2O_3						
30	10	60	515	4.209	1.748	4.322	1.640	0.652
40	10	50	499	4.529	1.784	4.203	1.722	0.700
50	10	40	471	4.842	1.815	4.087	1.822	0.754
60	10	30	449	5.152	1.851	4.004	1.963	0.819
65	10	25	440	5.292	1.870	3.981	2.057	0.858
30	20	50	482	5.445	1.895	5.145	1.858	0.771
40	20	40	462	5.688	1.933	5.109	1.997	0.834
50	20	30	435	5.964	1.968	5.035	2.146	0.892
20	30	50	440	5.993	1.986	6.256	2.037	0.850
30	30	40	421	6.192	2.021	6.235	2.187	0.906
40	30	30	398	6.497	2.059	6.124	2.322	0.951
20	40	40	397	6.798	2.095	7.130	2.259	0.931
10	50	40	376	7.151	2.155	8.182	2.374	0.967

(x = 10–65, y = 10–50, z = 25–60 mol%) are synthesized by using the conventional melt-quenching method [40]. The glasses show the values of T_g = 376–515 °C (see Table 11.3), and it is seen that the value of T_g decreases with increasing Bi_2O_3 content. It is obvious that Bi_2O_3 has a stronger effect for lowering the glass transition temperature, that is, weakening the glass network structure, of ternary ZnO–Bi_2O_3–B_2O_3 glasses. As seen in Table 11.3, ZnO–Bi_2O_3–B_2O_3 glasses have large densities of d = 4.209–7.151 g/cm³ and large refractive indices of n = 1.748–2.155, and thus, it is of interest to clarify the electronic polarizabilities of the glasses. The electronic polarizabilities and optical basicities of ZnO–Bi_2O_3–B_2O_3 glasses were estimated using the following equations [40]:

$$R_m = \left[\frac{(n^2-1)}{(n^2+2)}\right]\left(\frac{M}{d}\right) = \left[\frac{(n^2-1)}{(n^2+2)}\right]V_m = \frac{4\pi\alpha_m N_A}{3}, \tag{11.4}$$

$$\alpha_{O^{2-}}(n) = \left[\frac{R_m}{2.52} - \Sigma\alpha_i\right](N_{O^{2-}})^{-1}, \tag{11.5}$$

$$\Lambda = 1.67\left(1 - \frac{1}{\alpha_{O^{2-}}}\right), \tag{11.6}$$

where R_m is the molar refraction, M is the molecular weight, V_m is the molar volume, α_m is the molar polarizability, N_A is the Avogadro's number, $\alpha_{O2-}(n)$ is the average electronic polarizability of oxide ions, $\Sigma\alpha_i$ denotes molar cation polarizability, N_{O2-} denotes the number of oxide ions in the chemical formula, and Λ is the so-called optical basicity of the oxide medium. Equation (11.4) is the so-called Lorentz–Lorenz equation giving the relationship between molar refraction (R_m) and refractive index (n). Equation (11.5) gives the electronic polarizability of oxide ions ($\alpha_{O2-}(n)$) in oxide materials by subtracting the cation polarizability ($\Sigma\alpha_i$) from the molar polarizability (α_m). Equation (11.6) proposed by Duffy [41] gives an intrinsic relationship between electronic polarizability of the oxide ions and so-called optical basicity of the oxide medium (Λ). The optical basicity of an oxide medium is a numerical expression of the average electron donor power of the oxide species constituting the medium. Because increased oxide ion polarizability means stronger electron donor ability of oxide ions, the physical background of the oxide ion polarizability and optical basicity is naturally the same. Using Eqs. (11.4)–(11.6), the values of α_{O2-} and Λ of ZnO–Bi_2O_3–B_2O_3 glasses prepared in the present study were estimated, and the results are shown in Table 11.3. The values of $\alpha_{Zn} = 0.283$ Å^3 for Zn^{2+} ions, $\alpha_{Bi} = 1.508$ Å^3 for Bi^{3+} ions and $\alpha_B = 0.002$ Å^3 for B^{3+} ions are used [42]. As can be seen in Table 11.3, the average electronic polarizability of oxide ions and optical basicity in ZnO–Bi_2O_3–B_2O_3 glasses increase with increasing Bi_2O_3 content. As reported by Dimitrov and Sakka [42], the simple oxides of ZnO, Bi_2O_3, and B_2O_3 have the following values for optical basicity: $\Lambda = 1.03$ for ZnO, $\Lambda = 1.19$ for Bi_2O_3, and $\Lambda = 0.43$ for B_2O_3, respectively. It is, therefore, expected that the degree of basicity (electron donor ability of oxide ions) in these simple oxides is in the order: $B_2O_3 \ll ZnO < Bi_2O_3$. The general trend that the electronic polarizability of oxide ions in ZnO–Bi_2O_3–B_2O_3 glasses increases with the substitution of ZnO or Bi_2O_3 for B_2O_3 seems to be reasonable. The formation of B–O–Bi and B–O–Zn bridging bonds in the glass structure has been suggested from Raman and XPS spectra, and such bridging bonds would be closely related to the glass formation in the ZnO–Bi_2O_3–B_2O_3 system.

It is noted that $Bi_2ZnB_2O_7$ crystal exhibits excellent nonlinear optical properties and is crystallized from 33.3Bi_2O_3–33.3ZnO–33.3B_2O_3 glass corresponding to the stoichiometric composition of $Bi_2ZnB_2O_7$ [43]. Figure 11.8 shows the Raman scattering spectra at room temperature for the base glass and crystallized (530 °C, 3 h) samples. The glass shows very broad peaks being typical for glassy materials in the Raman scattering spectrum. The crystallized sample shows several sharp peaks, confirming the formation of the $Bi_2ZnB_2O_7$ crystalline phase constructed from the stacking of BiO_6, ZnO_4, BO_3, and BO_4 units. Furthermore, nonlinear optical $Bi_2ZnB_2O_7$ crystal lines with a high orientation were patterned in 3Sm_2O_3–30.3Bi_2O_3–33.3ZnO–33.3B_2O_3 glass by using a laser-induced crystallization technique [43], proposing a new potential for optical device applications in Bi_2O_3-based glasses.

11.5.3
Gd_2O_3–MoO_3/WO_3–B_2O_3-Based Glasses

The glass formation in the Gd_2O_3–MoO_3–B_2O_3 system is noted, because multiferroic (ferroelasticity and ferroelectricity) β′-$Gd_2(MoO_4)_3$ crystals with a unique

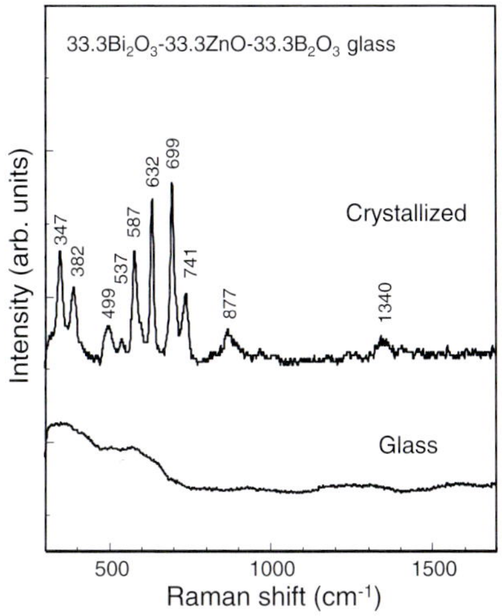

Figure 11.8 Raman scattering spectra at room temperature for the base glass and crystallized (530 °C, 3 h) samples in $33.3Bi_2O_3$–33.3ZnO–$33.3Bi_2O_3$ [43].

morphology are formed through the crystallization of Gd_2O_3–MoO_3–B_2O_3 glasses. For example, a glass with the composition of $21.25Gd_2O_3$–$63.75MoO_3$–$15B_2O_3$ (mol%) ($T_g = 540\,°C$, $T_p = 585\,°C$) (designated here as GM15B glass) was prepared by using the conventional melt-quenching method [44]. The temperature dependence of the magnetic susceptibility (χ_M) and inverse susceptibility (χ_M^{-1}) for the GM15B glass is shown in Figure 11.9, in which the magnetization is

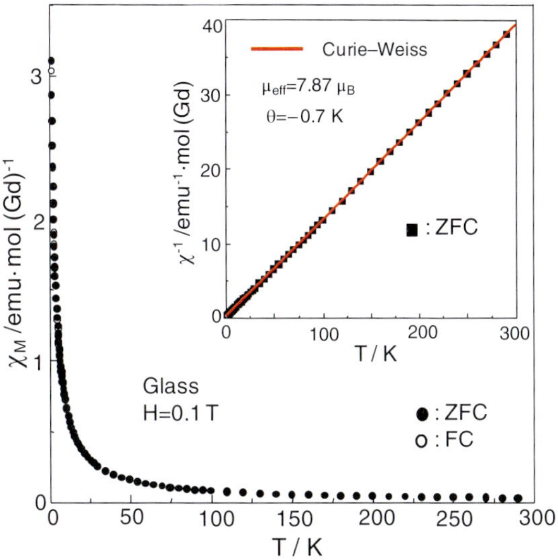

Figure 11.9 Temperature dependence of the magnetic susceptibility (χ_M) and inverse susceptibility (χ_M^{-1}) for $21.25Gd_2O_3$–$63.75MoO_3$–$15B_2O_3$ glass under the zero-field cooled (ZFC) and field-cooled (FC) conditions [44].

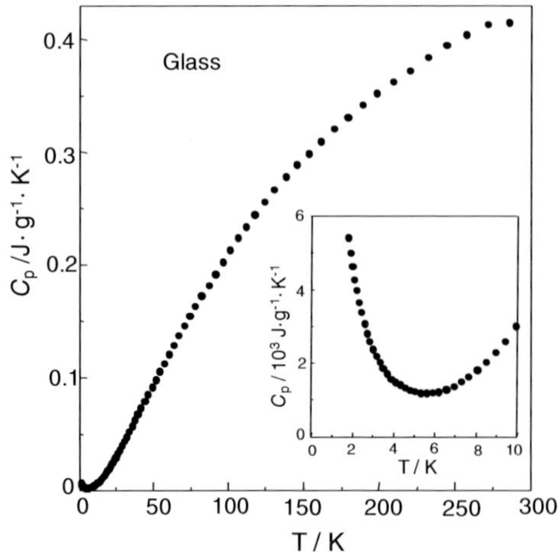

Figure 11.10 Temperature dependence of specific heat (C_p) in the zero magnetic field for $21.25Gd_2O_3$–$63.75MoO_3$–$15B_2O_3$ glass. The temperature region is 1.8–300 K [44].

reduced against the amount of 1 mol Gd ions. As can be seen in Figure 11.9, zero-field cooled (ZFC) and field-cooled (FC) magnetic susceptibilities have no differences and any anomaly is not observed. The magnetic susceptibility obeys the Curie–Weiss law, giving the effective magnetic moment of $\mu_{eff} = 7.87\,\mu_B$ and the Weiss constant of $\theta = -0.7$ K. It should be pointed out that the experimental value of $\mu_{eff} = 7.87\,\mu_B$ is very close to the value of Gd^{3+} ions with the 4f electron configuration of $^8S_{7/2}$, that is, $7.94\,\mu_B$. The temperature dependence of specific heat (C_p) in a zero magnetic field for GM15B glass is shown in Figure 11.10. It is seen that the specific heat decreases with decreasing temperature, indicating obviously the contribution of lattice vibrations in constituent ions and structural units of GM15B glass. In the temperature below $T = 5$ K, the specific heat increases with decreasing temperature. This excess specific heat would be arising from the magnetic entropy of Gd^{3+} ions. Gd^{3+} has the largest spin of $S = 7/2$ among all RE^{3+} ions and thus is expected to have a large magnetic entropy, that is, $S_{mag} = R\ln(2S+1) = 17.29\,\mathrm{J/(mol\ K)}$,, where R is the gas constant. The values of C_p/T were plotted as a function of temperature, and information on the magnetic entropy is obtained through the equation of $S_{mag} = \int(C_p/T)\mathrm{d}T$. In the temperature range of 1.8–5 K for these data, any peak such as λ-type anomaly was not observed, that is, it is considered that Gd^{3+} ions in GM15B glass are paramagnetic at least down to $T = 1.8$ K. These results shown in Figures 11.9 and 11.10 clearly demonstrate that Gd^{3+} ions in $21.25Gd_2O_3$–$63.75MoO_3$–$15B_2O_3$ glass exist as paramagnetic ions and basically magnetic clusters consisting of Gd^{3+} ions are not formed. In other words, Gd^{3+} ions might be distributed homogeneously and randomly in GM15B glass, even though a large amount (21.25 mol%) of Gd_2O_3 is included. The glass formation in the Gd_2O_3–MoO_3–B_2O_3 system indicates that MoO_3–B_2O_3-based glasses are good hosts for the homogeneous solubility of large amounts of Gd_2O_3. It has been also reported that La_2O_3 plays an important role in the glass formation of a homogeneous network in La_2O_3–MoO_3–B_2O_3 glasses (e.g., $25La_2O_3$–$50MoO_3$–$25B_2O_3$) and Mo–O–La linkages might be formed [45].

A clear glass is obtained for the composition of $25Gd_2O_3$–$65WO_3$–$15B_2O_3$ in which the content of B_2O_3 is only 15 mol% as similar to the case of Gd_2O_3–MoO_3–B_2O_3 system [46]. WO_3 is not classified into the so-called glass-forming oxide, and thus basically glasses containing WO_3 are prepared through the combination with glass-forming oxides such as B_2O_3 and P_2O_5. Furthermore, because there is no glass-forming region in the binary WO_3–B_2O_3 system, at least the third component must be added to the binary WO_3–B_2O_3 system in order to synthesize WO_3–B_2O_3 based glasses. The addition of both WO_3 and Gd_2O_3 to B_2O_3 enables to form glasses, that is, the glass formation in the ternary Gd_2O_3–WO_3–B_2O_3 system. However, it should be also pointed out that glasses are not formed in the binary Gd_2O_3–B_2O_3 system. The information on the glass-forming tendency suggests that it might be difficult to create rigid network structure with strong chemical bonds in the ternary Gd_2O_3–WO_3–B_2O_3 system, in particular with small B_2O_3 contents, that is, strong chemical bonds of W–O–B and Gd–O–B would not be expected in the Gd_2O_3–WO_3–B_2O_3 glasses. In other words, it would be difficult to create strong network structures with a long distance scale in the combination of Gd_2O_3, WO_3, and B_2O_3. However, as reported by Taki *et al.* [46], a clear glass is obtained for the composition of $25Gd_2O_3$–$65WO_3$–$15B_2O_3$ in which the content of B_2O_3 is only 15 mol%. Recently, Wright [47] pointed out that a floppy network also increases the probability of glass formation, via a smaller difference in strain energy between the crystalline and vitreous phases than would be the case for a stressed-rigid network. The glass formation in the ternary Gd_2O_3–WO_3–B_2O_3 system might be such a case, that is, a floppy network system. Furthermore, it should be pointed out that the nanoscale (2–3 nm) phase separation, that is, WO_3-rich phase and B_2O_3-rich phase, has been observed in the initial stage of crystallization in $25Gd_2O_3$–$40WO_3$–$35B_2O_3$ glass and $Gd_2(WO_4)_3$ crystals are formed easily through the heat treatment at temperatures above the glass transition temperature in Gd_2O_3–WO_3–B_2O_3 glasses [46].

The stability of borate glass structures depends on the delocalization of formal charges of B_nO_m anions as a result of their polymerization. Generally, the glass formation is realized through the one-, two-, or three-dimensional network structure, that is, polymerization of structural units and delocalization of formal charges of structural units. The formation of nonbridging oxygen means the increase in the localization of charges of structural units and thus the decrease in the polymerization (or delocalization). The formation and breakdown of the polymerization (network structure) among structural units provide the glass-forming region, for example, glasses for xLi$_2$O–(100–x)B_2O_3 with x(mol%) = 0–57. It is recognized that, according to the Lux-Flood acid–base concept, the polymerized borates are often less stable for high-valence and small-sized cations. Cations of higher valence such as MoO_3, WO_3, Nb_2O_5, and Ta_2O_5 can form rather rigid coordination polyphedra with oxygen atoms. Indeed, as reported by Imaoka [48], any wide glass-forming region has not been reported in these binary borate systems.

11.5.4
Li_2O–Nb_2O_5–P_2O_5 Glasses

Lithium phosphates such as $LiFePO_4$, $LiVOPO_4$, and $Li_{1+x}Al_xTi_{2-x}(PO_4)_3$ have been proposed to be potential candidates for use as cathode materials or electrolytes for the next generation of rechargeable lithium ion batteries (LIB) and are commonly synthesized via solid-state reaction, coprecipitation, hydrothermal method, and so on. Very recently, some lithium phosphates have been synthesized successfully through the crystallization processing of corresponding precursor glasses, and Li^+ ion conductivities have been examined [49–51]. On the other hand, Sakurai and Yamaki [52] reported the recharge-ability and good cycle-ability of V_2O_5–P_2O_5 glasses as cathodes for LIBs, which is the first report on the glass-based cathodes for LIBs. And then, various glasses having a possibility for LIBs such as Li_2O–FeO–P_2O_5 have been synthesized. It is of interest to study the glass formation of oxide glasses containing a large amount of Li_2O and their electrical properties.

Here, the glass formation in the Li_2O–Nb_2O_5–P_2O_5 system and electrical properties of the glasses obtained are described [53]. The combination of Nb_2O_5 and P_2O_5 is known to result in the formation of thermally stable glasses, and structure, optical properties, and crystallization behaviors of various Nb_2O_5–P_2O_5-based glasses have been studied. Okada *et al.* [53] synthesized the glasses with the compositions of $x$$Li_2O$–$(70-x)$$Nb_2O_5$–30$P_2O_5$, $x = 30$–60, by using the conventional melt-quenching method and examined Li^+ ion conductivities of glasses and glass-ceramics to clarify whether the glasses and glass-ceramics prepared have a potential as Li^+ conductive electrolytes or not. All melt-quenched samples with x = 30–60 show only broad halo patterns in XRD patterns. The DTA patterns for bulk and powdered samples of 60Li_2O–10Nb_2O_5–30P_2O_5 glass are shown in Figure 11.11, as an example. Endothermic peaks corresponding to the glass transition and exothermic peaks due to the crystallization are clearly observed in both samples, providing the values of $T_g = 429\,°C$, $T_{p1} = 541$ °C, and $T_{p2} = 607\,°C$ in the glass.

The temperature dependence (25–200 °C) of the electrical conductivity (σ) for 60Li_2O–10Nb_2O_5–30P_2O_5 glass containing a large Li_2O content of 60 mol% is shown in Figure 11.12 together with the data for the crystallized (544 and 612 °C) samples. It is found that the glass shows the value of $\sigma = 2.35 \times 10^{-6}$ S/cm at room temperature and the activation energy (E_a) of 0.48 eV for Li^+ ion mobility. It is noted that the data shown in Figure 11.12 indicate that the electrical conductivity of Li^+ ions decreases due to the crystallization and the activation energy tends to increase slightly due to the crystallization. The values of σ and E_a for $x$$Li_2O$–$(70-x)$$Nb_2O_5$–30$P_2O_5$ glasses are summarized in Table 11.4. It is seen that the electrical conductivity of the glasses increases monotonously with

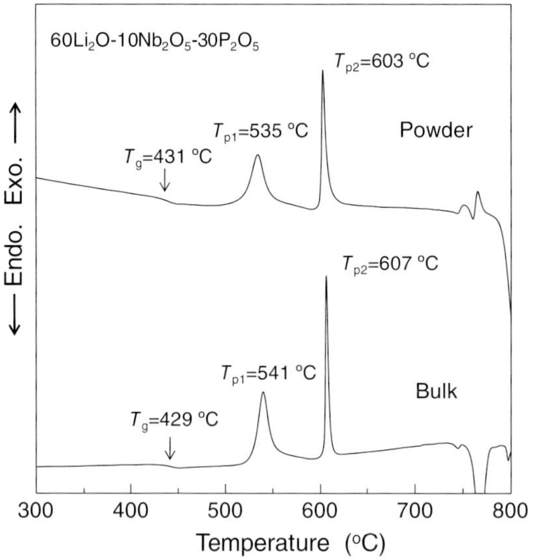

Figure 11.11 DTA patterns for the bulk and powdered samples of $60Li_2O–10Nb_2O_5–30P_2O_5$ glass. T_g and T_p are the glass transition and crystallization peak temperatures, respectively. Heating rate was 10 K/min [53].

increasing Li_2O content and the activation energy for Li^+ ion mobility is $E_a = 0.48–0.58$ eV, being almost independent on the Li^+ content.

Recently, Honma *et al.* [54] synthesized $xNa_2O–(70–x)Nb_2O_5–30P_2O_5$ glasses and reported the values of $\sigma = 5.6 \times 10^{-8}$ S/cm at room temperature and $E_a = 0.54$ eV for Na^+ ion mobility. The glass of $Na_2O–FeO–P_2O_5$ having the

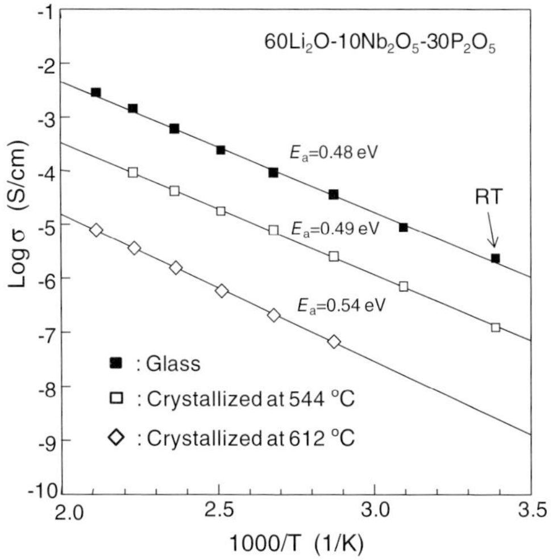

Figure 11.12 Temperature dependence of electrical conductivities (σ) for the base glass and crystallized (at 544 and 612 °C for 3 h) samples in $60Li_2O–10Nb_2O_5–30P_2O_5$ glass [53].

Table 11.4 Values of glass transition temperature T_g, the first crystallization peak temperature T_{p1}, density *d*, refractive index *n*, electrical conductivity σ, and activation energy of electrical conductivity E_a for $xLi_2O-(70-x)Nb_2O_5-30P_2O_5$ bulk glasses. *d*, *n*, and σ are the values at room temperature [53].

Glass	T_g (°C)	T_{p1} (°C)	d (g/cm^3)	n	σ (S/cm)	E_a (eV)
$x = 30$	626	791	3.57	1.92	1.27×10^{-7}	0.49
$x = 35$	592	839	3.48	1.89	3.38×10^{-7}	0.54
$x = 40$	559	830	3.36	1.84	5.52×10^{-7}	0.52
$x = 45$	524	836	3.26	1.80	9.56×10^{-7}	0.47
$x = 50$	499	775	3.11	1.75	1.38×10^{-6}	0.57
$x = 55$	468	736	2.99	1.71	2.71×10^{-6}	0.58
$x = 60$	429	541	2.83	1.65	2.35×10^{-6}	0.48

composition of $Na_2FeP_2O_7$ was synthesized, and the glass-ceramics with $Na_2FeP_2O_7$ crystals were proposed to be new cathode candidates for rechargeable sodium ion second battery (NaBs) [55]. The glasses in the $SnO-P_2O_5$ were also synthesized, and it has been proposed that $72SnO-28P_2O_5$ glass is a promising candidate as new anode materials that realizes LIBs with a high energy density, which can be used over a wide temperature range [56]. It should be pointed out that the total positive charge of Sn^{2+} ions in $72SnO-28P_2O_5$ glass anode is compensated by a reaction with lithium during the first charge by forming nanoscale Sn metal crystals [56]. Furthermore, it has been proposed that $72SnO.28P_2O_5$ glass is a promising candidate of new anode active materials that realizes high energy density NaBs [57].

11.5.5
TeO_2-Based Glasses

Tellurium oxide (TeO_2)-based glasses have attracted much attention due to their unique combination of properties such as chemical durability and stability, low phonon energy, low melting temperature, broad optical transmission window including mid infrared range, high dielectric constant, and high linear and nonlinear refractive indices. They are considered for application in photonic devices such as optical switches, tunable filters, optical sensors, and broadband optical amplifiers. The structure of TeO_2-based glasses is also of interest, because there are two types of basic structural units, that is, TeO_4 trigonal bipyramid (tbp) and TeO_3 trigonal pyramid (tp). It is known that pure TeO_2 does not become a glass under usual quenching rates and the addition of other elements is needed to form TeO_2-based bulk glasses. So far, many binary glasses such as Li_2O-TeO_2 or $ZnO-TeO_2$ have been synthesized. Even for the ternary systems, many glasses having excellent thermal resistance against crystallization such as $Li_2O-Na_2O-TeO_2$ and $K_2O-WO_3-TeO_2$ have been reported. Wang *et al.* [58] reported the

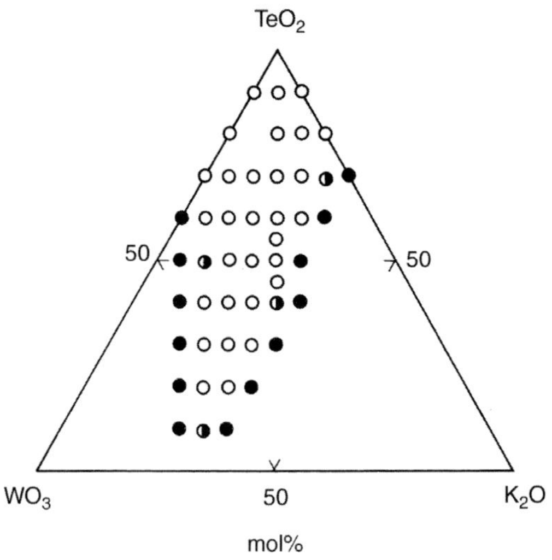

Figure 11.13 Glass-forming region in the K_2O–WO_3–TeO_2 system. ○: glass, ◑: partially crystallized, ●: crystallized [59].

system of Na_2O–ZnO–TeO_2 with rare earth ions is a good candidate for optical amplifier glasses.

The glass-forming region in the K_2O–WO_3–TeO_2 system is shown in Figure 11.13 [59]. The glass formation is observed in the compositions with 20–90 mol% TeO_2. It is proposed from Raman scattering spectra that the network structure is basically composed of TeO_4, TeO_3, and WO_4 (and WO_6) units, and the breaking of Te—O—Te bonds occurs due to the addition of K_2O and further Te—O—W bonds are formed due to the substitution of W^{6+} for Te^{4+}. The melting temperatures of TeO_2 and WO_3 crystals are 452 °C and 1473 °C, respectively, implying that W—O bonds in K_2O–WO_3–TeO_2 glasses are much stronger than Te—O bonds. These structural and bonding features would cause the compositional dependence of T_g in these glasses, that is, the substitution of K_2O for TeO_2 gives a rapid decrease in T_g, but the substitution of WO_3 for TeO_2 provides the increase in T_g. Furthermore, the formation of Te—O—W bonds might retard the rearrangements of Te^{4+} or W^{6+} structures necessary for crystallization. Indeed, high thermal stable glasses against crystallization are obtained in the composition of 60–70 mol% TeO_2. The glass of $10K_2O$–$10WO_3$–$80TeO_2$ shows the values of $T_g = 308$ °C, crystallization onset temperature $T_x = 456$ °C, $d = 5.31$ g/cm^3, and $n = 2.04$, and the glass of $10K_2O$–$30WO_3$–$60TeO_2$ exhibits the values of $T_g = 361$ °C, $d = 5.61$ g/cm^3, and $n = 2.05$.

The heat capacity C_p, in the glass transition region for the $10K_2O$–$10WO_3$–$80TeO_2$ sample is shown in Figure 11.14 [59]. A jump in C_p is clearly observed at the transformation from the glassy state to the super-cooled liquid state. The relaxation overshoot in the heat capacity is also clearly observed. From the C_p versus temperature curve, the heat capacities of glass C_{pg}, and of super-cooled

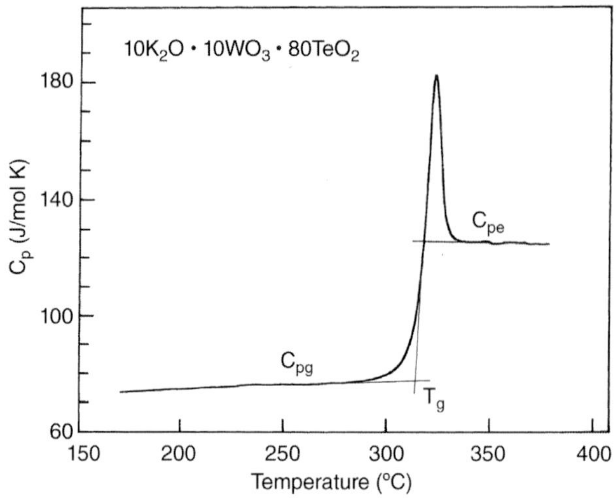

Figure 11.14 Heat capacity C_p, in the glass transition region for the 10 K_2O–10WO_3–80TeO_2 sample. C_{pg} and C_{pe} are the heat capacities of the glasses and supercooled liquids, respectively [59].

liquids C_{pe}, are estimated to be $C_{pg}(200\,°C) = 73$ J/(mol K) and $C_{pe}(350\,°C) =$ 121 J/(mol K). It should be pointed out that the value of $C_{pe}/C_{pg} = 1.66$ obtained in the 10 K_2O–10WO_3–80TeO_2 sample is extremely large. The value of $C_{pg} = 73$ J/(mol K) for the glass was compared with the classical Dulong–Petit value of 3Rn, and the value of $n = 2.93$ is obtained. This value is very close to the number of atoms (3.01) in the constituent oxides of K_2O, WO_3, and TeO_2. As proposed by Angell [8], glass-forming liquids having small C_{pe}/C_{pg} are called strong liquid, while those showing large C_{pe}/C_{pg} are called fragile liquids. It is well recognized that SiO_2 and GeO_2 having small $C_{pe}/C_{pg} = 1.1$ are strong glass-forming liquids. SiO_2 and GeO_2 glasses have generally tetrahedrally coordinated network structures with strong covalent bonds that are expected to experience relatively little disruption during heating. It is obvious that the 10 K_2O–10WO_3–80TeO_2 sample having the value of $C_{pe}/C_{pg} = 1.66$ is included in the category of fragile liquids, implying that an inferred rapid breakdown of their configurational structure occurs with increasing temperature. The temperature dependence of heat capacity for paratellurite TeO_2 crystal, annealed glass, and a crystallized sample of 10Li_2O–10Na_2O–80TeO_2 is shown in Figure 11.15 [60]. The C_p value becomes constant at temperatures above ~400 K, which would be the reason why TeO_2-based glasses hold a classical Dulong–Petit heat capacity at temperatures that are similar to T_g.

Because TeO_2-based glasses exhibit high dielectric constant and high linear and nonlinear refractive indices, it is of importance to clarify the temperature dependence of refractive index. The values of refractive indices for RO–TeO_2 glasses (RO = MgO, BaO, and ZnO) glasses were measured as a function of temperature, and as an example, the data for ZnO–TeO_2 glasses are shown in Figure 11.16 [61]. All glasses show the almost linear increase in the refractive

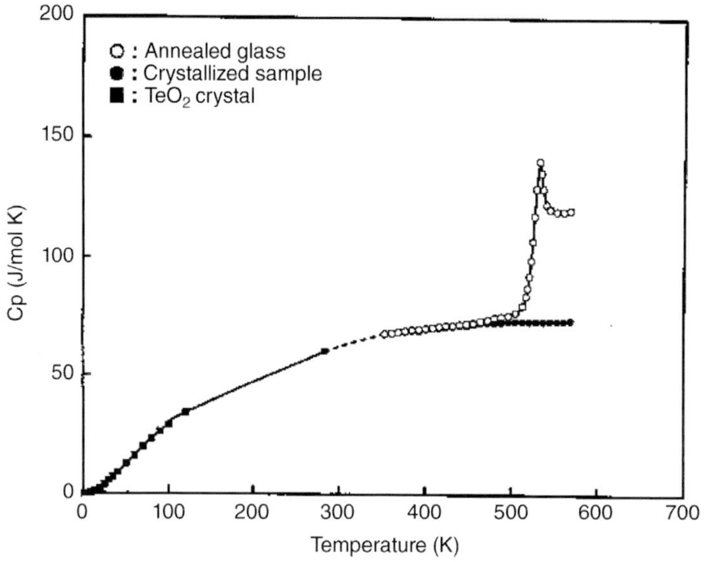

Figure 11.15 Temperature dependence of heat capacity for (■) paratellurite TeO_2 crystal, (○) annealed glass, and (●) a crystallized sample of $10Li_2O$–$10Na_2O$–$80TeO_2$ [60].

index with increasing temperature (30–150 °C). The temperature dependence of the refractive indices for these glasses, dn/dT, is evaluated using the least square fitting method and the values obtained are given in Table 11.5. It is noted that the glasses show large temperature dependences of $dn/dT = 8.90$–$9.69 \times 10^{-5}\,K^{-1}$ for MgO–TeO_2, $dn/dT = 7.32$–$8.54 \times 10^{-5}\,K^{-1}$ for BaO–TeO_2, and $dn/dT = 8.75$–$8.97 \times 10^{-5}\,K^{-1}$ for ZnO–TeO_2. Prod'homme [62] tried to explain the

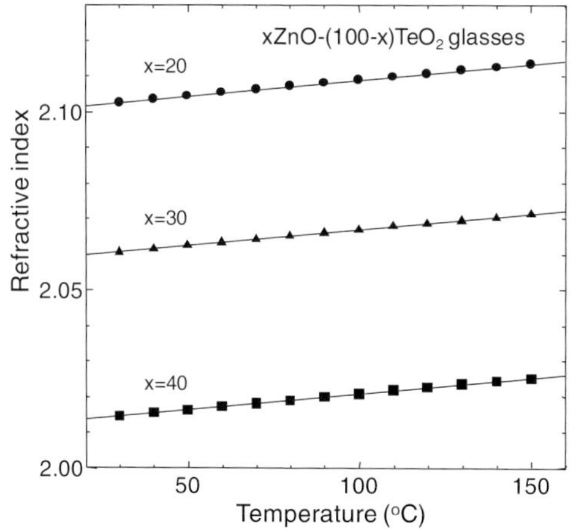

Figure 11.16 Temperature dependence of refractive indices (n) for xZnO–(100–x)TeO_2 glasses in the temperature range of 30–150 °C. The uncertainty of n is $\pm 3 \times 10^{-4}$ [61].

Table 11.5 Values of temperature dependences of refractive index (dn/dT), mean volume thermal expansion coefficient (β), and electronic polarizability (ϕ) calculated by Eq. (11.7) in the temperature range of 50–150 °C for xRO–(100–x)TeO_2 glasses [61].

x (mol%)	$\frac{dn}{dT}$ (1/K)	β (1/K)	ϕ (1/K)
MgO			
10	9.18×10^{-5}	3.30×10^{-5}	8.40×10^{-5}
15	8.90×10^{-5}	3.39×10^{-5}	8.56×10^{-5}
20	9.68×10^{-5}	3.09×10^{-5}	8.70×10^{-5}
BaO			
10	8.54×10^{-5}	3.87×10^{-5}	8.60×10^{-5}
15	7.86×10^{-5}	4.09×10^{-5}	8.63×10^{-5}
20	7.32×10^{-5}	4.36×10^{-5}	8.81×10^{-5}
ZnO			
20	8.97×10^{-5}	3.44×10^{-5}	8.58×10^{-5}
30	8.75×10^{-5}	3.35×10^{-5}	8.67×10^{-5}
40	8.84×10^{-5}	3.21×10^{-5}	8.96×10^{-5}

temperature dependence of refractive index of glasses, that is, thermo-optic coefficient, by using the following equation obtained by differentiating the Lorentz–Lorenz equation:

$$\frac{dn}{dT} = \frac{(n^2-1)(n^2+2)}{6n}(\phi - \beta), \qquad (11.7)$$

where the term ϕ is the polarizability coefficient indicating the temperature dependence of the electronic polarizability α and the term β is the volume expansion coefficient.

$$\phi = \frac{1}{\alpha}\frac{d\alpha}{dT} \qquad (11.8)$$

From Eq. (11.7), if the contribution of volume expansion β is large compared with the contribution of polarizability, that is, $\phi < \beta$, the sign of dn/dT would be negative. Contrary, if $\phi > \beta$, the sign of dn/dT would be positive. For RO–TeO_2 glasses, the contribution of polarizability ϕ on thermo-optic coefficient was calculated using Eq. (11.7), where the values of the refractive indices at 50 °C were used as the values of n in Eq. (11.7), and the results are given in Table 11.5. The values of $\phi = 8.40–8.70\times10^{-5}\,K^{-1}$ for MgO–TeO_2 glasses, $\phi = 8.60–8.81\times10^{-5}$ K^{-1} for BaO–TeO_2 glasses, and $\phi = 8.58–8.96\times10^{-5}\,K^{-1}$ for ZnO–TeO_2 glasses are obtained. It is noted that the contribution of polarizability is almost the same, being independent on the kind and content of RO. Furthermore, it should be pointed out that the condition of $\phi > \beta$ is held in all glasses examined in this study. It is suggested that the electronic polarizability has an important contribution on thermo-optic coefficients, but the contribution of volume thermal expansion on thermo-optic coefficients is not ignored.

11.5.6 Oxyfluoride Glasses

Fluoride glasses and crystals are desirable hosts for optically active ions because of their transparencies in the range from the near ultra violet (UV) to the middle of infrared (IR), low phonon energy, and large RE^{3+} ion solubility. Generally, fluoride materials have, however, drawbacks to chemical and thermal stabilities, limiting in the use of optical devices. For this problem, Wang and Ohwaki [63] gave a solution. That is, they proposed oxyfluoride-based glass-ceramics consisting of fluoride nanocrystals. RE^{3+} ions are incorporated into fluoride nanocrystals being embedded in oxide-based glass matrices, and thus, oxyfluoride-based glass-ceramics can maintain good chemical and thermal stabilities. After their first proposal, numerous researches on the fabrication and optical properties of transparent RE^{3+}-doped oxyfluoride-based crystallized glasses have been reported [64–66]. Usually, oxyfluoride glass-ceramics are fabricated through the crystallization of oxyfluoride glasses, and the size and dispersion state of fluoride nanocrystals are controlled by glass composition and heat treatment (i.e., crystal nucleation and growth) condition. The synthesis of oxyfluoride glasses has received, therefore, much attention. In case of oxyfluoride glass-ceramics with CaF_2 nanocrystals, a typical glass system is CaF_2–CaO–Al_2O_3–SiO_2 and the content of CaF_2 is usually limited to around 30 mol% [64]. In aluminosilicate-based oxyfluoride glasses such as $25CaF_2$–5CaO–$20Al_2O_3$–$50SiO_2$, the network structure constructed by Al_2O_3 and SiO_2 is rigid and stable thermally and consequently depresses the large size growth of CaF_2 crystals [26]. In particular, it is well recognized that the increase in the viscosity of the residual glassy phase (i.e., Al_2O_3–SiO_2-based oxide glasses) hinders the diffusion of Ca^{2+} and F^- ions, consequently keeping nanoscale sizes of CaF_2. For transparent oxyfluoride glass-ceramics with LaF_3 nanocrystals, $15LaF_3$–$20Na_2O$–$25Al_2O_3$–$40SiO_2$ glass is synthesized [67].

An oxyfluoride glass of $50BaF_2$–$25Al_2O_3$–$25B_2O_3$ (mol%) with a large fraction of fluorine, that is, $F/(F+O)=0.4$, was synthesized by using the conventional melt-quenching method in order to develop new glass-ceramics containing nonlinear optical $BaAlBO_3F_2$ crystals [68]. This glass has the values of $T_g = 488\,°C$, $T_p = 566\,°C$, $d = 4.02\,g/cm^3$, and $n = 1.564$ at $\lambda = 632.8\,nm$. It is noted that the density of the glass is much smaller (i.e., about 10%) compared with the value of $d = 4.47\,g/cm^3$ for $BaAlBO_3F_2$ single crystal, indicating that the glass has a large open structure compared with $BaAlBO_3F_2$ crystal. The optical transmittance spectrum for the glass with a thickness of 1 mm is shown in Figure 11.17. The UV cutoff wavelength is ~200 nm and the absorption edge is ~290 nm. The glasses of xEu_2O_3–$50BaF_2$–$25Al_2O_3$–$25B_2O_3$ ($x = 0$–10) (mol) and yEu_2O_3–$1Tb_4O_7$–$50BaF_2$–$25Al_2O_3$–$25B_2O_3$ ($y = 0$, 0.5) (mol) were synthesized, and photoluminescence properties such as quantum yield (η), lifetime (τ), and concentration quenching effect of Eu^{3+} ions were clarified [69]. The data on the quantum yield are shown in Figure 11.18 as a function of Eu_2O_3 content. The glasses show an excellent red luminescence with extremely high quantum yield of $\eta = 97\%$ in the visible region at the excitation of $\lambda = 396\,nm$ and a long lifetime of $\tau = 3.29$

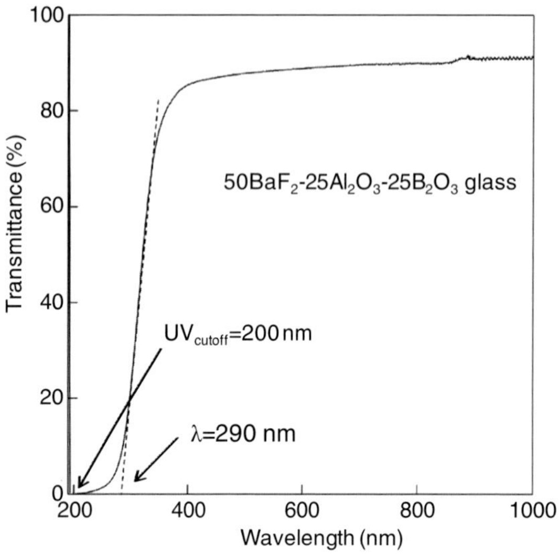

Figure 11.17 Optical absorption spectrum at room temperature for $50BaF_2$–$25Al_2O_3$–$25B_2O_3$ glass [68].

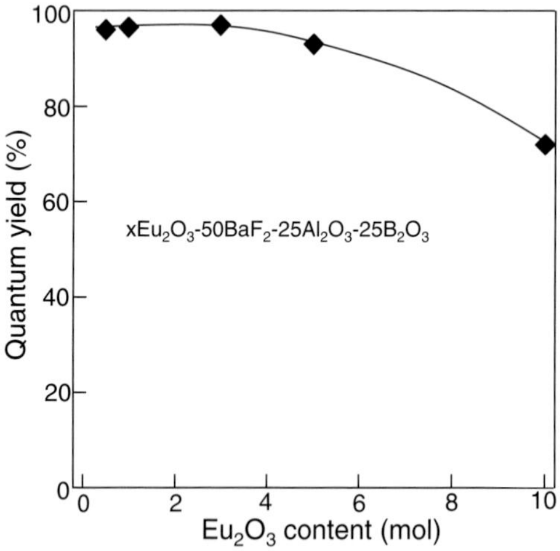

Figure 11.18 Total quantum yield for Eu^{3+} in the visible region as a function of content of Eu_2O_3 in xEu_3O_3–$50BaF_2$-$25Al_2O_3$-$25B_2O_3$ ($x = 0.5$–10) glasses. The wavelength of the excitation light was 396 nm. The line is drawn as a guide for the eye [69].

ms for the emission ($\lambda = 612$ nm) of Eu^{3+} ions at the ^5D level. The effect of concentration quenching in these glasses is very small, for example, $\eta = 72\%$ even for the glass with $10Eu_2O_3$. The values of $\eta = 52\%$ for the green emission of Tb^{3+} ions and $\eta = 81\%$ for the yellow emission of Eu^{3+}–Tb^{3+} codoped ions were also achieved in the glasses. It is proposed that the structure of $1Er_2O_3$–$50BaF_2$–$25Al_2O_3$–$25B_2O_3$ glass is composed of BO_3 and $Al(O,F)_x$ units and its nanoscale structure is similar to the $BaAlBO_3F_2$ crystalline phase [70]. In the studies on PL properties for Eu_2O_3–Tb_4O_7–BaF_2–Al_2O_3–B_2O_3 glasses, Shinozaki *et al.* [69] have proposed that Eu^{3+} and Tb^{3+} ions are dispersed homogeneously in these oxyfluriode glasses and the degree of their site asymmetry (i.e., the electric field gradient) is widely distributed due to the coordination of both F^- and O^{2-} ions. Furthermore, the emission spectra of Er^{3+} ions at 1.5 μm in the glass and glass-ceramics with $BaAlBO_3F_2$ crystals show very broad peaks. The oxyfluoride glasses with high BaF_2 and Al_2O_3 contents have a high potential as hosts for optical device applications such as phosphors and broadband optical amplifiers.

Microfabrication of glass materials has found increasingly more applications in optoelectronics, telecommunications, and photonic devices such as optical gratings and waveguides, and laser irradiation to glass has received considerable attention as a new tool of spatially selected microfabrications in glasses. The spatially selected patterning of crystals in glasses by laser irradiations, that is, laser-induced crystallization, is also extremely interesting and important in glasses. This is really a new challenge in the science and technology of glasses and glass-ceramics [1,71,72]. The laser-induced crystallization was applied to the patterning of CaF_2 crystals in oxyfluoride glasses [66]. Figure 11.19 shows the polarized optical microscope (POM) (a) and confocal scanning laser microscope (CSLM) (b) photographs and micro-PL spectrum (c) for the line obtained by irradiations of continuous-wave Yb: YVO_4 fiber lasers (wavelength $\lambda = 1080$ nm, laser power $P = 1.0$ W, laser scanning speed $S = 2$ μm/s) in $0.5ErF_3$–$3NiO$–$20CaF_2$–$16NaF$–$4CaO$–$20Al_2O_3$–$10B_2O_3$–$30SiO_2$ glass with $T_g = 541\,°C$ and $T_p = 613\,°C$ [66]. A homogeneous line with the width of 4 μm and the height of 0.3 μm is patterned at the glass surface. The morphology and dispersion state of CaF_2 nanocrystals in the line were examined from transmission electron microscope (TEM) observations combined with focused ion beam (FIB) sample preparations, and it was found that the size of CaF_2 nanocrystals present in the region close to the center of lines was larger (~200 nm) than that (~50 nm) in the region close to the edge of lines. The PL spectrum for the line part shows the clear emissions corresponding to the f–f transitions of $^2H_{11/2} \rightarrow {}^4I_{15/2}$ at ~525 nm, $^4S_{3/2} \rightarrow {}^4I_{15/2}$ at 540–570 nm, and $^4F_{9/2} \rightarrow {}^4I_{15/2}$ at ~660 nm for Er^{3+} ions. The glassy part, that is, nonlaser irradiated part, shows broad PL peaks as shown in Figure 11.19. These results indicate that Er^{3+} ions are incorporated into CaF_2 crystals formed by laser irradiations. Shinozaki *et al.* [68] succeeded in patterning of nonlinear optical $BaAlBO_3F_2$ single crystal lines by laser irradiations in $3NiO$–$50BaO$–$25Al_2O_3$–$25B_2O_3$ oxyfuoride glass. It should be pointed out that $50BaO$–$25Al_2O_3$–$25B_2O_3$ oxyfuoride glass shows the bulk nanocrstallization in the crystallization in an electric furnace. Their results, therefore, suggest that the laser

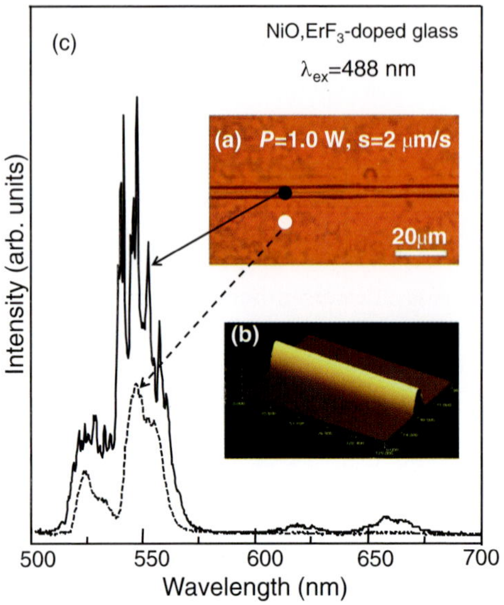

Figure 11.19 Polarized optical microscope (a) and confocal scanning laser microscope (b) photographs and micro-photoluminescence spectrum (c) of Er^{3+} ions (λ_{ex} = 488 nm) for the line patterned by Yb:YVO_4 fiber lasers in $0.5ErF_3$–3NiO–$20CaF_2$–16NaF–4CaO–$20Al_2O_3$–$10B_2O_3$–$30SiO_2$ glass [66].

patterning of CaF_2 single crystal lines might be possible through a well design of glass system and composition in oxyfluoride glasses, in particular through the design of oxide component parts.

11.6 Conclusion

Numerous oxide glasses (glass formation) with different systems and compositions, that is, the design of chemical, physical, thermal, and mechanical properties, have been synthesized from the scientific interests and practical applications. For the glass formation, crystallization in super-cooled liquids must be avoided kinetically, and thus, it is of importance to understand the relationship between the glass composition and crystallization behavior. From this point of view, the basic scenario for glass formation, the fragility concept, and the nanoscale structure in oxide glasses were described. Some interesting glass-forming oxide systems such as BaO–TiO_2–SiO_2/GeO_2, ZnO–Bi_2O_3–B_2O_3, Gd_2O_3–MoO_3/WO_3–B_2O_3, K_2O–WO_3–TeO_2, and oxyfluoride glasses were focused, and their features relating to the glass formation were discussed. The glass formation and crystallization are becoming more and more important as a novel technique for the development of new functional hybrid materials and also as a step for a deep understanding of nanoscale heterogeneous structure in glasses.

Acknowledgments

The author thanks Prof. R. Sato, Prof. Y. Benino, Prof. T. Fujiwara, Prof. T. Honma, Prof. Y. Takahashi, Dr R. Ihara, Dr K. Shinozaki, and Prof. V. Dimitrov for their guides and supports in our research on glasses and crystallization. The author also thanks some financial supports from the Grant-in-Aid for Scientific Research from the Ministry of Education, Science, Sports, Culture, and Technology, Japan (e.g., No. 23246114), and from the scientific programs in Nagaoka University of Technology such as Program for High Reliable Materials Design and Manufacturing.

References

1 Komatsu, T. (2015) *J. Non Cryst. Solids*, **428**, 156.
2 Zanotto, E.D., Tsuchida, J.E., Schneider, J.F., and Eckert, H. (2015) *Int. Mater. Rev.*, **60**, 376.
3 Turnbull, D. and Cech, R.E. (1950) *J. Appl. Phys.*, **21**, 804.
4 Spaepen, F. (1975) *Acta Metal.*, **23**, 729.
5 Kuribayashi, K. and Ozawa, S. (2006) *J. Alloys Compd.*, **408–412**, 266.
6 Fokin, V.M., Zanotto, E.D., Yuritsyn, N.S., and Schmelzer, J.W.P. (2006) *J. Non Cryst. Solids*, **352**, 2681.
7 Burger, L.L. and Weinberg, M.C. (2001) *Phys. Chem. Glasses*, **42**, 184.
8 Angell, C.A. (1991) *J. Non Cryst. Solids*, **131–133**, 13.
9 Komatsu, T., Aida, K., Honma, T., Benino, Y., and Sato, R. (2002) *J. Am. Ceram. Soc.*, **85**, 193.
10 Sun, K.H. (1947) *J. Am. Ceram. Soc.*, **30**, 277.
11 Dimitrov, V. and Komatsu, T. (2005) *Phys. Chem. Glasses*, **46**, 521.
12 Dimitrov, V. and Komatsu, T. (2008) *Phys. Chem. Glasses: Eur. J. Glass Sci. Technol. B*, **49**, 97.
13 Dimitrov, V. and Komatsu, T. (2012) *J. Solid State Chem.*, **196**, 574.
14 Greaves, G.N. (1985) *J. Non Cryst. Solids*, **71**, 203.
15 Elliot, S.R. (1992) *Europhys. Lett.*, **19**, 201.
16 Sokolov, P., Kisliuk, A., Soltwisch, M., and Quitmann, D. (1992) *Phys. Rev. Lett.*, **69**, 1540.
17 Rouxel, T. (2007) *J. Am. Ceram. Soc.*, **90**, 3019.
18 Dargaud, O., Cormier, L., Menguy, N., and Patriarche, G. (2012) *J. Non Cryst. Solids*, **358**, 1257.
19 Tokuda, Y., Oka, T., Takahashi, M., and Yoko, T. (2011) *J. Ceram. Soc. Jpn.*, **119**, 909.
20 Vargheese, K.D., Tandia, A., and Mauro, J.C. (2010) *J. Chem. Phys.*, **132**, 194501.
21 Zhang, Y., Yang, G., and Yue, Y. (2013) *J. Am. Ceram. Soc.*, **96**, 3035.
22 Takahashi, Y., Osada, M., Masai, H., and Fujiwara, T. (2009) *Phys. Rev. B*, **79**, 214204.
23 Phillips, J.C. (1979) *J. Non Cryst. Solids*, **34**, 153.
24 Thorpe, M.F. (1983) *J. Non Cryst. Solids*, **57**, 355.
25 Bocker, C., Rüssel, C., and Avramov, I. (2012) *Chem. Phys.*, **406**, 50.
26 Rüssel, C. (2005) *Chem. Mater.*, **17**, 5843.
27 Wright, A.C. (2014) *Int. J. Appl. Glass Sci.*, **5**, 31.
28 Maruyama, N., Honma, T., and Komatsu, T. (2009) *J. Solid State Chem.*, **182**, 246.
29 Maruyama, N., Honma, T., and Komatsu, T. (2011) *Mater. Res. Bull.*, **46**, 922.
30 Schneider, M., Richter, W., Keding, R., and Rüssel, C. (1998) *J. Non Cryst. Solids*, **226**, 273.
31 Cabral, A.A., Fokin, V.M., Zanotto, E.D., and Chinaglia, C.R. (2003) *J. Non Cryst. Solids*, **330**, 174.
32 Imaoka, M. and Yamazaki, T. (1964) *J. Ceram. Soc. Jpn.*, **72**, 182.
33 Takahashi, Y., Benino, Y., Fujiwara, T., and Komatsu, T. (2004) *J. Appl. Phys.*, **95**, 3503.

34 Takahashi, Y., Benino, Y., Fujiwara, T., and Komatsu, T. (2002) *Jpn. J. Appl. Phys.*, **41**, L1455.
35 Fukushima, T., Benino, Y., Fujiwara, T., Dimitrov, V., and Komatsu, T. (2006) *J. Solid State Chem.*, **179**, 3949.
36 Torres, F., Narita, K., Benino, Y., Fujiwara, T., and Komatsu, T. (2003) *J. Appl. Phys.*, **94**, 5265.
37 Dumbaugh, W.H. (1986) *Phys. Chem. Glasses*, **27**, 119.
38 Maeder, T. (2013) *Int. Mater. Rev.*, **58**, 3.
39 Terashima, K., Shimoto, T., and Yoko, T. (1997) *Phys. Chem. Glasses*, **38**, 211.
40 Inoue, T., Honma, T., Dimitrov, V., and Komatsu, T. (2010) *J. Solid State Chem.*, **183**, 3078.
41 Duffy, J.A. (1989) *Phys. Chem. Glasses*, **30**, 1.
42 Dimitrov, V. and Sakka, S. (1996) *J. Appl. Phys.*, **79**, 1736.
43 Inoue, T., Gao, X., Shinozaki, K., Honma, T., and Komatsu, T. (2015) *Front. Mater. Glass Sci.*, **2**, 42.
44 Suzuki, F., Honma, T., Doi, Y., Hinatsu, Y., and Komatsu, T. (2012) *Mater. Res. Bull.*, **47**, 3403.
45 Alexsandrov, L., Iordanova, R., and Dimitriev, Y. (2009) *J. Non Cryst. Solids*, **355**, 2023.
46 Taki, Y., Shinozaki, K., Honma, T., Komatsu, T., Aleksandrov, L., and Iordanova, R. (2013) *J. Non Cryst. Solids*, **381**, 17.
47 Wright, A.C. (2013) *Int. J. Appl. Glass Sci.*, **4**, 214.
48 Imaoka, M. (1961) *J. Ceram. Soc. Jpn.*, **69**, 282.
49 Fu, J. (1997) *Solid State Ion.*, **104**, 191.
50 Hirose, K., Honma, T., Benino, Y., and Komatsu, T. (2007) *Solid State Ion.*, **178**, 801.
51 Nagamine, K., Hirose, K., and Komatsu, T. (2008) *Solid State Ion.*, **179**, 508.
52 Sakurai, Y. and Yamaki, J. (1985) *J. Electrochem. Soc.*, **132**, 512.
53 Okada, T., Honma, T., and Komatsu, T. (2010) *Mater. Res. Bull.*, **45**, 1443.
54 Honma, T., Okamoto, M., Togashi, T., Ito, N., Shinozaki, K., and Komatsu, T. (2015) *Solid State Ion.*, **269**, 19.
55 Honma, T., Togashi, T., Ito, N., and Komatsu, T. (2012) *J. Ceram. Soc. Jpn*, **120**, 344.
56 Yamauchi, H., Park, G., Nagakane, T., Honma, T., Komatsu, T., Sakai, T., and Sakamoto, A. (2013) *J. Electrochem. Soc.*, **160**, A1725.
57 Honma, T., Togashi, T., Kondo, H., Komatsu, T., Yamauchi, H., Sakamoto, A., and Sakai, T. (2013) *Appl. Phys. Lett. Mater.*, **1**, 052101.
58 Wang, J.S., Vogel, E.M., and Snitzer, E. (1994) *Opt. Mater.*, **3**, 187.
59 Kosuge, T., Benino, Y., Dimitrov, V., Sato, R., and Komatsu, T. (1998) *J. Non Cryst. Solids*, **242**, 154.
60 Komatsu, T., Noguchi, T., and Sato, R. (1997) *J. Am. Ceram. Soc.*, **80**, 1327.
61 Komatsu, T., Ito, N., Honma, T., and Dimitrov, V. (2012) *Solid State Sci.*, **14**, 1419.
62 Prod'homme, L. (1960) *Phys. Chem. Glasses*, **1**, 119.
63 Wang, Y. and Ohwaki, J. (1993) *Appl. Phys. Lett.*, **63**, 3268.
64 Shinozaki, K., Honma, T., Oh-ishi, K., and Komatsu, T. (2011) *Opt. Mater.*, **33**, 1350.
65 Pablos-Martin, A., Duran, A., and Pascual, M.J. (2012) *Int. Mater. Rev.*, **57**, 165.
66 Shinozaki, K., Noji, A., Honma, T., and Komatsu, T. (2013) *J. Fluor. Chem.*, **145**, 81.
67 Kusatsugu, M., Kanno, M., Honma, T., and Komatsu, T. (2008) *J. Solid State Chem.*, **181**, 1176.
68 Shinozaki, K., Honma, T., and Komatsu, T. (2012) *J. Appl. Phys.*, **112**, 093506.
69 Shinozaki, K., Honma, T., and Komatsu, T. (2014) *Opt. Mater.*, **36**, 1384.
70 Shinozaki, K., Pisarski, W., Affatigato, M., Honma, T., and Komatsu, T. (2015) *Opt. Mater.*, **50**, 238.
71 Komatsu, T., Ihara, R., Honma, T., Benino, Y., Sato, R., Kim, H.G., and Fujiwara, T. (2007) *J. Am. Ceram. Soc.*, **90**, 699.
72 Komatsu, T. and Honma, T. (2013) *J. Asian Ceram. Soc.*, **1**, 9.

12
Glass-Forming Ability, Recent Trends, and Synthesis Methods of Metallic Glasses

Hidemi Kato,[1] *Takeshi Wada,*[1] *Rui Yamada,*[2] *and Junji Saida*[2]

[1]*Tohoku University, Institute for Materials Research, Katahira 2-1-1, 980-8577 Sendai, Japan*
[2]*Tohoku University, Frontier Research Institute of Interdisciplinary Sciences, Aramaki aza Aoba 6-3, 980-8578 Sendai, Japan*

12.1
Glass Formation and Glass-Forming Ability

A liquid is sometimes metastably maintained even below the melting point (T_m) at which the free energy of the liquid is equal to that of the crystalline solid during the cooling process. This phenomenon is called undercooling. For a liquid to transform to a crystalline solid, nucleation and growth processes are usually required. The theory of which was developed by Becker *et al.* [1] and applied to the liquid–solid phase transformation by Turnbull *et al.* [2]. Generation and growth of a homogeneous solid nucleus causes a decrease in the volume energy rather than an increase in the surface energy, which is expressed as

$$\Delta G_n = \frac{4}{3}\pi r^3 \Delta G + 4\pi r^2 \sigma, \tag{12.1}$$

where ΔG_n is the energy balance during nucleation, r is the radius of the nucleus, ΔG is the free energy difference between the liquid and solid, and σ is the interface energy. This energy balance contributes to the critical nucleus size r^* and the energy barrier for its stable growth ΔG^*, which are expressed at $d\Delta G_n/dr = 0$ as

$$r^* = -\frac{2\sigma}{\Delta G} \tag{12.2}$$

and

$$\Delta G^* = \frac{16\pi\sigma^3}{3\Delta G^2}. \tag{12.3}$$

Considering the frequencies of the contact of a required atom with the nucleus and of its diffusion into the nucleus, the steady-state nucleation frequency I_{ss} can

Handbook of Solid State Chemistry, First Edition. Edited by Richard Dronskowski, Shinichi Kikkawa, and Andreas Stein.

be expressed with viscosity $\eta(T)$ as

$$I_{ss} = \frac{A}{\eta(T)} \exp\left(-\frac{\Delta G^*}{kT}\right), \tag{12.4}$$

where A is a constant ($\sim 10^{36}$) [3] related to the Zeldovich factor [4]. The temperature dependence of the viscosity of the undercooled liquid can be expressed by the Vogel–Fulcher–Tammann (VFT) formula [5–7]:

$$\eta = \eta_0 \exp\left(\frac{Q}{T - T_0}\right), \tag{12.5}$$

where η_0 is the prefactor, Q is the VFT temperature, and T_0 is the ideal glass transition temperature where the viscosity diverges. If there is an impurity that wets the liquid well with an angle of θ, heterogeneous nucleation should be taken into account. The nucleation barrier decreases by the factor $0 < f(\theta) < 1$ to be [8]

$$\Delta G^*_{wet} < \Delta G^* f(\theta). \tag{12.6}$$

The growth rate of the nucleus is related to the difference between the liquid–solid and solid–liquid transition states. Thus, it can be expressed by the Arrhenius law [9] with the free energy difference between the liquid and solid ΔG as

$$U = U_d\left[1 - \exp\left(-\frac{\Delta G}{kT}\right)\right], \tag{12.7}$$

where U_d is atomic diffusive speed.

Using I_{ss} and U, the fraction of the crystal x for small x at time t can be expressed as [10,11]

$$x = \frac{1}{3}\pi I_{ss} U^3 t^4. \tag{12.8}$$

Here, we consider that crystalline precipitates can be detected at x_c (e.g., $\sim 10^{-6}$). From Eq. (12.8), the incubation time τ for devitrification detected at x_c is expressed as

$$\tau = \left(\frac{3x_c}{\pi I_{ss} U^3}\right)^{1/4}. \tag{12.9}$$

Figure 12.1 shows the typical temperature dependence of τ, that is, the time temperature transformation (TTT) curve. In the high-temperature region of the undercooled liquid when the undercooling state is shallow, the nucleation frequency is low because of the low driving force, while the growth rate is high because of high atomic diffusivity, so the incubation time tends to be long. In contrast, in the lower temperature region of the undercooled liquid when the undercooling state is deep, the nucleation frequency is high because of the high driving force, while the growth rate is low because of low atomic diffusivity, so the life time of the undercooled liquid is long. Therefore, the TTT curve has a "C" shape whose tip is called the "nose." If the thermal history does not pass the

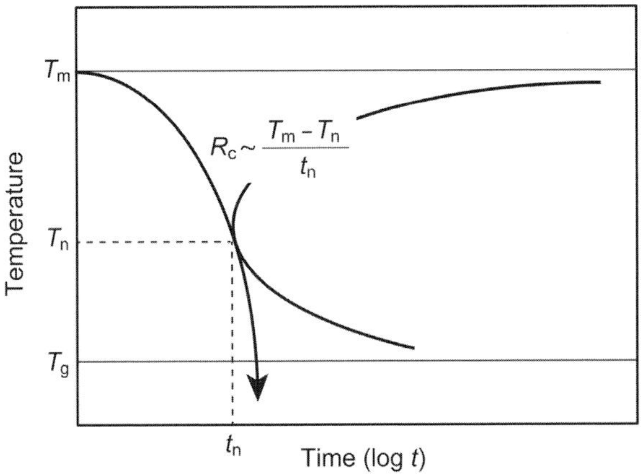

Figure 12.1 Schematic illustration of TTT curves from the melting point (T_m) to the glass transition temperature (T_g).

C curve during the cooling process, with decreasing temperature, the viscosity of the undercooled liquid increases by many orders of magnitude and finally ideally reaches $\sim10^{12}$ Pa s at >0.6 T_m [3], where the undercooled liquid behaves like an amorphous solid. This kinetic freezing phenomenon is called the glass transition. The constant cooling rate that can contact just the nose is defined as the critical cooling rate of the alloy: $R_c \approx (T_m - T_n)/t_n$, where T_m is the melting temperature, T_n is the nose temperature, and t_n is incubation time at the nose.

Polymer and inorganic glasses consist of long chain structures with C—C bonds and networks with highly anisotropic Si—O covalent bonds, respectively, which disturb the undercooled liquid to form crystalline quartz phases. Therefore, these glassy materials are easily obtained and have a long history of human usage. Metallic glasses are simple atomic glassy structures (or contain atomic clusters) with metallic bonds that are much more isotropic than the Si—O covalent bond. Therefore, the incubation time for devitrification of the undercooled metallic liquid is short, and thus preparation of metallic glasses is much more difficult than other types of glassy materials, as shown schematically in Figure 12.1. For example, the nose point is estimated to be at $t_n \approx 10^{-2}$ s for a Pd–Si metallic glass and $t_n \approx 10^0$ s for a Pd–Cu–Si bulk metallic glass [12], but $t_n \approx 10^2$ s for a SiO_2 glass [13].

The first reported metallic glass was a small and thin piece of a Au–Si eutectic alloy prepared by a special quenching technique called the gun method, where the cooling rate was estimated to be $\sim10^{6-7}$ K/s by Duwez in 1960 [14]. A series of glass-forming alloys was subsequently found in binary eutectic systems using various quenching techniques. Eutectic systems tend to be obtained when the lattice structure of the two components is different and/or the atom size mismatch between the two components is large. In particular, near the eutectic point composition, the solid solution (or intermetallic compound) is difficult to precipitate from the undercooled liquid as the primary crystalline phase, because

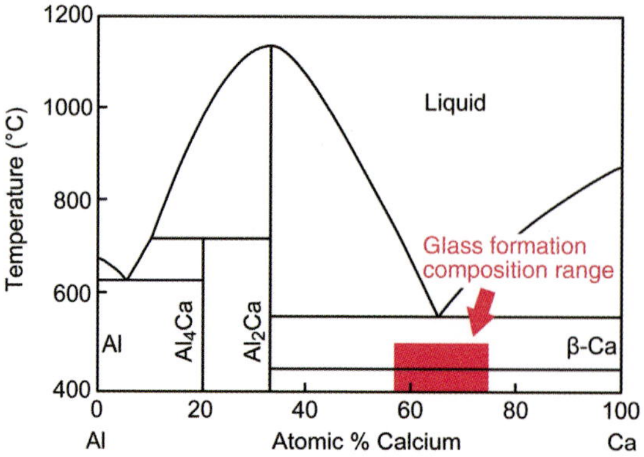

Figure 12.2 Equilibrium phase diagram of the Al–Ca binary alloy. The region near the eutectic point (57–75 at.% Ca) indicates the glass-forming composition obtained by the single-roller melt-spinning method [16].

there is no driving force for precipitation, so the undercooled liquid has to separate into two phases with the characteristic lamella structure [15]. This phase separation is achieved by relatively large-scale atomic diffusion. Therefore, a large driving force is required, making the undercooled liquid more stable to lower temperature and thus, undercooled eutectic alloys tend to be frozen to glassy solids. Figure 12.2 shows the equilibrium phase diagram for an Al–Ca typical binary eutectic system. In the diagram, the composition range for glass formation by the single-roller melt-quenching technique is indicated near the eutectic point of the Al_2Ca intermetallic compound and β-Ca solid solution [16]. Table 12.1 summarizes a series of binary glass-forming systems and their composition ranges obtained by the single-roller melt-quenching technique [16].

Except for the Nb–Ni (maximum bulk glass diameter $D_{max} \approx 2$ mm) [17] and Zr–Cu ($D_{max} \approx 2$ mm) [18] systems, glass formation of binary alloys of the eutectic system is achieved using the hyperquenching technique because the incubation time for devitrification of their undercooled liquid is very short ($\sim 10^{-4}$–10^{-2} s) [12]. Therefore, only small pieces, powders, ribbons, and thin films of their glass can be obtained. In decreasing order of the estimated cooling rate (R), typical hyperquenching techniques, which are currently used in this research field, are sputtering ($R \approx 10^7$ K/s), single-roller melt spinning ($R \approx 10^6$ K/s), and gas or water atomization ($R \approx 10^{3-4}$ K/s) that will be explained in detail in Section 12.4. It is worth pointing out that solid-state reaction is an alternative route to prepare metallic glasses [19]. Mixtures of pure metallic powder components that are balanced to be near the eutectic composition and eutectic crystalline powders are glassy solids by the ball milling technique. The former is known as mechanical alloying (MA) and the latter is known as mechanical grinding (MG), which will be explained in detail in Section 12.4. Ball-milled Pd–Ni–P amorphous powder has been confirmed to exhibit the glass transition phenomenon by differential scanning calorimetry [20].

Table 12.1 Binary eutectic systems with glass-forming composition ranges obtained by the single-roller melt-quenching technique [16].

System	Amorphous ribbon formation range (at.%)	System	Amorphous ribbon formation range (at.%)	System	Amorphous ribbon formation range (at.%)
Ag–Ca	Ca: 58–88	Cu–Gd	Gd: 66–76	Hf–Ni	Ni: 20–90
Ag–Dy	Dy: 25–85	Cu–Hf	Hf: 30–60	Hf–Si	Si: 13–18
Ag–La	La: 20–85	Cu–Mg	Mg: 55–85	Ir–Zr	Zr: 70–80
Ag–Pr	Pr: 20–80	Cu–Ti	Cu: 30–70	La–Ni	Ni: 28–42
Al–Ca	Ca: 57–75	Cu–Y	Y: 25–70	La–Pb	Pb: 7–10
Al–Gd	Gd: 78	Cu–Zr	Zr: 30–75	La–Si	Si: 15–28
Al–La	La: 35–85	Dy–Fe	Fe: 40–80	La–Sn	Sn: 15–20
Al–Sm	Sm: 35–85	Fe–Gd	Gd: 20–60	Mg–Sr	Sr: 65
Al–Sr	Sr: 65	Fe–Hf	Hf: 9–10	Mg–Zn	Zn: 23–35
Al–Y	Y: 50–75	Fe–Hf	Hf: 60–80	Mn–P	P: 23–26
Au–La	La: 75–82	Fe–Nd	Nd: 8–70	Mn–Si	Si: 25–29
Au–Zr	Zr: 65–75	Fe–P	P: 16–20	Nb–Ni	Nb: 34–78
B–Co	B: 15–30	Fe–Tb	Tb: 60	Nb–Si	Si: 17–21
Fe–B	B: 12–35	Fe–Th	Th: 28–80	Ni–P	P: 17–22
B–Mn	Mn: 73–85	Fe–Ti	Ti: 68–71	Ni–Pr	Pr: 47–72
B–Ni	B: 33–43	Fe–Y	Y: 20–75	Ni–Sm	Sm: 37–72
B–Ta	B: 15–25	Fe–Zr	Zr: 65–80		
Ca–Mg	Ca: 60–75	Fe–Zr	Zr: 8–11	Ni–Ta	Ta: 30–50
Ca–Cu	Cu: 13–63	Ga–Gd	Gd: 75–79	Ni–Ti	Ni: 25–65
Ca–Zn	Zn: 18–63	Ge–Hf	Ge: 14–16	Ni–Y	Y: 45–75
Co–Dy	Dy: 65	Ge–La	La: 76–82	Ni–Zr	Zr: 20–80
Co–Gd	Co: 45–75	Ge–Mn	Ge: 23–26	Os–Zr	Zr: 70–80
Co–Er	Er: 65	Ge–Nb	Ge: 24–25	Pd–P	P: 18–21
Co–Hf	Hf: 40–80	Ge–Pd	Pd: 75–82	Pd–Si	Si: 14–26
Co–Nd	Nd: 65	Ge–Zr	Ge: 13–21	Pd–Zr	Pd: 20–35
Co–Pr	Pr: 45–70	Gd–Mn	Mn: 40	Pt–Zr	Zr: 75–80
Co–Sm	Sm: 45–70	Gd–Ni	Ni: 31–40	Rh–Zr	Zr: 70–80
Co–Ti	Ti: 21–23	Gd–Pd	Pd: 24	Ru–Zr	Zr: 75
Co–Y	Co: 30 – 55	Gd–Pt	Pt: 17	Si–Ta	Si: 15–25
Co–Zr	Zr: 9–16	Gd–Rh	Rh: 18	Si–Ti	Ti: 77–85
Co–Zr	Zr: 40–80	Gd–Ru	Ru: 15–40	Si–Zr	Zr: 76–88

12.2
Improving the Glass-Forming Ability for Bulk Metallic Glasses

Binary eutectic glass-forming liquids that do not have sufficient incubation time before crystals precipitate can be frozen into glasses with small dimensions, such as thin firms and powders. However, a small amount of a third element additive can drastically enhance their glass-forming ability, sometimes enough for bulk glass formation, by improving the thermal stability by enhancing the life time of the undercooled liquid by many orders of magnitude. Figure 12.3a shows a schematic illustration of the temperature dependence of the lifetime of the undercooled liquid of a multicomponent alloy with improved thermal stability compared with that of the binary glass former. If the thermal stability of the undercooled liquid can be improved and the C curve thus shifted to longer time, the glassy solid of the alloy can be obtained with a lower cooling rate, and thus with larger dimensions. Figure 12.3b shows experimental TTT curves of typical Zr-based multicomponent bulk glass formers [21]. The nose points of these bulk metallic glass formers are located at several seconds to ~100 s, which are much longer times than typical times for binary glass formers ((10^{-4}–10^{-2} s) [12].

The question is how can the glass-forming ability of the alloy be improved? Unfortunately, a perfect rule for preparing alloys with high glass-forming ability has not been determined, although some criteria based on the thermal properties have been suggested [3,22–25]. Some empirical rules have been proposed. Inoue's group at Tohoku University suggested three empirical rules [26] for eutectic alloy systems based on their extensive experience in bulk metallic glass preparation:

1) Multicomponent systems with three or more elements
2) Negative heats of mixing among the three main constituent elements
3) Significant atomic size mismatch among the three main constituent elements (>12%)

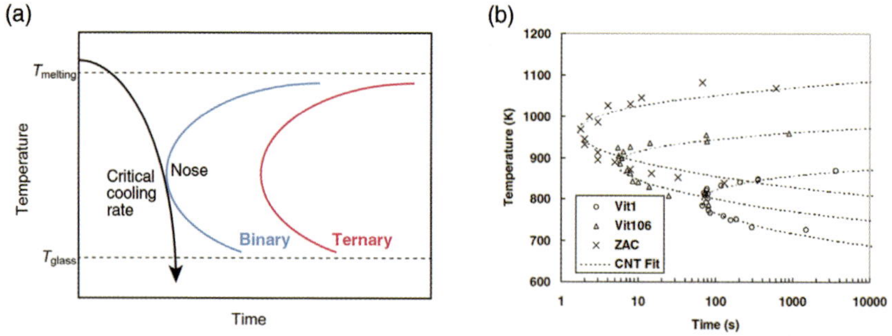

Figure 12.3 (a) Schematic illustration of TTT curves of a binary alloy with low glass-forming ability and a ternary (or multicomponent) alloy with improved glass-forming ability. (b) Experimental curves of the Zr-based multicomponent bulk glass formers $Zr_{41.2}Ti_{13.8}Cu_{12.5}Ni_{10}Be_{22.5}$ (Vit1), $Zr_{57}Cu_{15.4}Ni_{12.6}Al_{10}Nb_5$ (Vit106) and $Zr_{55}Al_{22.5}Co_{22.5}$(ZAC). The dotted curves show fitting using classical nucleation theory. (Reprinted with permission from Ref. [21].)

The second rule indicates that the homogeneous undercooled liquid is more stable than the phase separated state. The third rule indicates that the high strain energy arising from large size mismatch among the atoms, especially in the low-temperature region, is sufficient to disturb precipitation of the solid solution, which usually does not require a large atomic diffusion process as the primary phase, resulting in deep undercooling until the eutectic reaction. However, in the deep undercooling state, atomic diffusion is sufficiently inhibited, so the eutectic reaction requiring large atomic diffusion becomes difficult. Finally, the deep undercooled liquid is frozen into the glass solid at T_g, where η reaches $\sim 10^{12}$ Pa s. In addition, a undercooled liquid that satisfies the three empirical rules tends to develop local short-range order that decreases the free energy difference between the undercooled liquid and the corresponding crystalline phase (ΔG), resulting in an increase in nucleation energy barrier ΔG^*, as explained by Eq. (12.3). It has been reported that Zr–Cu–Al metallic glasses, which are known to be one of the best bulk glass formers, contain a high proportion of clusters with icosahedral-like short-range ordered atomic configuration from X-ray structural analysis [27]. Pd–TM (transition metal)–metalloid metallic glasses (e.g., Pd–Cu–Ni–P) contain highly dense-packed atomic configurations of two types of polyhedra of Pd–Cu–P and Pd–Ni–P pairs [28]. Other metal–metalloid types such as Fe–Ln (lanthanoid)–B and Fe–(Zr, Hf, Nb)–B ternary systems contain medium-range network-like atomic configurations in which a distorted trigonal prism and an anti-Archimedean prism of Fe and B are connected in face- and edge-shared configuration modes through glue atoms of Ln and the early transition metal Zr, Hf, or Nb [29]. These orderings are shown in Figure 12.4 [30]. In Table 12.2 [30], typical bulk metallic glasses are summarized with the year that they were reported and the preparation method: water quenching, copper mold casting, arc-melt tilt casting, or arc-melt suction casting, which are explained in detail in Section 12.4.

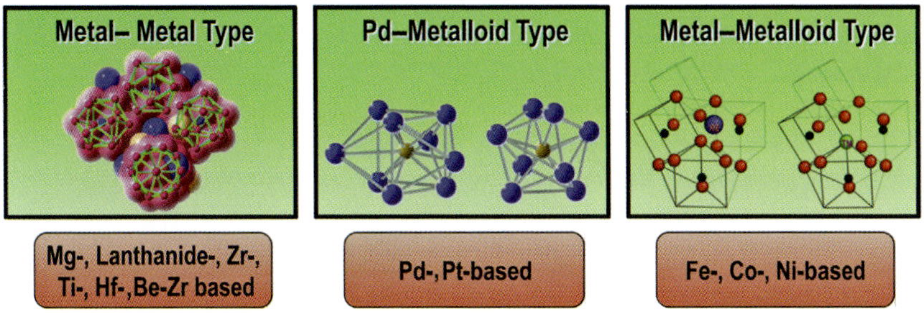

Figure 12.4 Atomic clusters (short-range ordering) of metal–metal eutectic system based metallic glasses (e.g., Zr–Cu–Ni–Al), Pd–TM–metalloid eutectic systems (e.g., Pd–Cu–Ni–P), and other metal–metalloid type systems (e.g., Fe–Ln–B and Fe–(Zr. Hf. Nb)–B) determined by X-ray structural analysis. Formation of these clusters is considered to decrease the free energy of the undercooled liquid, resulting in an increase of the nucleation energy barrier to crystallization. (Reprinted with permission from Ref. [30].)

Table 12.2 Typical multicomponent bulk metallic glass-forming systems with the year that they were reported.

Nonferromagnetic alloy systems	Year	Ferromagnetic alloy systems	Year
Mg–Ln–M (Ln = lanthanide metal, M = Ni. Cu, Zn)	1988	Fe–(Al,Ga)–(P,C,B,Si,Ge)	1995
Ln–Al–TM (TM = Fe, Co, Ni, Co)	1989	Fe–(Nb, Mo)–(Al, Ga)–(P, B, Si)	1995
Ln–Ga–TM	1989	Co–(Al, Ga)–(P, B, Si)	1996
Zr–Al–TM	1990	Fe–(Zr, Hf, Nb)–B	1996
Zr–Ln–Al–TM	1992	Co–(Zr, Hf, Nb)–B	1996
Ti–Zr–TM	1993	Fe–Co–Ln–B	1998
Zr–Ti–TM–Be	1993	Fe–Ga–(Cr, Mo)–(P, C, B)	1999
Zr–(Ti, Nb, Pd)–Al–TM	1995	Fe–(Cr, Mo)–(C, B)	1999
Pd–Cu–Ni–P	1996	Ni–(Nb, Cr, Mo)–(P, B)	1999
Pd–Ni–Fe–P	1996	Co–Ta–B	1999
Ti–Ni–Cu–Sn	1998	Fe–Ga–(P, B)	2000
Ca–Cu–Ag–Mg	2000	Ni–Zr–Ti–Sn–Si	2001
Cu–Zr, Cu–Hf	2001	Ni–(Nb, Ta)–Zr–Ti	2002
Cu(Zr, Hf)–Ti	2001	Fe–Si–B–Nb	2002
Cu–(Zr, Hf)–Al	2003	Co–Fe–Si–B–Nb	2002
Cu–(Zr, Hf)–Al–(Ag, Pd)	2004	Ni–Nb–Sn	2003
Pt–Cu–Ni–P	2004	Co–Fe–Ta–B–Si	2003
Ti–Cu–(Zr, Hf)–(Co, Ni)	2004	Ni–Pd–P	2004
Au–Ag–Pd–Cu–Si	2005	Fe–(Cr, Mo)–(C, B)–Ln (Ln = Y, Er, Tm)	2004
Ce–Cu–Al–Si–Fe	2005	Co–(Cr, Mo)–(C,B)–Ln (Ln = Y, Tm)	2005
Cu–(Zr, Hf)–Ag	2005	Ni–(Nb, Ta)–Ti–Zr–Pd	2006
Pd–Pt–Cu–P	2007	Ni–Pd–P–B	2009
Zr–Cu–Al–Ag–Pd	2007	Fe–(Nb, Cr)–(P, B, Si)	2010
Ti–Zr–Cu–Pd	2007		
Ti–Zr–Cu–Pd–Sn	2007		

Source: Reprinted with permission from Ref. [30].

12.3 Heterometallic Glasses

One of the recent researches trending in metallic glasses is development of heterometallic glasses, that is, metallic glasses with nanoscale heterogeneity such as phase separation and precipitation, to improve the mechanical or soft magnetic properties.

12.3.1 High-Plasticity Zr–Al–Ni–Pd Bulk Metallic Glasses

Most metallic glasses exhibit high strength close to the ideal strength, in contrast to other crystalline metals. However, their amorphous structure cannot generate

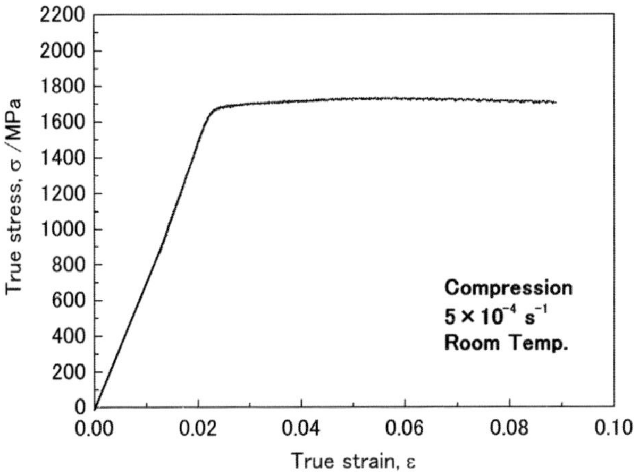

Figure 12.5 True compressive stress–strain curve of Zr–Al–Ni–Pd bulk metallic glass (2 mm in diameter) at a strain rate of $5 \times 10^{-4}\,s^{-1}$ with significant plastic strain at room temperature (~6%). (Reprinted with permission from Ref. [32].)

dislocations that cause plastic deformation and/or work-hardening behavior in ordinary crystalline metals. Therefore, metallic glasses usually do not exhibit work-hardening behavior and catastrophically fracture along with highly localized slipping at the maximum shear planes without apparent plastic deformation, especially under uniaxial tensile/compressive modes. This shortcoming in the mechanical properties limits applications of metallic glasses and much effort has been devoted to solving this critical problem. It is worth mentioning that Saida *et al.* successfully improved the plastic deformability by addition of noble elements, such as Au, Pd, and Pt, into Zr–Al–Ni(-Cu) metallic glasses. Figure 12.5 shows the compressive stress–strain curve of Zr–Al–Ni–Pd bulk metallic glass [31,32]. The noble metals have higher chemical affinity with Zr that can be confirmed by the corresponding heat of mixing ($\Delta H_{mix} = -91$ kJ/mol (Pd–Zr)) being higher than that of the other metals (e.g., $\Delta H_{mix} = -23$ kJ/mol (Cu–Zr) and -49 kJ/mol (Ni–Zr)) [33]. This makes the metallic glass unstable, and nanoscale quasicrystals or the distorted phase of metastable fcc-Zr_2Ni precipitate in the shear bands, as shown in Figure 12.6 [32]. The mechanism for developing this nanostep-like fcc-Zr_2Ni zone is schematically shown in Figure 12.7 [32]. Dynamic precipitation of fcc-Zr_2Ni is considered to suppress fast shear slipping by increasing the viscosity by many orders of magnitude, leading to significant improvement of plastic deformation by developing multiple shear bands. Yamada *et al.* [34] reported that minor addition of Au (3 at.%) to $Zr_{60}Cu_{22}Ni_5Al_{10}Au_3$ bulk metallic glass causes drastic improvement of the room temperature plasticity because of dynamic precipitation of the nanoscale icosahedral crystalline phase (I phase). In addition, thermal treatment at $T_g = 50$ or 100 K prior to mechanical testing significantly improved the plasticity because the treatment generated nuclei of the I phase. Recently, such inhomogeneity in

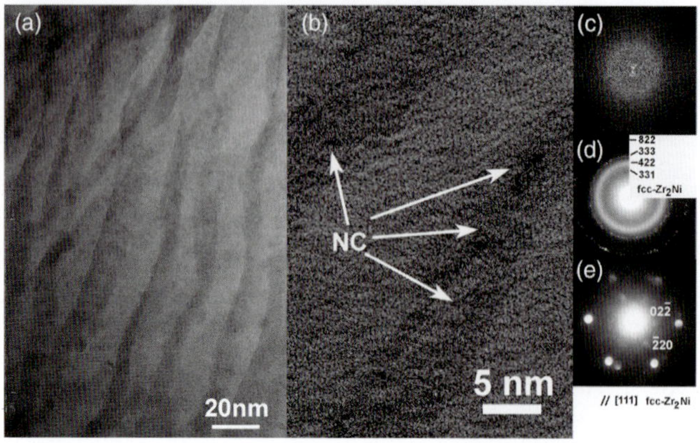

Figure 12.6 TEM images of the shear band developed in a $Zr_{65}Al_{7.5}Ni_{10}Pd_{17.5}$ bulk metallic glass that exhibited 6% compressive strain. (a) and (b) High-resolution images. (c) Fast Fourier transform image obtained from the bright region. (d) Selected area electron diffraction image taken from the whole area. (e) Nanobeam electron diffraction pattern taken from the dark region in the high-resolution image in part (a). (Reprinted with permission from Ref. [32].)

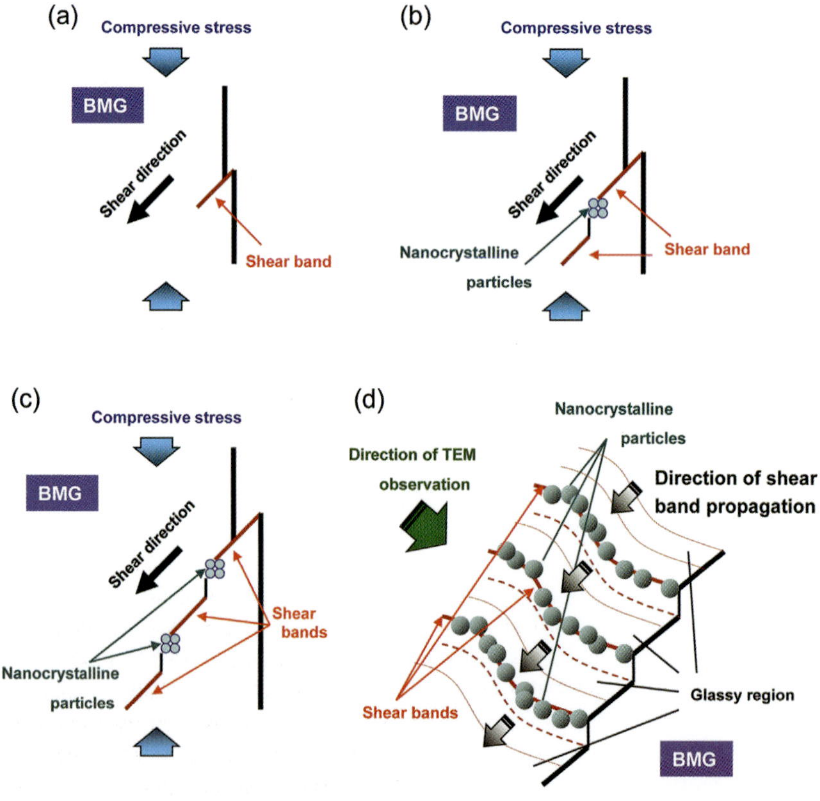

Figure 12.7 Schematic illustration of dynamic nanoprecipitation of metastable fcc-Zr_2Ni in the shear band, resulting in slow down of the shear slipping process to enhance the multiple shear bands [32].

metallic glass has been widely studied to improve the ductility. These findings are expected to bring a novel concept of structure and a solution to the problem of brittleness in metallic glasses [35–39].

12.3.2
Fe-Based Heterometallic Glasses for Excellent Soft Magnetic Nanocrystalline Alloys

Fe-based metallic glasses are known to have excellent soft magnetic properties. To obtain high saturation magnetization, a high Fe content is required. However, when the Fe concentration exceeds ~80 at.%, coarse α-Fe grains begin to disperse in the matrix phase. These α-Fe grains inevitably degrade the excellent magnetic softness of Fe-based metallic glasses. Makino *et al.* drastically improved the soft magnetic properties in terms of the saturated magnetization (J_s), coercivity (H_c), and permeability (μ) by minor addition of Cu and P into Fe–Si–B metallic glass [40–42]. Cu has a repulsive chemical interaction with Fe ($\Delta H_{mix} = 14$ kJ/mol (Fe–Cu)), whereas P has an attractive interaction with Cu ($\Delta H_{mix} = -17.5$ kJ/mol (Cu–P)) [33]. Simultaneous addition of Cu and P makes Fe–Si–B metallic glass unstable and Cu-rich and P-rich regions are generated, as shown in Figure 12.8a [41]. This heterogeneity may cause homogeneous precipitates of nanoscale α-Fe phase (2–3 nm) in the Fe-based metallic glass matrix. By controlling the crystallization process, mainly in terms of the heating rate, a homogeneous nanocrystalline structure with α-Fe nanocrystalline alloy with 10–17 nm successfully develops, as shown in Figure 12.8b [41]. This nanocrystallized $Fe_{83.3-84.3}Si_4B_8P_{3-4}Cu_{0.7}$ (at.%) alloy is found to exhibit excellent soft magnetic properties, such as high $J_s = 1.88$–1.94 T, which is close to that of Fe-3.5 mass% Si crystalline alloys, low $H_c < 10$ A/m, and higher effective permeability $\mu = 16\,000$–$25\,000$ at 1 kHz.

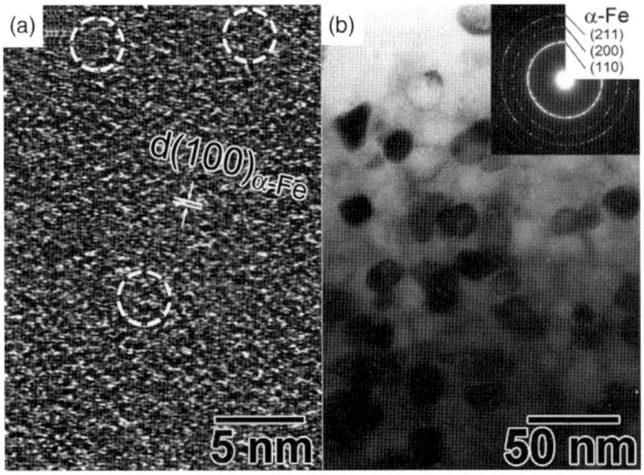

Figure 12.8 (a) High-resolution TEM image of as-melted spun $Fe_{83.3}Si_4B_8P_4Cu_{0.7}$ alloy. (b) Bright field TEM image and selected area electron diffraction pattern of the crystallized counterpart at 748 K. (Reprinted with permission from Ref. [41].)

12.4
Preparation Methods for Metallic Glasses

12.4.1
Melt Spinning

The melt-spinning technique was developed to prepare long continuous thin ribbon-shaped metallic glasses that are widely used in the metallic glass research field because the cooling rate is fast (~10^6 K/s), a small amount of alloy is required, and the ribbon shape of the specimen allows the mechanical properties of metallic glasses to be characterized. Figure 12.9 shows a schematic illustration of the single-roller melt-spinning technique. Crushed ingots are inserted in a quartz nozzle that is then inductively heated above the melting point in an inert atmosphere. The molten metal is ejected through an orifice (several hundred micrometers in diameter) to a fast rotating copper wheel (~25 cm in diameter, ~1–5 krpm). The molten metal is continuously spread on the surface of the wheel by the rapid shear action and quenched. The cooling rate of the roll-contacted side is higher than the other "free solidified" side. However, the free solidified side is sometimes influenced by the wheel material. This difference in the quenching conditions sometimes causes differences in the density, hardness, and thermal stability between the two sides. The controlling factors, such as the molten metal temperature, copper wheel rotating speed, atmosphere gas, and ejection gas pressure, should be controlled to prepare metallic glass ribbons with desirable shapes, and thus experience using this equipment is usually required.

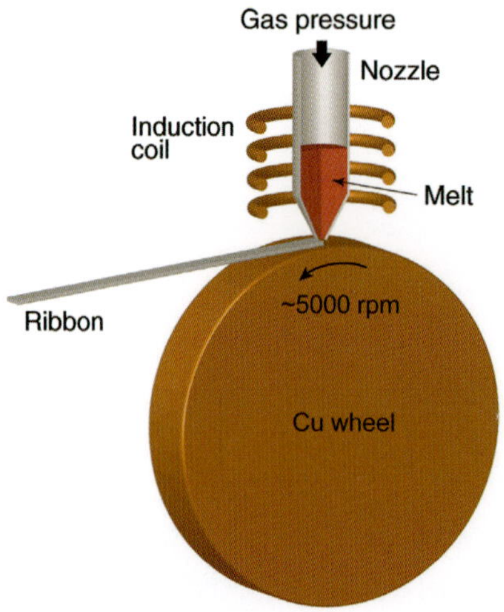

Figure 12.9 Schematic illustration of the typical single-roller melt-quenching technique.

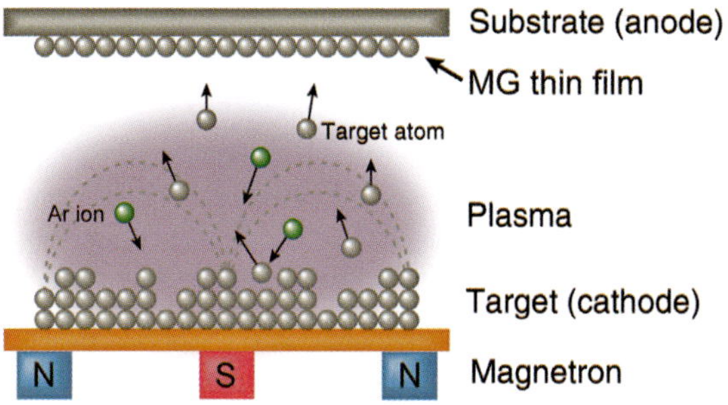

Figure 12.10 Schematic illustration of the typical magnetron sputtering technique.

12.4.2
Sputtering

Sputtering is a common technique to prepare thin film materials, and it is not unique to metallic glass preparation. The main advantage of this technique is that the achievable cooling rate is the highest of the metallic glass production methods. There are several types of sputtering equipment, but the principle is that accelerated gas ions are irradiated on the alloy target and the target components are forced out and deposit on the substrate. Figure 12.10 shows a schematic illustration of the magnetron sputtering method. In this process, the gas phase is considered to be directly frozen to the solid phase and the achievable cooling rate is very high. Using this technique, alloys with low glass-forming ability can be frozen into the glassy state.

12.4.3
Gas (Water) Atomization

Gas atomization is a method to obtain fine metal powders using a high-pressure gas jet. The typical setup of the gas atomization machine is shown in Figure 12.11a. The ingot is inductively melted in a crucible with a small nozzle. The melt stream is continuously ejected from the tip of the nozzle by either applying pressure or removing the stopper inserted at the bottom of the crucible. The melt stream is instantly broken up into fine metal droplets by the high-pressure (up to ~10 MPa) gas jet and the droplets are simultaneously quenched by heat transfer between the gas and the melt. Generally, the size of the gas-atomized powder follows a log-normal distribution function, and higher atomization pressure results in finer average powder size [43]. The cooling rate of the melt by gas atomization depends on various factors, such as the gas species, gas pressure, elements in the melt, and size of the powder. Measurement of the secondary dendrite arm spacing of the gas-atomized crystalline metal powder reveals that the cooling rate is approximately 10^3–10^4 K/s [43]. If this cooling rate is higher than the critical cooling rate for glass formation of the alloy, the atomized

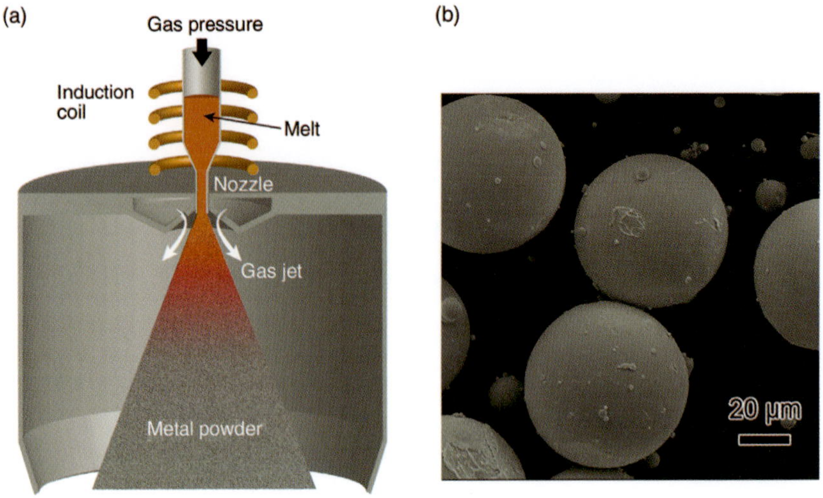

Figure 12.11 (a) Typical setup of the gas atomizer. (b) SEM micrograph of gas-atomized Zr–Al–Ni–Cu metallic glass powder.

powder can be glassy. Figure 12.11b shows a representative image of gas-atomized Zr-based metallic glass powder. The powder has a spherical shape with an extremely smooth surface because of a lack of crystalline grains. The as-atomized powder often has a broad size distribution and large crystalline particles that cannot be vitrified owing to an insufficient cooling rate. In this case, the large crystalline particles should be removed by sieving and only a glassy powder with a particular size range should be used. Recently, the atomization technique has been applied to prepare metallic glass nanowires [44]. Water atomization is an alternative technique using a water jet in place of the gas jet, and it is sometimes used to prepare Fe-based metallic glass powders.

12.4.4 Pulsated Orifice Ejection Method

The pulsated orifice ejection method (POEM) is a novel method to prepare very large particles with high sphericity. The obtained particles are so monodispersed that their microstructures can be unified. This unique technique is attracting much attention as a new method to prepare metallic glassy particles.

Figure 12.12a shows a schematic illustration of the POEM apparatus. This apparatus is roughly divided into three parts: (A) a droplet ejection part, (B) a solidification part, and (C) a particle collection part. Figure 12.12b shows the outline of the droplet ejection part. There are four steps to create the droplet. The first step is to apply gas pressure to the melt to fix the liquid surface (step 1). A pulse voltage is then applied to the actuator. The rod, which is directly attached to the actuator, is reciprocating and droplets are continuously created

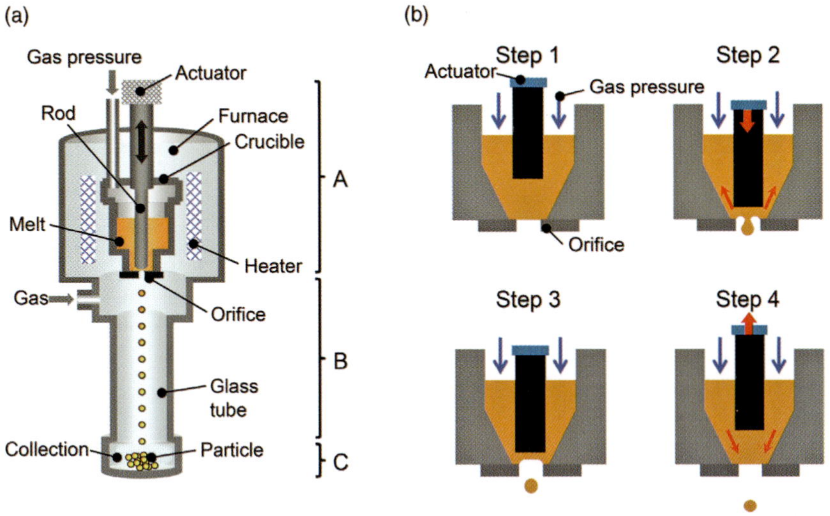

Figure 12.12 (a) Schematic illustration of the POEM. (b) Outline of the droplet ejection part of the apparatus.

and ejected from the orifice to the glass tube (steps 2–4). By tuning the pulse voltage or changing the diameter of the orifice, various particles with different sizes can be easily prepared. In addition, by changing the frequency of the reciprocating cycle, more than 100 particles can be obtained per second. The ejected droplet free falls in the glass tube, which is filled with a gas with high thermal conductivity (e.g., Ar or He), and is finally cooled. This cooling system enables the droplet to be quenched and a cooling rate of up to several thousand Kelvin per second can be obtained. During this process, the droplet is not solidified in the same way as conventional methods (e.g., the melt quench method and copper mold casting technique), but it is solidified in a specific environment. The droplet is solidified under microgravity. This prevents segregation and a homogeneous microstructure of the inside of the particles is attained. Moreover, this is a container-less process and the negative factors of crystallization (e.g., impurities contamination, gas entrainment, and impact) can be completely avoided. From this point of view, the POEM is an advantageous and favorable process for preparing metallic glasses.

In previous studies, metallic and nonmetallic particles, such as Cu, solder, Si, Ge, and TiO_2, have been successfully obtained using this method [45–48]. Recently, metallic glassy particles with diameters of around 200–700 μm have been successfully prepared in the Fe-based metallic glass system [49,50]. Figure 12.13a shows the outer appearance of $[(Fe_{0.5}Co_{0.5})_{0.75}Si_{0.05}B_{0.2}]_{96}Nb_4$ metallic glassy particles obtained by the POEM method. In this image, there are no traces of solidification shrinkage and cracks, and the particles maintain a clean and smooth surface. The particle size distribution shows that the obtained particles are almost monodispersed and their average diameter is around 470 μm

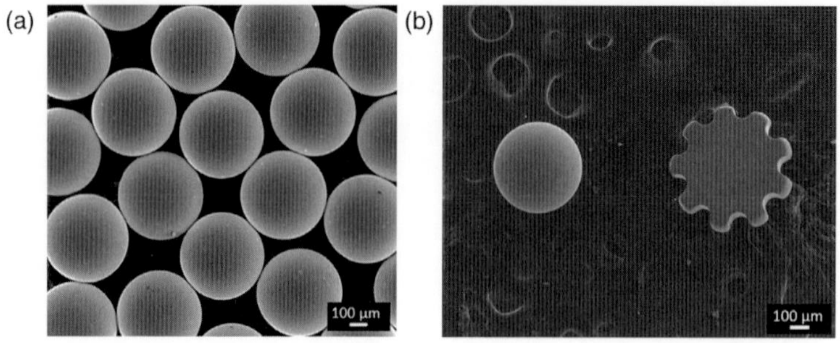

Figure 12.13 (a) Outer appearance of $[(Fe_{0.5}Co_{0.5})_{0.75}Si_{0.05}B_{0.2}]_{96}Nb_4$ metallic glassy particles. (b) A metallic glassy gear fabricated from a single particle.

in a sharp size distribution. TEM indicates that the microstructure of the particles is in the fully amorphous state. Other metallic glassy particles with different compositions have also been successfully prepared [51].

This process provides useful information for fundamental research of metallic glasses. For example, the shape of the particles obtained by this method is quite simple (i.e., highly spherical) that is advantageous for modeling or calculating heat transfer between the gas and the droplet during solidification. This enables easy estimation of the critical cooling rate [49,51]. In addition, the crystallization behavior is much easier to discuss [49] because the initial microstructures of the particles are clean and sound compared with bulk or ribbon samples. Moreover, the obtained particles are expected to be used as raw materials for several applications. Some unique applications have already been demonstrated, such as three-dimensional terahertz photonic crystals and microparts (e.g., gears (Figure 12.13b) and relay contacts) [46,50], and they are expected to be used as engineering products in the future.

12.4.5
Mechanical Alloying and Mechanical Grinding

Mixtures of pure metallic powder components that are balanced to be near the eutectic composition and eutectic crystalline powders form glassy solids by the ball milling technique [19]. The former is known as mechanical alloying and the latter is known as mechanical grinding. In these techniques, balanced powder mixtures of pure metal elements (MA) or near eutectic crystalline powders (MG) are added with hard balls to a jar containing an inert gas. The jar is then rotated at a desired rotation rate for a suitable length of time. In the rotating jar, the powders are plastically deformed, cold welded, and fractured by balls repeatedly hitting balls or the jar wall. These actions make the powder structure disordered by inducing dislocations and defects, resulting in amorphization [52]. These techniques can generate a metallic glass even for alloy systems that cannot be synthesized by liquid quenching techniques. Especially, it is very effective to produce the metallic glasses containing the elements with huge different melting

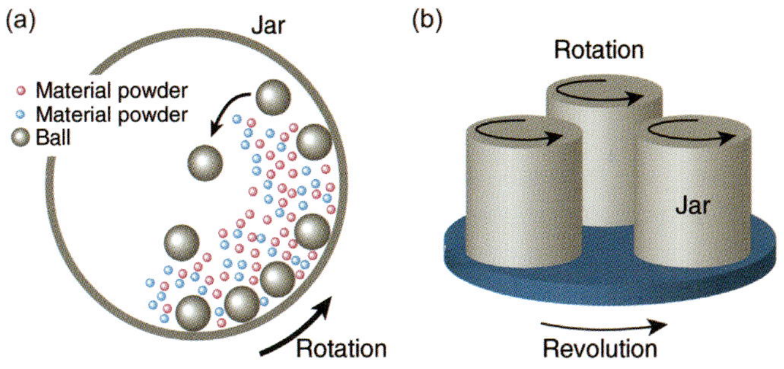

Figure 12.14 Schematic illustrations of (a) simple ball milling and (b) high-energy planetary ball milling techniques.

temperature or immiscible tendency [53,54]. However, the shortcoming of this method is contamination of the alloy with undesirable elements from the ball, jar, or atmosphere. Figure 12.14 shows the typical ball milling techniques: simple ball milling (Figure 12.14a) and planetary ball milling (Figure 12.14b). In the planetary ball milling technique, the jar is revolved by rotating it on its own axis, and higher mechanical energy can be induced in powders than the simple counterpart. Actually, several metallic glass systems have been successfully produced by mechanical alloying or grinding [20,55–59].

12.4.6 Water Quenching

In the water quenching method, master alloy ingots with bulk glass-forming ability are encapsulated with an inert gas in a quartz tube, as shown in Figure 12.15. The tube is heated by a furnace above the melting point of the alloy and then quenched in a water bath (sometimes with ice to enhance the cooling rate). Considering the relatively low thermal conductivity of quartz, the cooling rate obtained by the method is not very high ($\sim 10^2$ K/s). In addition, silicon and oxygen contamination from quartz sometimes decreases the glass-forming ability of the alloy when the melt reacts with the quartz container. Therefore, this

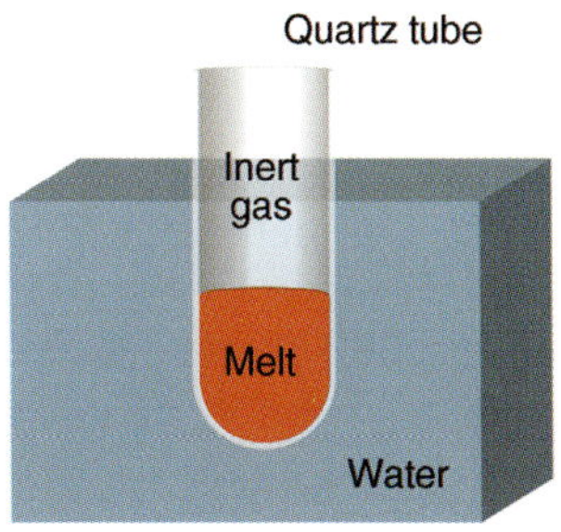

Figure 12.15 Schematic illustration of the water quenching technique.

method is usually used to prepare Pd–P-, Pt–P-, Ni–P-, and Fe–metalloid-based multicomponent bulk metallic glasses owing to their high bulk glass-forming ability and no reaction with quartz.

12.4.7 Metallic Mold Casting

Metallic mold casting is the most frequently used method to prepare metallic glasses. The alloy melt is poured into a cavity in a metallic mold and is rapidly cooled by heat transfer between the melt and the mold. Copper is used as a mold material because of its high thermal conductivity. The typical setup of copper mold casting is shown in Figure 12.16. The prealloyed ingot is inserted into a nozzle with a small orifice. The ingot is inductively heated and melted, and then ejected into the cavity of the mold by applying gas pressure [60]. The material of the nozzle must be selected considering the reactivity with the melt because impurities originating from the nozzle decrease the glass-forming ability of the alloy. Quartz, alumina, and boron nitride are commonly used as the nozzle material. The mold cavity is generally cylindrical, but rectangular, plate-, wedge-, cone-, and dog bone-shaped mold cavities can be used depending on the purpose of the experiment. In evaluating the glass-forming ability of an alloy, the melt is cast into cylindrical cavities with various diameters, and each cast sample is examined by XRD to determine if the casts are glassy or crystalline. By preloading a skeleton reinforcement, such as an open-cell metal foam or spring, in the mold cavity and infiltrating the metallic glass melt into the open space of the reinforcement, a metallic glass matrix composite can be prepared [61,62]. By rotating the mold at high speed, the centrifugal force accelerates the flow of the melt into the small complicated shaped cavity. This enables almost net-shaped metallic glasses with complex features, such as rings [63] and tooth roots, to be fabricated [64].

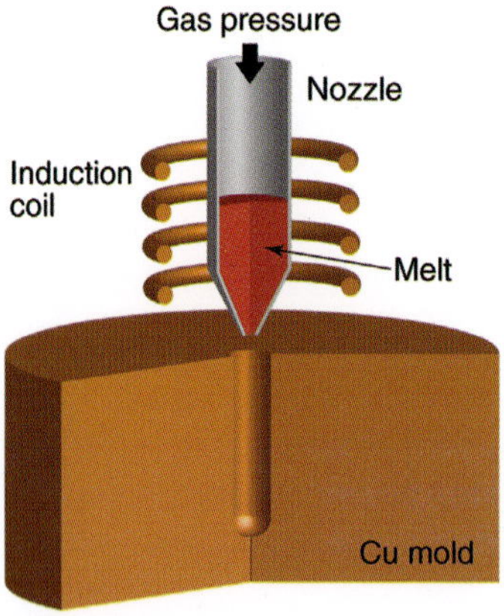

Figure 12.16 Typical setup of copper mold casting.

The cooling rate during metallic mold casting is strongly affected by various factors, such as the thermal conductivity and diffusivity of the materials of the mold and melt, heat transfer coefficient between the mold and the melt interface, shape of the mold cavity, and atmospheric gas and its pressure. The mold is kept at room temperature or cooled by liquid nitrogen. The wall of the mold cavity should be smooth for good contact between the mold and the melt to obtain rapid heat transfer. Casting of an alloy with active elements must be performed in an atmospheric chamber filled with an inert gas such as Ar or He because oxidization enhances crystallization. However, some alloys with high oxidation resistance, such as the Mg–Gd–Cu alloy, can be cast even in air. According to measurements using a thermocouple directly inserted into the melt, the cooling curve of the melt cast into the mold follows Newton's law of cooling. It has been reported that the cooling rate of a Zr–Al–Ni–Cu–Pd alloy melt cast into a cylindrical cavity in a Cu mold exceeds 10^3 K/s at the liquid temperature and 10^{1-2} K/s near the glass transition temperature [65]. This method can contaminate the alloy melt with silicon or oxygen, and turbulence of the ejected alloy melt sometimes inserts inert gas into the cast sample. To overcome the mold and gas contamination problems of the copper-cold casting method, arc-melt tilt casting or suction casting methods can be used.

12.4.8
Arc-Melt Tilt Casting and Suction Casting

Tilt casting is the method where the molten alloy is poured into a mold cavity by tilting. Tilt casting is often used to prepare larger metallic glass samples than those produced by the copper mold casting method. Because tilt casting only relies on gravity, it is not suitable to produce a few millimeter-sized metallic glasses, in which the effect of surface tension is significant. Formation of centimeter-sized metallic glasses by tilt casting has been reported for various systems, such as Zr- [66], Mg- [67], Cu- [68], and La–Ce-based [69] alloys. There are two main types of tilt casting: one uses induction melting with a crucible [67,69] and the other uses arc melting without a crucible [66,68]. Arc-melt tilt casting has the advantage of reducing contamination from the melting container, so it is suitable for casting melts with active elements. Here, arc-melt tilt casting, which is often used to prepare large-sized Zr-based metallic glass, is discussed.

When an alloy ingot is arc-melted on a water-cooled Cu hearth, the bottom of the ingot in contact with the Cu hearth always has a semisolid region even though the top of the ingot looks like the liquid state [70]. While this provides the advantage of preventing contamination of the alloy melt by the Cu hearth, it also means that the completely liquid state cannot be achieved by the normal arc-melting system. If such a semisolid is cast into the mold, a glassy phase hardly forms because the preexisting solid acts as a heterogeneous nucleation site that enhances crystallization of the supercooled liquid. Another problem is that such crystalline inclusions sometimes seriously decrease the mechanical

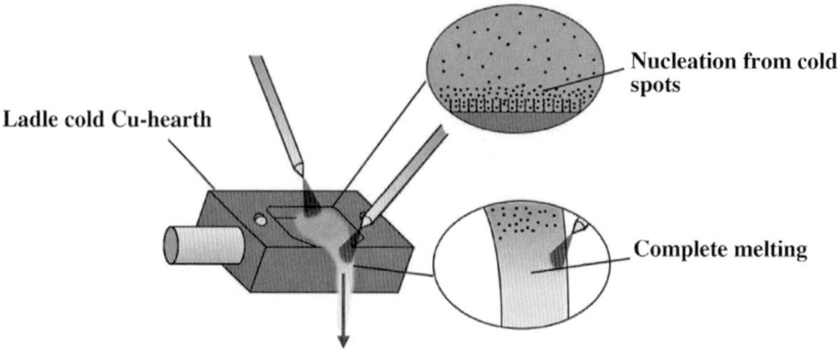

Figure 12.17 Typical setup of arc-melt tilt casting. (Reprinted with permission from Ref. [71].)

properties of the metallic glass. Thus, attention should be paid to ensure that such a crystalline solid is not included in the melt.

The setup of the typical arc-melt tilt casting machine is shown in Figure 12.17 [71]. A ladle-shaped copper hearth is connected to the shaft. This shaft is connected to the handle outside the chamber and the Cu hearth can be freely tilted by moving the handle. The ingot placed on the Cu hearth is melted by arc discharge. To prevent inclusion of unmelted particles in the melt, the tilt casting machine has more than one arc torch. One torch heats the center of the melt bath while the other heats the melt at the pouring gate to achieve complete melting. Figure 12.18 shows a representative picture of a Zr-based bulk metallic glass prepared by tilt casting [66]. The surface of the cast looks very smooth like liquid metal and no defects are visible. In the copper mold casting method, a stream of melt is strongly ejected from the orifice of the nozzle. The turbulent flow of the melt causes formation of defects such as voids and cold shuts in the

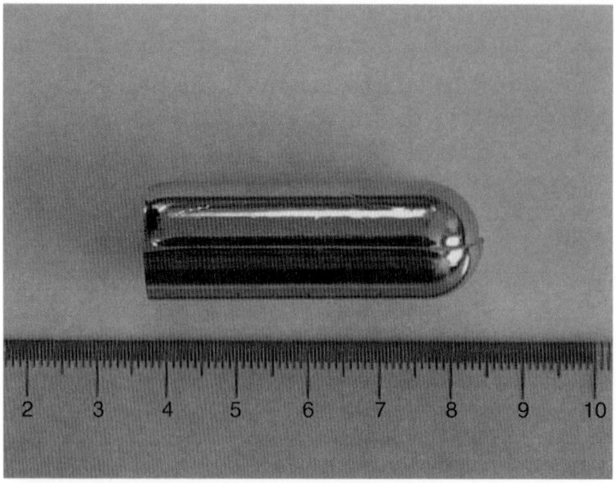

Figure 12.18 A 16 mm diameter Zr-based metallic glass prepared by arc-melt tilt casting. (Reprinted with permission from Ref. [66].)

cast product [72]. Such defects are reported to cause large variation in the mechanical properties even if the sample is completely glassy [72]. In contrast, in tilt casting the melt can be slowly poured to make laminar flow without significant increase of the specific surface area. This enables formation of defect-free metallic glasses with less variation in the mechanical properties.

After the melt is poured into the mold, the heat of the melt rapidly dissipates by heat transfer to the bottom and sidewall of the mold. However, the top of the cast is not quenched because of the lack of a heat sink that sometimes causes the top of cast to crystallize. To overcome this problem, the cap casting method was developed, in which a metallic cap is placed on the melt just after pouring the melt into the cavity, and formation of larger bulk metallic glass has been reported using this method [73].

The setup of the typical arc-melt suction casting machine is shown in Figure 12.19 [71]. An alloy ingot is melted by arc plasma on a water-cooled copper hearth. The molten alloy is then sucked into a mold cavity under the hearth through a small orifice (<3 mm diameter) set at the center of the hearth by applying a pressure difference between the chamber and the mold cavity. During melting, the molten alloy does not drop into the cavity because of the large surface tension of the alloy.

Arc melting tends to heat the melt above 1000 °C. Thus, these casting methods are unsuitable for alloys containing elements with high vapor pressure or low boiling point.

12.4.9
Metallic Glass Powder Consolidation

Powder metallurgy is an alternative method to prepare bulk metallic glasses. Metallic glass powders are prepared by atomizing with water or a high-pressure gas. Consolidation is then performed by various techniques such as hot pressing [74], warm extrusion [75–77], and spark-plasma sintering [78]. Schematic

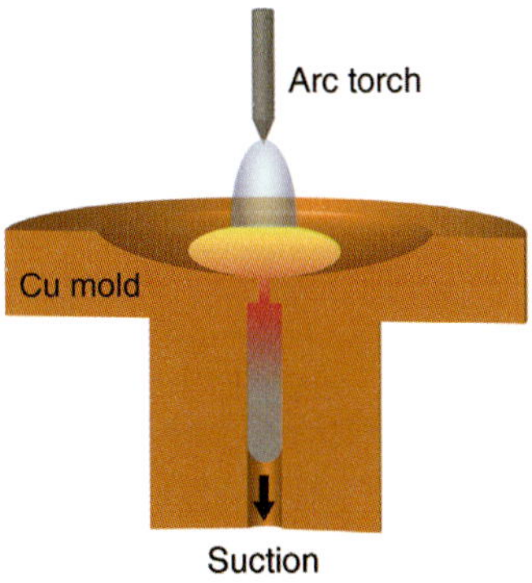

Figure 12.19 Typical setup of the arc-melt suction casting technique.

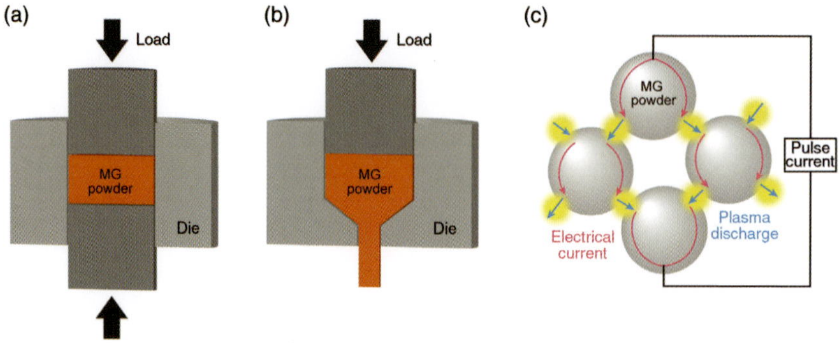

Figure 12.20 Schematic illustrations of metallic glass powder consolidation techniques. (a) Hot pressing. (b) Warm extrusion. (c) Spark plasma sintering.

illustrations of these techniques are shown in Figure 12.20. A metallic glass becomes an undercooled liquid when heated above its glass transition temperature, at which the viscosity of the glass powder decreases to 10^{5-9} Pa s. The relative density of the compact is almost 100%, although the bonding strength between powders varies with the method. The surface oxide layer on the metallic glass powder disturbs the strong bonding between the powder, resulting in lower tensile strength of the powder compacts compared with the mold cast metallic glass counterpart. To overcome this problem, Kawamura *et al.* applied large shear deformation to Zr-based metallic glass powders by the warm extrusion technique in the undercooled liquid state to directly contact the metallic surfaces by breaking the surface oxide layer [75]. Using this approach, metallic glass powder compacts (10 mm in diameter) with full tensile strength were successfully prepared.

12.4.10
Severe Plastic Deformation

In the above-described preparation methods, the glassy phase is formed by rapidly cooling liquid alloys. Severe plastic deformation (SPD) is another approach to form an amorphous (glassy) phase by solid-state reaction. Among the various SPD methods, ball milling is a convenient method to obtain amorphous metal powders. However, the ball-milled powder needs additional steps for consolidation to form bulk samples.

The SPD process can produce bulk samples. One method is accumulative roll bonding (ARB) [79]. Figure 12.21a shows a schematic illustration of ARB. In ARB, pure metal sheets are stacked and rolled (e.g., 50% rolled), and the obtained multilayer sheet can be cut into two pieces and stacked again. Because the cross-sectional dimensions of the stack do not change, this process can be infinitely repeated and very large plastic strain of more than 10 in equivalent strain can be accumulated in the material. This method was originally developed to prepare bulk ultrafine grained material, but it has also been applied to

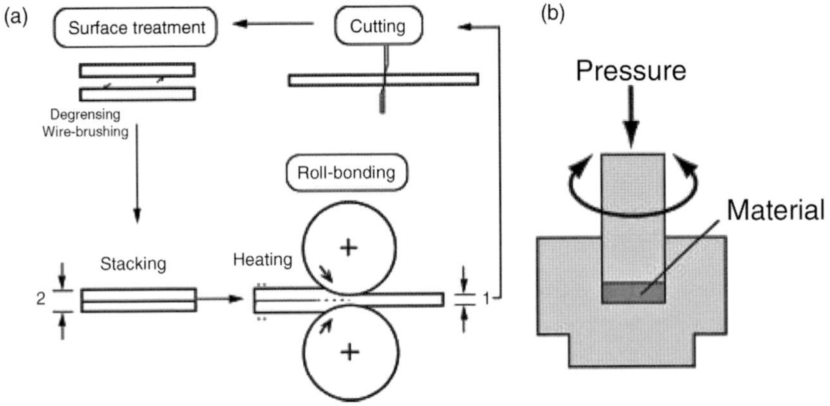

Figure 12.21 Schematic illustrations of (a) ARB and (b) HPT processes. (Reprinted with permission from Ref. [79].)

produce Zr–Cu bulk metallic glass sheets [80]. Seven layers of Cu and six layers of Zr subjected to ten cycles of the ARB process showed a nanoscale Zr/Cu lamellar structure with an amorphous phase at the interface. Subsequent annealing of the processed sample at 673 K induced solid-state amorphization in most of the samples.

High-pressure torsion (HPT) is another SPD method to produce bulk materials. Figure 12.21b shows a schematic illustration of HPT. In HPT, a disk-shaped sample is deformed by torsion under high pressure and a large plastic strain can be induced in the sample. Because the sample dimensions do not change with this process, very large shear strain can be achieved. However, it should be noted that the amount of strain differs depending on the radial distance [79]. A pure Zr and Cu multilayer have been deformed at 0.2 rpm under 5 GPa pressure [81]. With increasing rotation number, atomic mixing proceeded and after 20 rotations the sample became fully glassy. Similarly, using crystalline Zr–Al–Cu alloy as a precursor, HPT has been performed at 1 rpm under 5 GPa pressure [82]. After rotating 50 times, the processed sample became almost glassy and exhibited glass transition and crystallization at the same temperature as those of melt-quenched metallic glass. HPT has been recently applied to modify the structure of bulk metallic glass. By applying a large amount of shear strain to the glassy phase, a less relaxed glassy structure (more free volume) is formed that improves the room temperature plasticity of metallic glasses [83,84].

12.5 Important Factors for Metallic Glass Preparation

Several techniques have been proposed for synthesis of metallic glasses [12]. It is well known that metallic glasses are in a nonequilibrium state, so techniques need to be developed to avoid nucleation of the stable crystalline phase during their production. From this viewpoint, two different approaches have been suggested. One is a way to suppress heterogeneous nucleation in the supercooled liquid, because nucleation plays a dominant role in precipitation or formation of primary crystalline phases. If heterogeneous nucleation can be suppressed, glass formation can be achieved at a much lower cooling rate. The other approach is acceleration of the cooling rate by changing the production conditions. Saida *et al.* found that the atmosphere and pressure in the casting chamber affects the cooling rate, especially in the temperature region just above the glass transition temperature [65]. Because such conditions are very easy to control in conventional Cu mold casting for bulk metallic glass production, an effective cooling process can be obtained for various alloy systems. In this section, two key methods are explained that provide useful information for production of high-quality (bulk) metallic glasses.

12.5.1 Impurities

Heterogeneous nucleation is dominant for nucleating crystalline phases. In other words, if the liquid is supercooled to a temperature below the glass transition temperature without heterogeneous nucleation, a metallic glass can be easily produced. Heterogeneous nucleation generally occurs on the surface of the molten alloy or at the interface with an impurity, such as oxide, because the energy barrier for nucleation decreases according to Eq. (12.6). In the $Pd_{40}Ni_{40}P_{20}$ alloy, Drehman *et al.* found that the glass-forming ability is enhanced by etching the alloy ingot to eliminate surface heterogeneities [85]. Using the purified alloy, they estimated the critical cooling rate to be around 1 K/s, which is almost the same as that to avoid homogeneous nucleation. Kui *et al.* proposed another way to eliminate oxide impurities in a molten alloy using a flux [86]. They succeeded in removing impurities from the $Pd_{40}Ni_{40}P_{20}$ alloy using boron oxide (B_2O_3) flux that resulted in production of bulk metallic glass with diameter of 10 mm. When the molten alloy was heated at high temperature (~1273 K) while immersed in molten B_2O_3 flux, most of the heterogeneous impurities dissolved, leading to separation from the molten alloy because of the different specific gravity. After it was cooled, the purified alloy spontaneously separated from the flux containing impurities from the initial mother alloy ingot. Repeating this process, the alloy was significantly purified. Because the B_2O_3 flux is still in the liquid state near the glass transition temperature, the supercooled alloy liquid can be solidified because it is in a purified state, that is, heterogeneous nucleation is strongly suppressed. This method is regarded as a flux-melting technique and it is widely used to avoid heterogeneous nucleation during solidification.

Nishiyama *et al.* investigated the effect of flux treatment on nucleation and the glass-forming ability [87]. Figure 12.22 shows the relationship between the incubation time and temperature for crystallization of a highly purified $Pd_{40}Cu_{30}$-$Ni_{10}P_{20}$ melt under continuous cooling. They investigated glass formation in the fluxed alloy using a high-purity phosphorus polycrystal (up to 6 N), and

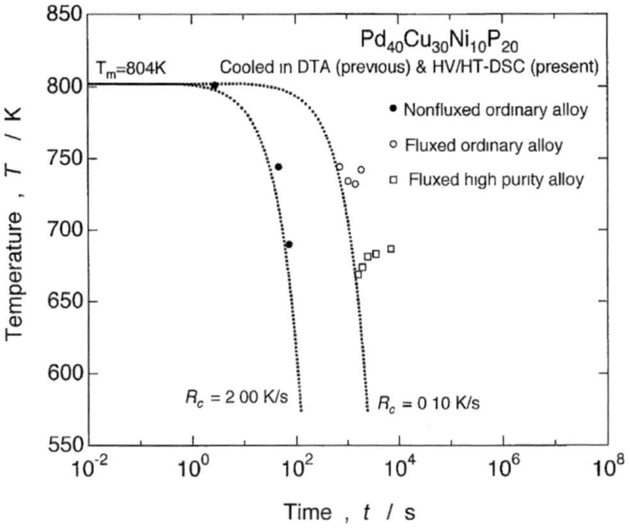

Figure 12.22 Relationship between incubation time and temperature for crystallization of a highly purified $Pd_{40}Cu_{30}Ni_{10}P_{20}$ melt under a continuous cooling rate. Previous data for the nonfluxed and fluxed ordinary alloy melts are also shown for comparison. (Reprinted with permission from Ref. [87].)

compared the nonfluxed and fluxed ordinary alloys using 98% phosphorus. The crystalline phase precipitated in a shorter time and at a higher supercooled temperature in the nonfluxed alloy. The critical cooling rate for glass formation was about 2 K/s. In the fluxed alloy, precipitation of the crystalline phase was suppressed, that is, a longer incubation time and lower supercooled temperature were exhibited that results in an extremely low critical cooling rate of 0.1 K/s. Moreover, the magnitude of supercooling for the fluxed high-purity alloy was enhanced by approximately 80 K compared with that for the fluxed ordinary melt. Thus, purification and flux treatment of the alloy is very effective to enhance the stability of the supercooled melt against crystallization.

Although flux treatment is a very useful way to suppress heterogeneous nucleation, it has only been reported for Pd-based alloys and it cannot be applied to many other alloy systems, such as the Zr-based system, because of the strong chemical affinity between the dominant element and oxygen. In this case, impurities need to be reduced using high-purity raw elements. For example, because the primary crystalline phase during solidification of the $Zr_{55}Al_{10}Ni_5Cu_{30}$ alloy from the melt is Zr_4Cu_2O, de Oliveira *et al.* concluded that oxygen is responsible for the presence of the crystalline phase [88]. Yokoyama *et al.* reported enhanced glass-forming ability using high-purity crystal Zr instead of sponge Zr as the raw material in $Zr_{50}Cu_{40}Al_{10}$ [89]. The measured cooling curves of $Zr_{50}Cu_{40}Al_{10}$ ingots using sponge and crystal Zr are shown in Figure 12.23. Distinct recalescence corresponding to precipitation of the crystalline phase is observed in only the cooling curve with sponge Zr. The oxygen

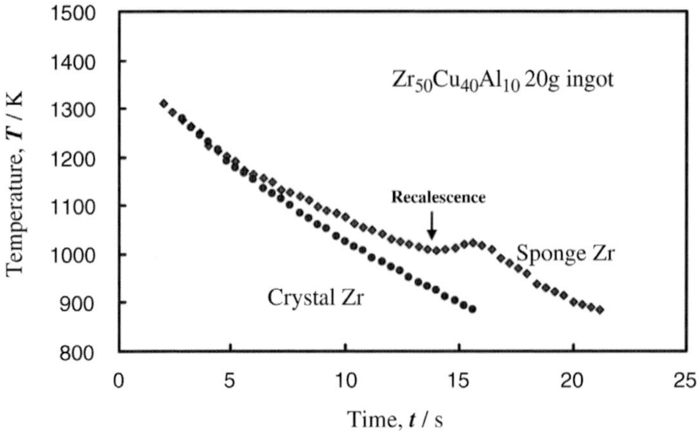

Figure 12.23 Measured cooling curves of $Zr_{50}Cu_{40}Al_{10}$ ingots using sponge Zr and crystal Zr. (Reprinted with permission from Ref. [89].)

concentration in sponge Zr is 1400 ppm, which is much larger than that in crystal Zr (45 ppm). They concluded that the glass-forming ability is enhanced by reduction of oxygen impurities introduced by the Zr raw element. Similar results of the influence of oxygen on glass formation have also been reported [90,91].

These studies indicate that suppression of heterogeneous nucleation can be achieved by purification of the raw elements of the mother alloy as well as by flux treatment during casting.

12.5.2 Atmosphere Gas and Its Pressure

Kato *et al.* found that the glass-forming ability is correlated with the hydrostatic pressure in Cu mold casting of $Zr_{65}Al_{7.5}Ni_{10}Cu_{7.5}Pd_{10}$ and $Zr_{65}Al_{7.5}Ni_{10}Pd_{17.5}$ alloys [92]. They reported that in both alloys, a single glassy phase is obtained in the sample cast in Ar atmosphere with a pressure of 0.1 MPa, while the crystalline structure is mainly formed in alloys cast in vacuum. Setyawan *et al.* investigated the effects of the hydrostatic pressure during casting and the gas species on glass formation in Zr–Al–Ni–Cu–Pd alloys with various Pd contents using structural analysis and experimental measurements of the cooling process [93]. They found that introduction of Ar or He gas results in superior glass-forming ability in the high Pd content alloys, whereas the glass-forming ability does not change in the low Pd content alloy. This different atmosphere dependence originates from the cooling process and precipitation behavior of the crystalline phase during casting.

Figure 12.24 shows measured cooling curves of the $Zr_{65}Al_{7.5}Ni_{10}Pd_{17.5}$ alloy cast into a 3 mm diameter Cu mold in vacuum (Z-17.5Pd3-V), 1 kPa Ar (Z-17.5Pd3-1 k), 10^5 Pa Ar (Z-17.5Pd3-A), and 10^5 Pa He (Z-17.5Pd3-H) atmospheres [94]. A high-purity Zr crystal was used as the raw material. Although

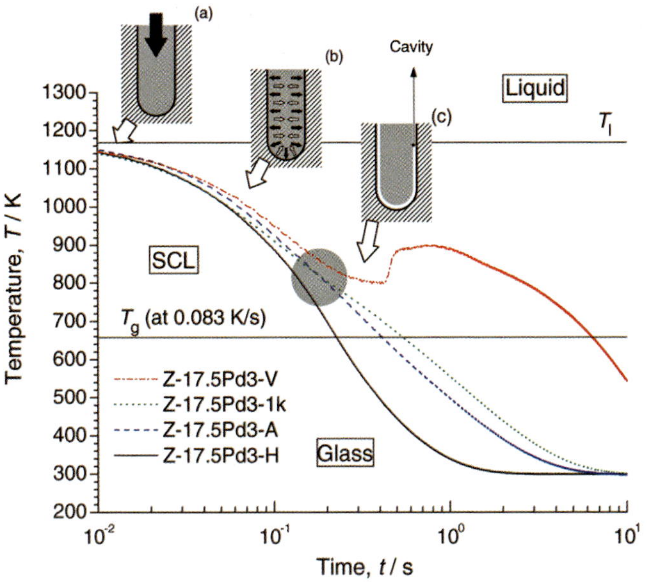

Figure 12.24 Measured cooling curves of the $Zr_{65}Al_{7.5}Ni_{10}Pd_{17.5}$ alloy cast into a 3 mm diameter Cu mold in vacuum (Z-17.5Pd3-V), 1 kPa Ar (Z-17.5Pd3-1 k), 10^5 Pa Ar (Z-17.5Pd3-A), and 10^5 Pa He (Z-17.5Pd3-H) atmospheres. (Reprinted with permission from Ref. [94].)

recalescence corresponding to precipitation of the crystalline phase (tetragonal Zr_2Ni) is observed in the curve in vacuum (Z-17.5Pd3-V), the other curves exhibit a continuous and smooth cooling process that indicates glassy solidification. There are no obvious differences in the curves in the initial stage of cooling just below the liquidus temperature, although they significantly differ in the final stage just above the glass transition temperature. The cast alloy in He atmosphere cools much faster than in the other atmospheres in the low-temperature range. Formation of a cavity between the cast alloy and the wall of the Cu mold affects the cooling effect. The cast alloy is directly cooled by heat transfer with the Cu mold in the initial stage, leading to cooling being independent of the atmosphere. The cavity then appears with shrinkage of the cast alloy by cooling. The cavity changes the cooling mode to indirect heat transfer via the atmosphere between the alloy and the mold, which indicates that the cooling characteristics strongly depend on the gas species and pressure in the chamber.

The glass-forming ability depending on the composition is also explained by the change of the cooling behavior. Figure 12.25 shows a schematic illustration of the different dependence of the glass-forming ability on the casting atmosphere for low Pd content ($x < 12.5$, Figure 12.25a) and high Pd content ($x \geq 12.5$, Figure 12.25b) alloys in the $Zr_{65}Al_{7.5}Ni_{10}Cu_{17.5-x}Pd_x$ system [95]. The continuous cooling transformation (CCT) diagram was constructed based on experimental data. The low Pd content alloys possess a CCT curve corresponding to the crystalline phase transformation located at the high-temperature side

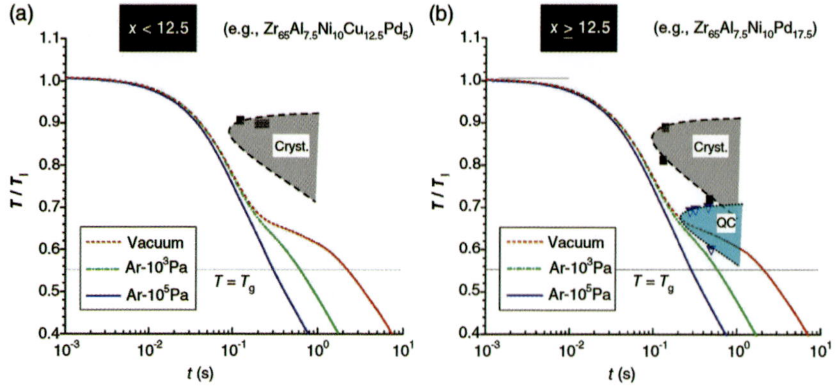

Figure 12.25 Schematic illustration of the different dependence of the glass-forming ability on the casting atmosphere for (a) low Pd content ($x < 12.5$) and (b) high Pd content ($x \geq 12.5$) $Zr_{65}Al_{7.5}Ni_{10}Cu_{17.5-x}Pd_x$ alloys. (Reprinted with permission from Ref. [95].)

in the supercooled liquid region, in which the cooling rate does not significantly depend on the atmosphere. An additional CCT curve that exhibits precipitation of the quasicrystalline (QC) phase exists in the low-temperature region in the high Pd content alloy. In this case, the glass-forming ability strongly depends on the atmosphere, that is, the cast alloy avoids precipitation of the QC phase because of the higher cooling rate in Ar or He atmosphere.

Thus, controlling the atmosphere during casting is very effective in accelerating the cooling rate in the low-temperature side of the supercooled liquid region. This should be regarded as one of the most important parameters for metallic glass production.

The atmosphere conditions have also been investigated for relaxation control of metallic glasses [96]. It is well known that the mechanical properties depend on the relaxation state, even if the glassy structure is maintained. The relaxation state is fixed by the cooling rate just above the glass transition temperature [65], which leads to control by changing the atmosphere during casting. Figure 12.26 shows specific heat curves of the $Zr_{65}Al_{7.5}Ni_{10}Cu_{12.5}Pd_5$ glassy alloy in the almost fully relaxed ($C_{p,r}$) and as-cast states ($C_{p,q}$) after casting in vacuum and ambient Ar pressure. The enthalpy of relaxation evaluated by $\int \Delta C_p dT$, where $\Delta C_p = C_{p,r} - C_{p,q}$, for the as-cast state in ambient Ar is much larger than that in vacuum that indicates production of metallic glass in a less relaxed state by casting in Ar atmosphere. The results show the possibility of obtaining metallic glasses with improved mechanical properties using this method. Setyawan *et al.* also reported increased plasticity in the less relaxed Zr-based metallic glass cast in Ar or He [97].

Thus, the synthesis conditions of metallic glass production should be carefully considered because they strongly affect the various properties and structure of the metallic glass. We expect further improvement of the synthesis techniques for metallic glass production.

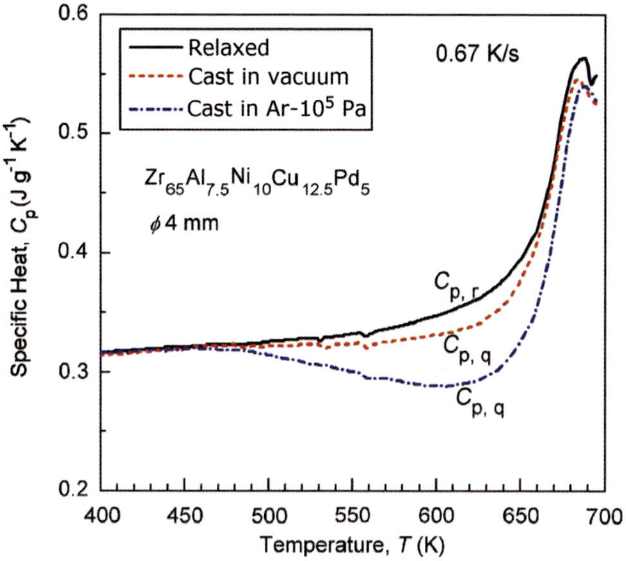

Figure 12.26 Specific heat curves of the $Zr_{65}Al_{7.5}Ni_{10}Cu_{12.5}Pd_5$ glassy alloy in the almost fully relaxed ($C_{p,r}$) and the as-cast states ($C_{p,q}$) cast under vacuum and ambient Ar. (Reprinted with permission from Ref. [96].)

12.6 Conclusions

Because of the isotropic nature of metallic bonds and the strong chemical affinity between metals and oxygen to form oxides, it took longer for glass formation in metals to be reported than formation of other types of glassy materials, such as inorganic/organic glasses. After the first report of a small thin piece of metallic glass of the Au–Si alloy in 1960, a lot of glass-forming alloys have been systematically discovered in eutectic binary alloy systems. Bulk metallic glasses have been successfully prepared by modifying binary glass formers, developing new equipment and treatments, and optimizing the preparation conditions. These bulk metallic glasses contribute to the fundamental understanding of most important research topics in metallic glasses, such as glass transition/formation, the rule of the alloy system/composition, local/medium-range structures, and the mechanical, chemical, and magnetic properties, as well as industrial applications.

References

1 Becker, R. and Doring, W. (1935) *Annu. Phys.*, **24**, 719–752.

2 Turnbull, D. and Fisher, J.C. (1949) *J. Chem. Phys.*, **17**, 71–73.

3 Turnbull, D. (1969) *Contemp. Phys.*, **10**, 473–488.

4 Zeldovich, J.B. (1943) *Acta Physicochim.*, **18**, 1–22.

5 Vogel, H. (1921) *Phys. Z.*, **22**, 645–646.
6 Fulcher, G.S. (1925) *J. Am. Ceram. Soc.*, **8**, 339–355.
7 Tammann, G. and Hesse, W. (1926) *Z. Anorg. Allg. Chem.*, **156**, 245–257.
8 Volmer, M.Z. (1929) *Z. Elektrochemie*, **35**, 555–561.
9 Uhlmann, D.R. (1969) *Materials Science Research*, vol. **4**, Plenum, New York.
10 Avrami, M. (1941) *J. Chem. Phys.*, **9**, 177–184.
11 Christian, J. (1965) *The Theory of Phase Transformations in Metals and Alloys*, Pergamon, Oxford.
12 Suryanarayana, C. and Inoue, A. (2010) *Bulk Metallic Glasses*, CRC Press, Boca Raton.
13 Uhlmann, D.R. (1972) *J. Non Cryst. Solids*, 7, 337–348.
14 Klement, W., Willens, R.H., and Duwez, P.O.L. (1960) *Nature*, **187**, 869–870.
15 Herlach, D.M. (1994) *Mater. Sci. Eng. R*, **12**, 177–272.
16 Mizutani, U., Hoshino, Y., and Yamada, H. (1986) *Handbook for Preparing Binary Amorphous Metals by Rapid Quenching Methods*, Agne Gijutu Center, Tokyo.
17 Xia, L., Li, W.H., Fang, S.S., Wei, B.C., and Dong, Y.D. (2006) *J. Appl. Phys.*, **99**, 026103.
18 Wang, D., Li, Y., Sun, B.B., Sui, M.L., Lu, K., and Ma, E. (2004) *Appl. Phys. Lett.*, **84**, 4029–4031.
19 Koch, C.C., Cavin, O.B., McKamey, C.G., and Scarbrough, J.O. (1983) *Appl. Phys. Lett.*, **43**, 1017–1019.
20 Inoue, A., Matsuki, K., and Masumoto, T. (1990) *Mater. Trans.*, **31**, 148–151.
21 Mukherjee, S., Schroers, J., Johnson, W.L., and Rhim, W.K. (2005) *Phys. Rev. Lett.*, **94**, 245501.
22 Donald, I.W. and Davies, H.A. (1978) *J. Non Cryst. Solids*, **30**, 77–85.
23 Hrubý, A. (1972) *Czech. J. Phys. B*, **22**, 1187–1193.
24 Lu, Z.P. and Liu, C.T. (2002) *Acta Mater.*, **50**, 3501–3512.
25 Inoue, A. (1995) *Mater. Trans.*, **36**, 866–875.
26 Inoue, A. (2000) *Acta Mater.*, **48**, 279–306.
27 Miracle, D.B., Greer, A.L., and Kelton, K.F. (2008) *J. Non Cryst. Solids*, **354**, 4049–4055.
28 Park, C., Saito, M., Waseda, Y., Nishiyama, N., and Inoue, A. (1999) *Mater. Trans.*, **40**, 491–497.
29 Nakamura, T., Matsubara, E., Imafuku, M., Koshiba, H., Inoue, A., and Waseda, Y. (2001) *Mater. Trans.*, **42**, 1530–1534.
30 Inoue, A. and Takeuchi, A. (2011) *Acta Mater.*, **59**, 2243–2267.
31 Kato, H., Saida, J., and Inoue, A. (2004) *Scr. Mater.*, **51**, 1063–1068.
32 Saida, J., Setyawan, A.D.H., Kato, H., and Inoue, A. (2005) *Appl. Phys. Lett.*, **87**, 151907.
33 Takeuchi, A. and Inoue, A. (2005) *Mater. Trans.*, **46**, 2817–2829.
34 Yamada, M., Yamasaki, T., Fujita, K., Yokoyama, Y., and Kim, D.H. (2014) *J. Jpn. Inst. Met.*, **78**, 449–458.
35 Liu, Y.H., Wang, G., Wang, R.J., Zhao, D.Q., Pan, M.X., and Wang, W.H. (2007) *Science*, **315**, 1385–1388.
36 Chen, L.Y., Fu, Z.D., Zhang, G.Q., Hao, X.P., Jiang, Q.K., Wang, X.D., Cao, Q.P., Franz, H., Liu, Y.G., Xie, H.S., Zhang, S.L., Wang, B.Y., Zeng, Y.W., and Jiang, J.Z. (2008) *Phys. Rev. Lett.*, **100**, 075501.
37 Sun, Y.L., Qu, D.D., Sun, Y.J., Liss, K.D., and Shen, J. (2010) *J. Non Cryst. Solids*, **356**, 39–45.
38 Peng, H.L., Li, M.Z., Sun, B.A., and Wang, W.H. (2012) *J. Appl. Phys.*, **112**, 023516.
39 Li, W., Bei, H., Tong, Y., Dmowski, W., and Gao, Y.F. (2013) *Appl. Phys. Lett.*, **103**, 171910.
40 He, M., Cui, L., Kubota, T., Yubuta, K., Makino, A., and Inoue, A. (2009) *Mater. Trans.*, **50**, 1330–1333.
41 Makino, A., Men, H., Kubota, T., Yubuta, K., and Inoue, A. (2009) *J. Appl. Phys.*, **105**, 7A308.
42 Makino, A., Men, H., Kubota, T., Yubuta, K., and Inoue, A. (2009) *Mater. Trans.*, **50**, 204–209.
43 German, R.M. (1984) *Powder Metallurgy Science*, Metal Powder Industries Federation, Princeton.
44 Nakayama, K.S., Yokoyama, Y., Wada, T., Chen, N., and Inoue, A. (2012) *Nano. Lett.*, **12**, 2404–2407.

45 Takagi, K., Masuda, S., Suzuki, H., and Kawasaki, A. (2006) *Mater. Trans.*, **47**, 1380–1385.

46 Takagi, K., Seno, K., and Kawasaki, A. (2004) *Appl. Phys. Lett.*, **85**, 3681–3683.

47 Masuda, S., Takagi, K., Dong, W., Yamanaka, K., and Kawasaki, A. (2008) *J. Cryst. Growth*, **310**, 2915–2922.

48 Dong, W., Takagi, K., Masuda, S., and Kawasaki, A. (2006) *J. Jpn. Soc. Powder Powder Metall.*, **53**, 346–351.

49 Miura, A., Dong, W., Fukue, M., Yodoshi, N., Takagi, K., and Kawasaki, A. (2011) *J. Alloys Compd.*, **509**, 5581–5586.

50 Yodoshi, N., Yamada, R., Kawasaki, A., and Makino, A. (2014) *J. Alloys Compd.*, **615**, S61–S66.

51 Yodoshi, N., Yamada, R., Kawasaki, A., and Makino, A. (2015) *J. Alloys Compd.*, **643**, S2–S7.

52 Eckert, J. (1997) *Mater. Sci. Eng. A*, **226**, 364–373.

53 El-Eskandarany, M.S., Zhang, W., and Inoue, A. (2002) *Mater. Trans.*, **43**, 1422–1425.

54 Zhang, X.Q., Ma, E., and Xu, J. (2006) *J. Non Cryst. Solids*, **352**, 3985–3994.

55 Sagel, A., Wunderlich, R.K., Perepezko, J.H., and Fecht, H.-J. (1997) *Appl. Phys. Lett.*, **70**, 580–582.

56 Seidel, M., Eckert, J., Bächer, I., Reibold, M., and Schultz, L. (2000) *Acta Mater.*, **48**, 3657–3670.

57 Zhang, L.C., Shen, Z.Q., and Xu, J. (2003) *J. Mater. Res.*, **18**, 2141–2149.

58 El-Eskandarany, M.S., Saida, J., and Inoue, A. (2003) *J. Mater. Res.*, **18**, 250–253.

59 El-Eskandarany, M.S., Saida, J., and Inoue, A. (2003) *Acta Mater.*, **51**, 1481–1492.

60 Inoue, A., Shinohara, Y., Yokoyama, Y., and Masumoto, T. (1995) *Mater. Trans.*, **36**, 1276–1281.

61 Wang, H., Li, R., Wu, Y., Chu, X.M., Liu, X.J., Nieh, T.G., and Lu, Z.P. (2013) *Compos. Sci. Technol.*, **75**, 49–54.

62 Shakur Shahabi, H., Scudino, S., Kühn, U., and Eckert, J. (2014) *Mater. Des.*, **59**, 241–245.

63 Zhang, Q.S., Guo, D.Y., Wang, A.M., Zhang, H.F., Ding, B.Z., and Hu, Z.Q. (2002) *Intermetallics*, **10**, 1197–1201.

64 Furuya, Y., Kudou, H., Kodaira, K., Hashimoto, K., Kimura, H., and Inoue, A. (2007) The 15th materials and processing conference (M&P2007), The Japan Society of Mechanical Engineers, Tokyo (in Japanese), 371.

65 Saida, J., Setyawan, A.D., Kato, H., and Inoue, A. (2011) *Metall. Mater. Trans. A*, **42**, 1450–1455.

66 Zhang, Q.S., Zhang, W., Wang, X.M., Yokoyama, Y., Yubuta, K., and Inoue, A. (2008) *Mater. Trans.*, **49**, 2141–2146.

67 Ma, H., Shi, L.L., Xu, J., Li, Y., and Ma, E. (2005) *Appl. Phys. Lett.*, **87**, 181915.

68 Dai, C.-L., Guo, H., Shen, Y., Li, Y., Ma, E., and Xu, J. (2006) *Scr. Mater.*, **54**, 1403–1408.

69 Zhang, T., Li, R., and Pang, S. (2009) *J. Alloys Compd.*, **483**, 60–63.

70 Yokoyama, Y., Inoue, K., and Fukaura, K. (2002) *Mater. Trans.*, **43**, 2316–2319.

71 Yokoyama, Y., Fukaura, K., and Inoue, A. (2002) *Intermetallics*, **10**, 1113–1124.

72 Yokoyama, Y., Gary Harlow, D., Liaw, P.K., and Inoue, A. (2010) *Metall. Mater. Trans. A*, **41**, 1780–1786.

73 Yokoyama, Y., Mund, E., Inoue, A., and Schulz, L. (2009) *J. Phys. Conf. Ser.*, **144**, 012043.

74 Itoi, T., Takamizawa, T., Kawamura, Y., and Inoue, A. (2001) *Scr. Mater.*, **45**, 1131–1137.

75 Kawamura, Y., Inoue, A., and Masumoto, T. (1993) *Scr. Mater.*, **29**, 25–30.

76 Kawamura, Y., Kato, H., Inoue, A., and Masumoto, T. (1995) *Appl. Phys. Lett.*, **67**, 2008–2010.

77 Bae, D.H., Lee, M.H., Kim, D.H., and Sordelet, D.J. (2003) *Appl. Phys. Lett.*, **83**, 2312–2314.

78 Xie, G., Louzguine-Luzgin, D.V., and Inoue, A. (2011) *J. Alloys Compd.*, **509** (Suppl. 1), S214–S218.

79 Tsuji, N., Saito, Y., Lee, S.H., and Minamino, Y. (2003) *Adv. Eng. Mater.*, **5**, 338–344.

80 Sun, Y., Tsuji, N., Kato, S., Ohsaki, S., and Hono, K. (2007) *Mater. Trans.*, **48**, 1605–1609.

81 Sun, Y.F., Nakamura, T., Todaka, Y., Umemoto, M., and Tsuji, N. (2009) *Intermetallics*, **17**, 256–261.

82 Meng, F.Q., Tsuchiya, K., and Yokoyama, Y. (2013) *Intermetallics*, **37**, 52–58.

83 Dmowski, W., Yokoyama, Y., Chuang, A., Ren, Y., Umemoto, M., Tsuchiya, K., Inoue, A., and Egami, T. (2010) *Acta Mater.*, **58**, 429–438.

84 Joo, S.H., Pi, D.H., Setyawan, A.D.H., Kato, H., Janecek, M., Kim, Y.C., Lee, S., and Kim, H.S. (2015) *Sci Rep.*, **5**. doi: 10.1038/srep09660.

85 Drehman, A.J., Greer, A.L., and Turnbull, D. (1982) *Appl. Phys. Lett.*, **41**, 716–717.

86 Kui, H.W., Greer, A.L., and Turnbull, D. (1984) *Appl. Phys. Lett.*, **45**, 615–616.

87 Nishiyama, N., Matsushita, M., and Inoue, A. (2001) *Mater. Trans.*, **42**, 1068–1073.

88 de Oliveira, M.F., Botta F, W.J., Kaufman, M.J., and Kiminami, C.S. (2002) *J. Non Cryst. Solids*, **304**, 51–55.

89 Yokoyama, Y., Fredriksson, H., Yasuda, H., Nishijima, M., and Inoue, A. (2007) *Mater. Trans.*, **48**, 1363–1372.

90 Lin, X.H., Johnson, W.L., and Rhim, W.K. (1997) *Mater. Trans.*, **38**, 473–477.

91 Gebert, A., Eckert, J., and Schultz, L. (1998) *Acta Mater.*, **46**, 5475–5482.

92 Kato, H., Inoue, A., and Saida, J. (2004) *Appl. Phys. Lett.*, **85**, 2205–2207.

93 Setyawan, A.D., Kato, H., Saida, J., and Inoue, A. (2007) *Mater. Sci. Eng. A*, **449–451**, 903–906.

94 Setyawan, A.D., Kato, H., Saida, J., and Inoue, A. (2007) *Mater. Trans.*, **48**, 1266–1271.

95 Setyawan, A.D., Kato, H., Saida, J., and Inoue, A. (2008) *J. Appl. Phys.*, **103**, 044907.

96 Saida, J., Setyawan, A. D., and Matsubara, E. (2011) *Appl. Phys. Lett.*, **99**, 061903.

97 Setyawan, A.D., Kato, H., Saida, J., and Inoue, A. (2008) *Philos. Mag.*, **88**, 1125–1136.

13
Crystal Growth Via the Gas Phase by Chemical Vapor Transport Reactions

Michael Binnewies,[1] Robert Glaum,[2] Marcus Schmidt,[1] and Peer Schmidt[3]

[1]Max-Planck-Institut für Chemische Physik fester Stoffe, Nöthnitzer Str. 40, 01187 Dresden, Germany
[2]Institut für Anorganische Chemie der Universität Bonn, Gerhard-Domagk-Str.1, 53121 Bonn, Germany
[3]Brandenburgische Technische Universität Cottbus – Senftenberg, Anorganische Chemie, Universitätsplatz 1, 01968 Senftenberg, Germany

13.1
Chemical Vapor Transport – Principles

The term chemical vapor transport (CVT) summarizes a variety of reactions that show one common feature: a condensed phase, typically a metallic or salt like solid, is volatilized in the presence of a gaseous reactant, the so-called transport agent, and deposits elsewhere, usually in the form of crystals. The deposition will take place if the site of volatilization and the site of crystallization have different temperatures. In many cases, chemical vapor transport is associated with a purification effect. CVT-reactions of elements, oxides, sulfides, selenides, tellurides, phosphates, sulfates, halides, oxide halides, phosphides, arsenides, intermetallics, and so on, are known.

A vast number of heterogeneous reactions involving the gas phase hardly differ from each other. If a condensed substance encounters a temperature gradient, it may move from the place of dissolution via the gas phase to the place of deposition. One observes the transport of a solid to another place. The transfer of a solid into the gas phase and vice versa may occur by different processes.

Sublimation

A well-known example is aluminum(III) chloride which is present in the gas phase in a large fraction as the dimeric molecule Al_2Cl_6.

$$2\,AlCl_3(s) \leftrightharpoons Al_2Cl_6(g)$$

Decomposition Sublimation

A solid can also decompose into various gaseous products during heating. While cooling down, the initial solid can be recrystallized from the gas phase. This is

Handbook of Solid State Chemistry, First Edition. Edited by Richard Dronskowski, Shinichi Kikkawa, and Andreas Stein.

called decomposition sublimation. Ammonium chloride is an example of this. It decomposes into ammonia and hydrogen chloride in the vapor phase.

$$NH_4Cl(s) \leftrightharpoons NH_3(g) + HCl(g)$$

During cooling, solid ammonium chloride is formed again.

Chemical Vapor Transport Reaction

A chemical vapor transport reaction is characterized by the fact that another substance, the transport agent, is required for the dissolution of a solid in the gas phase. The transport of iron(III) oxide with hydrogen chloride is an example.

$$Fe_2O_3(s) + 6\,HCl(g) \leftrightharpoons 2\,FeCl_3(g) + 3\,H_2O(g)$$

Due to the *forward reaction*, Fe_2O_3 is transferred into the gas phase, this is usually called dissolution. Due to the *reverse reaction*, it is deposited from the gas phase. The site of the volatilization is called the source and the site of deposition the crystallization zone or sink. The chemical equation describing dissolution and deposition is called transport equation. Reaction enthalpy and entropy of a transport reaction always refer to the dissolution reaction.

van Arkel and de Boer [1] were the first scientists to carry out specific transport reactions in the laboratory from 1925 onwards. van Arkel and de Boer used the so-called *hot wire method*, often called *iodide process*. In this process, the crude metal M transforms during an exothermic reaction into a gaseous metal iodide (MI_n) in the presence of iodine as transport agent. The iodide reaching a hot wire, heated up electrically to high temperatures, decomposes under formation of the purified metal and iodine. Due to this much higher temperature, the reverse reaction of the exothermic dissolution reaction is favored by le Chatelier's principle. Such a vapor transport reaction can be described by the following equation:

$$M(s) + n\,I(g) \leftrightharpoons MI_n(g) \quad \text{exothermic}$$

The source and deposition temperatures are commonly given as T_1 and T_2, respectively, T_1 representing the lower temperature. Therefore, exothermic reactions always lead to a transport from T_1 to T_2 ($T_1 \rightarrow T_2$), endothermic reactions from T_2 to T_1 ($T_2 \rightarrow T_1$). A chemical vapor transport reaction is typically expressed as follows:

$$i\,A(s) + kB(g) \leftrightharpoons jC(g) + \ldots$$

There may be further gaseous products in addition to $C(g)$.

13.1.1 Transport via a Single Equilibrium Reaction

Let us consider a well-known example, the vapor transport of zinc sulfide with iodine as transport agent applying a temperature gradient of $1000 \rightarrow 900\,°C$. Here, we assume for simplicity that iodine is present as $I_2(g)$ and sulfur as $S_2(g)$

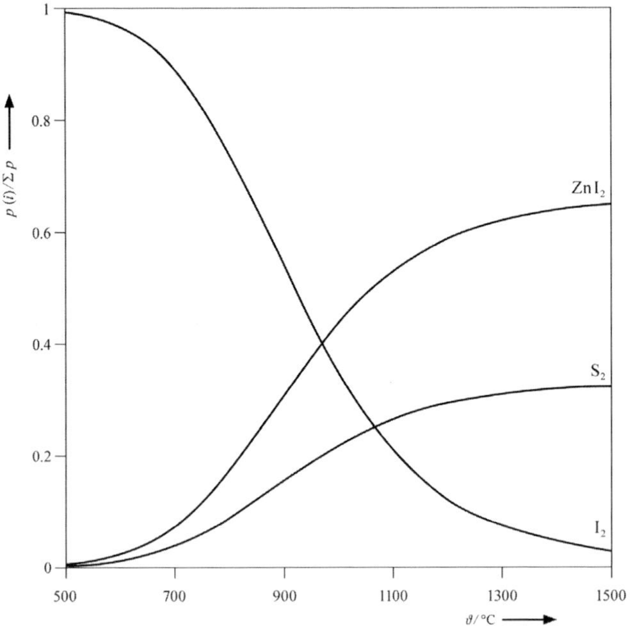

Figure 13.1 Temperature dependency of the normalized partial pressures of I_2, ZnI_2, and S_2 in the reaction of zinc sulfide with iodine. (Ref. [2], p. 5.)

at the reaction conditions. Thus, the transport equation follows:

$$ZnS(s) + I_2(g) \rightleftharpoons ZnI_2(g) + 1/2S_2(g)$$

$$\Delta_r H^0_{298} = 144\,\text{kJ/mol}, \quad \Delta_r S^0_{298} = 124\,\text{J/(mol·K)}$$

Figure 13.1 emerges if the partial pressures of iodine, zinc iodide, and sulfur (S_2) are calculated from the thermodynamic data of the transport reaction and plotted as a function of the temperature, taking into account the experimental conditions.

As expected, the amount of iodine in the gas phase decreases with rising temperatures and the amounts of zinc iodide and sulfur are increasing. Zinc iodide and sulfur are the so-called transport-effective species, because they contain the atoms that constitute the solid zinc sulfide. Furthermore, the partial pressure of S_2 is exactly half that of ZnI_2 in the entire temperature range. This corresponds to the ratio of the stoichiometric factors of S_2 and ZnI_2 in the transport equation. The transport proceeds from T_2 to T_1. The expected transport effect is largest where the quotient $\Delta p/\sum p$ at a defined temperature interval ΔT reaches the highest value for the transport-effective species. In this case, it is close to the inflection point of the illustrated graph for ZnI_2 (or S_2), that is, at temperatures around 900 °C. Calculating the equilibrium constant K_p for three chosen temperatures, 700, 900, and 1100 °C, one obtains the following values: 0.03 $\text{bar}^{0.5}$, 0.5 $\text{bar}^{0.5}$, and 3.8 $\text{bar}^{0.5}$. These values show exemplarily that the expected transport effect is particularly large if the numerical value of the equilibrium constant for the transport reaction is close to unity.

The Optimum Transport Temperature

The temperature at which the numerical value of the equilibrium constant K_p equals 1 is referred to as optimum transport temperature T_{opt}. The equilibrium constant is easily calculated from the thermodynamic data of the transport reaction. For this purpose, it is sufficient to use the reaction enthalpy and reaction entropy at 298 K. van't Hoffs equation establishes the link between K and the thermodynamic data of the reaction enthalpy and entropy.

$$\ln K = -\frac{\Delta_r H^0}{R \cdot T} + \frac{\Delta_r S^0}{R}. \tag{13.1}$$

For $K = 1$, the following expression results:

$$T_{opt} = \frac{\Delta_r H^0}{\Delta_r S^0}. \tag{13.2}$$

It is not always necessary to consider thermodynamics prior to the experiment in order to choose a suitable transport agent. Often, it is sufficient to check whether similar solids have already been transported, which transport agents have been used and at which temperatures the transport has occurred.

The transport rate, the amount of deposited substance per time in the sink, is highest in a transport gradient at the optimum transport temperature. However, transport reactions are often carried out at higher temperatures than the calculated optimum transport temperature. Usually, such conditions are chosen because at the calculated temperature, condensation of unwanted reaction products would occur or crystal growth would be kinetically inhibited.

According to Schäfer [2,3] the transport rate $\dot{n}(A)$ can be calculated in simple cases with the help of the following equation, which is based on a diffusion approach:

$$\dot{n}(A) = \frac{n(A)}{t'} = \frac{i}{j} \cdot \frac{\Delta p(C)}{\Sigma p} \cdot \frac{\overline{T}^{0.75} \cdot q}{s} \cdot 0.6 \cdot 10^{-4}\ \text{mol/h}, \tag{13.3}$$

$\dot{n}(A)$ transport rate/(mol/h)

i, j stoichiometric coefficients in the transport equation

$$i\,A(s) + k\,B(g) \rightleftharpoons j\,C(g) + \ldots,$$

$\Delta p(C)$ partial pressure difference of the *transport effective* species C/bar

Σp total pressure/bar

$\overline{T}$ mean temperature along the diffusion path/K (practically, $\overline{T}$ can be taken as average of T_1 and T_2)

q cross-section of the diffusion path/cm^2

s length of the diffusion path/cm

t' duration of the transport experiments/h

13.1.2
The Gas Phase Solubility λ

According to van Arkel, iron can be transported exothermically with iodine from 800 to 1000 °C. At first, the following transport equation comes into consideration:

$$\begin{array}{l} \mathrm{Fe(s)} + \mathrm{I_2(g)} \leftrightharpoons \mathrm{FeI_2(g)} \\ \Delta_r H^0_{298} = 24\,\mathrm{kJ/mol} \end{array}.$$

This reaction, however, is endothermic. According to le Chatelier's principle, a transport from T_2 to T_1 is expected. Because of the strict validity of this principle, one has to assume that the transport obviously cannot be described, or at least not alone by this reaction. In contrast, the following exothermic reaction could be describing the observed transport direction:

$$\begin{array}{l} \mathrm{Fe(s)} + 2\,\mathrm{I(g)} \leftrightharpoons \mathrm{FeI_2(g)} \\ \Delta_r H^0_{298} = -128\,\mathrm{kJ/mol} \end{array}.$$

A detailed investigation showed that other reactions take place as well. Accordingly, iron(II) iodide forms monomeric and dimeric molecules, FeI_2 and Fe_2I_4, in the vapor. The reaction of iron with iodine under formation of gaseous Fe_2I_4 molecules can be described as

$$\begin{array}{l} \mathrm{Fe(s)} + 2\,\mathrm{I(g)} \leftrightharpoons 1/2\,\mathrm{Fe_2I_4(g)} \\ \Delta_r H^0_{298} = -209\,\mathrm{kJ/mol} \end{array}.$$

Again, this reaction equation has the character of a transport equation. The reaction is exothermic. Therefore, one expects a transport from T_1 to T_2. The situation becomes more complicated because of the temperature dependent dissociation of iodine. Iodine molecules are dominantly present at low temperatures while iodine atoms react with iron by formation of $FeI_2(g)$ and $Fe_2I_4(g)$ at high temperatures. Different transport equations result, even if the molecules FeI_2 and Fe_2I_4 act as transport-effective species in both cases.

Figure 13.2 shows the molar fractions of all molecules involved in the transport reaction as a function of the temperature.

Figure 13.2 clearly shows the following:

- The amounts of iodine atoms and iodine molecules in the gas phase are small compared to those of iron iodides. The equilibria, which lead to the formation of FeI_2 and Fe_2I_4, are apparently positioned on the product side.
- The fraction of FeI_2 initially increases with rising temperatures. This corresponds to the expectation of the formation of FeI_2 from iron and iodine molecules by an endothermic equilibrium (le chatelier's principle). If this equilibrium was the only determinant of the transport of iron, it would lead to migration of iron from the higher to the lower temperature in the considered

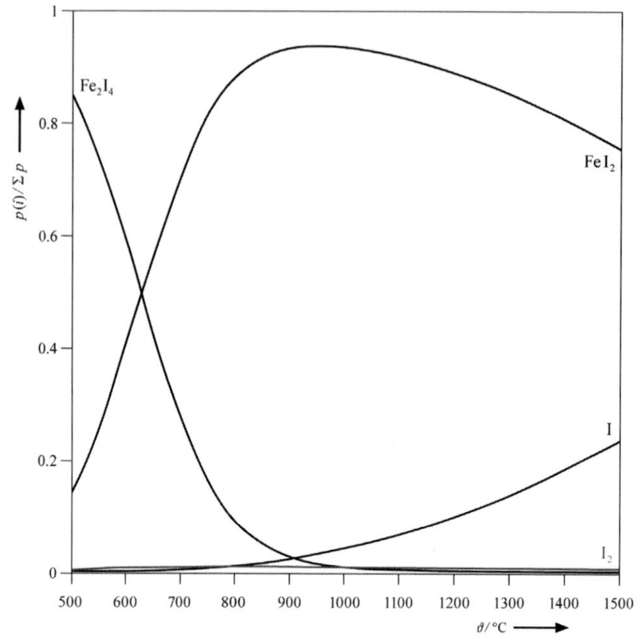

Figure 13.2 Temperature dependency of the molar fractions of I, I_2, FeI_2, and Fe_2I_4 in the gas phase formed by the reaction of iron with iodine. (Ref. [2], p. 7.)

interval (800 → 1000 °C). Above 1000 °C, the fraction of FeI_2 decreases with rising temperatures. This is due to the presence of mainly atomic iodine in this temperature range. The reaction of iron with iodine atoms is exothermic, thus the amount of FeI_2 decreases with rising temperature and the amount of iodine atoms increases with rising temperature, the fraction of Fe_2I_4 decreases, too.

Below 1000 °C, we therefore deal with two opposing processes – the increasing formation of FeI_2 and the decreasing formation of Fe_2I_4, both due to rising temperature. The first process lets us expect transport toward the zone of lower temperature, the second one to the higher temperature zone. Based on the so far used considerations, it is not predictable which process will dominate. A new term – the gas phase solubility [4] – is helpful in answering this question.

The term gas phase solubility λ is used with reference to the solubility of a substance in a liquid. Liquid solutions are used for the purification of the dissolved substance through recrystallization. One uses the temperature dependency of the solubility, respectively of the solubility equilibrium. From a solution saturated at higher temperature, recrystallization of the solute is achieved by cooling. Generally, this is associated with a purifying effect. A chemical vapor transport reaction basically works the same way. Here, too, one uses the temperature dependence of the gas phase solubility in order to crystallize and to purify. In both cases, one deals with heterogeneous equilibria; in the first case between a solid and a liquid, in the second between a solid and a gas phase.

The example of the transport of iron shows the advantage of the concept of the solubility (of a solid) in the gas phase for the description of complicated transport reactions. According to the transport equations, iron is volatilized as FeI_2 and Fe_2I_4, and thus dissolved in the gas phase. The gas phase is the solvent, that is, all gaseous species together. On the basis of this idea, the description of the solubility of iron in the gas phase in a quantitative way has to take into consideration that one molecule Fe_2I_4 includes two Fe-atoms, whereas the FeI_2 molecule includes only one. This is done by multiplying the partial pressure of Fe_2I_4 by the factor 2. The same applies to the solvent gas phase. If one primarily assumes iodine atoms, one has to multiply the partial pressures of the other gaseous molecules with the corresponding factors. The solubility of iron in the gas phase can be described as

$$\lambda(\mathrm{Fe}) = \frac{p(\mathrm{FeI_2}) + 2 \cdot p(\mathrm{Fe_2I_4})}{p(\mathrm{I}) + 2 \cdot p(\mathrm{I_2}) + 2 \cdot p(\mathrm{FeI_2}) + 4 \cdot p(\mathrm{Fe_2I_4})}. \tag{13.4}$$

The temperature dependency of the so defined solubility of iron in the gas phase takes both ferrous molecules FeI_2 and Fe_2I_4 into consideration. Figure 13.3 presents the solubility of iron in the gas phase as a function of the temperature.

The solubility decreases with rising temperatures. If less iron dissolves at higher temperatures in the gas phase than at lower temperatures, iron must be transported from lower to higher temperatures. This is in line with experimental observations. With the aid of the *solubility of a solid in the gas phase*, it is

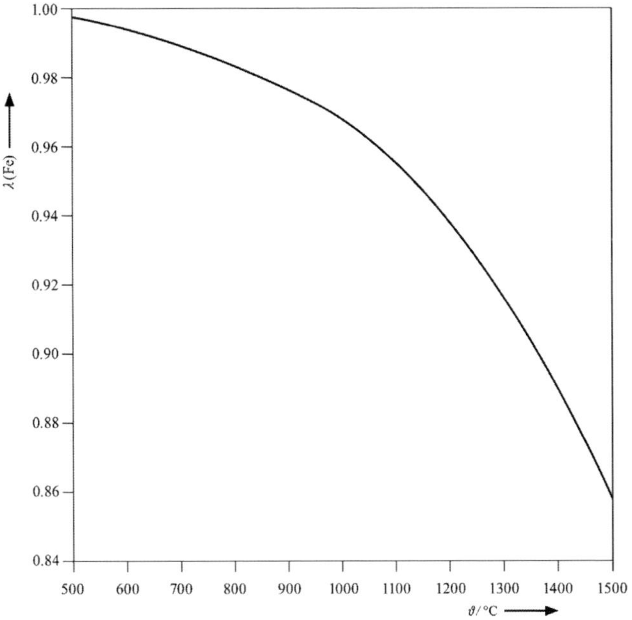

Figure 13.3 Temperature dependency of the gas phase solubility of iron. (Ref. [2], p. 8.)

possible to describe the transport direction correctly even when several transport reactions are occurring simultaneously.

$$\dot{n}(A) = \frac{n(A)}{t'} = \lambda \cdot \frac{\overline{T}^{0.75} \cdot q}{s} \cdot 0.6 \cdot 10^{-4}\ \mathrm{mol/h}. \qquad (13.5)$$

13.2 Working Techniques

In most cases, transport reactions are conducted in closed ampoules of suitable glass. The selection of the kind of glass is defined by the transport temperatures. Today, borosilicate glass that is suitable up to 600 °C is frequently used. Glass made from pure silicon dioxide, quartz, or silica glass is appropriate at higher temperatures and for more corrosive fillings. Silica tubes are stable at temperatures up to 1100 °C. It is important to note that water is released during heating of silica glass (water content up to 50 ppm [5]).

Vapor transport reactions take place in a temperature gradient. In order to set up the gradient in a controlled manner, tube furnaces with two independent heating coils are used. The transport furnace should be in horizontal position in order to keep convection as part of the gas motion as small as possible. However, if the aim of the transport is the preparation of large amounts of substance by an endothermic transport, the furnace can be tilted so that the sink side is higher than the source side. This increases the transport rate. One has to make sure that the thermocouples measure the temperatures of the source and sink zone as accurately as possible. It is important that the tip of the thermocouple, the junction, is as close as possible to the source and the sink zone of the ampoule, respectively. Further, the other end of the thermocouple sticks out of the furnace so that the connection point between the actual thermocouple and the connection line to the temperature regulator is at room temperature; if not, measurement errors can occur. Figure 13.4 schematically shows the experimental set up for a chemical vapor transport in a two-zone furnace. In Figure 13.5 a typical temperature characteristic in a two-zone furnace is given.

There are various experimental procedures for preparing transport ampoules. Above all, they depend on the physical and chemical properties of the transport agent. Following three cases can be distinguished:

1) The transport agent is solid or liquid at ambient conditions but has a considerable vapor pressure (e.g., iodine).
2) The transport agent is gaseous at ambient conditions (e.g., HCl, Cl_2).
3) The transport agent does not have a vapor pressure at room temperature, but has to be prepared in a prereaction (e.g., $TeCl_4$, $AlCl_3$).

For 1:A tube closed at one side is prepared (diameter 10–20 mm). Approximately 15–20 cm from the closed end, the tube is narrowed so that the reactants can be filled in without any problem; the narrow part should be as narrow as

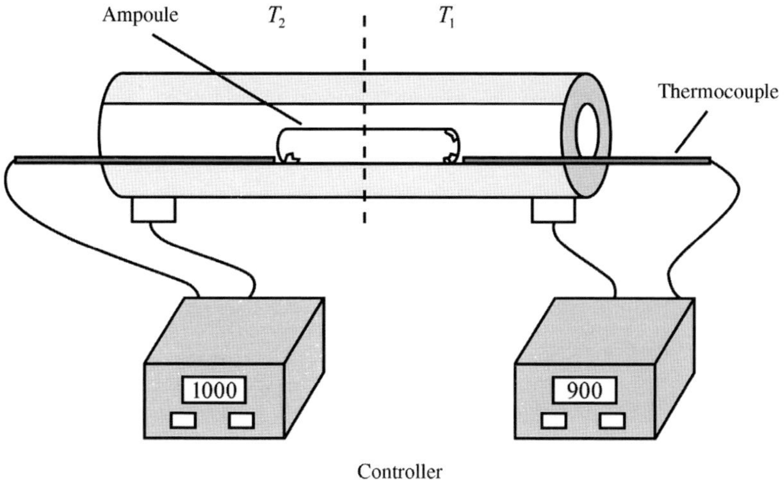

Figure 13.4 Experimental set up for chemical vapor transport in a two-zone furnace. (Ref. [2], p. 557.)

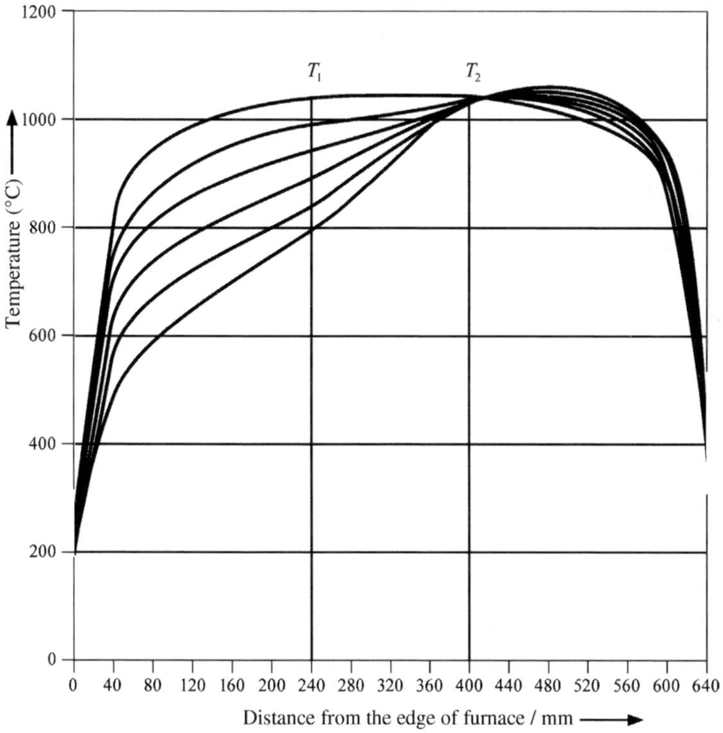

Figure 13.5 Typical temperature characteristic of a two-zone furnace (T_2 = constant, T_1 = variable). (Ref. [2], p. 557.)

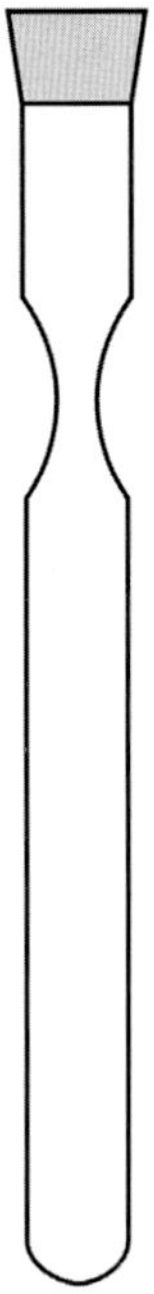

Figure 13.6 Transport ampoule with a ground joint. (Ref. [2], p. 558.)

possible in order to simplify the sealing but still wide enough to offer sufficient space for a long funnel. The open end of the tube is to be arranged in a way to enable a smooth and vacuum-tight junction to the vacuum line. Thus prepared, the ampoules typically look like illustrated in Figure 13.6.

First, the tubes prepared as above are filled with approximately 1 g of the solid that is to be transported. For this purpose, a funnel long enough for the outlet to be near the ampoule bottom is used. Thus, small particles of the solid can be prevented from sticking to the ampoule wall and later functioning as crystal seeds during deposition. This would lead to an unwanted formation of many small crystals. In the same way, the transport agent can be added. Its amount is often selected so that the pressure (approximately, expressed by the initial pressure of the transport agent) in the ampoule is 1 bar at the experiment temperature (gas law).

The ampoule is attached to the vacuum line in vertical arrangement, the valve stays shut. Usually, the content of the ampoule must be cooled before evacuation in order to avoid vaporization or sublimation. For cooling, the bottom end of the vertically arranged ampoule is dipped 5 cm in the cooling medium. After approximately 3 min, the valve to the vacuum apparatus is opened; the reaction tube is evacuated and sealed. The cooling of the ampoule is necessary when the vapor pressure of one or more components of the filling is as high at room temperature for there to be a danger of vaporization or sublimation. *If iodine is used as transport agent, cooling is obligatory.*

For 2: Another way to introduce a certain amount of gaseous transport agent into the ampoule is via condensation. Often, chlorine is used as transport agent. If one wants to avoid the handling of a steel cylinder filled with liquid chlorine, the element can be evolved easily by thermal decomposition of platinum(II) chloride at 525 °C.

$$PtCl_2(s) \rightarrow Pt(s) + Cl_2(g)$$
(above 525 °C)

The amount of $PtCl_2$ needed depends on the setting of the initial pressure, the temperature, and the volume of the ampoule. It can be calculated with the help of the ideal gas law. In order to set a pressure of 1 bar at room temperature in a volume of 1 ml, about 0.01 g $PtCl_2$ is needed.

$$p \cdot V = n \cdot R \cdot T = \frac{m}{M} R \cdot T \tag{13.6}$$

$$m = \frac{p \cdot V \cdot M}{R \cdot T} \tag{13.7}$$

$$m = \frac{1\,\text{bar} \cdot 10^{-3}\, l \cdot 266\,\text{g/mol}}{0.083141 \cdot \text{bar}/(\text{K} \cdot \text{mol}) \cdot 298\,\text{K}} = 1.073 \cdot 10^{-2}\,\text{g}. \tag{13.8}$$

Equally, $PtBr_2$ can serve as a bromine source. It decomposes to the elements above 475 °C.

Mostly, the according ammonium halides are used as source for hydrogen halides, since they are particularly easy to dose. One has to keep in mind, however, that the ammonia which is formed besides HX(g) decomposes into nitrogen and hydrogen at temperatures above 700 °C and a reducing atmosphere is generated. If this is not wanted, the hydrogen halides have to be inserted with a special working technique.

A certain amount of water can easily be provided by heating $BaCl_2{\cdot}2\,H_2O$ to 150 °C. Oxygen forms during heating of gold(III) oxide to approximately 200 °C.

For 3: It is particularly easy to feed the transport ampoule, if the transport agent has practically no vapor pressure at room temperature, for example, aluminum(III) chloride and tellurium(IV) chloride. However, these additives, which are frequently used as transport agents, are very moisture-sensitive. In order to insert them into transport ampoules, different working techniques are used. These are described in detail in Ref. [2].

The Transport Experiment

The prepared transport ampoule is placed in the furnace so that the middle of the furnace and the middle of the ampoule are at the same position. Before the actual transport temperatures are set, the ampoule is exposed to a reversed temperature gradient for approximately 1 day. This way, the ampoule walls of the sink side are freed of small crystallization seeds. This approach is called back transport or a transport in a reverse temperature gradient. The ampoule is taken out carefully after terminating the experiment. In order to maintain crystals without being contaminated by the condensed gas phase, one has to make sure that the gas phase condenses on the source side. Different approaches have been established to achieve this.

The transport ampoule is removed from the furnace on its source side as far for the ampoule to stick out approximately 5 cm. Within a few minutes, the gas phase condenses there and the ampoule can be taken out. Opening of the ampoule can be done in different ways. Either the ampoule is scarified at a suitable spot, wrapped with a firm cloth and broken open or an approximate 1 mm slot is cut in the ampoule with a suitable tool. After securing with a cloth, the ampoule is then opened by levering with a screwdriver. Finally, the sink side is rinsed out with a suitable solvent.

13.3 Examples

In the following, the vapor transport of some selected solids is described. In particular, the crystallization of elements, intermetallics, halides, oxides, sulfides, selenides, tellurides, and pnictides is discussed exemplarily regarding typical representatives.

13.3.1 Chemical Vapor Transport of Elements

The most commonly used transport agent for elements is iodine. Iodine is unsuitable if an element forms an extremely stable or an extremely unstable gaseous iodide. Examples for such type of behavior are the metals of group I and II, and the platinum metals respectively. If a gaseous metal iodide is highly unstable, transport may be possible with bromine or chlorine as transport agents, as in the case of some platinum metals. The solubility of a metal in the presence of a halogen is often enhanced by addition of aluminum halides. Thus, so-called "gas complexes" can be formed (see Section 13.3.3). These can be useful in the case of the transport of noble metals [6].

van Arkel–de Boer Process

The van Arkel–de Boer process is always exothermic. The deposition temperature exceeds 1000 °C in all cases. Therefore, special experimental equipment is needed [7].

$$\mathrm{Zr(s)} + 4\,\mathrm{I(g)} \leftrightharpoons \mathrm{ZrI_4(g)} \quad (400 \rightarrow 1100 \ldots 2000\,^\circ\mathrm{C})$$

Conproportionation Reactions

Germanium can be transported, too, using iodine as transport additive. The reaction, however, is endothermic and therefore obviously not of the van Arkel–de Boer type. Detailed investigations of the solid/gas-reactions have shown the formation of $GeI_4(g)$ from the elements in a first reaction step. However, the endothermic transport of solid germanium is caused by the formation and decomposition of $GeI_2(g)$ at 500 °C and 400 °C, respectively. The following equation describes the transport process. In this case, iodine is a *transport additive*, the *transport agent* however is $GeI_4(g)$.

$$Ge(s) + GeI_4(g) \rightleftharpoons 2\,GeI_2(g) \quad (500 \rightarrow 400\,°C)$$

In the case of the transport of silicon with iodine as transport additive, endothermic and exothermic transport processes have been described depending on pressure and temperature conditions [8]. This is a special case with an inversion of the transport direction. The exothermic transport is of the van Arkel–de Boer – the endothermic one of the conproportionation – type.

Transport with Oxygen

Some noble metals can be transported with oxygen as transport agent. The noble metals are volatilized at temperatures above 1000 °C and deposited at lower temperatures. The transport of platinum as an example can be described by the following equation:

$$Pt(s) + O_2(g) \rightleftharpoons PtO_2(g) \quad (1500\,°C \rightarrow T_1)$$

13.3.2
Chemical Vapor Transport of Intermetallics

In principle, the same transport agents are suitable for metals and intermetallics. The transport mechanisms and conditions are very similar, too. The challenge in the selection of the transport agent suitable for a particular intermetallic system is that the transport agent must react with all components of the intermetallic phase forming gaseous reaction products of similar stability at the transport temperature and pressure. Additionally, the reaction enthalpies must have the same sign. In many cases, the transport agent causes an enrichment of one component of the solid in the gas phase. In these cases the compositions in source and sink often differ from each other. In many intermetallic systems several intermetallic compounds occur. During the transport process, often neighboring compounds in the phase diagram crystallize in the sink. The composition of the deposited solid can be influenced by the following parameters:

1) The composition of the source solid
2) The kind of transport agent and its concentration
3) The temperatures of source and sink as well as the resulting temperature gradient

The transport of intermetallic compounds is well known in binary systems and in some ternary ones.

Intermetallic Phases with Wide Homogeneity Range

Molybdenum and tungsten are two metals with very high melting points. They are isotypic and show complete miscibility in the solid and liquid state. It requires great experimental effort to grow molybdenum/tungsten mixed-crystals.

With the help of chemical vapor transport reactions, this succeeds far below the liquidus curve at temperatures around 1000 °C. Both metals can be transported at the same conditions when mercury(II) bromide is added.

$$Mo_{1-x}W_x(s) + 2\,HgBr_2(g) \leftrightharpoons (1-x)MoBr_4(g) + x\,WBr_4(g) + 2\,Hg(g)$$
$$(1000 \rightarrow 900\,°C)$$

In the following binary systems, mixed crystals are transportable: cobalt/nickel, iron/nickel, silver/copper, gold/copper, copper/nickel, gold/nickel, and copper/gallium [2]. The composition of the deposited mixed crystals can be influenced by the experimental conditions, in particular by the composition of the source solid. One cannot always assume that the composition of source and sink solid are identical. Enrichment effects by transport actions can also lead to concentration gradients within individual crystals.

Stoichiometrically Composed Intermetallic Phases

As a typical example, we consider the system nickel/tantalum. The respective phase diagram shows the existence of binary compounds Ni_8Ta, Ni_3Ta, Ni_2Ta, $NiTa$, and $NiTa_2$, Figure 13.7. All compounds in this system can be transported

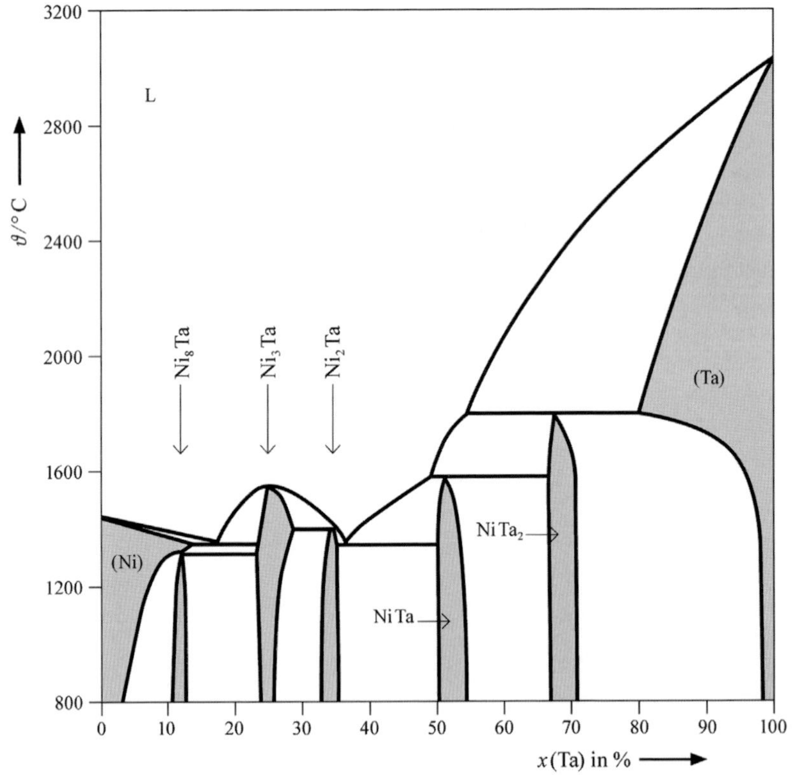

Figure 13.7 Phase diagram of the system Ni/Ta. (Ref. [10].)

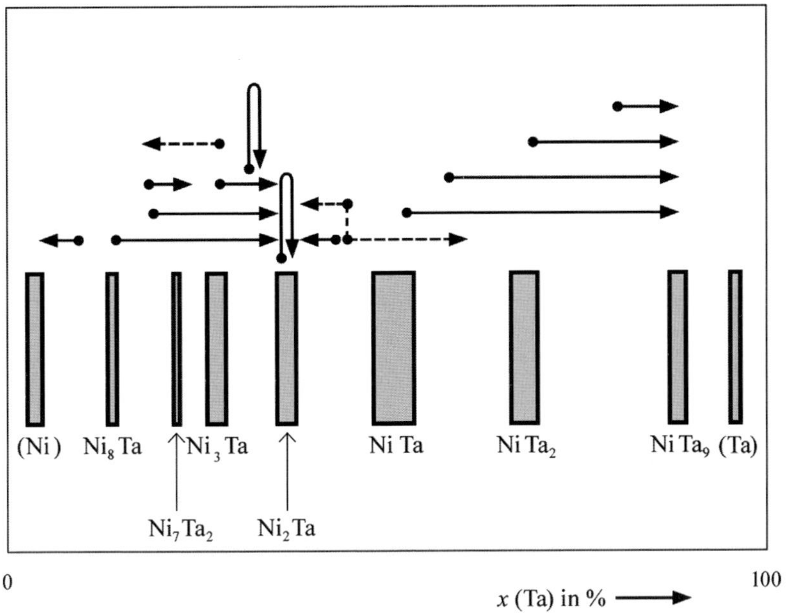

Figure 13.8 "Transport shed" in the system Ni/Ta. (Ref. [9])

from 800 to 950 °C using iodine as transport additive. In the majority of the cases, one observes an incongruent transport, the compositions of source and sink solid are different. This is shown in Figure 13.8. The arrows show the different compositions in source and sink, for example, starting with NiTa, $NiTa_9$ is deposited in the sink. The activities of the two components in the source solid are the main parameters that determine the composition of the deposited solid. In particular, the composition in the source determines the composition in the sink. Additionally, chemical composition and amount of the transport agent, and the temperature gradient have an influence as well; however, they are of minor importance.

13.3.3
Halides

Many metal halides can be obtained by chemical vapor transport reactions. In the process, four different types of solid/gas-reactions play a central role. These are discussed in the following:

- Halogens as transport agent, formation of higher halides

$$CrCl_3(s) + 1/2\,Cl_2(g) \rightleftharpoons CrCl_4(g)$$

- Conproportionation reactions

$$MoCl_3(s) + MoCl_5(g) \rightleftharpoons 2\,MoCl_4(g)$$

- Formation of gas complexes

$$CoCl_2(s) + Al_2Cl_6(g) \rightleftharpoons CoAl_2Cl_8(g)$$

Formation of a Higher Halide

An example of a vapor transport of a metal halide well known to chemists is the preparation of pure chromium(III) chloride. First, solid chromium(III) chloride is synthesized from the elements as raw material. Afterwards it is "sublimated" at high temperature in a stream of chlorine via gaseous chromium(IV) chloride. In fact, this is not a sublimation but a chemical vapor transport reaction according to the following equation:

$$CrCl_3(s) + 1/2\,Cl_2(g) \rightleftharpoons CrCl_4(g) \quad (T_2 \rightarrow 400\,°C)$$

Conproportionation Reactions

Transition metals can appear in their binary halides in more than two oxidation states stable under transport conditions, which particularly applies to metals of group 5 and 6. This can be used in order to transport a solid metal halide with a low oxidation number of the metal through a gaseous metal halide in which the metal has an oxidation number higher by two units or more. The transport of niobium(III) chloride with niobium(V) chloride as transport agent is an example.

$$NbCl_3(s) + NbCl_5(g) \rightleftharpoons 2\,NbCl_4(g) \quad (400 \rightarrow 300\,°C)$$

Gaseous niobium(IV) chloride is formed that disproportionates in the sink to solid niobium(III) chloride and gaseous niobium(V) chloride. Usually, the according halogen is used as transport additive instead of the one formulated in the transport equation. The actually effective transport agent forms through the reaction of the transport additive with the solid. Additional examinations and/or thermodynamic modal calculations are necessary in order to decide whether the added halogen or a higher halide formed by halogen is the actual transport agent.

Formation of Gas Complexes

The term gas complex refers to a gaseous metal halide in which several metal atoms are connected to each other by halogen bridges. Gas complexes are important for many chemical vapor transports [11], in particular for the chemical vapor transports of halides; in these cases, the solid to be transported is a metal halide with a high boiling temperature, the transport agent is a highly volatile halide, particularly often an aluminum halide. The aluminum halides have low boiling temperatures and form stable gas complexes with a number of metal halides. Gallium(III) halides, indium(III) halides, and iron(III) halides are used as transport agents as well. At relatively low temperatures around 300 °C, the transport of dihalides, for example, $CoCl_2$, with aluminum(III) chloride takes place via MAl_2Cl_8 as transport effective species.

$$MCl_2(s) + Al_2Cl_6(g) \rightleftharpoons MAl_2Cl_8(g) \quad (400 \rightarrow 300\,°C)$$

The formation of gas complexes has also been used for the volatilization of halides of the lanthanoids. The solubility of lanthanoid chlorides is enhanced by about 10 orders of magnitude in the presence of aluminum chloride.

It is important to note that the addition of the very moisture sensitive aluminum(III) chloride as transport agent requires special experimental care. The same applies to iron(III) halides and indium(III) halides, and particularly to gallium(III) halides.

13.3.4
Oxides

Chlorinating equilibria proved suitable for the chemical vapor transport of oxides. Apart from chlorine and hydrogen chloride, tellurium(IV) chloride is an important transport agent. Tellurium(IV) chloride is especially useful when the oxygen partial pressure in the system varies and the setting of the oxygen partial pressure is of essential importance for the transport and the composition of the deposited solid. Due to the generally unfavorable equilibrium position, brominating and iodinating equilibria are of minor importance. Some further transport agents, respectively transporting effective additives are oxygen and water. In some cases, the solid oxides can form gaseous oxide halides, transport effective species which contain both oxygen and halogen atoms. There are many metal/oxygen systems in which the oxygen partial pressure varies depending on the composition of the solid phase. The transport agent must allow for this fact.

The oxides are the largest group among all compounds that have been crystallized by chemical vapor transport reactions with more than 600 examples [2]. Among these are binary oxides, oxides with complex anions, such as phosphates or sulfates, as well as oxides with several cations, such as $CoFe_2O_4$ or $Zn_{1-x}Mn_xO$. Ternary oxides, to which belong the majority of the compounds with complex anions (e.g. $SrMoO_4$), are most common. There are more than 100 examples of binary oxides, most of them containing transition metals [2]. The smallest group is the one of quaternary and multinary oxides with slightly more than 50 examples [2].

The question whether an oxide is transportable not only depends on its own thermodynamic stability but also on the stabilities of the gas species that are formed. Thus, zirconium(IV) oxide, which is thermodynamically very stable, is suited for the transport reactions in contrast to less stable oxides, for example, rubidium oxide.

Chlorine as Transport Agent

Chlorine has been used as transport agent in the case of many metal oxides. One can distinguish between chlorinating equilibria under formation of gaseous metal chlorides and metal oxide chlorides, respectively.

$$Ga_2O_3(s) + 3\,Cl_2(g) \leftrightarrows 2\,GaCl_3(g) + 3/2\,O_2(g) \qquad (1000 \rightarrow 800\,^\circ C)$$
$$Cr_2O_3(s) + 5/2\,Cl_2(g) \leftrightarrows 3/2\,CrO_2Cl_2(g) + 1/2\,CrCl_4(g) \qquad (1000 \rightarrow 900\,^\circ C)$$

The transport of metal oxides with chlorine as transport additive is always endothermic ($T_2 \rightarrow T_1$).

Hydrogen Halides as Transport Agent

Sometimes, hydrogen halides are preferred as transport agents, especially in the case of thermodynamically very stable oxides like MgO. The formation of H_2O shifts the equilibrium to the product side.

$$MgO(s) + 2\,HCl(g) \rightleftharpoons MgCl_2(g) + H_2O(g) \quad (1000 \rightarrow 800\,°C)$$

Often, ammonium halides (NH_4X, X=Cl, Br) are used as hydrogen halide source. These solids are easy to handle and to dose. They decompose to ammonia and hydrogen halide at increased temperature. However, the formation of ammonia leads to an increase of the total pressure in the system and thus to lower transport rates. It is important to consider that ammonia and hydrogen, which is formed by decomposition above 700 °C, create a reducing atmosphere. This can lead to a reduction of the gas species and/or the solid phase.

The transport of metal oxides with hydrogen halides as transport additive is almost always endothermic ($T_2 \rightarrow T_1$).

Tellurium(IV) Halides as Transport Agents

Tellurium(IV) chloride is a flexible transport agent, which can especially be used for oxides of the transition metals and compounds with complex anions. Tellurium(IV) chloride is specially suited as transport additive for oxide systems with a wide range of oxygen partial pressures $p(O_2)$ between 10^{-25} and 1 bar. This is due to the creation of a complex redox system under formation of different tellurium containing gas species, mainly TeO_2, $TeOCl_2$, $TeCl_2$, TeO, and Te_2. Tellurium(IV) chloride proves an ideal transport additive for those oxides which differ only slightly in their composition and stability and thus are thermodynamically stable only in narrow ranges of the oxygen partial pressure, such as Magnéli-phases like V_nO_{2n-1} (n=4 to 8) [12,13]. Tellurium(IV) chloride is also suitable for the transport of oxide phases which show homogeneity ranges dependent on the oxygen partial pressure, such as "VO_2."

Combination of Transport Additives

The combination of two transport additives is often used, for example, the combination of sulfur + chlorine, sulfur + bromine, sulfur + iodine, selenium + chlorine, selenium + bromine, carbon + chlorine, and carbon + bromine. These also form complex redox systems and can be treated similarly to tellurium(IV) chloride. The mechanism of the combination of sulfur + iodine is described exemplarily by the following equation:

$$2\,Ga_2O_3(s) + 3/2\,S_2(g) + 6\,I_2(g) \rightleftharpoons 4\,GaI_3(g) + 3\,SO_2(g) \quad (1000 \rightarrow 900\,°C)$$

Other Transport Additives

Sometimes water is used as transport agent. It can always be used when a volatile hydroxide, respectively a gaseous acid (e.g., H_2WO_4) will be formed.

$$WO_3(s) + H_2O(g) \rightleftharpoons H_2WO_4(g) \quad (1050 \rightarrow 900\,°C)$$

If oxygen is used as transport agent, the transport will always occur via a higher, more volatile oxide, which is formed in the transport equilibrium.

$$IrO_2(s) + 1/2\,O_2(g) \rightleftharpoons IrO_3(g) \quad (1050 \rightarrow 850\,°C)$$

13.3.5
Sulfides, Selenides, and Tellurides

The chemical vapor transport of metal sulfides, metal selenides, and metal tellurides has been examined in detail. Due to the lower thermodynamic stability of the metal sulfides, selenides, and tellurides in comparison to the oxides, other transport agents are often used, for example, iodine or iodine compounds, which are almost unsuitable for oxides.

All metals form stable binary sulfides, selenides, and tellurides. Most of them, and in addition many ternary and quaternary compounds can be crystallized by chemical vapor transport. The transport of these compounds clearly differs from the ones of the oxides.

Iodine as Transport Agent

The transport of a binary sulfide can be described by the following simple equation:

$$ZnS(s) + I_2(g) \rightleftharpoons ZnI_2(g) + 1/2\,S_2(g) \quad (900 \rightarrow 800\,°C).$$

This holds for ternary sulfides (selenides and tellurides), too. One example of this is the transport of $CuGaS_2$ [14]. Materials like this are used in so called CIGS solar cells.

$$CuGaS_2(s) + 2\,I_2(g) \rightleftharpoons 1/3\,Cu_3I_3(g) + GaI_3(g) + S_2(g) \quad (850 \rightarrow 800\,°C).$$

Hydrogen Chloride as Transport Agent

In analogy to the transport of oxides, some sulfides can be transported with hydrogen chloride.

$$FeS(s) + 2\,HCl(g) \rightleftharpoons FeCl_2(g) + H_2S(g) \quad (850 \rightarrow 800\,°C)$$

The behavior of selenides is nearly identical to that of the sulfides in contrast to that of the tellurides: H_2Te is too unstable to be formed under these conditions.

$$PbTe(s) + 2\,HCl(g) \rightleftharpoons PbCl_2(g) + H_2(g) + 1/2\,Te_2(g) \quad (830 \rightarrow 370\,°C)$$

Hydrogen as Transport Agent

A very special way is the transport with hydrogen. The transport becomes possible in some cases if the metal has a high vapor pressure.

$$ZnSe(s) + H_2(g) \rightleftharpoons Zn(g) + H_2Se(g) \quad (850 \rightarrow 800\,°C)$$

13.3.6
Pnictides

Chemical bonding in metal pnictides is very variable and ranges from metallic, ionic, or covalent nitrides and phosphides via the rather covalent or metallic arsenides and antimonides to the typical metallic bismutides.

Elemental halogens, in particular iodine, and in some cases halogen compounds are the preferred transport additives. The stabilities of pnictogen halides increase from nitrogen to bismuth. The formation of phosphorus halides is important during the transport of phosphides. The main difference between the transport of a chalcogenide and a pnictide becomes obvious by comparing the following two transport equations:

$$CoS(s) + I_2(g) \rightleftharpoons CoI_2(g) + 1/2\, S_2(g)$$
$$CoP(s) + 5/2 I_2(g) \rightleftharpoons CoI_2(g) + PI_3(g)$$

The stability of phosphorus halides, arsenic halides, and antimony halides is so strong in some cases that metal halides with transport effective partial pressures are not formed.

The boiling temperatures of the pnictogens increase from nitrogen to bismuth. While nitrogen, phosphorus, and arsenic have sufficiently high saturation pressures to be transport effective in elemental form, it is necessary to generate transport effective compounds for the antimonides and bismutides.

Concerning the transport of phosphides, in the gas phase mostly phosphorus(III) halides occur. During the transport of arsenides and antimonides, one has to expect, at rising temperatures, the formation of monohalides. This applies in particular for the heavy halogens. An example of this is the transport of GaAs with the transport additive I_2. The transport is due to the following transport equation:

$$GaAs(s) + AsI_3(g) \rightleftharpoons 3/2\, GaI(g) + 1/4\, As_4(g)$$

In this case, the transport agent is AsI_3 but not the transport additive I_2.

The high volability of GaAs in arsenic vapor is explained by the formation of the peculiar gas species $GaAs_5$ [2]. This example presents an interesting link between the high-temperature chemistry of vapor phase transport and the coordination chemistry of cyclopentadienide analogue anions at ambient conditions.

$$GaAs(s) + As_4(g) \rightleftharpoons GaAs_5(g)$$

Another example of the formation of an unusual gas species is a pnictogen chalcogenide halide, such as AsSeI [15].

13.4
Modeling of Chemical Vapor Transport Experiments

Modeling the simplest case of a transport experiment (which comprises identical and single-phase solids *A* at source and sink) requires calculation of the equilibrium gas phase for the two sub-spaces of a transport ampoule and eventually estimation of the transport rate $\dot{n}(A)$ using Eq. (13.3) as was already described in Section 13.1.1. However, frequently this simple situation is not met by the experiment. Multiphase solids at the source as well as their time dependent (sequential) migration to and deposition at the sink are observed. Such a nonstationary transport behavior is typically caused by complicated reactions between the starting solid and the gaseous transport agent and/or a strong influence of temperature on the chemical composition of the starting solid. The transport systems CoP/I_2 (formation of CoI_2(l) [16]), Rh_2O_3/Cl_2 (formation of metallic Rh and/or $RhCl_3$(s) [17]), VO_x/HCl (HCl by hydrolysis of $TeCl_4$;$1.5 < x \leq 2.0$; formation of neighboring oxides [18]), and eventually $NiSO_4/PbCl_2$ (formation of $PbSO_4$(s), $NiCl_2$(s), NiO(s) [19]) have been described in detail in literature and may serve to illustrate deviations from the simple transport behavior. Figure 13.9 shows the mass transfer versus time for the chemical vapor transport of Rh_2O_3 using chlorine as transport agent. The shown data were recorded using a so-called transport balance [2,19] (Figure 13.10).

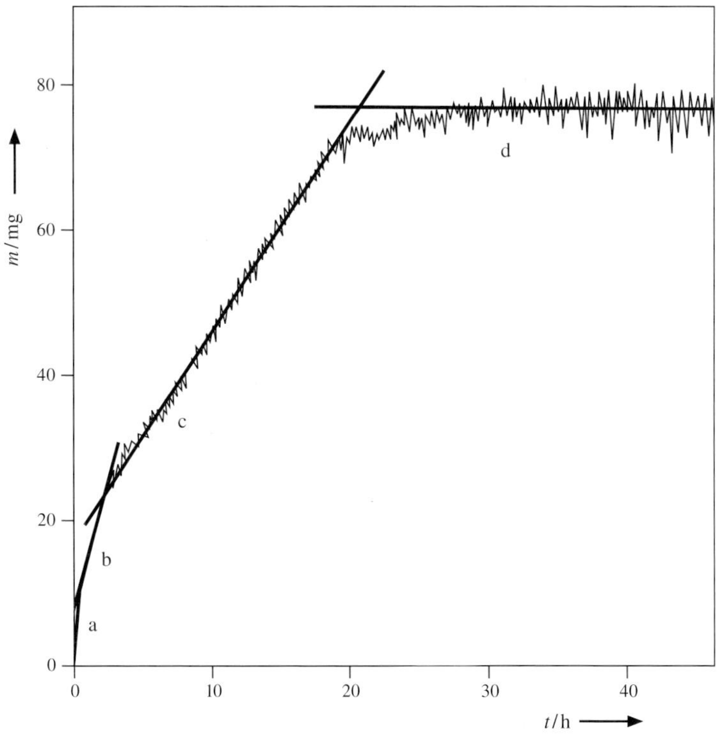

Figure 13.9 Diagram showing the course of mass versus time for a typical transport of Rh_2O_3 by chlorine according to [15] (a) setting of the chemical equilibrium between the starting solid and the transport agent, (b) sublimation of $RhCl_3$, (c) transport of Rh_2O_3, (d) constant mass achieved after complete transport of Rh_2O_3 (1075 → 975 °C; starting amounts: 75.6 mg Rh_2O_3, 56.7 mg Cl_2).

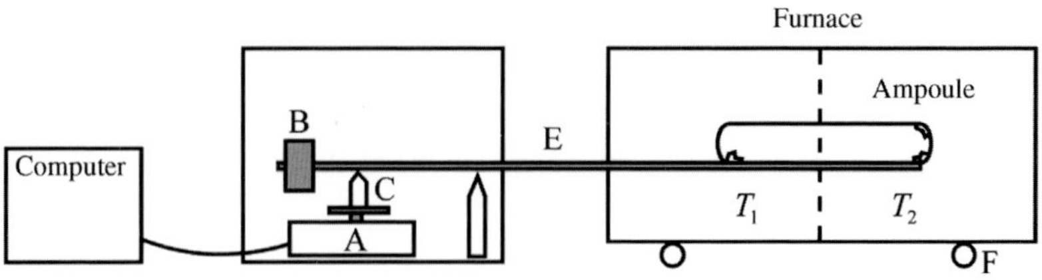

Figure 13.10 Schematic set up of the transport balance (A: scale; B: counterweight; C: plug; D: edge; E: lever; F: two-zone furnace).

Monitoring the course of a nonstationary chemical vapor transport experiment (by using a transport balance) provides detailed information on the masses of all solids deposited at the sink as well as on their individual transport rates. Describing and understanding all these observations on a thermodynamic basis requires a more general model than that introduced by Schäfer [3] (Section 13.1). Such a model was developed by Gruehn and Schweizer [20]. It is called "model of co-operative transport" and is based on the idea of two co-operating equilibrium zones (source and sink). Thus, the restriction of two identical solids at source and sink in the theoretical treatment of transport experiments has been overcome. The main idea of the model as well as the course of such a calculation are visualized in Figure 13.11.

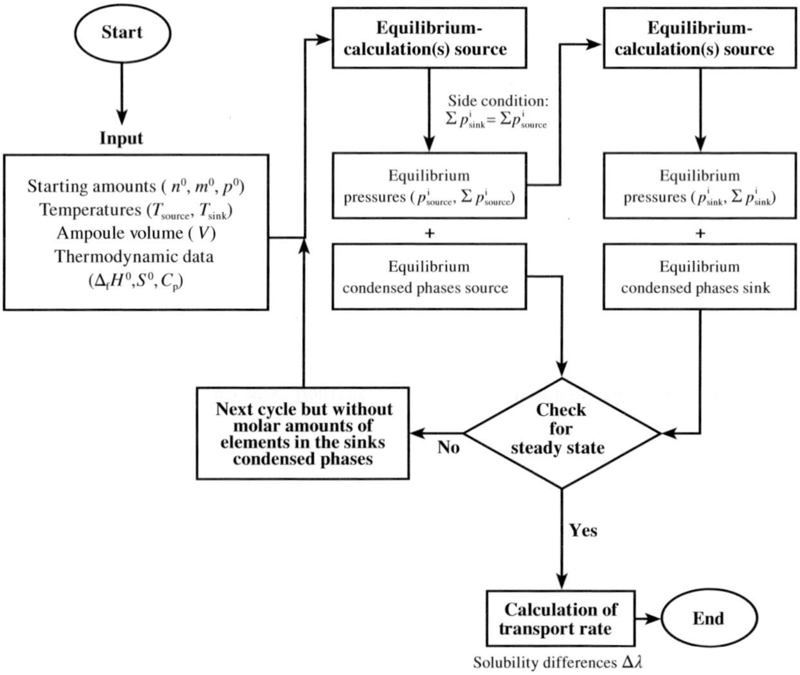

Figure 13.11 Flow chart for calculations on the basis of the model of co-operative transport.

In the first step, equilibrium condensed phase(s) and gas phase at the source (T_{source}, V_{source}) are calculated using the G_{min} method [21]. The elemental balance of the sources gas phase is then used for the equilibrium calculation at the sink (T_{sink}) with $\Sigma p_{source} = \Sigma p_{sink}$ as only restriction. In case of deposition of a condensed phase at the sink ("chemical transport"), the molar amounts of the deposited elements are subtracted from the elemental balance of the source, followed by a new equilibrium calculation for the source and subsequently for the sink. By repetition of this cycle of calculations, eventually all condensed phases from the source are transferred to the sink. The calculated deposition sequence (as well as the amounts and individual transport rates) of condensed phases in the sink can be compared to the experiment. The aforementioned calculations can be routinely performed using computer program CVTRANS [22]. Due to its facile use and short computing times, CVTRANS allows even fitting of the calculations to the observations in cases where crucial thermodynamic data is missing. Thus, the detailed observations during nonstationary transport experiments may be used to narrow down thermodynamic data for one of the condensed phases or a crucial gaseous species.

In this way, a comprehensive thermodynamic understanding of a transport system can be achieved. In turn, such an understanding is beneficial for the choice of optimum experimental parameters to achieve a particular goal (high transport rates, growth of crystals for structure analysis, deposition of mixed-crystals with set chemical composition).

References

1 van Arkel, A.E. and J. H., deBoer. (1925) Darstellung von reinem titanium-, zirkonium-, hafnium und thoriummetall. *Z. Anorg. Allg. Chem.*, **148**, 345–350.

2 Binnewies, M., Glaum, R., Schmidt, M., and Schmidt, P. (2012) *Chemical Vapor Transport Reactions*, De Gruyter, Berlin.

3 Schäfer, H. (1962) *Chemische Transportreaktionen*, Verlag Chemie, Weinheim.

4 Schäfer, H. (1973) Der chemische transport und die Löslichkeit des Bodenkörpers in der Gasphase. *Z. Anorg. Allg. Chem.*, **400**, 242–152.

5 Schmidt, G. and Gruehn, R. (1981) Zum chemischen Transport von Ge mit H_2O. *J. Cryst. Growth*, **55**, 599–610.

6 Schäfer, H. and Trenkel, M. (1975) Der chemische Transport von Cu, Ag, Au, Ru, Rh, Pd, Os, Ir, Pt unter Mitwirkung von Gaskomplexen. Al_2Cl_6, Fe_2Cl_6 oder Al_2I_6 als Komplexbildner. *Z. Anorg. Allg. Chem.*, **414**, 137–150.

7 van Arkel, A.E. (1934) Über die Herstellung von hochschmelzenden Metallen durch thermische Dissoziation ihrer Verbindungen. *Metallwirtschaft*, **13**, 405–408.

8 Schäfer, H. and Morcher, B. (1957) Über den Transport von Silicium im Temperaturgefälle unter Mitwirkung der Silicium(ll)-halogenide und über die Druckabhängigkeit der Transportrichtung. *Z. Anorg. Allg. Chem.*, **290**, 279–291.

9 Neddermann, R. (1997) Ph.D. thesis, University of Hannover.

10 Massalski, H. (1990) *Binary Alloy Phase Diagrams*, 2nd edn, ASN International, p. 2867.

11 Schäfer, H. (1975) Complexes in the vapour and their implications for the vapour growth. *J. Cryst. Growth*, **31**, 31–35.

12 Oppermann, H., Reichelt, W., Krabbes, G., and Wolf, E. (1977) Zu dem Transportsystem V-0-Te-CI und dem

Transportverhalten der Vanadinoxide mit $TeCl_4$, (I). Der Transport von Vanadindioxid. *Krist. Tech.*, **12**, 717–728.

13 Oppermann, H., Reichelt, W., Krabbes, G., and Wolf, E. (1977) Zu dem Transportsystem V-0-Te-Cl und dem Transportverhalten der Vanadinoxide mit $TeCl_4$ (II). Der chemische Transport der Magnéliphasen V_nO_{2n-1}, des V_6O_{13} und des V_20_5. *Krist. Tech.*, **12**, 919–928.

14 Prabukanthan, P. and Dhanasekaran, R. (2007) Growth of $CuGaS_2$ single crystals by chemical vapor transport and characterization. *Cryst. Growth Des.*, **7**, 618–623.

15 Brünig, C., Locmelis, S., Milke, E., and Binnewies, M. (2006) Mischphasenbildung und Chemischer Transport im System ZnSe/GaAs. *Z. Anorg. Allg. Chem.*, **632**, 1067–1072.

16 Schmidt, A. (1999) Ph.D. thesis, University of Giessen.

17 Görzel, H. and Glaum, R. (1996) Untersuchungen zur Kristallisation von Rhodium(III)-Oxoverbindungen unter Beteiligung der Gasphase –Der chemische Transport von Rh_2O_3 mit Chlor. *Z. Anorg. Allg. Chem.*, **622**, 1773–1780.

18 Wenzel, M. and Gruehn, R. (1989) Untersuchungen zum chemischen Transport der Vanadiumoxide V_2O_5, V_3O_7 und V_6O_{13}. *Z. Anorg. Allg. Chem.*, **568**, 95–105.

19 Plies, V., Kohlmann, T., and Gruehn, R. (1989) Eine Methode zur kontinuierlichen Bestimmung von Transportraten. Experimente im System $GeO_2/WO_2/H_2O$ und $NiSO_4/PbSO_4/PbCl_2$. *Z. Anorg. Allg. Chem.*, **568**, 62–72.

20 Gruehn, R. and Schweizer, H.J. (1983) Feststoffpräparation durch chemischen Transport – Interpretation und Steuerung mit dem Kooperativen Transportmodell. *Angew. Chem., Int. Ed.*, **22**, 82–95.

21 Eriksson, G. (1971) Thermodynamic studies of high temperature equilibria. *Acta Chem. Scand.*, **25**, 2651–2658.

22 Trappe, O., Glaum, R., and Gruehn, R. (1999) *Das Computerprogramm CVTRANS*, University of Giessen.

14
Thermodynamic and Kinetic Aspects of Crystal Growth

Detlef Klimm

Leibniz Institute for Crystal Growth, Simulation & Characterization, Max-Born-Str. 2, 12489 Berlin, Germany

14.1
Introduction

In the year 1857, William Thomson (Lord Kelvin) introduced thermodynamics as "the subject of the relation of heat to forces acting between contiguous parts of bodies, and the relation of heat to electrical agency" [1]. In this time, the focus of interest was on the quantitative understanding and improvement of steam engines and other technical equipment. Later the qualified terms "statistical thermodynamics" and "chemical thermodynamics" were introduced – the first refers to the description of systems in terms of atomic movements, and the latter is focused on the description of equilibrium states with several substances, where equilibrium can be reached by chemical reactions and/or physical changes of state, respectively. It will be demonstrated in the course of this chapter that chemical and physical processes are often closely correlated, for example, during melting or evaporation processes that are accompanied by dissociation reactions.

During crystal growth processes the desired crystalline solid phase can be created by a physical transition from a nutrient phase. This is often a melt (e.g., Czochralski or Bridgman methods for many technologically relevant crystals) or the gas phase (sublimation or "physical vapor transport" PVT). Alternatively, the crystalline phase can be obtained via chemical reactions. For fine-grained dielectric lead and alkaline earth titanates and niobates, Pechini prepared mixtures of organic salts of the constituents. The desired oxides were prepared by combustion of the organic parts [2]. Alternatively, the growth of salt crystals with low solubility is possible by separating solutions containing the cation or the anion, respectively, with a semipermeable gel that slows down diffusion

Handbook of Solid State Chemistry, First Edition. Edited by Richard Dronskowski, Shinichi Kikkawa, and Andreas Stein.

processes. As an example, calcium oxalate single crystals with several millimeter diameter can be grown within several weeks if aqueous solutions of oxalic acid and cadmium chloride are interdiffusing through a metasilicate gel [3].

The driving force for most crystal growth processes, irrespective whether they rely on physical or chemical transformations, is the change in Gibbs free energy $G = H - TS$. The Gibbs energy aspires to a minimum, and this means equilibrium. The growth of bulk crystals is typically performed over time spans ranging from hours to months, and over such long periods equilibrium is usually almost obtained. Consequently, such processes can be described quite well by calculations of the thermodynamic equilibrium. This is not always so for atomistic processes occurring for example, at interfaces during epitaxial or bulk growth, where kinetic effects play a larger role. However, also then equilibrium thermodynamics is often a useful concept that helps to indicate the direction the system wants to go.

Phase diagrams are one major outcome of thermodynamic equilibrium considerations, and the focus of this chapter will be on the description of typical situations of crystal growth in terms of several types of phase diagrams. Only to some extend kinetic phenomena will be included in the discussion. Purely kinetic phenomena, such as diffusion, grain growth, and adsorption at epitaxial layers, are not discussed here.

14.2 Equilibrium Thermodynamics

The thermodynamic approach is purely macroscopic, and the behavior of a system (definition in Section 14.2.1) is described in terms of properties that can be directly measured, such as temperature T, pressure p, volume V, or the concentration of the ith chemical component x_i. These macroscopic quantities are interrelated without making any microscopic assumptions [4]. Actually, this restriction is often a fortune because microscopic assumptions and theories are developing fast – in contrast to the basics of thermodynamics that rely mainly on physical and chemical properties of the chemical elements and on statistics. Especially in bulk growth processes information on the atomistic behavior is typically not available, and is often not necessary to describe the experiment. In such cases equilibrium thermodynamics may give an even better quantitative picture, compared to the atomistic description.

Situations where a phenomenological formalism can give an appropriate quantitative description of reality can be found also in other fields of physics: Ohm's law combines accurately the current flowing through an electric circuit with the applied voltage and the resistivity of the circuit. Of course, Ohm's law does not consider the movement of single electric carriers; and so thermodynamics says nothing about the behavior of single atoms on a crystals surface – nevertheless both theories give extremely accurate predictions in their specific field.

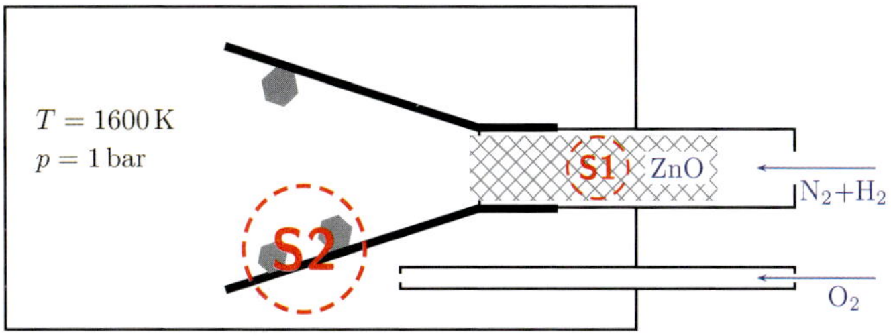

Figure 14.1 Helbig's [5] apparatus for the growth of ZnO crystals by isothermal chemical transport of ZnO feedstock with flowing N_2/H_2 from "S1" to "S2" where reaction with O_2 occurs.

14.2.1 Basic Terms

14.2.1.1 System

A thermodynamic system is the content of a macroscopic volume that can be described by a set of variables, such as temperature T, pressure p, and internal energy U. The universe itself is the largest possible system, but for crystal growers beyond the scope of interest. Then also smaller portions can be chosen as system to be studied. Nevertheless, the boundaries of the system should be defined with care, to allow a useful description: The flow of matter or energy through the boundaries should be either prohibited, or should proceed in a controlled manner. In thermodynamic equilibrium, thermal gradients do not occur by definition, because heat flow would level them. Nevertheless, crystal growth processes rely often on mass transport that is driven by thermal gradients. Then it can be useful to consider partial systems: a Bridgman ampoule is often exposed to low thermal gradients, and all relevant parts (ampoule wall, solid, melt, gas) can then be treated as one single system. In contrast, in setups for crystal growth from the gas phase, significant temperature or concentration gradients are a precondition for mass transport. Then dividing the setup in partial systems may be helpful. This shall be demonstrated for the gas-phase transport of zinc oxide that was performed by Helbig in a setup that is depicted in Figure 14.1.

For the growth, an excess (600–1000 g) of polycrystalline annealed ZnO is rinsed inside a ceramic rod by flowing N_2 (80 l/h) + H_2 (2 l/h) at 1600 K and 1 bar. One can consider this part of the reactor as system "S1." Under these conditions a "gas 1" phase is formed that flows to the left, where it reacts with another gas flow of oxygen (0.5–2 l/h). This is demonstrated in Eq. (14.1) for the gas amounts flowing during 1 h.

$$\left.\begin{array}{r}3.27\text{ mol }N_2(g)\\0.082\text{ mol }H_2(g)\\9.83\text{ mol }ZnO(sol)\end{array}\right\}\xrightarrow{1600\text{ K, S1}}\begin{array}{c}\left.\begin{array}{c}3.27\text{ mol }N_2(g)\\0.081\text{ mol }Zn(g)\\0.081\text{ mol }H_2O(g)\end{array}\right\}3.43\text{ mol}\left[\text{gas }1\right]\\9.74\text{ mol }ZnO(sol)\end{array}\tag{14.1}$$

In system S2, the gas leaving S1 reacts with the separate oxygen flow (bottom in Figure 14.1). For this calculation, an average flow of 1 l (0.041 mol) per hour was used.

$$\left.\begin{array}{l}3.43\text{ mol }\left[\text{gas }1\right]\\ 0.041\text{ mol }O_2(g)\end{array}\right\}\xrightarrow{1600\text{ K, S2}}\begin{array}{c}3.27\text{ mol }N_2(g)\\ 0.082\text{ mol }H_2O(g)\\ 0.003\text{ mol }Zn(g)\\ 0.078\text{ mol }ZnO(sol)\end{array} \tag{14.2}$$

The 0.078 mol ZnO(sol) resulting from Eq. (14.2) are 6.4 g of ZnO that can be transported during 1 h, which should be considered as upper limit. If the flow rate is too high, or if the gas in S1 is not in perfect contact with ZnO feedstock, somewhat lower transport rates must be expected. Helbig [5] did not mention in his paper the total amount of ZnO that was transported during the long (up to 200 h) growth runs, but usually several single crystals with masses up 20 g each were obtained. It must be assumed that besides single crystals also micro-crystalline debris crystallized in the conical part around S1.

Taking this into account, the estimation given above seems reasonable. Practically, the processes in both systems S1 and S2 can be simplified to the subsequent chemical reactions

$$\text{S1}: \quad ZnO(sol) + H_2(gas) \rightarrow Zn(gas) + H_2O(gas) \tag{14.3}$$

$$\text{S2}: \quad Zn(gas) + \frac{1}{2}O_2(gas) \rightarrow ZnO(sol) \tag{14.4}$$

where in Eq. (14.3) ZnO(sol) is always available in excess, whereas in Eq. (14.4) Zn(gas) can be either in excess or in undersupply, depending on the oxygen flow. This is shown in Figure 14.2. There the calculations of Eqs. (14.1) and (14.2) are repeated for a range of oxygen flows exceeding the experimental data given by Helbig [5]. If the flow is small, all O_2 (gas) reacts to ZnO (sol).

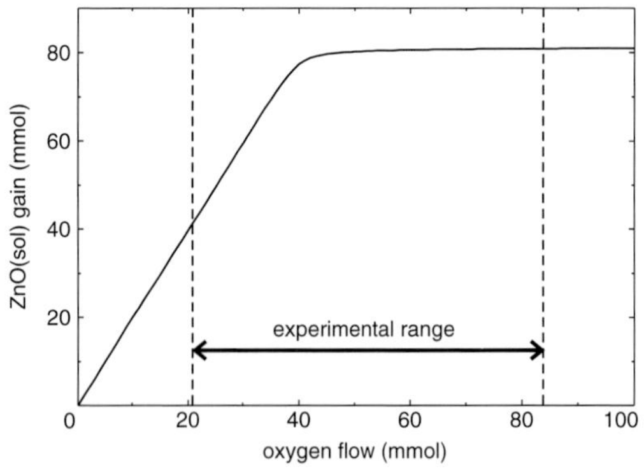

Figure 14.2 Amount of ZnO(sol) which can be transported during 1 h in Helbig's apparatus, in dependency of oxygen flow. Vertical dashed lines mark the experimental limits that are given in the paper [5].

Under such conditions excess Zn(gas) remains, leading to reducing conditions and possibly to Zn excess in the growing crystals [6]. For larger flows, excess oxygen is available and the effective ZnO growth rate is no longer dependent on O_2 flow.

14.2.1.2 **Component and Phase**

Component

A thermodynamic system contains substances that can undergo chemical or physical transformations. Often chemical equilibria are more complex than expressed by simple equations such as Eq. (14.3) or (14.4). For instance, "normal" diatomic oxygen molecules are always in equilibrium with atoms and ozone: $3\,O \rightleftarrows 1.5\,O_2 \rightleftarrows O_3$. A full calculation of the equilibrium reaction Eq. (14.1) with FactSage [7] results in a "gas 1" that contains, besides the three species given above, minor quantities of H_2, NO, O_2, O, and some others. All of them are present in amounts $< 7 \times 10^{-4}$ mol, and can often be neglected. The question arises, how many basic substances, or *components*, must be taken into account to be sure that all species can be created. In the general case, this is simply the number of chemical elements that occur in the corresponding system. In the case of ZnO transport, these are the components Zn, O, H, and N. Systems with C components are chemically defined by $C-1$ mole fractions $x_i (i = 1 \cdots C-1)$, because $\sum x_i = 1$.

The calculation and especially the graphical representation of systems with many components is difficult, and it is often desirable to find a description with the lowest possible number of components. This is possible and useful, if chemical compounds are chosen as component, rather than elements. On this basis, large collections of compound phase diagrams have been created, where for example, pseudobinary (two compounds as components) or more complex systems are critically reviewed and presented [8].

It must not be forgotten, however, that the selection of chemical compounds for components of phase diagrams must be done with care. The construction of a phase diagrams relies on the assumption that all compounds appearing in the system can be constructed by chemical reaction of the components. Indeed, all compounds that are labeled in the Figure 14.3a can be constructed in this way: $CaFe_4O_7 = CaO \cdot 2\,Fe_2O_3$, $CaFe_2O_4 = CaO \cdot Fe_2O_3$, and $Ca_2Fe_2O_5 = 2\,CaO \cdot Fe_2O_3$. This diagram coincides basically with the results of Hillert *et al.* [9], except that there the small rim solubility of Fe in CaO was neglected. However, the left phase diagram is valid only for oxygen fugacities p_{O_2} exceeding several bars, where Fe^{3+} is stable.

In Figure 14.3b an additional phase field "melt + Fe_3O_4" appears, but Fe_3O_4 cannot be constructed by linear combination of the components CaO and Fe_2O_3. Besides, the melting point of "pure" Fe_2O_3 is shifted because Fe_2O_3 decomposes at high $T \gtrsim 1390\,°C$ to Fe_3O_4, which has a higher melting temperature. Generally, all temperatures resulting from equilibria between solid and liquid phases are altered, because the melt contains some Fe^{2+} at $p_{O_2} = 0.21$ bar.

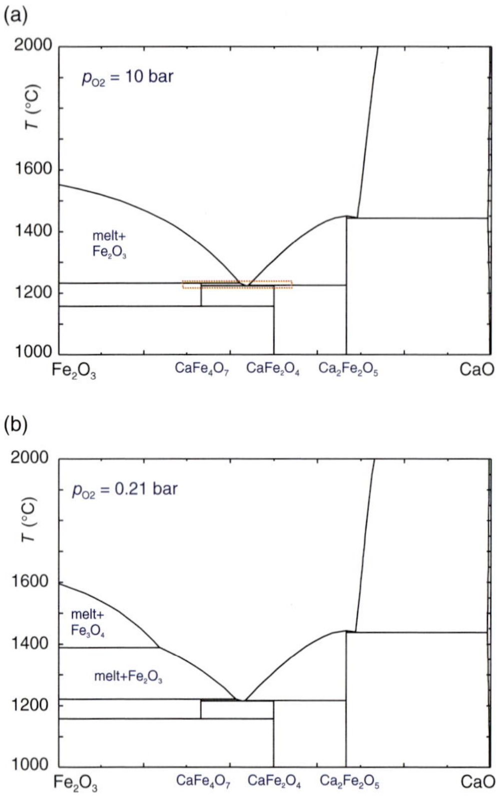

Figure 14.3 (a) The system $CaO–Fe_2O_3$, under the high oxygen pressure p_{O_2} = 10 bar shows only phases with Fe^{3+}. (b) Under ambient p_{O_2} = 0.21 bar iron(III) oxide is reduced to iron(II,III) oxide Fe_3O_4 at elevated *T*. (The region inside the dotted box (a) is reproduced in Figure 14.8.)

It is obvious that the situation in Figure 14.3b cannot be described correctly as binary system. Alternatives would be Ca–Fe–O or $CaO–Fe_3O_4–Fe_2O_3$. For lower p_{O_2}, even FeO or Fe should be taken into account.

Phase

A phase is a part of a thermodynamic system that is homogeneous at least down to a submicrometer scale. This means that an optical microscope and even a normal electron microscope (without atomic resolution) cannot observe chemical inhomogeneity. If different substances are intermixing, they form one phase. This is typically the case for all gases, which can mix in arbitrary ratio, except for extreme conditions. Liquids tend to mix if they are chemically and physically not too different: Most molten metals are intermixing in arbitrary ratio, but Al and Pb do not. Molten salts with the same anion are often intermixing, but with much different anions (e.g., chlorides and iodides) this is not always the case. For solids, complete miscibility is rather

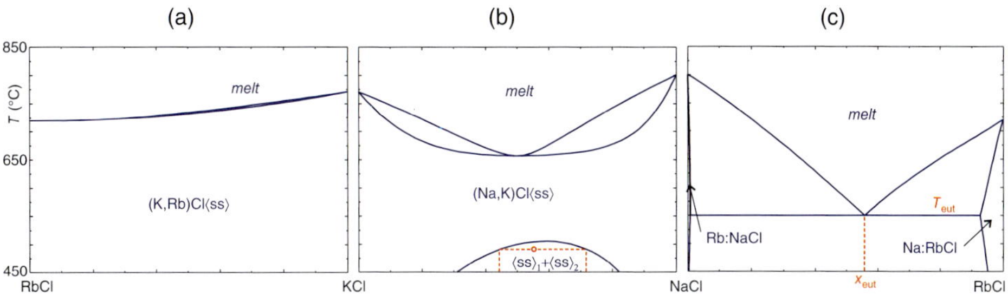

Figure 14.4 Binary phase diagrams of systems where all components crystallize in the halite (NaCl) structure: KCl–RbCl (4.6%), NaCl–KCl (10.3%), and RbCl–NaCl (14.4%) (from left to right, lattice parameter misfits $(a_{large} - a_{small})/a_{large}$ given in brackets). The horizontal red dashed "tie" line in the NaCl–KCl system is explained in Section 14.2.2.1.

seldom. Necessary, but not always sufficient preconditions for complete miscibility are the following:

1) Identical crystal structure, because only then the first end member of the system can be changed into the second in a sequence without interruption.
2) Similar lattice parameters, although it is hard to give accurate numbers for "similar": Some examples for different lattice parameter misfits, where nevertheless the previous condition is always fulfilled, are given in Figure 14.4.
3) Similar bonding types, which is often not a significant constraint, because typically similar substances are brought into relation. Nevertheless: solid helium and solid silver are both crystallizing in the cubic close-packed structure with space symmetry group $Fm\overline{3}m$, with lattice constants of 4.242 Å and 4.085 Å, respectively. Irrespectively of the small lattice parameter difference of −3.7% of Ag with respect to He, solid solutions are not known, and not expected to exist.

Gibb's Phase Rule

In the 1870s, Josiah Willard Gibbs published a seminal paper on heterogeneous equilibria that was soon considered as landmark in the foundation of chemical thermodynamics [10]. In this paper (page 152ff), he derived that in a thermodynamic system consisting of C components with P phases the number of degrees of freedom F equals their difference in equilibrium, increased by the number of independent intensive variables. Often pressure p and temperature T are considered as independent, and one has

$$F = C - P + 2, \tag{14.5}$$

or for the isobar case (e.g., for ambient pressure, $p = 1$ bar)

$$F = C - P + 1 \qquad (p = \text{const.}). \tag{14.6}$$

The number of independent variables is not necessarily limited to the two cases Eq. (14.5) or (14.6) if additional contributions to the Gibbs free energy G of the system play a significant role, such as large magnetic or electric [11] fields, or

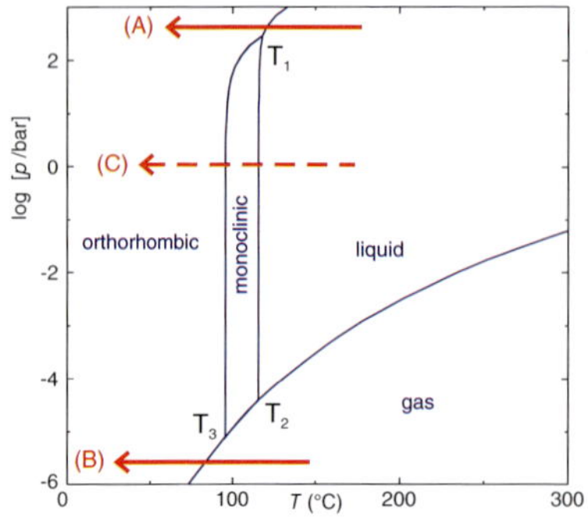

Figure 14.5 The pressure–temperature ($p - T$) phase diagram of sulfur shows three triple points T_i. (A) and (B) represent isobar direct paths from a fluid phase to the orthorhombic solid; (C) traverses the monoclinic phase field.

mechanical stress ("stain engineering" [12]). In such cases, it must be raised for example, to 3.

For systems with only p and T as independent variables, the phase rule Eq. (14.5) limits the maximum number of coexisting phases for one component to $P = 3$. These are triple points that are labeled "T_i" ($i = 1 \cdots 3$) in Figure 14.5 for sulfur. This chalcogen crystallizes under ambient conditions in the typical yellow and brittle orthorhombic modification ($\alpha - \mathrm{S}$). Before melting (for 1 bar at 115°C) it transforms at 95°C to monoclinic $\beta - \mathrm{S}$, which is darker and very soft. Each phase field in Figure 14.5 contains only one of the four phases orthorhombic/monoclinic/liquid/gas. Thus one has $P = 1, C = 1$ in Eq. (14.5) and hence $F = 2$ degrees of freedom. This means T as well as p can be changed independently without leaving the phase field. The lines in Figure 14.5 are phase boundaries between two coexisting phases. With $P = 2, C = 1$, one has $F = 1$ that means that only one of the intensive variables can be changed independently, but the other is defined then by the system. For example, the upper, almost parabolic boundary of the "gas" field represents the vapor pressure of solid or liquid sulfur as a function of T.

During crystal growth processes the corresponding substance usually undergoes a phase transition from a fluid phase, where transport of matter is fast, to the desired solid phase. Under ambient conditions, only orthorhombic sulfur ($\alpha - \mathrm{S}$) is stable and can be crystallized by isobaric cooling following several different paths that are shown in Figure 14.5:

A: Crystallization of a melt at elevated pressure above the triple point T_1. Only there, for $p > 283$ bar, $\alpha - \mathrm{S}$ is in equilibrium with molten sulfur.
B: Sublimation from the gas phase below T_3. Only there, for $p < 8 \times 10^{-6}$ bar, $\alpha - \mathrm{S}$ is in equilibrium with sulfur gas.

Under normal pressure $p = 1$ bar, which is path (C), $\beta - \mathrm{S}$ is formed first that prevents crystallization of large $\alpha - \mathrm{S}$ crystals even under low T. It is a general rule that for crystal growth experiments the conditions should be adjusted in such a way that the fluid mother phase (gas, liquid melt, or solution) is in equilibrium with the desired crystalline (solid) phase.

14.2.2
Phase Diagrams

Crystal growth is often performed in systems with more than one component. But even in one-component systems, the stability of the desired substance in the corresponding phase states with respect to crucible/ampoule materials and/or atmosphere must be considered. For instance, during the Czochralski growth of silicon, the Si melt is in contact with the silica (SiO_2) crucible that results in the introduction of oxygen atoms into the growing crystal. Besides, semiconductors are often doped by the addition of doping ions into the melt feedstock.

Most crystal growth processes are performed under ambient pressure, $p = 1$ bar = const., but temperature T changes during the crystallization process. If T is used as one coordinate in 2D diagrams (typically as ordinate), just one concentration parameter x can be used as abscissa. This gives a two-component system, and the concentration of the second component is $1 - x$. Resulting $x - T$ phase diagrams are very often used in materials science.

14.2.2.1 Solid–Liquid Equilibria

No Intermediate Compounds

The topology of binary phase diagrams depends basically on the miscibility of their components in the low-T and high-T state. These are often solid and liquid, but from the thermodynamic point of view it is only relevant that both states can be changed into another by a first-order phase transition. This condition is also fulfilled for liquid/gas, solid/gas, and for many solid/solid equilibria. For many crystal growth processes, however, solid/liquid equilibria are most relevant; and very often the molten states of all components form one common liquid phase, which means unlimited miscibility.

If also the solid phases exhibit unlimited solubility, they form another common phase that is called "solid solution" or "mixed crystal." The system KCl–RbCl in Figure 14.4 gave an example where the "melt" field and the solid solution field (K,Rb)Cl⟨ss⟩ are separated by a narrow lens-shaped 2-phase field where both phases coexist. (The abbreviation ⟨ss⟩ stands for "solid solution.") The lower and upper boundaries of this phase field are called solidus and liquidus, respectively, and meet in the corresponding melting points T_f^{KCl} and T_f^{RbCl} for the pure components on both sides. The lens becomes broader on the side of the corresponding component for larger heat of fusion ΔH_f. If ΔH_f is significantly different for both components, the 2-phase field can bend slightly downwards or upwards.

For the next system, in Figure 14.4 NaCl–KCl, solidus and liquidus show a local minimum. If such minimum appears, it must be common for both phase boundaries, and is called "azeotropic point." Such points are interesting for crystal growth of mixed crystals [13], because only there no segregation occurs (see Section 14.3.1). Azeotropic points are possible only if significant excess energies occur at least in one phase. Usually these are positive (repulsive) interactions in the solid phase that is demonstrated by the lowest phase boundary in the NaCl–KCl system. Only above this concave line (Na,K)Cl⟨ss⟩ = $Na_xK_{1-x}Cl$ is stable for all compositions x. For points below the boundary, only limited amounts of K can be introduced into NaCl, and vice versa. Consequently, one homogeneously mixed crystal decomposes to separate Na-rich and K-rich solid solutions. Their compositions are found on the intersections of a horizontal (= isothermal) "tie line" through the corresponding point inside the $\langle ss \rangle_1 + \langle ss \rangle_2$ field in Figure 14.4b. It should be noted that such processes require diffusion in the solid state and are usually slow; consequently they are for experiments often not easily accessible.

If both components are miscible in the liquid, but immiscible in the solid phase, a eutectic system such as RbCl–NaCl in Figure 14.4 is formed. In such systems a specific composition x_{eut} melts at a fixed temperature T_{eut} suddenly to a homogeneous melt, and vice versa a melt x_{eut} crystallizes completely at T_{eut} under the formation of two solid phases. Also other compositions start to form a small portion of melt with composition x_{eut} at T_{eut}. The horizontal eutectic line is the solidus. Then, the melting process continues until the liquidus of one of the components, right or left, is reached.

For thermodynamic reasons, some finite mutual solubility of components appears always at $T > 0$ K. If this solubility is so small that it cannot be presented in the x scale of the phase diagram, it is usually neglected and one speaks of a "line compound" or "stoichiometric compound" (daltonide). It should not be forgotten, however, that also minor additive ("dopant") concentrations can have a significant influence, for example, on the electrical properties of semiconductors. In Figure 14.4, Rb^+ can substitute for about 1% of Na^+ in NaCl, and Na^+ can substitute for about 7% of Rb^+ in RbCl. This results in the small 1-phase fields at the rims of the diagram. Below the eutectic line, a mixture of the two Na-rich and Rb-rich solid solutions is formed.

The upper part of Figure 14.6 is similar to the KCl–RbCl system discussed above (Figure 14.4), because both components K_2SO_4 and K_2CO_3 show complete miscibility in the hexagonal high-T phase and in the melt. Below 584 °C (K_2SO_4) or 422 °C (K_2CO_3), respectively, both end members of the system undergo phase transitions to orthorhombic [14] or monoclinic [15] crystal structures. Different crystal structures cannot intermix at arbitrary compositions; and actually only about 6% sulfate can substitute for carbonate on the left side of the diagram. K_2SO_4, in contrast, can be substituted by $\leq 88\%$ K_2CO_3, which is an extremely vast amount. Interestingly such mixed crystals that contain more carbonate than sulfate have still the orthorhombic structure of K_2SO_4.

If a hexagonal solid solution $xK_2SO_4 + (1-x)K_2CO_3$ with $x = 0.07$ is cooled below 389 °C, it decomposes to a mixture of the two (orthorhombic and monoclinic) low-T modifications. Like in a eutectic reaction, one high-T phase decomposes under the formation of two low-T phases, with the difference that the

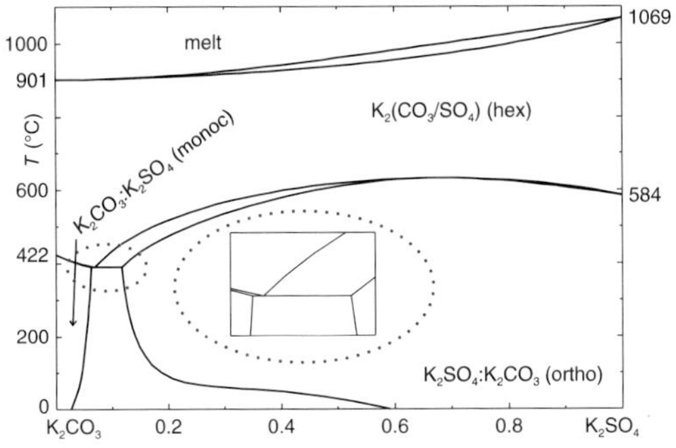

Figure 14.6 Potassium sulfate K_2SO_4 and potassium carbonate K_2CO_3 undergo solid-state transitions from low-T orthorhombic or monoclinic modifications to a high-T hexagonal modification. Only the high-T modification exhibits miscibility in arbitrary ratio, whereas both low-T phases show limited mutual solubility.

high-T phase is solid. Such process is called "eutectoid reaction," and 389 °C is the eutectoid decomposition temperature in this case. Like melting, transitions between the K_2SO_4:K_2CO_3 (ortho) or K_2CO_3:K_2SO_4 (monoc) phase fields and $K_2(CO_3/SO_4)$ (hex) are first-order phase transitions. Consequently, upper and lower phase boundaries separating the 2-phase field between them are subject to the same rules as liquidus and solidus: Both are either monotonously rising or falling, like shown on the K_2CO_3 side; or have a common extremum, like shown at $x = 0.67$ on the K_2SO_4 side.

Intermediate Compounds

In the previous paragraph systems were presented that contain exclusively pure or mixed components in different phase states (solid phases/liquid/gas). However, often components A, B form intermediate compounds with fixed stoichiometry (daltonides), such as AB, AB_2, and A_2B_3. Sometimes, the composition may vary (bertollide), and the composition can be written for example, as $A_{1\pm\delta}B$. Daltonides appear as vertical lines in the $x - T$ phase diagrams, such as Cu_5Sr, CuSr, and Mg_2Cu in Figure 14.7, whereas the stability (= homogeneity) field of the bertollide Laves phase $MgCu_2$ extends over some finite region on the concentration (x) axis. Consequently, the formula for this phase would more accurately be written for example, $Mg_{1\pm\delta}Cu_2(\delta \approx 0.11)$, but even such expression does not account for the T dependence of δ. Not always the stability field of daltonides extends nearly symmetrically to both sides: magnesium aluminum spinel $MgAl_2O_4$ is almost stoichiometric only below about 1000 °C but extends from $0.38 \lesssim x_{Al_2O_3} \lesssim 0.88$ around 2000 °C [16]. "δ-MoNi" [17] shows its nominal 1:1 composition only near the decomposition temperature 1360 °C and extends its homogeneity range to both sides by 2% about 60 K below that temperature. Near room temperature the homogeneity range becomes narrower again, centered around 43% Ni, and the 1 : 1 composition is no longer stable.

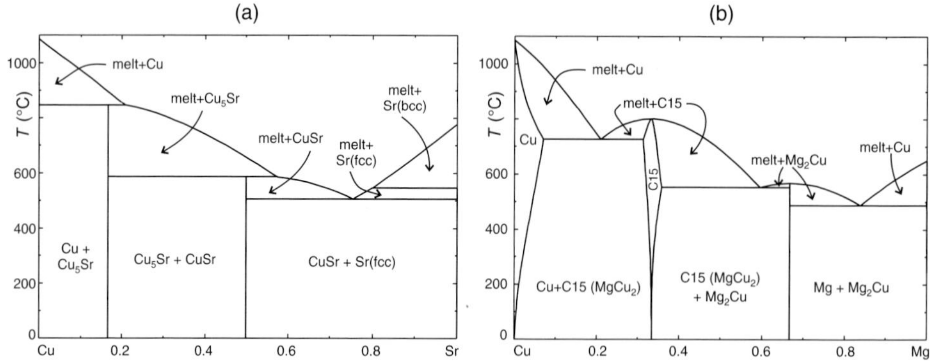

Figure 14.7 (a) Copper forms with strontium the intermediate compounds Cu_5Sr and CuSr, both show peritectic melting. (b) Copper forms with magnesium the intermediate compounds "C15" and Mg_2Cu, both are congruent melting. C15 is the Laves phase $MgCu_2$ with variable composition $0.31 \lesssim x_{Mg} \lesssim 0.36$ [18].

Only the two intermediate compounds in the Mg–Cu system, $MgCu_2$ and Mg_2Cu, have stability regions that extend up to the "melt" phase field. The dome-shaped liquidus has a local maximum at the point where it meets the solidus of the compound, and the slope of the liquidus is almost the same on both sides if appropriate (equimolar) concentration scaling is chosen. Even for berthollides the homogeneity range degenerates to a point on the apex of liquidus. At the liquidus maximum, the solid phase coexists with a melt of identical composition, and such behavior is called "congruent melting."

A different behavior is shown by the intermediate compounds of the Sr–Cu system in Figure 14.7a. CuSr decomposes upon heating at the horizontal "peritectic line" (586 °C) to solid Cu_5Sr (Cu-rich) and a Sr-rich melt, $x = 0.58$. This is a kind of "incongruent melting," because the corresponding solid CuSr and the melt that is in equilibrium have different composition. Upon further heating to 845 °C also solid Cu_5Sr melts peritectically to a melt with $x \approx 0.21$ under the release of solid Cu.

It should be noted that peritectic melting and incongruent melting are no synonyms. "Peritectic" means that one solid is in equilibrium with another solid and a melt, whereas "incongruent" means that a solid and a liquid that are in equilibrium have different composition. Such behavior, however, is already shown by typical mixed-crystal systems (see Section 14.3.1).

Out of the three intermediate compounds of the $CaO–Fe_2O_3$ system in Figure 14.3 only $Ca_2Fe_2O_5$ melts congruently. Such compounds can be grown from stoichiometric melts by oriented crystallization, for example, with the Czochralski or Bridgman methods. Figure 14.8 demonstrates more detailed that the eutectic in this system is formed from two compounds with peritectic melting. For the left one, $CaFe_4O_7$, the peritectic point is at $x_{per} = 0.422$, and at 1231 °C $CaFe_2O_7$ crystallizes from melts between $x_{per} < x < x_{eut}$, but the crystals decompose during cooling below 1157 °C (outside the scale of this figure). The other

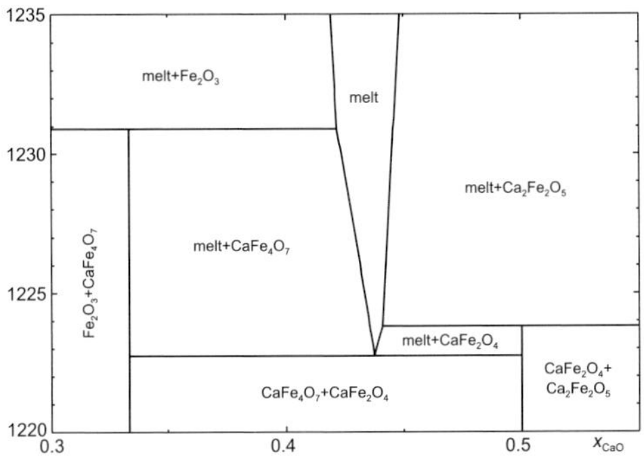

Figure 14.8 Central part of the CaO–Fe_2O_2 system (cf. Figure 14.3a): Both $CaFe_4O_7$ and $CaFe_2O_4$ show peritectic melting and form the eutectic at $x_{eut} = 0.437$.

component of the eutectic, $CaFe_2O_4$, is stable down to room temperature. Here, however, the peritectic composition $x'_{per} = 0.440$ is extremely close to x_{eut} and the peritectic decomposition temperature is only about 1 K above T_{eut}, which makes crystal growth not feasible also for this phase. Generally, realistic options for crystal growth from melts require usually that the liquidus for the corresponding phase extends over a significant concentration range, and that T_{per} is at least so much above T_{eut}, that a useful supercooling can be reached. For instance, small crystals of Cu_5Sr that melts also peritectically (Figure 14.7a) could be grown from melts with $x_{Sr} \approx 0.29$ by cooling from 850 °C [19], because there these conditions are fulfilled. It is not possible to give here useful numbers with general validity, because the necessary supercooling depends on several parameters such as the system itself, and the type of the growing surface (flat or rough) [20].

Miscibility Gaps

The preconditions of miscibility were already discussed in Section 14.2.1.2, in the context of the term "phase." If these conditions are fulfilled, the formation of mixture phases, either gaseous, liquid, or solid, is energetically profitable. This results mainly from the Gibbs free energy of mixing, which is for an ideal mixture (without chemical interaction) $G_{mix,id} = RT[x\ln x + (1-x)\ln(1-x)]$ (R – gas constant, x – molar fraction of a binary mixture). The factor T causes that $G_{mix,id}$ becomes smaller upon cooling, and consequently the benefit from mixing disappears at low T. Practically, many systems with unlimited mutual solubility are not totally ideal, because at least in the solid phase some repulsive interaction exists. This leads to a slight destabilization of the corresponding phase, compared to ideal mixing, and is responsible, for example, for the small downward bend of solidus and liquidus in the KCl–RbCl system (Figure 14.4). The interaction is more repulsive in the (Na,K)Cl solid solution, where solidus and liquidus have a minimum. For even lower T, in such cases often a miscibility gap appears where the solid solution (one phase) decomposes in two solid solutions with different composition.

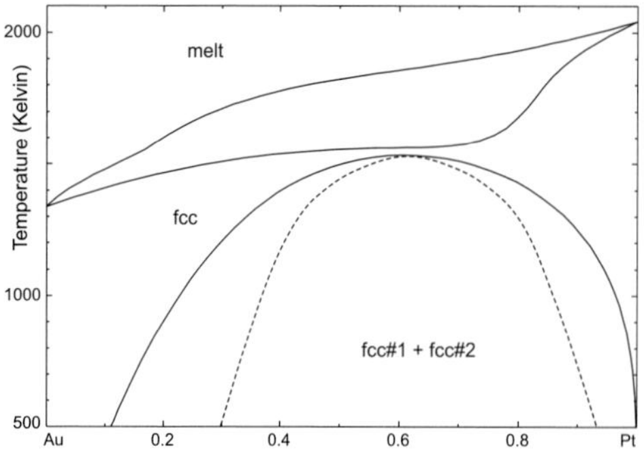

Figure 14.9 Pt and Au (both face-centered cubic) show complete miscibility at high *T* and below a miscibility gap (dashed line: spinodal curve). (Data are compiled from [7,21].)

Very often in phase diagrams only the stability limits of the corresponding phases are shown, like in the middle panel of Figure 14.4 for NaCl–KCl. Typically this equilibrium ("coexistence") curve envelopes a region were the system is metastable with respect to microscopic concentration fluctuations. Spontaneous decomposition appears then inside the spinodale, which is shown in Figure 14.9 as dashed line. It should be noted that the shape of both curves, equilibrium and spinodal, may be influenced by other fields, such as mechanical stress of the demixing phases [21]. During epitaxial growth in this way phases can be stabilized that would be, under equilibrium bulk conditions, significantly outside their stability fields.

Also liquid phases can show total or partial immiscibility. Figure 14.10 demonstrates this for the system Pb–Cu where around 1000 °C a Pb-rich and a Cu-rich melt coexist. Only above the "melt#1 + melt#2" phase field one homogeneous liquid phase is formed. Such behavior is called monotectic and is found not only for some intermetallic systems (especially with metals from the first or second

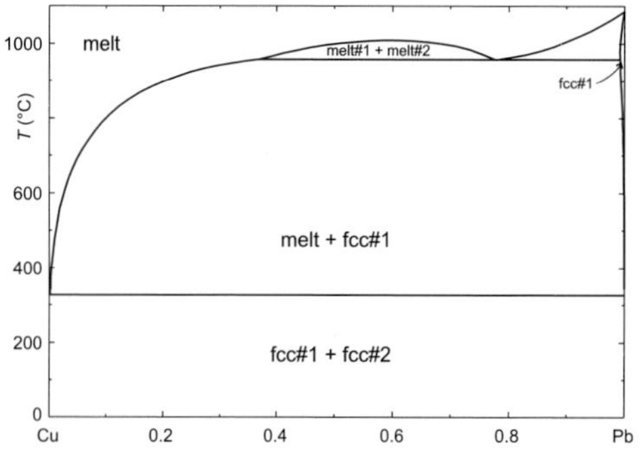

Figure 14.10 The face centered solid phases of Pb and Cu (fcc#1 and fcc#2) are almost immiscible, except a minor solubility of Cu in Pb near 1000 °C. Besides, immiscibility appears for melts with intermediate concentrations.

group of the periodic system, or lead, as one component), but also quite often for systems where the components are compounds with substantially different anions. It should be remarked that some systems with complex anions, such as silicate and phosphate, together with oxides, show also miscibility gaps in the liquid state. The system I_2–Pb is reported to show a monotectic miscibility gap close to pure lead. Besides, the intermediate phase PbI_2 decomposes at 406 °C under formation of two immiscible liquids [22], which is called a syntectic reaction. Monotectic or syntectic behavior does not necessarily impede crystal growth in the corresponding system [23,24].

14.2.2.2 Solid–Gas Equilibria

All phase diagrams in Section 14.2.2.1 described equilibria between solid and liquid phases, and indeed such diagrams are sufficient for many applications where the vapor pressure of all components can be neglected. However, the vapor pressure p_A of a chemical substance A over a condensed phase never vanishes for $T > 0$ K and increases with temperature T. Typically the gas phase cannot be neglected in phase diagrams if a significant fraction of the relevant species appears in the gas phase, which is for sure the case in the vicinity of "gas" phase fields in diagrams such as Figure 14.5. Other conditions apply for the CaO–Fe_2O_3 (or better: Ca–Fe–O_2) systems in Figure 14.3 where the metal oxide phases remain under all conditions in condensed states. Nevertheless the equilibrium between Fe_2O_3 and Fe_3O_4 is influenced by the oxygen fugacity in the whole system. Both cases will be handled separately in the following sections.

Volatility of Main Components

Binary compounds $A^{III}B^{V}$ between a metal from the third group of the periodic system (Al, Ga, In) and one of the elements N, P, As, Sb from the fifth group are semiconductors with great relevance for example, for optoelectronics. They have in common that all components exhibit low melting points (e.g., for gallium: 29.8 °C = 302.9 K; for red phosphorus: 582 °C = 855 K). In contrast, the melting point of the binary compounds is high; for GaP values of 1740 K or even more were reported [25]. The solubility of excess A or B in the $A^{III}B^{V}$ compounds is small and obviously ≪ 1% [26]. In $x - T$ phase diagrams of the whole system, the $A^{III}B^{V}$ phase appears as a vertical line, such as shown in Figure 14.11a for GaP. The much higher melting point T_f of the intermediate binary, compared to the components, shifts the eutectics so close to the rim of the phase diagrams that they may almost coincide with the components.

The $x - T$ phase diagram in Figure 14.11a describes only the equilibrium between solid GaP and the Ga–P melt. Irrespective of the circumstance that the vapor pressure especially of phosphorus is large (boiling point at 1 bar around 280 °C, the gas phase is neglected. Diagrams like in Figure 14.11b are consequently a useful complement of the $x - T$ diagram. The fugacity (≈ vapor pressure p_X of the most important species (X = P_4, P_2, P, or Ga, respectively) is given for conditions along the liquidus curve of the $x - T$ diagram. Typically T^{-1}

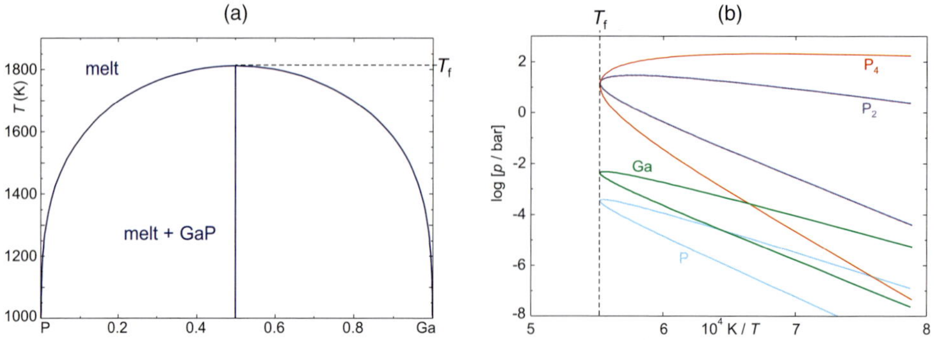

Figure 14.11 (a) Binary phase diagram Ga–P with intermediate GaP and degenerate eutectics on both sides. (b) Vapor pressure of several gaseous species along the GaP liquidus.

rather than T is chosen as abscissa, resulting in nearly linear dependencies $\log[p] - (T^{-1})$ for low T. The apex of the almost hyperbolic curves corresponds to the melting temperature T_f. From these diagrams, one can read that P_4 has the highest and almost constant fugacity in the order of 100 bar along the left side of the GaP liquidus. (Bass [27] reported 35 bar.) The fugacity of P_2 is smaller, because under such phosphorus-rich conditions the equilibrium $P_4 \rightleftarrows 2\,P_2$ is shifted to the educt side. The lower branches represent both phosphorus species on the Ga-rich side, and there P_2 prevails over P_4. The fugacity of Ga does never exceed several mbar, and for this species the upper branch represents the Ga-rich side of the $x - T$ phase diagram.

Volatile Anions

Many chemical compounds such as oxides, halides, or sulfides are of high technological or scientific relevance. Such substances are typically composed of a cation (mainly a metal) and the corresponding anion (here oxygen, a halogen, or sulfur). Many metals (especially transition metals) can take different oxidation states and the corresponding ions are connected via redox reactions of the kind

$$2\,\mathrm{MeO}_{m/2} + \frac{1}{2}\mathrm{O}_2 + 2\,\mathrm{MeO}_{(m+1)/2} \tag{14.7}$$

where m is the valency of the corresponding ion. Often the metal oxide fugacity, compared to the anion fugacity, can be neglected. Then, the equilibrium constant K of reaction Eq. (14.7) is described by the relation

$$-RT\ln K = \frac{1}{2}RT\ln p_{\mathrm{O}_2} = \Delta G^0 = \Delta H^0 - TS^0, \tag{14.8}$$

which means that plots of $2\Delta G^0 = RT\ln p_{\mathrm{O}_2}$ versus T are linear for each reaction of type Eq. (14.7) [28,29]. Such plots are called Ellingham diagrams. Figure 14.12a shows such plots for the different oxidation states of iron, ranging from Fe^0 (metal) to Fe^{3+} (hematite). The blue solid inclined curves represent the

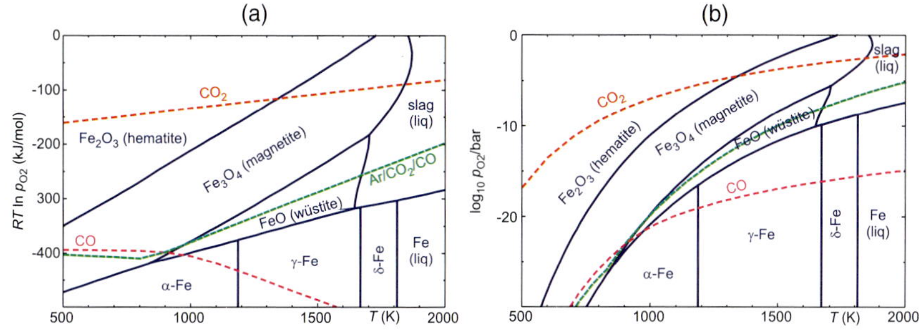

Figure 14.12 (a) Ellingham type predominance diagram for the system Fe–O_2 (blue solid lines). The dashed lines represent the $RT\ln p_{O_2}(T)$ that are created by thermolysis of CO_2, CO, or by a mixture 85% Ar + 10% CO_2 + 5% CO. (b) The same data plotted with the alternative scaling $\log_{10} p_{O_2}(T)$. Total pressure always $p = 1$ bar.

Gibbs free energy changes ΔG^0 for reactions of type Eq. (14.7) between subsequent iron oxidation states. Vertical or almost vertical lines represent first-order phase transitions without change of the oxidation state. These are the phase transitions between different structures of solid iron, and the melting of iron of iron oxides. It is obvious that the melting temperature of iron oxides depends significantly on p_{O_2} and that all iron oxides melt to one "slag" phase with variable composition FeO_x.

Such Ellingham type predominance phase diagrams have the drawback that the fugacity of the volatile (often oxygen) cannot be read directly. From the experimentalist's viewpoint the alternative scaling shown in Figure 14.12b is sometimes more convenient. There a constant fugacity of a volatile (e.g., a fixed concentration of oxygen in a noble gas) would be represented by a horizontal line. The right panel makes obvious, that Fe_3O_4 and FeO have a very limited stability range on the T axis for any fixed p_{O_2}. In contrast, several gaseous oxides such as CO_2 and CO can deliver by thermolysis $CO_2 \rightleftarrows CO + 0.5\,O_2 \rightleftarrows C + O_2$ a $p_{O_2}(T)$ that changes approximately in the same way with T like the stability ranges of different metal oxides. Even if the predominance field of wüstite is narrow, a suitable mixture of Ar, CO_2, and CO is able to stabilize Fe^{2+} over a wide T angle [30]. This is remarkable because the equilibrium oxygen fugacity at the eutectoid stability limit of FeO near 850 K is so small ($p_{O_2} \ll 10^{-20}$ bar), that even the oxygen impurity traces of argon with 99.999% purity are sufficient to oxidize FeO completely to Fe_3O_4. Here the gas mixture containing CO_2 and CO is superior because it acts as a smoothly adjustable buffer.

In Figure 14.12b, the $p_{O_2}(T)$ curves are monotonously rising for all gases, whereas the corresponding curve for CO is falling beyond 900 K in Figure 14.12a. This behavior results from the very low oxygen concentration at these conditions, where the equilibrium $O_2 \rightleftarrows 2\,O$ is shifted to the right-hand side, but only p_{O_2} contributes to the ordinate value. Together with the rising T values, this results in a negative slope of the curve.

14.3
Kinetic Aspects

14.3.1
Segregation

Crystallization processes are usually performed in such a way that an intensive variable describing the system (often the temperature T) is changed. This was demonstrated in Figure 14.5 with the cases (A) and (B) where the system moved from a fluid state (liquid or gaseous, respectively) to the crystalline orthorhombic state. Both, start and end of the reaction arrow are here inside "single-phase" fields. The situation is more complicated if the system moves from a "1-phase" field (a single phase) to a "2-phase" field. In Figure 14.7a from Cu-rich melts ($x_{Sr} < 0.21$) pure Cu crystallizes first. Consequently, upon further cooling the melt is depleted in Cu and the melt composition moves along the liquidus curve until the peritectic point of Cu_5Sr is reached ($x_{Sr} = 0.21$; $T = 845\,°C$). Here crystallization of Cu_5Sr starts, under further Cu depletion of the melt. This phase again crystallizes down to its peritectic point ($x_{Sr} = 0.58$; $T = 586\,°C$), and is followed by the crystallization of CuSr down to the eutectic point. Such preferred transformation of higher melting components or phases toward the solid state, which goes along with an enrichment of lower melting components in the liquid state, is called segregation.

The process of segregation is more complicated if the composition of the solid phase is not fixed, which is the case for example, for "Cu" and "C15" in the Mg–Cu system Figure 14.7b. From melts left of the $MgCu_2$/Cu eutectic, $x_{Mg} < 0.21$, a Cu-rich solid solution crystallizes first. The concentration x of Mg in this solid solution (sometimes written as Cu:Mg or $Cu_{1-x}Mg_x$) can be obtained by drawing an isothermal (= horizontal) "tie line" through the diagram. This procedure allows also to obtain quantitative data on the concentration and on the amount of different phases during segregation and is discussed in the next paragraph. Besides it will be shown that often the thermodynamic equilibrium state under practical conditions cannot be reached, and that kinetically governed diffusion processes must be taken into account.

14.3.2
Lever Rule

Horizontal lines in $x - T$ phase diagrams are isotherms. If such lines are drawn inside 2-phase fields, like in Figure 14.13, they connect compositions two phases that coexist under the given conditions in equilibrium. These are "tie lines." For each of the three cases in Figure 14.13, cooling of a homogeneous melt starts inside the "liquid"-phase field. The formation of the solid phase begins when the liquidus line is touched. In Figure 14.13a this is at T_1. Type and composition of the solid phase are found by following the horizontal (coarse dashed) tie line to the next phase boundary. For the three cases, these are (a) the intermediate

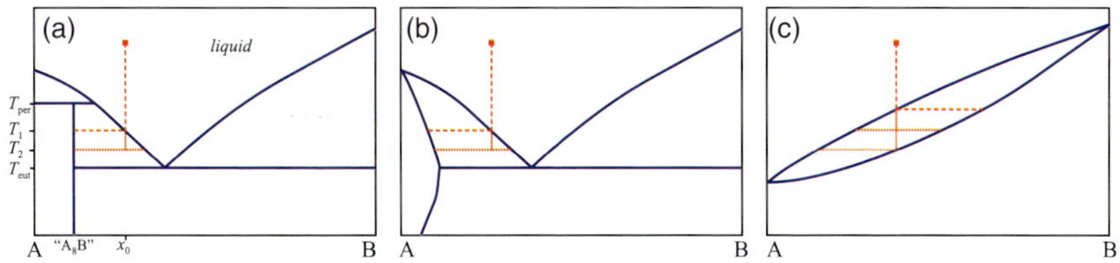

Figure 14.13 Segregation and lever rule for different types of binary phase diagrams. (a) peritectic, (b) eutectic with partial miscibility in the solid, (c) total miscibility in the liquid and solid phases.

compound "A_8B," (b) a solid solution of some B component in an A-rich mixed crystal. The composition of this mixed crystal can be found by dropping a line perpendicular from the point, where the tie line meets the solidus of the mixed crystal. (c) A solid solution of A and B where again the concentration can be found from the perpendicular at the end of the tie line.

Upon further cooling to the somewhat lower temperature T_2 in Figure 14.13a still solid A_8B is in equilibrium with the melt, but the melt contains now more B than the initial melt x_0. This segregation results from the continued crystallization of the A-rich compound A_8B and the resulting depletion of A in the melt. The relative quantities of A_8B and melt can be obtained by comparing the lengths of the "levers" that are spanned from x_0 to the phase boundaries. The right lever between x_0 and liquidus gives the share of A_8B and the left lever between x_0 and the vertical A_8B solidus at $x = 1/(1+8) = 0.111$ gives the share of remaining melt. The crystallization can be continued down to T_{eut} where the melt reaches the eutectic composition.

From Figure 14.13a it is obvious why crystals of substances with peritectic melting behavior cannot be grown from melts of the same chemical composition: For all melt compositions left from the intersection between the peritectic line at T_{per} and the liquidus,"A" crystallizes first. "A_8B" can only be obtained from melt compositions between the peritectic and eutectic points, x_{per} or x_{eut}, respectively. With the lever rule one can derive that the yield of crystalline material with composition x_{sol} crystallizing from a melt x_0 is

$$Y_{max} = \frac{x_0 - x_{eut}}{x_{sol} - x_{eut}} \tag{14.9}$$

and the maximum yield for a peritectic material is obtained for $x_0 = x_{per}$ [31]. Growth processes where a crystal is grown from a melt with different composition are often called flux growth processes, or melt solution growth. If, like described here, the flux consists of the same components like the crystal in just a different concentration, it is often called a "self flux" [20].

The lever rule can also be used to describe growth processes where the solid phase has a variable composition. In Figure 14.13b, this is done for the case where the crystal (here: A) has a limited but significant solubility for the minor component B. For most practical cases, this situation leads to an overlay of kinetic and equilibrium effects during crystallization that was discussed already

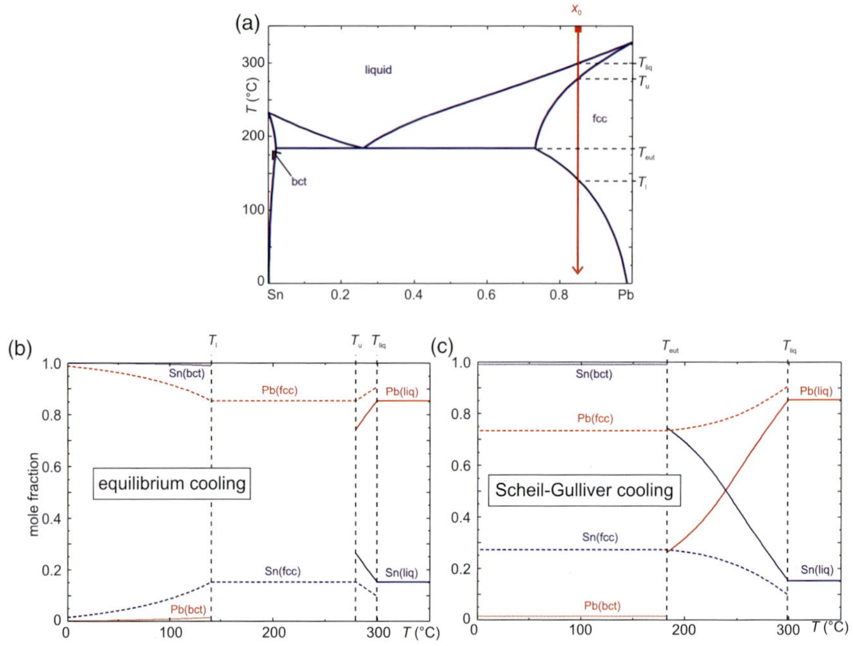

Figure 14.14 The phase diagram Pb–Sn shows a significant solubility of Sn in the face-centered cubic (fcc) Pb structure and a small solubility of Pb in the body-centered tetragonal (bct) Sn structure. The lower diagrams demonstrate from right to left the cooling process under thermodynamic equilibrium or Scheil–Gulliver conditions, respectively, for the melt composition x_0 indicated in the phase diagram.

one century ago by Gulliver [32]. An example is given in Figure 14.14 where a lead-rich melt x_0 is cooled from the liquid state down to room temperature.

Figure 14.14b shows from right to left the sequence of processes occurring upon cooling of a melt x_0 if the system is allowed to reach equilibrium at any instant of time ("equilibrium cooling"):

1) The melt composition is constant down to T_{liq}, where crystallization of the fcc solid starts.
2) The lead concentration [Pb] of fcc is larger than x_0, consequently [Pb] drops in the melt, and [Sn] rises. The application of the lever rule in the phase diagram makes obvious, that between T_{liq} and the upper boundary of the fcc field, T_{u}, the concentration of tin rises in the liquid as well as in the fcc phase.
3) Between upper and lower boundary of the fcc field ($T_{\text{u}} \geq T \geq T_{\text{l}}$) nothing is changed. Only the fcc phase is stable with the same constant composition x_0 like the initial melt.
4) Below T_l the system enters the large phase field under the eutectic where fcc and bcc coexist. Both phase fields are narrowing, however, with lower T. Consequently fcc becomes almost pure Pb, and bct becomes almost pure Sn.

The fulfillment of all processes given above requires, that also solid phases adopt their composition if T is lowered. The necessary diffusion processes might be possible at least partially as long as the temperature is still high, and a part of the system is still liquid. Besides it is necessary that the liquid phase is in close contact to the already crystallized phase(s). This could be the case just below the start of crystallization, for $T \lesssim T_{\text{liq}}$, but even there the time that is required for diffusion processes often exceeds experimentally accessible periods. This problem is enforced in many crystal growth processes, such as Czochralski or Bridgman. There the bulk of the grown crystal is no longer in direct contact with the remaining melt, and the pulling (growth) rates in the order of millimeters per hour are usually faster than the speed of atoms diffusing in solids. Consequently, the alternative scenario in Figure 14.14c is usually observed in practice:

1) Like under equilibrium conditions, the melt composition is constant down to T_{liq} where the crystallization of fcc starts.
2) The first crystallizing fcc phase has the same composition like the first crystallizing solid in the equilibrium case. However, this material is no longer in exchange with the liquid phase, for example, because it is removed from the melt by pulling in a Czochralski process. As [Pb] in the fcc phase is higher than x_0, the melt composition shifts toward the eutectic.
3) The crystallization of fcc from the melt continues until the melt reaches the eutectic point. But as the melt is continuously enriched in tin, also [Sn] in the fcc phase grows. Practically the melt composition moves along the liquidus line down toward the eutectic, and the fcc composition can be read from the lever rule inside the "liquid + fcc" field.
4) The crystallization terminates at T_{eut} with the first formation of bct (Sn rich mixed crystal) and the phase state is frozen in – the compositions of fcc and bct remain on the level which was reached at T_{eut}.

It can be summarized that under equilibrium cooling the resulting solid is a mixture of nearly pure Pb (fcc) and nearly pure Sn (bct). In contrast, Scheil–Gulliver cooling leads to a gradient crystal with fcc structure, where the Pb concentration drops from the first crystallized part to the last crystallized part. At the very end of the cooling process, a two-phase mixture with eutectic composition follows.

14.3.3
Constitutional Supercooling

The previous Section 14.3.2 discussed the case where homogeneous equilibrium distribution of components could not be obtained in a solid phase due to kinetic effects. However, analogous effects might appear also inside a fluid phase like the melt. Whether this occurs or not depends on parameters such as the growth rate of a crystal, on temperature gradients, or on the melt viscosity. Tiller [33] showed that segregation during crystallization, together with incomplete mixing inside the melt, leads to an instability that is called "constitutional supercooling."

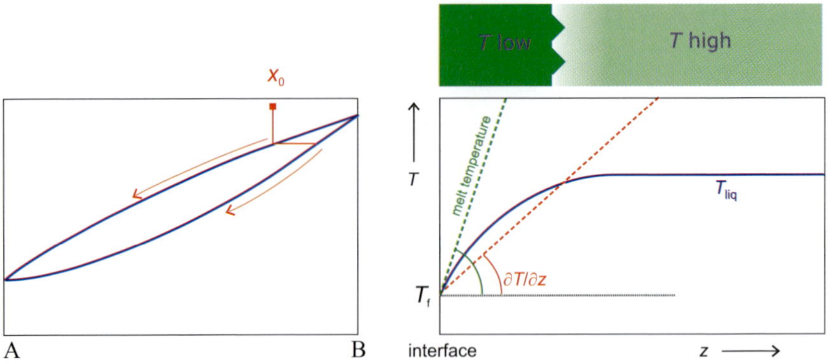

Figure 14.15 (a) Segregation in a mixed crystal system (see also Figure 14.13) enriches the lower melting component A in front of the interface. (b) Due to the high A concentration, T_{liq} drops and stronger cooling is required for crystallization. If $\partial T/\partial z < \partial T_{liq}/\partial z$, random fluctuations of the growing interface are amplified.

Figure 14.15 demonstrates this for a mixed crystal system such as in Figure 14.13c, but adoption to eutectic systems like shown in Figure 14.13a or b is possible too. If a melt with composition x_0, close to a higher melting compound (here: solid B), is cooled, the solid crystallizing at the corresponding T_{liq} is enriched in B. In the course of the crystallization process, both solid and melt become richer in A, but always the solid will incorporate more B than the current melt composition. In the top right drawing this is symbolized by darker green color of the solid (left) than the melt. The preferred incorporation of B in the solid leads to a depletion of this component in front of the solid/liquid interface, which is symbolized by an even lighter color there. According to the phase diagram, this local depletion in B results in a drop of T_{liq}. The right diagram shows this local variation by a blue full line.

Besides, the real temperature of the melt rises from T_f at the solid/liquid interface to a somewhat higher value in the melt volume. If this almost linear temperature gradient $\partial T/\partial z$ (z – coordinate pointing into the melt volume) is sufficiently large, the situation is stable (green dashed line in Figure 14.15). In contrast, a too small temperature gradient (red dashed line) brings melt volumes in front of the interface below the corresponding T_f. This means that solidification is enhanced on top of random surface fluctuations, and unstable "cell formation" or even dendritic growth occurs [34]. Since melt mixing is able to reduce the depletion of B, enhancing convection is a way to stabilize the interface.

14.4
Conclusion

Most crystal growth processes are performed with low growth rate, and close to thermodynamic equilibrium. The growth itself is a phase transition from a nutrient phase (melt, solution, gas) to the desired crystalline phase, and can be described qualitatively and often even quantitatively by equilibrium phase diagrams. It should be taken into account that also for crystallization processes from a melt or solution the liquid phase, the crystal, and if applicable also the crucible are in contact with a surrounding gas phase that might be relevant (Section 14.2.2.2). Such phase diagrams are constructed under the condition that time plays no role, which means that concentration gradients inside one phase are totally leveled out by diffusion. Even if this condition is usually not fully satisfied and solid phases with variable composition (mixed crystals) are formed, a realistic description of the crystallization process can be given by Scheil–Gulliver solidification (Section 14.3.2). Farer from equilibrium, also inhomogeneities in the fluid mother phase can occur, leading to unstable growth (Section 14.3.3).

References

1 Thomson, W. (1857) IX. – On the dynamical theory of heat. Part V. Thermo-electric currents. *Earth Environ. Sci. Trans. R Soc. Edinb.*, **21**, 123–171.

2 Pechini, M.P. (1967) Method of preparing lead and alkaline earth titanates and niobates and coating method using the same to form a capacitor. US Patent 3,330,697.

3 Raj, A.M.E., Jayanthi, D.D., and Jothy, V.B. (2008) Optimized growth and characterization of cadmium oxalate single crystals in silica gel. *Solid State Sci.*, **10** (5), 557–562.

4 Adkins, C.J. (1983) *Equilibrium Thermodynamics*, Cambridge University Press.

5 Helbig, R. (1972) Über die Züchtung von grösseren reinen und dotierten ZnO-Kristallen aus der Gasphase. *J. Crystal Growth*, **15**, 25–31 (in German).

6 Hagemark, K.I. and Toren, P.E. (1975) Determination of excess Zn in ZnO. The phase boundary Zn–$Zn_{1+x}O$. *J. Electrochem. Soc.*, **122**, 992–994.

7 CRCT (2015) Factsage thermodynamic software and databases, version 7.0. Available at www.factsage.com (accessed Nov. 24, 2016).

8 The American Ceramic Society, 600N (2014) Phase equilibria diagrams, version 4.0. Available at http://ceramics.org/publications-and-resources/phase-equilibria (accessed Nov. 24, 2016).

9 Hillert, M., Selleby, M., and Sundman, B. (1990) An assessment of the Ca–Fe–O system. *Metall. Trans. A*, **21** (10), 2759–2776.

10 Gibbs, J.W. (1875 ff) On the equilibrium of heterogeneous substances. *Trans. Conn. Acad. Arts Sci.*, **II**, 108–524. Online available at: http://web.mit.edu/jwk/www/docs/Gibbs1875-1878-Equilibrium_of_Heterogeneous_Substances.pdf.

11 Ye, Z.G. and Schmid, H. (1993) Optical, dielectric and polarization studies of the electric field-induced phase transition in $Pb(Mg_{1/3}Nb_{2/3})O_3$ [PMN]. *Ferroelectrics*, **145** (1), 83–108.

12 Haeni, J.H., Irvin, P., Chang, W., Uecker, R., Reiche, P., Li, Y.L., Choudhury, S., Tian, W., Hawley, M.E., Craigo, B., Tagantsev, A.K., Pan, X.Q., Streiffer, S.K., Chen, L.Q., Kirchoefer, S.W., Levy, J., and Schlom, D.G. (2004) Room-temperature ferroelectricity in strained $SrTiO_3$. *Nature*, **430**, 758–761.

13 Klimm, D., Rabe, M., Bertram, R., Uecker, R., and Parthier, L. (2008) Phase diagram analysis and crystal growth of solid solutions $Ca_{1-x}Sr_xF_2$. *J. Crystal Growth*, **310**, 152–155.

14 McGinnety, J.A. (1972) Redetermination of the structures of potassium sulphate and potassium chromate: the effect of electrostatic crystal forces upon observed bond lengths. *Acta Crystallogr. B*, **28** (9), 2845–2852.

15 Gatehouse, B.M. and Lloyd, D.J. (1973) Crystal structure of anhydrous potassium carbonate. *J. Chem. Soc., Dalton Trans.*, 70–72.

16 Hallstedt, B. (1992) Thermodynamic assessment of the system $MgO–Al_2O_3$. *J. Am. Ceram. Soc.*, **75**, 1497–1507.

17 Shoemaker, C.B. and Shoemaker, D.P. (1963) The crystal structure of the δ phase, Mo–Ni. *Acta Crystallogr.*, **16** (10), 997–1009.

18 Stein, F., Palm, M., and Sauthoff, G. (2004) Structure and stability of laves phases. Part I. Critical assessment of factors controlling laves phase stability. *Intermetallics*, **12**, 713–720.

19 Snyder, G.J. and Simon, A. (1994) Crystal-structure of copper strontium (5/1), Cu_5Sr. *Z. Krist.*, **209** (4), 384.

20 Elwell, D. and Scheel, H.J. (1975) *Crystal Growth from High Temperature Solutions*, Academic Press, London.

21 Xu, X.N., Qin, G.W., Ren, Y.P., Shen, B., and Pei, W.L. (2009) Experimental study of the miscibility gap and calculation of the spinodal curves of the Au-Pt system. *Scripta Mater.*, **61** (9), 859–862.

22 Zhu, X.H., Zhao, B.J., Zhu, S.F., Jin, Y.R., He, Z.Y., Zhang, J.J., and Huang, Y. (2006) Synthesis and characterization of PbI_2 polycrystals. *Cryst. Res. Technol.*, **41** (3), 239–242.

23 Delves, R.T. (1965) Constitutional supercooling and two-liquid growth of HgTe alloys. *Brit. J. Appl. Phys.*, **16**, 343–351.

24 Klimm, D., Ganschow, S., Bertram, R., Doerschel, J., Bermúdez, V., and Kłos, A. (2002) Phase separation during the melting of oxide borates $LnCa_4O(BO_3)_3$ (Ln = Y, Gd). *Mat. Res. Bull.*, **37**, 1737–1747.

25 Richman, D. (1963) Dissociation pressures of GaAs, GaP and InP and the nature of III–V melts. *J. Phys. Chem. Solids*, **24**, 1131–1139.

26 Van Vechten, J.A. (1975) Nonstoichiometry and nonradiative recombination in GaP. *J. Electron. Mater.*, **4** (5), 1159–1169.

27 Bass, S. and Oliver, P. (1968) Pulling of gallium phosphide crystals by liquid encapsulation. *J. Cryst. Growth*, **3–4**, 286–290.

28 Cahn, R.W., Haasen, P., and Kramer, E.J. (1991) *Materials Science and Technology*, vol. **5**, VCH, Weinheim, Thermodynamics and Phase Diagrams of Materials (Arthur D. Pelton), pp. 1–73.

29 Klimm, D., Ganschow, S., Schulz, D., Bertram, R., Uecker, R., Reiche, P., and Fornari, R. (2009) Growth of oxide compounds under dynamic atmosphere composition. *J. Crystal Growth*, **311**, 534–536.

30 Klimm, D. and Ganschow, S. (2005) The control of iron oxidation state during FeO and olivine crystal growth. *J. Crystal Growth*, **275**, e849–e854.

31 Klimm, D. (2014) Phase equilibria, in *Handbook of Crystal Growth*, (ed T. Nishinaga), Elsevier, pp. 85–136.

32 Gulliver, G.H. (1913) The quantitative effect of rapid cooling upon the constitution of binary alloys. *J. Inst. Met.*, **9**, 120–157.

33 Tiller, W., Jackson, K., Rutter, J., and Chalmers, B. (1953) The redistribution of solute atoms during the solidification of metals. *Acta Metall. Mater.*, **1** (4), 428–437.

34 Jackson, K.A. (2004) Constitutional supercooling surface roughening. *J. Cryst. Growth*, **264** (4), 519–529.

15
Chemical Vapor Deposition

Takashi Goto and Hirokazu Katsui

Institute for Materials Research, Tohoku University, 2-1-1, Katahira, Aoba-ku, Sendai 980-8577, Japan

15.1
Introduction

Chemical vapor deposition (CVD) is a versatile technique for mainly preparing thin films typically used in semiconductor devices and protective coatings [1]. Highly pure and dense self-standing materials can also be prepared by CVD followed by removing the substrate after deposition. Without the substrate, CVD can form powder or particles in a gas phase. In addition, by changing the wide-ranged deposition condition of CVD, it can be employed to discover new materials. CVD can also be used to prepare highly oriented, anisotropic or porous materials.

As CVD generally adopts thermochemical reactions to prepare various forms of materials, it is fundamentally a high-temperature process; this is sometimes disadvantageous for practical applications. In such cases, auxiliary energy such as plasma and photo (laser) can be employed to accelerate chemical reactions and enable low-temperature deposition. This chapter mainly describes the preparation of films and various forms of materials by CVD and auxiliary energy-assisted CVD.

15.2
Basics of CVD

Gas-phase deposition process can be generally categorized into two types: physical vapor deposition (PVD) and CVD. In PVD, a target material is evaporated or sputtered to atoms, molecules, and clusters and deposited as a film on a substrate. PVD is characterized by the absence of chemical reactions associated with the deposition process and an insignificant difference between the compositions of the source materials (target) and resultant film. Therefore, PVD readily controls the composition of the deposited films. Conversely, CVD includes

Handbook of Solid State Chemistry, First Edition. Edited by Richard Dronskowski, Shinichi Kikkawa, and Andreas Stein.

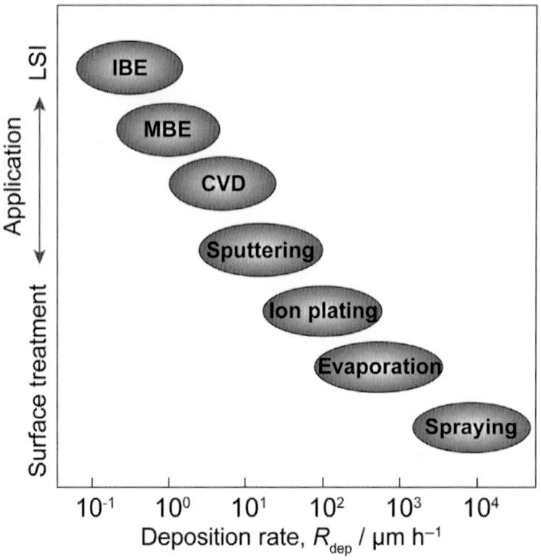

Figure 15.1 Typical deposition rates of various film deposition processes.

several chemical reactions, and the composition of the resultant film may be different from that of the gases (precursors). Thus, the composition of films grown by CVD might depend on the concurrent chemical reaction rates.

CVD is generally a slow deposition process. Figure 15.1 shows the typical deposition rates of various film deposition processes. In semiconductor, large-scale integration (LSI) fabrication, very low-speed processes such as ion-beam epitaxy (IBE) and molecular beam epitaxy (MBE), and sputtering are employed. For refractory or antioxidation coating, thick coatings of approximately several hundred micrometer thickness are often required, and then high-speed deposition processes such as plasma spray or vacuum evaporation are used. The deposition rate of CVD is usually several tens micrometers per hour. CVD can produce highly pure and dense films even on complex-shaped or rough surfaces with high adherence and good conformal coverage. In addition, CVD may necessitate a high temperature depending on the precursors used, while the deposition in CVD proceeds under relatively low vacuum as compared to that in PVD.

In CVD, films are formed by chemical reactions at the gas/substrate interface (heterogeneous reaction) or in a gas phase (homogeneous reaction) and various forms of materials can be obtained by controlling deposition conditions [2]. Figure 15.2 depicts the general trend of the effects of deposition conditions (temperature and supersaturation of source gases) on the morphology of CVD materials. Supersaturation is defined as the ratio of vapor pressure of source gases to the vapor pressure at thermal equilibrium. At low temperature and high supersaturation, powder-like deposits are formed. With increasing temperature and decreasing supersaturation, the deposits would change from amorphous to fine crystals, columnar crystals, dendritic crystals, whiskers, plate-like crystals, and epitaxial single-crystal films.

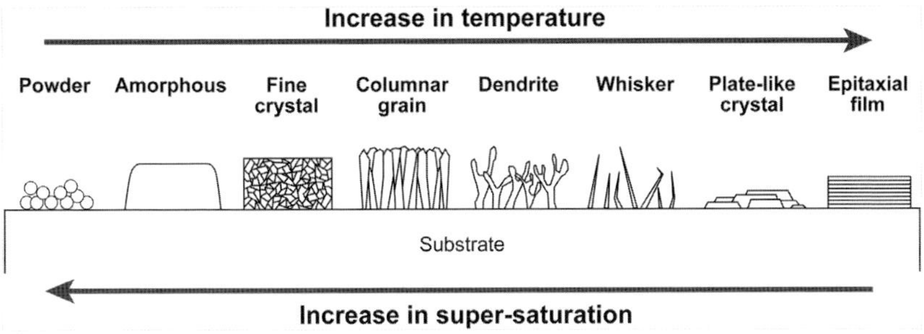

Figure 15.2 Effects of deposition temperature and supersaturation of source gases on the morphology of CVD materials.

CVD uses thermal energy (heat) for chemical reactions, such as electric current heating and radio frequency (RF) and microwave induction heating. Therefore, conventional CVD is called thermal CVD, and can be further categorized into cold-wall CVD and hot-wall CVD depending on the type of CVD chamber [3]. Figure 15.3 depicts schematics of hot- and cold-wall CVD. In hot-wall CVD, the wall of the CVD chamber is heated, which indirectly heats the substrate. Since hot-wall CVD has a uniform temperature distribution in the CVD chamber, many substrates can be coated simultaneously. For example, in an industrial setting, several thousands of cutting tools can be coated by hot-wall CVD simultaneously. As the chemical reactions take place on hot walls in a wide area and as the source gases may be depleted and thus deposition rate are

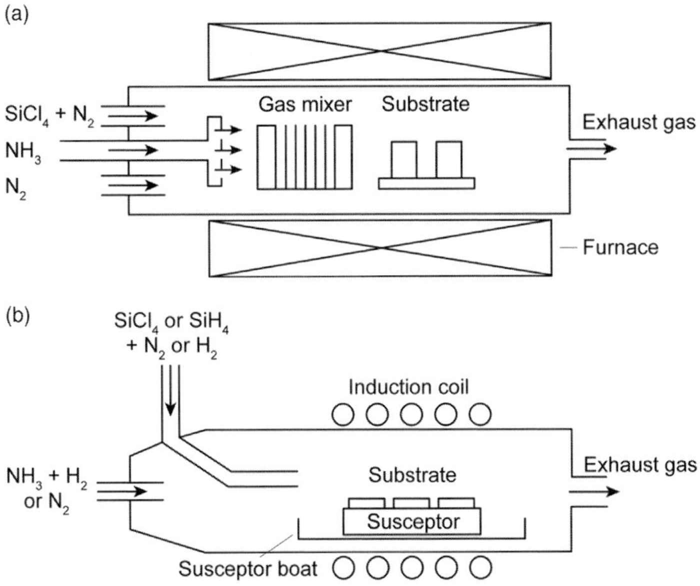

Figure 15.3 Schematics of (a) hot- and (b) cold-wall CVD.

usually low. In a cold-wall CVD, on the other hand, the substrate is locally and directly heated by RF induction current or laser irradiation. The chemical reactions on a cold wall are usually insignificant, and thus, the deposition rate is high. However, the uniform temperature region in cold-wall CVD may be small, and thus, only one or a few substrates can be coated at a time.

CVD can be also categorized according to the source materials used. CVD using metal halogens (fluoride, chloride, and bromide) is called halide CVD and that using metal-organic (MO) compounds is called MOCVD. MO precursors are generally more reactive than halides, and thus, films developed by MOCVD are generally deposited at a lower temperature than those developed by halide CVD. Several types of MO precursors have been developed to prepare various oxide and nonoxide films [4]. Since MO precursors are usually neither corrosive nor explosive, MOCVD is currently more common than halide CVD. In addition, MOCVD can be widely employed for practical applications such as preparations of superconducting [5] or ferroelectric films [6].

15.3 Kinetics of CVD

Chemical reactions, diffusion, and grain growth are associated with CVD, and hence, many parameters need to be controlled to prepare films. CVD uses various reactions, for example, thermal decomposition, reduction, oxidation, nitridation, hydrolysis, chemical transport, disproportion, catalytic reaction, and photodecomposition. Intermediate chemical species form in a gas phase, and nucleation occurs in a gas phase or substrate surface. Next, grain growth takes place on the substrate. A schematic of the film-formation processes is shown in Figure 15.4 [7–9].

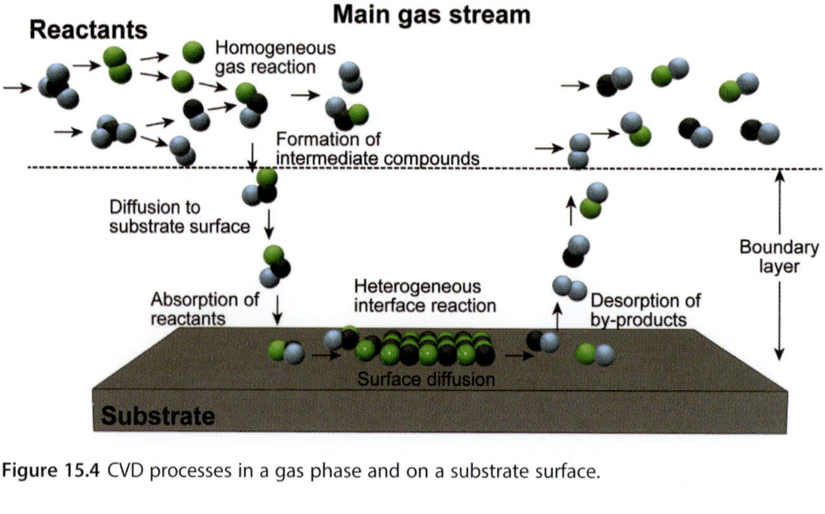

Figure 15.4 CVD processes in a gas phase and on a substrate surface.

i) From gas phase to substrate
   a) homogeneous gas reaction,
   b) formation of intermediate chemical species, and
   c) diffusion/transport of reaction gas/intermediate species to the substrate.
ii) On the substrate surface
   a) absorption of reaction species,
   b) heterogeneous interface reaction, and
   c) diffusion/transport of by-product gas to the main gas stream.

The film formation process in CVD is the consequence of the above-mentioned successive processes, and the slowest process is rate determining of deposition rate. The rate-determining process generally changes depending on temperature, as shown in Figure 15.5. The deposition rate is limited by chemical reaction in a low-temperature region; this is called the chemical reaction-limited process. The chemical reaction rate in this temperature range is generally low. Since the chemical reaction rates are low and the consumption of source gases is small, a sufficient amount of source gases is supplied to the substrate surface. Hence, the deposition rate or thickness distribution of films depends on the substrate temperature. The temperature difference (distribution) causes the inhomogeneity of the film thickness (or deposition rate). The deposition rate significantly increases with increasing temperature in the chemical reaction-limited process. With increasing temperature, the chemical reaction rate increases exponentially, and is sufficiently high at high temperature. In a high-temperature region, the deposition rate is determined by the supply of source gases or the diffusion of by-product gases, instead of the chemical reaction rate. This is called

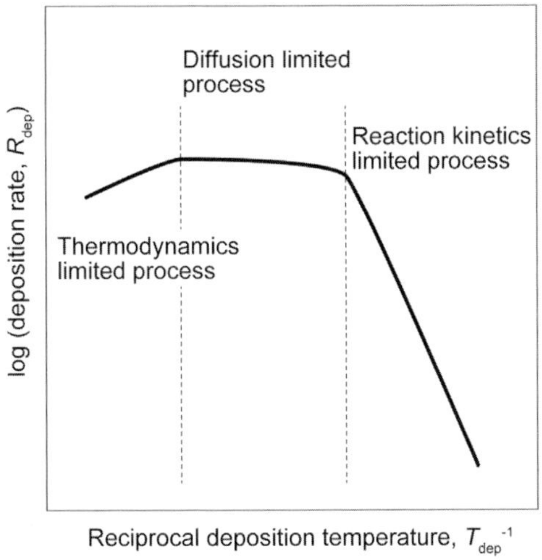

Figure 15.5 Typical temperature dependence of deposition rate by CVD in the Arrhenius format.

diffusion (or mass transport)-limited process. In the diffusion-limited process, since the temperature dependence of the diffusion coefficient of gas species is small, that of the deposition rate is also small. Thus, the temperature difference in a substrate does not cause significant thickness inhomogeneity, whereas the gas flow or gas supply affects the difference in the film thickness on the substrate. Since the deposition rate is high in the diffusion-limited process, a thick film or bulky material can be prepared in this process. At higher temperatures, the deposition rate may decrease because of premature chemical reactions (homogeneous reactions) in a gas phase usually forming powder. The deposition rate can be determined by a thermodynamic equilibrium at further higher temperature. In general, the Gibbs free energy ($-\Delta G^{\circ}$) decreases with increasing temperature, and then, the deposition rate may decrease at high temperature. Moreover, the deposited film may evaporate at high temperature, which in turn will decrease the deposition rate.

Silicon carbide (SiC) is the most widely studied material by thermal CVD. Thin and thick SiC films have been prepared on account of the wide-ranged applications of CVD [10,11]. SiC films can be prepared by thermal CVD using various precursors, such as $SiCl_4 + CH_4 + H_2$, $CH_3SiCl_3 + H_2$, and $SiH_4 + CH_3$ [12–18]. Methyltrichlorosilane (MTS; CH_3SiCl_3) is widely used to prepare SiC films in laboratory and industry because it has a fixed composition, Si/C = 1/1, and appropriate vapor pressure at room temperature. Highly pure and dense SiC thin and thick films and bulky SiC have been prepared by using MTS and H_2. By changing the H_2/MTS ratio and temperature, the morphology of SiC film can be widely changed. Figure 15.6 shows the effects of supersaturation (H_2/MTS ratio) and temperature on the morphology of the CVD SiC films [12]. At high temperature and low supersaturation (high H_2/MTS ratio), whiskers or faceted crystals are formed, while at low-temperature and high supersaturation, smooth nodular crystals are formed. Schematics of faceted crystals and smooth nodular are illustrated in Figure 15.7 [12]. The cross section of a faceted crystal

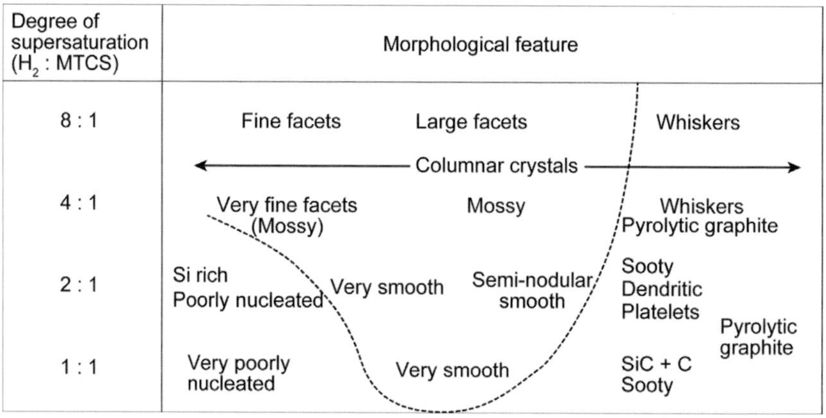

Figure 15.6 Effects of supersaturation and temperature on the morphology of CVD SiC films.

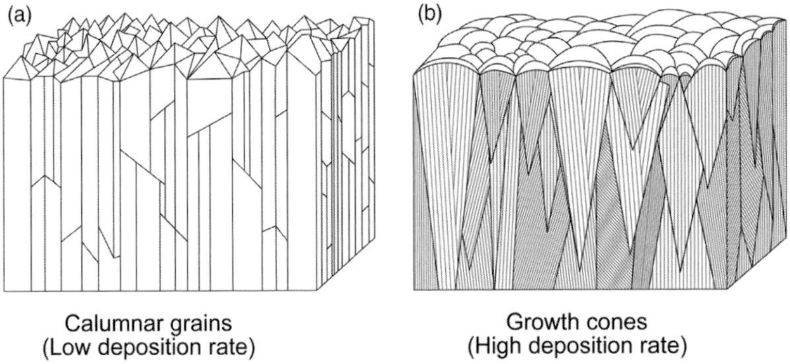

Figure 15.7 Schematic of faceted crystals (a) and smooth nodular films (b) by CVD.

is columnar, while that of a smooth nodular is cone-like. Hence, they are often called cone and columnar structures, respectively. These microstructures are commonly observed in thermal CVD. A Si_3N_4 thick film was prepared by thermal CVD using $SiCl_4$, NH_3, and H_2 gases [19]. The morphology at low temperature and high supersaturation was cone, and that at high temperature and low supersaturation was faceted. AlN was prepared by thermal CVD using $AlCl_3$, NH_3, and H_2 gases, exhibiting typical cone and faceted structures as shown in Figure 15.8 [20]. These effects of temperature and supersaturation on the microstructure of films prepared by CVD are in accordance with those summarized in Figure 15.2.

15.4 Thermodynamics of CVD

Although CVD does not proceed under a thermal equilibrium, the driving force of chemical reactions is related to $-\Delta G^\circ$. Therefore, CVD can be thermodynamically analyzed [21–24].

The thermal equilibrium of CVD can be calculated by an optimization method. The nondimensional Gibbs free energy (G/RT) of the CVD system is given by Eq. (15.1).

$$G/RT = \sum_{i}^{m} x_i^g[(g^0/RT)_i^g + \ln P + \ln(x_i^g/X)] + \sum_{i}^{S} x_i^c(g^0/RT)_i^c, \qquad (15.1)$$

where R refers to the ideal gas construct; T refers to the absolute temperature; x_i refers to the mole number of chemical species i; g^0 refers to the chemical potential at the standard state; P refers to the total pressure; X refers to the total molar amount; g and c refer to the gas and solid phases, respectively; m and s are the numbers of gas and solid species, respectively. The thermal equilibrium state is defined as the case of G/RT minimized

Figure 15.8 Cone (a) and faceted (b) microstructures of CVD AlN films.

under the mass balance condition of Eq. (15.2), as follows:

$$\sum_{i}^{m} a_{ij} x_i^g + \sum_{i}^{S} a_{ij} x_i^c = b_j, \tag{15.2}$$

where, b_j is the total amount of element j. The thermodynamic calculation based on SOLGASMIX-PV [24] and FactSage [25,26] can be used for the calculation. This calculation is useful for a multicomponent system such as Si–Ti–C–Cl–H as several CVD parameters are required to be considered. This system has many gas and solid species, as shown in Table 15.1 [27,28]. By using thermodynamic data in the JANF table [27], the thermodynamic equilibrium can be calculated by solving Eqs. (15.1) and (15.2). Figure 15.9 presents the equilibrium gas pressures under deposition conditions of 1773 K, $CCl_4/H_2 = 1.7 \times 10^{-2}$, and $P = 40$ kPa as a function of $SiCl_4/(SiCl_4 + TiCl_4)$ (molar ratio) in the source gases. The broken lines in the figure represent partial pressures of H_2, $SiCl_4$, $TiCl_4$, and CCl_4 in the source gases. In the equilibrium, H_2 and HCl are major gases, and $SiCl_4$ is decomposed to $SiCl_2$. $TiCl_4$ decomposes to $TiCl_3$ and CCl_4 is converted

Table 15.1 Solid and gas species of the Si–Ti–C–Cl–H system.

<Gas species>						
SiC	C	Si	H_2	Cl_2	Si_2	Si_3
C_2	C_3	Si_2C	SiC_2	SiH	SiH_4	CH
CH_2	CH_3	CH_4	C_2H	C_2H_2	C_2H_4	$Si(CH_3)_4$
SiCl	$SiCl_2$	$SiCl_3$	$SiCl_4$	CCl	CCl_2	CCl_3
CCl_4	C_2Cl	C_2Cl_4	C_2Cl_6	HCl	CHCl	$CHCl_3$
CH_2Cl	CH_3Cl	C_2HCl	SiH_3Cl	SiH_2Cl_2	$SiHCl_3$	$SiCl_3CH_3$
H	Cl	Ti	TiCl	$TiCl_2$	$TiCl_3$	$TiCl_4$
<Solid species>						
Si	Ti	C	SiC	TiC	TiSi	$TiSi_2$
Ti_5Si_3	Ti_3SiC_2					

to various hydrocarbons and HCl. Figure 15.10 depicts the solid phases by the calculation and experiments at $CCl_4/H_2 = 1.7 \times 10^{-2}$ and $P = 40$ kPa as functions of temperature and $SiCl_4/(SiCl_4 + TiCl_4)$ in the source gases. The calculation coincides well with the experimental results, that is, the formation of TiC, SiC and TiC mixture, and SiC. The TiC formation in the experiments is more significant than in the calculation. This implies that at low temperature, TiC is formed more readily than SiC. In experiments, well-faceted and crystallized TiC can be prepared as compared to SiC at the same temperature and pressure. Thus, TiC is kinetically easier to form than SiC. Monolytic Ti_3SiC_2 was first synthesized by thermal CVD using $SiCl_4$, $TiCl_4$, CCl_4, and H_2 as source gases [29].

15.5 Rotary CVD

Although CVD is typically employed to prepare films, it can also synthesize powders by homogeneous nucleation in a gas phase or can deposit particles on particulate substrates. As shown in Figure 15.2, powders are formed at high supersaturation and low deposition temperature. Nanometer-sized particles dispersed on particulate supports with large surface area are considerably important in the development of highly active catalysts. There are various techniques to synthesize catalysts, such as solid-state reaction and wet process (e.g., impregnation and sol–gel). Among these techniques, CVD is especially advantageous in fabricating highly active nanoparticles as it avoids the aggregation and growth of particle sizes. Catalytic nanoparticles such as Ni [30], Ir–C [31], Pr-Ir [32], and Ru-MgO [33] have been synthesized by thermal CVD, as nucleation in a gas phase and growth of particles occur in a hot-wall CVD. Various CVD techniques

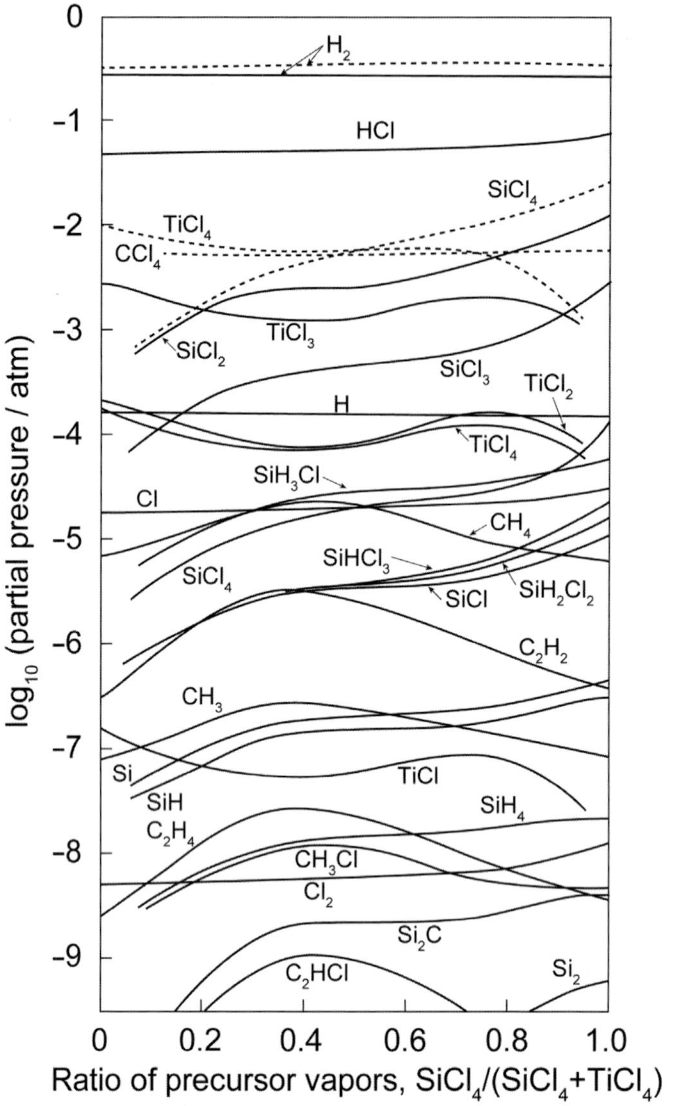

Figure 15.9 Equilibrium gas pressures under deposition conditions of 1773 K, $CCl_4/H_2 = 1.7 \times 10^{-2}$, and $P = 40$ kPa as a function of $SiCl_4/(SiCl_4 + TiCl_4)$ in source gases. The dotted lines indicate source gas pressures.

such as catalytic CVD [34], fluidized-bed CVD (FBCVD) [35], and two-step CVD [36,37] have been developed for the synthesis of catalytic nanoparticles, nanotubes, nanofibers, and nanocomposites. FBCVD is commonly used to prepare nanoparticles on particulate substrates. Figure 15.11 shows a schematic of FBCVD [38,39]. In a vertical-type reactor, the upper stream of source gases with

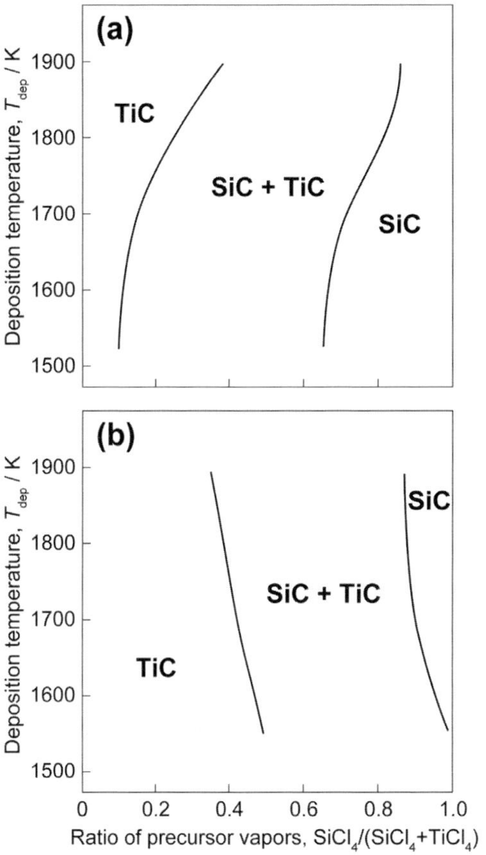

Figure 15.10 Solid phases by calculation (a) and experiments (b) at $CCl_4/H_2 = 1.7 \times 10^{-2}$ and $P = 40$ kPa as functions of temperature and $SiCl_4/(SiCl_4 + TiCl_4)$ in source gases.

a carrier gas induces the fluidization of powder in a gas phase, where the source gases chemically react and form nanoparticles on the surface of the particulate substrates. Figure 15.12 depicts SiO_2 powder deposited with Rh nanoparticles by FBCVD using $[Rh(\mu\text{-}Cl)CCO_2]_2$ as a source precursor [35]. Rh nanoparticles with a size of 1.7 nm were uniformly deposited on the SiO_2 powder surface. However, the SiO_2 particles agglomerated and formed secondary particles, which were several micrometers in size, as seen in Figure 15.12. Particle sizes less than several hundred nanometers are often agglomerated, and the agglomerated particles behave like big particles; they cannot be fluidized. If the nanometer-sized powder is not agglomerated, the powder can be easily blown away by the gas flow in FBCVD. If the powder size exceeds several hundred micrometers, the powder cannot fluidize through the gas flow owing to its heavy weight. Generally, the particle sizes of powders for fluidization range from several tens to several hundred micrometers in FBCVD [40].

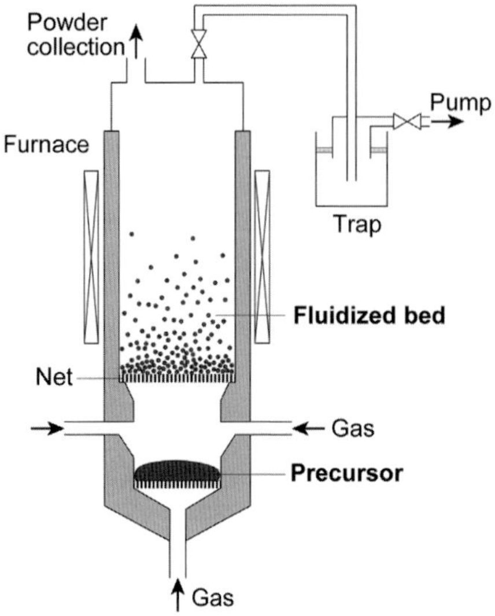

Figure 15.11 Schematic of typical FBCVD [38,39].

Rotary CVD has been developed to prepare nanoparticles on the surfaces of particulate substrates [41]. Figure 15.13 shows a schematic of rotary CVD apparatus. Drum-shaped CVD reactor with blades on the inner wall is rotated. A powder substrate is alternatively pulled up and dropped in the rotated reactor. During its fall in the CVD reactor, a particulate substrate can be suspended in a gas phase. Source gases are carried into the rotated reactor. Nanoparticles form

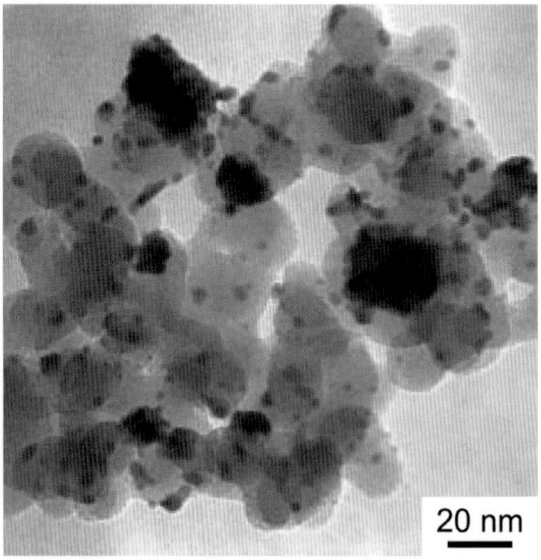

Figure 15.12 Rh-precipitated SiO_2 powder by FBCVD [35].

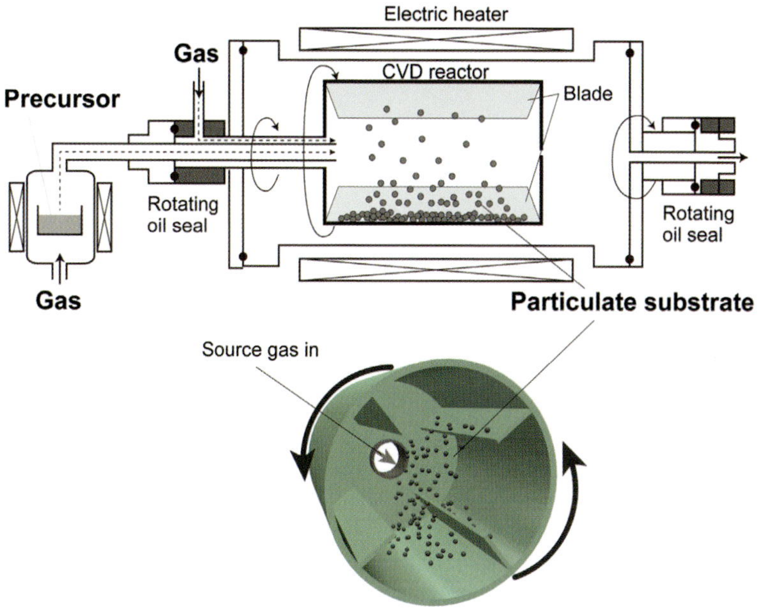

Figure 15.13 Schematic of rotary CVD.

on the surface of the suspended particulate substrate. There is no limit on the powder size to suspend the powder in the rotary CVD. Figure 15.14 shows a TEM bright field image of Ni nanoparticles precipitated on hexagonal BN powder by rotary CVD using nickelocene as a precursor [41]. Ni nanoparticles with

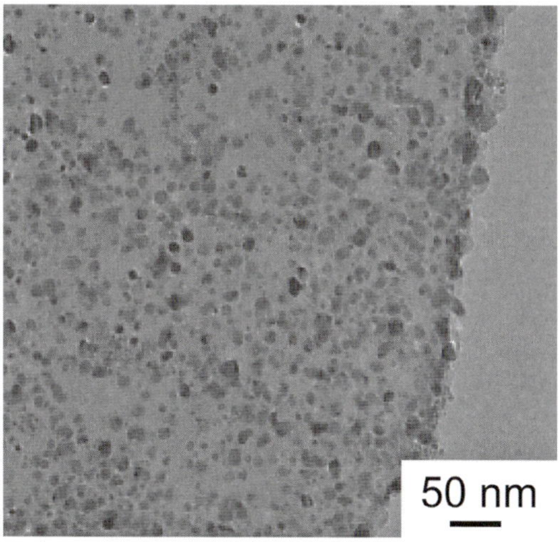

Figure 15.14 TEM bright field images of Ni nanoparticles precipitated on hexagonal BN powders by rotary CVD.

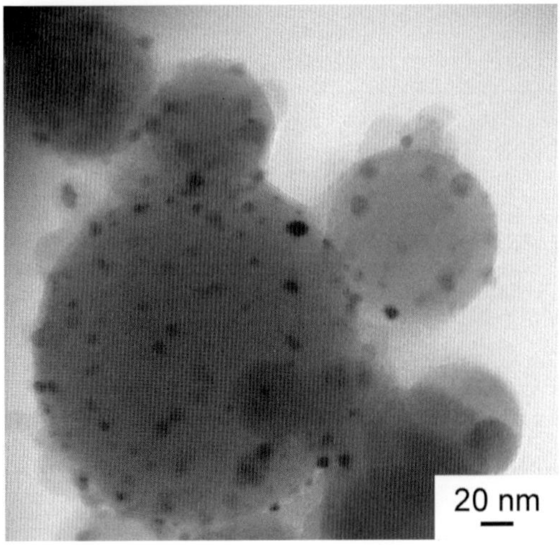

Figure 15.15 TEM bright field image of Ni nanoparticles precipitated on Al_2O_3 powder with the mean particle size of 0.08 μm by rotary CVD.

several tens nanometer particle sizes were uniformly and densely precipitated on the surface of the hexagonal BN powder. As the Ni nanoparticles possess excellent catalytic activity, Al_2O_3-supported Ni catalysts have been extensively studied using various processes, such as heterogeneous precipitation method [42], impregnation [43], solution-spray plasma method [44], sol–gel [45], electro-less deposition [46], ball milling [47], and FBCVD [48]. Wet processes, such as the solution-spray plasma method and sol–gel, require treatments to remove liquids and for hydrogen reduction. This results in the grain growth of Ni particles, and thus, particle sizes range from 10 nm to 1 μm. Ni particles were prepared on Al_2O_3 powder with a mean particle size of 250 μm by FBCVD [48]. By rotary CVD, catalytic nanoparticles can be precipitated on submicron-sized powder substrates. Figure 15.15 shows a TEM bright field image of Ni nanoparticles precipitated on Al_2O_3 powder with particle sizes less than 200 nm by rotary CVD [49]. The size of the Ni nanoparticles was less than 10 nm. Mesoporous silica can be an attractive support for catalysts due to its high-specific surface area. Figure 15.16 shows TEM bright field images of mesoporous silica (a) and Ni-precipitated mesoporous silica by rotary CVD (b) [50]. A stripped pattern was observed and the spacing of the stripes corresponded to the mesopore channel (Figure 15.16a). In addition to the striped pattern, several Ni nanoparticles (particle sizes: approximately 5 nm) with a dark contrast were precipitated by rotary CVD, as shown in Figure 15.16b. The Ni nanoparticles precipitated in the mesopore channels (in addition to those on the cluster surface), as the expansion of *d*-spacing and disorder of the (100) mesoporous lattice were identified by low-angle XRD. Figure 15.17 shows the rate of hydrogen production by the steam methane reforming reaction of Ni nanoparticles precipitated Al_2O_3, mesoporous

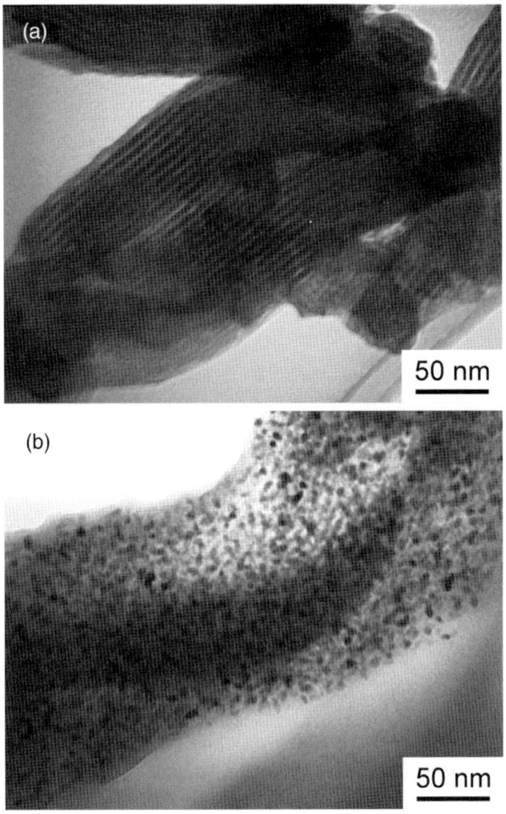

Figure 15.16 TEM bright field images of mesoporous silica (a) and Ni-precipitated mesoporous silica by rotary CVD (b).

silica, and zeolite powders prepared by rotary CVD. The hydrogen production rates of typical Ni-based catalysts in the literature [51–53] are also included for comparison. The pore sizes of zeolite powder were too small to be deposited inside the zeolite. Raney Ni is the most common Ni catalyst. It has sponge-like grains produced from Ni–Al alloy by selectively leaching Al in a concentrated NaOH solution. The catalytic activities of Ni-Al_2O_3 and Ni-mesoporous silica prepared by rotary CVD were approximately three times higher than those of the Raney Ni catalysts. Although the BET surface area of Ni-Al_2O_3 is lower than that of NiSn/$MgAl_2O_3$, the Ni-Al_2O_3 catalyst prepared by rotary CVD exhibited a significantly higher hydrogen production rate. The hydrogen production rate of Ni-mesoporous silica was six times higher than that of layered (Ni, Al) double hydroxide catalysts.

The wettability of Ni to Al_2O_3 powder surface is low, and thus, discontinuous island-like nanoparticles were formed (Figures 15.14 and 15.15). Continuous nanolayers of SiO_2 can be deposited on the powder surface by rotary CVD. This

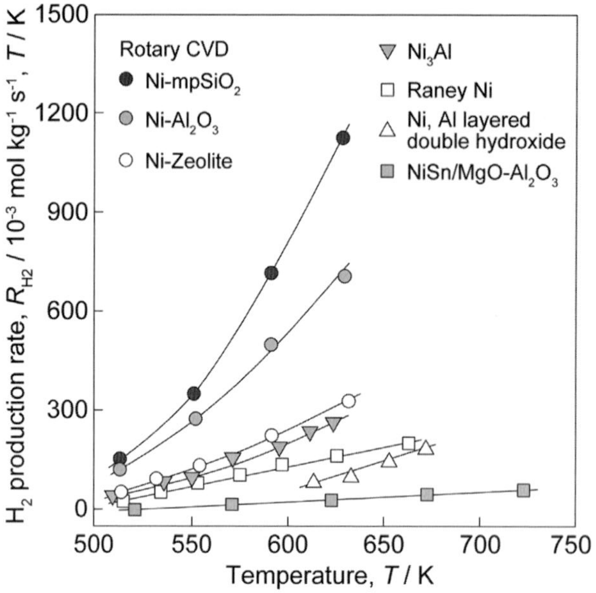

Figure 15.17 Hydrogen production rate by steam methane reforming reaction of Ni nanoparticles precipitated Al_2O_3, mesoporous silica and zeolite powders prepared by rotary CVD. The hydrogen production rates of typical Ni-based catalysts in the literature [51–53] are also included for comparison.

was performed by using tetraethyl orthosilicate (TEOS) as a precursor to give the high wettability of SiO_2 to various materials [54–57]. Figure 15.18 shows the TEM bright field image of SiO_2-coated SiC powder prepared by rotary CVD using TEOS. An 80∼100-nm thick SiO_2 layer was uniformly deposited on SiC powder with a particle size of ∼ 2 μm. This continuous thin SiO_2 layer covering the entire surface of SiC powder prevented the self-contact of SiC. The surface of the SiO_2 layer can be rapidly densified via viscous sintering at a relatively low sintering temperature by spark plasma sintering (SPS). As a result, the SiC–SiO_2 composite sintered body has small-sized mosaic microstructures with no grain growth during sintering, as shown in Figure 15.19, exhibiting high hardness and fracture toughness [56]. The SiC layer can be uniformly formed on diamond powder by rotary CVD [58]. Diamond powder is generally consolidated under ultrahigh pressure that exceeds several giga Pascals, because diamond readily transforms to graphite at high temperatures under moderate pressure below several hundred mega Pascals. The SiC-coated diamond composite powder prepared by rotary CVD, and mixed with SiO_2 powder can be fully densified with no transformation to graphite by SPS under 100 MPa. However, noncoated monolithic diamond powder and a mixture powder of diamond and SiO_2 cannot be densified.

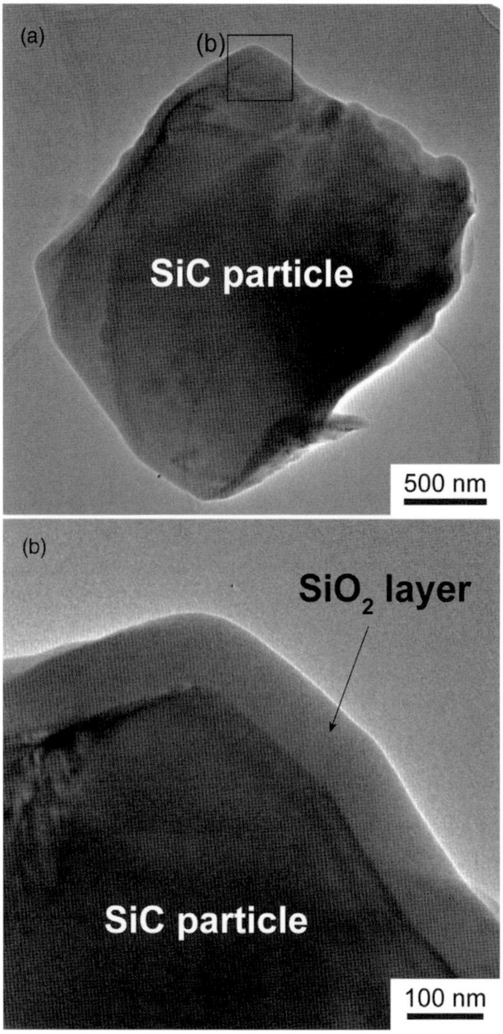

Figure 15.18 TEM bright field image of SiO_2-coated SiC powder prepared by rotary CVD using a TEOS precursor. (b) is the enlarged image of the SiO_2-coated SiC powder edge designated in (a).

15.6
Plasma CVD

Thermal CVD requires thermal energy (heat) and high temperature to deposit films. Therefore, it is difficult to prepare films on nonrefractory metals and polymers by thermal CVD. Thus, the high-temperature deposition is a disadvantage of thermal CVD. Thermal CVD is widely employed to prepare thin films on Si wafer. An example of this is semiconductor fabrication. However, the high-

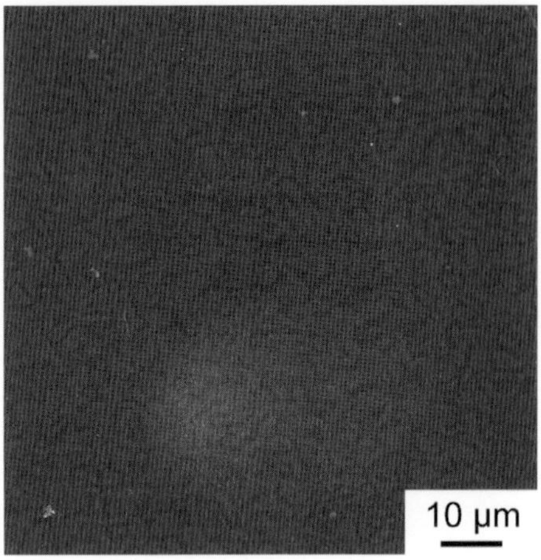

Figure 15.19 Mosaic microstructure resulting from the sintering of SiO_2-coated SiC powder prepared by rotary CVD.

temperature deposition may often degrade the impurity profile of the Si wafer. Therefore, low-temperature deposition is required in various applications. By applying plasma to CVD, chemical reactions in CVD are accelerated and films can be deposited at lower temperature [59–61]. This is called plasma-enhanced CVD (PECVD).

Plasma can be categorized into two types: thermal equilibrium high-temperature plasma and nonequilibrium low-temperature plasma [62]. In PECVD, low-temperature plasma is employed to prepare films. Plasma is formed by applying high voltage to electrodes in an evacuated chamber. At a total pressure of approximately 10 kPa and several killovolts of voltage, a streak-like plasma is formed. At a total pressure of 0.1 kPa, the plasma becomes stable and is uniformly disturbed throughout the chamber. This stable plasma (glow discharge low-temperature plasma) is used for PECVD. Figure 15.20 depicts a schematic of dc (direct current) plasma CVD [63]. Plasma is formed between the electrodes by the application of a high dc voltage to electrodes. Source gases are introduced in the chamber and a film forms on the cathode. Since delaminated films or debris are often dropped on the substrate, the substrate can be set up or side of the chamber. In general, the glow discharge plasma damages the film by ion bombardments or sputtering. RF and microwave electricity can form plasma without electrodes; they are called nonelectrode plasma. In dc plasma, the film and target (electrode) should be conductive, and only metallic materials can be prepared. On the other hand, in RF and microwave plasma, nonconductive materials can also be prepared. Thus, nonelectrode plasma is widely employed in PECVD. Several types of films have been prepared by introducing source gases in the plasma. PECVD can be used to prepare metastable or nonequilibrium films such as diamond and diamond-like carbon films [64–66] and amorphous Si–H semiconductor films [59,60]. PECVD was used to prepare an yttria-

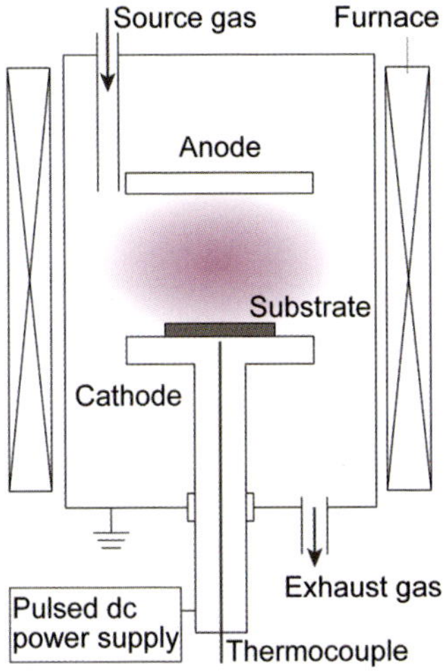

Figure 15.20 Schematic of d.c. plasma CVD.

stabilized zirconia (YSZ) thermal barrier coating that was several hundred micrometers in thickness [67].

15.7 Laser CVD

Laser is an energy source of heat and light (photon), and has been widely employed for several types of material processing such as welding, surface treatment, melt-solidification, and laser ablation. Laser has been also used for CVD and applied to prepare films mainly for semiconductor devices. This is called laser CVD. Figure 15.21 shows schematics of laser CVD. Laser CVD can be categorized into two types: photolytic laser CVD (Figure 15.21a), in which the laser is used for photochemical reactions, and pyrolytic laser CVD (Figure 15.21b), in which the laser is used for pyrolytic (thermochemical) reactions [68,69]. In photolytic laser CVD, a high-energy (short wave length) laser, typically an ultra-violet laser, is used. The laser can decompose the source gases and a film forms without heating. Since a laser beam cannot simultaneously decompose a large amount of precursors, it is assumed that photolytic laser CVD cannot prepare films at high deposition rate on a wide-area substrate. Photolytic laser CVD can only prepare very thin films, nanodots and nanofibers (whiskers). Conversely, pyrolytic laser CVD uses CO_2 laser (infra-red laser) as the heat source. A laser with several hundred watts or several hundred kilowatts laser can be used to heat a substrate by focusing the laser beam. The chemical reaction occurs very quickly due to the introduction of source gases to the laser-irradiated area. Figure 15.22 presents a schematic diagram of pyrolytic laser CVD. Thin rods

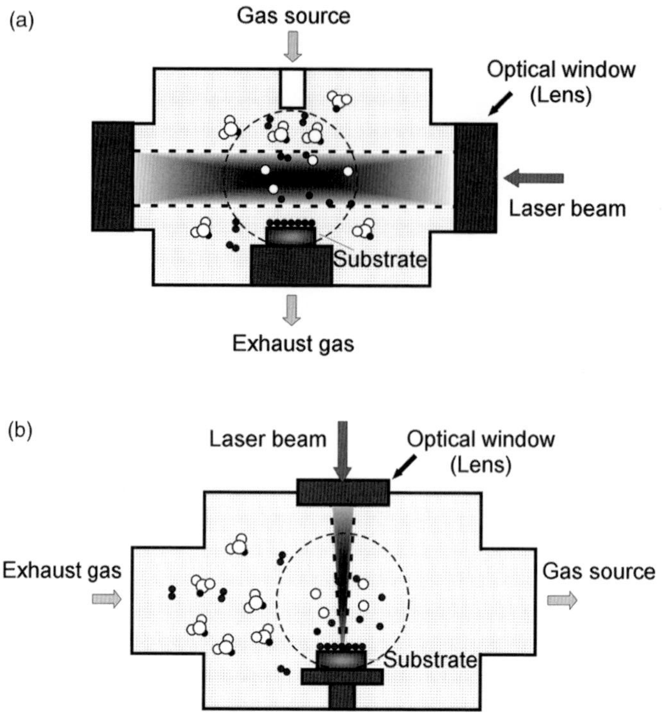

Figure 15.21 Schematics of photolytic laser CVD (a) and pyrolytic laser CVD (b).

and whiskers have been prepared by pyrolytic laser CVD at high deposition rate of several hundred meters per hour.

Figure 15.23 shows the relationship between laser density and the deposition rates of various films in pyrolytic laser CVD. The deposition rate is defined as an increase rate of thickness per unit time (thickness deposition rate) [70–100]. As the laser is focused on a small area (less than a few millimeters), the thickness deposition rate of pyrolytic laser CVD can be more than 10^6 times that of a conventional thermal CVD. However, the volume deposition rate (defined as an increase in volume per unit time) is very small because the laser beam is thin and the area is very small. Even at high temperature, the deposition rate of pyrolytic laser CVD cannot be limited by diffusion (mass transport) because the source gases can easily access the deposition area, as shown in Figure 15.22. However, on a wide-area substrate, the deposition rate is limited by the diffusion of gases to the substrate similar to that in thermal CVD. Thus, it is assumed that pyrolytic laser CVD cannot prepare thick and wide-area films at high deposition rate. Conversely, we found that laser CVD can prepare thick films at high deposition rate equal to approximately several hundred micrometers per hour on a wide-area substrate approximately 20 mm in diameter. Figure 15.24 presents the comparison between the volume deposition rate obtained by a conventional

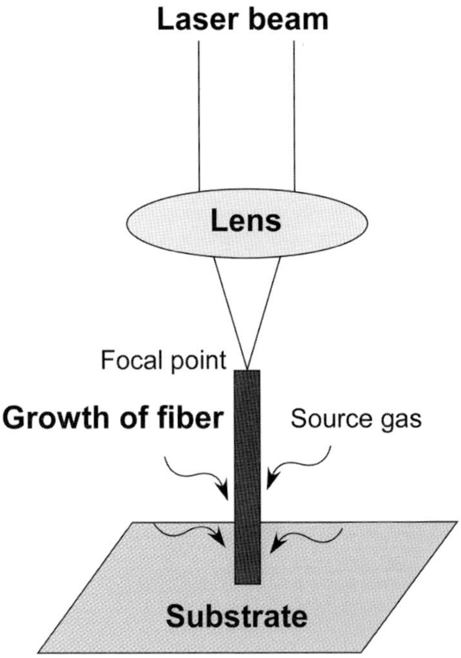

Figure 15.22 Schematic of the growth of fiber by pyrolytic laser CVD.

pyrolytic laser CVD and that by the laser CVD proposed by us (hereafter LCVD). Since the laser beam in (LCVD) is expanded and irradiated to a wide area of the substrate surface, the laser densities of LCVD are lower than those of a conventional pyrolytic laser CVD. Nevertheless, LCVD can prepare various films with a

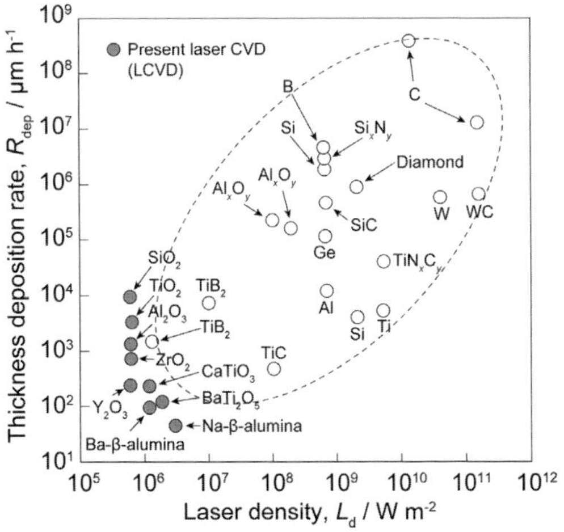

Figure 15.23 Comparison of deposition rates in thickness obtained by conventional pyrolytic laser CVD and present laser CVD (LCVD).

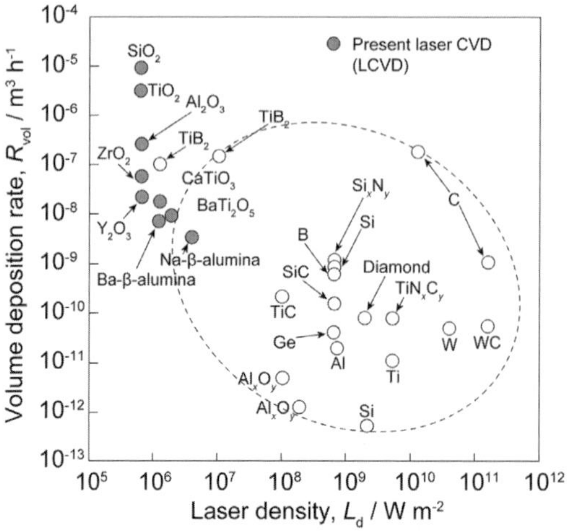

Figure 15.24 Comparison of deposition rates in volume obtained by conventional pyrolytic laser CVD and present laser CVD (LCVD).

volume deposition rate that is several thousand times higher than that obtained by a conventional pyrolytic laser CVD, as shown in Figure 15.24 [92–101], while a conventional pyrolytic laser CVD can deposit films at a high thickness deposition rate in a small area (Figure 15.23).

In a conventional pyrolytic CVD, the deposits often possess donut-like or volcano-like shapes as shown in Figure 15.25 [70,102]. This may be attributed to (i) the excessively high temperature in the center part, which evaporates the deposits, (ii) the difficulty in the diffusion of source gases into the center part, (iii) the occurrence of an upward convection flow at the center part. However, in LCVD, the temperature distribution in a substrate using Nd:YAG laser is uniform approximately 15 mm in diameter within ±5 K at 1200 K as shown in

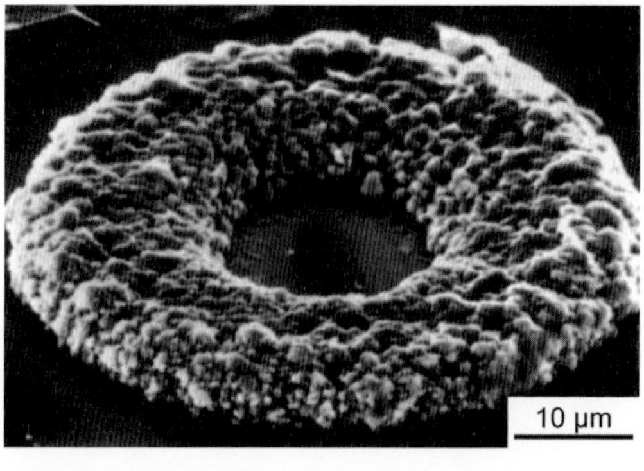

Figure 15.25 Donut-like or volcano-like morphology by conventional pyrolytic CVD.

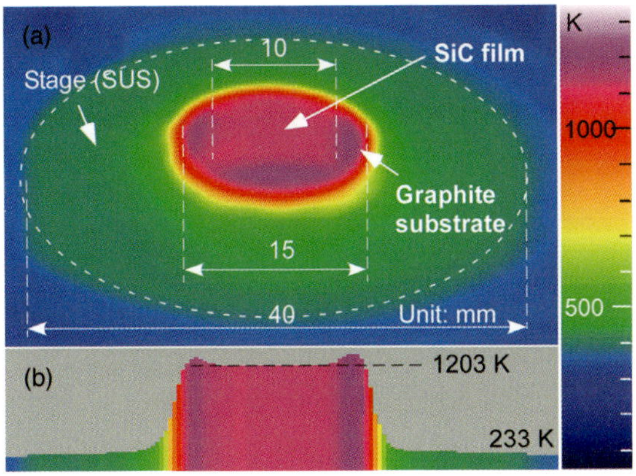

Figure 15.26 Temperature distribution on substrate under irradiation with recent Nd:YAG laser.

Figure 15.26 [103]. By using a scanning laser or by moving the substrate, a much wider substrate can be coated by LCVD. Figure 15.27 demonstrates a YSZ-coated Ni-base super alloy gas turbine blade as a thermal barrier coating (TBC). TBC is generally implemented by PVD, that is, plasma spray or electron-beam physical vapor deposition (EBPVD). LCVD can also be a candidate method for TBC. Figure 15.28 presents the microstructure of the YSZ film prepared by LCVD at a deposition rate of 230–660 μm/h [104,105]. A typical columnar structure (Figure 15.28a) and nanopores (Figure 15.28b) at the columnar boundary can be observed. The nanopores at the columnar boundary can relax the stress between the YSZ film and the metal substrate. This microstructure of LCVD is similar to that of EBPVD [106]. Since the laser can significantly accelerate the chemical reactions, the deposition rate is mainly dependent on the supply of

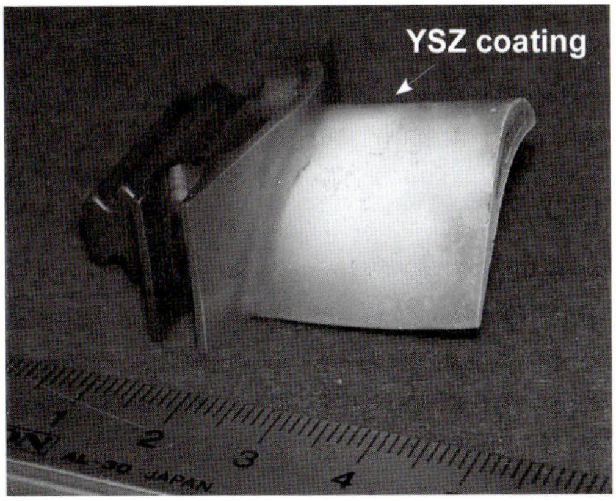

Figure 15.27 YSZ coating on Ni-base super alloy gas turbine blade by laser CVD.

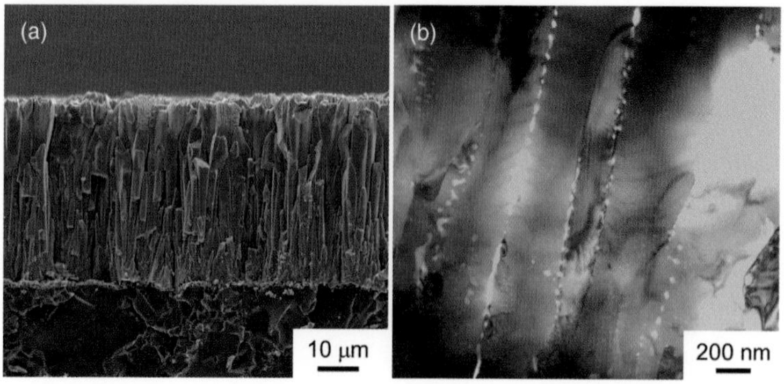

Figure 15.28 Cross-sectional microstructure of YSZ film prepared by LCVD; (a) SEM and (b) TEM bright field images.

source gases. The use of a high-vapor pressure source gas such as TEOS can increase the deposition rate. Figure 15.29 shows the relationship between the deposition rate of SiO_2 film and laser power by using TEOS as a precursor [96]. The microstructure of the SiO_2 film changes from dense to dendritic to porous with an increasing deposition rate. A porous SiO_2 film can be prepared at 27.5 mm/h. The deposition rate increases significantly above the threshold laser power of 150 W for the SiO_2 film and 70 W for the YSZ film. Plasma (or a flame) forms around the substrate when the threshold laser power is crossed. Although the formation mechanism of plasma in LCVD is not clearly understood, it is hypothesized that the plasma formation is associated with the high deposition rate.

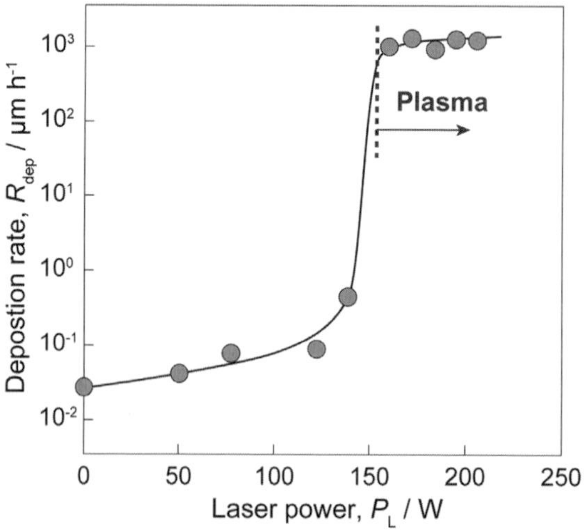

Figure 15.29 Effect of laser power on the deposition rate of SiO_2 film prepared by LCVD.

15.8 Summary

CVD is a key technology for preparing thin and thick films. Highly pure and dense materials can also be prepared by CVD. Many precursors have been developed, following which CVD has become more useful for preparing various forms of materials. CVD is commonly used as thermal CVD, which requires high temperature to deposit films. By applying the axial energy of plasma and/or laser, CVD can become more useful and available across various applications.

References

1 Choy, K.L. (2003) Chemical vapour deposition of coatings. *Prog. Mater. Sci.*, **48** (2), 57–170.

2 Park, J.-H. and Sudarshan, T.S. (2001) *Chemical Vapor Deposition*, vol. **2**, Surface Engineering Seies, AMS International.

3 Blocher, J.M., Jr. (1974) Structure/property/process relationships in chemical vapor deposition CVD. *J. Vac. Sci. Technol.*, **11** (4), 680–686.

4 Pierson, H.O. (1999) CVD Processes and Equipment, *Handbook of Chemical Vapor Deposition (CVD) (Second Edition), Principles, Technology, and Applications*, William Andrew, 108–146.

5 Berry, A.D., Gaskill, D.K., Holm, R.T., Cukauskas, E.J., Kaplan, R., and Henry, R.L. (1988) Formation of high T_C superconducting films by organometallic chemical vapor deposition. *Appl. Phys. Lett.*, **52** (20), 1743–1745.

6 Brierley, C.J., Trundle, C., Considine, L., Whatmore, R.W., and Ainger, F.W. (1989) The growth of ferroelectric oxides by MOCVD. *Ferroelectrics*, **91** (1), 181–192.

7 Jensen, K.F. (1989) Chemical vapor deposition, in *Microelectronics Processing: Chemical Engineering Aspects* (eds D.W. Hess and K.F. Jensen), American Chemical Society, Washington, DC, pp. 199–263.

8 Holstein, W.L., Fitzjohn, J.L., Fahy, E.J., Gilmour, P.W., and Schmelzer, E.R. (1989) Mathematical modeling of cold-wall channel CVD reactors. *J. Cryst. Growth*, **94**, 131–144.

9 Sarin, V.K. (1995) Systematic development of customized CVD coatings. *Surf. Coat. Technol.*, **72** (1–2), 23–33.

10 Schlichting, J. (1980) Chemical vapor deposition of silicon carbide. *Intern. J. Powder Metallurgy*, **12** (3), 141–147.

11 Schlichting, J. (1980) Chemical vapor deposition of silicon carbide. *Int. J. Powder Metall.*, **12** (4), 196–200.

12 Weiss, J.R. and Diefendorf, R.J. (1973) Chemically vapor deposited SiC for high temperature and structural applications, Silicon Carbide 1973 (eds R.C. Marshall, J.W. Faust Jr., and C.E. Ryan), Proceedings of the Third International Conference on Silicon Carbide, Florida.

13 Burk, A.A. (2006) Development of multiwafer warm-wall planetary VPE reactors for SiC device production. *Chem. Vap. Deposition*, **12** (8–9), 465–473.

14 Pedersen, H., Leone, S., Kordina, O., Henry, A., Nishizawa, S., Koshka, Y., and Janzén, E. (2012) Chloride-based CVD growth of silicon carbide for electronic applications. *Chem. Rev.*, **112** (4), 2434–2453.

15 Davis, R.F., Kelner, G., Shur, M., Palmour, J.W., and Edmond, J.A. (1991) Thin film deposition and microelectronic and optoelectronic device fabrication and characterization in monocrystalline alpha and beta silicon carbide. *Proc. IEEE*, **79** (5), 677–701.

16 Chin, J., Gantzel, P.K., and Hudson, R.G. (1977) The structure of chemical vapor deposited silicon carbide. *Thin Solid Films*, **40**, 57–72.

17 Kim, H.J. and Davis, R.F. (1986) Theoretically predicted and experimentally determined effects of the Si/(Si+C) gas phase ratio on the growth and character of monocrystalline beta silicon carbide films. *J. Appl. Phys.*, **60**, 2897–2903.

18 Stinespring, C.D. and Wormhoudt, J.C. (1988) Gas phase kinetics analysis and implications for silicon carbide chemical vapor deposition. *J. Cryst. Growth*, **87** (4), 481–493.

19 Nihara, K. and Hirai, T. (1976) Chemical vapour-deposited silicon nitride. *J. Mater. Sci.*, **11**, 593–603.

20 Goto, T., Tsuneyoshi, J., Kaya, K., and Hirai, T. (1992) Preferred orientation of AlN plates prepared by chemical vapour deposition of $AlCl_3$+NH_3 system. *J. Mater. Sci.*, **27** (1), 247–254.

21 van Zeggeren, F. and Storey, S.H. (1970) *The Computation of Chemical Equlibria*, Cambridge Univ. Press, London.

22 White, W.B., Johnson, S.M., and Dantzig, G.B. (1958) Chemical equilibrium in complex mixtures. *J. Chem. Phys.*, **28**, 751–755.

23 Eriksson, G. (1971) Thermodynamic studies of high temperature equilibria. III. SOLGAS, a computer program for calculating the composition and heat condition of an equilibrium mixture. *Acta. Chem. Scand.*, **25**, 2651–2658.

24 Besmann, T.M. (April (1977)) ORNL/TM-5775.

25 Bale, C.W., Chartrand, P., Degterov, S.A., Eriksson, G., Hack, K., Ben Mahfoud, R., Melançon, J., Pelton, A.D., and Petersen, S. (2002) FactSage thermochemical software and databases. *Calphad*, **26** (2), 189–228.

26 Dinsdale, A.T. (1991) SGTE data for pure elements. *Calphad*, **15** (4), 317–425.

27 Chase, M.W., Davis, C.A., Downey, J.R., Frurip, D.J., McDonald, R.A., and Syverud, A.N. (1986) *NIST JANAF Thermochemical Tables*, 3rd edn, American Institute of Physics, New York.

28 Barin, I. and Knacke, O. (1973) *Thermochemical Properties of Inorganic Substances*, Heidelberg, Supplments (1977).

29 Goto, T. and Hirai, T. (1987) Chemically vapor deposited Ti_3SiC_2. *Mat. Res. Bull.*, **22** (9), 1195–1201.

30 Omata, K., Mazaki, H., Yagita, H., and Fujimoto, K. (1990) Preparation of nickel-on-active carbon catalyst by CVD method for methanol carbonylation. *Catal. Lett.*, **4** (2), 123–127.

31 Goto, T., Vargas, J.R., and Hirai, T. (1996) Preparation of iridium clusters by MOCVD and their electrochemical properties. *Mat. Sci. Eng. A-Struct.*, **217–218**, 223–226.

32 Colindres, S.C., García, J.R.V., Antonio, J.A.T., and Chavez, C.A. (2009) Preparation of platinum-iridium nanoparticles on titania nanotubes by MOCVD and their catalytic evaluation: barium-promoted oxide-supported ruthenium. *J. Alloy. Compd.*, **483** (1–2), 406–409.

33 Bielawa, H., Hinrichsen, O., Birkner, A., and Muhler, M. (2001) The ammonia-synthesis catalyst of the next generation: barium-promoted oxide-supported ruthenium. *Angew. Chem., Int. Ed.*, **40** (6), 1061–1063.

34 Escobar, M., Rubiolo, G., Candal, R., and Goyanes, S. (2009) Effect of catalyst preparation on the yield of carbon nanotube growth. *Physica B*, **404** (18), 2795–2798.

35 Hierso1, J.-C., Serp, P., Feurer, R., and Kalck, P. (1998) MOCVD of rhodium, palladium and platinum complexes on fluidized divided substrates: novel process for one-step preparation of noble-metal catalysts. *Appl. Organomet. Chem.*, **12** (3), 161–172.

36 Dossi, C., Psaro, R., Ugo, R., Zhang, Z.C., and Sachtler, W.M.H. (1994) Non-acidic Pd/Y zeolite catalysts from organopalladium precursors: preparation and catalytic activity in MCP reforming. *J. Catal.*, **149** (1), 92–99.

37 Xu, C. and Zhu, J. (2004) One-step preparation of highly dispersed metal-supported catalysts by fluidized-bed MOCVD for carbon nanotube synthesis. *Nanotechnology*, **15** (11), 1671–1681.

38 Serp, P., Feurer, R., Morancho, R., and Kalck, P. (1995) One-step preparation of highly dispersed supported rhodium

catalysts by low-temperature organometallic chemical-vapor-deposition. *J. Catal.*, **157** (2), 294–300.

39 Vahlas, C., Caussat, B., Serp, P., and Angelopoulos, G.N. (2006) Principles and applications of CVD powder technology. *Mat. Sci. Eng. R*, **53** (1–2), 1–72.

40 Geldart, D. (1973) Types of gas fluidization. *Powder Technol.*, 7 (5), 285–292.

41 Zhang, J., Tu, R., and Goto, T. (2010) Preparation of Ni-precipitated hBN powder by rotary chemical vapor deposition and its consolidation by spark plasma sintering. *J. Alloy. Compd.*, **502** (2), 371–375.

42 Li, G.-J., Huang, X.-X., and Guo, J.-K. (2003) Fabrication and mechanical properties of Al_2O_3–Ni composite from two different powder mixtures. *Mater. Sci. Eng. A*, **352** (1–2), 23–28.

43 Fujiyama, T., Otsuka, M., Tsuiki, H., and Ueno, A. (1987) Control of the impregnation profile of Ni in an Al_2O_3 sphere. *J. Catal.*, **104** (2), 323–330.

44 Watanabe, M., Yamashita, H., Chen, X., Yamanaka, J., Kotobuki, M., Suzuki, H., and Uchida, H. (2007) Nano-sized Ni particles on hollow alumina ball: catalysts for hydrogen production. *Appl. Catal. B*, **71** (3–4), 237–245.

45 Rodeghiero, E.D., Tse, O.K., Chisaki, J., and Giannelis, E.P. (1995) Synthesis and properties of Ni-α-Al_2O_3 composites via sol-gel. *Mater. Sci. Eng. A*, **195** (1), 151–161.

46 Leon, C.A. and Drew, R.A.L. (2000) Preparation of nickel-coated powders as precursors to reinforce MMCs. *J. Mater. Sci.*, **35** (19), 4763–4768.

47 Li, J., Li, F., and Hu, K. (2004) Preparation of Ni/Al_2O_3 nanocomposite powder by high-energy ball milling and subsequent heat treatment. *J. Mater. Process. Tech.*, **147** (2), 236–240.

48 Chen, C.-C. and Chen, S.-W. (1997) Nickel and copper deposition on Al_2O_3 and SiC particulates by using the chemical vapour deposition–fluidized bed reactor technique. *J. Mater. Sci.*, **32** (16), 4429–4435.

49 Zhang, J., Tu, R., and Goto, T. (2013) Precipitation of Ni nanoparticle on Al_2O_3 powders by novel rotary chemical vapor deposition. *J. Ceram. Soc. Jpn.*, **121** (2), 226–229.

50 Zhang, J., Tu, R., and Goto, T. (2013) Precipitation of Ni and NiO nanoparticle catalysts on zeolite and mesoporous silica by rotary chemical vapor deposition. *J. Ceram. Soc. Jpn.*, **121** (10), 891–894.

51 Xu, Y., Kameoka, S., Kishida, K., Demura, M., Tsai, A., and Hirano, T. (2005) Catalytic properties of alkali-leached Ni_3Al for hydrogen production from methanol. *Intermetallics*, **13** (2), 151–155.

52 Qi, C., Amphlett, J.C., and Peppley, B.A. (2007) K (Na)-promoted Ni, Al layered double hydroxide catalysts for the steam reforming of methanol. *J. Power Sources*, **171** (2), 842–849.

53 Zhang, J., Tu, R., and Goto, T. (2011) Hydrogen production by methanol steam reforming on NiSn/MgO–Al_2O_3 catalysts: the role of MgO addition. *Appl. Catal. A Gen.*, **392** (1–2), 184–191.

54 He, Z., Tu, R., Katsui, H., and Goto, T. (2013) Synthesis of SiC/SiO_2 core–shell powder by rotary chemical vapor deposition and its consolidation by spark plasma sintering. *Ceram. Int.*, **39** (3), 2605–2610.

55 He, Z., Katsui, H., Tu, R., and Goto, T. (2014) Surface modification of silicon carbide powder with silica coating by rotary chemical vapor deposition. *Key. Eng. Mater.*, **616**, 232–236.

56 He, Z., Katsui, H., Tu, R., and Goto, T. (2014) High-hardness and ductile mosaic SiC/SiO_2 composite by spark plasma sintering. *J. Am. Ceram. Soc.*, **97** (3), 681–683.

57 Katsui, H., He, Z., and Goto, T. (2016) Preparation and sintering of silica-coated silicon carbide composite powder. *J. Jpn. Soc. Powder Powder Metall.*, **63** (3), 1–6.

58 He, Z., Katsui, H., and Goto, T. (2016) High-hardness diamond composite consolidated by spark plasma sintering. *J. Am. Ceram. Soc.*, 1862–1865.

59 Chittick, R.C., Alexander, J.H., and Sterling, H.F. (1969) The preparation and properties of amorphous silicon. *J. Electrochem. Soc.*, **116** (1), 77–81.

60 Spear, W.E. and Le Comber, P.G. (1975) Substitutional doping of amorphous silicon. *Solid State Commun.*, **17** (9), 1193–1196.

61 Matsuda, A. (2004) Thin-film silicon –growth process and solar cell application. *Jpn. J. Appl. Phys.*, **43** (12), 7909–7920.

62 Bárdos, L. and Baránková, H. (2010) Cold atmospheric plasma: sources, processes, and applications. *Thin Solid Films*, **518** (23), 6705–6713.

63 Leonhardt, A. (1995) Deposition of adherent hard coatings on steel by plasma-enhanced CVD. *Ceram. Int.*, **21** (6), 421–425.

64 Kamo, M., Sato, Y., Matsumoto, S., and Setaka, N. (1983) Diamond synthesis from gas phase in microwave plasma. *J. Cryst. Growth*, **62** (3), 642–644.

65 Schwander, M. and Partes, K. (2011) A review of diamond synthesis by CVD processes. *Diam. Relat. Mater.*, **20** (9), 1287–1301.

66 Silva, S.R.P., Clay, K.J., Speakman, S.P., and Amaratunga, G.A.J. (1995) Diamond-like carbon thin film deposition using a magnetically confined r.f. PECVD system. *Diam. Relat. Mater.*, **4** (7), 977–983.

67 Préauchat, B. and Drawin, S. (2001) Properties of PECVD-deposited thermal barrier coatings. *Surf. Coat. Technol.*, **142–144**, 835–842.

68 Bauerle, D. (2000) Laser-CVD of microstructures, in *Laser Processing and Chemistry*, Springer, pp. 337–360.

69 Goto, T. (2007) Development of highly-functional ceramic materials by chemical vapor deposition. *J. Jpn. Soc. Powder Powder Metallugy*, **54** (12), 863–872.

70 Duty, C., Jean, D., and Lackey, W.J. (2001) Laser chemical vapour deposition: materials, modelling, and process control. *Int. Mater. Rev.*, **46** (6), 271–287.

71 Lehmann, O. and Stuke, M. (1994) Three-dimensional laser direct writing of electrically conducting and isolating microstructures. *Mater. Lett.*, **21** (2), 131–136.

72 Maxwell, J., Shah, J., Webster, T., and Mock, J. (1998) Rapid prototyping of titanium nitride using three-dimensional laser chemical vapor deposition, Solid Freeform Fabrication Proceedings, Austin, Texas, pp. 575–580.

73 Tonneau, D., Bourée, J.E., and Pauleau, Y. (1995) Kinetics of laser thermal decomposition of trimethylamine alane. *Appl. Surf. Sci.*, **86** (1–4), 488–493.

74 Baum, T.H., Larson, C.E., and Jackson, R.L. (1989) Laser-induced chemical vapor deposition of aluminum. *Appl. Phys. Lett.*, **55**, 1264–1266.

75 Wanke, M.C., Lehmann, O., Müller, K., Wen, Q., and Stuke, M. (1997) Laser rapid prototyping of photonic band-gap microstructures. *Science*, **275**, 1284–1286.

76 Wallenberger, F.T. (1997) Inorganic fibres and microfabricated parts by laser assisted chemical vapour deposition (LCVD): structures and properties. *Ceram. Int.*, **23** (2), 119–126.

77 Oliveira, M.N., Botelho do Rego, A.M., and Conde, O. (1998) XPS investigation of $B_xN_yC_z$ coatings deposited by laser assisted chemical vapour desposition. *Surf. Coat. Technol.*, **100–101**, 398–403.

78 Maxwell, J.L., Boman, M., Williams, K., Larsson, K., Jaikumar, N., and Saiprasanna, G. (1999) High-speed laser chemical vapor deposition of amorphous carbon fibers, stacked conductive coils, and folded helical springs, Proceedings of the SPIE, Volume 3874, Micromachining and Microfabrication Process Technology V 227, pp. 227–235.

79 Moylan, C.R., Baum, T.H., and Jones, C.R. (1986) LCVD of copper: deposition rates and deposit shapes. *Appl. Phys. A*, **40** (1), 1–5.

80 Maxwell, J.L., Pegna, J., and Messia, D.V. (1998) Real-time volumetric growth rate measurements and feedback control of three-dimensional laser chemical vapor deposition. *Appl. Phys. A*, **67** (3), 323–329.

81 Petzoldt, F., Piglmayer, K., Kräuter, W., and Bäuerle, D. (1984) Lateral growth rates in laser CVD of microstructures. *Appl. Phys. A*, **35** (3), 155–159.

82 Westberg, H., Boman, M., Johansson, S., and Schweitz, J. (1993) Free-standing silicon microstructures fabricated by laser chemical processing. *J. Appl. Phys.*, **73** (11), 7864–7871.

83 Park, S.I. and Lee, S.S. (1990) Growth kinetics of microscopic silicon rods grown on silicon substrates by the pyrolytic laser-induced chemical vapor deposition process. *Jpn. J. Appl. Phys.*, **29** (1), L129–L132.

84 Chang, H., Lee, L.J., Hwang, R.L., Yeh, C.T., Lin, M.S., Lou, J.C., Hseu, T.H., Wu, T.B., Chen, Y., and Tang, C. (1996) Physical and chemical properties of the cylindrical rods SiC_x (x=0.3–1.2) grown from $Si(CH_3)_2C_{12}$ by laser pyrolysis. *Mater. Chem. Phys.*, **44** (1), 59–66.

85 Maxwell, J., Krishnan, R., and Haridas, S. (1997) High pressure convectively-enhanced laser chemical vapor disposition of titanium, Solid Freeform Fabrication Proceedings, Austin, Texas, pp. 497–504.

86 Hopfe, V., Tehel, A., Baier, A., and Scharsig, J. (1992) IR-laser CVD of TiB_2, TiC_x and TiC_xN_y coatings on carbon fibres. *Appl. Surf. Sci.*, **54** (1), 78–83.

87 Elders, J., Quist, P.A., Rooswijk, B., van Voorst, J.D.W., and van Nieuwkoop, J. (1991) CO_2-laser-induced chemical vapour deposition of TiB_2. *Surf. Coat. Technol.*, **45** (1–3), 105–113.

88 Elders, J. and van Voorst, J.D.W. (1993) Laser-induced CVD of titanium diboride and the influence of atomic hydrogen. *Appl. Surf. Sci.*, **69** (1–4), 267–271.

89 Zergioti, I., Hatziapostolou, A., Hontzopoulos, E., Zervaki, A., and Haidemenopoulos, G.N. (1995) Pyrolytic laser-based chemical vapour deposition of TIC coatings. *Thin Solid Films*, **271** (1–2), 96–100.

90 Croonen, Y.H. and Verspui, G. (1993) Laser induced chemical vapour deposition of TiN coatings at atmospheric pressure. *J. Phys. IV France*, **3**, C3-209–C3-215.

91 Chen, X. and Mazumder, J. (1995) Laser chemical-vapor deposition of titanium nitride. *Phys. Rev. B*, **52** (8), 5947–5952.

92 Chi, C., Katsui, H., and Goto, T. (2015) Preparation of Na-beta-alumina films by laser chamical deposition. *Surf. Coat. Technol.*, **276** (25), 534–538.

93 Ito, A., You, Y., Katsui, H., and Goto, T. (2013) Growth and microstructure of Ba β-alumina films by laser chemical vapor deposition. *J. Eur. Ceram. Soc.*, **33** (13–14), 2655–2661.

94 Ito, A., Guo, D., Tu, R., and Goto, T. (2012) Preparation of (0 2 0)-oriented $BaTi_2O_5$ thick films and their dielectric responses. *J. Eur. Ceram. Soc.*, **32** (10), 2459–2467.

95 Katsui, H., Kumagai, Y., and Goto, T. (2016) High-speed deposition of highly-oriented calcium titanate film by laser CVD. *J. Jpn. Powder Powder Metallugy*, **6** 3(3), 123–127.

96 Endo, J., Ito, A., Kimura, T., and Goto, T. (2010) High-speed deposition of dense, dendritic and porous SiO_2 films by Nd: YAG laser chemical vapor deposition. *Mat. Sci. Eng. B-Solid*, **166** (3), 225–229.

97 Guo, D., Ito, A., Goto, T., Tu, R., Wang, C., Shen, Q., and Zhang, L. (2013) Effect of laser power on orientation and microstructure of TiO_2 films prepared by laser chemical vapor deposition method. *Mater. Lett.*, **93** (15), 179–182.

98 Katokura, H., Ito, A., Kimura, T., and Goto, T. (2010) Moderate temperature and high-speed synthesis of α-Al_2O_3 films by laser chemical vapor deposition using Nd:YAG laser. *Surf. Coat. Technol.*, **204** (14), 2302–2306.

99 Banal, R., Kimura, T., and Goto, T. (2005) High speed deposition of Y_2O_3 films by laser-assisted chemical vapor deposition. *Mater. Trans.*, **46** (9), 2114–2116.

100 Goto, T. (2004) High-speed deposition of zirconia films by laser-induced plasma CVD. *Solid State Ionics*, **172** (1–4), 225–229.

101 Fujie, K., Ito, A., Tu, R., and Goto, T. (2010) Laser chemical vapor deposition of SiC films with CO_2 laser. *J. Alloy Compd.*, **502** (1), 238–242.

102 Suzuki, Y. (1989) Tungsten-carbon X-ray multilayered mirror prepared by photo-chemical vapor deposition. *Jpn J. Appl. Phys.*, **28** (5), 920–924.

103 Zhang, S., Tu, R., and Goto, T. (2012) High-speed epitaxial growth of β-SiC film on Si(111) single crystal by laser chemical vapor deposition. *J. Am. Ceram. Soc.*, **95** (9), 2782–2784.

104 Kimura, T. and Goto, T. (2003) Rapid synthesis of yttria-stabilized zirconia films by laser chemical vapor deposition. *Mater. Trans.*, **44** (3), 421–424.

105 Goto, T. (2005) Thermal barrier coatings deposited by laser CVD. *Surf. Coat. Technol.*, **198** (1–3), 367–371.

106 Lu, T.J., Levi, C.G., Wadley, H.N.G., and Evans, A.G. (2001) Distributed porosity as a control parameter for oxide thermal barriers made by physical vapor deposition. *J. Am. Ceram. Soc.*, **84** (12), 2937–2946.

16
Growth of Wide Bandgap Semiconductors by Halide Vapor Phase Epitaxy

Yuichi Oshima, Encarnación G. Víllora, and Kiyoshi Shimamura

National Institute for Materials Science, Optical Single Crystals Group, 1-1 Namiki, Tsukuba, Ibaraki 305-0044, Japan

16.1
Introduction

Halide vapor phase epitaxy (HVPE), which is also called *hydride* vapor phase epitaxy, is a type of chemical vapor deposition (CVD). HVPE was first reported by a group RCA Corporation as an epitaxial growth technique for $GaAs_{1-x}P_x$ films [1]. Before the invention of HVPE, $GaAs_{1-x}P_x$ was grown by vaporizing premixed raw materials, which were transported with a single agent such as iodine or water to the low temperature zone in the reactor [2,3]. In contrast, HVPE utilizes hydride gases (this is the origin of the name "Hydride" VPE) such as AsH_3 and PH_3 as V-element sources. The Ga source is GaCl, which is produced upstream in the reactor by the chemical reaction between Ga and HCl. These V-hydrides and GaCl react with each other to form $GaAs_{1-x}P_x$, and as a result, the composition controllability is dramatically improved. Nowadays, CVD techniques that utilize hydride gases are not limited to HVPE. Therefore, although the name "Halide" VPE is preferred, since the expression describes precisely the fact that III-elements are transported in their halide form, both names are used conventionally.

Today, there are many well-developed epitaxial growth techniques in addition to HVPE, such as molecular beam epitaxy (MBE), metal-organic chemical vapor deposition (MOCVD), and so on. Among them, one of the most important features of HVPE is the possibility to achieve a high deposition rate. For example, an ultrahigh growth rate of GaN about 2000 μm/h is possible without any increase in the dislocation density [4]. In addition, HVPE can be carried out under atmospheric pressure, and large-scale multiwafer HVPE equipment is available for mass production. These features make HVPE an indispensable growth technique of thick layers of various III–V semiconductors, such as $GaAs_{1-x}P_x$, InP, GaN, and AlN. HVPE is especially important for the fabrication of nitride semiconductor bulk crystals, which cannot be produced from melt.

Handbook of Solid State Chemistry, First Edition. Edited by Richard Dronskowski, Shinichi Kikkawa, and Andreas Stein.

Actually, HVPE is currently the only way for the mass production of freestanding GaN wafers (hereinafter simply referred to as GaN wafers). HVPE is also advantageous for the growth of high-purity crystals. For example, HVPE-grown AlN is promising for UV-LED applications because of the high transparency in the UV region, thanks to the low carbon impurity concentration, in contrast to AlN crystals grown by the sublimation method [5]. Furthermore, HVPE is also applicable to the growth of oxide wide bandgap semiconductors such as ZnO and Ga_2O_3.

In this chapter, we first present the outline of HVPE technique, including the principle and technical basics of the equipment. Second, we review the production technologies of GaN wafers, which is one of the most important applications of HVPE. Finally, we overview the recent progress in the HVPE of Ga_2O_3, a new emerging oxide with a great potential to realize high-performance power devices, even better than those based on GaN and SiC.

16.2 Outline of HVPE Technique

16.2.1 Principle of HVPE

In this section, the principle of HVPE is explained, using the important GaN by the way of example. HVPE of GaN was first reported by Maruska and Tietjen [6]. Figure 16.1 shows a schematic of a typical HVPE reactor for GaN. The growth is carried out in a hot-wall quartz tube reactor in most of the cases. GaCl and NH_3 are used as reactant materials to produce GaN. GaCl is formed *in situ*, in the Ga container located upstream in the reactor, through the chemical reaction between liquid Ga and HCl around 850 °C:

$$Ga + HCl \rightarrow GaCl + 1/2\,H_2 \qquad (16.1)$$

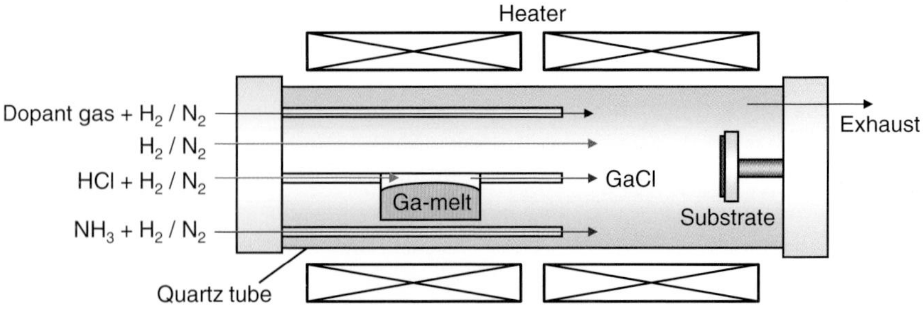

Figure 16.1 Schematic of an HVPE reactor for GaN.

GaN is grown on a substrate placed in the downstream region, through the reaction between GaCl and NH_3 at ~1000 °C:

$$GaCl + NH_3 \rightarrow GaN + HCl + H_2 \tag{16.2}$$

Each reactant is transported by a carrier gas such as H_2, N_2, Ar, and so on, and its partial pressure is adjusted in order to achieve an appropriate reactant concentration and flow rate. The selection of the carrier gas has a great impact on the effective reaction temperature and the gas flow pattern, since the gas properties, such as thermal conductivity, mass density, and viscosity, play a fundamental role.

16.2.2
What Enables the High-Speed Growth?

As already stated, HVPE is characterized by a rapid growth. In order to understand this fact, it is necessary to explain at first that the growth rate of CVD is sometimes limited by a parasitic gas phase reaction at high reactant concentrations. From this point of view, we will see the reason why HVPE enables large growth rates compared with those of other CVD techniques, such as MOCVD.

16.2.2.1 Determinant Factors of the Growth Rate

Figure 16.2 shows schematically the general trend of the CVD growth rate as a function of temperature under a constant reactant feed rate. The region A is the surface reaction limited regime, in which the reaction at the crystal surface is limited by the temperature, and the growth rate increases with increasing the temperature. The region B is the mass transport limited regime. In this regime, the surface reaction process is sufficiently quick, and the growth rate is dominated by the feed rate of the reactants to the surface. The region C starts at higher temperatures, when the absorption process to the surface is counteracted by the opposing one of desorption, so that the growth rate decreases with the temperature rise. CVD is usually carried out in the mass transport limited regime (region B), because this is generally believed to be advantageous to achieve high-quality crystals.

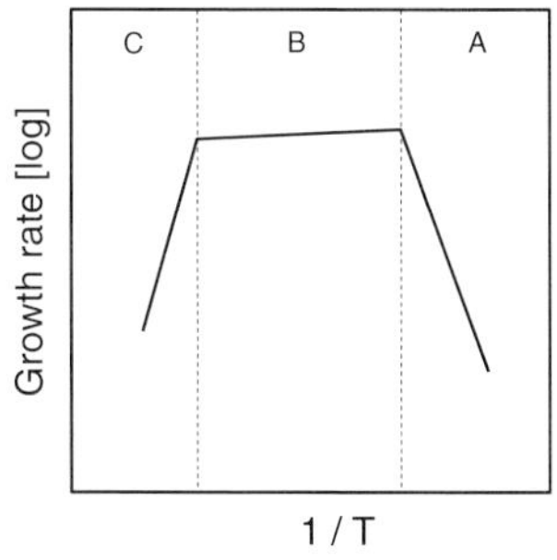

Figure 16.2 Growth rate as a function of temperature under constant reactant supply.

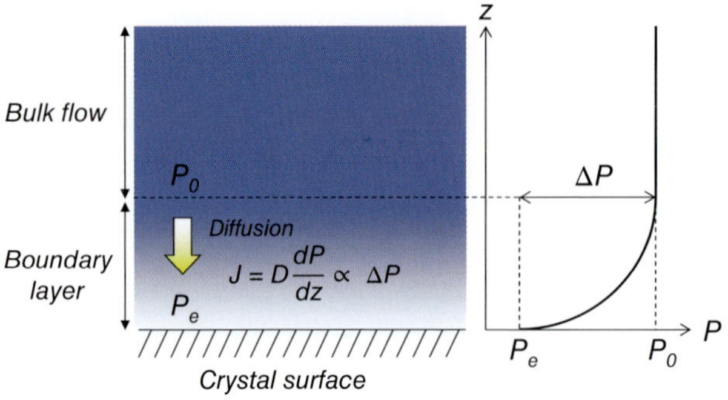

Figure 16.3 Distribution of reactant concentration in the vicinity of crystal surface under mass-transport limited regime.

The distribution of the reactant concentration in the vicinity of the crystal surface, assuming the mass transport limited regime, is described in Figure 16.3. In the bulk flow, which is sufficiently apart from the surface, the partial pressure P of a reactant is equal to the supply partial pressure P_0. On the other hand, P is equal to the equilibrium partial pressure P_e on the surface, where a thermodynamic equilibrium is always established. The reactant is transported from the bulk flow to the surface by diffusion through the boundary layer. The growth rate is then proportional to the difference in pressure, $\Delta P = P_0 - P_e$, because the feed rate J of the reactant to the surface is proportional to the concentration gradient. Therefore, ΔP needs to be large in order to achieve a large growth rate.

16.2.2.2 **Spontaneous Nucleation by a Parasitic Gas-Phase Reaction**

In most cases, a large ΔP can be realized by simply increasing P_0, as we already know well. However, the growth rate is sometimes limited by a parasitic gas-phase reaction, a spontaneous nucleation that takes place upstream in the growth region. Such gas phase reaction leads not only to the decrease in the growth rate but also to the degradation of the film quality due to the nucleated particles.

Here, we discuss the conditions for such gas phase reaction leading to the spontaneous nucleation in terms of the classical homogeneous nucleation theory [7]. When a nucleus of radius r precipitates, the free energy change ΔG can be expressed as follows:

$$\Delta G = -\frac{4\pi r^3}{3v}\Delta\mu + 4\pi r^2\gamma, \tag{16.3}$$

where v is the volume of the molecule, $\Delta\mu$ is the chemical potential change, and γ is the surface free energy density. The first term shows the decrease in the bulk free energy, and the second term is the increase of surface free energy. Figure 16.4 shows ΔG as a function of r. When r is smaller than the critical value r_c, ΔG increases with increasing r, due to the large contribution of the surface free energy. In another words, even if a nucleus with $r < r_c$ is formed by thermal

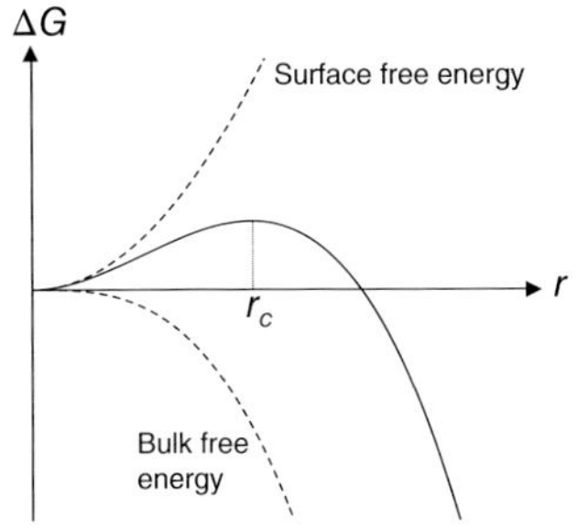

Figure 16.4 Free energy change by nucleation as a function of nucleus radius.

fluctuation, the nucleus is unstable and cannot keep growing. On the other hand, when $r > r_c$, ΔG decreases with increasing r, due to the large contribution of the bulk free energy, and the nucleus can initiate a stable growth. Here, r_c is expressed as follows:

$$r_c = \frac{2v\gamma}{\Delta\mu} = \frac{2v\gamma}{k_B T ln(P_0/P_e)} \tag{16.4}$$

where k_B is the Boltzmann constant and T is the temperature in Kelvin. We can see that r_c decreases with increasing the super saturation ratio P_0/P_e. As the result, the probability for the formation of stable nuclei will increase.

In conclusion, the key to achieve a rapid growth is to increase the driving force for the growth, that is, a large ΔP, with keeping the ratio P_0/P_e sufficiently small in order to avoid the gas phase reaction. Based on this consideration, in Section 16.2.2.3 we will see the reason why a rapid growth is possible in HVPE. For it, the growth of GaN by HVPE is compared with the one by MOCVD, which is widely used for the mass production of GaN-based devices.

16.2.2.3 **Comparison Between HVPE and MOCVD**

In this consideration, we assume the mass transport limited regime, in which the reaction rate is sufficiently quick, and the thermodynamic equilibrium is always established on the crystal surface. In this regime, the growth rate is dominated by the diffusion rate of the reactant from the bulk flow to the surface. The diffusion rate is proportional to ΔP, therefore the growth rate is also proportional to ΔP, as described in Section 16.2.2.1.

For each growth method, the precursors and the main reaction to synthesize GaN are as follows:

(HVPE)
$GaCl(g) + NH_3(g) \rightarrow GaN(s) + H_2(g) + HCl(g)$

(MOCVD)
$Ga(g) + NH_3(g) \rightarrow GaN(s) + 3/2\, H_2(g)$

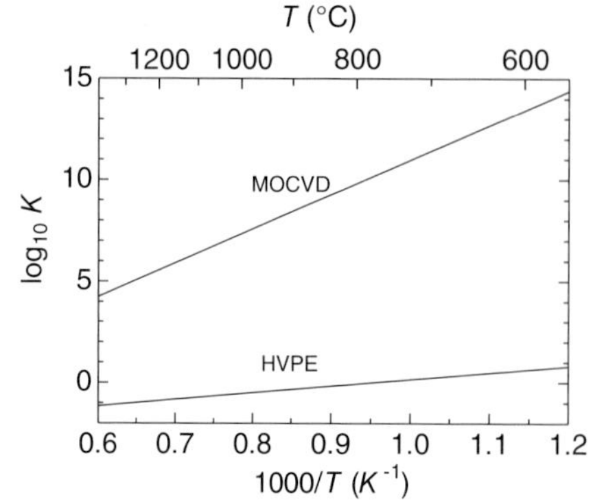

Figure 16.5 Equilibrium constants of GaN growth reactions as a function of temperature.

where Ga vapor is produced by the thermal decomposition of trimethylgallium (TMGa) or triethylgallium (TEGa). The equilibrium constants K as a function of temperature are shown in Figure 16.5 for both reactions. As can be seen, the K for HVPE (K_{HVPE}) is over five orders of magnitude smaller than the K for MOCVD (K_{MOCVD}). For example, at the typical growth temperature of 1000 °C the K_{HVPE} is ~0.3, while K_{MOCVD} is $\sim 2.2 \times 10^7$. This means that partial pressures of the precursors under thermal equilibrium, is very large in the case of HVPE. Accordingly, P_0/P_e will be overwhelmingly small in the case of HVPE compared to that of MOCVD under the same P_0. Consequently, under the assumption of the same P_0, a larger P_e means smaller ΔP, and thus the growth rate of HVPE will be smaller than that of MOCVD. However, since in the case of HVPE P_0 can be increased without causing any parasitic gas-phase reaction, thanks to the small P_0/P_e, the growth rate can be far greater than that of MOCVD.

For the better understanding of the explanation above, it might be helpful to see a concrete result of thermodynamic analysis. Figure 16.6a–c illustrates the calculated ΔP and P_0/P_e as a function of P_0. Note that P_0 and P_e are the values given just for the III-element source materials. The calculation assumed H_2 as the carrier gas, a growth temperature of 1000 °C, and a partial pressure ratio of V/III = 10 under atmospheric pressure. Figure 16.6a indicates that both ΔP_{HVPE} and ΔP_{MOCVD} increase with increasing P_0, and that ΔP_{HVPE} is smaller than ΔP_{MOCVD} due to the large P_e of HVPE. On the other hand, Figure 16.6b and c demonstrates that the increase of P_0 causes only a small increase in $(P_0/P_e)_{HVPE}$, while $(P_0/P_e)_{MOCVD}$ exhibits a remarkable rise over several order of magnitude. For example, $(P_0/P_e)_{HVPE}$ is on the order of 10^0 when $\Delta P = 5 \times 10^{-1}$ kPa, while $(P_0/P_e)_{MOCVD}$ is far greater, that is, on the order of 10^3 at the same ΔP. Thus, in the case of HVPE, we can supply a high concentration of reactants with a low risk of parasitic gas-phase reaction, and therefore large growth rates can be achieved.

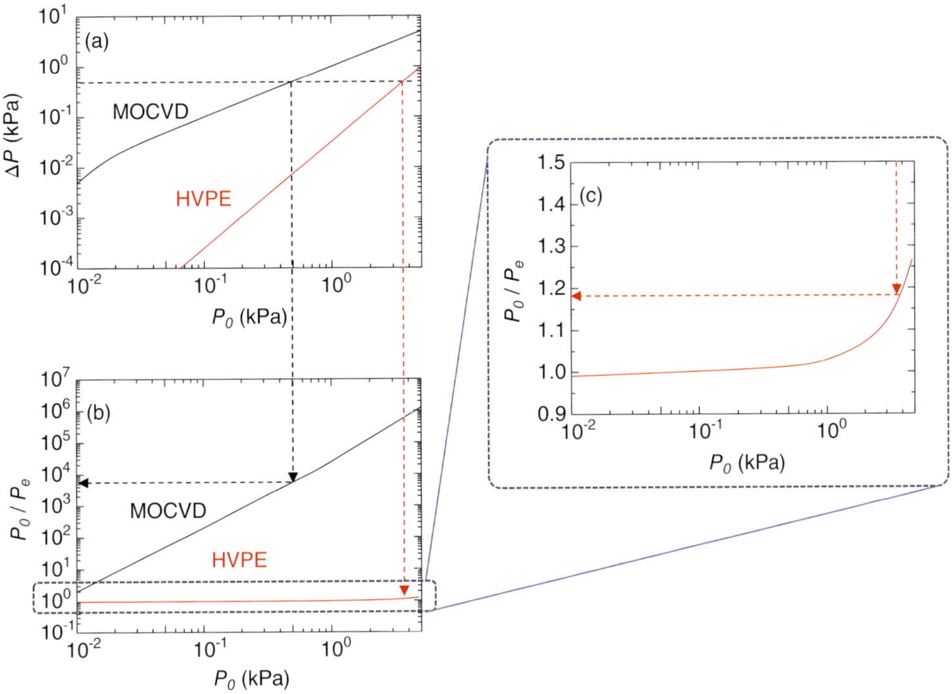

Figure 16.6 (a) Growth driving force of GaN (ΔP) and (b and c) super saturation ratio (P/P_e) as functions of supply partial pressure (P_0) (results of thermodynamic calculation).

16.2.3
Technical Characteristics and Functionality of HVPE Equipment

The availability of ready-made HVPE equipment is quite limited, in contrast to the case of MOCVD and MBE. Therefore, a home-built HVPE equipment is frequently used. In this section, we overview the technical features of the HVPE equipment, including some technical issues to which special attention should be paid when designing or operating the apparatus. The explanation is mainly intended for persons who have experience in MOCVD.

16.2.3.1 Outline of the Entire System

Figure 16.7 shows a schematic diagram of a typical HVPE system. This is basically similar to that of MOCVD equipment. Gas purifiers and cylinder cabinets provide high-purity carrier gases and precursor gases, respectively. The gas mixer, which is composed of many air-operated valves and mass flow controllers (MFCs), transfers the gases to the reactor with controlled composition and flow rate. The exhaust system is equipped with filters to remove fine particles from the exhaust gas, and a vacuum pump to control the growth pressure in the reactor. The exhaust gases from the reactor and the vent lines of the gas mixer

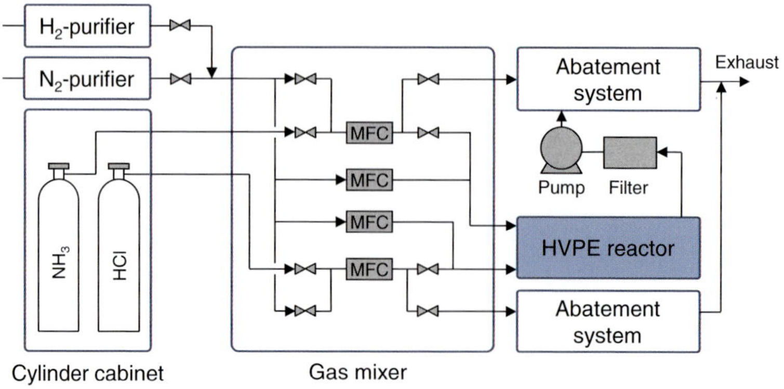

Figure 16.7 Schematic diagram of a typical HVPE system for GaN.

contain toxic and/or flammable components. Therefore, these gases need to be processed by abatement systems before being released in the air. Solid absorbents are often used to remove HCl in the exhaust gas from the vent line. A wet scrubber or a catalytic oxidation abatement system is useful to remove both HCl and NH_3. Obviously, the HVPE system must be equipped with gas detectors and an interlock system, which guarantee an automatic safe shut down upon gas leakage.

16.2.3.2 **Fundamentals of the Reactor Design**

Materials for HVPE Reactor

Even under a dry environment, most of metals, with a few exceptions such as W and Ir, are corroded by HCl at high temperatures. Therefore, the HVPE reactor should basically be made of nonmetallic components such as quartz, alumina, carbon, SiC, BN, and so on. Furthermore, in order to avoid contamination during crystal growth, these materials should not contain impurities whose chloride form exhibits a high vapor pressure.

Basic Design Policies of the HVPE Reactor

The reactor needs to be designed so that vortex generation in the gas stream is minimized, because this sometimes leads to a large spatial variation of the growth rate or to a serious parasitic deposition in the vicinity of the gas inlets. Therefore, an angular-shaped structure should be avoided as much as possible to enable a smooth gas flow. The influence of thermal convection and difference in the mass density between different gas species should also be considered to realize an ideal gas flow. To simulate the flow using a thermal fluid analysis software is sometimes helpful to design the reactor. For GaN, such software is commercially available, so that the internal shape of the reactor can be designed taking into consideration both the chemical reaction and the parasitic deposition.

Design of the Source-Metal Container

The source-metal container, in which the source metal and HCl react with each other to produce the metal chloride reactant (e.g., GaCl), should be designed so that the injected HCl fully reacts with the metal. An incomplete reaction would lead to a decrease in the growth rate due to the diminished amount of metal chloride supply and the increase of unreacted HCl. Additionally, the source-metal container has to be designed in a way that the conversion efficiency varies easily neither with the HCl feed rate nor the quantity of the metal in the container. One simple way to meet such requirement would be to use a large container, so that the injected HCl encounters a large surface area to react. However, the turn-on and turn-off time of the metal chloride concentration increase with increasing the container size. Therefore, rather than that, it is preferable to minimize the container volume and design the internal structure of the container so that HCl and the metal can interact effectively.

How to Hold Substrates

Substrates can be held in face up or down style. In the face-up style, substrates of any shape, size, and number, are conveniently just placed on a susceptor. In contrast, although the face down style needs certain measures to hold the substrates, this location has the advantage to avoid the fall off of polycrystalline flakes on the substrate surface. Anyway, in both configurations the periphery of the substrate should be covered in order to prevent from the exposure to the reactant gases and the subsequent polycrystalline deposition, which can even stick on the susceptor in the case of thick layer growth.

Counter Measures Against the Parasitic Deposition

During the HVPE growth, the growth takes place not only on the substrate but also on other high temperature parts, such as the inner wall of the reactor, the gas inlets, and the substrate holder. Such depositions are especially severe at the locations where high-concentration reactant gases are present (Figure 16.8a). In the case of thick film growth, these thick depositions can easily break quartz parts by thermal stress. One convenient way to prevent this problem is to cover the quartz parts with, for example, high-purity carbon flexible sheets. On the other hand, thick depositions on gas inlets influence the gas flow pattern and lead to the drift of growth conditions even in a single growth run. These can be reduced effectively by introducing a separation gas that inhibits the mixing of the reactant gases in the vicinity of the gas inlets (Figure 16.8b).

16.2.3.3 Counter-Measures Against the Deposition of By-Products

Sometimes, HVPE is accompanied by the generation of solid-phase by-products. In the case of GaN, for example, $GaCl_3$ and NH_4Cl, un-reacted GaCl transforms into $GaCl_3$, a compound with low boiling and melting points (201 and 78 °C, respectively) that deposits as a white solid at the low temperature part in the downstream of the reactor. NH_4Cl results from the reaction between HCl on the right hand side of Eq. (16.2) and the residual NH_3. It deposits in the exhaust

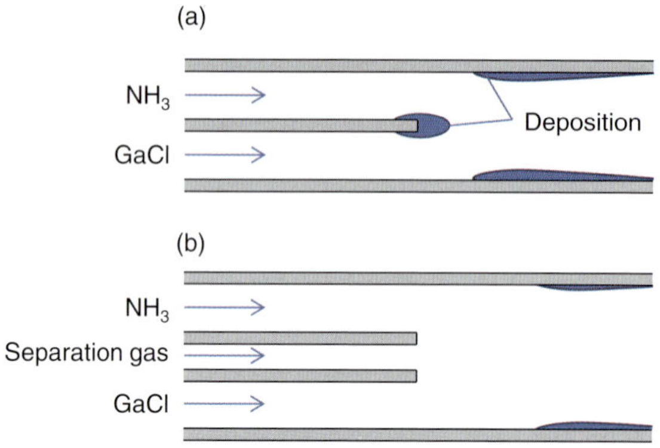

Figure 16.8 Schematic of gas-inlet structure (a) without separation gas line, (b) with separation gas line.

piping as a whitish powder and can even clog up the piping during a thick layer growth. In order to eliminate these by-products, sufficiently large filters or traps should be located at the place where these deposit.

16.2.3.4 Counter-Measures Against the Corrosion

Much attention should be paid to the corrosion of the metal parts by HCl and chlorides, as well as in the case of MOCVD, HVPE equipment utilizes a lot of stainless steel parts such as flanges, gas pipes, valves, pressure regulators, and MFCs. Stainless steel is resistant to dry HCl at room temperature, however, is markedly corroded by chloride ion under the presence of water. Therefore, HCl needs to be purged out before exposing the inner structure of the HVPE equipment to the air. Analogously, after the exposure, the air moistures must be purged out before the use of HCl. Furthermore, chlorine ions are also released by the deposition of the by-products such as $GaCl_3$ and NH_4Cl, which are sometimes sticking on the stainless steel flange. Accordingly, all the tasks that require the exposure to the air of inner metallic parts should be finished as soon as possible. In addition, a leak check of the gas piping should be conducted on a regular basis in order to find any corroded part. It is noteworthy that due to this corrosion of stainless steel, the grown crystals are generally contaminated with Fe, Cr, and Ni. Therefore, an unexpected increase of these impurities in chemical analysis of the grown crystals can be a good indicator of the presence of corrosion in the HVPE system.

16.3 Fabrication of GaN Wafers

In this section, we review the production technologies of GaN wafers. The fabrication of these substrates is one of the most important applications of HVPE. First, the basic properties of GaN and important applications of GaN wafers are

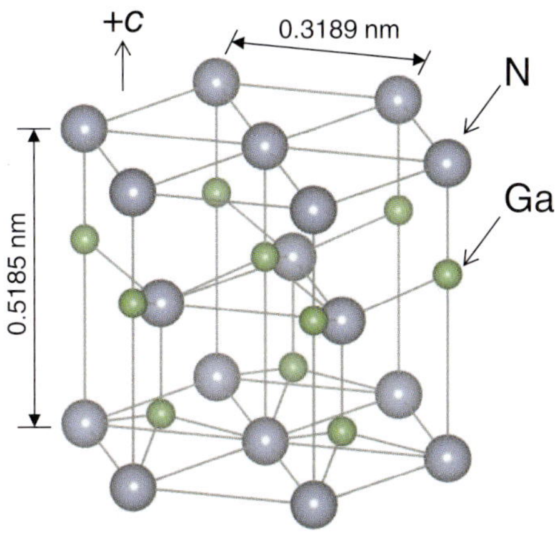

Figure 16.9 Crystal structure of GaN.

described. We then overview the technical issues and related technologies for the fabrication of GaN substrates.

16.3.1 Features and Applications of GaN

GaN is a III–V semiconductor that can crystallize in both the cubic zinc-blende and the hexagonal wurtzite structures. From the technological point of view, the latter is the desired phase for the fabrication of efficient semiconductor devices, and therefore, from now on we will refer just to the wurtzite phase simply as GaN. This compound is characterized by a direct bandgap of 3.4 eV, and its crystal structure is illustrated in Figure 16.9. The lattice parameters are $a = 0.3189$ nm and $c = 0.5185$ nm, and the Ga-polar (0001) plane is usually utilized for device fabrication. Figure 16.10 shows the relationship between the bandgap energy and the bond length of III-nitrides, together with the same data for other

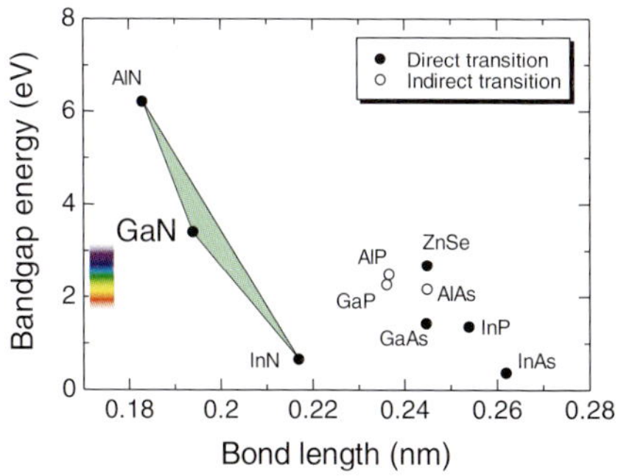

Figure 16.10 Relationship of band gap energy and bond length of III-nitrides and other III–V and II–IV compound semiconductors.

representative compound semiconductors. GaN can make solid solutions with the isostructural AlN and InN, which are direct bandgap semiconductors as well, so that the solid solution $Al_xIn_yGa_{1-x-y}N$ covers a very wide range of bandgap energes, extending from 0.65 to 6.2 eV. Therefore, these mixed nitrides are suitable for optical devices such as sensors, LEDs and LDs, which can operate in a wide wavelength range from IR to UV. In addition, GaN is also promising for applications in not only high-power but also high-frequency electronic devices, thanks to its large bandgap and related high breakdown electric field (~3 MV/cm) [8], and to its high saturation-drift velocity ($\sim 2.5 \times 10^7$ cm/s) [9], respectively.

16.3.2
Necessity and Applications of GaN Wafers

Unlike other conventional semiconductors such as GaAs and Si, large-size single crystalline GaN cannot be grown from the melt, since the formation of GaN melt requires a super high pressure, as high as 6 GPa or more in order to suppress its decomposition into Ga and 1/2 N_2. Accordingly, single crystal GaN wafers were not available, and GaN-based devices were fabricated only on foreign substrates, mainly sapphire and SiC. However, such heteroepitaxial GaN layers contained a high density of dislocations, mainly caused by the large lattice mismatch. The quality of those heteroepitaxial GaN films was not sufficient for the mass production of reliable blue-violet GaN-based LDs, which were indispensable for the commercialization of the Blu-ray system. Therefore, the realization of high-quality GaN wafers was strongly desired.

Accordingly, various growth methods, including the high-pressure high-temperature method [10], the ammonothermal method [11], the Na-flux method [12,13], and the HVPE method, have been investigated. Among them, large-area and high-quality GaN wafers were realized firstly by HVPE around 2001, and lead to the commercialization of the Blu-ray system in 2003, and even to date (2016), HVPE is the only mass production technique of GaN wafers.

Nowadays, GaN-based LDs are also used for other applications, for example, projectors (2010–). In addition, super-high-luminance white light sources, consisting of blue GaN-based LDs plus yellow phosphors, has been developed intensively, and some of them have already been installed in headlight units of commercial vehicles (2014–). Besides, due to the recent drastic price drop of GaN wafers, the commercialization of high-brightness LEDs is rapidly growing, so that the market size of GaN wafers for LEDs is now comparable or even greater than that for LDs. Moreover, the development of GaN power devices on GaN wafers is now being intensively pursued [14–17].

16.3.3
Basic Strategies and Technical Issues for the Fabrication of GaN Wafers

Figure 16.11 exemplifies the basic concept of the fabrication of GaN wafers by HVPE. In principle, a freestanding GaN crystal can be obtained by growing a thick GaN layer (several hundred micrometers or more) on a base substrate

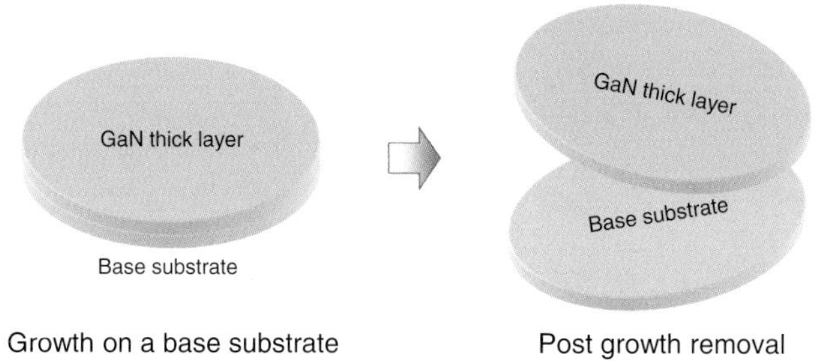

Figure 16.11 Schematic of the fabrication of freestanding GaN wafers by HVPE.

(e.g., sapphire, SiC, and GaAs) and its subsequent removal. In reality, however, some technical difficulties need to be overcome. In the following, these technical issues and the countermeasures are discussed.

16.3.3.1 **Reduction of the Dislocation Density**

Dislocations in GaN and Their Adverse Effects

So far, no foreign substrate with a matching lattice to GaN has been found. Heteroepitaxial GaN layers therefore include threading dislocations with concentrations around $10^9\,cm^{-2}$ or higher, if no measures are taken. Dislocations are nonradiative recombination centers in GaN, and therefore lower the efficiency of optical devices [18]. In addition, the dislocations scatter the carriers, especially when the carrier density is low, thus reducing the mobility [19]. Furthermore, these are also detrimental to the reliability of devices [20]. The degradation mechanisms of GaN-based devices by dislocations have not been fully clarified yet, although the possibility of abnormal dopant diffusion via dislocations has been pointed out [20].

The required crystal quality is specific to each device structure and operating condition. Therefore, it is difficult to predict the tolerable dislocation density precisely. However, roughly speaking, a dislocation density below $10^7\,cm^{-2}$ is required in the case of blue-violet LDs for Blu-ray. A further reduction of the dislocation density would be needed depending on the emission wavelength and output power. For GaN-based power devices, the influence of dislocations on the performance is still under the discussion, and devices that include a certain amount still show an excellent performance. For example, Ohta *et al.* [14] have demonstrated a pn-junction diode with a record breakdown voltage as high as 4.7 kV, which was fabricated on a GaN substrate with a dislocation density of $3 \times 10^6\,cm^{-2}$. In any case, more detailed investigations have to be carried out in order to elucidate the role of dislocations in the performance and long-term durability of GaN-based power devices.

Basic Strategies for the Dislocation Reduction

Since HVPE-GaN crystals are heteroepitaxially grown on highly lattice-mismatched substrates, these include a high density of threading dislocations if no

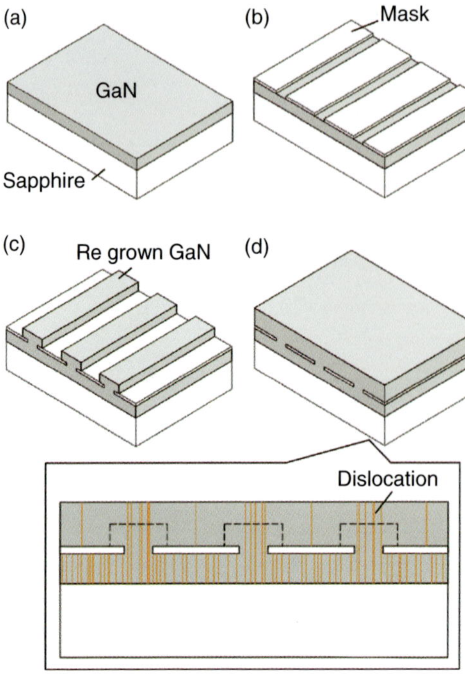

Figure 16.12 Process flow of ELO technique.

countermeasures are taken. These dislocations must be reduced, since they propagate into the epitaxial layers grown on the GaN substrate and thereby compromise the device performance.

Low dislocation density GaN wafers have been achieved through the combination of several primary technologies, namely, epitaxial lateral over-growth (ELO) [21], dislocation bending by large inclined facets [22,23], and thick film growth [24].

Figure 16.12 illustrates the sequential processes for standard ELO. First, a GaN thin film template is prepared. Second, a periodic mask pattern is formed on the template by photolithography. SiO_2 or SiN_x are often used as masking materials, and a simple stripe pattern (with a few micrometers period) is frequently employed. The regrowth of GaN is then carried out on the GaN template with the mask. It starts on the openings and spreads laterally till the GaN stripes coalesce forming a compact layer. During the regrowth process, the dislocations in the template layer under the mask do not propagate into the laterally growth region, thus diminishing the dislocation density. However, the dislocation density does not decrease above the openings, and moreover, new dislocations appear at the coalescence interfaces.

The propagation of dislocations at the openings can be suppressed by facet-initiated ELO (FIELO [22]) or facet-controlled ELO (FACELO) [23]. Figure 16.13 shows schematically the processing sequence of the FIELO

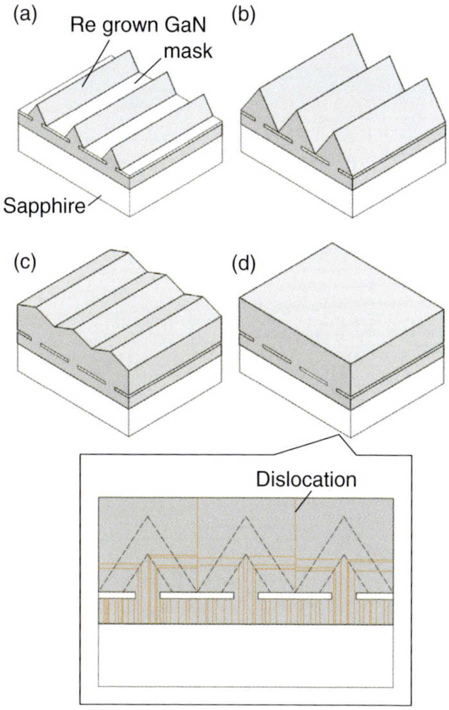

Figure 16.13 Process flow of FIELO technique.

technique. The process is similar to that of ELO, but it differentiates in the regrowth step. In this case, the conditions are controlled in such a way that the regrowth takes place at inclined facets such as $(10\bar{1}1)$, $(11\bar{2}2)$, and so on. In this case, even though the dislocations propagate into the regrown GaN at the openings, they change the direction of propagation at the faceted interfaces in order to minimize the elastic energy. As a result, the dislocations do not propagate along the growth direction anymore, and a top-end surface with a diminished dislocation density is achieved.

The dislocation density can also be reduced by simply increasing the thickness of a flat GaN layer. During the thick layer growth, dislocations having Burger's vectors with opposite signs attract each other, then they counteract and finally disappear by forming a dislocation loop. For example, Fujito *et al.* [24] demonstrated the linear decrease of the dislocation density from $5 \times 10^6\,cm^{-2}$ to $1 \times 10^6\,cm^{-2}$ during the growth of a 5-mm-thick layer.

16.3.3.2 **Removal of the Base Substrate**

Cracking Problem Upon the Removal of the Base Substrate

One of the most important requirements for GaN wafers is a sufficient wafer size for practical use. The diameter should be at least 2 inch or more in order to

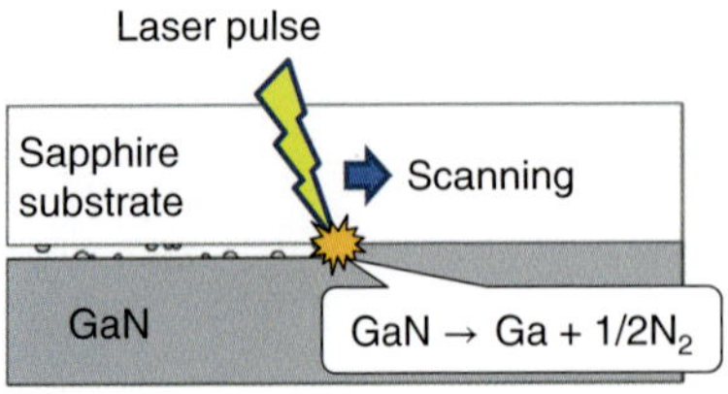

Figure 16.14 Concept of laser separation technique.

obtain a large amount of devices in every run, and so satisfy the cost requirements. Homogeneous heteroepitaxial growth on large-area substrates was achieved; however, serious cracking occurred during the removal process of the base substrate. This was mainly caused by the large thermal stress of grown crystals, originating from the difference in the thermal expansion coefficients of GaN and the base substrate.

Basic Strategies to Remove the Base Substrate

So far, many approaches have been made to remove the base substrate, namely, chemical etching [25], mechanical lapping [21], laser lift-off [26,27], spontaneous separation [28,29], and so on. Among them, the chemical etching and the mechanical lapping do not work well because of the severe cracking during the process. The laser lift-off consists in the thermal decomposition of GaN in the vicinity of the base substrate by irradiation with a high power laser, for example, an excimer laser or a pulsed Nd:YAG laser (Figure 16.14). Due to the high temperature gradients, caused by the rapid heating and cooling at the irradiated point, the crystals also tend to crack and the window for a satisfactory removal is quite narrow [27]. Accordingly, it is difficult to produce large-area GaN wafers with sufficiently high reproducibility for mass production. Instead, spontaneous separation is the most successful and has been implemented industrially. In this method, a mechanically flimsy sacrificial layer is formed between the GaN layer and the base substrate. The mechanical stress of the heteroepitaxial layer, which arises during the cooling process after the growth, is enough to break the brittle sacrificial layer, so that the whole GaN layer is automatically separated from the base substrate without cracking (Figure 16.15). In contrast to the other separation methods, the spontaneous separation is cost-effective since the separation

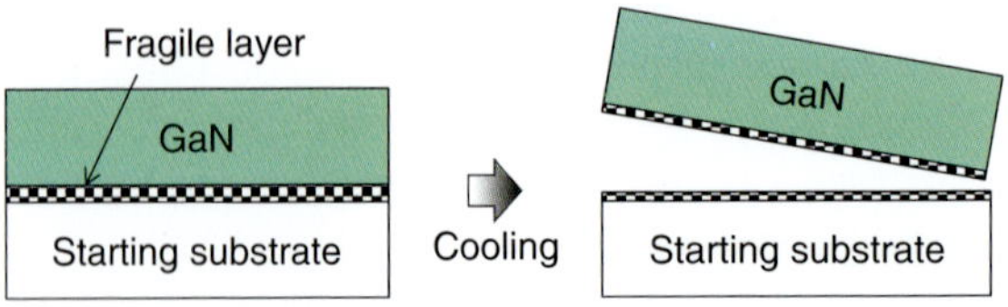

Figure 16.15 Concept of spontaneous separation technique.

process automatically completes within the growth apparatus. There are two basic trends among the different types of spontaneous separation methods, the ones that utilize a fragile interface layer with numerous small voids [28], and the ones with base substrates that are chemically unstable in the growth atmosphere of GaN [29]. The concrete description of the former method is given in Section 3.4.

16.3.3.3 Control of Electrical Properties

Necessity of Impurity Doping

The resistivity of GaN wafers should be as low as possible for vertical device applications to decrease the operating voltage, and to establish a stable ohmic back contact. For this purpose, the carrier concentration should be approximately $1 \times 10^{18}\,cm^{-3}$ or higher.

On the other hand, it is difficult to obtain a low-resistive p-type GaN due to the large ionization energy of Mg in GaN (~170 meV) [30]. Accordingly, the resistivity of GaN wafers is controlled with n-type dopants, such as Si, O, and Ge [28,31–33]. The ionization energies of these dopants are below 30 meV [34,35], and therefore most of the donors are ionized at room temperature.

Doping Techniques in HVPE

Doping control in HVPE is possible by introducing an additional dopant gas together with the precursors. This should not decompose easily at the growth temperature, because a thermally unstable gas cannot reach the growth zone in a hot wall reactor. The simplest way to control the doping concentration is to use a dopant source whose gas phase is stable at room temperature, for example, SiH_2Cl_2, so that it can be supplied from a gas cylinder [28]. Another possible way is the bubbling of a liquid-phase dopant source, for example, $GeCl_4$ [32,33]. Furthermore, it is also possible to put the doping element directly in the reactor and to produce the chloride gas in a similar way as the Ga-precursor, that is, GaCl gas. However, the structure of the reactor becomes complicated, and it is not easy to achieve a sharp doping profile.

On the other hand, it should be noted that Si is automatically doped from quartz in the case of HVPE [33], most probably due to the reaction with HCl. The Si concentration, which depends on the growth conditions and the internal structure of the reactor, can reach a value as high as $10^{18}\,cm^{-2}$. Therefore, to achieve a precise doping control, such background donor concentration needs to be low enough compared to the target concentration. In order to suppress such unintentional doping, it is effective to use alternative materials other than quartz for the parts that are in contact with HCl at high temperature.

Problems Associated with the Impurity Doping

In general, impurity doping has a negative impact on material properties. Therefore, countermeasures should be taken in order to minimize the impurity concentrations as much as possible. In this section, the most common problems associated with the n-type doping are described.

First, impurities lower the transparency below the bandgap energy, which is unfavorable for LEDs [32,33]. Second, the dislocation density increases by heavy Si doping. This degradation can be avoided by using Ge instead of Si, since the ionic radius of Ge is closer to that of Ga [32,33]. Furthermore, the bowing of as-grown GaN wafers is promoted by the increase in doping concentration, regardless of the doping element [33]. The mechanism governing this phenomenon will be described in more details in Section 16.3.3.4. Moreover, it should also be noted that the thermal conductivity of high-quality GaN does not significantly decrease up to very high doping concentrations, as high as $1 \times 10^{19}\,\text{cm}^{-3}$, while the thermal conductivity of low-quality GaN presents a remarkable decrease [32].

For practical purposes, the doping concentration is optimized so as to maximize the performance and the yield of the devices, thus finding a compromise between the advantages of a low resistivity and the demerits of a high impurity concentration.

16.3.3.4 Reduction of the Off-Angle Variation

Off-Angle Variation of GaN Wafers and Its Undesired Effect

In general, as-grown heteroepitaxial freestanding GaN crystals have a tendency to concave bowing, as can be seen in Figure 16.16. It is important to keep in mind that this bowing has no relation with the difference in the thermal expansion coefficient between the GaN epilayer and the foreign substrate. Although depending on the substrate a tensile or compressive strain unavoidably appears upon cooling, such strain can be released after the removal of the substrate. The different origin and permanent nature of the actual concave bowing of freestanding GaN wafers will be explained in Section "Mechanism of Wafer Bowing by Island Coalescence." GaN wafers are fabricated by lapping and polishing the as-grown crystals without flattening the bow. Even if a geometrically flat shape is achieved during the processing, the crystal planes in the GaN wafers are still

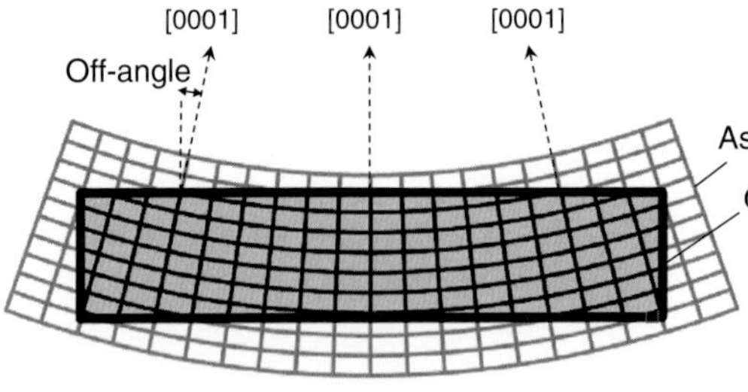

Figure 16.16 Cross-section of an as-grown GaN wafer with concave bending, and creation of off-angle distribution in a final GaN wafer.

bowed, that is, the GaN wafers have an off-angle variation. One of the undesired effects of this variation is that epilayers deposited on such wafers present a compositional gradient, which causes a continuous shift in the emission wavelength of the light emitting devices. It is noteworthy that the off-angle variation proportionally increases with the wafer diameter, if the radius of curvature is constant. Thus, at present, the mainstream of GaN wafer size is till 2 inch, though recently 4-inch wafers are starting to be used. Therefore, even though an increase in diameter is readily achievable by state-of-the-art HVPE technology, it needs to be complemented by a correspondingly restrained off-angle variation. The reduction of the bowing is thus one of the most important technical issues for the practical implementation of large diameter GaN wafers and is considered in the following. The wafer bowing can be explained by two correlative phenomena, on one side the tensile strain caused during the island coalescence, and on the other hand the inclination of edge dislocations after the coalescence. Each mechanism is briefly explained in the following sections.

Mechanism of Wafer Bowing by Island Coalescence

In most cases, heteroepitaxy of GaN takes place through the Volmer–Weber growth mode, according to which isolated GaN islands are formed on the substrate at the early stage, these grow and then coalesce to form a compact layer. The ELO process of GaN can also be considered to be a type of Volmer–Weber growth mode. The coalescence of two adjacent small islands is accompanied by a decrease in the surface free energy [36]. During the "zipping" process of the islands, the sidewalls are displaced, inducing an in-plane tensile strain that causes the bowing. Nix and Clemens [37] estimated the stress evolution during this process assuming a regular array of small hexagonal islands with isotropic surface free energy γ_{sv}, as shown in Figure 16.17. The free energy of this array per unit film area just before the coalescence ($\Delta \sim 0$) is expressed as

$$E_1 = E_0 + \frac{2h\gamma_{sv}}{a}, \tag{16.5}$$

where E_0 is the free energy per unit film area associated with both the top surface of the film and the film/substrate boundary. The second term is the free energy of the sidewalls per unit film area. When the gap Δ is closed by the elastic deformation of each crystallite along the radial direction, a biaxial

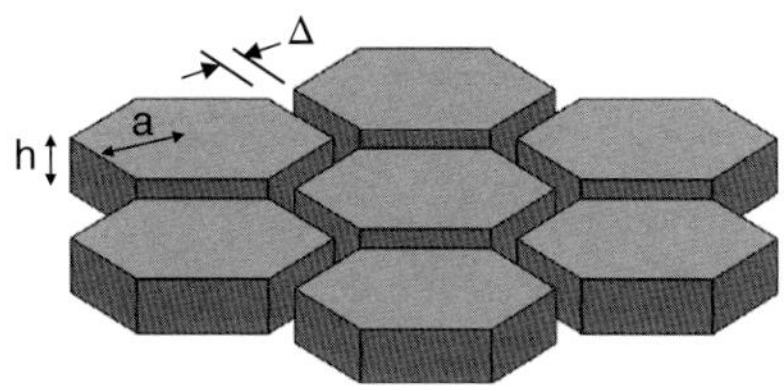

Figure 16.17 Calculation model for tensile strain produced by island coalescence.

strain $\varepsilon = \Delta/(2a)$ is introduced in each island. Therefore, the free energy density of this array per unit film area just after the coalescence is given as

$$E_2 = E_0 + \frac{h\gamma_{gb}}{a} + \frac{E}{1-\nu}h\left(\frac{\Delta}{2a}\right)^2, \tag{16.6}$$

where the second term represents the contribution from the grain boundary free energy γ_{gb}. The third term corresponds to the strain energy per unit film area. E and ν are the Young's modulus and Poisson's ratio, respectively. By setting $E_1 = E_2$, we can deduce the maximum gap Δ_{max} which can be closed spontaneously, as well as the associated maximum stress σ_{max} as follows:

$$\Delta_{max} = \sqrt{4a\left(2\gamma_{sv} - \gamma_{gb}\right)\frac{1-\nu}{E}}, \tag{16.7a}$$

$$\sigma_{max} = \sqrt{\frac{2\gamma_{sv} - \gamma_{gb}}{a}\frac{E}{1-\nu}} \tag{16.7b}$$

In the case of the ELO process of GaN, the crystal orientations of the islands do not differ largely, and therefore the grain boundaries after the coalescence are barely distinguishable. Nevertheless, the development of the tensile strain can be explained by this mechanism. Figure 16.18 shows the Δ_{max} and σ_{max} as a function of the island diameter assuming common parameter values as $E = 150$ GPa, $\nu = 0.38$, $\gamma_{sv} = 5.3\,\mathrm{J/m^2}$, and $\gamma_{gb} = 0$ in Eqs. 16.7a and b. For example, for a typical diameter $a = 10\,\mu\mathrm{m}$ we obtain a $\Delta_{max} = 4.2$ nm and a $\sigma_{max} = 0.5$ GPa. The actual measured values, evaluated by Raman scattering, are far smaller by a factor of 10 or more. However, this discrepancy is reasonable if we consider the large model simplification, which does not take into account variable parameters such as islands size, shape, and position, as well as strain relaxation by the generation and movement of crystal defects. Böttcher *et al.* obtained a more realistic value of Δ for MOVPE-grown GaN on a *c*-plane sapphire substrate by considering these factors [38]. The estimated value of $\Delta = 0.29 \pm 0.16$ nm is close to the length of the Burger's vector of threading edge dislocations in *c*-plane GaN.

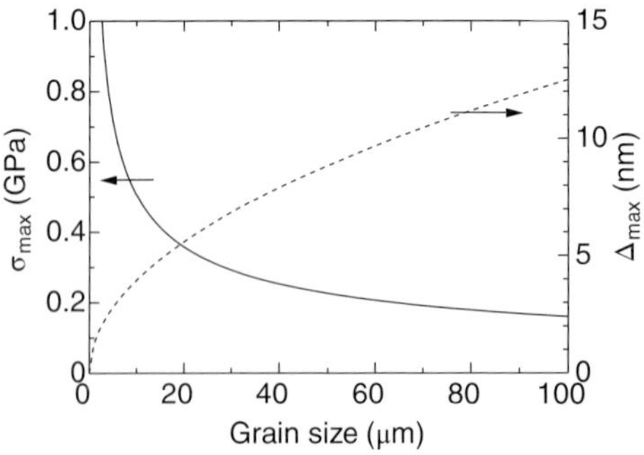

Figure 16.18 Calculated maximum stress σ_{max} and gap size Δ_{max} as a function of island radius.

Mechanism of Wafer Bowing by Inclination of Edge Dislocations

As described in Section 16.3.3.3, wafer bowing is enhanced by donor doping even after the coalescence. Interestingly, the degree of the bowing is approximately the same for both Si and Ge if their concentrations are the same [33]. This fact suggested that the bowing originates in the inclination of dislocations [39] assisted by the Fermi level effect [40]. In n-type GaN, the concentration of gallium vacancies increases with the carrier concentration, because the formation energy of a negatively charged vacancy decreases through the rise of the Fermi level. Such increase in the vacancy concentration would facilitate the surface-mediate edge dislocation climb, which would result in the decrease of the area of the extra half plains. The resulting volume shrinkage would cause the tensile stress and concave bowing.

Basic Strategies for the Reduction of Wafer Bowing

According to Eq. (16.7b), σ_{max} is inversely proportional to the square root of the island diameter. We can therefore suppress the wafer bowing by increasing the island diameter through the reduction of the island density at the early stage of the growth. At the same time, a smaller island density is also helpful to diminish the dislocation density, because the coalescence front itself is a dislocation source, and in turn also the bowing caused by the dislocation inclination mechanism [39,40], which is enhanced by donor doping, as described in Section "Mechanism of Wafer Bowing by Inclination of Edge Dislocations." However, with the decrease in the island density, the deposition time and the layer thickness required for the formation of a compact layer increase, and with it the probability for the formation of inversion domains (N-polar GaN), which lead to inverse pyramidal pits on the surface due to the difference in the growth rate of the Ga-polar matrix. At the end, the actual growth conditions are chosen not only taking into account the technical requirements concerning the bowing and the surface morphology but also guaranteeing a cost-effective production.

On the other hand, the enhancement of the film thickness is also effective to decrease the bowing. As described above, the island coalescence and edge dislocations climb promote a concave bowing during the growth at high temperature, as shown in Figure 16.16. After the full coalescence, only the dislocation climb remains and, although its effect diminishes gradually with the film thickness as the dislocation density decreases, the bowing should increase continuously. However, on such bowing crystal surface, a regular homogeneous epitaxy is not possible. The own bowing becomes the origin for an in-plane compressive stress, since the number of unit cells per area increases with the thickness. This compression counteracts the dislocation climb mechanism and the concave bowing itself, thus the thicker the grown layer the smaller the final bowing. However, the layer thickness should be as small as possible from the viewpoint of the production cost, if only one GaN wafer is produced from one base substrate. In this sense, it is preferable to produce multiple GaN wafers at one time by slicing a thick GaN boule grown by HVPE. The technical issues to realize this manufacture are described in Section 16.3.5.

16.3.4
Concrete Examples of the Production Technology of GaN Wafers

Commercial GaN wafers need to be produced so as to meet various technical requirements (wafer size, dislocation density, conductivity, optical transmittance, off-angle variation, etc.) for the fabrication of the target devices. At the same time, the production cost must be sufficiently low for a successful business. So far, a lot of trials have been made to realize the mass production of GaN wafers. Among them, however, only a handful of technologies have been put into practical use. In this section, the two most representative methods are reviewed as concrete examples.

The DEEP Method [41,42]
The DEEP (Dislocation Elimination by the Epitaxial-growth with inverse-pyramidal Pits) method, developed by Sumitomo Electric Industries, Ltd, is the one that realized the world-first mass production of GaN wafers. Figure 16.19 shows schematically the process steps of this method. It utilizes GaAs (111) as base substrate for the HVPE growth of a thick Ga-polar GaN (0001) layer. As the name of the method indicates, N-polar inversion domains (so-called "cores") are

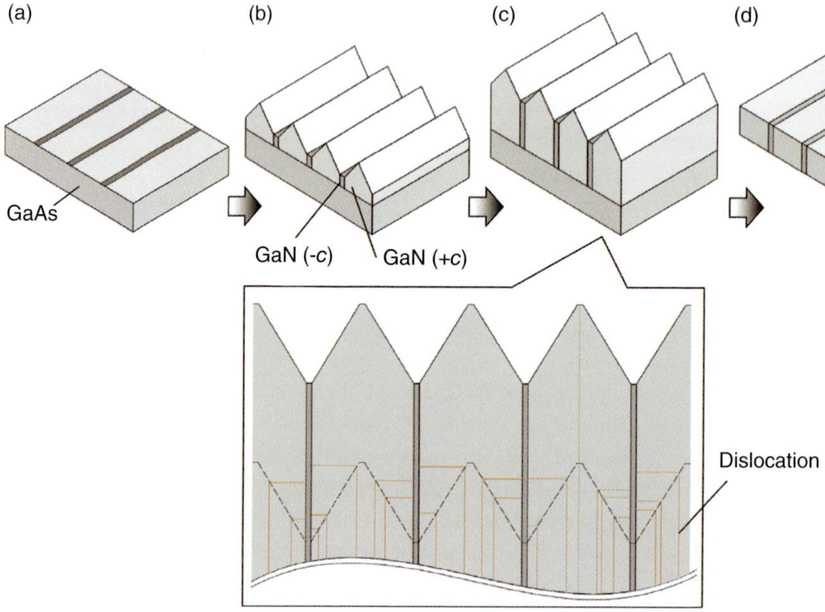

Figure 16.19 Process flow of DEEP technique.

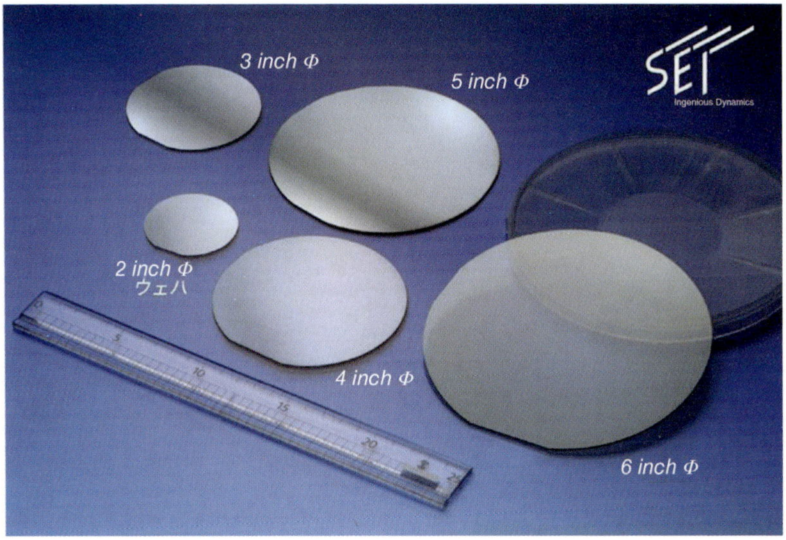

Figure 16.20 Photograph of GaN wafers produced by HVPE with DEEP. (By courtesy of Sumitomo Electric Industries, Ltd.)

intentionally formed as strips with typical periodicity of 0.5 mm during the HVPE growth by conducting some undisclosed processing on the GaAs substrate prior to the deposition. As the result, the wide Ga-polar surface grows along two oblique facets, like in the case of the FIELO technique, with N-polar cores in between. Although a high density of dislocations is generated at the GaN/GaAs interface and these dislocations propagate along the *c*-direction of GaN, they bend when they come across the inclined facets, as described in Section "Basic Strategies for the Dislocation Reduction." In contrast to the FIELO technique, the faceted surface structure is kept throughout the growth, thanks to the presence of the cores. Therefore, the dislocations are gathered thoroughly toward the cores. As the result, the dislocation density is decreased down to remarkably low values, namely, $10^5\,cm^{-2}$ or less between the cores. A GaN wafer is obtained by removing the GaAs substrate and lapping/polishing both sides of the freestanding GaN crystal (Figure 16.20). The obtained GaN wafer consists of high-quality regions and defective cores, alternately aligned with a typical period of about 0.5 mm. Accordingly, the device fabrication should be done avoiding the defective regions.

The VAS Method [28]

The VAS (void-assisted separation) method, developed by Hitachi-cable, Ltd, (the name was changed to SCIOCS, Ltd), realized the world-first mass production of GaN wafers with a homogeneous dislocation density. The VAS method utilizes porous GaN templates with a TiN nanomask on the top as base substrates for the HVPE growth of a thick Ga-polar GaN (0001) layer. The

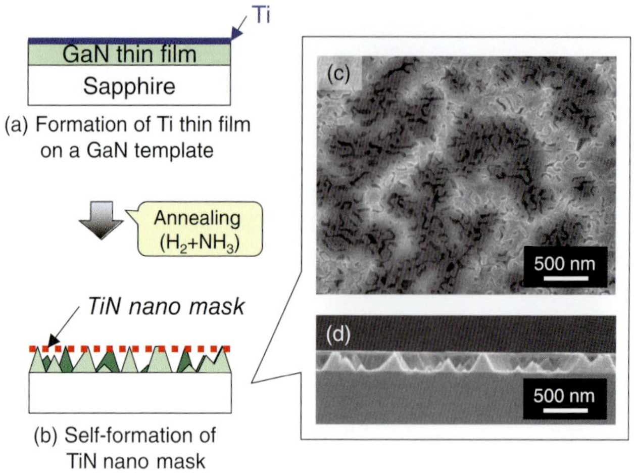

Figure 16.21 (a and b) Schematic diagram of the fabrication of porous templates, (c and d) surface and cross-sectional SEM of the porous template.

nanomask is made through a self-formation process without any costly photolithography. Figure 16.21 describes the process sequence to produce the porous GaN templates. First, a Ti film with a thickness around 20 nm is deposited on a GaN/sapphire template. The template is then annealed in a mixture gas stream of H_2 and NH_3 at around 1000 °C. As a result, the Ti layer turns into a TiN "nanomask," having a mesh-like structure with many nanoscale holes due to a thermal agglomeration, while the GaN template layer is heavily attacked leading to a void structure.

Figure 16.22 shows how the HVPE growth of GaN proceeds on the porous template. It is initiated selectively on the GaN template layer under the nanomask, forming numerous GaN islands that appear on the surface through the holes of the nanomask. These islands coalesce with each other during growth till a compact layer is formed. During the growth, many voids remain around the nanomask, resulting in a fragile porous layer that is used for the base substrate

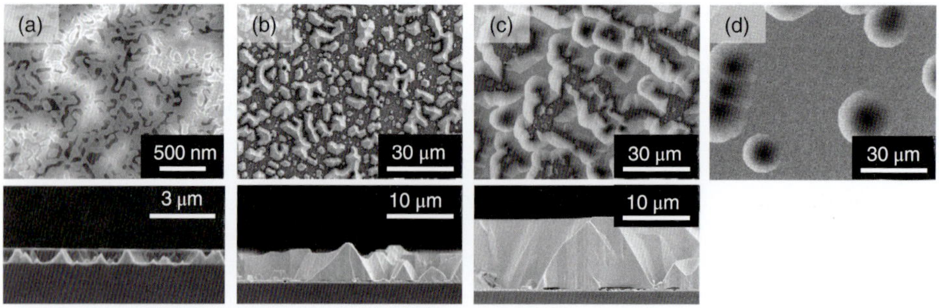

Figure 16.22 Surface and cross-sectional SEM images of GaN grown on a porous template at different growth stages.

Figure 16.23 Photograph of a GaN wafer produced by HVPE with VAS method. (By courtesy of SCIOCS, Ltd.)

separation. This sacrificial layer breaks by the thermal stress created during the cooling process after the deposition of a thick GaN layer, so that the base substrate is spontaneously separated and a freestanding GaN crystal without any cracking is obtained. Furthermore, the base sapphire substrate can be reused after surface polishing.

In this method, the dislocation density is remarkably reduced during the island growth process through the FIELO mechanism described in Section 16.3.3.1. Although the dislocation density is not uniform just after the coalescence, the distribution is homogenized during the thick layer deposition and at the end an uniform dislocation density as low as $10^6\,cm^{-2}$ is achieved. Figure 16.23 shows the photograph of a 4-inch GaN wafer produced by the VAS method. Compared to DEEP-GaN wafers, the VAS-GaN ones are very uniform, thus, there is no limitation in the device size and the position where the devices are fabricated.

16.3.5 Future Prospectives

Recently, the application field of GaN wafers is not restricted anymore to LDs, the use of these wafers for LEDs is rapidly augmenting. Also, power device applications are being investigated and developed intensively. In order to suit these applications, GaN wafers need to be further improved, especially from the viewpoint of the crystal quality, wafer size, and production cost. However, it is not easy to largely reduce the production cost as long as only one GaN wafer is produced from one base substrate. In addition, it is difficult to significantly improve

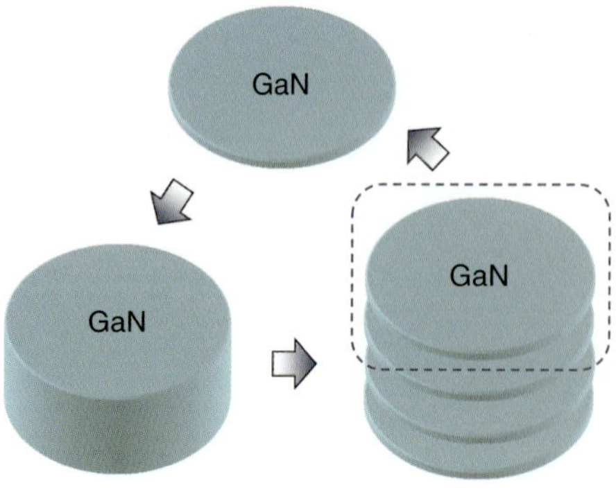

Figure 16.24 Schematic of the fabrication of freestanding GaN wafers through boule growth.

multiple properties at the same time, such as the dislocation density, off-angle variation, and homogeneity, while securing a limited thickness and growth time.

As mentioned above, such difficulties can be overcome if multiple GaN wafers can be produced at the same time by slicing a GaN boule, as shown in Figure 16.24. Foreign base substrates do not have to be prepared anymore, once one GaN wafer is available as seed crystal for the next boule growth. In addition, the dislocation density could be largely improved by repeating the cycle, since it would be equivalent to the growth of a super long boule. It would be also possible to use an ultrahigh-quality GaN seed crystal made by other methods, such as the ammonothermal [11] or the Na-flux method [12,13], whose productivity is still insufficient to be applied for mass production independently.

In order to realize such production method, however, some technical problems need to be solved. One of the critical ones is the shrinkage of the *c*-plane. When the thickness of GaN increases during the HVPE, inclined $\{10\bar{1}1\}$ facets gradually develop at the periphery, so that the area of *c*-plane decreases (Figure 16.25). In order to maintain the production cycle of the boule method, the diameter of the boule should not diminish during the

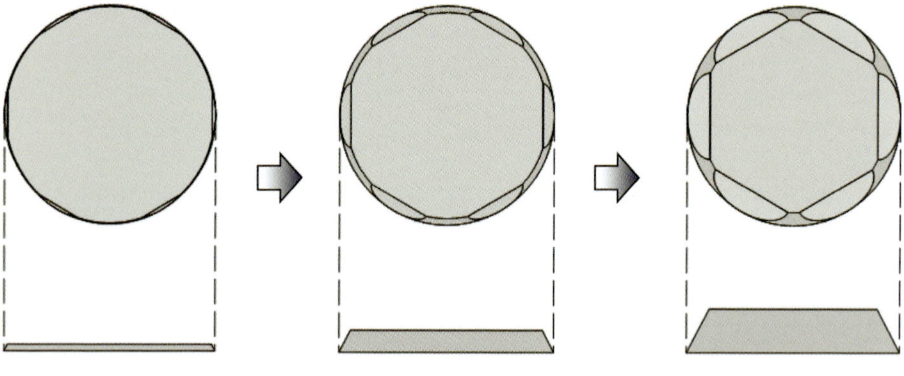

Figure 16.25 Shrinkage of top surface area during very thick growth of *c*-plane GaN.

growth. Even if the boule is completely surrounded by $\{10\bar{1}1\}$ facets, in principle it would be possible to maintain the production cycle if the $\{10\bar{1}1\}$ facets would continue growing. However, when the $\{10\bar{1}1\}$ planes grow to a macroscopic scale, $\{10\bar{1}0\}$ facets start to develop on the $\{10\bar{1}1\}$ facets. The growth rate of $\{10\bar{1}0\}$ facets is very small, and polycrystalline GaN easily deposit on the $\{10\bar{1}0\}$ facets. As a result, the boule growth is interrupted. Even if the boule growth is carried out using a GaN seed crystal with a principal plane other than (0001), the growth is finally disturbed by the development of $\{10\bar{1}0\}$ facets as well. In conclusion, the key to the successful boule growth is the suppression of the $\{10\bar{1}0\}$ facets growth and the subsequent polycrystallization. If this problem could be solved, it would revolutionize the GaN wafer industry.

At last, it should be mentioned that boule growth of GaN has also been tried intensively by the ammonothermal and Na-flux methods, and even the shipment of sample products has started recently. However, these solution growth techniques need a drastic breakthrough to improve the productivity and so meet the cost requirements. Accordingly, there is no doubt that HVPE leads the GaN wafer industry for the present and in the near future.

16.4 HVPE of Ga_2O_3

Ga_2O_3 is reported to possess five different polymorphs, namely, α-, β-, δ-, ε-, and γ-phase [43,44]. Among them, the β- and the α-phase have been confirmed to be promising wide bandgap semiconductors.

β-Ga_2O_3 is known to be thermodynamically the most stable phase among the five polymorphs under atmospheric pressure. Its crystal structure belongs to monoclinic system, space group *C*2/*m*. The unit cell and lattice parameters are shown in Figure 16.26. The bandgap energy is as large as 4.7–4.9 eV [45–47],

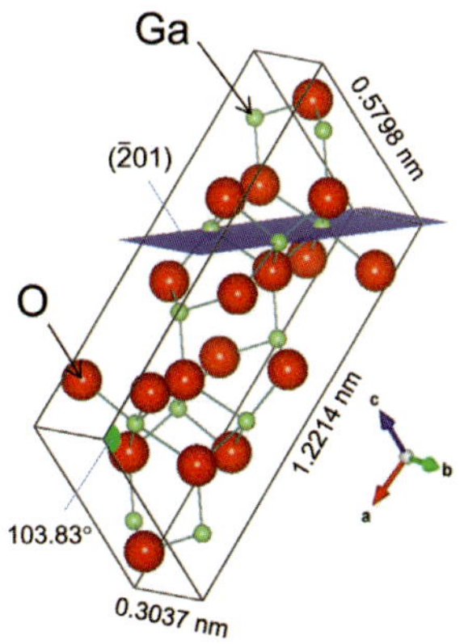

Figure 16.26 Crystal structure of β-Ga_2O_3.

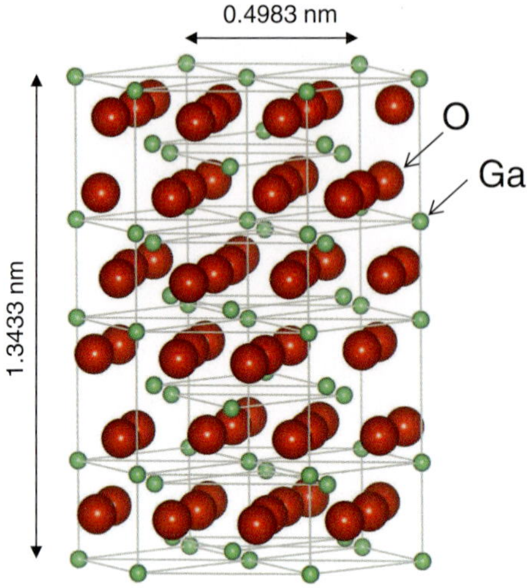

Figure 16.27 Crystal structure of α-Ga_2O_3.

well above that of GaN, and the electrical conductivity can be controlled by doping donor impurities such as Si or Sn [48,49]. In contrast to GaN, high-quality bulk single crystals can be grown from the melt by standard techniques, such as edge-defined film-fed growth (EFG) [50], floating zone (FZ) [51], and Czochralski (CZ) [52], and therefore single crystal wafers are commercially available. These unique characteristics make this material a promising candidate for various applications, such as transparent/conductive substrates for GaN-based high-brightness LEDs [53,54] and solar-blind UV sensors [55]. In addition, β-Ga_2O_3 is attracting remarkable attention as a promising material to realize high-performance power devices, such as Schottky barrier diodes (SBDs) [56], metal–semiconductor field effect transistors (MESFETs) [57], and metal–oxide–semiconductor field effect transistors (MOSFETs) [58].

On the other hand, corundum α-Ga_2O_3 (Figure 16.27) is a metastable phase of Ga_2O_3, which is reported to be stable up to 550–700 °C [43,44,59,60]. Its bandgap energy is 5.2–5.3 eV, which is even greater than that of β-Ga_2O_3 [61,62]. Its electrical conductivity can be controlled by doping donor impurities such as Sn [63,64]. Thus, α-Ga_2O_3 is a promising wide bandgap semiconductor as well as β-Ga_2O_3. In contrast to the β-phase, the α-one has the potential to realize bandgap engineering through the formation of heterostructures and solid solutions with other corundum-structured oxides, so that new novel functional materials can be fabricated [65]. For this purpose, the formation of α-$(Al_xGa_{1-x})_2O_3$ and α-$(In_xGa_{1-x})_2O_3$ films, and the consequent bandgap control have already been demonstrated [66,67]. Moreover, it has been also proven that the solid solutions of α-Ga_2O_3 and α-Fe_2O_3 exhibit a remarkably large magnetization, which is far greater than that of α-Fe_2O_3 alone [68]. Accordingly, we can expect the realization of high-performance, multifunctional devices exploiting such variety of solid solutions and heterostructures.

In contrast to the remarkable achievements of HVPE in the field of III–V semiconductors, there were only a few reports about the HVPE of Ga_2O_3. However, promising results have been reported that demonstrate rapid growth of high quality Ga_2O_3.

In this section, the general characteristics of Ga_2O_3 are introduced, and then the cutting edge technologies developed for HVPE of β-Ga_2O_3 and α-Ga_2O_3 are described.

16.4.1 Recent Progress in HVPE of β-Ga_2O_3

16.4.1.1 Technical Issues for the Epitaxial Growth of β-Ga_2O_3

So far, the study of β-Ga_2O_3 power devices [56–58] has been conducted mainly using homoepitaxial layers grown by MBE, whose typical growth rate is ~1 μm/h or less. However, it is desirable to develop an epitaxial growth technique that enables a high growth rate, and thereby a high productivity, especially if we take into account that power devices need sometimes a thick film growth in order to guarantee a sufficiently high break down voltage.

At present, 2-inch β-Ga_2O_3 wafers produced by EFG method are available, and larger wafers up to 6 in. are under development. However, the EFG method requires the use of dies and crucibles made of the very expensive noble metal iridium, and therefore the production cost of β-Ga_2O_3 wafers rapidly increases with the diameter. One of the most practical alternatives is to utilize heteroepitaxial β-Ga_2O_3 films deposited at a high growth rate on large-area, mass-producible substrates such as sapphire. It could be also possible to grow a very thick β-Ga_2O_3 layer to obtain freestanding β-Ga_2O_3 wafers in a similar way to that of GaN. Unfortunately, at present, there is no report on the successful growth of single crystalline β-Ga_2O_3 layers on foreign substrates. In the following, we introduce the recent achievements by HVPE, which could solve all these technical issues.

16.4.1.2 Rapid Growth of β-Ga_2O_3 by HVPE

The first report related to HVPE of β-Ga_2O_3 was made by Matsumoto *et al.* [69]. They did not use any substrates, but they successfully synthesized small needles and platelets of β-Ga_2O_3 on the inner wall of their quartz reactor tube by using GaCl and O_2 as precursors. After this, no report on HVPE of β-Ga_2O_3 using substrates was made until 2014.

Recently, a research group of Tokyo University of Agriculture and Technology has proposed some guidelines for the rapid growth of β-Ga_2O_3 based on their thermodynamic analysis of HVPE using GaCl and O_2 as precursors [70]. According to these, for the rapid growth, it is important to feed the precursors so that the partial pressure of O_2 is greater than that of GaCl. In addition, it is also desirable to keep the partial pressure of H_2 as small as possible, because H_2

significantly reduces the driving force of the growth. The same group has confirmed that these estimations are in good agreement with the experimental observations. They deposited homoepitaxially (001) β-Ga_2O_3 under atmospheric pressure using O_2 and GaCl, which formed *in situ* upstream in the reactor by the reaction of Ga with Cl_2. The growth rate increased proportionally with the partial pressures of the precursors, reaching a large value of about 30 μm/h. They also demonstrated that the growth rate was virtually independent of the growth temperature in the range of 900–1050 °C, thus proving that the growth in the mass transport limited regime is possible [71].

On the other hand, Oshima *et al.* demonstrated the rapid growth of heteroepitaxial ($\bar{2}01$) β-Ga_2O_3 on sapphire substrates by HVPE. They carried out the growth under atmospheric pressure at 1050 °C using O_2 and GaCl, the latter formed in this case by the reaction between Ga and HCl in a way similar to that of GaN-HVPE [72]. Figure 16.28 shows the growth rate as a function of the partial pressures of (a) HCl and (b) O_2. It increased monotonically with both partial pressures, achieving a very large growth rate, over 250 μm/h, which is comparable to those of GaN. These experimental results demonstrate that HVPE of β-Ga_2O_3 can achieve faster growth rates than other epitaxial growth techniques by a factor of 10–100. High growth rates, however, have been related to the degradation of surface morphology and crystal quality. Therefore, further detailed investigations are required in order to find out optimal growth conditions for the deposition of high quality of epilayers at high speed.

16.4.1.3 Homoepitaxy of β-Ga_2O_3 by HVPE

Evidently, homoepitaxy is the best way to grow high-quality β-Ga_2O_3 layers for high-performance devices, if sufficiently high-quality and large-area substrates are available for a reasonable price. The crystal quality of the homoepitaxial layer should be the same or better than that of the substrate. In addition, the background carrier concentration needs to be low enough for the intentional doping control. Furthermore, a smooth surface is preferable so that the as-grown film can be used for the device fabrication without any additional polishing process. Of course, the growth rate should be as large as possible.

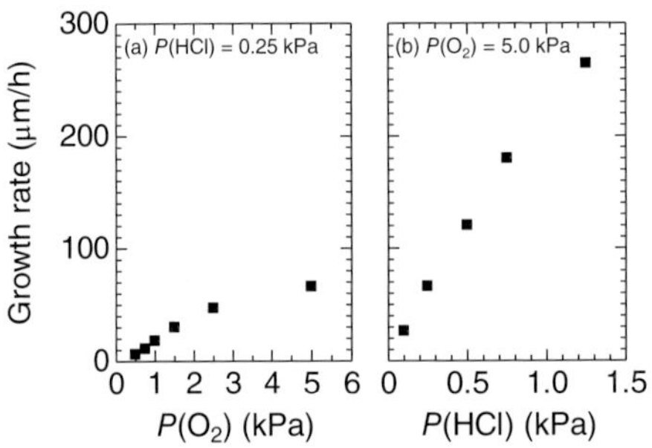

Figure 16.28 Growth rate of ($\bar{2}01$) β-Ga_2O_3 as a function of (a) $P(O_2)$ and (b) $P(HCl)$.

Murakami *et al.* demonstrated the homoepitaxial growth of (001) β-Ga_2O_3 by HVPE [71]. They have shown that relatively smooth homoepitaxial layers can be obtained at 1000 °C without degradation of the crystal quality, while a rough surface and a significant broadening of the X-ray rocking curve were observed below 900 °C. Secondary ion mass spectrometry (SIMS) evidenced the high purity of the film, with impurity concentrations of [H], [C], [Si], and [Sn] below the detection limits, and [Cl] as low as $1 \times 10^{16}\,cm^{-3}$. Their *C–V* measurement indicated that the residual carrier concentration was less than $10^{13}\,cm^{-3}$, that is, low enough for the doping control at the low levels required to guarantee high-breakdown voltages in power devices.

16.4.1.4 HVPE of β-Ga_2O_3 on Foreign Substrates

Heteroepitaxy of β-Ga_2O_3 has been pursued by various growth methods such as MBE [73–75], MOCVD [76–78], and HVPE [72]. Their common problem is the formation of in-plane rotational domains, which reflect the in-plane rotational symmetry of the substrate [73–78]. For example, a $(\bar{2}01)$ β-Ga_2O_3 film grown on a *c*-plane sapphire substrate comprise six kinds of domains with different in-plane orientations [73–75], reflecting the symmetry of the in-plane oxygen arrangement of the substrate. In principle, this problem can be solved simply by using a substrate plane with rotational symmetry, which is not higher than that of the target plane of β-Ga_2O_3. However, β-Ga_2O_3 does not have any rotational symmetry axes except the [010], and therefore, a substrate with twofold or nonrotational symmetry is required to deposit single crystalline β-Ga_2O_3 film.

Oshima *et al.* have demonstrated the remarkable suppression of such in-plane rotational domains in $(\bar{2}01)$ β-Ga_2O_3 films by using off-angled *c*-plane sapphire substrates [72]. Figure 16.29 shows the X-ray *002* pole figures of β-Ga_2O_3 films grown on off-angled *c*-plane substrates. The reference pole figure from a single crystal $(\bar{2}01)$ β-Ga_2O_3 wafer grown by the EFG method is shown together for comparison (Figure 16.29a). In this case, in accordance with the symmetry, only one *002* peak is observed because the $(\bar{2}01)$ plane does not have any rotational symmetry. Note that the $\bar{2}02$ peak also appears because the (001) and $(\bar{1}01)$ planes have almost the same spacing and thus the same Bragg angle. In contrast, six equivalent *002* peaks appear in the case of the HVPE-grown $(\bar{2}01)$ β-Ga_2O_3 film on a nonoff-angled *c*-plane sapphire substrate (Figure 16.29b), in accordance with the substrate symmetry. The use of off-angled substrates remarkably changes the amount and predominance of the domains. Even a small off-angle of $\Delta = 2°$ leads to a decrease in the number of *002* peaks from six to three (Figure 16.29c), reflecting the bulk symmetry of sapphire. At the same time, the intensity of the *002* peak along the off-direction increased, while those of other two peaks decreased, indicating that the growth of the first domain is favored. This tendency is strongly enhanced with increasing the off-angle (Figure 16.29d and e), yielding to a pole figure which is very similar to that of single crystalline β-Ga_2O_3 when $\Delta = 5°$ or more. Although a trace amount of in-plane rotational domains still exist and the mosaicity also needs to be improved, this technology

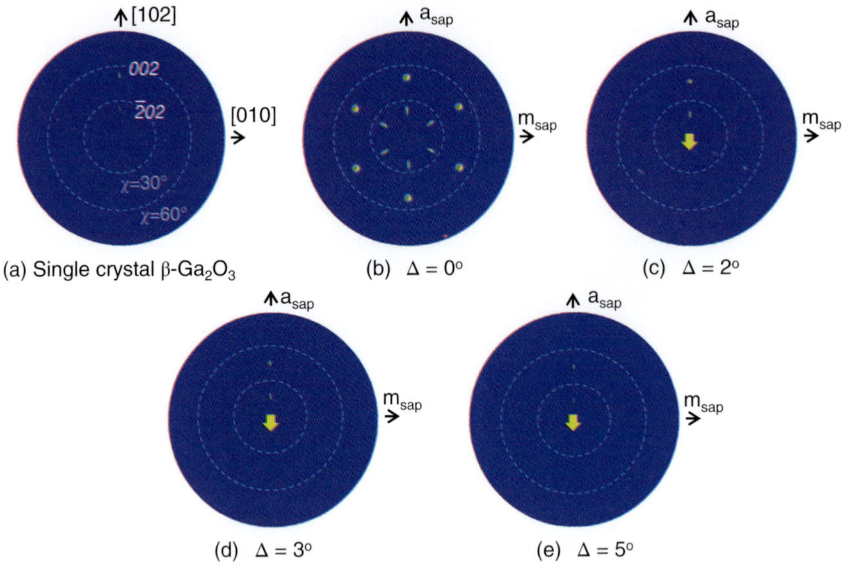

Figure 16.29 *002* Pole figures of (a) single crystal β-Ga_2O_3 wafer, (b)–(e) β-Ga_2O_3 layers grown on sapphire (0001) substrates with various off-angles Δ_a. Yellow arrows show the direction of off-angle.

represents a first breakthrough toward the realization of high-quality β-Ga_2O_3 templates and wafers for cost-competitive β-Ga_2O_3 devices.

16.4.2 Recent Progress in HVPE of α-Ga_2O_3

16.4.2.1 Technical Issues for the Epitaxial Growth of α-Ga_2O_3

As α-Ga_2O_3 is a metastable phase, its growth from the melt is not possible. Accordingly, the fabrication of single crystalline α-Ga_2O_3 can be carried out only through heteroepitaxy. Sapphire is one of the most promising substrates for this heteroepitaxy, since both α-Ga_2O_3 and sapphire have the same corundum structure, and large-area substrates are available at reasonable prices. The successful deposition of α-Ga_2O_3 films has been demonstrated already by mist-CVD [79,80], MBE [81], and HVPE [62]. Mist-CVD realized the heteroepitaxy of α-Ga_2O_3 for the first time. However, the mist-CVD-grown films were found to include a small fraction of 180° rotational domains [79] and to contain a high concentration of impurities ([H] $= 3 \times 10^{19}\,cm^{-3}$, [C] $= 1 \times 10^{19}\,cm^{-3}$, [Si] $= 9 \times 10^{18}\,cm^{-3}$) [64]. Kumaran *et al.* reported the successful MBE heteroepitaxy of Nd-doped *a*-plane α-Ga_2O_3 on *a*-plane sapphire at 500 °C, while β-Ga_2O_3 was dominant in a film grown on *c*-plane sapphire at the same temperature [81]. In both mist-CVD and MBE, the typical growth rates are quite low with a value about 1 μm/h or less. Therefore, there was a necessity to establish a more productive epitaxial growth technique, which enables a faster growth under atmospheric pressure, especially when α-Ga_2O_3 is deposited to manufacture power devices. Furthermore, if a very large growth rate is realized, the fabrication of freestanding α-Ga_2O_3 wafers, which enable vertical-structured devices, becomes feasible through the thick layer growth and the subsequent removal of the base substrate.

Rapid Growth of α-Ga_2O_3 by HVPE [62]

Recently, the rapid HVPE growth of twin-free α-Ga_2O_3 has been reported for the first time. The films were grown on *c*-plane sapphire substrates using gallium chloride and O_2 as precursors and a deposition temperature around 525–600 °C under atmospheric pressure. These growth conditions are virtually the same as those for β-Ga_2O_3 except the growth temperature. Figure 16.30 shows an X-ray *ω-2θ* scan profile of an α-Ga_2O_3 film grown by HVPE. It exhibits only diffraction peaks from the (0001) plane of α-Ga_2O_3 together with the substrate peaks, indicating the successful synthesis of phase-pure *c*-plane α-Ga_2O_3. Further, Figure 16.31 shows the X-ray *$10\bar{1}2$* pole figures of the α-Ga_2O_3 layer and the substrate. Only three peaks appeared at the positions that are expected for single crystalline α-Ga_2O_3, thus proving the successful growth of twin-free single-crystalline α-Ga_2O_3. The epitaxial relationships between the α-Ga_2O_3 layer and the sapphire substrate were determined to be $[10\bar{1}0]$ α-Ga_2O_3 || $[10\bar{1}0]$ sapphire and (0001) α-Ga_2O_3 || (0001) sapphire. On the other hand, the tilt and twist angles of a typical HVPE-grown α-Ga_2O_3 film are relatively large, that is, around 600 and 1300 arcsec, respectively, probably because of the large in-plane lattice mismatch

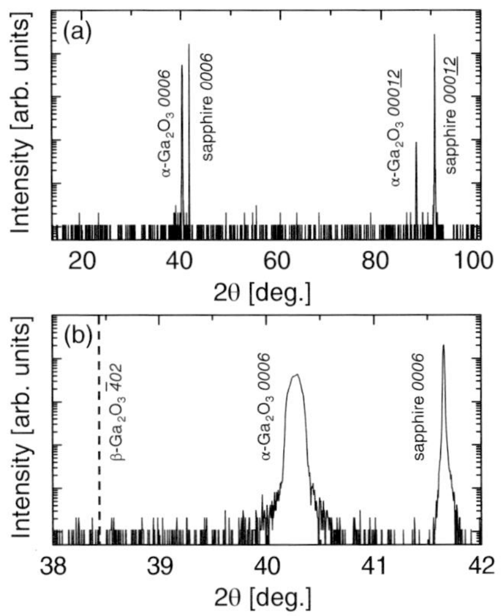

Figure 16.30 XRD *ω-2θ* profile of an α-Ga_2O_3 layer grown on a sapphire (0001) substrate. (a) Wide-scan profile, (b) narrow-scan profile around *0006*.

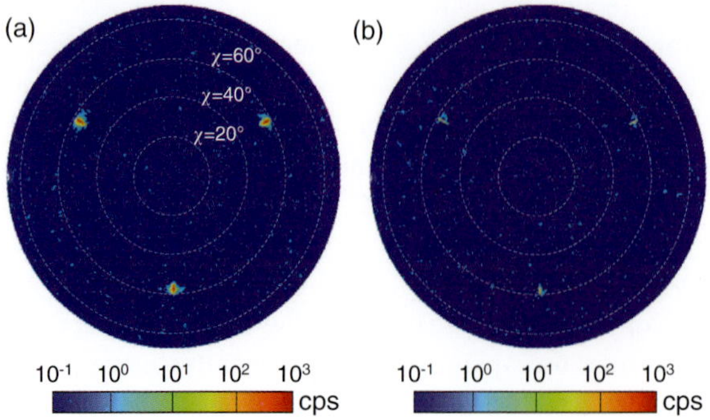

Figure 16.31 X-ray *10$\bar{1}$2* pole figures (log-scale) of (a) α-Ga_2O_3 layer and (b) sapphire substrate.

(~4.5%). Therefore, in order to drastically improve the crystalline quality, novel techniques, such as ELO, need to be developed in addition to the optimization of the growth conditions.

The impurity concentrations in a typical HVPE-grown α-Ga_2O_3 film measured by SIMS are shown in Table 16.1. These concentrations were much lower than the reported values for α-Ga_2O_3 grown by mist-CVD. [H], [C], and [Al] were below the detection limits, while [Si] was much lower than the reported value, which is necessary in order to control the carrier concentration in a wide range. [Cl] cannot be compared because there has been no other report, however, it is noteworthy that Murakami *et al.* have reported the incorporation of Cl impurity with the same concentration order as in the HVPE-grown β-Ga_2O_3, and the Cl impurity do not act as an effective donor [71].

The growth rate of α-Ga_2O_3 as a function of the partial pressures of HCl and O_2 is illustrated in Figure 16.32. This increases monotonically with both partial pressures, reaching over 100 μm/h, which would be sufficient for the fabrication of very thick layers and freestanding wafers. The surface of such rapidly grown α-Ga_2O_3 is still mirror-like, and confirmed to be twin-free single crystal through pole figure measurements.

Table 16.1 Impurity concentrations in α-Ga_2O_3 measured by SIMS.

Element	Detection limit (D. L.) (cm^{-3})	Concentration (cm^{-3})
H	4×10^{17}	<D. L.
C	6×10^{17}	<D. L.
Al	4×10^{15}	<D. L.
Si	1×10^{16}	2×10^{16}
Cl	1×10^{16}	7×10^{16}

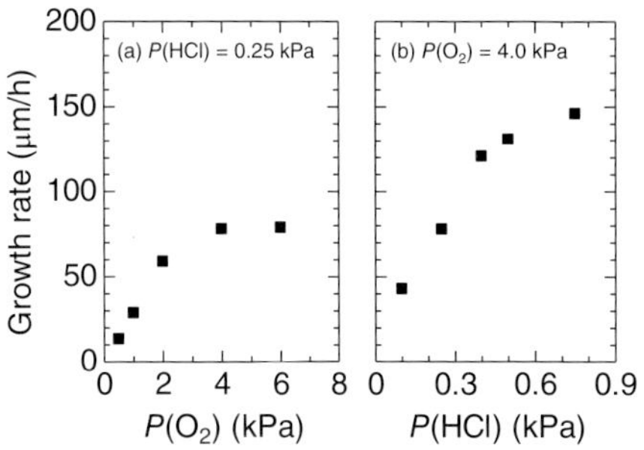

Figure 16.32 Growth rate of (0001) α-Ga_2O_3 as a function of (a) $P(O_2)$ and (b) $P(HCl)$.

16.4.3
Summary and Future Prospectives

Recent progress of the HVPE technologies of β- and α-Ga_2O_3 is reviewed. Concerning β-Ga_2O_3, the rapid growth and the low background donor concentration have already been demonstrated. Hereafter, the key issue is how to maintain the crystalline quality and surface morphology during the rapid growth. For this purpose, total optimization of the growth conditions will be required, including not only the growth parameters such as the partial pressures of the precursors, the growth temperature but also the crystal orientations. In the case of heteroepitaxy, the improvement of the crystal quality is also essential. Concerning α-Ga_2O_3, the rapid growth of single crystalline films has also been demonstrated, however, the improvement of the crystalline quality remains as one of the most important technical issues. In the case of the heteroepitaxial growth of both β- and α-Ga_2O_3, similar problems with those in the case of GaN, such as bowing, cracking upon the substrate removal, would need to be solved as well.

The chemical reaction between gallium chloride and oxygen is very reactive and therefore, the growth rate variation along the gas stream tends to be steep. In the worst case, the obtained thickness cannot be uniform even if the substrate is rotated during the growth. Accordingly, a more precise control of the chemical reaction and gas flow will be required in order to realize a rapid and uniform growth on large-area substrates.

16.5
Conclusion

HVPE of wide bandgap semiconductors was overviewed through two representative examples, namely, the production technologies of GaN wafers and the recent progress in Ga_2O_3 growth. Although, more than 10 years have passed since the beginning of the mass-production of GaN wafers, the technologies are still immature compared to those of other highly sophisticated semiconductors, such as GaAs and Si. GaN growth technologies will continue improving in order to be suitable for new applications such as power devices. Concerning Ga_2O_3, the application of HVPE to the device development is still on the way, however, the establishment of technological basis, such as rapid growth of high-purity epitaxial films, is steadily advancing. Furthermore, the development of freestanding AlN wafers by HVPE is also in progress, though it is not described in detail in this chapter. Advantages of HVPE-grown AlN wafers have already been demonstrated in the fabrication of high-performance UV-LEDs, which can replace conventional low-efficiency and nonenvironment-friendly (contain Hg) mercury lamps [82,83].

Under current circumstances, in which the development of high-efficiency devices is fundamentally important to meet the global requirements for energy-saving and environmental conservation, the HVPE technologies for the deposition of wide bandgap semiconductors will keep increasing in importance.

References

1 Ku, S.M. (1963) *J. Electrochem. Soc.*, **110**, 991.

2 Gottlieb, G.E. (1965) *J. Electrochem. Soc.*, **112**, 192.

3 Tietjen, J.J. and Amic, J.A. (1966) *J. Electrochem. Soc.*, **113**, 724.

4 Yoshida, T., Oshima, Y., Watanabe, K., Tsuchiya, T., and Mishima, T. (2011) *Phys. Status. Solidi C*, **8**, 2110.

5 Kumagai, Y., Kubota, Y., Nagashima, T., Kinoshita, T., Dalmau, R., Schlesser, R., Moody, B., Xie, J., Murakami, H., Koukitu, A., and Sitar, Z. (2012) *Appl. Phys. Express*, **5**, 055504.

6 Marusuka, H.P. and Tietjen, J.J. (1969) *Appl. Phys. Lett.*, **15**, 327.

7 Abraham, F.F. (1974) *Homogeneous Nucleation Theory*, Academic Press, NY.

8 Chow, T.P. and Ghezzo, M.J. (1996) SiC power devices, in *III-Nitride, SiC, and Diamond Materials for Electronic Devices*, vol. **423** (eds. D.K. Gaskill, C.D. Brandt, and R.J. Nemanich), Material Research Society Symposium Proceedings, Pittsburgh, PA, pp. 69–73.

9 Eastman, L., Chu, K., Schaff, W., Murphy, M., and Wiemann, N.G. (1997) *MRS Internet J. Nitride Semicond. Res.*, **2**, 2.

10 Grzegory, I., Jun, J., Bockowski, M., Wroblewski, M., Lucznik, B., and Porowski, S. (1995) *J. Phys. Chem. Solids*, **56**, 639.

11 Dwiliński, R., Doradziński, R., Garczyński, J., Sierzputowski, L.P., Puchalski, A., Kanbara, Y., Yagi, K., Minakuchi, H., and Hayashi, H. (2008) *J. Cryst. Growth*, **310**, 3911.

12 Yamane, H., Shimada, M., Clarke, S.J., and DiSalvo, F.J. (1997) *Chem. Mater.*, **9**, 413.

13 Kawamura, F., Iwahashi, T., Omae, K., Morishita, M., Yoshimura, M., Mori, Y., and Sasaki, T. (2003) *Jpn. J. Appl. Phys.*, **42**, L4.

14 Ohta, H., Kaneda, N., Horikiri, F., Narita, Y., Yoshida, T., Mishima, T., and Nakamura, T. (2015) *IEEE Electron Device Lett.*, **36**, 1180.

15 Kizilyalli, I.C., Edwards, A.P., Nie, H., Bour, D., Prunty, T., and Disney, D. (2014) *IEEE Electron Device Lett.*, **35**, 247.

16 Oka, T., Ueno, Y., Ina, T., and Hasegawa, K. (2014) *Appl. Phys. Express*, **7**, 021002.

17 Nie, H., Diduck, Q., Alvarez, B., Edwards, A.P., Kayes, B.M., Zhang, M., Ye, G., Prunty, T., Bour, D., and Kizilyalli, I.C. (2014) *IEEE Electron Device Lett.*, **35**, 939.

18 Sugahara, T., Sato, H., Hao, M., Naoi, Y., Kurai, S., Tottori, S., Yamashita, K., Nishino, K., Romano, L.T., and Sakai, S. (1998) *Jpn. J. Appl. Phys.*, **37**, L398.

19 Look, D.C. and Sizelove, J.R. (1999) *Phys. Rev. Lett.*, **82**, 1237.

20 Takeya, M., Mizuno, T., Sasaki, T., Ikeda, S., Fujimoto, T., Ohfuji, Y., Oikawa, K.,

Yabuki, Y., Uchida, S., and Ikeda, M. (2003) *Phys. Status Solidi C*, **0**, 2292.
21 Nakamura, S., Senoh, M., Nagahama, S., Iwasa, N., Yamada, T., Matsushita, T., Kiyoku, H., Sugimoto, Y., Kozaki, T., Umemoto, H., Sano, M., and Chocho, K. (1998) *Appl. Phys. Lett.*, **72**, 2014.
22 Usui, A., Sunakawa, H., Sakai, A., and Yamaguchi, A.A. (1997) *Jpn. J. Appl. Phys.*, **36**, L899.
23 Hiramatsu, K., Nishiyama, K., Onishi, M., Mizutani, H., Narukawa, M., Motogaito, A., Miyake, H., Iyechika, Y., and Maed, T. (2000) *J. Cryst. Growth*, **221**, 316.
24 Fujito, K., Kubo, S., Nagaoka, H., Mochizuki, T., Namita, H., and Nagao, S. (2009) *J Cryst. Growth*, **311**, 3011.
25 Melnik, Yu.V., Vassilevski, K.V., Nikitina, I.P., Babanin, A.I., Davydov, V.Yu., and Dmitriev, V.A. (1997) *MRS Internet J. Nitride Semicond. Res.*, **2**, 39.
26 Kelly, M.K., Ambacher, O., Dimitrov, R., Handschuh, R., and Stutzmann, M. (1997) *Phys. Status Solidi A*, **159**, R3.
27 Tavernier, P.R. and Clarke, D.R. (2001) *J. Appl. Phys.*, **89**, 1527.
28 Oshima, Y., Eri, T., Shibata, M., Sunakawa, H., Kobayashi, K., Ichihashi, T., and Usui, A. (2003) *Jpn. J. Appl. Phys*, **42**, L1.
29 Wakahara, A., Yamamoto, T., Ishio, K., Yoshida, A., Seki, Y., Kainosho, K., and Oda, O. (2000) *Jpn. J. Appl. Phys.*, **39**, 2399.
30 Götz, W., Johnson, N.M., Walker, J., Bour, D.P., and Street, R.A. (1996) *Appl. Phys. Lett.*, **68**, 667.
31 van de Walle, C.G., Stampfl, C., and Neugebauer, J. (1998) *J. Cryst. Growth*, **189**, 505.
32 Oshima, Y., Yoshida, T., Watanabe, K., and Mishima, T. (2010) *J. Cryst. Growth*, **312**, 3569.
33 Yoshida, T., Suzuki, T., Kitamura, T., Abe, Y., Fujikura, H., Shibata, M., and Saito, T. (2015) Proceedings of the 39th International Conference on Advanced Ceramics and Composites, Daytona beach, FL, USA.
34 Götz, W., Johnson, N.M., Chen, C., Liu, H., Kuo, C., and Imler, W. (1996) *Appl. Phys. Lett.*, **68**, 3144.
35 Wang, H. and Chen, A.B. (2000) *J. Appl. Phys.*, **87**, 7859.
36 Hoffman, R.W. (1981) *Surf. Interface Anal.*, **3**, 62.
37 Nix, W.D. and Clemens, B.M. (1999) *J. Mater. Res.*, **14**, 3467.
38 Böttcher, T., Einfeldt, S., Figge, S., Chierchia, R., Heinke, H., and Hommel, D. (2001) *Appl. Phys. Lett.*, **78**, 1976.
39 Romanov, A.E. and Speck, J.S. (2003) *Appl. Phys. Lett.*, **83**, 2569.
40 Xie, J., Mita, S., Rice, A., Tweedie, J., Hussey, L., Collazo, R., and Sitar, Z. (2011) *Appl. Phys. Lett.*, **98**, 202101.
41 Motoki, K., Okahisa, T., Matsumoto, N., Matsushima, M., Kimura, H., Kasai, H., Takemoto, K., Uematsu, K., Hirano, T., Nakayama, M., Nakahata, S., Ueno, M., Hara, D., Kumagai, Y., Koukitu, A., and Seki, H. (2001) *Jpn. J. Appl. Phys.*, **40**, L140.
42 Motoki, K., Okahisa, T., Hirota, R., Nakahata, S., Uematsu, K., and Matsumoto, N. (2007) *J. Cryst. Growth*, **305**, 377.
43 Roy, R., Hill, V.G., and Osborn, E.F. (1952) *J. Am. Chem. Soc.*, **74**, 719.
44 Playford, H.Y., Hannon, A.C., Barney, E.R., and Walton, R.I. (2013) *Chem. Eur. J*, **19**, 2803.
45 Tippins, H.H. (1965) *Phys. Rev.*, **140**, A316.
46 Lorenz, M.R., Woods, J.F., and Gambino, R.J. (1967) *J. Phys. Chem. Solids*, **28**, 403.
47 Orita, M., Ohta, H., Hirano, M., and Hosono, H. (2000) *Appl. Phys. Lett.*, **77**, 4166.
48 Suzuki, N., Ohira, S., Tanaka, M., Sugawara, T., Nakajima, K., and Shishido, T. (2007) *Phys. Status Solidi C*, **4**, 2310.
49 Víllora, E.G., Shimamura, K., Yoshikawa, Y., Ujiie, T., and Aoki, K. (2008) *Appl. Phys. Lett*, **92**, 202120.
50 Aida, H., Nishiguchi, K., Takeda, H., Aota, N., Sunakawa, K., and Yaguchi, Y. (2008) *Jpn. J. Appl. Phys.*, **47**, 8506.
51 Víllora, E.G., Shimamura, K., Yoshikawa, Y., Aoki, K., and Ichinose, N. (2004) *J. Cryst. Growth*, **270**, 420.
52 Galazka, Z., Irmscher, K., Uecker, R., Bertram, R., Pietsch, M., Kwasniewski, A., Naumann, M., Schulz, T., Schewski, R., Klimm, D., and Bickermann, M. (2014) *J. Cryst. Growth*, **404**, 184.

53 Shimamura, K., Víllora, E.G., Domen, K., Yui, K., Aoki, K., and Ichinose, N. (2005) *Jpn. J. Appl. Phys.*, **44**, L7.

54 Víllora, E.G., Arjoca, S., Shimamura, K., Inomata, D., and Aoki, K. (2014) *Proc. SPIE*, **8987**. DOI: 10.1117/12.2063892

55 Oshima, T., Okuno, T., Arai, N., Suzuki, N., Ohira, S., and Fujita, S. (2008) *Appl. Phys. Express*, **1**, 011202.

56 Sasaki, K., Kuramata, A., Masui, T., Víllora, E.G., Shimamura, K., and Yamakoshi, S. (2012) *Appl. Phys. Express*, **5**, 035502.

57 Higashiwaki, M., Sasaki, K., Kuramata, A., Masui, T., and Yamakoshi, S. (2012) *Appl. Phys. Lett.*, **100**, 013504.

58 Higashiwaki, M., Sasaki, K., Kamimura, T., Wong, M.H., Krishnamurthy, D., Kuramata, A., Masui, T., and Yamakoshi, S. (2013) *Appl. Phys. Lett.*, **103**, 123511.

59 Eckert, L.J. and Bradt, R.C. (1973) *J. Am. Ceram. Soc.*, **56**, 229.

60 Lee, S.D., Akikawa, A., and Fujita, S. (2013) *Phys. Status Solidi C*, **10**, 1592.

61 Shinohara, D. and Fujita, S. (2008) *Jpn. J. Appl. Phys.*, **47**, 7311.

62 Oshima, Y., Víllora, E.G., and Shimamura, K. (2015) *Appl. Phys. Express*, **8**, 055501.

63 Kawaharamura, T., Dang, G.T., and Furuta, M. (2012) *Jpn. J. Appl. Phys.*, **51**, 040207.

64 Akaiwa, K. and Fujita, S. (2012) *Jpn. J. Appl. Phys.*, **51**, 070203.

65 Fujita, S. and Kaneko, K. (2014) *J. Cryst. Growth*, **401**, 588.

66 Ito, H., Kaneko, K., and Fujita, S. (2012) *Jpn. J. Appl. Phys.*, **51**, 100207.

67 Suzuki, N., Kaneko, K., and Fujita, S. (2014) *J. Cryst. Growth*, **401**, 670.

68 Kaneko, K., Kakeya, I., Komori, S., and Fujita, S. (2013) *J. Appl. Phys.*, **113**, 233901.

69 Matsumoto, T., Aoki, M., Kinoshita, A., and Aono, T. (1974) *Jpn. J. Appl. Phys.*, **13**, 1578.

70 Nomura, K., Goto, K., Togashi, R., Murakami, H., Kumagai, Y., Kuramata, A., Yamakoshi, S., and Koukitu, A. (2014) *J. Cryst. Growth*, **405**, 19.

71 Murakami, H., Nomura, K., Goto, K., Sasaki, K., Kawara, K., Thieu, Q.T., Togashi, R., Kumagai, Y., Higashiwaki, M., Kuramata, A., Yamakoshi, S., Monemar, B., and Koukitu, A. (2015) *Appl. Phys. Express*, **8**, 015503.

72 Oshima, Y., Víllora, E.G., and Shimamura, K. (2015) *J. Cryst. Growth*, **410**, 53.

73 Víllora, E.G., Shimamura, K., Kitamura, K., and Aoki, K. (2006) *Appl. Phys. Lett.*, **88**, 031105.

74 Oshima, T., Okuno, T., and Fujita, S. (2007) *Jpn. J. Appl. Phys.*, **46**, 7217.

75 Tsai, M.Y., Bierwagen, O., White, M.E., and Speck, J.S. (2010) *J. Vac. Sci. Technol. A*, **28**, 354.

76 Mi, W., Ma, J., Zhu, Z., Luan, C., Lv, Y., and Xiao, H. (2012) *J. Cryst. Growth*, **354**, 93.

77 Mi, W., Ma, J., Luan, C., Lv, Y., Xiao, H., and Li, Z. (2012) *Mater. Lett.*, **87**, 109.

78 Mi, W., Luan, C., Li, Z., Zhao, C., Xiao, H., and Ma, J. (2013) *Mater. Lett.*, **107**, 83.

79 Shinohara, D. and Fujita, S. (2008) *Jpn. J. Appl. Phys.*, **47**, 7311.

80 Kaneko, K., Kawanowa, H., Ito, H., and Fujita, S. (2012) *Jpn. J. Appl. Phys.*, **51**, 020201.

81 Kumaran, R., Tiedje, T., Webster, S.E., Penson, S., and Li, W. (2010) *Opt. Lett.*, **35**, 3793.

82 Kinoshita, T., Obata, T., Nagashima, T., Yanagi, H., Moody, B., Mita, S., Inoue, S., Kumagai, Y., Koukitu, A., and Sitar, Z. (2013) *Appl. Phys. Express*, **6**, 092103.

83 Inoue, S., Naoki, T., Kinoshita, T., Obata, T., and Yanagi, H. (2015) *Appl. Phys. Lett.*, **106**, 131104.

17
Growth of Silicon Nanowires

Fengji Li and Sam Zhang

Nanyang Technological University, School of Mechanical & Aerospace Engineering, 50 Nanyang Avenue, Singapore 639798, Singapore

17.1
Growth Techniques for Silicon Nanowires

17.1.1
Introduction

Silicon (Si) is the eighth most common element in the universe by mass, but very rarely occurs as the pure free element in nature. It is most widely distributed in dusts, sands, planetoids, and planets as various forms of Si dioxide or silicates. The atomic number of Si is 14. It is a diamond cubic crystal lattice with a lattice parameter of 5.43 Å (cf. Figure 17.1) [1]. Each Si atom is surrounded by four other Si atoms arranged at the corner of a tetrahedron.

Si nanowires are important building blocks in optical devices [2–4], field effect transistors [5–7], lithium batteries [8,9], and power generators [10,11]. They have the substantial quantum size effects as long as the diameter reaches 3 nm and below [12], which could result in the increase of the bandgaps from 1.1 to 3.3 eV. The large amount of available surface area to molecular adsorption greatly increases the sensitivity of the nanowire electrical properties to the environment. The small size of Si nanowires can be used to prepare dense arrays of sensors. Si nanowires are semiconducting, and the dopant type and concentration of Si nanowire are controllable, which enables the sensitivity to be tuned in the absence of an external gate. Si nanowires allow the combination of one-dimensional transport and self-assembled processing with the well-known Si technology. Si nanowires can circumvent the pulverization and capacity fading in the application of lithium battery due to the great volume changes of Si upon insertion and extraction of lithium. To date, however, no production-scale Si nanowire applications have shown in the marketplace, but many simple device structures have been demonstrated, such as optical devices [2–4], biosensors [5],

Handbook of Solid State Chemistry, First Edition. Edited by Richard Dronskowski, Shinichi Kikkawa, and Andreas Stein.

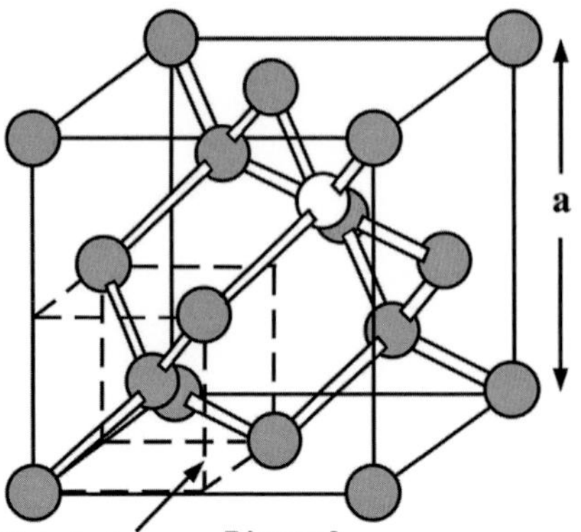

Figure 17.1 Face-centered cubic Si crystal structure [1].

field effect transistors [6,7,13,14], and lithium battery [8,9,15], and power generators [10,11,16], exhibiting the great possibility that may be marketable in future.

"Top-down" and "bottom-up" methods are widely applied to synthesize Si nanowires [17]. In the top-down methods, bulk materials of desired composition are carved into the desired nanoscale structure by lithographic techniques. Such methods dominated the material processing over the last century. They are still of principal importance for production of electronic components. However, as the desired length scales of devices and applications shrink, these techniques become more problematic. This is partly a practical problem: Techniques to carve out ever smaller structures are difficult to find. More importantly, uniformity of bulk crystals on nanometer length scales is not very high, and quality of structures becomes difficult to control.

Bottom-up methods mimic nature's way of self-assembling atoms to form increasingly larger structures. Such techniques involve controlled crystallization of materials from vapor, liquid, or solid sources, typically yielding uniform and highly ordered nanometer-scale structures. Si is provided by distinct sources in different growth mechanisms [18], either bulk Si or Si-containing compounds in the form of vapor, gases, or fluid. For instance, Si nanowires could grow by laser ablation of Si—Fe powder mixture [19], chemical vapor deposition (CVD) of $SiCl_4/H_2$ [20–22] or SiH_4 gases [23], thermal degradation of diphenylsilane in a supercritical hexane fluid using the size-monodispersed Au nanocrystals as the seeds [24,25], thermal evaporation of bulk Si onto metal-covered substrates under ultrahigh vacuum [26,27], thermal evaporation of Si-containing powders in quartz or alumina tube [28–30], or thermal annealing of metal-covered Si wafers [31,32]. In the growth, Si atoms come from vaporization of powder mixture [19], decomposition of gaseous molecules [20–22], degradation of diphenylsilane [24,25], sublimation of bulk Si [26,27], Si-containing powders [28–30], or Si wafers [31,32], and then melt into the metal–Si eutectic droplet. Upon supersaturation of Si, Si precipitates out to seed a nanowire, followed by the growth of the nanowire with continuous supply of precipitating Si atoms from the liquid droplet.

Among these methods, however, laser ablation suffers from the low production yield; chemical vapor deposition has the risk of flammable or toxic precursor gases; thermal evaporation of Si-containing powder gives rise to high level of defects; thermal degradation in a supercritical hexane fluid suffers from the surface contamination of hydrocarbon or limited choice of metal catalyst and Si reactant; evaporation of bulk Si has limited growth rate and considerable Si film growth. Si nanowires grown from Si wafers via thermal annealing metal-coated wafers could get rid of not only the metallic contamination at the tip of the nanowire but also the Si-containing flammable and toxic gases.

17.1.2
Metal-Assisted Chemical Etching

As one of the top-down methods, metal-assisted chemical etching produces vertical Si nanowires out of Si wafers, where the diameter is defined by a lithography step preceded by reactive ion etching. During metal-assisted chemical etching, Si is wet-chemically etched, with the Si dissolution reaction being catalyzed by the presence of a noble metal that is added as a salt to the etching solution [33]. Alternatively, a continuous but perforated noble metal film can be used. During etching, this film could etch down into the Si wafer to produce vertical Si nanowires at the locations of the holes in the film [34,35]. In a typical metal-assisted chemical etching procedure [36], a Si wafer partly covered by a noble metal is subjected to an etchant composed of hydrofluoric (HF) acid and an oxidative agent. Typically, the Si beneath the noble metal is etched much faster than the Si without noble metal coverage. As a result, the noble metal sinks into the Si wafer, generating pores in the Si wafer or, additionally, Si nanowires. The detailed geometries of the resulting Si structures depend mostly on the initial morphology of the noble metal coverage. Under certain conditions, microporous structures could form in the regions without noble metal coverage.

Figure 17.2 describes the metal-assisted chemical etching process: (1) The oxidant is preferentially reduced at the surface of the noble metal due to the

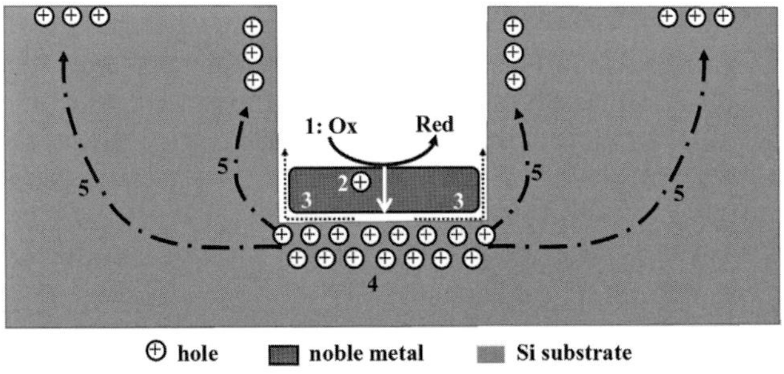

Figure 17.2 Schematic diagram of the metal-assisted chemical etching process [36].

catalytic activity of the noble metal on the reduction of the oxidant. (2) The holes are generated due to the reduction of the oxidant diffuse through the noble metal. They are injected into the Si that is in contact with the noble metal. (3) The Si is oxidized by the injected holes and dissolved at the Si–metal interface by HF. The reactant (HF) and the by-products diffuse along the interface between the Si and the noble metal. (4) The concentration of holes has the maximum at the Si–metal interface. Therefore, the Si that is in contact with the metal is etched much faster by HF than a bare Si surface without metal coverage would be. (5) The holes diffuse from the Si under the noble metal to off-metal areas or to the wall of the pore if the rate of hole consumption at the Si–metal interface is smaller than the rate of hole injection. Accordingly, the off-metal areas or sidewalls of the pore may be etched and form microporous Si.

There are some unsolved technical problems. First, the injection process of the hole has not been systematically studied. The influence of the intrinsic properties of the Si wafer (i.e., doping type and doping level), surface state of the Si wafer, and the type and morphologies of the noble metal catalyst on the hole injection process remains incompletely understood. Second, the proper explanation of some currently poorly understood phenomena, such as different etching behaviors of Si wafers with different doping type and doping level and the different etching behaviors induced by different type of metal catalysts, relies upon the full understanding of the hole injection process. Third, the origin of the driving force of the movement of the noble metal particles is simply the preferred etching of Si under the particles. The etching path of the noble metal particles and the possible role of the shape of the noble metal particle (sphere, half sphere, or disk) have yet been investigated. The driving force for aligning the movement directions of adjacent particles for longer etching times, which appears to be some kind of cooperative effect, is not presently understood. Fourth, there is not a systematic study of the relation between the etching rate, etching direction, and orientation of Si wafers. Fifth, a possible influence of the specific dopant element, doping levels, etchants, and noble metals on the morphologies of the etched structures has not been studied. Last but not the least, the etching morphologies of these non-Si semiconductors have been well characterized, whereas the detailed etching mechanisms are yet to be explored.

17.1.3
Vapor–Liquid–Solid Growth

In 1964, Wagner and Ellis developed the vapor–liquid–solid (VLS) mechanism to explain the growth process of Si whiskers [20]. The growth takes place as follows. A small Au particle is placed on the {111}surface of a Si wafer and heated up to 950 °C, forming a small droplet of Au–Si alloy (cf. Figure 17.3a). A mixture of H_2 and $SiCl_4$ is introduced. The liquid alloy acts as a preferred sink for arriving Si atoms or, perhaps more likely, as a catalyst for the chemical process involved. Si atoms enter the liquid and freeze out, with a very small concentration of Au in the solid solution, at the interface between the solid Si and the

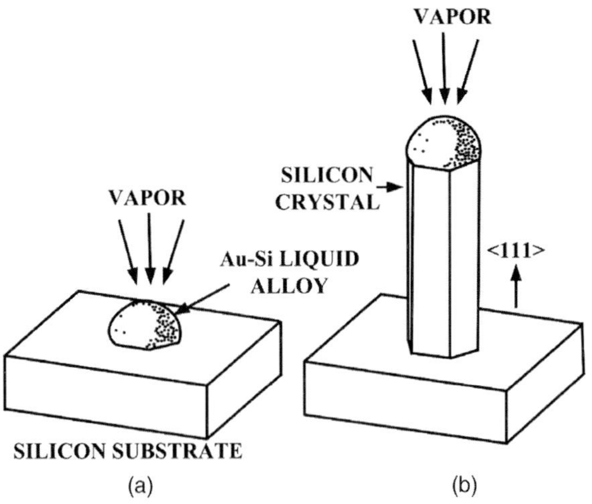

Figure 17.3 VLS growth of the Si whisker (see also Ref. [20]). (a) Initial condition with a liquid Au—Si droplet formed on the Si substrate. (b) Growing Si crystal with the liquid droplet at the tip. (From Ref. [20] with the permission of AIP Publishing.)

liquid alloy. By a continuation of this process, the alloy droplet becomes displaced from the substrate crystal and "rides" atop the growing whisker (cf. Figure 17.3b). Since the eutectic point of Au–Si alloy is about 363 °C, the droplet could form during heating the Au-covered Si wafer at 950 °C. The VLS mechanism reflects the pathway of Si, which coming from the vapor phase diffuses through the liquid droplet and ends up as a solid Si nanowire. It represents the core of semiconductor nanowire research, since it works not only for Si but also for a much broader range of nanowire materials.

Si nanowires grown via CVD of gaseous Si precursors follow the VLS mechanism. Silane (SiH_4), disilane (Si_2H_6), dichlorosilane (Si dichloride, SiH_2Cl_2), and tetrachlorosilane (Si tetrachloride, $SiCl_4$) are frequently adopted as the precursors. When $SiCl_4$ is used, the growth temperature ranges from about 800 °C [37–40] to well beyond 1000 °C [22,41]. In the presence of SiH_4, the growth temperature is about 400–600 °C [42–44]. The use of a chlorinated silane precursor in the presence of hydrogen would lead to the creation of hydrochloric acid during nanowire growth, causing some desirable or undesirable etching to the substrate, the nanowires, and the equipment. Additionally, chlorinated silanes are chemically more stable than their nonchlorinated counterparts. Thus, higher temperature is required to thermally crack the precursor.

High-temperature CVD (>700 °C) is often carried out in tubular hot wall reactors [38,40,41]. A gas flow, typically hydrogen or a hydrogen–inert gas mixture, is directed through an externally heated quartz tube held at about atmospheric pressure. Prior to entering the reactor, a part of the gas is led through a bubbler filled with $SiCl_4$ ($SiCl_4$ is liquid at room temperature and atmospheric

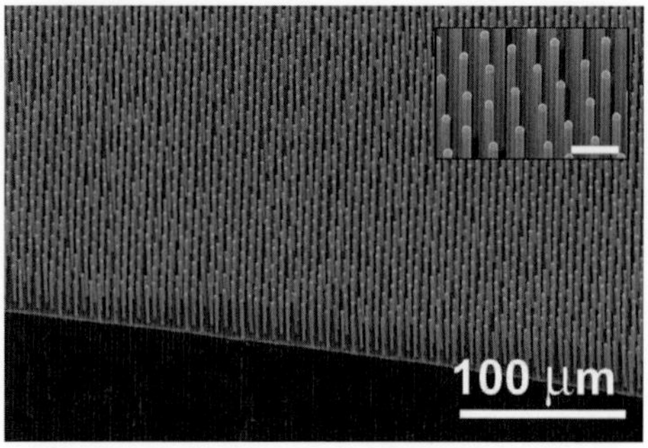

Figure 17.4 Cu-catalyzed Si nanowire array with nearly 100% fidelity over a large area (>1 cm^2). The scale bar in the inset is 10 µm. (From Ref. [38] with the permission of AIP Publishing.)

pressure), thereby supplying $SiCl_4$ to the reactor. If the Si substrate deposited with some amount of the metallic catalyst is placed in the hot zone of the reactor beforehand, Si nanowires would start to grow.

As shown in Figure 17.4 [38], Cu-catalyzed Si nanowire array with nearly 100% fidelity over a large (>1 cm^2) area is grown through an oxide buffer layer to confine the Cu catalyst to the desired areas in the pattern. The nanowires are grown for up to 30 min at 850–1100 °C under 1 atm of H_2 and $SiCl_4$ at flow rates of 1000 and 20 sccm, respectively.

High-temperature CVD is also preceded in a closed reaction vessel. Wagner *et al.* [45], for instance, obtained their early results using evacuated and sealed quartz ampules into which iodine was placed together with Si and the metallic catalyst materials. Upon heating, iodine reacts with Si to form gaseous silicon iodide products, which then serve as a locally produced CVD precursor. And then, Si nanowire grows at the colder parts of the quartz ampule.

The advantage of high-temperature CVD is the much broader choice of the metallic catalyst materials. However, the high growth rates (on the order of micrometers per minute [38], or even micrometers per second [41]) and the diffusion of the metallic catalysts (Ostwald ripening effect [46,47]) at elevated temperatures often result in the difficulty to control the length and to grow nanowires with well-defined diameters. Therefore, high-temperature CVD suits for the growth of microscopic Si wires rather than nanowires. Independent of the catalyst material used, the main crystallographic growth direction of the nanowires is the ⟨111⟩ direction, probably due to the large diameter of the wires. Wagner *et al.* pointed out that ⟨112⟩-oriented wires showing a twin defect parallel to the wire axis can also be found [45]. Besides, the dopant precursor could influence the wire morphology. Givargizov pointed out that the periodic instability of the wires (a periodic variation of the wire diameter observed at high temperatures and pressures) disappears once $AsCl_3$ is added to the gas mixture [21,22], presumably because $AsCl_3$ changed the surface tension configuration of the wire and/or the catalyst droplet.

Silane and disilane are often used in low-temperature CVD (<700 °C). SiH_4 could decompose at about 350 °C. Si_2H_6 is more reactive than silane, as it allows Si nanowire growth at much lower pressures. Both SiH_4 and Si_2H_6 are self-igniting gases that are potentially explosive if they are in contact with air. The dominant orientation of nanowires with diameters larger than about 50 nm is ⟨111⟩. When Au is used as the catalyst, high-vacuum equipment with a base pressure around 10^{-6} mbar is sufficient for the nanowire growth. This is, however, not the case when more sensitive metallic catalysts are used. For instance, as Al is sensitive to oxidation, the use of an ultrahigh-vacuum reactor with a base pressure lower than about 10^{-9} mbar is recommended. As long as oxidation of Al is prevented, excellent nanowires can be obtained [48].

The silane partial pressure typically ranges from 0.1 mbar [42,44] to 0.5 mbar [49], resulting in growth rates on the order of nanometers per second. The growth rate increases approximately linearly with the partial pressure of silane [49]. Au-catalyzed Si nanowires by using silane as source are usually slightly tapered (cf. Figure 17.5), indicating that the radial growth is slow compared to axial growth. Radial growth rates are about two orders of magnitude lower than the axial growth rates. The activation energies for Au-catalyzed Si nanowire growth are about half as large as those for Si thin film growth: 19 versus 35 kcal/mol for silane [49] and 12 versus 28 kcal/mol for disilane [50]. This reduction by about a factor of 2 makes the Au droplet a catalyst droplet. Plasma-enhanced CVD makes use of the plasma in the reactor [51], thus the Si precursor, silane in most cases, is partially precracked. Such a precracking could facilitate the supply of Si to the catalyst droplet.

The advantages of the low-temperature CVD are that nanowires with a large variety of diameters and length could be grown epitaxially on Si substrates. With

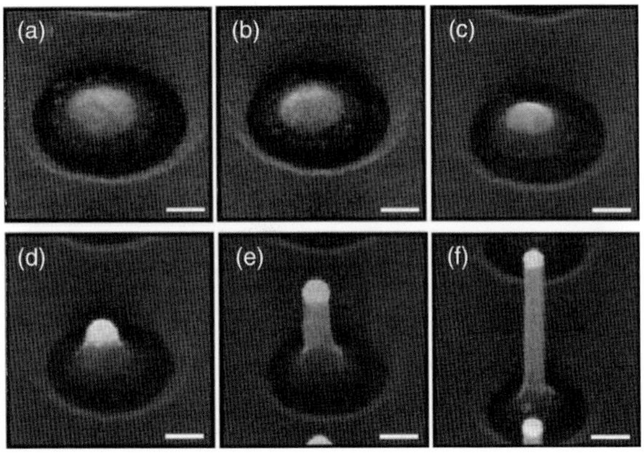

Figure 17.5 Diameter-dependent nucleation of Si nanowires. (a–f) The evolution of a flat Au disk into a hemispherical Au droplet on top of a Si nanowire. Diameters range from 250 nm (a) to 80 nm (f). Scale bars are 100 nm. (From Ref. [49] with the permission of AIP Publishing.)

the lengths of the nanowires being essentially proportional to the process time, they can be easily adjusted. Nanowire growth at predefined positions on the substrate is possible. The properties of the nanowires could be tuned directly by doping from the gas phase, allowing also for modulated doping profiles. One major problem is that they exhibit a certain variation of the growth direction, especially for diameters smaller than about 50 nm. Another problem is that a certain percentage of the nanowires tend to change their growth direction during growth, resulting in the appearance of a kink [52].

17.1.4
Vapor–Solid–Solid Growth

Vapor–solid–solid (VSS) growth is of reduced processing temperature and more abrupt heterojunction [53]. The interface between the Si nanowire and the metallic catalyst provides a preferential interface for the incorporation of adatoms into the growing crystal [54]. The phase of the metallic catalyst may be the same both during and after the growth. The growth could be broken down into four steps (cf. Figure 17.6): (a) delivery of precursor species to the substrate; (b) precursor decomposition; (c) diffusion in or on the catalyst particle; and (d) incorporation of adatoms into the growing nanowire. Each step introduces specific requirements for appreciable VSS growth to be realized. First of all, sufficient source materials must be delivered to the substrate for growth to occur. Then, selective precursor decomposition at the catalyst can facilitate the anisotropic growth. Next, Si adatoms must be delivered to the growth interface by

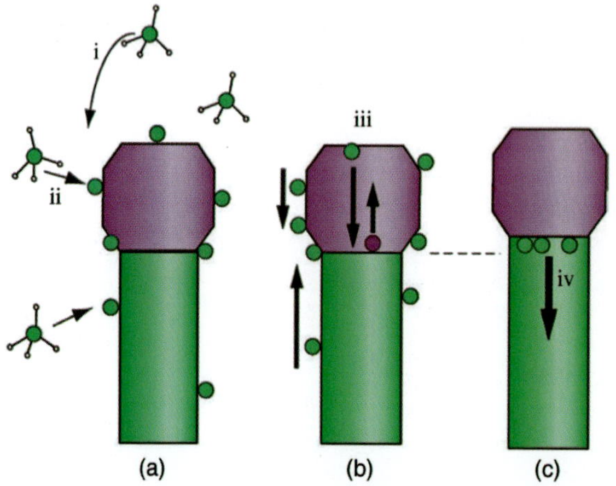

Figure 17.6 VSS mechanism of the nanowire growth [53]. (a) Delivery of the precursor to the nanowire surface, followed by the decomposition of this precursor. (b) Atoms are delivered to the metal catalyst–nanowire interface by surface or bulk diffusion processes. (c) Semiconductor atoms are incorporated at the growth interface (metal catalyst–nanowire interface) leading to anisotropic growth.

bulk or surface diffusion. Finally, anisotropic growth requires the preferential incorporation of adatoms at one interface.

Compared to the VLS growth, VSS growth is of the following advantages. First, in the same metal system, VSS growth takes place at reduced temperatures because the metallic catalyst remains in a solid state. Lower temperatures could reduce the thermal budgets and slow diffusion in the grown structure. Second, particle ripening by diffusion could be mitigated if the solid nanoparticles are used, particularly at lower temperatures, resulting in narrow diameter distributions. Third, the crystal structure and orientation of the solid catalyst may provide a stronger influence on the orientation of the nanowire [37,55–58], leading to opportunities to control orientation and crystallinity independent of the substrate. Fourth, the incorporation of unintentional impurities, particularly from the metal catalyst, may be reduced due to the reduction in atom diffusivity and solid solubility associated with nanowire growth at lower temperatures. Finally, the growth may yield more abrupt interfaces in axial nanowire heterostructures that are generally difficult to realize with the VLS mechanism.

As shown in Figure 17.7, Si nanowires are grown by chemical vapor deposition on a Cu-coated Si (111) substrate placed through the VSS mechanism in a TEM chamber [56]. Si nanowires start to grow after exposure to disilane, Si_2H_6 (20% diluted in helium (He)). The growth temperature ranges between 470 and 550 °C and the disilane partial pressure between 1×10^{-7} and 8×10^{-6} Torr. The Cu_3Si

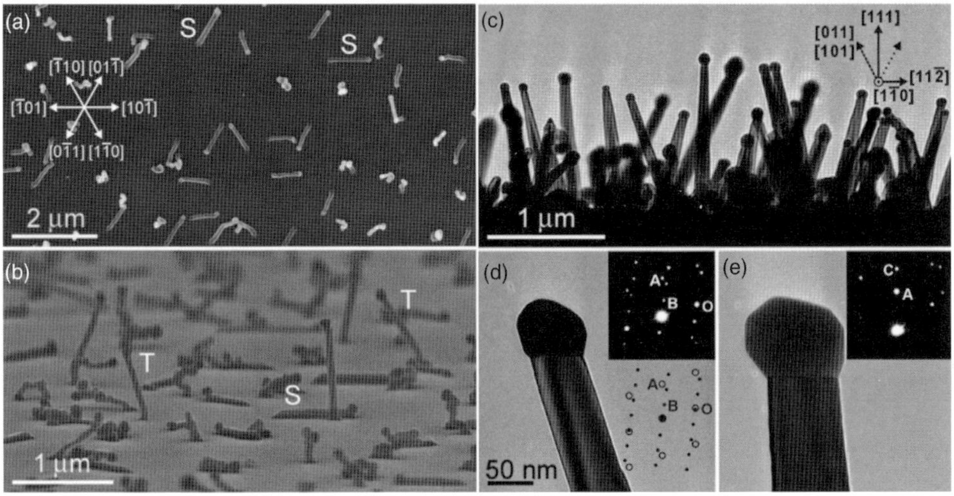

Figure 17.7 (a and b) SEM plan-view and grazing angle (10°) images of Cu_3Si-catalyzed Si nanowires after 3 h of growth on Si (111) at 510 °C and 1×10^{-6} Torr Si_2H_6 [56]. (c) Bright-field TEM image of Si nanowires recorded during growth at 530 °C and 1×10^{-6} Torr Si_2H_6. (d) TEM image and electron diffraction patterns of a [101] nanowire. (e) TEM image and electron diffraction patterns showing a nanowire and catalyst with another orientation relation, Si (111)//$Cu_3Si(1\bar{1}03)$, labeled A and C, respectively. The growth direction is close to Si [111].

phase serves as the catalyst and the epitaxial growth orientation is predominantly ⟨110⟩, but depends on the relative orientation of the catalyst and substrate. The real-time observations show that the nanowire growth involves rigid rotations of the catalyst particles and is by repeated ledge nucleation and flow at the Cu_3Si–Si interface, with the ledges propagating in a jumpy manner due to pinning by interfacial dislocations at the growth interface. In Figure 17.7a and b, S indicates the nanowires that grow along the surface in ⟨110⟩ directions. T indicates wires growing off the surface, like those in Figure 17.7c, with growth directions ⟨110⟩. In Figure 17.7c, the dashed arrow indicates the projection of the vector in the [011] or [101] orientation. The dotted arrow indicates the mirror symmetric directions. In Figure 17.7d, the growth plane of the nanowire is Si (111), which is labeled as A in the diffraction patterns. And the Cu_3Si (0001) plane is labeled as B. The schematic illustration shows the electron diffraction patterns of Si $[1\bar{1}0]$ and Cu_3Si $[11\bar{2}0]$ zone axes, open circles and solid circles, respectively, and the superposition of Si $(11\bar{1})$ and Cu_3Si $(1\bar{1}01)$ reflections, labeled O. Averaged over several nanowires, the Cu_3Si [0001] orientation is 8.3 (0.7° to Si [111] in the diffraction patterns), while the nanowire growth orientation is 18.4 (0.7° off the Si [111] axis), as expected for a nanowire growing out of plane in the Si [011] or Si [101] orientation.

The key features of the VSS growth are as follows. Prior to the initiation of growth, the solid seed particles must be formed. Upon exposure of the metal particles to a vapor-phase flux of Si, silicides begin to form. The solid silicide phases are expected to form from the metal particles at temperatures as low as $\sim 0.35 \times T_{eut}$. When excess Si is present, solid Si would crystallize at an interface with the silicide because crystallization of the semiconductor at this metal/semiconductor interface is energetically favorable. The atom diffusion through the particle and the subsequent atom incorporation at the catalyst–semiconductor interface result in the nanowire growth. However, the VSS growth rates of nanometer/minute are 10–100 times slower than that of the VLS growth, depending on the catalyst, growth temperature, and precursor partial pressure [53]. The reduced rates can be attributed to slower precursor decomposition kinetics, diffusion through a solid rather than a liquid, and/or slower atomic incorporation at subeutectic temperatures. Besides, in comparison with the VLS growth, the VSS growth may require more stringent process control, because of the dependence of the nanowire orientation and growth rate on the details of the catalyst shape and orientation relation.

17.1.5 Supercritical–Fluid–Liquid–Solid Growth

Supercritical fluid–liquid–solid (SFLS) growth utilizes highly pressurized supercritical organic fluids enriched with a liquid Si precursor, such as diphenylsilane, and metal catalyst particles [24] to produce high yield of Si nanowires. Diphenylsilane, $SiH_2(C_6H_5)_2$, is used as the Si precursor, which is mixed with hexane and sterically stabilized gold nanoparticles in a high-pressure reactor. High pressures

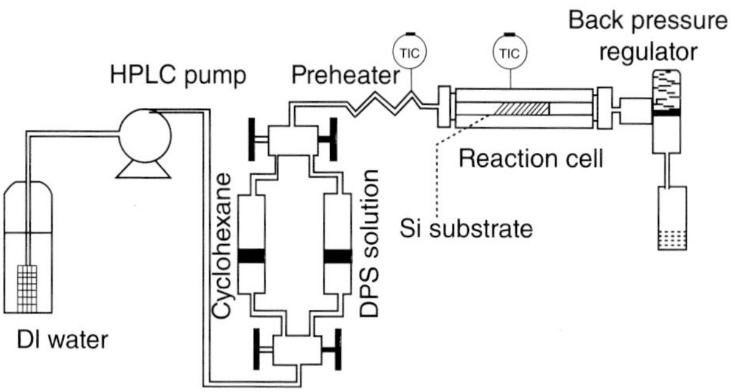

Figure 17.8 Schematic of the flow reactors used to grow Si nanowires [60].

(200–270 bar) and a temperature of 500 °C are applied to the reaction vessel. Under such conditions, hexane becomes supercritical. As the reaction temperature is above the metal–Si eutectic point, the Si atoms decomposed from the Si precursor form an alloy with gold. Once the Au–Si alloy gets supersaturated with Si, the alloy droplet starts to precipitate into a Si nanowire.

As shown in Figure 17.8, the growth takes place in a flow reactor instead of a closed reaction vessel. The colloid gold particles and Si precursor, for example, diphenylsilane, are fed into the solution as dispersions and transferred into the heated and pressurized reactor. Monophenylsilane [59], that is, SiC_6H_8, could lead to a higher product yield, however, additionally to a smaller amount of carbonaceous by-product. Supercritical fluids such as hexane and cyclohexane disperse high concentrations of sterically stabilized nanocrystals, and precursor concentrations can range from 0.1 to 1.0 mol/l [60], in contrast with typical values of 4.0×10^{-3} mol/l in the CVD [61]. The high concentrations together with the high diffusion coefficients that approach gas-phase values afford fast growth rates and efficient nanowire production. Similar to the VSS growth, Cu-, Ni-, and Co-catalyzed Si nanowires grown at temperatures below the eutectic point of the corresponding metal–Si alloy have been demonstrated [62]. Au- or Bi-catalyzed Si nanowires are demonstrated at atmospheric pressure [25]. Trisilane, Si_3H_8, which is even more reactive than disilane, acts as the Si precursor. The growth takes place in a vessel filled with a long-chain, low-vapor pressure hydrocarbon. The nanowire growth takes place at temperatures higher than the eutectic points through the SFLS mechanism.

In the SFLS growth, large quantities of thin nanowires with good crystalline quality and uniform size could be synthesized. The variability in the choice of the precursors offers an additional degree of freedom for optimization. The ability to manipulate the precursor concentration, the concentration and size of the metal seed, and the solvent strength of the supercritical fluid via pressure and temperature provides flexibility in controlling the nanowire composition, morphology, and orientation. The flow reactor provides the kinetic tenability necessary to minimize undesirable Si particle deposition and to optimize the production of straight nanowires. Si nanowire growth in the supercritical fluid could be mediated by a solid catalyst particle. However, a controlled, in-place, epitaxial growth is hardly to realize. The elements with alloy eutectic temperatures above the boiling point and even the critical temperature of most solvents could not use the SFLS technique.

17.1.6 Solid–Liquid–Solid Growth

17.1.6.1 Molecular Beam Epitaxy

Solid–liquid–solid growth includes molecular beam epitaxy (MBE), laser ablation, thermal evaporation, and thermal annealing using solid Si source, such as bulk Si, Si-containing powders, or Si wafers. Molecular beam epitaxy growth takes place by evaporating bulk Si onto the substrate that is covered with metallic catalysts [27]. An ultrahigh vacuum system with a base pressure in the 10^{-10} mbar range is required to satisfy the mean free path for the vaporized Si atoms to reach the substrate. To maintain such low pressures even during the nanowire growth, parts of the system are, therefore, often cooled with liquid nitrogen. For Au-catalyzed Si nanowires, Au is deposited onto the substrate. Annealing the substrate at temperatures above the Au—Si eutectic temperature causes the Au film to break up. Au mixes with Si to form Au—Si eutectic droplets, followed by the subsequent nanowire growth. Si atoms are only from the bulk Si. The metallic catalyst merely facilitates the crystallization of Si. Chemical reaction is not involved, therefore, the metal–Si droplets are not real catalysts from a chemical point of view. However, Si nanowires grow faster than the substrate by about a factor of 2, indicating that the droplets have an effect on Si crystallization. An example of the Si nanowhiskers grown by MBE is presented in Figure 17.9 [27]. In the experiment, ⟨111⟩-oriented 5 in. Si wafers are used as substrates. The experiments are initiated in the MBE chamber with a desorption of the oxide layer by thermal annealing at 850 °C. Afterward, a thin Au film is deposited at a substrate temperature of 525 °C with thicknesses between 1.5 and 2 nm. The size of the thus generated droplets could be chosen between 70 and 200 nm. During the subsequent whisker growth, the constant Si flux amounted to 0.5 Å/s. A set of samples were grown at 525 °C. A total growth time of 60 min (Figure 17.9a) and 120 min (Figure 17.9b and c) is chosen. Both experiments are characterized by a homogeneous distribution of ⟨111⟩-oriented whiskers of cylindrical morphology, partly hexagonal faceted. They have a diameter in the range of 100–160 nm, which is determined by the Au droplet sizes.

Si nanowires are typically grown at a substrate temperature of 500–700 °C. The evaporated Si diffuses on the substrate surface until it either crystallizes directly or finds a Au particle. Because the nanowire growth strongly relies on Si surface diffusion, the growth takes place at a temperature higher than 500 °C. However, at these temperatures, it is difficult to form small Au—Si droplet due to the effect of Ostwald ripening. Therefore, the diameter of the nanowires usually exceeds 40 nm. Gibbs–Thomson effect is another potential cause of the fact

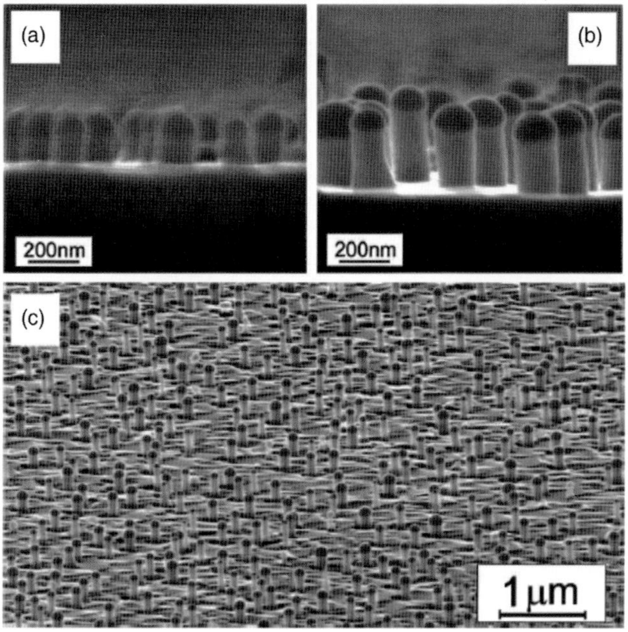

Figure 17.9 Si nanowires grown on a ⟨111⟩ Si substrate at 0.5 Å/s at 525 °C for (a) 60 min and (b) 120 min. (c) An overview of the Si whiskers in part (b). (From Ref. [27] with the permission of AIP Publishing.)

that only nanowires with diameters larger than about 40 nm can be obtained. The slow surface diffusion results in a comparably small growth rate, typically at 1–10 nm/min. The growth rate is inversely proportional to the nanowire diameter. Single-crystalline ⟨111⟩-oriented Si nanowires can be homoepitaxially grown on Si substrate at predefined positions. The incoming particle fluxes can be accurately controlled, which is particularly important for a doping of the Si nanowires. If additional evaporation sources, for example, Ge, exist, axial heterostructures could be formed by switching sources. Therefore, Si nanowires with well controlled doping profiles could be realized. The main disadvantages are the considerable Si film growth on the substrate and the limited flexibility concerning nanowire diameters and aspect ratios.

17.1.6.2 **Laser Ablation**

Si nanowire growth takes place in the vapor phase during laser ablation of the catalyst/Si disk. Lieber *et al.* pioneered this nanowire synthesis method, using a mixed Si/Fe target containing about 90% Si and 10% Fe [19]. They heated the tube furnace to 1200 °C, which is close to the minimum eutectic point to form Fe—Si droplet. Under such condition, they ablated Fe and Si from the laser disk by shooting at it with a pulsed, frequency-doubled Nd:YAG laser (wavelength 532 nm). The ablated material collides with inert gas molecules and condenses

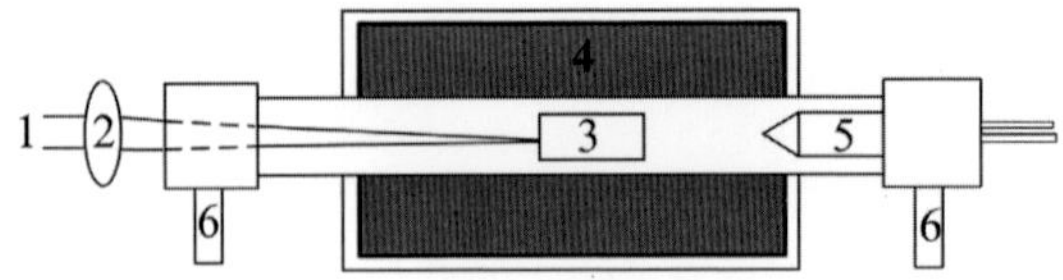

Figure 17.10 Schematic diagram of the nanowire growth apparatus (see also Ref. [19]). The output from a pulsed laser (1) is focused (2) onto a target (3) located within a quartz tube; the reaction temperature is controlled by a tube furnace (4). A cold finger (5) is used to collect the product as it is carried in the gas flow that is introduced (6, left) through a flow controller and exits (6, right) into a pumping system.

in the gas phase, resulting in Fe—Si nanodroplets, which act as seeds for Si nanowire growth (cf. Figure 17.10). Formation of Fe—Si eutectic droplet takes place at 1207 °C: although this temperature is 7 °C higher than that at the disk, laser ablation may have produced vapor Fe or Si atoms or nanosized particles, thus the size effect operated here [19]. As such, Si dissolves into the liquid eutectic droplet. Upon supersaturation, Si precipitates and supplies to the growth of the Si nanowire.

For laser ablation, sophisticated gas installations and substrate are not needed. The composition of the nanowires can be varied by changing the composition of the target. The high temperature allows the use of non-gold catalyst materials such as Fe and Ni. The size of the nanowires depends on both the type of metal catalysts and the gases that are streamed through the furnace, such as H_2, He, Ar, or N_2 [63]. However, it is not suitable for the in-place epitaxial growth of Si nanowires and the product yield is low.

In Sections 17.2 and 17.3, thermal evaporation and thermal annealing will be discussed, respectively.

17.1.7
Summary

Silicon nanowires could be synthesized by top-down or bottom-up methods. Top-down method carves bulk materials of desired composition into the desired nanoscale structure by lithographic techniques. To date, some technical problems, such as injection process, different etching behaviors, driving force, and dopant effect, are still unsolved. Bottom-up methods mimic nature's way of self-assembling atoms to form increasingly larger structures. Such techniques involve controlled crystallization of materials from vapor, liquid, or solid sources, typically yielding uniform and highly ordered nanometer-scale structures. Till now, VLS, VSS, and SFLS growth have been developed comprehensively. In the SLS growth, molecular beam epitaxy and laser ablation are studied systematically. However, some controversies or problems in thermal evaporation and thermal annealing are yet to be resolved.

17.2 Thermal Evaporation Growth of Silicon Nanowire

17.2.1 Introduction

Thermal evaporation of commercially available Si-containing powders, for example, Si, SiO_2, or SiO, is a cost-effective method to produce Si nanowires [64]. A two-zone tube furnace connected to an inert gas supply and the above-mentioned powders are the basic requirements for the synthesis of Si nanowires. The high-temperature tube has a temperature gradient from about 1350 to 900 °C. It is widely explained by the "oxide-assisted growth" theory, that is, Si nanoparticles precipitated through the disproportionation of SiO seed and grow into Si nanowires [64], and the Si oxide produced by the disproportionation reaction directly forms the sheath. As such, there is no need of metallic catalyst. The key experiments were (i) thermal evaporation of Si/Fe, Si disk, or Si/SiO_2 powder mixture in a quartz tube at 1200 °C under Ar atmosphere [28,29,65–67] and (ii) thermal evaporation of SiO powders in an alumina tube at 1200–1320 °C under Ar + H_2 atmosphere (cf. Table 17.1) [30,68–70]. Interestingly, the need for H_2 atmosphere comes whenever the alumina tube is used and not the quartz tube. The commercial quartz tube is, however, not made of 100% SiO_2, but contains certain amount of metallic elements, including Al and Cu [71–74]. Could it be possible that these metals actually acted as catalysts? Scrutiny of the nanowire sampling procedures in the papers related to the "oxide-assisted growth" revealed that the nanowires were never sampled at the tube wall. Li *et al.* examined the composition of the quartz tube through careful studies of microscopy, X-ray photoelectron spectroscopy, and X-ray diffraction [75]. In conjunction with the analysis of the vapor content, growth atmosphere, and grown nanowires, a metal-catalyzed solid–liquid–solid mechanism is developed to better explain the seeding and growth of Si nanowires during thermal evaporation of Si-containing powders (presented in detail in later sections).

17.2.2 Growth of Si Nanowires in Quartz Tube

17.2.2.1 Metallic Al and Cu in Quartz Tube

In 1996 [71], 2002 [72], 2003 [73], and 2004 [74], it was already reported that quartz tube contains metallic impurities, including Al and Cu. In 2014, Li *et al.* carefully examined a quartz tube (UFO Labglass, Singapore) to further confirm the existence of metallic impurities on the tube wall. Upon heating and dwelling, a lot of grain boundaries shown up on the tube wall (cf. Figure 17.11, dark arrows). This is the indication of the crystallization of the fused quartz.

The crystallization is further confirmed by the XRD peaks (cf. Figure 17.12), exhibiting the typical crystalline structure of the crystal quartz.

Table 17.1 Key experimental condition to grow Si nanowires via thermal evaporation of Si-containing powders in quartz tube and alumina tube [75].

Reference/ Year	High-temperature tube	Atmosphere/flow rate (sccm)/pressure ($\times 10^4$ Pa)	Furnace temperature (°C)	Growth temperature (°C)	Source materials (supplier, purity)	Process (duration)
[65]/1998	Quartz	Ar/50/6.67	1200	Unknown	Si/Fe disk made by pressing Si powder mixed with 0.5% iron	Laser ablation (2 h)
[66]/1999	Quartz	Ar/50/6.67	1200	Unknown	Si disk made by pressing the Si powder at 150 °C for 48 h under a hydraulic press	Laser ablation (2 h)
[28]/1998	Quartz	Ar/unknown/6.67	1200	930	Si powder mixed with approximately 70 wt% SiO_2 (Goodfellow, purity 99.99%)	Laser ablation or thermal evaporation (12 h)
[29]/1999	Quartz	Ar/unknown/6.67	1200	930	Si powder mixed with about 70 wt% SiO_2 (Goodfellow, 99.99%)	Thermal evaporation (12 h)
[67]/1999	Quartz	Ar/50/6.67	1200	920–950	Mixed disk of Si (purity: 99.998%, particle size: ~45 μm) and SiO_2 powder (purity: 99.995%, particle size: ~50 μm) with a hydraulic press	Laser ablation (unknown)
[68]/2000	Alumina	5%H_2 + Ar/50/5.33	1300	930	SiO powders (Goodfellow, 99.95%)	Thermal evaporation (7 h)
[69]/2001	Alumina	5%H_2 + Ar/50/5.33	1250	950	SiO powders (unknown)	Thermal evaporation (2 h)
[70]/2003	Alumina	5%H_2 + Ar/50/5.33	1320	Unknown	SiO powders (Aldrich, 325 mesh, 99.9%)	Thermal evaporation (7 h)
[30]/2003	Alumina	4%H_2 + Ar/100/5.33	1200	900	SiO powders (unknown)	Thermal evaporation (unknown)

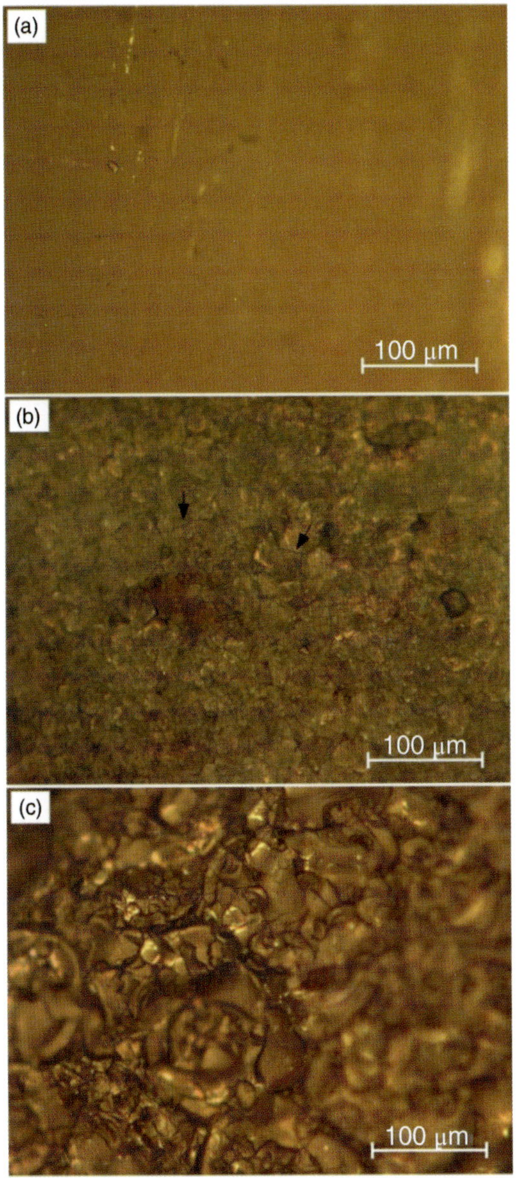

Figure 17.11 Surface morphology observed at (a) the unheated fused tube wall, (b) 3 cm away from the heat center, and (c) heat center [75].

The XPS core-level spectra detected from the outside tube wall reveals the dramatic increase of the atomic concentration of Zn, Al, and Cu (cf. Figures 17.13–17.15). It implies the diffusion of the metallic impurities from the quartz tube, although the spectra exhibit their oxide states, that is, 1022.4 eV for

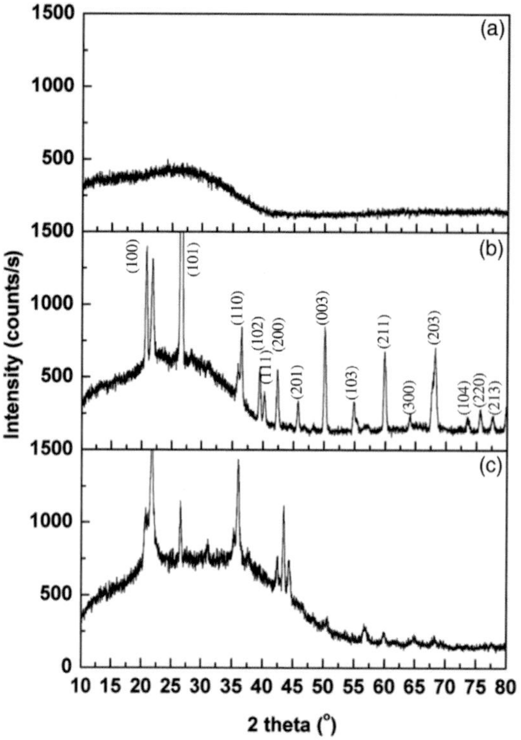

Figure 17.12 XRD curves detected at (a) the unheated fused tube wall, (b) 3 cm away from the heat center, and (c) heat center [75].

Zn $2p_{3/2}$ [76], 934.0 eV for Cu $2p_{3/2}$ [77], and 74.6 eV for Al $2p_{3/2}$ [78]. Understandably, Zn, Al, and Cu could be oxidized in air at high temperatures, but could still reserve the metallic state in the quartz tube under the Ar atmosphere.

The mass concentration or atomic concentration of the metal at the tube surface could be estimated by the XPS results. The amount of Cu could be calculated, given the mass of Si at the X-ray spot (detection area: $700 \times 300\,\mu m^2$, detection depth: 10 nm). If the density of SiO_2 is 2.30 g/cm^3, the amount of SiO_2 in this spot is about 4.83×10^{-9} g. Thus, the amount of Si is about 2.42×10^{-9} g. XPS shows that at the heat center, the mass concentration of Si is 26.39 wt%, that of Cu is 1.89 wt%, and that of Al is 3.59 wt%. In this area, therefore, the amount of Cu is about 1.73×10^{-10} g, and that of Al is about 3.29×10^{-10} g.

Figure 17.16a shows the backscattering electron image (BEI) of the ion-milled cross section of the quartz tube at the heat center, where small white nanoparticles are observed on the inner tube wall and the crack path. The particles are of diameters ranging from about 20 to 60 nm, matching with the nanowire diameter of 43 nm in Ref. [65]. In Figure 17.16b, the EDS spectrum is generated from the white nanoparticle that is pointed by a dark arrow in Figure 17.16a. It

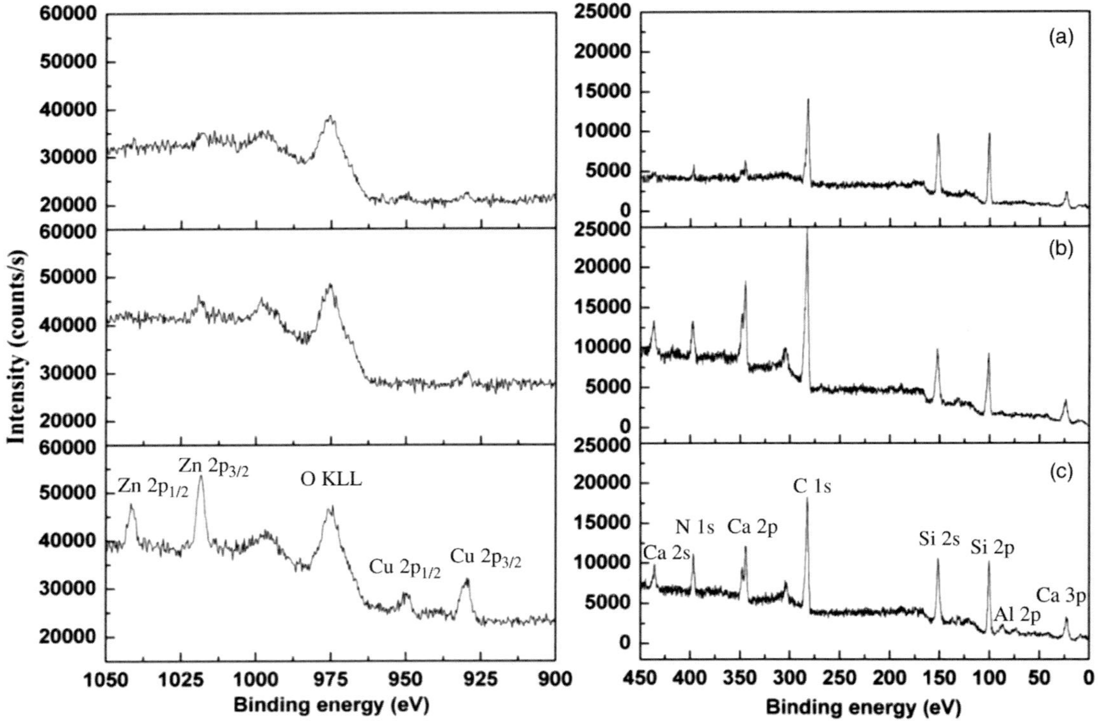

Figure 17.13 XPS survey scan spectra detected at (a) the unheated fused tube wall, (b) 3 cm away from the heat center, and (c) heat center [75].

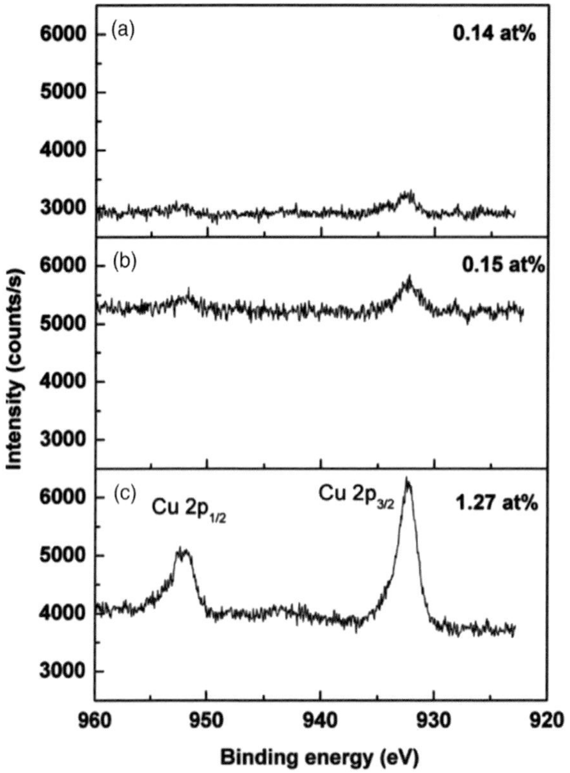

Figure 17.14 Cu 2p XPS core-level spectra detected at (a) the unheated fused tube wall, (b) 3 cm away from the heat center, and (c) heat center [75].

confirms that the white nanoparticle is Al, where C (54.3 at.%) comes from the atmosphere and the flowing gases, O (13.2 at.%) comes from the oxidation and the quartz tube, and Si (1.7 at.%) comes from the quartz tube. However, Cu is not detected by EDS in this cross section observation, possibly due to the small quantity, that is, 1.73×10^{-10} g on the area of 700 μm × 300 μm.

To further confirm the crystallization of the fused quartz occurring in the first heating, a brand new fused quartz tube is heated up to 1150 °C at 5 °C/min, held at this temperature for 45 min (cf. Figure 17.17a), and then cooled down to room temperature at 5 °C/min. Upon heating, the grain boundary does appear on the quartz tube wall, again confirming the crystallization of the fused quartz (cf. Figure 17.17b–e).

Three reasons explain the appearance of the grain boundary and the diffusion of the metallic impurities: (i) the crystallization of the amorphous fused quartz taking place in the first heating produces the grain boundaries (cf. Figure 17.11), (ii) the $\alpha \rightarrow \beta$ phase transition of the crystalline quartz occurs at 573 °C (in the succeeding heating) [79–82], resulting in the change of the quartz crystals in cell

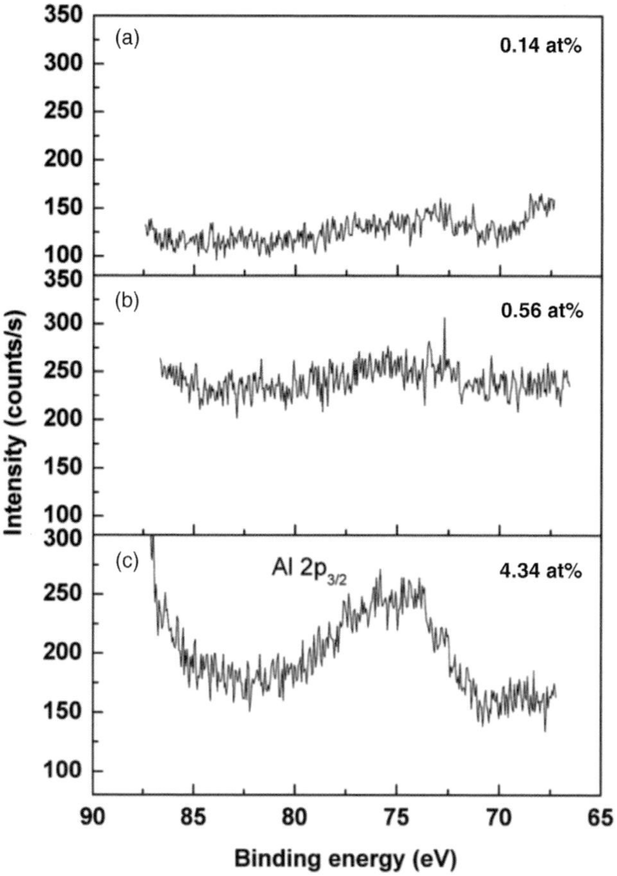

Figure 17.15 Al 2p core-level XPS spectra detected at (a) the unheated fused tube wall, (b) 3 cm away from the heat center, and (c) heat center [75].

dimension and volume and thus thickening the grain boundaries; (iii) the melting of the metallic impurities, for example, Zn (melting point, 420 °C) [83], Al (melting point, 660 °C), and Cu (melting point, 1083 °C). Though the impurity level may change from supplier to supplier, Al and Cu almost always exist in all the quartz tubes, and could effectively play the role of seeds to the growth of Si nanowires (cf. Table 17.2).

17.2.2.2 **Source of Si: Ejected Micrometer-Sized Si Particles**

Powders contain large quantities of absorbed or occluded gases. Upon heating, the spontaneous release of gases due to the volume expansion would cause violent ejection of the evaporant particles [84]. This is the reason why the baffled sources have been designed to inhibit direct line-of-sight transmission from the evaporant to the substrates in the technology of thin films (cf. Figure 17.18). During thermal evaporation of the Si–SiO_2 mixed powders in quartz tube at

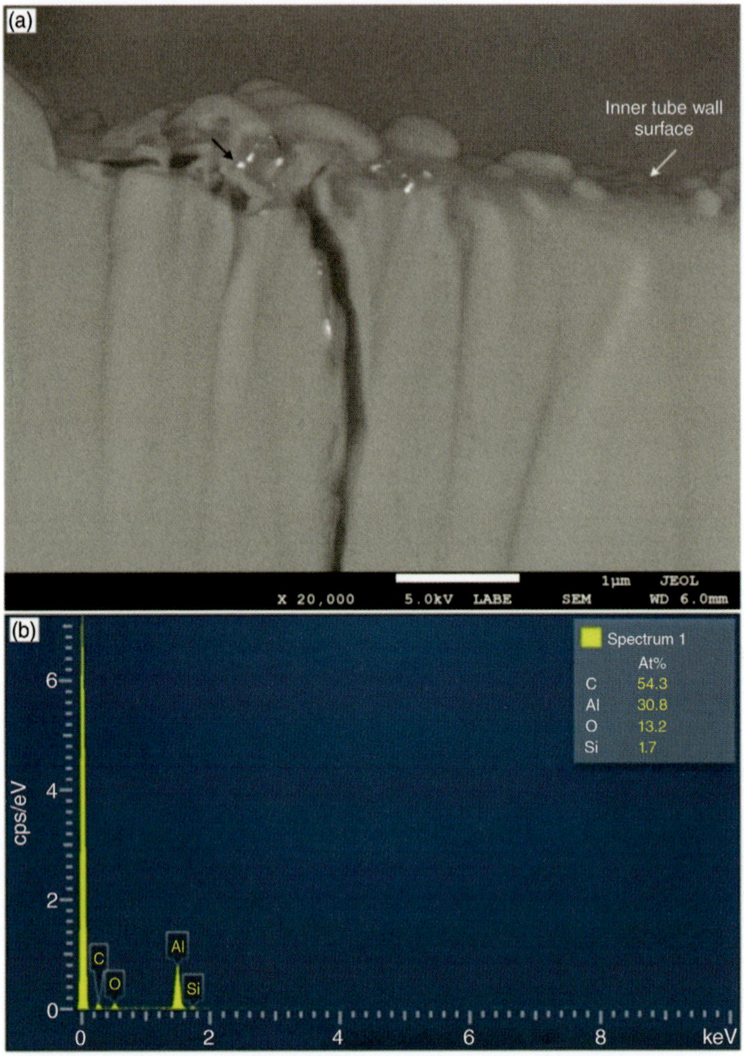

Figure 17.16 (a) Cross section of the quartz tube that is broken manually at the heat center, where white nanoparticles with diameters ranging from 20 to 60 nm are observed on the inner wall and the crack path. (b) EDS spectrum generated from the white nanoparticle that is pointed by an arrow in part (a) [75].

1200 °C [83], Si atoms and SiO_2 molecules are sublimated from the particle surface; meanwhile, the microparticles are also ejected into the atmosphere due to volume expansion of the residual gases.

It is known that commercial Si and SiO_2 powder particles are in the micrometer scale, for example, around 45 μm in diameter [67], at least five orders of magnitude larger than the diameter of Si atoms and Ar or SiO_2 molecules

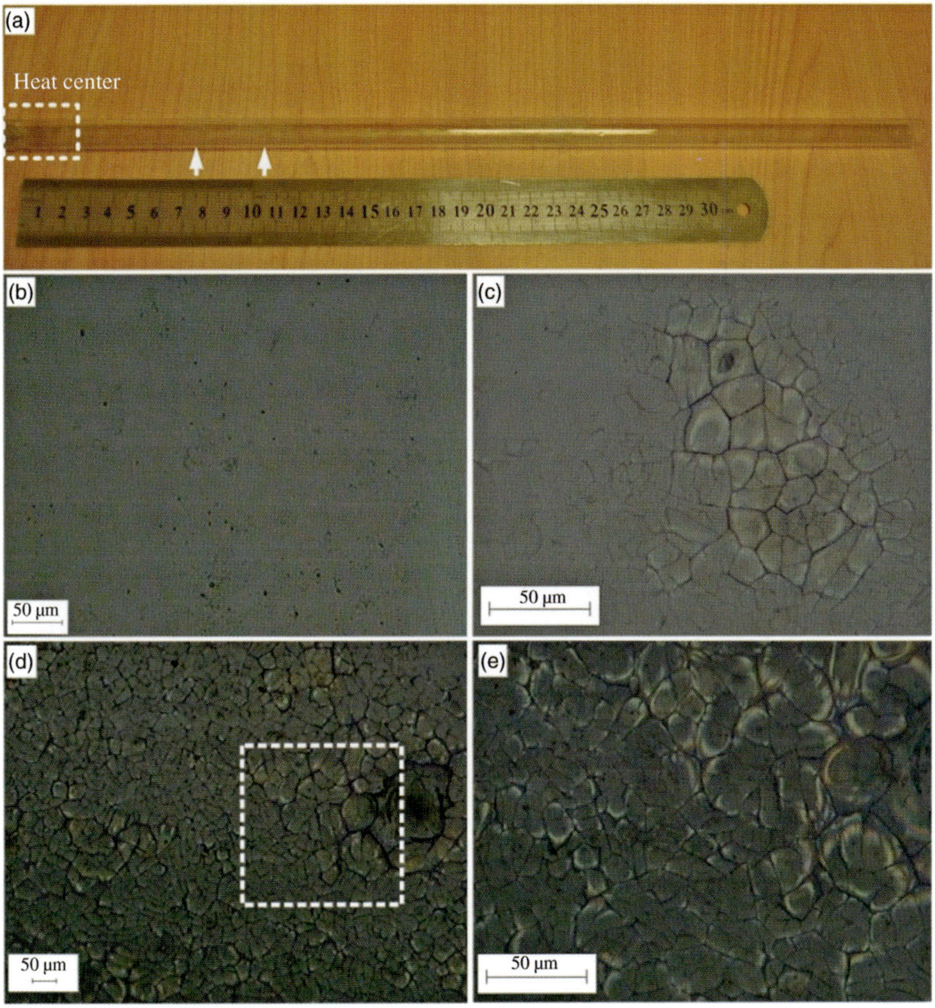

Figure 17.17 (a) The fused quartz tube after heating up to 1150 °C and holding for 45 min, (b) the unheated fused quartz tube wall, (c) 7.5 cm away from the heat center, (d) 10.5 cm away from the heat center, and (e) the magnified area in part (d). The rectangle in part (a) does not indicate the deformed area after heating, but the constriction designed by the supplier for holding the samples in the vertical tube furnace. The white arrows point at the observation locations of parts (c) and (d) [75].

(cf. Table 17.3, around 0.30 nm [85]). The mass of a Si atom (i.e., 4.67×10^{-23} g) is 0.7 times heavier than an Ar molecule, and the mass of a SiO_2 molecule (i.e., 9.98×10^{-23} g) is 1.5 times heavier than an Ar molecule, while the mass of a Si microparticle is around 1.12×10^{-7} g, 1.69×10^{15} times heavier than an Ar molecule. Si vapor pressure of around 111.57 Pa could be produced during thermal evaporation of bulk Si at 1200 °C [83], under 2.67×10^4–6.67×10^4 Pa

Table 17.2 Chemical composition and/or metallic impurities of quartz and alumina tubes from common suppliers [75].

Materials (supplier)	Purity (%) or grade	Chemical composition (percentage by weight) or trace element content (parts per million by weight)
Fused quartz tube (Quartz Scientific)	99.995	**Al (16)**, **Cu (<0.1)**, As (<0.4), B (<0.1), Ca (0.6), Cd (<0.01), Cr (0.05), Fe (0.3), K (0.7), Li (1), Mg (0.1), Mn (0.1), Na (1), Ni (<0.1), P (1.5), Sb (<0.4), Ti (1.1), Zr (1.5), OH (<5)
Fused quartz tube (Heraeus)	HQS grade	**Al (15)**, **Cu (<0.05)**, Ca (0.5), Cr (<0.05), Ca (0.6), Fe (0.1), K (0.4), Li (0.6), Mg (0.05), Mn (<0.05), Na (0.3),Ti (1.1), Zr (0.7), OH (<30)
Fused silica glass (Tosoh)	N grade	**Al (8)**, **Cu (<0.01)**, Ca (0.6), Fe (0.2), Na (0.6), K (0.1), Li (<0.07), Mg (0.4), OH (200)
Fused quartz tube (UFO Labglass)	Corning 7980 ?R?*R*	**Al (40), Cu (<13),** Ni (<7), Zn (<30), Ti (<40), Zr (<30), U (<1), Mg (<25), K (<21), Cr (<1), Fe (<15), Mo (<5), Li (<1), Ca(<20), B (<100), As (<5), (Na < 150), V (<10)
High-purity alumina tube (Vesuvius/McDanel)	99.8	**Al_2O_3 (99.82%),** Fe_2O_3 (0.025%), SiO_2 (0.06%), CaO (0.04%), MgO (0.035%), Na_2O (0.008%), TiO_2 (0.004%), Cr_2O_3 (0.003%), K_2O (0.001%), B_2O_3 (0.001%)
High-purity alumina tube (Graphtek LLC)	>99.6	**Al_2O_3 (>99.6%),** Fe_2O_3 (<0.05%), SiO_2 (<0.1%), Na_2O (<0.1%), K (<0.1%), L (<0.1%)

(cf. Table 17.1) [84,86,87]; however, the mean free path of these vaporized atoms or molecules is lower than 3.11 μm, that is, a galaxy away from the location where the nanowire grew (i.e., 15 cm downstream [65,67]).

To reach the growth location, the vaporized Si atoms would collide at least 4.8×10^4 times with the Ar molecules. These collisions would scatter the sublimated atoms or molecules, and attenuate or annihilate the momentum. As such, the vaporized atoms or molecules could not reach the growth location to supply the growth of Si nanowires. On the other hand, the collisions have negligible impact on the ejected microparticles due to the dramatic difference in diameter (5 orders of magnitude) and mass (15 orders of magnitude) with the Ar molecules. The collisions could not affect the kinetic energy of the ejected microparticles; thus, the micrometer-sized Si particles could reach the growth location to supply Si for the growth of Si nanowires.

17.2.2.3 Alternative Growth Mechanism

Al- or Cu-Catalyzed Growth

In the ejected micrometer-sized Si particles, the Si atoms are still confined in the particle, thus not free to move around. Growth of Si nanowires needs freely

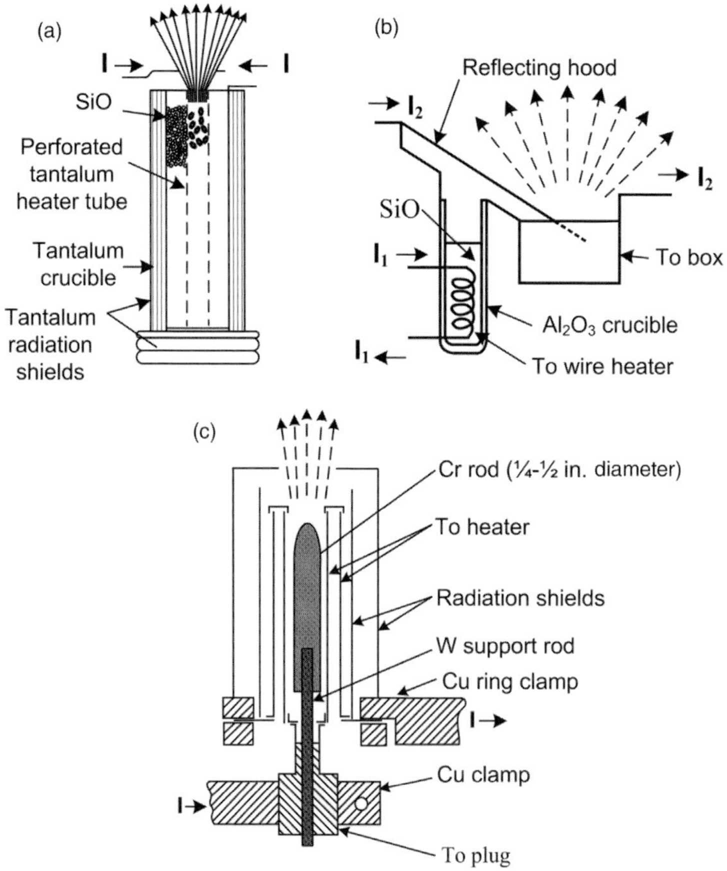

Figure 17.18 Thermal evaporation source [84]. (a) The Drumheller source. (b) Compartmentalized source. Powder source materials, such as SiO, would release large quantities of residual gases upon heating, thus causing the ejection of evaporant particles. In part (a), a ring-shaped lid prevents the ejection of the particles, and in part (b), a reflecting hood is placed on the top of the powder source to stop the ejected particles. (c) Chromium sublimation source, the electric current flows through the tantalum cylinder (heavy lines). Since the source is compact and well shielded, sublimation occurs from the entire surface area of the rod, which is uniformly heated by radiation. Therefore, the ejection of particles does not occur.

mobile Si atoms in vapor [20], or melt [19], or solution [24]. As the crystalline size reduces to certain degree, its melting temperature drastically reduces. This is called "size effect." But the Si particle has to be less than 20 nm in diameter before it can have any size effect on melting point [88,89]. And, to melt below 900 °C, Si particle must be less than 10 nm. The diameter of the as-grown Si nanowires could reach 43 nm [65]. So "size effect" does not work here, because commercial Si and SiO_2 powders are reported in micrometer scale [67], that is,

Table 17.3 Physical constants of the microparticles, atoms, and molecules [75].

Substance	Diameter (nm)	Mass (g)	Density (g/cm^3)	Mean free path (μm, at 1200 °C, under 2.67×10^4–6.67×10^4 Pa)	Reference
Si microparticle	4.4–4.5×10^4	1.11–1.12×10^{-7}	2.32–2.34	Nil	Aldrich, Goodfellow, [83]
SiO microparticle	4.4–4.5×10^4	0.95 –1.02×10^{-7}	2.13	Nil	Aldrich, Goodfellow, [83]
SiO_2 microparticle	4.5×10^4	1.04–1.11×10^{-7}	2.18–2.32	Nil	Goodfellow, [83]
Si atom	0.235	4.67×10^{-23}	Nil	1.24–3.11	[85]
SiO molecule	~0.30	7.32×10^{-23}	Nil	0.76–1.91	
SiO_2 molecule	~0.30	9.98×10^{-23}			
Ar molecule	0.394	6.64×10^{-23}	1.4 (87 K)	0.44–1.12	
H_2 molecule	0.240	3.35×10^{-24}	0.07 (20 K)	1.19–2.98	

45 μm, three orders of magnitude larger than the diameter of the nanowires. With metallic impurities of Al or Cu, however, melting point of Si can drastically decrease: Al—Si eutectic droplet forms at only 577 °C [90]; Cu—Si eutectic droplet forms at 802 °C. As examined in the previous section, metallic Al and Cu already exist as impurity in the quartz tube, at the growth temperature (900–950 °C), therefore, the ejected micrometer-sized Si particles could melt with the metallic Al or Cu on the quartz tube wall to form liquid Al—Si or Cu—Si eutectic. Different quartz tube contains different concentration of metal, resulting in different size of the droplet and the corresponding nanowire diameter, such as 3–43 nm in the work of Lee and coworkers [65]. The amount of the metal catalysts could be estimated by XPS calculation. It is further suggested that the amount of the metal catalysts on the surface is determined by the composition of the quartz tube, the annealing time, and the annealing temperature. Based on the nanowires grown in Ref. [65], "The sample consists almost entirely of nanowires with diameters ranging from 3 to 43 nm and length up to a few hundreds of mm." The amount of the metal catalyst should not determine whether the nanowires can grow, but control the size of the nanowires. During laser ablation of Si/Fe disk (cf. Figure 17.21a), vapor-phase Fe or Si atoms or nanosized particles could be produced. In this case, size effect works here, thus the formation of Fe—Si eutectic droplet at 1200 °C (7 °C lower than the eutectic point) is possible [19]. Once the liquid eutectic droplet forms, Si in the vapor phase or in the ejected microparticles dissolves easily into the droplet. Upon supersaturation, Si precipitates and supplies to the growth of the Si nanowire (cf. step 4 in Figure 17.21). Under the present growth condition, however, the formation of the droplet taking place on the inner wall of the quartz tube is almost technically impossible to directly observe by transmission electron microscope since the quartz tube is embedded in the furnace. Although it would be interesting to cut the tube after growth and then the actual growth would be observed by doing TEM on a piece of the quartz tube, the actual nanowire growth is difficult to carry out due to the harsh experiment conditions. However, Li and coworkers observed the ion-milled cross section of the quartz tube at the heating center area. As shown in Figure 17.16, lots of Al nanoparticles are observed on the inner quartz tube wall. They could effectively seed the growth of Si nanowires.

Seeding Location: The Wall of the Quartz Tube

As the tube wall is at 900–950 °C, a water-cooled "copper finger" points in the center of the tube, and the Ar gas flows in the tube, the tube center must be colder than the wall (thus creating a thermal gradient, the *driving force*); therefore, the eutectic droplet stays at the wall and the wire grows away from the wall toward the center of the tube (thus the droplet would always be found at the "root" of the wire instead of "atop"). This is confirmed in Ref. [65], where the nanowires were "formed on the inner wall of the quartz tube at the position surrounding the tip of the copper cold finger." As such, no metallic trace should be found if the "root" at the wall is not examined (cf. step 4 in Figure 17.21). In the meantime, as the nanowire freezes out from the liquid "foundation," it could not stand and thus lies either on the surface of the tube wall or on the Mo gird that is placed on the tube wall. The nanowires were either "taken off directly from the produced spongelike Web" [65], or sampled through placing the Mo grid "in the region of the quartz tube where nanowires grew" [28,29]; therefore, the observed particles or nanowires standing alone on the grid were not the Si nucleation sites (cf. Figure 1d in Ref. [29]), but the broken end sections of the nanosphere chains or smooth Si nanowires catalyzed at the beneath tube surface since the thickness of the Mo grid is only around 30 μm, but the length of the nanowires was reported to be "a few hundreds of μm" [65]. At the evaporation temperature (>1200 °C), liquid Al or Cu droplets are diffused onto the inner surface of the tube, including the upper side or the ceiling area of the tube. To overcome the sticking coefficient and surface tension of the droplet, however, the droplet on the ceiling is difficult to drop on the top side of the Mo grid. However, the bottom side of the Mo grid is still in contact with the quartz tube. Therefore, liquid metal droplets that are diffused from the quartz tube could still stick on the Mo grid, serving as the seeds for the Si nanowires. This explains why some nanowires look like they have not broken off (cf. Figure 1b and c in Ref. [29]). The layer warping around the entire wire is suggested to be SiO_x that is precipitated from the metal–Si–O droplet, rather than the sublimated SiO. The seeding of the liquid metal–Si eutectic droplet, that is, liquid foundation, at the tube wall explains why the Si nanowires are lying on the tube wall [28,29,65,66], rather than perpendicularly standing on the tube wall. As such, the Si nanowires could be in any growth orientations. It is different from the ones grown through vapor–liquid–solid mechanism on Si wafers with a specific crystal orientation.

Formation of the Si Oxide Sheath

The Si nanowires thus formed are usually wrapped within a layer of Si oxide sheath. The Si oxide sheath is formed via the precipitation of the supersaturated Si oxide from the oxidized metal–Si droplet, rather than directly formed by the SiO_2 molecules that are produced through the disproportionation of SiO. The metal–Si droplet is surrounded by residual oxygen and Ar gas in the tube. Thus, the surface of the droplet is easily oxidized. Supersaturation of Si in the liquid metal–Si droplet is known to be the key for the precipitation of a Si nanowire. Therefore, the precipitation of SiO_x from the liquid metal–Si–O droplet into the solid must satisfy supersaturation. In conjunction with the coaxial Si/SiO_x nanowire, it is reasonable to speculate that the metal–Si and metal–Si–O phases are coaxially distributed in the metal–Si–O droplet. It is noted that even in the presence of metallic impurities, Ref. [28] failed to produce any Si nanowires with pure SiO_2 powders in quartz tube at 930 °C, because here the SiO_2 powders did not contain any elemental Si, and 930 °C could not melt SiO_2 (melting point: 1703 °C [83]). It is not difficult to infer that the vaporized SiO_2 microparticle in Refs [28,29,67] (cf. Figure 17.21b) did not melt or participate in the growth of the Si nanowire. In the meantime, this also explains why the yield of the Si nanowires did not show any dependence on the fraction of Si oxides in the Si–SiO_2 mixed powders [28] (e.g., the nanowire yield was lower than 0.01 mg at 0 wt%

SiO_2, but it became zero when 100 wt% SiO_2 powder was used); however, it was dependent on the fraction of Si in the Si–SiO_2 powder mixture. It is also reported that the nanowire yield reached an optimum point at 50% Si/SiO_2 to support their oxide-assisted growth theory. However, the nanowires were only collected in the following way, "a Mo grid was placed in the region of the quartz tube where the nanowires grew." Therefore, the amount of the grown nanowires may not be accurate.

Trace of the Metal Impurities

Earlier in 2000, Ni-catalyzed Si nanowires were grown through thermal annealing of Ni-coated Si wafers [31]. In the recent Ni-catalyzed growth of Si nanowires via thermal annealing of a Ni-coated Si wafer beneath sputtered amorphous carbon [32], the Ni—Si—O nanoclusters were found seeding at the root of the grown Si nanowires (cf. Figure 17.19) [91]. These facts confirm the

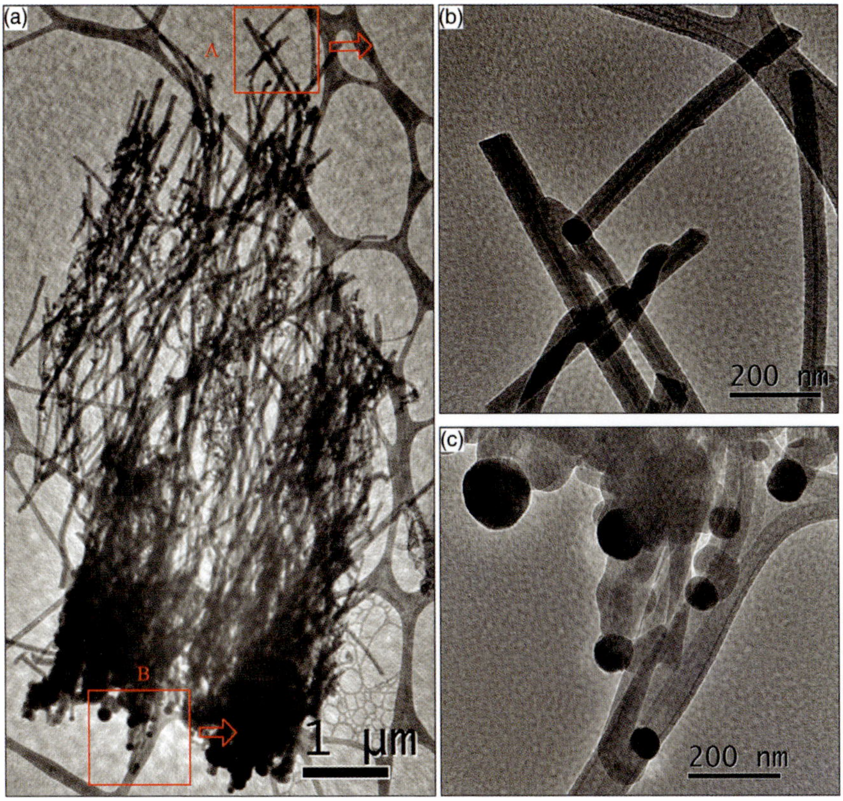

Figure 17.19 Morphology of the Ni-catalyzed Si nanowires grown through thermal annealing of the Ni-coated Si wafer [91]. (a) A bundle of Si nanowires, where the tip and root are respectively highlighted by the red squares A and B. (b) The enlarged tips of the nanowires without dark Ni nanoclusters. (c) The enlarged roots of the nanowires with the dark Ni nanoclusters.

possibility of seeding and growing Si nanowires at the metal-covered substrates (i.e., the quartz tube wall). A solid trace of the metallic impurity is readily found as the dark eggs embedded in the nanowire, that is, frog-egg-shaped Si nanowire (cf. Figure 3b in Ref. [66], Figure 17.20a) that was grown in quartz tube, although the dark "eggs" were declared to be the sites of Si nuclei (precipitated through the disproportionation of SiO) embedded in Si nanowires in Ref. [66]. In that case, there should be no contrast between the "eggs" and the nanowire since they are Si.

In Ref. [29], the c-Si particles did exist in the nanowire. But the nanowires were of a sinusoidal profile (cf. Figure 1 in Ref. [29]), different from the smooth one in Ref. [66]. It is known that the sinusoidal profile resulted from the vibration of the liquid metal–Si droplet during precipitation of Si. It is possible that the dark "eggs" are c-Si, where the a-SiO_2 (light region) has formed the disconnection in the nanowire growth since the composition was never characterized in Ref. [66]. However, it is more likely that these "eggs" are metallic Al or Cu because similar dark Au nanospheres (in or not in contact with the Si core) are observed entrapped in the smooth Si nanowires through the similar metal-catalyzed growth process: thermal annealing of a Au/a-Si/Si wafer at 1030 °C [92] (cf. Figure 1a and b in Ref. [92], Figure 17.20b and c), thermal annealing of Au/SiN_y/Si wafer sandwiched with a Au-coated Si wafer at 530 °C [93] (cf. Figure 2e and f in Ref. [93], Figure 17.20d and e), or thermal evaporation of SiO powders at 1050 °C on a Au-coated Si wafer in a quartz tube [94] (cf. Figures 2 and 4 in Ref. [94], and Figure 17.20f–j). Till now, however, why the catalyst gets "sucked" in the nanowire is still controversial. Some believe that the nanospheres form by neck breaking of two liquid Au—Si droplets supersaturated with Au [92], some think liquid Au flows into the silica nanotube due to capillary force [93], and other view it as a result of forced deformation of the Au—Si droplet due to faster growth of the Si oxide sheath [94]. The phase diagram of Al—Si and Cu—Si reveals that Al and Si are immiscible below 577 °C, while Cu and Si are immiscible below 802 °C. In this work, therefore, it is believed that the entrapment should result from the eutectic precipitation of metal and Si in the metal–Si–O droplet that is still supersaturated with Si during the cooling process. As such, the nanowire embedded with nanoparticles still has a smooth profile.

17.2.3 Growth of Si Nanowires in Alumina Tube

17.2.3.1 Reduction of Alumina Tube by H_2

Metallic Al could be reduced from alumina tube (cf. Table 17.2) when H_2 was introduced at 1200–1320 °C [30,68–70]. Thermodynamic data show that liquid or gaseous aluminum could be formed at 1600 K (i.e.,1327 °C) or above by direct reaction between molecular hydrogen and alumina [95]:

$$Al_2O_3(s) + 3H_2(g) \leftrightarrow 2Al(l, g) + 3H_2O(g) \qquad (17.1)$$

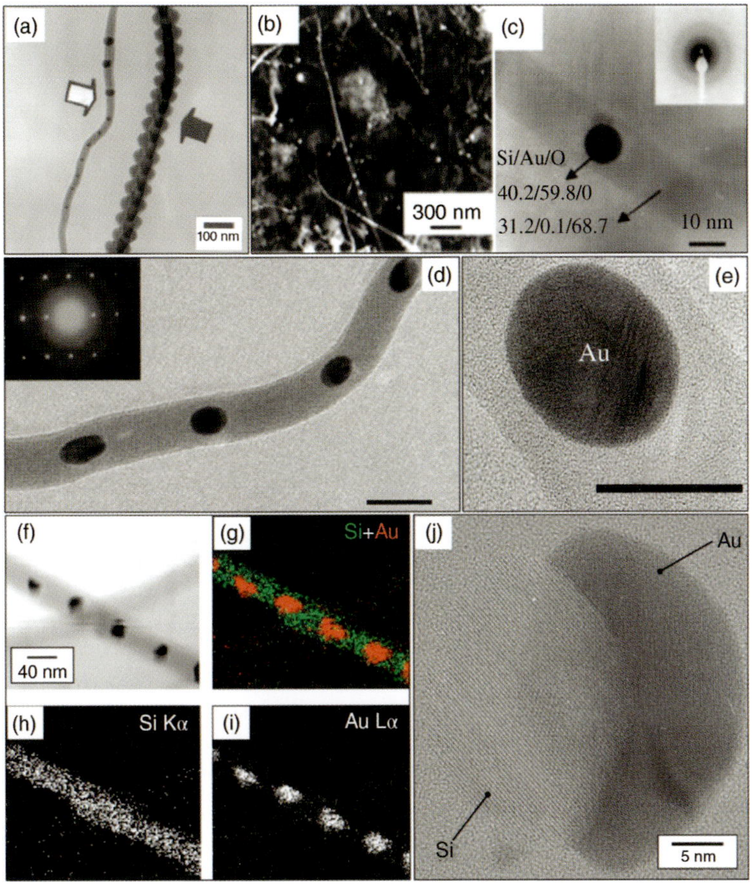

Figure 17.20 Nanosphere-entrapped Si nanowires grown through laser ablation of Si disk in a quartz tube under Ar atmosphere [66], or thermal annealing of Au/amorphous Si bilayers in a nitrogen ambient atmosphere [92], or heating the Au/SiN_y/Si wafer sandwiched with a Au-coated Si wafer in a microwave chamber [93], or thermal evaporation of Si monoxide powders on Au-coated Si wafer [94]. (a) Frog-egg-shaped and fishbone-shaped Si nanowires, where the eggs marked by the hollow arrow actually were the metallic Cu or Al nanoparticles, while the fishbone marked by the solid arrow was the wavy crystalline Si core. The wavy structure resulted from the periodic vibration of the metal–Si droplet seeded at the quartz tube wall [66]. (b) Morphology of the Au_2Si nanopea-podded Si oxide nanowires [92]. (c) Nanostructure and analytical electron microscopic analysis of the nanowire and the peapodded nanosphere [92]. The inset shows the amorphous patterns corresponding to the nanowire. (d) TEM images of the gold nanopeapodded silica nanowire [93]. The inset shows an electron-diffraction pattern recorded along the [102] zone axis of Au. (e) HRTEM image of the gold core in a silica nanowire. (f) The periodic chains of Au/Si nanoparticles entrapped in the SiO_2 nanowire [94]. (g) The colored and overlayed element mappings (silicon: green; gold: red). Energy-dispersive X-ray spectroscopy element mappings of (h) the Si Kα line and (i) the Au Lα line. (j) High-resolution TEM image of a nanoparticle. The two regions of the particle were identified as crystalline Si and Au.

Laboratory experiments demonstrate that hydrogen dissolved in molten aluminum could possibly reduce alumina to aluminum at temperature as low as 700 °C [95]:

$$Al_2O_3(s) + 6AlH(g) \leftrightarrow 8Al(l) + 3H_2O(g) \quad (17.2)$$

17.2.3.2 **Si in the Ejected Micrometer-Sized SiO Particles**

In 1996 [96], 2003 [97,98], and 2004 [99], it was reported that the commercially available SiO powders (cf. Table 17.3, around 44 μm in diameter) were not 100% SiO, but an inhomogeneous mixture of amorphous SiO_2 and amorphous Si with SiO phase at the interface (cf. Figure 17.21c, "starting SiO particle"). Upon thermal evaporation of the SiO powders, the micrometer-sized SiO particles could eject in space due to the expansion of the residual gas in the powder [84], and then land on the inner tube wall (cf. step 2 in Figure 17.21c). Disproportionation of SiO,

$$2SiO \rightarrow Si + SiO_2 \quad (17.3)$$

takes place at around 900 °C through the migration of electrons [97]. However, no physical movement of Si atoms or SiO_2 molecules occurs in this redox reaction. After the reaction, therefore, the particle becomes a microscopic mixture of Si and SiO_2 clusters. These clusters are not separable, but still confined in the particle, not able to move freely and separately in space (cf. Figure 17.21c, "ejected SiO particle"). Similarly, thermal annealing of the Si-rich SiO_2 thin film produces crystal Si particles embedded in amorphous SiO_x matrix [100]. The morphology of the thin film may change a little due to the rearrangement of the crystal and amorphous phases in the film. However, the produced Si particles are still confined in the 2D thin film, not able to move freely in space.

17.2.3.3 **Alternative Growth Mechanism: Al-Catalyzed Growth**

As metallic Al is reduced from alumina tube, however minute it could be, nano-sized Al—Si droplet could form at 577 °C with the Si in the SiO particle (either existed in the starting powder or produced from the disproportionation reaction but still confined in the SiO particle). As the tube wall was over 900 °C, the liquid Al—Si eutectic droplet could form and seed the growth of Si nanowires. As the tube temperature was over 900 °C, about 323 °C higher than the eutectic temperature of Al—Si, the seed should be in the liquid state during the growth [101]. However, temperature is supplied for the formation of the liquid droplet seed. The nanowire growth is driven by the supersaturation of Si in the droplet and the overcooling from the surrounding gases [31]. Since the SiO_2 portion is also confined in the microparticle (cf. Figure 17.21c) and could not melt into the droplet to participate in the growth of the Si nanowire, the Si oxide sheath is formed through the precipitation of the supersaturated Si oxide from the oxidized metal–Si droplet.

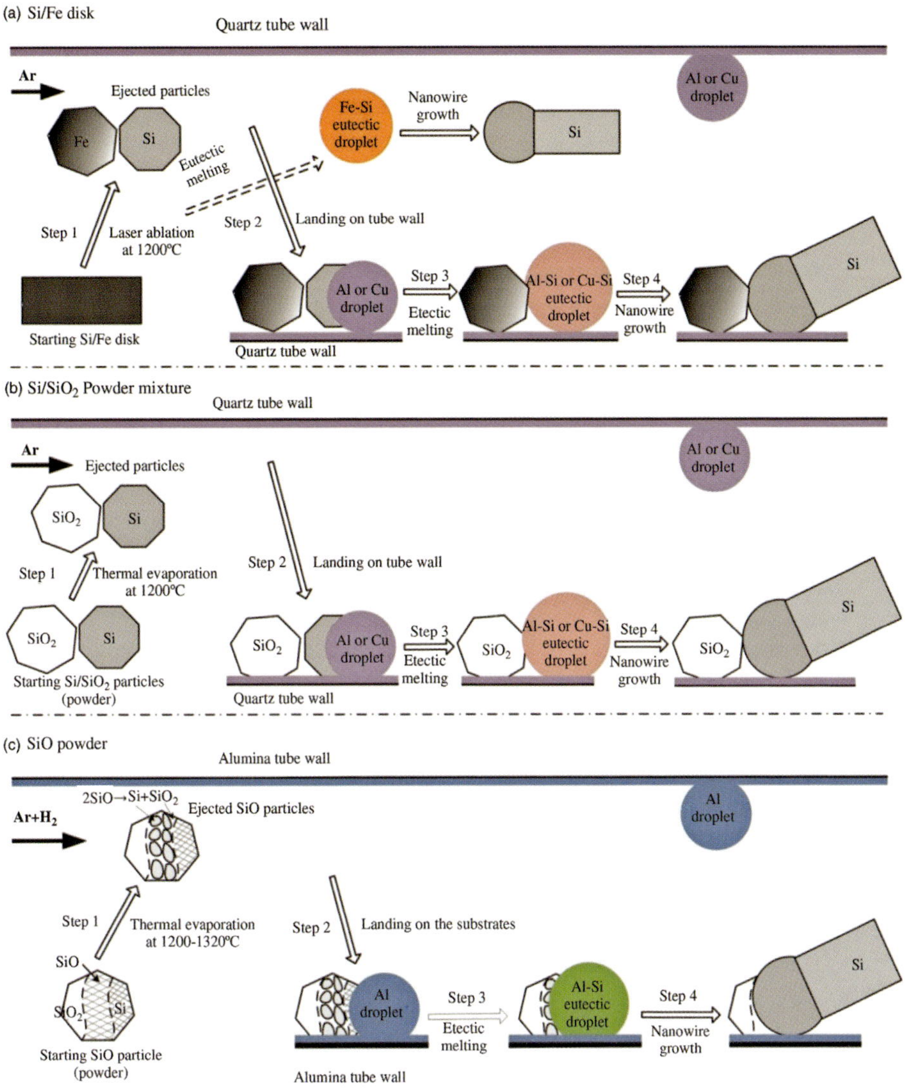

Figure 17.21 Schematic diagrams of the Al-, Cu-, and Fe-catalyzed growth of Si nanowires during (a) laser ablation of Si/Fe disk in quartz tube at 1200 °C under Ar atmosphere, (b) thermal evaporation of Si/SiO_2 powder mixture in quartz tube at 1200 °C under Ar atmosphere, and (c) thermal evaporation of SiO powders in alumina tube at 1200–1320 °C under $Ar + H_2$ atmosphere. To simplify the mechanism, the slight deformation of the ejected microparticle, the vaporized atoms or molecules, the size difference between the microparticles and the droplets, and the oxide sheath in the Si nanowires are ignored [75].

17.2.4
Conclusion

A metal-catalyzed solid–liquid–solid growth mechanism explains the seeding and growth of Si nanowires via thermal evaporation of Si-containing powders in quartz tube. The ejected Si microparticle (the *solid*) gets in touch with metallic Al or Cu that is inherited in the tube to form Al—Si or Cu—Si eutectic droplet (the *liquid*) on the tube wall. Si thus dissolves into the droplet. As it is supersaturated with Si, Si precipitates toward the cold tube center and adds into the Si nanowire (thus *solid* again). The oxide sheath is formed through the precipitation of the supersaturated Si oxide from the oxidized metal–Si droplet. Si nanowires grown via thermal evaporation of SiO powders in alumina tube should also follow the solid–liquid–solid mechanism, where the metallic Al comes from the reduction of alumina tube by H_2.

17.3
Growth of Silicon Nanowires through Thermal Annealing

17.3.1
Introduction

In 2000, Yu and coworkers reported the growth of Si nanowires via thermal annealing of Ni-covered Si wafers [31]. Thereafter, Au [102,103], Pt/Au [104], NiO [105], and Fe [106] coated Si wafers were used to grow amorphous Si, SiO_x, or crystalline Si/amorphous SiO_x nanowires, that is, c-Si/a-SiO_x. For instance, amorphous SiO_x nanowires can be grown directly by heating a NiO-catalyzed Si substrate without using WO_3/graphite powder in a reductive environment in a tube furnace [105], or by thermal heating Au or Ni thin film-coated Si substrates [107]. Park and Yong synthesized c-Si/a-SiO_x nanowires by heating a NiO-catalyzed Si substrate under a reductive environment of WO_3, C, and Ar atmosphere [105]. A growth mechanism based on chemical reaction was proposed: The CO gas from the carbothermal reaction of WO_3/C reacts with SiO_x to produce SiO and CO_2, where SiO_2 is formed from Si—Ni—O liquid droplets in the NiO–Si interface. Si and SiO_x are generated through the disproportionation of SiO in the film [105]. The c-Si/a-SiO_x nanowire is grown via the phase separation of Si and SiO_x from the supersaturated Si—Ni—O droplets by the cooling effects of the Ar flow. Considering the oxidation of the droplet at the beginning stage of the growth, Nie *et al.* developed a growth mechanism catalyzed by the temperature-dependent alloy, metallic and ionic Fe (Fe—FeO_x) to describe the growth of Si—SiO_x core–shell nanowires during heating the Fe-coated Si wafers at three different temperatures. In the mechanism, the formation of the 1D amorphous SiO_x nanowires is grown from the fully oxidized Fe–Si alloying droplets as a result of the continuously feeding Si into the alloying droplets, whereas partly oxidized large ones form Fe–Si alloying cores and

Fe—O—Si shells, which result in Si—SiO_x core–shell nanowires. However, most people overlooked the oxidation taking place at high temperatures. Besides, if the disproportionation reaction took place, the Si crystal would be embedded in the amorphous SiO_2 matrix in the form of 2D film [108].

In comparison with the conventional furnace, rapid thermal annealing offers precise control of the annealing temperature and duration, annealing profile, ambient purity, cycle time, and process flexibility. The fundamental flexibility in creating different types of thermal processes arises from the dynamic control of the source temperature, which permits fast heating combined with dynamic optimization of temperature uniformity [109,110]. Owing to the high cost, the annealing chamber is normally under a gas atmosphere rather than a vacuum environment. To grow crystalline Si nanowires, therefore, Li and coworkers sputtered a layer of carbon on the Ni catalyst that is deposited on the Si wafer substrate to retard the oxidation during rapid thermal annealing [32,111].

17.3.2
Growth of the Si Nanowires under a-C

17.3.2.1 Experimental Details

In the experiments of Li and coworkers [32], the Ni catalyst layer is sputtered on the N-type Si (100) wafers in an E303A magnetron sputtering system (Penta-Vacuum, Singapore). Amorphous-C films of around 130 and 570 nm in thickness are respectively sputtered for 15 and 60 min. As a comparison study, a layer of Ni film is deposited from the Ni target for 5 min. And then, the as-deposited a-C/Ni films underwent rapid thermal annealing (RTA, Jipelec Jetfirst 100 rapid thermal processor, France) in an Ar ambient at 1100 °C for 1 and 3 min, respectively.

17.3.2.2 Structural Evolution of the a-C/Ni Film

Figure 17.22 shows the transformation of carbon sp^2 bonding structure in the 570 nm film. The Raman spectra are deconvoluted into D-band (at ~1350 cm^{-1} A_{1g} mode) and G-band (at ~1580 cm^{-1}, E_{2g} mode) using a mixture of Lorentzian and Gaussian functions [112,113]. D peak is the breathing mode of those sp^2 sites only in rings [112]. Its intensity is directly dependent on the presence of the sixfold rings. G peak is attributed to the in-plane stretching vibration of any pair of sp^2 bonded carbon atoms, either in C=C chains or in aromatic rings [112]. I_D/I_G increases as the number of rings per cluster increases and the fraction of chain groups falls [114].

Upon annealing, G peak is upshifted from 1533 to 1588 cm^{-1}. The corresponding I_D/I_G increases from 0.98 to 1.77. The second-order bands at approximately 2643 and 2908 cm^{-1} are extruded. The band at 2643 cm^{-1} is the most intensive one and attributed to the first overtone of the D-band (around two times of 1350) [115,116]. An additional higher order band at 2908 cm^{-1} has been assigned to a combination of the G and D modes characteristic of disturbed graphitic structures. To summarize, once the temperature surpasses a certain point, the graphitization inclination and the crystallization are dramatically

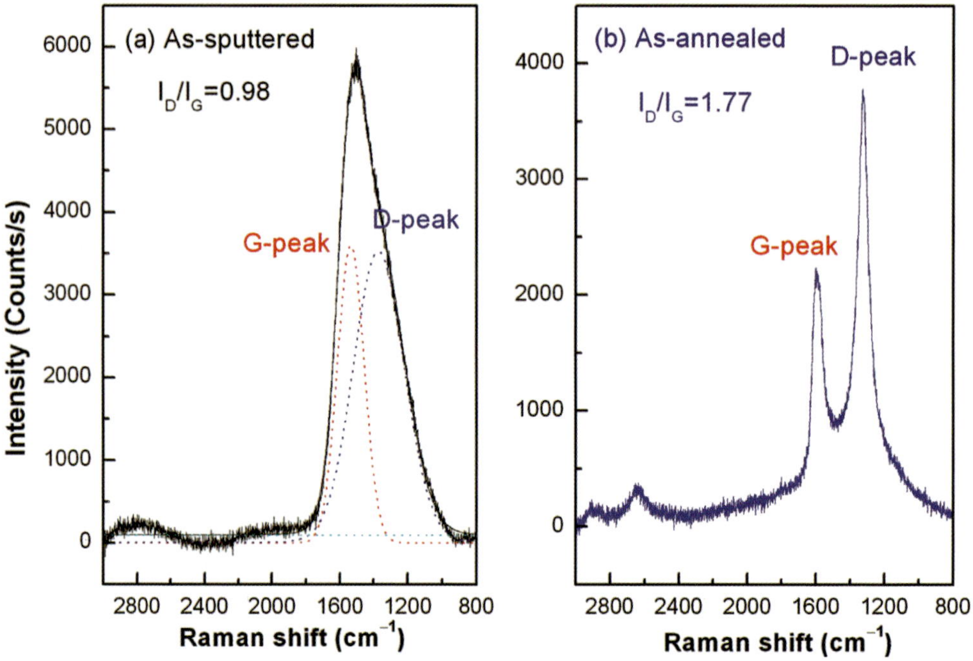

Figure 17.22 Carbon sp^2 bonding structure of the 600 nm a-C/Ni film (a) as-sputtered and (b) 3 min annealed [32].

accelerated and strengthened. The tremendous increase of I_D/I_G, the huge upshift of the G-band position, the appearance of the second-order bands at high Raman shift centers are attributed to growing sp^2 clusters and the conversion from sp^3 to sp^2 bonding configuration during the rapid thermal annealing.

17.3.2.3 Morphology of the Annealed a-C/Ni Film

One minute annealing of the 600 nm a-C/Ni film gives rise to white wire-like materials on the film surface (cf. Figure 17.23a and b). The wires are of around 70 nm in diameter and up to 10 μm in length, entangling with each other. They become inhomogeneous with bright and transparent outer sheaths and a white inner core indicating a heterogeneous structure. Some wires are growing from the cracked film. However, Ni particles do not appear at the tip of the wire.

Figure 17.24a shows a bundle of primarily wire-like nanostructures entangling into each other with remarkably uniform diameter on the order of 80 nm and with length longer than 10 μm. Statistical calculation on 20 randomly selected coaxial Si nanowires reveals that the mean diameter of the Si core is 17.0 ± 6.3 nm, and the mean thickness of SiO_x ($x = 2$) sheath is 19.7 ± 3.7 nm. The nanowire diameter equals that of the Si core plus two times of the SiO_x sheath thickness. The nanowires terminate at the root in dark ball-shaped nanoclusters with a diameter of 1.2–1.5 times that of the connected nanowires (cf. Figure 17.24b). But no nanocluster is observed at the tip (cf. Figure 17.24c).

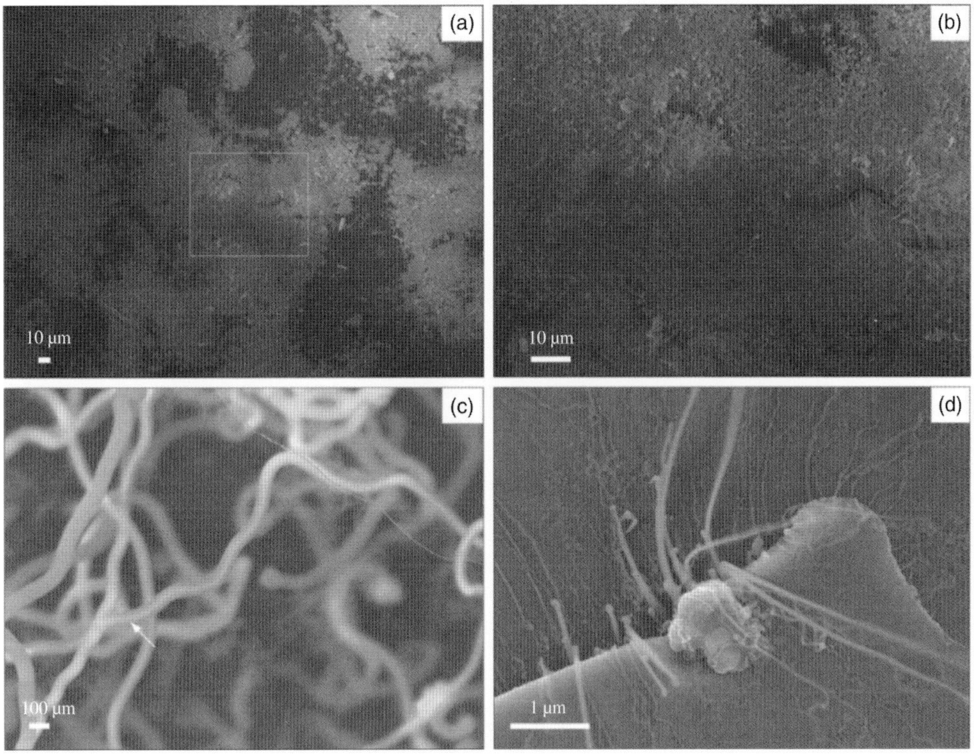

Figure 17.23 One minute annealed 600 nm a-C/Ni film surface. (a and b) Low magnification of the annealed film surface. (c) High magnification of the nanowires, showing the coaxial structure. (d) Nanowires growing from the cracked film.

Similar morphology is reported elsewhere [117]. This is suggestive of the Ni-catalyzed SLS growth similar to the VLS growth, except for the opposite seeding location of the nanoclusters. Upon annealing, the Ni layer collapses into particles due to surface tension. Ni particles melt with the beneath Si wafer to form Ni—Si eutectic, leading to the continuous diffusion of Si into the droplet. Upon supersaturation, Si is precipitated from the droplet to grow into nanowires. As Si is from the Si wafer and the liquid Ni—Si droplet acts as the medium, the nanowire always grows out from the liquid droplet while the droplet always stays on the wafer. Experimentally, Ni is not detected in the FESEM image of the annealed Si wafer surface [32]. Therefore, the Ni—Si eutectic droplet always stays at the root of the wire.

Figure 17.25 shows the structure of the nanowires grown via different duration of annealing of the a-C/Ni films. It is to be noted that by prolonging the annealing to 3 min, the ratio of the core diameter over the sheath thickness is 0.3 ± 0.1, which is dramatically decreased compared to the 1 min annealing. The end of the core is surrounded by a 12.1 nm thin sheath layer, as shown in the inset of Figure 17.25d. The details of the grown nanowires are summarized in Table 17.4.

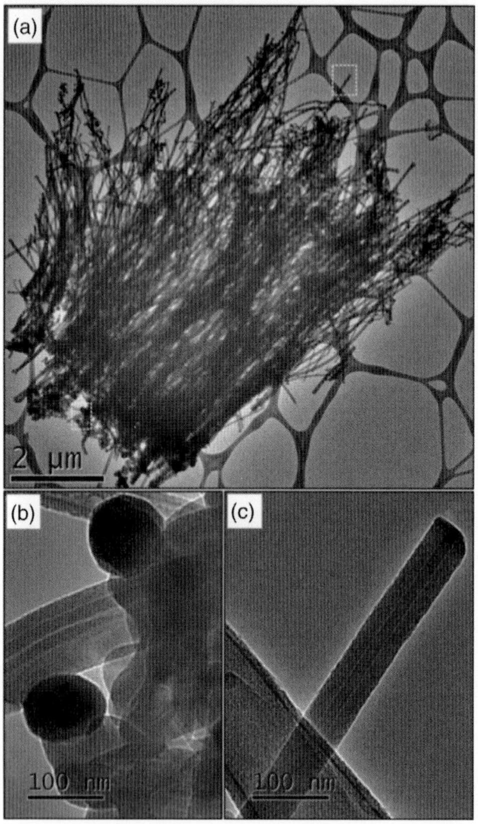

Figure 17.24 Morphology of Si nanowires. (a) A bundle of Si nanowires. (b) Dark nanoclusters are located at the root of Si nanowires. (c) Nanocluster-free nanowire tip, where the inner dark Si core is surrounded by the gray SiO_x outer sheath [118].

Obviously, oxidation takes place inward to consume the core. With or without the a-C film on the top of the Ni layer, homogeneous or heterogeneous nanowires are able to form. Si in the nanowire is from Si wafer. The a-C cover layer can retard the oxidation. Upon 3 min annealing, the ratio of the core diameter over the sheath thickness decreases from 0.3 ± 0.1 to 0 as the a-C layer thickness reduces from 570 to 130 nm (cf. Figure 17.25c–f). Besides, for the same a-C layer thickness of 570 nm, prolonging the annealing from 1 to 3 min gives rise to reduced ratio of the core diameter over the sheath thickness from 1.2 ± 0.1 to 0.3 ± 0.1 (cf. Figure 17.25a–d).

17.3.2.4 Composition of the Nanowire

The cross-sectional element mapping in Figure 17.26a reveals the coaxial nanowire consists of Si and O only. The intensity of Si and O exhibits almost the same parabolic function along the cross-sectional route, indicating the stable

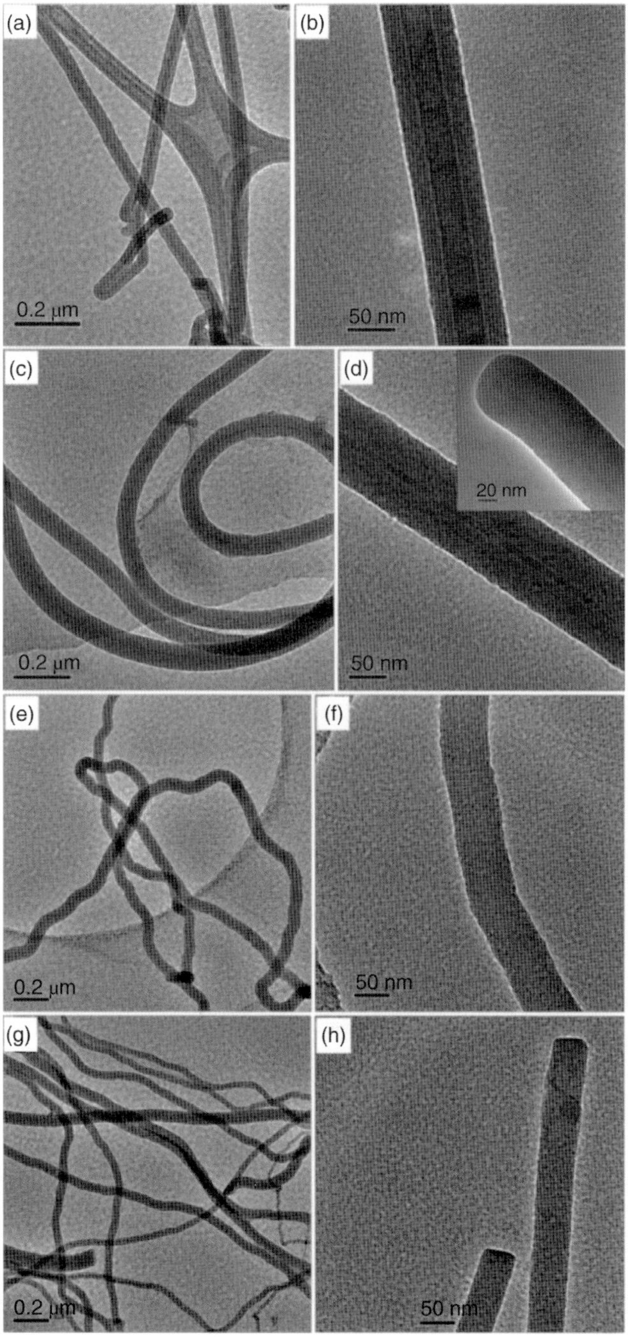

Figure 17.25 TEM images of the grown nanowires via (a and b) 1 min annealing of the 600 nm a-C/Ni film; (c and d) 3 min annealing of the 600 nm a-C/Ni film; (e and f) 3 min annealing of the 160 nm a-C/Ni film, and (g and h) 3 min annealing of the 50 nm Ni film.

Table 17.4 Details of the as-deposited a-C/Ni, Ni film, and the grown nanowires [32].

Film structure	Thickness of the a-C layer (nm)	Thickness of the Ni layer (nm)	Annealing duration (min)	Ratio of core diameter over sheath thickness
a-C/Ni	570	30	1	1.2 ± 0.1
a-C/Ni	570	30	3	0.3 ± 0.1
a-C/Ni	130	30	3	0
Ni	0	50	3	0

atomic concentration ratio (1 : 2) between Si and O along the route, that is, Si/SiO_x ($x = 2$). Particularly, the intensity of O has a small plateau, whereas that of the Si continues to increase while approaching the core of the nanowire, which is regarded as an implication of the Si core.

Figure 17.26b shows a triple-concentric nanowire constructing with a 33.3 nm-diameter core, a 14.1 nm-thick interlayer, and a 17.6 nm-thick outer sheath.

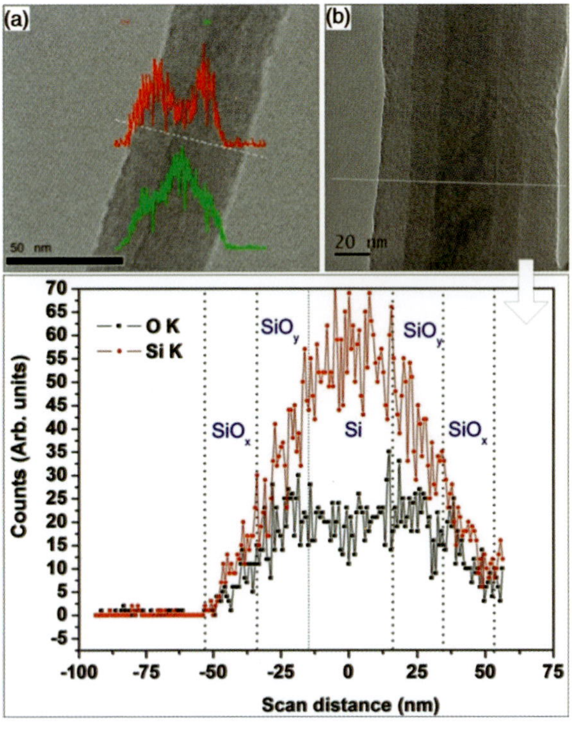

Figure 17.26 Composition of the nanowires via 1 min annealing of the 600 nm-thick a-C/Ni film, (a) Si/SiO_x ($x = 2$) coaxial nanowire [32], (b) $Si/SiO_y/SiO_x$ ($y = 1$) triple-concentric nanowire, through EDX cross-sectional element mapping [91].

The cross-sectional element mapping reveals the core is Si, the interlayer is SiO_y ($y = 1$), and the outer sheath is SiO_x, that is, Si/SiO_y/SiO_x. Due to the drift of the nanowire during EDX cross sectional element mapping, the signal is not sharp enough to clearly distinguish the interfaces of core–interlayer and the interlayer–sheath. However, it is known that the EDX intensity is related to the atomic number and the atomic weight of the element. In Figure 17.26b, the intensity of Si and O exhibits almost the same parabolic function as the signal proceeds into the outer sheath, indicating the atomic concentration of around 1 : 2 between Si and O since the ratio of the atomic weight between Si and O is 2 : 1. As the signal proceeds into the interlayer, the intensity of the O does not keep increasing, while that of the Si continues to increase. This can indicate the round cross section of the nanowires and also the decrease of the atomic concentration of O in the interlayer. It shows that the intensity ratio between Si and O is close to 2 : 1, therefore, the atomic concentration between Si and O in the interlayer is estimated to be 1 : 1. In the center area of the nanowire, the intensity of O only comes from the outer sheath and the interlayer, and thus almost keeps constant. The intensity of Si comes from the core, interlayer, and outer sheath, and thus it keeps increasing to the maximum of the parabolic curve in the center. In conjunction with the thickness of the core, interlayer, and outer sheath, the composition of the triple-concentric nanowire is estimated, that is, Si core, SiO_y ($y = 1$) interlayer, and SiO_x ($x = 2$) outer sheath, or Si/SiO_y/SiO_x structure.

17.3.2.5 Structure of the Nanowire

Figure 17.27a shows a coaxial nanowire consisting of a dark Si core surrounded by a gray SiO_x sheath. A unit cell of the cubic structure (lattice parameter = 5.43 Å) is consistent with the SAED pattern that is shown in the inset. The pattern is indexed to be from the Si[111] zone axis, suggesting the nanowire is growing along the $[\bar{1}10]$ orientation [119]. Another Si/SiO_y/SiO_x triple-concentric nanowire shown in Figure 17.27b and c is constructed with a 31.8 nm-diameter crystalline Si core, a 15.7 nm-thick amorphous SiO_y interlayer, and a 17.8 nm-thick amorphous SiO_x outer sheath. The atom-resolved crystalline Si core is surrounded by the amorphous SiO_y interlayer (cf. Figure 17.27d). The two-dimensional fast Fourier transform (FFT) of the atom-resolved Si core is indexed for Si [102] zone axis, determining the nanowire grows along the $[21\bar{1}]$ orientation. The orientation is further confirmed in the atom-resolved crystalline Si core, which clearly shows the $(2\bar{1}\bar{1})$ atomic planes (spacing, 2.22 Å), the (020) atomic planes (spacing, 2.72 Å), and their dihedral angle (65.9°). In short, the separation of the Si, SiO_y, and SiO_x phase does exist in Si nanowires, and follows the sequence of crystalline Si, amorphous SiO_y, and SiO_x from the core to the sheath.

17.3.2.6 Conclusion

One minute annealing of the 600 nm a-C/Ni film gives rise to wire-like materials on the film surface. The growing carbon sp^2 clusters and the conversion from sp^3 to sp^2 bonding configuration result in the increment of I_D/I_G from 0.98 to

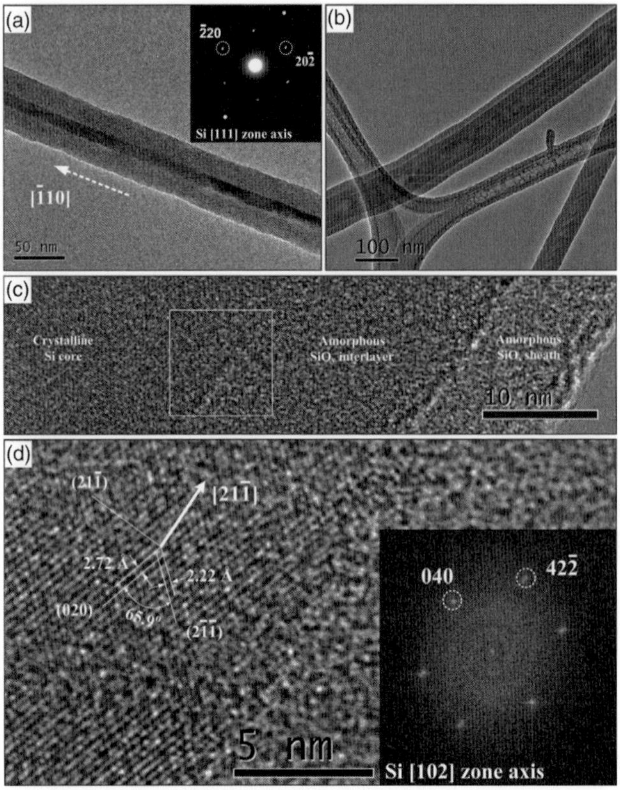

Figure 17.27 HRTEM images of the nanowires via 1 min annealing of the 600 nm a-C/Ni film. (a) A Si/SiO_x coaxial nanowire grows along the [$\bar{1}$10] orientation. The inset shows the SAED pattern of the Si core [32]. (b) A $Si/SiO_y/SiO_x$ triple-concentric nanowire. (c) The enlarged image showing the crystalline Si core, amorphous SiO_y interlayer and SiO_x outer sheath are separated by atomically sharp interfaces. (d) The atom-resolved crystalline Si core grows along the [21$\bar{1}$]orientation [91].

1.77, the upshift of G-band position from 1533 to 1588 cm^{-1}, and the appearance of the second-order bands at 2643 and 2908 cm^{-1}. The coaxial nanowire is constructed with crystalline Si surrounded by amorphous SiO_x ($x = 2$) sheath. The triple-concentric nanowire is $Si/SiO_y/SiO_x$ ($y = 1$), with a crystalline Si core surrounded by the amorphous SiO_y interlayer, and followed by the amorphous SiO_x outer sheath. The homogeneous nanowire is the amorphous SiO_x nanowires. The Ni particle does not appear at the tip of the nanowire. The Si source material comes from the Si wafer. The a-C covering layer reduced the oxidation. Within the same annealing duration of 3 min, the ratio of the core diameter over the sheath thickness decreased from 0.3 ± 0.1 to 0 as the a-C layer thickness reduces from 570 to 130 nm. For the 570 nm a-C covering layer, prolonging the annealing from 1 to 3 min gave rise to reduced ratio of the core diameter over the sheath thickness from 1.2 ± 0.1 to 0.3 ± 0.1.

17.3.3 Origin of the Growth Orientation

17.3.3.1 Introduction

Growth orientation is crucial in tailoring properties of Si nanowire. For instance, the bandgap, electrical conductance, and mechanical properties are dramatically different in nanowires with different growth orientations [12,120–125]. The ⟨100⟩-oriented Si nanowires exhibit a significantly higher exciton energy than the ⟨110⟩ counterparts [24]. The ⟨110⟩-oriented wire has the lowest total energy and the highest sensitivity to surface modification [121], whereas the ⟨111⟩-oriented wire is structurally stable to the change of diameter and has the lowest sensitivity to surface modification [12]. The ⟨110⟩-oriented one displays a direct transition, the ⟨111⟩-oriented one possesses a competitive indirect–direct gap character, and the ⟨112⟩-oriented one has an indirect bandgap [123]. Tetrahedral Si nanowires oriented in ⟨111⟩ are the most stable and the best suited for quantum confinement effects [124]. The ⟨112⟩ nanowires exhibit the highest tensile strength of 12 GPa, and the weakest link fracture is a by-product of orientation-dependent flaw populations [125]. To date, what really determines the growth orientation is still unclear, although lots of factors have been reported to have effect, including surface chemical treatment of substrate [126], nanowire diameter [23,61,127], composition of the initial alloy droplet [128,129], growth temperature [130–132], reaction pressure [24,133], and precursor molar ratios [134]. Most people believe the preferential orientation is controlled by the energetics at the solid–liquid interface of the nanocrystal emerging from the liquid droplet. It involves the surface energy of various nanocrystal planes [23,70,135]. As one of the metal-catalyzed Si nanowire growth methods, however, SLS growth involves (i) the formation of the metal–Si eutectic droplet from solid [31,32,105,106] and (ii) the subsequent Si precipitation or crystallization process. The first process requires thermal energy to melt the solid Si and Ni to liquid Si—Ni eutectic, that is, an energy-consuming process. The subsequent crystallization process solidifies the liquid eutectic to Si nanowires, that is, an energy-releasing process [136]. Therefore, energy may not be necessary in the second process. It is therefore speculated that there may be alternative determinant factors controlling the growth orientation.

In the growth of Si nanowires, Au was widely used as the catalyst; however, the incorporation of Au in Si nanowires would possibly induce a deep-level trap, causing a reduction in the efficiency of the solar cells [137]. In addition, the ionization energy level of Ni within Si is located far from the mid-bandgap, offering a higher threshold level ($5 \times 10^{15}\,cm^{-3}$) in impurity concentration compared with that ($3 \times 10^{13}\,cm^{-3}$) of Au [138]. With these concerns, in the work of Li [118], highly crystalline Si nanowires are grown via thermal annealing of Ni-coated Si wafers. The growth orientation and the relation to the seeding Ni catalyst particles are examined through HRTEM and SAED pattern. The determinant factors for the growth orientation are explored.

17.3.3.2 Growth Orientation

To date, three techniques have been applied to determine the growth orientation: (i) HRTEM image and the corresponding two-dimensional FFT pattern [19,25]; (ii) TEM image and the corresponding SAED pattern [23,105]; and (iii) geometric relation to the substrate surface by microscope, SEM or TEM [44]. Approaches (i) and (ii) are the most accurate ways since they completely eliminate the influence from the environment and instrument. Approach (iii) may be the most convenient, but of the biggest error since the observed orientations are likely changed during sampling, especially for the long and curved nanowires. In the work of Li, the growth orientation is determined by HRTEM image and the corresponding 2D FFT pattern, in conjunction with the TEM image and the corresponding SAED pattern.

Growth Orientation Determined by HRTEM Image and 2D FFT Pattern

As shown in Figure 17.28, HRTEM images captured of three individual Si nanowires along different growth orientations provide atomic insight into the structure of the nanowires. The nanowires consist of a uniform diameter crystalline Si core surrounded by an amorphous SiO_x sheath. The core and sheath are separated by an atomically sharp interface. The amorphous SiO_x sheath surrounding the crystalline Si core could be removed by hydrofluoric acid [19]. In Figure 17.28a, the reciprocal lattice peaks obtained from the 2D FFT of the atom-resolved image of the Si core are indexed for the [102] zone axis of crystalline Si, suggesting that the nanowire grows along the [2, 1, −1] orientation. This orientation is further confirmed in the atom-resolved TEM image of the crystalline Si core, which clearly shows the (2, −1, −1) atomic planes (spacing: 2.22 Å), the (020) atomic planes (spacing: 2.72 Å), and their dihedral angle 65.9°. Similarly, Figure 17.28b and c respectively indexes the [111] and [112] zone axis of the crystalline Si from the 2D FFT pattern of the atom-resolved Si core. In conjunction with the HRTEM images, the nanowires grow along [0, −1, 1] and [1, 1, −1] orientations, respectively. Figure 17.28b reveals the atomic planes of (−1, 1, 0) and (0, −1, 1) (spacing: 3.84 Å) and their dihedral angle 120.0°, where (0, −1, 1) plane is perpendicular to the nanowire axis. Figure 17.28b has been used in the previous work to prove the grown Si nanowires are highly crystalline [32]. In this work, this figure is cited to prove HRTEM and 2D FFT pattern can determine the different growth orientations. Figure 17.28c shows the atomic planes of (2, −2, 0) (spacing: 1.92 Å) and (1, 1, −1) (spacing: 3.14 Å), and their dihedral angle 90.0°, where (1, 1, −1) planes are perpendicular to the growth axis.

Growth Orientation Determined by TEM Image and SAED Pattern

As shown in Figure 17.29a, the SAED pattern in the inset reveals the crystalline Si core and the amorphous SiO_x sheath, that is, c-Si/a-SiO_x. A unit cell of face-centered cubic structure of Si (5.43 Å in lattice parameter) is consistent with the

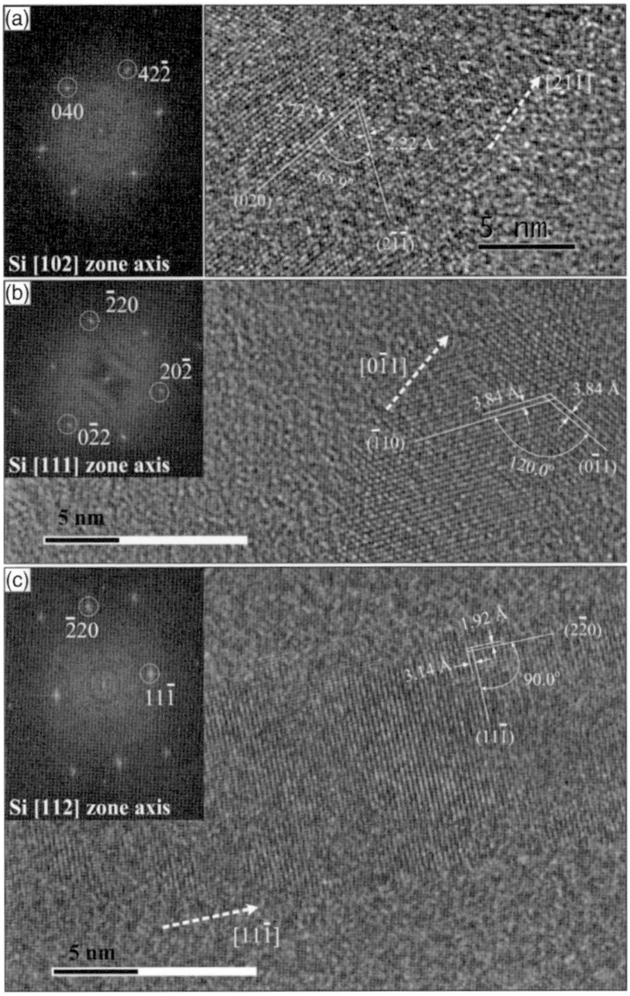

Figure 17.28 HRTEM images of Si nanowires growing along different orientations [118]. (a) [2, 1, −1]-oriented; (b) [0, 1, −1]-oriented; and (c) [1, 1, −1]-oriented Si nanowires. Inset shows the respective FFT pattern of the atom-resolved Si core generated from the zone axis perpendicular to the growth axis.

diffraction pattern. It is indexed for the Si [001] zone axis, indicating the Si nanowire grows along the [110] orientation. Likewise, the insets in Figure 17.29b–f respectively show the SAED pattern generated from the zone axis of Si [101], [101], [315], [323], and [116], determining the nanowires grow along [1, −1, 1], [0 1 0], [−2, 1, 1], [−1, 3, −1], and [−3, −3, 1] orientations.

As shown in Table 17.5, statistical results of 33 nanowires exhibit an occurrence preference of 30.3% for the nanowires growing along the ⟨112⟩ orientation,

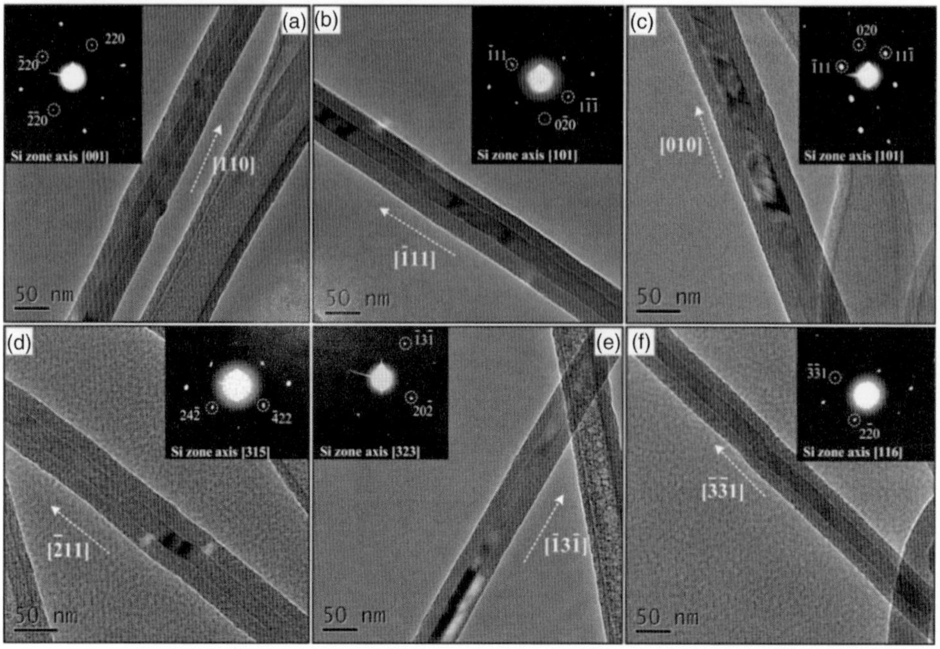

Figure 17.29 TEM images of Si nanowires growing along different orientations [118]. (a) [110], (b) [1, −1, 1], (c) [010], (d) [−2, 1, 1], (e) [−1, 3, −1], and (f) [−3, −3, 1]. Inset shows the respective SAED pattern generated from different zone axis.

followed by 27.3% nanowires growing along the ⟨110⟩ orientation, 24.3% along the ⟨111⟩ orientation, and 12.1% in the ⟨001⟩ orientation. Few nanowires grow in the ⟨113⟩ and ⟨133⟩ orientations.

Similar orientation distribution has been reported for nanowires (diameter ranging from 10 to 20 nm) grown by chemical vapor deposition of SiH_4 on Au-covered poly-L-lysine on Si wafer substrates [23,61], and those synthesized through thermal evaporation of SiO powders in an alumina tube under $H_2 + Ar$ atmosphere [70]. Both of them believe the orientation preference is determined by the surface energy of the various Si nanocrystal planes [23,70]. However, the energy argument made in Ref. [23] is under the assumption that the flat Si–Au interface is always the Si {111} plane (cf. Figure 3 in Ref. [23]). In Ref. [70] Si

Table 17.5 Statistical distribution of Si nanowires along different growth orientations [118].

Statistics	⟨112⟩	⟨110⟩	⟨111⟩	⟨001⟩	⟨113⟩	⟨133⟩
Quantity	10	9	8	4	1	1
Percentage (%)	30.3	27.3	24.3	12.1	3.0	3.0

nanoparticles that are precipitated through the disproportionation of SiO, that is, $2SiO \rightarrow Si + SiO_2$ [28,70], are considered to act as the seeds to the growth of Si nanowires without any supply of the metallic catalyst. Si nanowire nucleates from the Si{111} facets to reach the minimum surface energy [28]. However, in their experiments with SiO powder, reduction of Si^{2+} to Si^0 and oxidation of Si^{2+} to Si^{4+} (in formation of SiO_2) take place only with migration of electrons; thus, the process does not involve physical movement of Si atoms or SiO_2 molecules [75]. Recent study shows that metallic aluminum coming from the reduction of alumina tube by H_2 possibly acts as the catalyst for Si nanowire seeding and growth [75]. On the other hand, surface energy quantifies the disruption of intermolecular bonds that occur when a surface is created [139]. For a liquid, the surface energy density are identical everywhere. In the present growth process, surrounding residual oxygen may enter the surface of the droplet to form Ni—Si—O droplet. The growth of Si nanowires takes place as Si in the liquid Ni—Si—O eutectic droplet becomes supersaturated. Therefore, the surface energy density is identical everywhere in the droplet. For a solid, however, the molecules on the surface have more energy compared with the molecules in the bulk of the material. For a solid Si nanowire that is already grown from the liquid droplet, the Si oxide shell must have higher surface energy than the Si core because it is on the outer surface. But the growth orientation is already determined at the moment of precipitating Si from the droplet. Moreover, if the formation of Si oxide shell and Si core takes place simultaneously, the interfacial energy between Si oxide shell and Si core may be considered instead of the surface energy of a bare facet. The reasonably low Si/Si oxide interfacial energy may explain why few ⟨113⟩ and ⟨133⟩-oriented Si nanowires with relatively high surface energy are detected. It is therefore concluded that the theory of surface energy may not be enough to explain the orientation preference in the seeding and growth of Si nanowires. There might be other determinant factors related to the metallic catalyst controlling the growth orientation.

17.3.3.3 **Interfacial Structure between Si Nanowire and Ni Catalyst**

Figure 17.30a shows an elliptic Ni catalyst particle seeding a Si nanowire. The particle is of around 47.0 nm in semimajor axis and 39.3 nm in semiminor axis. The particle is surrounded by a 7.2 nm-thick amorphous SiO_x layer, forming the Ni—Si—O nanocluster. The EDX spectrum shown in Figure 17.30b is generated from the nanocluster, confirming the chemical composition of the core in the seeding nanocluster is Ni, where the signals of C and Cu are from the TEM grid, while Si and O are from the SiO_x layer surrounding the Ni core. The SAED systematic rows, generated by directing the electron beam on the area of both the seeding Ni particle and the connected grown Si nanowire, provide the crystalline information of both the catalyst and the nanowire (cf. Figure 17.30c). These systematic rows are indexed for the zone axis of Ni [112]. The dotted white squares are respectively indexed for Ni (−3, 1, 1), Ni (−1, −1, 1), and Ni (1, 1, −1). Though the information in the SAED spots is not enough to determine the growth orientation of the Si nanowire, the dotted white circle near

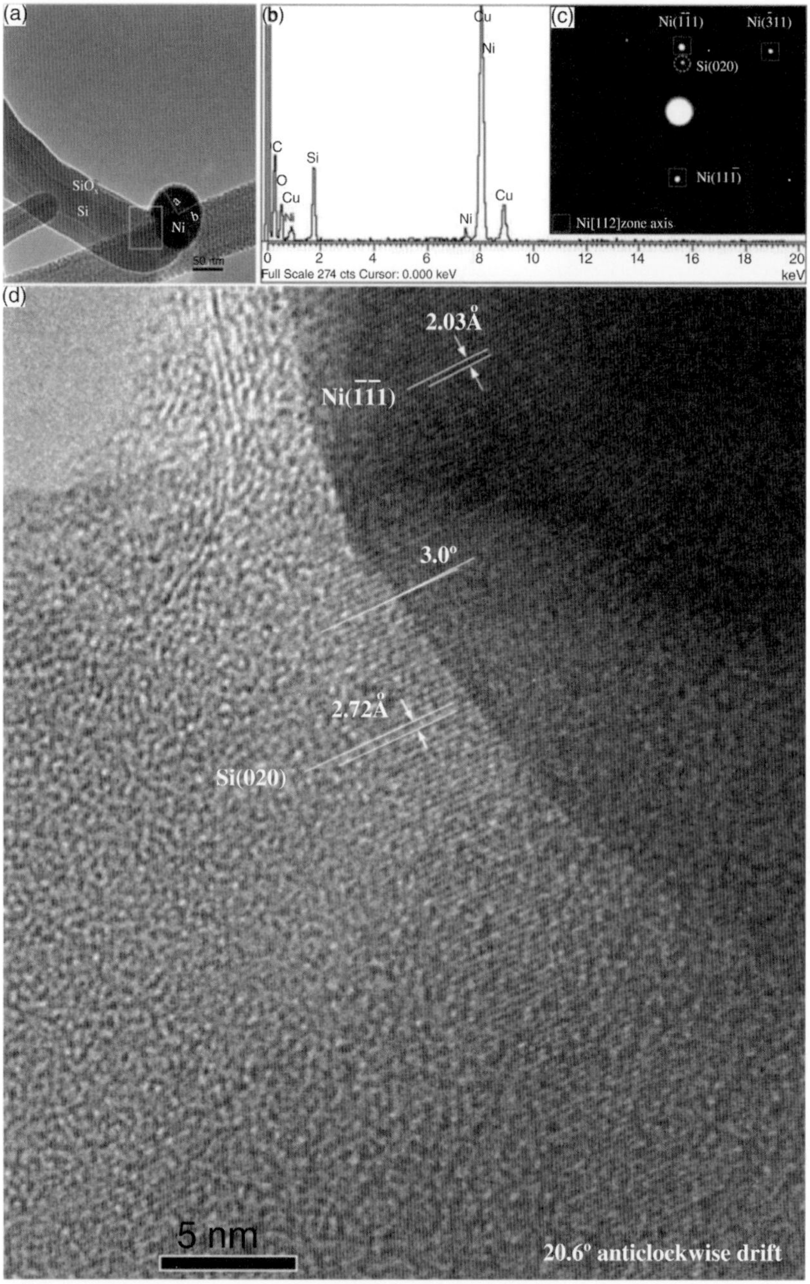

Figure 17.30 Interfacial structure between a Si nanowire and the seeding Ni catalyst particle [118]. (a) A Si nanowire is seeded by an elliptic Ni particle. (b) EDX spectrum generated from the nanocluster. (c) SAED systematic rows generated from the area covering both the ending nanocluster and the connected nanowire, which are indexed for the zone axis of Ni [112]. (d) High-resolution interfacial structure between the seeding Ni catalyst particle and the grown Si nanowire.

Ni (−1, −1, 1) still could be indexed for Si (020) based on the measured lattice spacing of around 2.70 Å (cf. Figure 17.30d) and the relationship to the spot reflected by the Ni (−1, −1, 1) plane. There is around 20.6° drift in anticlockwise during HRTEM observation. It is noted that the Si wire–Ni catalyst interface exhibits a round profile, rather than the flat ones reported in Ref. [23]. The round Ni/Si interfacial shape further reveals the formation of the liquid Ni—Si—O phase does take place during the growth process since the liquid Ni—Si eutectic droplet could form at only around 964 °C [90], at least 100 °C lower than the annealing temperature. Ni (−1, −1, 1) (spacing: 2.03 Å) and Si (020) (spacing: 2.72 Å) planes are almost parallel with a small dihedral angle of 3° (cf. Figure 17.30c and d), although there is about 25.4% difference in lattice spacing in reference to Si (cf. (2.72–2.03)/2.72). It is noted that Ni (−1, −1, 1) planes match with Si (020) planes, although the nanowire has a change in morphology along the growth orientation within 5–10 nm of the wire–catalyst interface. Growth orientation is determined during precipitation of Si from the liquid Ni—Si eutectic droplet. As such, images showing lattice match during growth between the catalyst and wire of different growth orientations would provide stronger evidence. This can be a topic for future study.

17.3.3.4 **Determinant Factors of Growth Orientation**

In the work of Li *et al.* [118], Si nanowires are of many growth orientations instead of certain preferred ones such as ⟨112⟩, ⟨110⟩, or ⟨111⟩ that are reported in Refs [23,61,70]. It was commonly believed that the preference of growth orientations was dictated by the surface energy of the various Si nanocrystal planes [23,70]. However, the growth of Si nanowires involves mass transfer of Si atoms from the Ni—Si—O droplet to a solid crystalline Si phase. It is an energy-releasing process. Therefore, surface energy may not be necessary in nucleation and growth of nanowires. As shown in Figure 17.30, a structure-sensitive seeding principle is found at the Ni catalyst–Si wire spherical interface, that is, Ni (−1, −1, 1) is parallel to Si (020) with only 3° in dihedral angle and 25.4% mismatch in lattice spacing. It indicates that the matching degree of the lattice planes at the Si wire–Ni catalyst interface may also dictate the nucleation and growth of Si nanowires. As such, it is reasonable to believe that the growth of nanowire follows certain structure-sensitive principle to achieve the most stable structure with the minimum mismatch in lattice spacing and dihedral angle at the wire–catalyst interface (in this study, Si and Ni).

This structure-sensitive principle works in other semiconductor nanostructures, such as ZnSe and ZnO [135,140]. For instance, in Ref. [135], besides the dominant (111) planes with the lowest energy (cf. Figure 17.31a), small fraction of (001) planes are also at the Au–ZnSe interfaces. It proves that the lowest surface energy may not be necessary in the nucleation and growth of nanowire. By performing the 2D FFT pattern on the ZnSe nanowires, it is further found that (i) (−1, 1, 1), (−1, −1, 1), and (0, −2, 0) planes of ZnSe and Au are parallel to each other in the [−1, 2, 1]-oriented nanowire (cf. Figure 17.31a); (ii) (1, −1, −1) and (1, 1, −1) planes of ZnSe and Au are parallel to each other in the [−1, 0, 1)-

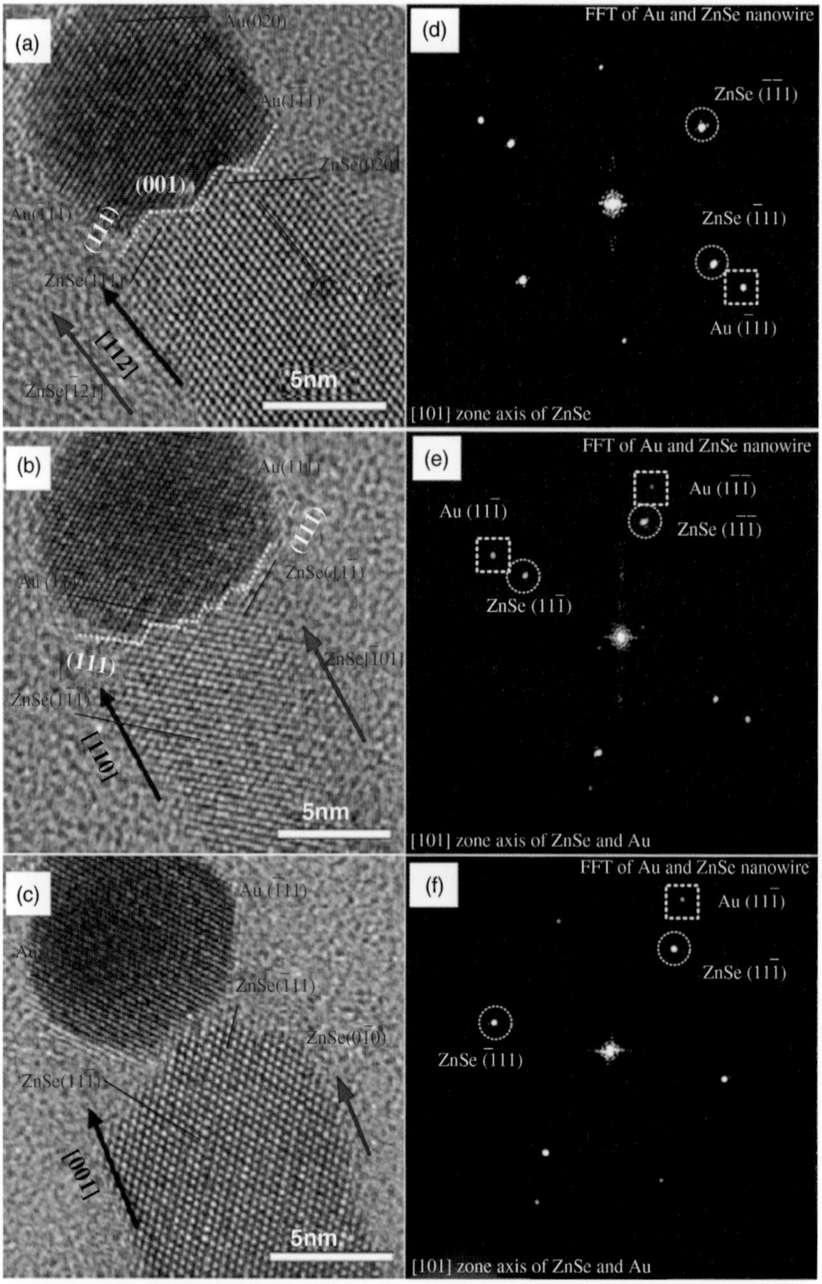

Figure 17.31 ZnSe nanowires growing along (a) ⟨112⟩, (b) ⟨110⟩, and (c) ⟨001⟩ orientations reported in Ref. [135]. (d–f) FFT pattern respectively generated from parts (a–c). *Notes:* (1) The FFT pattern is performed in this chapter. (2) The white and dark labels in parts (a–c) are indexed by the authors of Ref. [135]. (3) The blue labels are indexed by the authors of this chapter based on the FFT pattern in parts (d–f).

oriented nanowire (cf. Figure 17.31b); and (iii) (1, 1, −1) and (−1, 1, 1) planes of ZnSe and Au are almost parallel to each other in the [0, −1, 0]-oriented nanowire (cf. Figure 17.31c). Careful calculation finds that the $\{111\}_{ZnSe}$ and $\{002\}_{ZnSe}$ planes respectively match the $\{111\}_{Au}$ and $\{002\}_{Au}$ planes with a mismatch of 28.04% in lattice spacing in reference to ZnSe (cf. (5.67–4.08)/5.67).

Note: Reference [135] reports that the zigzag interfaces in the nanowires growing along the [112] direction mainly consist of (111) plus a small fraction of (001) facets (cf. Figure 17.31a). The zigzag interfaces in the [110] nanowires consist of (111) and $(1\bar{1}1)$ planes (Figure 17.31a). In this case, the flat (111) interface and the zigzag one have the same interface area. The reason that these (111) planes are dominant at the interfaces is attributed to its low interface (or surface) energy compared to other planes. However, in this work, two-dimensional FFT is respectively performed on Figure 17.31a–c. All patterns are indexed to be from the [101] zone axis of Au and ZnSe (cf. Figure 17.31d–f). It is found that in images (a) and (d), the $(\bar{1}\bar{1}1)$,$(11\bar{1})$, and $(0\bar{2}0)$ planes are respectively parallel; in images (b) and (e), $(1\bar{1}\bar{1})$ and $(11\bar{1})$ planes are respectively parallel; in images (c)–(f), $(11\bar{1})$ and $(\bar{1}11)$ planes are respectively parallel.

In Ref. [140], the interface of the [0001] growing ZnO nanowire is composed of $(0001)_{ZnO}$ and $(020)_{Sn}$ planes (cf. Figure 17.32a). The atoms in the ZnO

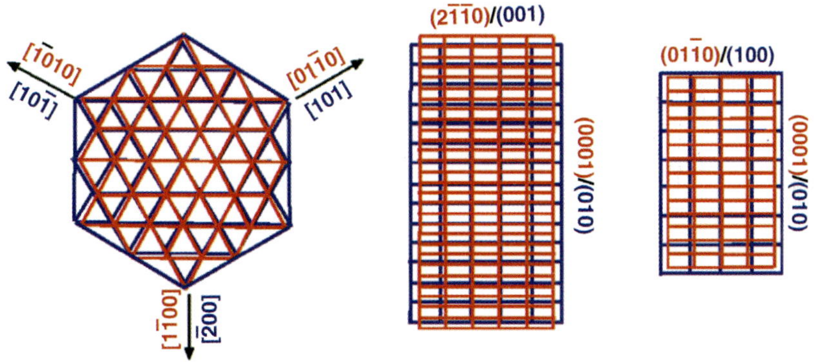

Figure 17.32 Schematics of atomic planes at the epitaxial interfaces between the Sn particle and the [0001], $[01\bar{1}0]$, and $[2\bar{1}\bar{1}0]$ growing ZnO nanostructures, respectively [140]. (a) The interface of the [0001] growing nanowire is composed of $(0001)_{ZnO}$ and $(020)_{Sn}$ planes. The atoms in the ZnO $(0001)_{ZnO}$ plane have sixfold symmetry. The angle between $(101)_{Sn}$ and $(\bar{1}01)_{Sn}$ is 57.1°. Therefore, the atoms in the $(020)_{Sn}$ plane have quasi-sixfold symmetry. Two of the $\{01\bar{1}0\}_{ZnO}$ planes match the $\{101\}_{Sn}$ planes with a lattice mismatch as small as 0.7% in reference to ZnO. The third one matches the $\{200\}_{Sn}$ with a lattice mismatch of 3.6%. (b) The interface of the $[01\bar{1}0]$growth ZnO nanobelts is composed of $(01\bar{1}0)_{ZnO}$ and $(200)_{Sn}$ planes. The lattice mismatches between $\{2\bar{1}\bar{1}0\}_{ZnO}$ and $(002)_{Sn}$ and between $\{01\bar{1}0\}_{ZnO}$ and $(101)_{Sn}$ are 2.1 and 0.7%, respectively. (c) The interface of the $[2\bar{1}\bar{1}0]$ growing ZnO nanobelt is composed of $(01\bar{1}0)_{ZnO}$ and $(100)_{Sn}$ planes. The lattice mismatches between $(0001)_{ZnO}$ and $(020)_{Sn}$ and between $(01\bar{1}0)_{ZnO}$ and $(200)_{Sn}$ planes are about 3.6 and 12.0%, respectively.

$(0001)_{ZnO}$ plane have sixfold symmetry. The angle between $(101)_{Sn}$ and $(-1, 0, 1)_{Sn}$ is 57.1°. Thus, the atoms in the $(020)_{Sn}$ plane have quasi-sixfold symmetry. Two of the $\{0, 1, -1, 0\}_{ZnO}$ planes match the $\{101\}_{Sn}$ planes with a lattice mismatch as small as 0.7% in reference to ZnO. The third one matches the $\{200\}_{Sn}$ with a 3.6% lattice mismatch. The interface of the [0, 1, −1, 0] growing ZnO nanobelts is composed of $(0, 1, -1, 0)_{ZnO}$ and $(200)_{Sn}$ planes (cf. Figure 17.32b). The lattice mismatches between $\{2, -1, -1, 0\}_{ZnO}$ and $(002)_{Sn}$ and $\{0, 1, -1, 0\}_{ZnO}$ and $(101)_{Sn}$ are 2.1 and 0.7%, respectively. The interface of the [2, −1, −1, 0] growing ZnO nanobelt is composed of $(0, 1, -1, 0)_{ZnO}$ and $(100)_{Sn}$ planes (cf. Figure 17.32c). The lattice mismatches between $(0001)_{ZnO}$ and $(020)_{Sn}$ and between $(0, 1, -1, 0)_{ZnO}$ and $(200)_{Sn}$ planes are respectively about 3.6 and 12.0%. It is also found that the ordered planes of the Sn catalyst at the ZnO nanostructure–Sn catalyst interface play an important role in initiating the nucleation and growth of the ZnO nanostructures [140]. As the temperature drops to room temperature, Sn preserves the orientation as defined by the interface with the ZnO nanostructures. It is in turn inferred that the ordered planes of the metallic catalyst at the wire–catalyst interface also control the growth orientation of the nanowire.

17.3.3.5 Conclusion

Thermal annealing of a Ni-coated Si wafer gives rise to highly crystalline Si nanowires in many growth orientations. The nanowires prefer to grow in the ⟨112⟩ orientation. Surface energy is not necessary in the nucleation and growth of nanowires. The growth of nanowire follows certain structure-sensitive seeding principle with the metal catalyst to minimize the mismatch in lattice spacing and dihedral angle at the wire–catalyst interface. The growth orientation of nanowires is determined by the ordered planes of the metallic catalyst at the wire–catalyst interface at the onset of the growth.

17.3.4 Origin of the Morphology

17.3.4.1 Introduction

Coaxial Si/Si oxide nanowires are reported to be grown by laser ablation of Si/Fe disk or Si disk in quartz tube [19,66], chemical vapor deposition of $SiCl_4/H_2$ [20–22], or SiH_4 [23], on metal-covered substrates; thermal degradation of diphenylsilane in a supercritical hexane fluid size-monodispersed with Au nanocrystals [24,25]; thermal evaporation of bulk Si onto metal-covered substrates under ultrahigh vacuum [26,27]; thermal evaporation of SiO powders in an alumina tube under H_2/Ar atmosphere [141]; or thermal annealing of metal-covered Si wafers [32,105,106,142]. In addition to the most abundant coaxial ones, diameter-oscillating [21,22,66,143,144], side-by-side biaxial [141], and nanosphere-entrapped nanowires (cf. Table 17.6) [66,92–94,145] may play more important roles in the devices of photonics [2,146], optoelectronics [93,143], quantum dots [147,148], and energy

Table 17.6 Key experimental condition in the reported growth of the various Si/SiO_x nanowires [91].

References	Growth process	Grown nanostructures
[21,22]	Chemical vapor deposition of $SiCl_4/H_2$ on Au-covered Si wafers at 900–1250 °C	Diameter-oscillating Si/SiO_x nanowires
[143,144]	Thermal annealing of Au-coated Si wafers up to 1250 °C in a quartz ampoule	Diameter-oscillating Si/SiO_x nanowires
[66]	Laser ablating the Si disk (made by pressing Si powder at 150 °C for 48 h under a hydraulic press) in a quartz tube under 50 sccm Ar atmosphere at 1200 °C	Diameter-oscillating and nanosphere-entrapped Si/SiO_x nanowires
[92]	Thermal annealing of a Au/a-Si/Si wafer at 1030 °C in nitrogen ambient flowing at 50 sccm and under 1 atmosphere pressure	Au_2Si nanosphere entrapped Si oxide nanowires
[93]	Thermal annealing of the Au/SiN_z/Si wafer sandwiched with a Au-coated Si wafer at 530 °C in a gas mixture of H_2 (200–250 sccm) and NH_3 (300–350 sccm) atmosphere under $6.7–9.3 \times 10^3$ Pa pressure in a microwave chamber	Au nanosphere entrapped Si oxide nanowires
[94]	Thermal evaporation of SiO powders at 1050 °C on a Au-coated Si wafer in a fused quartz tube at 2.5×10^3 Pa	Au/Si nanoparticle entrapped Si oxide nanowires
[145]	Thermal annealing of a Au/SiO_2/Si wafer in a quartz tube at 1200 °C in the presence of forming gas (95% N_2 + 5% H_2) for 30 min	Au nanosphere entrapped Si oxide nanowires
[141]	Thermal evaporation of SiO powders at 1250 °C in an alumina tube in 50 sccm 5% H_2 mixed with 95% Ar atmosphere at 5.3×10^4 Pa	Side-by-side biaxial Si/SiO_2 nanowires

storage [149,150], due to the unique quantum dots effect [151] and third-order nonlinear susceptibility [152,153]. So far, however, several doubts are still unclear in the reported morphologies: (i) Are the diameter-oscillating nanowires formed by the vibration of the metal–Si droplets [21,22,143,144], or by that of the Si nanoparticles [66]? (ii) Is the disproportionation of SiO (only involve the migration of electrons, but no rearrangement of the atoms and molecules produced) able to explain the growth of Si nanowires [66,75]? (iii) How do the Au_2Si nanospheres form through simultaneous precipitation of Au and Si, since they could not supersaturate at the same time in the Au—Si eutectic droplet [92]? (iv) How does the capillary force entrap Au nanopeas in the solid silica nanowires [93]? (v) How do the SiO_2 molecules entrap the Au/Si nanospheres in the silica nanowires, since SiO_2 molecules are still confined in the ejected SiO particles [94]?

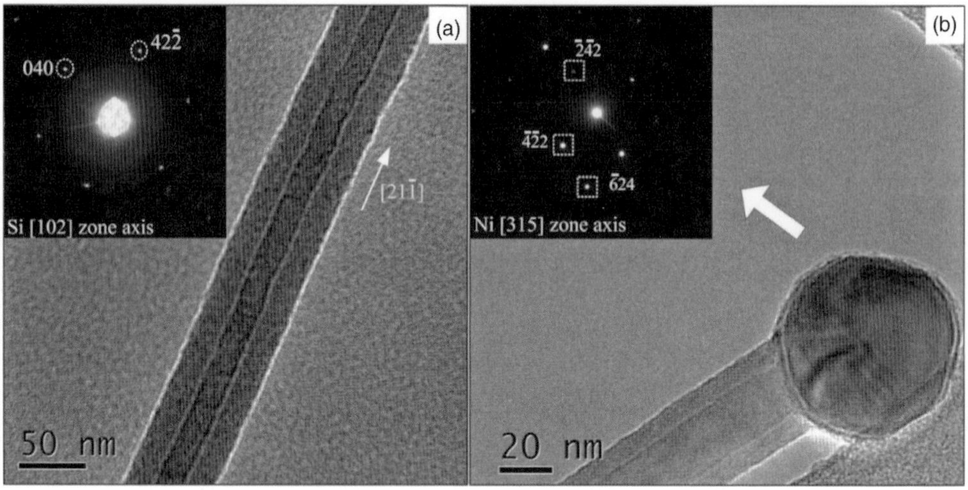

Figure 17.33 Coaxial Si/SiO$_x$ nanowires [91]. (a) A nanowire grows along the $[21\bar{1}]$ orientation. (b) Cross section of a Ni particle connected to a coaxial nanowire.

Through careful studies of transmission electron microscopy and selected area electron diffraction pattern, Li and coworkers explored the morphology of the nanowires and the relation to the seeding particles [91]. The origin for these morphologies is discussed in detail.

17.3.4.2 **Coaxial Nanowire**

Figure 17.33a shows a smooth Si/SiO$_x$ coaxial nanowire constructed with dark Si core surrounded by gray SiO$_x$ sheath. The SAED pattern is indexed for Si [102] zone axis, suggesting the wire is along the $[21\bar{1}]$ orientation. Figure 17.33b shows the cross section of a ball-shaped Ni particle with a diameter of 65.0 nm seeding a coaxial nanowire. The particle is surrounded by a 3.6 nm-thick SiO$_x$ layer, thus the Ni particle–Si nanowire interface is not observable. The SAED pattern is indexed for Ni [315] zone axis; however, it is not enough to exactly index the surrounding planes due to the low magnification of the Ni particle. The transition of the polycrystalline face-centered cubic structure of the sputtered Ni layer to the single-crystalline Ni particle reveals the formation of the liquid Ni—Si phase does take place during the growth process since the liquid Ni—Si eutectic droplet could form at only 800–964 °C [90], at least 100 °C lower than the annealing temperature.

17.3.4.3 **Side-by-Side Biaxial Nanowire**

Figure 17.34a shows a side-by-side Si/SiO$_x$ biaxial nanowire. This wire is connected with a 52.0 nm-diameter Ni particle. The SAED pattern generated from the nanowire is not a standard one, thus unable to be indexed to a specific Si zone axis (cf. Figure 17.34b, inset). However, it indicates the crystalline structure

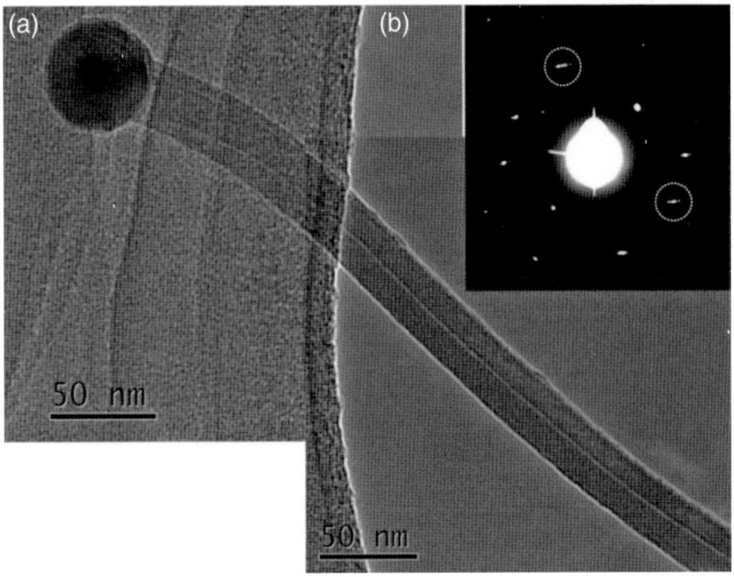

Figure 17.34 A ball-shaped Ni particle connects with a side-by-side Si/SiO_x biaxial nanowire. The inset is the SAED pattern generated from the wire [91].

of the nanowire. It is known that twins and stacking faults are important structural defects in Si nanowires [154–156]. They produce odd diffraction pattern and atom-resolved images due to the superposition of different crystal grains [157,158]. "Odd" means the pattern and images could not be obtained from classical face-centered cubic Si crystals. The line-shaped spots as pointed out by the dotted circles in Figure 17.34 imply the existence of the defects. They could be microtwins or stacking faults [156], requiring further atom-resolved observation to confirm.

17.3.4.4 **Triple-Concentric Nanowire**

As shown in Figure 17.35a, a smooth Si/SiO_y/SiO_x triple-concentric nanowire grows along the $[21\bar{1}]$ orientation. The structure has been carefully examined in Figure 17.27b and c. Figure 17.35b shows a triple-concentric sinusoidal nanowire grows along the $[2\bar{1}\bar{1}]$ orientation in a period of 65.5 nm. At the valley, the 24.7 nm-diameter Si core is surrounded by a 13.3 nm-thick SiO_y interlayer and an 11.7 nm-thick SiO_x outer sheath. At the peak, the 29.6 nm-diameter Si core is surrounded by a 13.5 nm-thick SiO_y interlayer and a 19.2 nm-thick SiO_x outer sheath. In Figure 17.35c, a fishbone-profiled nanowire is connected to a Ni particle with a width of 95.5 nm and length of 128.3 nm (cf. Figure 17.35c, bottom-left inset). The SAED pattern generated from the rectangle area of the Si nanowire is shown in the bottom-right inset. It is indexed for the Si [113] zone axis, determining the nanowire is along the$[30\bar{1}]$orientation. Figure 17.35d magnifies the

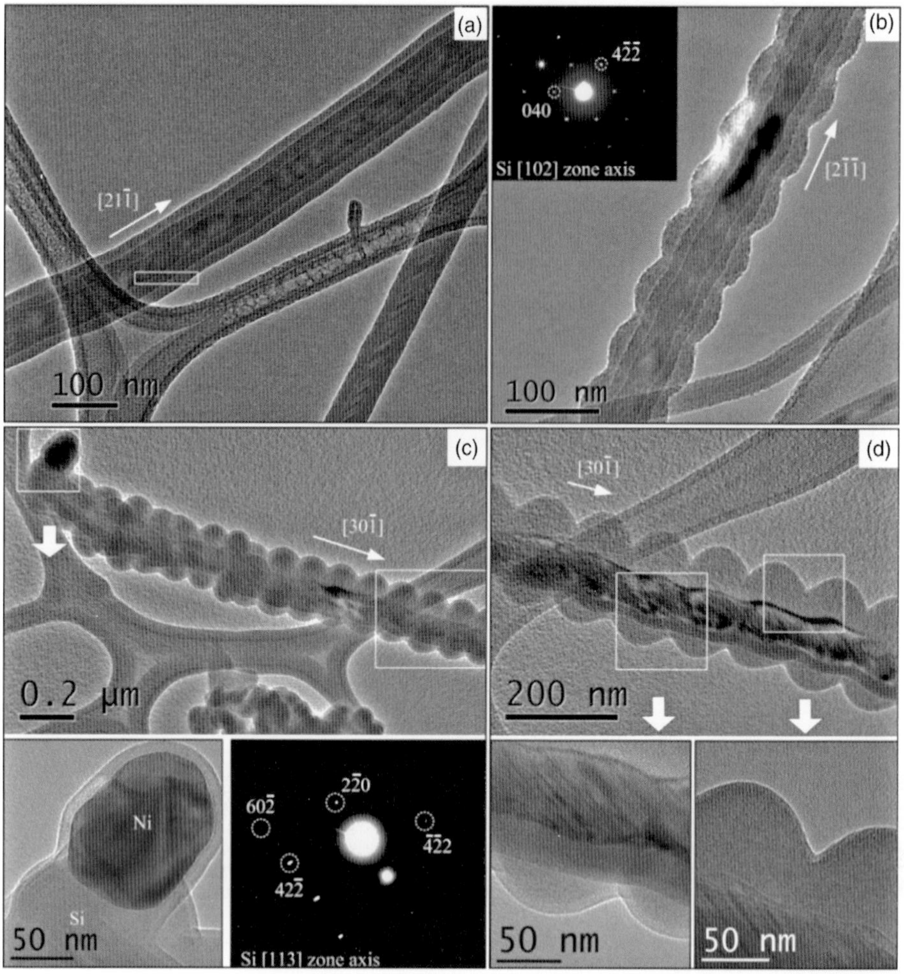

Figure 17.35 $Si/SiO_y/SiO_x$ triple-concentric nanowires [91]. (a) A smooth nanowire. (b) A sinusoidal nanowire grows along the [2$\bar{1}\bar{1}$] orientation. (c) A fishbone-profiled nanowire is along the [30$\bar{1}$] orientation. (d) Enlargement of the rectangle area in part (c). Insets in parts (c) and (d) are the respective SAED pattern and the enlarged structure of the wire.

rectangle areas of the fishbone-profiled nanowire. The nanowire is of a sinusoidal Si core with a valley diameter of 54.6 nm and a peak diameter of 68.4 nm, surrounded by a sinusoidal SiO_x outer sheath of 119.4 nm in period, 160.0 nm in valley diameter, and 199.8 nm in peak diameter. A 26.2 nm-thick SiO_y interlayer is between the bottom SiO_x sinusoidal outer sheath and the Si core (cf. Figure 17.35d, bottom-left inset). On the other side, an 8.6 nm-thick SiO_y interlayer is between the SiO_x outer sheath and the Si core (cf. Figure 17.35d, bottom-right inset).

17.3.4.5 Ni-Nanosphere Entrapment in Nanowires

Figure 17.36a shows four Ni nanospheres are entrapped in a triple-concentric Si/SiO_y/SiO_x nanowire. The Si core gradually disappears, followed by the shielding of the amorphous SiO_y interlayer by the SiO_x outer sheath. Two nanospheres are completely entrapped in the amorphous SiO_y core, that is, 45.1 nm-diameter nanosphere 1, and 48.0 nm-diameter nanosphere 2 (cf. Figure 17.36b). Figure 17.36c shows the SAED pattern generated from nanosphere 2 (cf. dotted white circle in Figure 17.36b), implying the crytalline structure. The EDX spectrum of nanosphere 2 further confirms the nanosphere is Ni, where C and Cu are from the TEM grid, while Si and O are from the wire. It is noted that all the nanospheres are not surrounded by the Si core, that is, nanosphere 3 is connected to the Si core (69.2 nm in length) only on the upper side, while nanosphere 4 is connected to the all-Si core only on the bottom side. This is consistent with the Au nanospheres reported in Ref. [94], where one side of the nanosphere is connected to the crystalline Si core (cf. Figure 4 in Ref. [94]), while the rest of the nanosphere is surrounded by Si oxides. It is reported that the

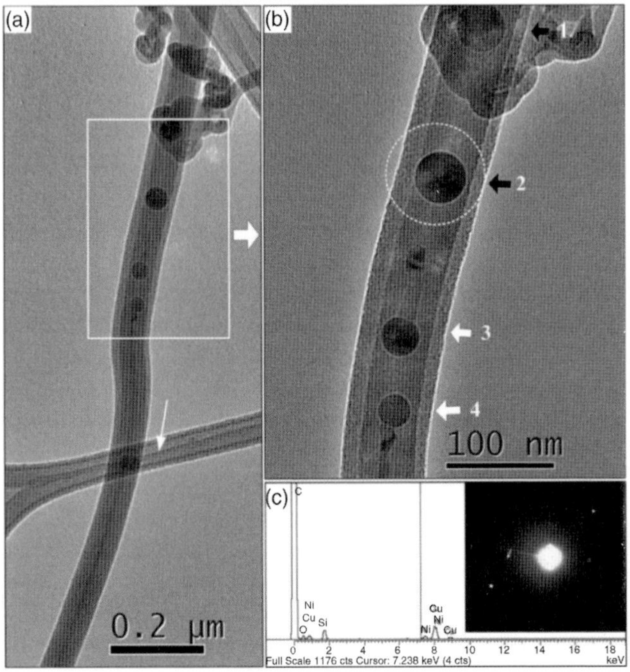

Figure 17.36 Entrapment of Ni nanospheres in a triple-concentric nanowire [91]. (a) Entrapment of four Ni nanospheres in the amorphous SiO_y core, where the triple-concentric nanowire transits to SiO_y/SiO_x coaxial one. (b) Magnification of the Ni nanosphere-entrapped Si nanowire shows the gradual disappearance of the Si core (white arrow) till the complete entrapping of the Ni nanospheres (dark arrow) in the SiO_y core. (c) SAED pattern and EDX spectrum generated from nanosphere 2.

addition of oxygen could increase the amount of the entrapped nanospheres. The continuous oxidation of the Au—Si phase in the Au—Si—O droplet could accelerate the precipitation of the Au nanospheres from the droplet (cf. Figure 6 in Ref. [94]).

Two distinct mechanisms are developed by the same group to explain the entrapping of the periodic chains of Au/Si [92] or Au nanospheres [93], formed during thermal annealing of a Au/a-Si/Si wafer [92] or during thermal annealing of the Au/SiN_z/Si wafer sandwiched with a Au-coated Si wafer [93], respectively. The first one believes the Au/Si nanospheres are entrapped by dislodging of the two necks of the supersaturated liquid Au—Si alloy during the extension of the alloy nanosphere [92]. In the liquid–solid interface, "consumption" of the constituents supersaturates the Au—Si alloy droplet, and there will be an increase in the molten curvature, thus forming neck "A". This is the growth of the nanospheres. Further extension of the nanosphere will reduce the supersaturation and thereby the curvature. As such, neck "B" forms. Finally, the formation of the nanosphere completes by dislodging from both the necks with the limited "supply" of material. The oxide sheath is produced by the oxidation in the atmosphere. In that case, however, the nanowire should have a sinusoidal sheath, rather than the smooth one. Moreover, Au and Si cannot be the solute in the Au—Si droplet at the same time to reach supersaturation.

The second theory believes the growth of Au-nanopeapodded nanowires starts with the formation of a silica nanotube with a gold attachment at the end [93]. With time, Au starts to flow into the hollow nanocavity due to the capillary force. With prolonged growth, the gold nanowires transform into a droplet-like chain owing to insufficient supply of gold, and finally evolve into a peapod. In fact, however, capillary force could not work because the silica nanotube has never been observed.

Lastly, it is reported that the Si core of the grown coaxial nanowire is formed by the phase separation of Si in the Au—Si droplet during thermal evaporation of SiO powders [94], or thermal annealing of a piece of Au/SiO_2/Si wafer [145]. In these processes, Si originates from the disproportionation of SiO; and the oxide sheath is directly formed by the disproportionate SiO_2 without entering the Au—Si droplet. Additional amount of oxygen changes the ratio of the disproportionate Si and SiO_2, that is, the amount of SiO_2 increases, and thus that of Si decreases. Therefore, the oxide sheath could grow faster than the Si core, resulting in the deformation of the Au—Si droplet into a cylindrical shape. Till a certain length, the liquid Au—Si droplet becomes unstable and pinches off periodically to entrap the Au/Si particles in the nanowire. However, commercially available SiO powders were not 100% SiO, but an inhomogeneous mixture of amorphous SiO_2 and amorphous Si with SiO phase at the interface [96–99]. Upon thermal evaporating the SiO powder at 1050 °C under 25 mbar nitrogen atmosphere, Si atoms and SiO and SiO_2 molecules could be sublimated by conversion of thermal energy to kinetic energy; meanwhile, the SiO microparticles could be ejected into space due to the volume expansion of the residual gas [84], and then land on the tube wall. The vaporized atoms or molecules have a limited mean free path of around 18.3 μm [84,86,87]. Therefore, they would collide with other atoms or molecules and thus lose energy. As such, they could not freely reach the location where the nanowire grows. Therefore, Si in the nanowire could only come from the Si portion in the ejected Si microparticles. As discussed, the disproportionation of SiO takes place through the migration of electrons [97]. There is no physical movement of Si atoms or SiO_2 molecules in this reaction. After the reaction, the particle becomes a microscopic mixture of Si and SiO_2 clusters. These clusters are not separable, but still confined in the particle, not able to move freely and separately in space [75]. Though Si has a high melting point of 1410 °C [83], Si and Au could form eutectic droplet at a low temperature of 363 °C to provide the mobile Si atoms (the growth temperature is 850–900 °C) [90]. The disproportionate SiO_2 molecules are still in the particle due to the high melting point (1703 °C) [83], thus unable to move around to form the oxide sheath. The growth takes place under nitrogen atmosphere (not a high-vacuum ambient). Therefore, the Au—Si droplet is easy to be partially oxidized into the Au—Si—O droplet. As such, the only source of the oxide sheath is the liquid Au—Si—O droplet, rather than the disproportionation of SiO. To summarize, what causes the entrapping of metal nanospheres is still unclear. To date, it is found that the entrapment takes place only in the metal-catalyzed growth, where the metallic catalyst seeds the growth at the root of the nanowire, and Si comes either from the Si-containing powder particles [66,94] or from the Si films [92,93,145].

17.3.4.6 **Transitional Nanowire**

Figure 17.37a shows a Si/SiO_x coaxial nanowire gradually transits to a side-by-side Si/SiO_x biaxial structure. The Si core diameter at the top is about 20.9 nm, decreasing to 18.7 nm at the bottom. Meanwhile, the SiO_x sheath thickness at the right-hand side gradually increases from 5.1 to 8.3 nm. The SAED pattern is indexed for the Si [111] zone axis, implying the growth occurs in the $[\bar{1}10]$ orientation. The dihedral angle 120.0° perfectly matches the theoretical dihedral angle between the atomic planes of $(2\bar{2}0)$ and $(20\bar{2})$. The white line-shaped pattern indicates microtwins or stacking faults [156], consistent with the white shining lines directly observed in the dark Si side. Note that the a-C layer that is sputtered on the Ni catalyst is only for retardation of the oxidation of the Ni—Si droplet. It is unable to melt with the Ni catalyst to form eutectic Ni—C droplet since the eutectic melting of Ni—C is at 1326.5 °C [90], 226.5 °C higher than the growth temperature of 1100 °C. SiC is impossible to form. Thus, the nanowire could not be SiC. As it is Si nanowire, the lattice parameter is 5.43 Å. From the odd diffraction patterns shown in Figure 17.37a, it is difficult to ensure that they are stacking faults or microtwins. However, the diffraction patterns could be indexed to Si [111] zone axis, thus revealing the stacking faults or twins take place on the plane of Si (110). As shown in Figure 17.37b, the wire is connected to a Ni particle with a horizontal width of 44.6 nm and a vertical height of 51.4 nm. The particle is surrounded by a 2.9 nm-thick SiO_x layer. The Si core offsets a little to the left, exhibiting a side-by-side biaxial morphology. As it

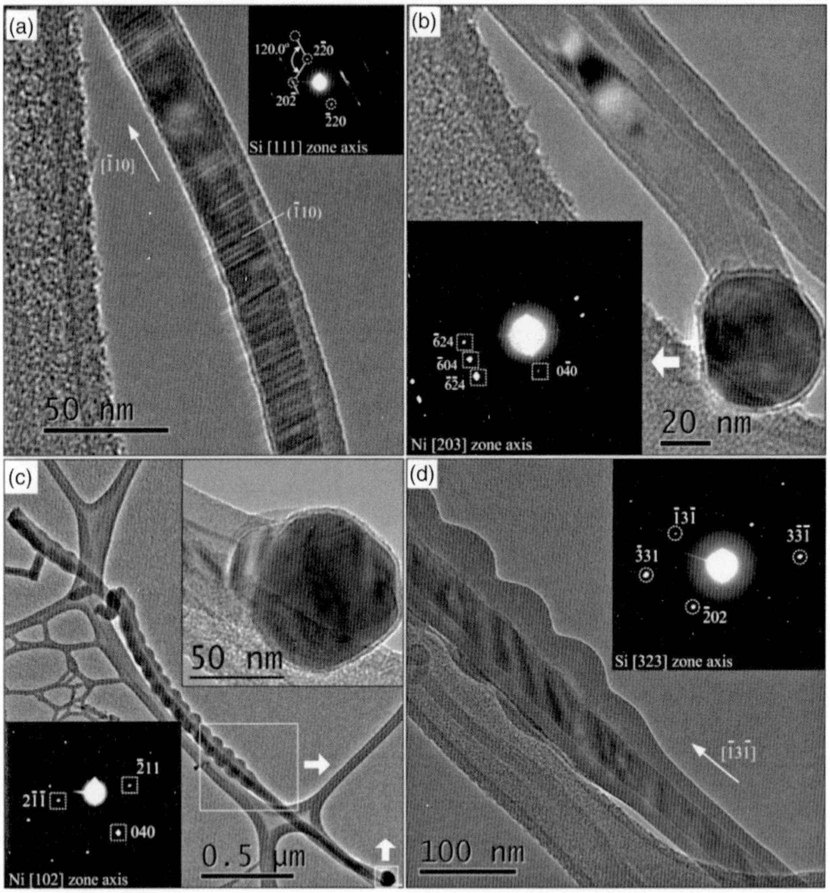

Figure 17.37 Si/Si oxide nanowires with transitional morphologies [91]. (a) Transition of a [$\bar{1}$10]-oriented Si/SiO_x nanowire from the coaxial structure (at the top) to the side-by-side Si/SiO_x biaxial structure (at the bottom). The SAED pattern in the inset is generated from the zone axis of Si [111]. (b) Cross-sectional view of the Ni particle connected with the wire in part (a). The SAED pattern in the inset is generated from the zone axis of Ni [203]. (c) Transition of a nanowire from the smooth surface profile (at the bottom) to the sinusoidal surface profile (at the top). The SAED pattern in the inset is generated from the zone axis of Ni [102]. (d) The wire grows along the [$\bar{1}$3$\bar{1}$]orientation. The SAED pattern is generated from the zone axis of Si [323].

proceeds, the wire becomes coaxial at the top of the nanowire. The SAED pattern shown in the bottom-left inset is indexed for the Ni [203] zone axis, surrounded by (020), (0$\bar{2}$0), ($\bar{3}$12), ($\bar{3}$02), ($\bar{3}\bar{1}$2), (31$\bar{2}$), (30$\bar{2}$), and (3$\bar{1}\bar{2}$) planes.

Figure 17.37c shows a nanowire transits from the sinusoidal surface profile (85.8 nm in period, 87.7 nm in valley diameter, 112.8 nm in peak diameter) to the smooth coaxial one (21.3 nm in core diameter, 15.6 nm in sheath thickness).

The seeding Ni particle (81.1 nm in diameter) exhibits a spherical Ni–Si interface. A 3.0 nm-thick SiO_x layer is surrounding the particle (cf. upper-right inset, Figure 17.37c). The SAED pattern shown at the bottom-left corner is indexed for the zone axis of Ni [102], indicating the single-crystalline structure of the Ni particle. Figure 17.37d enlarges the transition morphology of the Si nanowire, where the SAED pattern is indexed for the zone axis of Si [323], determining the wire is along the $[\bar{1}3\bar{1}]$ orientation.

17.3.4.7 Growth Mechanisms of the Various Morphologies

The nanowire growth in the work of Li *et al.* [32,91], Si vapor pressure of around 40.0 Pa, could be produced during thermal annealing the Si wafer at 1100 °C and under 1.08×10^5 Pa Ar atmosphere [83]. However, the mean free path of the vaporized Si atoms is only around 0.76 μm [84,86,87]. Moreover, the Si wafer is covered with a Ni layer, on top of which a thick a-C layer is sputtered. The Ni—Si—O nanoclusters are located at the root of the grown nanowires. Therefore, Si atoms could not come from the vapor, but only from the Si wafer substrate through the eutectic melting of Ni and Si at around 800–964 °C at the Ni–Si interface [90]. The actual eutectic temperature changes with the Si atomic concentration. The liquid droplet serves as a preferred site for melting of the surrounding solid-phase Si and Ni. Since Ar flow is provided by the gas cylinder placed at room temperature, Ar surrounds the droplet and exchanges energy and momentum with the atoms in the droplet, giving rise to overcooling effect of the droplet, and seeding the preferential growth of the Si nanowire at the root, once Si reaches supersaturation in the Ni—Si droplet. The driving force of the supersaturation of Si is attributed to the continuous dissolution of Si atoms from the beneath Si wafer into the Ni—Si droplet at 1100 °C. During the growth, the droplet directs the growth orientation and defines the diameter of the nanowire. Ultimately, the growth terminates when the temperature is below the eutectic melting point of the Ni—Si droplet or the reactants are run out. Though the residual oxygen could be partially retarded by the top carbon layer at the beginning of the growth, it could still oxidize the droplet into Ni—Si—O droplet when the a-C top layer becomes porous or cracked. Upon supersaturation of Si in the Ni—Si—O droplet, Si oxide is also precipitated from the droplet to form the oxide side of the Si nanowire. The single-crystalline Ni particle comes from the crystallization of the Ni—Si—O droplet in the final cooling stage of the annealing process. The above growth process is difficult to be observed *in situ.* It is well known that Si nanowires are grown via precipitation of the supersaturated Si from metal–Si droplet [20–22]. Based on the various morphologies, however, it is believed that the morphology of the nanowires is related to the seeding Ni—Si—O droplet.

Table 17.7 shows that the morphology of the grown Si nanowires is related to the diameter of the seeding Ni particles. The particles are embedded in the Ni—Si—O nanoclusters. They could be grouped into two categories according to the diameter: (i) smaller than 80 nm (seeding coaxial Si/SiO_x or side-by-side Si/SiO_x biaxial nanowires); (ii) larger than 80 nm (seeding smooth or sinusoidal

Table 17.7 Size distribution of the seeding Ni particles and the grown Si/Si oxide nanowires [91].

Morphology	Si thickness (nm)	SiO_y thickness (nm)	SiO_x thickness (nm)	Ni diameter (nm)	References
Coaxial smooth	19.1,20.1	0	23.1,8.6	65.0	Figure 17.33a and b
Side-by-side biaxial smooth	13.2	0	15.3	52.0	Figure 17.34
Side-by-side to coaxial	18.7–20.9	0	5.1–8.3	44.6–51.4	Figure 17.37a and b
Triple-concentric smooth	33.3, 31.8	14.1,15.7	17.6,17.8	—	Figure 17.27a and Figure 17.35a
Triple-concentric sinusoidal (valley)	24.7	13.3	11.7	—	Figure 17.35b
Triple-concentric sinusoidal (peak)	29.6	13.5	19.2		
Fishbone-profiled (valley)	54.6	8.6–26.2	10.0 (bottom), 39.9 (top)	95.5–128.3	Figure 17.35c and d
Fishbone-profiled (peak)	68.4		32.6 (bottom), 66.4 (top)		
Ni-nanosphere-entrapped	14.0	19.8–53.6	17.8	—	Figure 17.36a and b
Sinusoidal to smooth	21.3	0	15.6	81.3	Figure 17.37c and d

triple-phased nanowires). During thermal annealing, it is hard to control the diameter of the Ni particles that are broken from the Ni catalyst layer. On the other hand, the diameter of the Ni—Si—O eutectic droplet determines that of the seeding Ni particle that is crystallized at the end of the growth. As such, many phases in the nanowire are dictated by the size of the Ni—Si—O droplet. In the work of Li, the structure of the droplets is not depicted based on *in situ* observation, but according to that of the grown nanowires. It is believed that the Ni—Si—O droplet is constructed with Ni—Si and Ni—Si—O_y phases at the beginning of the growth process. With time, the Ni—Si—O_y phase is gradually oxidized due to the surrounding residual oxygen in the annealing chamber. This inward oxidation results in the interface at interlayer/outer sheath. The oxidation is related to the size of the droplet. In the case of the partial oxidation of the Ni—Si—O_y into Ni—Si—O_x phase, triple-concentric Si/SiO_y/SiO_x nanowire forms. The complete oxidation of the Ni—Si—O_y into Ni—Si—O_x phase results in the coaxial Si/SiO_x nanowire.

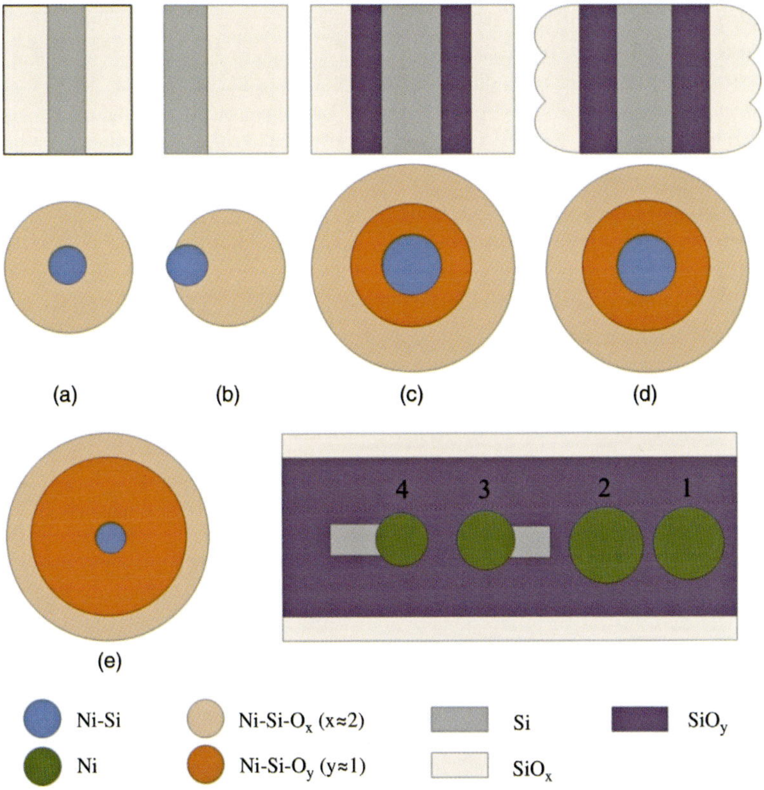

Figure 17.38 Schematic diagram of the Ni—Si—O droplets seeding the Si/Si oxide nanowires with various morphologies [91]. (a) Concentric Ni—Si—O droplet seeds the nanowire with Si core coaxially surrounded by SiO_x sheath. (b) Decentered Ni—Si—O droplet seeds the side-by-side Si/SiO_x biaxial nanowire. (c) Triple-concentric Ni—Si—O droplet seeds the smooth Si/SiO_y/SiO_x triple-concentric nanowire. (d) Periodically vibrated Ni—Si—O droplet seeds the Si/SiO_y/SiO_x sinusoidal and sometimes even fishbone-profiled nanowire. (e) Entrapment of Ni nanospheres in the nanowire via eutectic precipitation of Ni and Si in the Si supersaturated Ni—Si—O droplet. *Notes:* The droplet is related to the Ni particle that is embedded in the resultant Ni—Si—O nanocluster. The diameter of the Ni particles in parts (a and b) is smaller than 80 nm, while that in parts (c–e) is bigger than 80 nm.

Figure 17.38 schematically sketches the growth mechanism for the various morphologies. In Figure 17.38a and b, the Ni particles have a diameter smaller than 80 nm. In Figure 17.38c–e, the particles are bigger than 80 nm in diameter. The Si/SiO_x coaxial nanowire is grown through precipitating the supersaturated Si and SiO_x phases from the Ni—Si—O droplet, where the Ni—Si inner phase is concentrically surrounded by the Ni—Si—O_x outer phase in the Ni—Si—O droplet (cf. Figure 17.38a and Figure 17.33); while the side-by-side Si/SiO_x biaxial one is seeded by the droplet with decentered Ni—Si and Ni—Si—O_x phases in the Ni—Si—O droplet (cf. Figure 17.38b and Figure 17.34). The coaxial nanowire transits to side-by-side biaxial structure as the Ni—Si and Ni—Si—O phases in the Ni—Si—O droplet become decentered (cf. Figure 17.37a and b). The smooth Si/SiO_y/SiO_x triple-concentric nanowire is precipitated from the Si, SiO_y, and SiO_x supersaturated Ni—Si—O concentrical droplet (cf. Figure 17.38c and Figure 17.35a). The periodic vibration of the Ni—Si—O triple-concentric droplet gives rise to the sinusoidal or even fishbone-profiled Si/SiO_y/SiO_x nanowire (cf. Figure 17.38d and Figure 17.35b–d). As the droplet stays central symmetrically ball-shaped throughout the growth process, the grown nanowire is of a smooth triple-concentric structure (cf. Figure 17.35a). The periodical vibrating of the droplet is responsible for the sinusoidal triple-concentric ones (cf. Figure 17.35b). The fishbone-profiled Si nanowire is precipitated from the droplet with the most violent vibration (cf. Figure 17.35c and d). The sinusoidal nanowire becomes smooth, as the droplet stops the vibration (cf. Figure 17.37c and d). This work believes that the periodic vibration of the droplet comes from the surface energy effects and experimental conditions such as temperature, supersaturation, and impurities [21], the roughening transition at the liquid–solid interface, the area of the liquid–solid interface, and the contact angle of the droplet seeding the nanowire [92].

With time, Ni dissolves from the underneath Ni layer or surrounding Ni particles like the dissolution of Si from the underneath Si wafer. Therefore, both Ni and Si in the droplet are variables. In the Ni—Si eutectic phase diagram [90], different Ni phases would precipitate out if Si is lower than 41 at.% in the Ni—Si alloy droplet. If the droplet has more than 50 at.% Si, NiSi, $\alpha NiSi_2$, or $\beta NiSi_2$ alloy would precipitate during cooling. In this work, SAED pattern confirms the seeding particles are single-crystalline Ni, not NiSi or $NiSi_2$ (cf. Figure 17.33b and Figure 17.37b); therefore, Si concentration in the droplet should be lower than 41 at.% during the nanowire growth and the subsequent cooling process. As such, Ni and Si phases are immiscible in the Ni—Si—O droplet during cooling. They would separate via eutectic precipitation. Hence, both the seeding Ni particle and the entrapped Ni nanospheres are formed via the eutectic precipitation of Ni and Si in the cooling stage. Given supersaturated Si in the Ni—Si—O droplet in this stage, the precipitated Ni nanospheres would be entrapped out together with the precipitation of Si (cf. Figure 17.38e). As a result, Ni nanospheres are entrapped in the Si/Si oxide nanowires.

17.3.4.8 **Conclusion**

Thermal annealing of Ni-covered Si wafers gives rise to Si/Si oxide nanowires with various morphologies, that is, coaxial, side-by-side biaxial, smooth or sinusoidal triple-concentric, fishbone-profiled, Ni nanosphere entrapped, and the transitional morphologies. The morphology is controlled by the diameter, phase distribution, vibration, and eutectic precipitation of the seeding Ni—Si—O droplet. The oxidation of the droplet is the key in determining the phase number in the nanowire. The partial oxidation of the Ni—Si—O_y ($y = 1$) phase into Ni—Si—O_x ($x = 2$) phase in the bigger droplets (that is Ni particle larger than 80 nm in this work) results in triple-phased Si/SiO_y/SiO_x nanowires.

Double-phased Si/SiO_x nanowires result from the complete oxidation of the Ni—Si—O_y phase into Ni—Si—O_x phase in the smaller droplet. Vibration of the Ni—Si—O droplet results in nanowires of a sinusoidal profile. Extreme vibration gives rise to the fishbone morphology. At no vibration, the smooth surface profile is obtained. The side-by-side biaxial structure appears when the droplet is off-center. Given Si-supersaturated Ni—Si—O droplet during cooling, Ni nanospheres would be entrapped in the nanowire as a result of the eutectic precipitation of Ni and Si.

Acknowledgment

This chapter was supported by the Ministry of Education's Research Grant 208A1218 ARC4/08 and Nanyang Technological University Singapore.

References

1 Sze, S.M. and Ng, K.K. (2006) *Physics of Semiconductor Devices*, 3rd edn, John Wiley & Sons, Inc., Hoboken.

2 Xia, Y., Yang, P., Sun, Y., Wu, Y., Mayers, B., Gates, B. *et al.* (2003) One-dimensional nanostructures: synthesis, characterization, and applications. *Adv. Mater.*, **15**, 353–389.

3 Huang, Y., Duan, X., and Lieber, C.M. (2005) Nanowires for integrated multicolor nanophotonics. *Small*, **1**, 142–147.

4 Goodey, A.P., Eichfeld, S.M., Lew, K.K., Redwing, J.M., and Mallouk, T.E. (2007) Silicon nanowire array photoelectrochemical cells. *J. Am. Chem. Soc.*, **129**, 12344–12345.

5 Cui, Y., Wei, Q., Park, H., and Lieber, C.M. (2001) Nanowire nanosensors for highly sensitive and selective detection of biological and chemical species. *Science*, **293**, 1289–1292.

6 Bashouti, M.Y., Tung, R.T., and Haick, H. (2009) Tuning the electrical properties of Si nanowire field-effect transistors by molecular engineering. *Small*, **5**, 2761–2769.

7 Zheng, G., Gao, X.P.A., and Lieber, C.M. (2010) Frequency domain detection of biomolecules using silicon nanowire biosensors. *Nano Lett.*, **10**, 3179–3183.

8 Chan, C.K., Peng, H., Liu, G., McIlwrath, K., Zhang, X.F., Huggins, R.A. *et al.* (2008) High-performance lithium battery anodes using silicon nanowires. *Nat. Nanotechnol.*, **3**, 31–35.

9 Zhang, Q., Zhang, W., Wan, W., Cui, Y., and Wang, E. (2010) Lithium insertion in silicon nanowires: an *ab initio* study. *Nano Lett.*, **10**, 3243–3249.

10 He, L., Lai, D., Wang, H., Jiang, C., and Rusli (2012) High-efficiency Si/polymer hybrid solar cells based on synergistic surface texturing of Si nanowires on pyramids. *Small*, **8**, 1664–1668.

11 Boukai, A.I., Bunimovich, Y., Tahir-Kheli, J., Yu, J.K., Goddard, W.A. III, and Heath, J.R. (2008) Silicon nanowires as efficient thermoelectric materials. *Nature*, **451**, 168–171.

12 Ng, M.F., Zhou, L., Yang, S.W., Sim, L.Y., Tan, V.B.C., and Wu, P. (2007) Theoretical investigation of silicon nanowires: methodology, geometry, surface modification, and electrical conductivity using a multiscale approach. *Phys. Rev. B Condens. Matter Mater. Phys.*, **76**, 155435.

13 Schmidt, V., Riel, H., Senz, S., Karg, S., Riess, W., and Gösele, U. (2006) Realization of a silicon nanowire vertical surround-gate field-effect transistor. *Small*, **2**, 85–88.

14 Weber, W.M., Geelhaar, L., Graham, A.P., Unger, E., Duesberg, G.S., Liebau, M. *et al.* (2006) Silicon-nanowire transistors with intruded nickel-suicide contacts. *Nano Lett.*, **6**, 2660–2666.

15 Kang, K., Lee, H.S., Han, D.W., Kim, G.S., Lee, D., Lee, G. *et al.* (2010) Maximum Li storage in Si nanowires for the high capacity three-dimensional Li-ion battery. *Appl. Phys. Lett.*, **96**, 053110.

16 Peng, K., Xu, Y., Wu, Y., Yan, Y., Lee, S.T., and Zhu, J. (2005) Aligned single-crystalline Si nanowire arrays for photovoltaic applications. *Small*, **1**, 1062–1067.

17 Dick, K.A. (2008) A review of nanowire growth promoted by alloys and non-alloying elements with emphasis on Au-assisted III–V nanowires. *Prog. Cryst. Growth Charact. Mater.*, **54**, 138–173.

18 Schmidt, V., Wittemann, J.V., and Gösele, U. (2010) Growth, thermodynamics, and electrical properties of silicon nanowires. *Chem. Rev.*, **110**, 361–388.

19 Morales, A.M. and Lieber, C.M. (1998) A laser ablation method for the synthesis of crystalline semiconductor nanowires. *Science*, **279**, 208–211.

20 Wagner, R.S. and Ellis, W.C. (1964) Vapor–liquid–solid mechanism of single crystal growth. *Appl. Phys. Lett.*, **4**, 89–90.

21 Givargizov, E.I. (1973) Periodic instability in whisker growth. *J. Cryst. Growth*, **20**, 217–226.

22 Givargizov, E.I. (1975) Fundamental aspects of VLS growth. *J. Cryst. Growth*, **31**, 20–30.

23 Wu, Y., Cui, Y., Huynh, L., Barrelet, C.J., Bell, D.C., and Lieber, C.M. (2004) Controlled growth and structures of molecular-scale silicon nanowires. *Nano Lett.*, **4**, 433–436.

24 Holmes, J.D., Johnston, K.P., Doty, R.C., and Korgel, B.A. (2000) Control of thickness and orientation of solution-grown silicon nanowires. *Science*, **287**, 1471–1473.

25 Heitsch, A.T., Fanfair, D.D., Tuan, H.Y., and Korgel, B.A. (2008) Solution–liquid–solid (SLS) growth of silicon nanowires. *J. Am. Chem. Soc.*, **130**, 5436–5437.

26 Zakharov, N., Werner, P., Sokolov, L., and Gösele, U. (2007) Growth of Si whiskers by MBE: mechanism and peculiarities. *Physica E*, **37**, 148–152.

27 Schubert, L., Werner, P., Zakharov, N.D., Gerth, G., Kolb, F.M., Long, L. *et al.* (2004) Silicon nanowhiskers grown on ⟨111⟩Si substrates by molecular-beam epitaxy. *Appl. Phys. Lett.*, **84**, 4968–4970.

28 Wang, N., Tang, Y.H., Zhang, Y.F., Lee, C.S., and Lee, S.T. (1998) Nucleation and growth of Si nanowires from silicon oxide. *Phys. Rev. B Condens. Matter Mater. Phys.*, **58**, R16024–R16026.

29 Wang, N., Tang, Y.H., Zhang, Y.F., Lee, C.S., Bello, I., and Lee, S.T. (1999) Si nanowires grown from silicon oxide. *Chem. Phys. Lett.*, **299**, 237–242.

30 Ma, D.D.D., Lee, C.S., Au, F.C.K., Tong, S.Y., and Lee, S.T. (2003) Small-diameter silicon nanowire surfaces. *Science*, **299**, 1874–1877.

31 Yan, H.F., Xing, Y.J., Hang, Q.L., Yu, D.P., Wang, Y.P., Xu, J. *et al.* (2000) Growth of amorphous silicon nanowires via a solid–liquid–solid mechanism. *Chem. Phys. Lett.*, **323**, 224–228.

32 Li, F.J., Zhang, S., Kong, J.H., Guo, J., Cao, X.B., and Li, B. (2013) Growth of crystalline silicon nanowires on nickel-coated silicon wafer beneath sputtered amorphous carbon. *Thin Solid Films*, **534**, 90–99.

33 Hochbaum, A.I., Chen, R., Delgado, R.D., Liang, W., Garnett, E.C., Najarian, M. *et al.* (2008) Enhanced thermoelectric performance of rough silicon nanowires. *Nature*, **451**, 163–167.

34 Huang, Z., Fang, H., and Zhu, J. (2007) Fabrication of silicon nanowire arrays with controlled diameter, length, and density. *Adv. Mater.*, **19**, 744–748.

35 Huang, Z., Zhang, X., Reiche, M., Ltu, L., Lee, W., Shimizu, T. *et al.* (2008) Extended arrays of vertically aligned sub-10nm diameter [100] Si nanowires by metal-assisted chemical etching. *Nano Lett.*, **8**, 3046–3051.

36 Huang, Z., Geyer, N., Werner, P., Boor, J.D., and Gösele, U. (2011) Metal-assisted chemical etching of silicon: a review. *Adv. Mater.*, **23**, 285–308.

37 Garnett, E.C., Liang, W., and Yang, P. (2007) Growth and electrical characteristics of platinum-nanoparticle-catalyzed silicon nanowires. *Adv. Mater.*, **19**, 2946–2950.

38 Kayes, B.M., Filler, M.A., Putnam, M.C., Kelzenberg, M.D., Lewis, N.S., and Atwater, H.A. (2007) Growth of vertically aligned Si wire arrays over large areas (>1 cm^2) with Au and Cu catalysts. *Appl. Phys. Lett.*, **91**, 103110.

39 Hochbaum, A.I., Fan, R., He, R., and Yang, P. (2005) Controlled growth of Si nanowire arrays for device integration. *Nano Lett.*, **5**, 457–460.

40 Lombardi, I., Hochbaum, A.I., Yang, P., Carraro, C., and Maboudian, R. (2006) Synthesis of high density, size-controlled Si nanowire arrays via porous anodic alumina mask. *Chem. Mater.*, **18**, 988–991.

41 Nebol'sin, V.A., Shchetinin, A.A., Dolgachev, A.A., and Korneeva, V.V. (2005) Effect of the nature of the metal solvent on the vapor–liquid–solid growth rate of silicon whiskers. *Inorg. Mater.*, **41**, 1256–1259.

42 Lew, K.K. and Redwing, J.M. (2003) Growth characteristics of silicon nanowires synthesized by vapor–liquid–solid growth in nanoporous alumina templates. *J. Cryst. Growth*, **254**, 14–22.

43 Chung, S.W., Yu, J.Y., and Heath, J.R. (2000) Silicon nanowire devices. *Appl. Phys. Lett.*, **76**, 2068–2070.

44 Schmidt, V., Senz, S., and Gösele, U. (2005) Diameter-dependent growth direction of epitaxial silicon nanowires. *Nano Lett.*, **5**, 931–935.

45 Wagner, R.S., Ellis, W.C., Jackson, K.A., and Arnold, S.M. (1964) Study of the filamentary growth of silicon crystals from the vapor. *J. Appl. Phys.*, **35**, 2993–3000.

46 Chakraverty, B.K. (1967) Grain size distribution in thin films-1: conservative systems. *J. Phys. Chem. Solids*, **28**, 2401–2412.

47 Lifshitz, I.M. and Slyozov, V.V. (1961) The kinetics of precipitation from supersaturated solid solutions. *J. Phys. Chem. Solids*, **19**, 35–50.

48 Wang, Y., Schmidt, V., Senz, S., and Gösele, U. (2006) Epitaxial growth of silicon nanowires using an aluminium catalyst. *Nat. Nanotechnol.*, **1**, 186–189.

49 Schmid, H., Björk, M.T., Knoch, J., Riel, H., Riess, W., Rice, P. *et al.* (2008) Patterned epitaxial vapor–liquid–solid growth of silicon nanowires on Si(111) using silane. *J. Appl. Phys.*, **103**, 024304.

50 Kodambaka, S., Tersoff, J., Reuter, M.C., and Ross, F.M. (2006) Diameter-independent kinetics in the vapor–liquid–solid growth of Si nanowires. *Phys. Rev. Lett.*, **96**, 096105.

51 Sunkara, M.K., Sharma, S., Miranda, R., Lian, G., and Dickey, E.C. (2001) Bulk synthesis of silicon nanowires using a low-temperature vapor–liquid–solid method. *Appl. Phys. Lett.*, **79**, 1546–1548.

52 Zhang, X.Y., Zhang, L.D., Meng, G.W., Li, G.H., Jin-Phillipp, N.Y., and Phillipp, F. (2001) Synthesis of ordered single crystal silicon nanowire arrays. *Adv. Mater.*, **13**, 1238–1241.

53 Lensch-Falk, J.L., Hemesath, E.R., Perea, D.E., and Lauhon, L.J. (2009) Alternative catalysts for VSS growth of silicon and germanium nanowires. *J. Mater. Chem.*, **19**, 849–857.

54 Wacaser, B.A., Dick, K.A., Johansson, J., Borgström, M.T., Deppert, K., and Samuelson, L. (2009) Preferential interface nucleation: an expansion of the VLS growth mechanism for nanowires. *Adv. Mater.*, **21**, 153–165.

55 Hofmann, S., Sharma, R., Wirth, C.T., Cervantes-Sodi, F., Ducati, C., Kasama, T. *et al.* (2008) Ledge-flow-controlled catalyst interface dynamics during Si nanowire growth. *Nat. Mater.*, **7**, 372–375.

56 Wen, C.Y., Reuter, M.C., Tersoff, J., Stach, E.A., and Ross, F.M. (2010) Structure, growth kinetics, and ledge flow during vapor–solid–solid growth of copper-catalyzed silicon nanowires. *Nano Lett.*, **10**, 514–519.

57 Kamins, T.I., Williams, R.S., Basile, D.P., Hesjedal, T., and Harris, J.S. (2001) Ti-catalyzed Si nanowires by chemical vapor deposition: microscopy and growth mechanisms. *J Appl. Phys.*, **89**, 1008–1016.

58 Baron, T., Gordon, M., Dhalluin, F., Ternon, C., Ferret, P., and Gentile, P. (2006) Si nanowire growth and characterization using a microelectronics-compatible catalyst: PtSi. *Appl. Phys. Lett.*, **89**, 233111.

59 Lee, D.C., Hanrath, T., and Korgel, B.A. (2005) The role of precursor-decomposition kinetics in silicon-nanowire synthesis in organic solvents. *Angew. Chem., Int. Ed.*, **44**, 3573–3577.

60 Lu, X., Hanrath, T., Johnston, K.P., and Korgel, B.A. (2003) Growth of single crystal silicon nanowires in supercritical solution from tethered gold particles on a silicon substrate. *Nano Lett.*, **3**, 93–99.

61 Cui, Y., Lauhon, L.J., Gudiksen, M.S., Wang, J., and Lieber, C.M. (2001) Diameter-controlled synthesis of single-crystal silicon nanowires. *Appl. Phys. Lett.*, **78**, 2214–2216.

62 Tuan, H.Y., Lee, D.C., and Korgel, B.A. (2006) Nanocrystal-mediated crystallization of silicon and germanium nanowires in organic solvents: the role of catalysis and solid-phase seeding. *Angew. Chem., Int. Ed.*, **45**, 5184–5187.

63 Zhang, Y.F., Tang, Y.H., Peng, H.Y., Wang, N., Lee, C.S., Bello, I. *et al.* (1999) Diameter modification of silicon nanowires by ambient gas. *Appl. Phys. Lett.*, **75**, 1842–1844.

64 Zhang, R.Q., Lifshitz, Y., and Lee, S.T. (2003) Oxide-assisted growth of semiconducting nanowires. *Adv. Mater.*, **15**, 635–640.

65 Zhang, Y.F., Tang, Y.H., Wang, N., Yu, D.P., Lee, C.S., Bello, I. *et al.* (1998) Silicon nanowires prepared by laser ablation at high temperature. *Appl. Phys. Lett.*, **72**, 1835–1837.

66 Tang, Y.H., Zhang, Y.F., Wang, N., Lee, C.S., Han, X.D., Bello, I. *et al.* (1999) Morphology of Si nanowires synthesized by high-temperature laser ablation. *J. Appl. Phys.*, **85**, 7981–7983.

67 Zhang, Y.F., Tang, Y.H., Wang, N., Lee, C.S., Bello, I., and Lee, S.T. (1999) One-dimensional growth mechanism of crystalline silicon nanowires. *J. Cryst. Growth*, **197**, 136–140.

68 Shi, W.S., Peng, H.Y., Zheng, Y.F., Wang, N., Shang, N.G., Pan, Z.W. *et al.* (2000) Synthesis of large areas of highly oriented, very long silicon nanowires. *Adv. Mater.*, **12**, 1343–1345.

69 Shi, W., Peng, H., Wang, N., Chi Pui, L., Xu, L., Chun Sing, L. *et al.* (2001) Free-standing single crystal silicon nanoribbons. *J. Am. Chem. Soc.*, **123**, 11095–11096.

70 Li, C.P., Lee, C.S., Ma, X.L., Wang, N., Zhang, R.Q., and Lee, S.T. (2003) Growth direction and cross-sectional study of silicon nanowires. *Adv. Mater.*, **15**, 607–609.

71 Demars, C., Pagel, M., Deloule, E., and Blanc, P. (1996) Cathodoluminescence of quartz from sandstones: interpretation of the UV range by determination of trace element distributions and fluid-inclusion P-T-X properties in anthigenic quartz. *Am. Mineral.*, **81**, 891–901.

72 Flem, B., Larsen, R.B., Grimstvedt, A., and Mansfeld, J. (2002) *In situ* analysis of trace elements in quartz by using laser ablation inductively coupled plasma mass spectrometry. *Chem. Geol.*, **182**, 237–247.

73 Müller, A., Wiedenbeck, M., Van Den Kerkhof, A.M., Kronz, A., and Simon, K. (2003) Trace elements in quartz: a combined electron microprobe, secondary ion mass spectrometry, laser-ablation ICP-MS, and cathodoluminescene study. *Eur. J. Mineral.*, **15**, 747–763.

74 Götze, J., Plötze, M., Graupner, T., Hallbauer, D.K., and Bray, C.J. (2004) Trace element incorporation into quartz: a combined study by ICP-MS, electron spin resonance, cathodoluminescence, capillary ion analysis, and gas chromatography. *Geochim. Cosmochim. Acta*, **68**, 3741–3759.

75 Li, F.J., Zhang, S., and Lee, J.W. (2014) Rethinking of the silicon nanowire growth mechanism during thermal evaporation of Si-containing powders. *Thin Solid Films*, **558**, 75–85.

76 Islam, M.N., Ghosh, T.B., Chopra, K.L., and Acharya, H.N. (1996) XPS and X-ray diffraction studies of aluminum-doped zinc oxide transparent conducting films. *Thin Solid Films*, **280**, 20–25.

77 Ertl, G., Hierl, R., Knözinger, H., Thiele, N., and Urbach, H.P. (1980) XPS study of copper aluminate catalysts. *Appl. Surf. Sci.*, **5**, 49–64.

78 Chen, M., Wang, X., Yu, Y.H., Pei, Z.L., Bai, X.D., Sun, C. *et al.* (2000) X-ray photoelectron spectroscopy and auger electron spectroscopy studies of Al-doped ZnO films. *Appl. Surf. Sci.*, **158**, 134–140.

79 Watson, H.L. (1926) Some properties of fused quartz and other forms of silicon-dioxide. *J. Am. Ceram. Soc.*, **9**, 511–534.

80 Zarka, A. (1983) Observations on the phase transition in quartz by synchrotron-radiation X-ray topography. *J. Appl. Crystallogr.*, **16**, 354–356.

81 Kihara, K. (1990) An X-ray study of the temperature dependence of the quartz structure. *Eur. J. Mineral.*, **2**, 63–77.

82 Tucker, M.G., Keen, D.A., and Dove, M.T. (2001) A detailed structural characterization of quartz on heating through the α–β phase transition. *Mineral. Mag.*, **65**, 489–507.

83 Weast, R.C., Astle, M.J., and Beyer, W.H. (1986) *CRC Handbook of Chemistry and Physics*, 67th edn, CRC Press Inc., Boca Raton, FL.

84 Maissel, L.I. and Glang, R. (1970) *Handbook of Thin Film Technology* (eds L.I. Maissel and R. Glang), McGraw-Hill, New York.

85 Wallace, H.G., Stark, J.G., and McGlashan, M.L. (1982) *Chemistry Data Book*, 2nd edn in SI, International Ideas, London.

86 Hirschfelder, J.O., Curtiss, C.F., and Bird, R.B. (1965) *Molecular Theory of Gases and Liquids*, John Wiley & Sons, Inc., New York.

87 Jennings, S.G. (1988) The mean free path in air. *J. Aerosol Sci.*, **19**, 159–166.

88 Goldstein, A.N., Echer, C.M., and Alivisatos, A.P. (1992) Melting in semiconductor nanocrystals. *Science*, **256**, 1425–1427.

89 Goldstein, A.N. (1996) The melting of silicon nanocrystals: submicron thin-film structures derived from nanocrystal precursors. *Appl. Phys. A*, **62**, 33–37.

90 Okamoto, H. (2000) *Desk Handbook: Phase Diagrams for Binary Alloys*, ASM International, Materials Park, OH.

91 Li, F.J., Zhang, S., Guo, J., and Li, B. (2014) Morphology of silicon/silicon-oxide nanowires grown from nickel-coated silicon wafers. *Nanosci. Nanotechnol. Lett.*, **6**, 505–514.

92 Wu, J.S., Dhara, S., Wu, C.T., Chen, K.H., Chen, Y.F., and Chen, L.C. (2002) Growth and optical properties of self-organized Au_2Si nanospheres pea-podded in a silicon oxide nanowire. *Adv. Mater.*, **14**, 1847–1850.

93 Hu, M.S., Chen, H.L., Shen, C.H., Hong, L.S., Huang, B.R., Chen, K.H. *et al.* (2006) Photosensitive gold-nanoparticle-embedded dielectric nanowires. *Nat. Mater.*, **5**, 102–106.

94 Kolb, F.M., Berger, A., Hofmeister, H., Pippel, E., Gösele, U., and Zacharias, M. (2006) Periodic chains of gold nanoparticles and the role of oxygen during the growth of silicon nanowires. *Appl. Phys. Lett.*, **89**, 173111(1)–173111(3).

95 Braaten, O., Kjekshus, A., and Kvande, H. (2000) The possible reduction of alumina to aluminum using hydrogen. *JOM*, **52**, 47–54.

96 Friede, B. and Jansen, M. (1996) Some comments on so-called 'silicon monoxide'. *J. Non-Cryst. Solids*, **204**, 202–203.

97 Hohl, A., Wieder, T., Van Aken, P.A., Weirich, T.E., Denninger, G., Vidal, M. *et al.* (2003) An interface clusters mixture model for the structure of amorphous silicon monoxide (SiO). *J. Non-Cryst. Solids*, **320**, 255–280.

98 Schulmeister, K. and Mader, W. (2003) TEM investigation on the structure of amorphous silicon monoxide. *J. Non-Cryst. Solids*, **320**, 143–150.

99 Schnurre, S.M., Gröbner, J., and Schmid-Fetzer, R. (2004) Thermodynamics and phase stability in the Si–O system. *J. Non-Cryst. Solids*, **336**, 1–25.

100 Conibeer, G., Green, M., Cho, E.-C., König, D., Cho, Y.-H., Fangsuwannarak, T. *et al.* (2008) Silicon quantum dot nanostructures for tandem photovoltaic cells. *Thin Solid Films*, **516**, 6748–6756.

101 Ke, Y., Weng, X., Redwing, J.M., Eichfeld, C.M., Swisher, T.R., Mohney, S.E. *et al.* (2009) Fabrication and electrical properties of Si nanowires synthesized by Al catalyzed vapor–liquid–solid growth. *Nano Lett.*, **9**, 4494–4499.

102 Wong, Y.Y., Yahaya, M.M., Salleh, M., and Majlis, B.Y. (2005) Controlled growth of silicon nanowires synthesized via solid–liquid–solid mechanism. *Sci. Technol. Adv. Mater.*, **6**, 330–334.

103 Xing, Y.J., Yu, D.P., Xi, Z.H., and Xue, Z.Q. (2003) Silicon nanowires grown from Au-coated Si substrate. *Appl. Phys. A*, **76**, 551–553.

104 Elechiguerr, J.L., Manriquez, J.A., and Yacaman, M.J. (2004) Growth of amorphous SiO_2 nanowires on Si using a Pd/Au thin film as a catalyst. *Appl. Phys. A*, **79**, 461–467.

105 Park, B.T. and Yong, K. (2004) Controlled growth of core–shell Si-SiO_x and amorphous SiO_2 nanowires directly from NiO/Si. *Nanotechnology*, **15**, S365–S370

106 Nie, T.X., Chen, Z.G., Wu, Y.Q., Wang, J.L., Zhang, J.Z., Fan, Y.L. *et al.* (2010) Metallic and ionic Fe induced growth of Si-SiO_x core–shell nanowires. *J. Phys. Chem. C*, **114**, 15370–15376.

107 Ha, J.K. and Cho, K.K. (2012) Preparation of amorphous silicon oxide nanowires by the thermal heating of Ni or Au-coated Si substrates, **110–116**, *Appl. Mech. Mater.*, 1087–1093.

108 Zhang, W.L., Zhang, S., Yang, M., and Chen, T.P. (2010) Microstructure of magnetron sputtered amorphous SiO_x films: formation of amorphous Si core–shell nanoclusters. *J. Phys. Chem. C*, **114**, 2414–2420.

109 Timans, P.J. (1998) Rapid thermal processing technology for the 21st century. *Mater. Sci. Semicond. Process.*, **1**, 169–179.

110 Aitken, D., Mehta, S., Parisi, N., Russo, C.J., and Schwartz, V. (1987) A new VLSI compatible rapid thermal processing system. *Nucl. Instrum. Methods Phys. Res. B*, **21**, 622–626.

111 Li, F.J., Zhang, S., Kong, J.H., and Zhang, W.L. (2011) Study of silicon dioxide nanowires grown via rapid thermal annealing of sputtered amorphous carbon films doped with Si. *Nanosci. Nanotechnol. Lett.*, **3**, 240–245.

112 Ferrari, A.C. and Robertson, J. (2000) Interpretation of Raman spectra of disordered and amorphous carbon. *Phys. Rev. B Condens. Matter Mater. Phys.*, **61**, 14095–14107.

113 Irmer, G. and Dorner-Reisel, A. (2005) Micro-Raman studies on DLC coatings. *Adv. Eng. Mater.*, **7**, 694–705.

114 Tuinstra, F. and Koenig, J.L. (1970) Raman spectrum of graphite. *J. Chem. Phys.*, **53**, 1126–1130.

115 Wang, Y., Alsmeyer, D.C., and McCreery, R.L. (1990) Raman spectroscopy of carbon materials: structural basis of observed spectra. *Chem. Mater.*, **2**, 557–563.

116 Cuesta, A., Dhamelincourt, P., Laureyns, J., Martínez-Alonso, A., and Tascón, J.M.D. (1994) Raman microprobe studies on carbon materials. *Carbon*, **32**, 1523–1532.

117 Li, F.J., Zhang, S., Guo, J., and Li, B. (2014) Morphology of silicon/silicon-oxide nanowires grown from nickel-coated silicon wafers. *Nanosci. Nanotechnol. Lett.* doi: 10.1166/nnl.2014.1817.

118 Li, F.J., Zhang, S., Lee, J.-W., Guo, J., White, T.J., Li, B. *et al.* (2014) Orientation of silicon nanowires grown from nickel-coated silicon wafers. *J. Cryst. Growth*, **404**, 26–33.

119 Straumanis, M.E. and Aka, E.Z. (1952) Lattice parameters, coefficients of thermal expansion, and atomic weights of purest silicon and germanium. *J Appl. Phys.*, **23**, 330–334.

120 Rurali, R. (2010) Colloquium: structural, electronic, and transport properties of silicon nanowires. *Rev. Mod. Phys.*, **82**, 427–449.

121 Migas, D.B. and Borisenko, V.E. (2009) The role of morphology in stability of Si nanowires. *J. Appl. Phys.*, **105**, 104316.

122 Ng, M.F., Shen, L., Zhou, L., Yang, S.W., and Tan, V.B.C. (2008) Geometry dependent I–V characteristics of silicon nanowires. *Nano Lett.*, **8**, 3662–3667.

123 Migas, D.B. and Borisenko, V.E. (2007) Tailoring the character of the band-gap in ⟨011⟩-, ⟨111⟩- and ⟨112⟩-oriented

silicon nanowires. *Nanotechnology*, **18**, 375703.

124 Ponomareva, I., Menon, M., Richter, E., and Andriotis, A.N. (2006) Structural stability, electronic properties, and quantum conductivity of small-diameter silicon nanowires. *Phys. Rev. B Condens. Matter Mater. Phys.*, **74**, 125311.

125 Steighner, M.S., Snedeker, L.P., Boyce, B.L., Gall, K., Miller, D.C., and Muhlstein, C.L. (2011) Dependence on diameter and growth direction of apparent strain to failure of Si nanowires. *J Appl. Phys.*, **109**, 033503.

126 Mikkelsen, A., Eriksson, J., Lundgren, E., Andersen, J.N., Weissenreider, J., and Seifert, W. (2005) The influence of lysine on InP(001) surface ordering and nanowire growth. *Nanotechnology*, **16**, 2354–2359.

127 Wang, C.X., Hirano, M., and Hosono, H. (2006) Origin of diameter-dependent growth direction of silicon nanowires. *Nano Lett.*, **6**, 1552–1555.

128 Jagannathan, H., Deal, M., Nishi, Y., Woodruff, J., Chidsey, C., and McLntyre, P.C. (2006) Nature of germanium nanowire heteroepitaxy on silicon substrates. *J. Appl. Phys.*, **100**, 024318.

129 Song, M.S., Jung, J.H., Kim, Y., Wang, Y., Zou, J., Joyce, H.J. *et al.* (2008) Vertically standing Ge nanowires on GaAs(110) substrates. *Nanotechnology*, **19**, 125602.

130 Ihn, S.G., Song, J.I., Kim, T.W., Leem, D.S., Lee, T., Lee, S.G. *et al.* (2007) Morphology- and orientation-controlled gallium arsenide nanowires on silicon substrates. *Nano Lett.*, **7**, 39–44.

131 Shan, C.X., Liu, Z., and Hark, S.K. (2007) CdSe nanowires with controllable growth orientations. *Appl. Phys. Lett.*, **90**, 193123.

132 Fortuna, S.A., Wen, J., Chun, I.S., and Li, X. (2008) Planar GaAs nanowires on GaAs (100) substrates: self-aligned, nearly twin-defect free, and transfer-printable. *Nano Lett.*, **8**, 4421–4427.

133 Lugstein, A., Steinmair, M., Hyun, Y.J., Hauer, G., Pongratz, P., and Bertagnolli, E. (2008) Pressure-induced orientation control of the growth of epitaxial silicon nanowires. *Nano Lett.*, **8**, 2310–2314.

134 Dayeh, S.A., Yu, E.T., and Wang, D. (2007) III–V nanowire growth mechanism: V/III ratio and temperature effects. *Nano Lett.*, **7**, 2486–2490.

135 Wang, N., Cai, Y., and Zhang, R.Q. (2008) Growth of nanowires. *Mater. Sci. Eng. R*, **60**, 1–51.

136 Scheel, H.J. and Fukuda, T. (2003) *Crystal Growth Technology*, John Wiley & Sons, Inc., Chichester.

137 Chiang, C.H., Ci, J.W., Uen, W.Y., Lan, S.M., Liao, S.M., and Yang, T.N. (2012) Fabrication of silicon-submicron-wire-based solar cells on UMG-Si substrates using nickel catalyst. *J. Electrochem. Soc.*, **159**, H112–H116

138 Jee, S.W., Kim, J., Jung, J.Y., Um, H.D., Moiz, S.A., Yoo, B. *et al.* (2010) Ni-catalyzed growth of silicon wire arrays for a Schottky diode. *Appl. Phys. Lett.*, **97**, 042103.

139 Hartland, S. (2004) *Surface and Interfacial Tension: Measurement, Theory, and Applications* (ed. S. Hartland), Marcel Dekker, New York.

140 Ding, Y., Gao, P.X., and Wang, Z.L. (2004) Catalyst–nanostructure interfacial lattice mismatch in determining the shape of VLS grown nanowires and nanobelts: a case of Sn/ZnO. *J. Am. Chem. Soc.*, **126**, 2066–2072.

141 Teo, B.K., Li, C.P., Sun, X.H., Wong, N.B., and Lee, S.T. (2003) Silicon–silica nanowires, nanotubes, and biaxial nanowires: inside, outside, and side-by-side growth of silicon versus silica on zeolite. *Bioinorg. Chem.*, **42**, 6723–6728.

142 Li, F.J., Zhang, S., Lee, J.W., and Zhao, D. (2013) Wire or no wire: depends on the catalyst layer thickness. *J. Cryst. Growth*, **381**, 87–92.

143 Kohno, H. and Takeda, S. (1998) Self-organized chain of crystalline-silicon nanospheres. *Appl. Phys. Lett.*, **73**, 3144–3146.

144 Kohno, H. and Takeda, S. (2005) Chains of crystalline-Si nanospheres: growth and properties. *e-J. Surf. Sci. Nanotechnol.*, **3**, 131–140.

145 Nie, T., Chen, Z.G., Wu, Y., Guo, Y., Zhang, J., Fan, Y. *et al.* (2012) Fabrication of crystal α-Si_3N_4/Si-SiO_x core–shell/Au–SiO_x peapod-like axial double

heterostructures for optoelectronic applications. *Nanotechnology*, **23**, 305603.
146 Lu, W. and Lieber, C.M. (2006) Semiconductor nanowires. *J. Phys. D*, **39**, R387–R396.
147 Yang, C., Zhong, Z., and Lieber, C.M. (2005) Materials science: encoding electronic properties by synthesis of axial modulation-doped silicon nanowires. *Science*, **310**, 1304–1307.
148 Tallury, P., Malhotra, A., Byrne, L.M., and Santra, S. (2010) Nanobioimaging and sensing of infectious diseases. *Adv. Drug Deliv. Rev.*, **62**, 424–437.
149 McDowell, M.T., Lee, S.W., Ryu, I., Wu, H., Nix, W.D., Choi, J.W. *et al.* (2011) Novel size and surface oxide effects in silicon nanowires as lithium battery anodes. *Nano Lett.*, **11**, 4018–4025.
150 Yang, Y., McDowell, M.T., Jackson, A., Cha, J.J., Hong, S.S., and Cui, Y. (2010) New nanostructured Li_2S/silicon rechargeable battery with high specific energy. *Nano Lett.*, **10**, 1486–1491.
151 Nozik, A.J. (2002) Quantum dot solar cells. *Physica E*, **14**, 115–120.
152 Hache, F., Ricard, D., and Flytzanis, C. (1986) Optical nonlinearities of small metal particles: surface-mediated resonance and quantum size effects. *J. Opt. Soc. Am. B*, **3**, 1647–1655.
153 Haglund, R.F., Jr., Yang, L., Magruder, R.H. III, Witting, J.E., Becker, K., and Zuhr, R.A. (1993) Picosecond nonlinear optical response of a Cu:silica nanocluster composite. *Opt. Lett.*, **18**, 373–375.
154 Arbiol, J., Fontcuberta, A., Morral, I., Estradé, S., Peiró, F., Kalache, B., Roca, P., Cabarrocas, I. *et al.* (2008) Influence of the (111) twinning on the formation of diamond cubic/diamond hexagonal heterostructures in Cu-catalyzed Si nanowires. *J. Appl. Phys.*, **104**, 064312.
155 Lopez, F.J., Givan, U., Connell, J.G., and Lauhon, L.J. (2011) Silicon nanowire polytypes: identification by Raman spectroscopy, generation mechanism, and misfit strain in homostructures. *ACS Nano*, **5**, 8958–8966.
156 Hemesath, E.R., Schreiber, D.K., Kisielowski, C.F., Petford-Long, A.K., and Lauhon, L.J. (2012) Atomic structural analysis of nanowire defects and polytypes enabled through cross-sectional lattice imaging. *Small*, **8**, 1717–1724.
157 M. I., DenHertog., Cayron, C., Gentile, P., Dhalluin, F., Oehler, F., Baron, T. *et al.* (2012) Hidden defects in silicon nanowires. *Nanotechnology*, **23**, 025701.
158 Cayron, C., Hertog, M. Den, Latu-Romain, L., Mouchet, C., Secouard, C., Rouviere, J.L. *et al.* (2009) Odd electron diffraction patterns in silicon nanowires and silicon thin films explained by microtwins and nanotwins. *J. Appl. Crystallogr.*, **42**, 242–252.

18
Chemical Patterning on Surfaces and in Bulk Gels

Olaf Karthaus

Chitose Institute of Science and Technology, Department of Applied Chemistry and Bioscience, Bibi 65-758, 066-8655 Chitose City, Japan

18.1
Background

The term "chemical patterning" was first used in a patent in 1988 describing the wet chemical patterning of Hafnium boride layers [1]. It took seven years until the term appeared again, this time for describing the etching of potassium titanyl phosphate (KTP) in a short communication [2]. The next publication in 1996 was already concerned with biological applications – the modification of polymer surfaces by photolithography and embossing [3] – to produce topological features such as grooves in a thin film. The following years saw the firm establishment of the term, with a special emphasis for tissue engineering [4]. In this context, chemical patterning is used to give a "chemical" cue, instead of a physical one, like topography, to pattern cells on biodegradable substrates. It can also be more specifically referred to a spatial difference in van der Waals forces to support artificial bilayer membranes of lipids [5]. The term was also used to describe the reaction of sputtered inorganic $SiCl_x$ species with a self-assembled monolayer [6]. The authors opined that "such processes are potentially useful for the construction of novel surfaces from a monolayer substrate and for chemical patterning of surfaces with functional groups," for patterning in the plane of a substrate.

In more general terms, "chemical patterning" also occurs when a chemical reaction is used to make a pattern perpendicular to the layer plane. It can even be applied to patterns that arise from a chemical reaction in a three-dimensional matrix, for example, in a gel.

A pattern in the chemical sense is thus a regular structure that comprises at least two different moieties. The wavelength of the pattern is considerably larger than the size of its constituents; if not, a crystal formed by two chemical species, for example, a charge transfer complex, could be named "a pattern." Also, in

Handbook of Solid State Chemistry, First Edition. Edited by Richard Dronskowski, Shinichi Kikkawa, and Andreas Stein.

most cases, the symmetry of the pattern is not related to the crystal structure of the constituents.

There are two fundamentally different ways to make such patterns: *Top-down* and bottom-up. Top-down techniques reproduce a predesigned pattern by using tools. *Bottom-up* approaches use the inherent characteristics of molecules to form pattern via self-assembly or self-organization phenomena [7]. For the latter approach, molecular level interactions (e.g., autocatalysis, supersaturation, phase separation) and long-range phenomena (e.g., diffusion and electrostatic interaction) play an important role.

18.2
Examples of Chemical Processes that Lead to the Chemical Modification of Surfaces

Chemical modification on surfaces can be carried out by either etching into a surface or by grafting onto a surface. "Etching-into" can be used for metals, semiconductors, or oxide surfaces. The etchants react with surface atoms and molecules and remove them, leading to a different surface functionality than before the reaction, for example, treating an SiO_2 surface with alkali or acid will hydrolyze the Si—O—Si bond and form Si—OH surface groups that are more hydrophilic.

The "grafting-onto" approach has a wider applicability. Many surfaces can be modified by adding functionalities on top of the substrate material: metals, semiconductors, metal oxides, glasses, and even plastics.

The most well-known examples are adsorption of fatty acids on alumina [8], adsorption of alkylphosphates on metal oxides via electrostatic interaction (Al_2O_3, Ta_2O_5, Nb_2O_5, ZrO_2, and TiO_2) [9], chemisorption of primary alkanethiols on gold surfaces [10], the chemical reaction of alkyltrichlorosilanes [11] and trialkoxysilanes [12] with silicon oxide surfaces [13], and the chemical reaction of 1-alkenes with hydrogen-terminated silicon surfaces [14] (Figure 18.1). The first cue that influences the chemical properties of those self-assembled monolayers is the length of the alkyl chain [15]. Further, and more influential modification of surfaces is possible by adding a functionality at the opposite terminal of the alkyl chain. CH_3 and CF_3 groups change the lubrication and hydrophobicity [16]. Alkenes, carboxylic acids, esters [17], amines [18], thiols [19], and hydroxy groups can be used to further modify the chemical nature of the surface. Primary alkylthiols can be functionalized at the opposite end of the alkane chain with such functional groups, and even with peptides [20].

Reacting thiols on a gold substrate and alkenes on a hydrogen-terminated silicon substrate produce nothing but monomolecular films, since the terminal reacting group can only undergo a reaction in a single step. The single thiol group in the molecule reacts with gold atoms on the surface, one alkene group reacts with one Si-H, and so on. In the case of silanes, the situation is more complicated since alkoxysilanes and chlorosilanes are reactive species that

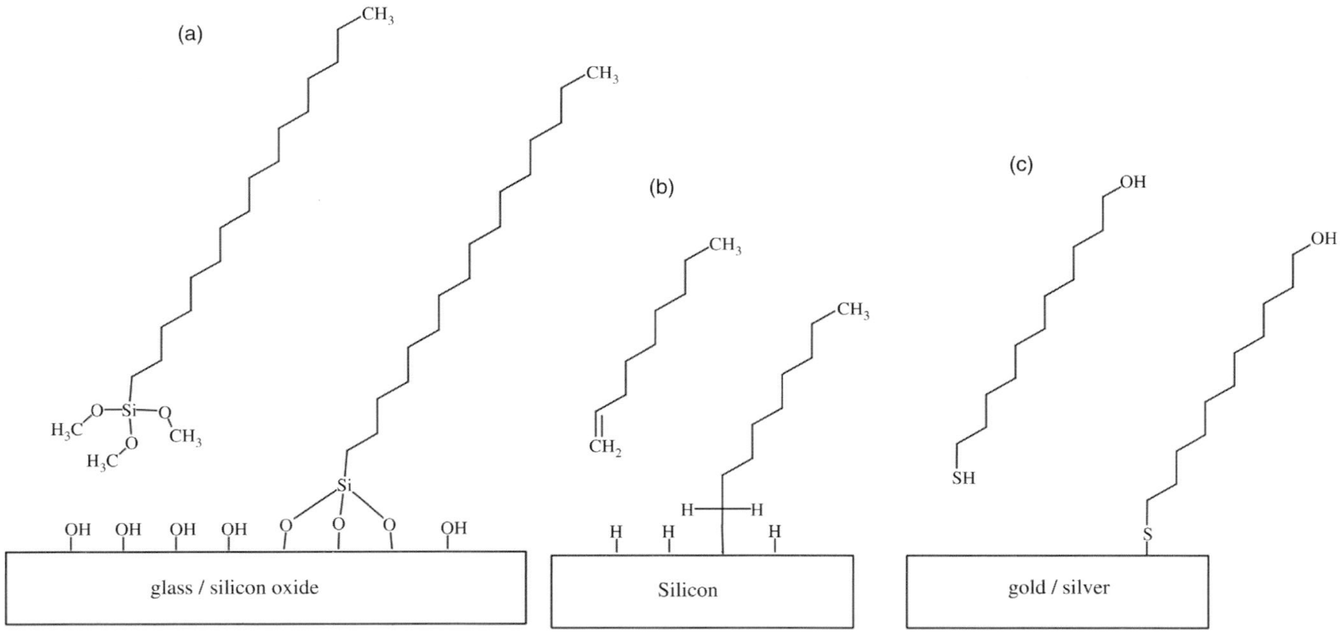

Figure 18.1 Surface modification by using reactive alkanes. Modification of glass or silicon oxide surfaces with trimethoxysilanes (a), modification of hydrogen-terminated silicon surfaces with alkenes (b), and modification of gold or silver surfaces with alkanethiols, in this case inclusing a terminal OH group (c).

hydrolyze in the presence of water to form hydroxysilanes [21]. In fact, this reaction is necessary to form a monolayer, because those hydroxyl species then react with hydroxyl groups on the surface of a SiO_2 substrate. But bi- or trifunctional silanes (bisalkyldichlorosilanes, alkyltrichlorosilanes, trialkoxysilanes, and bisalkoxysilanes) hydrolyze and form molecules with more than one hydroxyl group that then can condense with each other to form oligomers or polymers in solution. A reaction of those polymeric species with surface hydroxyl groups then will not lead to a homogeneous monomolecular layer [22,23]. Thus, silane solutions should be prepared from freshly distilled reactants and should not be stored for a prolonged time. Another possibility to form homogeneous films is to subject the substrate to a vapor of the reactant [24], since the problematic oligomers do not evaporate.

Besides using well defined, small molecules as reacting species, polymers can also change the chemical nature at surface. Here, the grafting of polymers, either *from* or *to* a surface is notable [25–27]. In the usual grafting-to approach, the substrate is immersed in a solution that contains a polymer that forms chemical bonds with a reactive surface [28], or a polymer that has a highly reactive end group such as a radical [29]. The grafting-from approach anchors an initiator to the surface that starts a polymerization reaction, most notably an atom transfer radical polymerization (ATRP) [30]. An exciting extension of this concept is the reversible switching of chemical contrast. Polymer brushes can swell and collapse depending on environmental clues such as temperature, pH, or solvent, and thus a mixed brush surface can be used to switch chemical contrast [31,32].

Another method to modify a surface with polymers is by the layer-by-layer alternating electrostatic adsorption of cationic and anionic polymeric compounds [33]. A charged (e.g., cationic) surface is immersed in a solution that contains a polymer with opposite charge (here, anionic). After the adsorption by electrostatic force, the substrate is taken out and rinsed to remove weakly adsorbed polymer. The surface now has the charge of the polymer (anionic). This is then immersed in a solution containing an oppositely charged polymer (cationic), leading to the adsorption of this polymer onto the anionic layer. The charge on the substrate thus can be switched from cationic to anionic, with an increase of the polymer multilayer thickness with increasing adsorption cycles (Figure 18.2).

18.3 Surface Patterning

The above-described methods by themselves do not produce a lateral pattern *per se*. Thus, in order to make a chemical pattern on the substrate, a patterned substrate is needed to begin with, on which selective chemical modification takes place, or the surfaces need to be patterned after modification. Both avenues can be taken to realize chemical patterns by the following top-down methods.

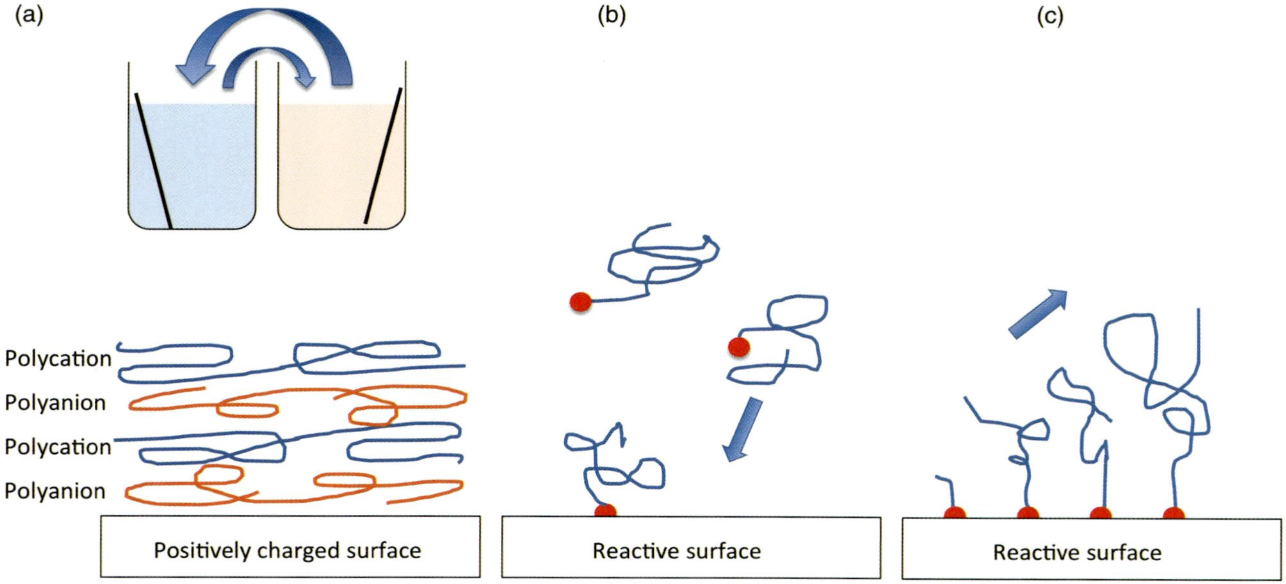

Figure 18.2 Chemical modification by using polymers. Layer-by-layer adsorption of polycations and polyanions (a), grafting-to with reactive polymers (b), and grafting-from a reactive surface (c).

- Using lasers and multiphoton excitation, surfaces can be ablated [34].
- Inkjet printing can be used to make patterns through reactive molecules and even proteins or DNA onto a substrate [35].
- Modified atomic force microscopes (AFMs) can be used to make chemical patterns. In the earliest example, the AFM tip is used to scratch a surface [36] or to remove a thin resist film [37], while the uncovered surface is *in situ* coated with a thiol compound. Around the same time, dip-pen lithography had been developed in which an AFM tip is first dipped into a solution of surface reactive agent and then scanned over the surface, much like a nanometer-sized dip pen [38]. Later on, it was shown that an AFM tip can be used to locally induce an electrochemical reaction at the surface, the so-called "electro pen" nanolithography [39].
- Microcontact printing was first reported in 1994 [40]. Liquid poly(dimethylsiloxane) (PDMS) is used to make a thick coat over a 3D pattern that can be produced by usual lithography methods on a stainless steel master mold. PDMS has a very low surface energy and thus, after gelation, this stamp can easily be removed from the master and coated with a reactive compound, often a thiol, and placed on the substrate. Only the protruding parts of the stamp come into contact with the substrate, and thus only there the transfer of the compound occurs, producing a chemical pattern on the substrate. Because there is the chance of contamination by accidentally transferred oligomeric siloxanes from the stamp material, polyolefine plastics (POP) has been used, too [41].
- Nanoimprinting is the transfer of a nanopattern by placing the steel master on a polymer substrate. Heating above the glass transition then imprints the 3D relief of the master into the polymer that then acts as a mask for further modification of the surface and finally the preparation of a chemical pattern [42].

The chemical modification of a patterned substrate in a single step depends on the orthogonality of the adsorption of two species – chemical A1 adsorbs on substrate part S1, and chemical A2 adsorbs on substrate part S2, but A1 does not adsorb on S2 (or even if it does, it is replaced by the much stronger A1–S1 interaction), and A2 does not adsorb on S1. Whitesides and coworkers made use of this concept in their seminal 1989 paper. An alumina-on-gold pattern was selectively coated with long-chain carboxylic acid and alkyl thiol in one step from a common solvent [43]. The thiol reacts with gold, while the carboxylate adsorbs on alumina. Even though they did not use the phrase "chemical patterning," this can be considered a very early example for it (Figure 18.3).

Another orthogonal system is poly(l-lysine)-poly(ethylene glycol) (PLL-PEG) block copolymer and alkyl phosphates. The PLL-PEG adsorbs preferentially on SiO_2, while alkyl phosphates adsorb on TiO_2 areas of a patterned SiO_2/TiO_2 substrate [44].

Another possibility to produce a pattern is the bottom-up approach in which the intrinsic properties of materials to form patterns are utilized.

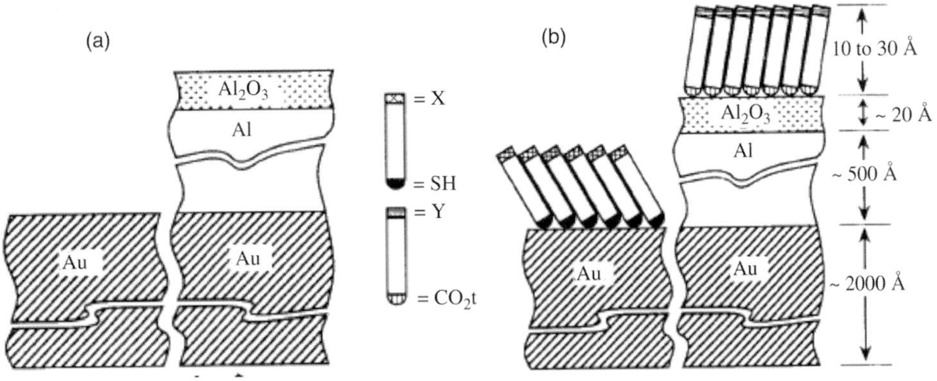

Figure 18.3 Orthogonal self-assembly on a gold/alumina pattern. Part (a) shows the cross section of a gold surface that is patterned with an aluminum oxide layer. Part (b) is the cross section of such a surface after the orthogonal self-assembly of alkane thiols on the gold surface and carboxylic acids on the alumina surface. (Adapted with permission from Ref. [43].)

Colloidal particles form monolayers on surfaces. The progress in the synthesis of nano- and microparticles with narrow size distribution leads to the formation of regular colloidal crystals of such particles with hexagonal arranged voids between the particles. Thus, the particles can act as a mask that allows the formation of a chemical pattern by adsorption of molecules into the interparticle spaces [45,46].

Block copolymers are prime examples for this, because they phase-separate in nanometer-sized domains whose structure (spherical, gyroid, columnar, or lamellar) and dimensions depend on the relative and absolute length of the blocks in the polymer [47]. Cast on substrates, the phase-separated structure is likely to be different from the bulk structure, because of interfacial energy and surface tension effects [48,49]. The nanometer-scale polymer pattern may show a chemical contrast by itself, depending on the chemical nature of the blocks, or the pattern can be used as a mask to etch a chemical pattern into the substrate. For example, spin-coated diblock copolymer thin films with well-ordered spherical or cylindrical microdomains were used as templates for silicon nitride-coated silicon wafers [50]. A cylindrical pattern was formed after etching in which the holes are 20 nm across, 40 nm apart, and hexagonally ordered.

One elegant method to control the size, spacing, and regularity of a chemical pattern by using block copolymers has been reported by Spatz *et al.* by micelle nanolithography [20,51]. A polystyrene-poly(vinyl pyrrolidone) (PS-PVP) block copolymer, in which the PVP block contains $AuCl_3$ forms micelles in toluene solution. The $AuCl_3$/PVP block forms a sphere that is surrounded by the PS block. By slow dip coating, these micelles are transferred to a solid substrate. Subsequent plasma treatment ashes the polymer and reduces the gold salt to gold, resulting in a regular gold pattern on the silicon substrate. By mixing

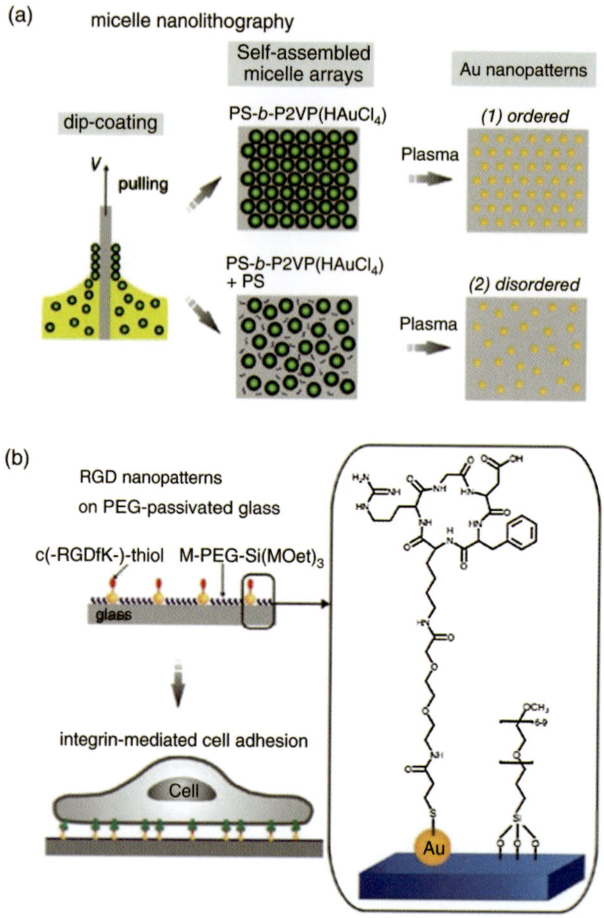

Figure 18.4 (a) Preparation of ordered and disordered nanopatterns by self-assembling of Au-containing polymer micelles and (b) their chemical modification by peptide-terminated alkane thiols. (Adapted with permission from Ref. [20]. Copyright 2009, American Chemical Society.)

polystyrene into the dip-coating solution, the pattern becomes irregular. A chemical reaction of the Au island with a peptide-functionalized thiol leads to a nanoscale chemical pattern that can be used to control cell attachment (Figure 18.4).

18.4
Application of Surface Patterns and Combination with Other Methods

From the very beginning of the research on chemical patterns, possible applications were envisioned, ranging from pure and applied physics (wetting, friction) to nanoelectronics (transistors, light emitting diodes, sensing) and biology (antifouling, cell adhesion, cell proliferation, and cell positioning).

By using micron-sized stripes with a hydrophilic/hydrophobic chemical contrast, it is possible to control the location [52], size, and polydispersity of liquid drops on a surface [53]. One application of such pinned liquid droplets is protein crystallization [54], because the drops are confined to the patterned area and stable over an extended time. Combined with heating elements, chemical patterns on a SiO_2 substrate can be used to control the flow of liquids [55].

Sirringhaus and coworkers have reported that a pattern of hydrophilic/hydrophobic surfaces can be used to self-align a solution of an organic semiconductor to produce an organic field-effect transistor with a channel length of 500 nm [56].

A chemical pattern on surfaces can also be used to direct the nanometer-size phase separation of a block copolymer [57,58]. This "directed self-assembly" (DSA) technique has been receiving much interest recently, because it can also be used for the so-called "pattern multiplication," in which the pattern of the polymer phase-separated film is smaller than the underlying chemical pattern itself (Figure 18.5). The polymer molecules align at the pattern edges and then fill the space between the edges with their natural phase separated structure. Usually, a topographical pattern is used to guide the polymer phase-separation, but chemical patterns have been reported, too [59].

Chemical patterns control cell adhesion and function [60], for example, by direct integrin patterning [61] or protein adsorption [62]. Silicon substrates covered with patterns of aminofunctionalized silanes can be used to guide the adsorption of bone cells and finally can direct the biomineralization of inorganic matter [63].

In vitro investigation of neural architectures requires cell positioning. For that purpose, Delacour *et al.* have developed micromagnets and combined with chemical patterning to attract cells to adhesive sites on silicon substrates and keep them there during incubation [64].

Chemical patterning can also occur in three dimensions, for example, in a gel that may be used to control cell behavior. An agarose gel was modified with a photoexcitable sulfide-containing coumarin derivative. Chemical contrast can be written in such a transparent 3D gel by confocal laser microscopy [65].

18.5 In Bulk (Liesegang patterns)

Periodic band or ring structures are chemical patterns that are ubiquitous in nature, studied by geologists, chemists, physicists, and biologists.

Chemically generated precipitation patterns of inorganic salts were extensively researched and described by Raphael Eduard Liesegang, a German chemist and photographer [66]. His first paper on the subject was published in 1896 [67] in which he described the formation of silver chromate rings after dropping a silver

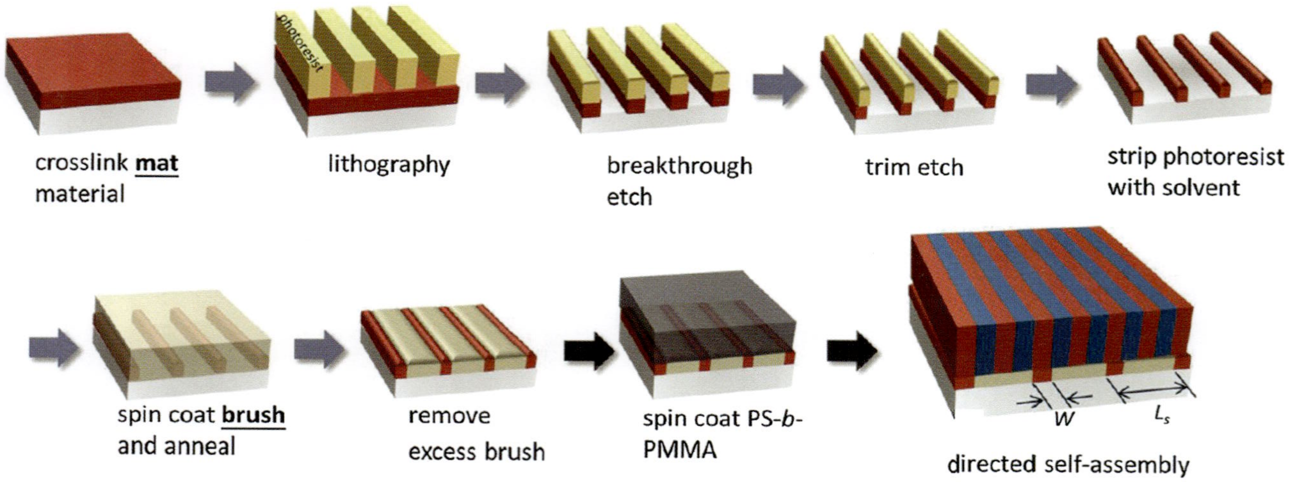

Figure 18.5 Preparation of a chemical pattern by conventional lithography. The pattern is then covered with a polystyrene PMMA block copolymer that has a smaller layer spacing (*W*) than the underlying chemical contrast (*L*). The PS-PMMA lamellae orient along the chemical pattern and produce a pattern in which *L* is a multiple of *W*. ((Reprinted with permission from Ref. [59]. Copyright 2013, American Chemical Society.)

nitrate solution onto a sheet of gel that contained chromate ions. He also found that not only concentric rings in such a quasi-two-dimensional gel sheet were formed, but also banded patterns in a quasi-one-dimensional column of gel in a test tube [68]. This serendipitous discovery intrigued not only him but also many other researchers, not at least because of the beauty of these patterns (Figure 18.6).

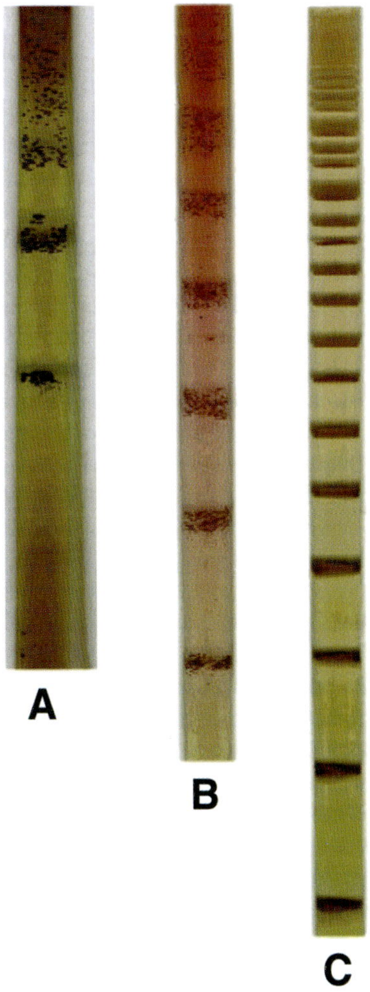

Figure 18.6 Photographs of Liesegang rings in glass test tubes. The top of the gel phase is at the top of the pictures. The mobile phase diffuses downward into the gel, leading to bands of precipitates that have an increasing spacing as the mobile phase continues to diffuse downward. The diameter of the test tubes is around 1 cm. ((Reprinted with permission from Ref. [77]. Copyright 2004, American Chemical Society.)

Generally, these patterns are formed by the precipitation of oppositely charged ions when their concentration exceeds the solubility constant of the salt. Usually, the distance between the precipitation rings increase with the distance from the point of origin. Also, the ring thickness increases with the number of rings. Thus, patterns become more spaced-out with increasing distance from the origin.

Patterns are best observed with salts that have a low solubility product such as halides [69], hydroxides [70–72], phosphates [73,74], oxalates [75,76], and 8-hydroxyquinoline salts [77]. Hausmann reported early already in 1904 on a large number of different salts and observed that while most precipitates were layered colloids, precipitates made of calcium, strontium, and barium sulfates, barium and silver oxalates, and thallium halides did not form patterns [78]. Ostwald also reported that other barely soluble salts such as $BaSO_4$ do not yield ring structures [79] either. Silver chromate has been reported to form Liesegang rings [80,81], but this has later been attributed to chloride contamination of the gelatin gel [82]. $Ag_2Cr_2O_7$ in well-washed gelatin does not form Liesegang rings, but an unbroken deposit free from lines. The addition of NaCl of up to 0.1% to the solution of washed gelatin causes a reappearance of Liesegang rings. The rings, therefore, do not form unless a chloride impurity is present in the gelatin. Thus, even though Liesegang rings seem to be a general phenomenon, there are some boundary conditions to be observed.

External stimuli can be used to control the precipitation bands such as light [83] or electrostatic potential [84,85]. Bena *et al.* found that the width of consecutive bands increases with increasing static electric field, and that above a certain field the bands merge and form a continuous deposition [86]. Examples of current dynamics yielding periodic bands of prescribed wavelength, as well as more complicated structures are given by the same authors [87] who theoretically as well as experimentally could show that a quasi-periodical field gave narrow nearly equidistant stripes of silver dichromate. Furthermore, the type of gel, gelatin or agarose [88,89], plays a role. The influence of light sometimes leads to an abnormal banding pattern in which the band distance decreases with increasing distance from the gel surface. Hatschek quantitatively studied Liesegang rings formed by the diffusion of $Pb(NO_3)_2$ in agar gel containing KI [90]. In addition to the normal narrow bands whose separation increased with depth, wider bands whose separation decreased with depth appeared. He could show that the cause of the abnormal bands were diurnal phenomena by using identical pairs of tubes except that one was wrapped in metal foil and the other exposed to the daily change in light. The wrapped tubes gave normal bands and all the exposed tubes showed anomalies. Unfortunately, Hatschek was not able to give a definite explanation for that behavior. A similar work was published by Karam *et al.* [91]. They found that the parallel one-dimensional bands obey either one of two known spacing trends. The overwhelming trend is an increase in spacing away from the interface, as shown in Figure 18.6, while few systems display a decrease in spacing as the bands get further away from the interface, for example, the $PbCrO_4$ system. They proposed a mechanism of revert spacing that is governed by the adsorption of the diffusing CrO_4^{2-} ions on the formed $PbCrO_4$ crystallites in the Liesegang bands. Experimenting with different reaction conditions, they found that this adsorption increases as the band number increases in revert spacing systems, while it decreases as the band number increases in normal spacing systems. The attraction between the CrO_4^{2-} and Pb^{2+} in the gel seems to cause the bands to form gradually closer and closer. Under some conditions, they could also observe thinner bands that formed within the main ones that are discussed in view of the light sensitivity of the chromate ion and the stability of the lead chromate sol.

Interestingly, there was little information about the size and shape of the precipitated particles for the first few decades of research. In 1982, Müller *et al.* used optical microscopy to elucidate the particle size distribution in a PbI_2 system [92] that was strictly speaking not a Liesegang system, because they started out with a uniform distribution of colloids suspended in a thin gel sheet with a thickness of 1 mm. Even though the initial distribution of the PbI_2 colloidal particles was homogeneous, the authors found that the Ostwald ripening process lead to the formation of macroscopic inhomogeneities. From photographs taken with an optical microscope to a resolution of 1 μm, they found that the distribution of micrometer-sized spherical particles in the gel was in such a way that millimeter-sized void structures were formed in which no colloidal particles could be observed. Their findings led to the hypothesis that the formation of such macroscopic inhomogeneities and periodic precipitation processes are the result of a chemical instability due to the coupling of autocatalytic growth of colloidal particles by Ostwald ripening and diffusion of the involved chemicals.

Al-Ghoul *et al.* have investigated the crystal morphology and shape of the metal hydroxide precipitation–resolution system that is a true Liesegang system. They found a polymorphic transition between alpha- and beta-$Ni(OH)_2$ during the transition of the precipitate from a propagating pulse to a stable Liesegang ring regime [89]. The same group also reported an Ostwald ripening front in cobalt hydroxide system. In agar as the gel, no Liesegang rings were formed, but the broad precipitation band showed a transformation from the blue alpha-Co$(OH)_2$ into the pink, crystalline, beta-$Co(OH)_2$. On the other hand, gelatin gels did give Liesegang banded structures under otherwise identical conditions [89]. The authors attribute this to the different nucleation rates in both gels [93]. Gelatin allows for a high nucleation rate and many small crystals, and thus Liesegang banding, while agar favors less nucleation and thus larger crystals behind a single reaction front.

One difficulty for determining the size distribution of particles is the tedious measurement: preparing Liesegang rings, removing the gel out of the reactor tube, collecting the particles in each band, and finally analyzing them by either X-ray scattering or electron microscopy. Mandalian *et al.* circumvented those problems by investigating a system that produces large crystals that can be observed under an optical microscope, the cobalt oxinate system [77]. They found that the average size of their millimeter-sized particles increases with the band number.

Recently, Walliser *et al.* have reported that a 2D Liesegang ring system can be used to prepare nanoparticles with various sizes that are spatially sorted into nearly monodisperse regions [95]. Silver chromate particle size increases with ring number in a similar fashion as Mandalian had reported for the cobalt oxinate system [77]. Walliser's finding opens the possibility to prepare colloidal particles with a narrow, defined diameter by using the Liesegang rings method. Such a synthesis of nano- and microparticles of controlled size has gained importance because they are important in chemistry, physics, materials science, and medicine. The size and shape of particles often influence their physical and chemical properties, for example, catalytic activity [96,97], toxicity [98,99], photonic properties [100,101], or even magnetic susceptibility [102].

The conventional procedure to produce particles with a narrow size distribution rely on the presence of additives [103], the use of micelles [104], or microfluidic devices [105]. The actual size can then be adjusted by the reaction conditions. But for each new particle size, a new reaction mixture has to be found. Using Liesegang ring systems, the diffusion of the two reactants lead automatically to a whole library of different reaction mixture, depending on their position in the gel, and thus Walliser *et al.* succeeded to produce a whole library of batches of particles with narrow size in just one experiment.

The precipitation of colloidal metal particles instead of ions into bands also had been investigated from the very early stages of the research [80] (Figure 18.7), a topic that has been picked up again in recent years [93,94]. The majority of the Liesegang ring systems are ionic in nature, and Lagzi *et al.* investigated the possibility to use charged entities that were larger than atoms and ions and made ring structures out of them. The precipitation of two different types of nanoparticles that are functionalized with oppositely charged capping reagents should depend on charge neutrality and not on the solubility product as for molecular ions. The charge density on the surface of nanoparticles can be adjusted by changing the density and nature of the capping reagent. Lagzi *et al.* could demonstrate that the coupling of the diffusion of the nanoparticles with charge neutralization lead to precipitation bands. As a result, Liesegang rings that differ in chemical composition by using different nanoparticle cores can be prepared. They used nanoparticles with metal core diameters of 7–10 nm capped with either mercaptoundecyl ammonium salts or mercaptoundecanoic acid. No rings were observed with uncharged particles or with sizes larger than 12–15 nm, because aggregation was due to van der Waals interaction. The precipitation reaction is fundamentally different from Liesegang patterns: Mixtures of oppositely charged nanoparticles are stable in solution if one particle type is in excess [106], or if the concentration is below a critical one [107]. If these conditions are not met, precipitation occurs. Since the diffusion is much slower than the precipitation, a nucleation and growth model can be used to explain the patterns. The authors also found an unexpected sorting mechanism. In contrast to traditional pattern-forming systems, in which all of the ions or molecules of a given type have the same sizes, nanoparticles are inherently polydisperse. An

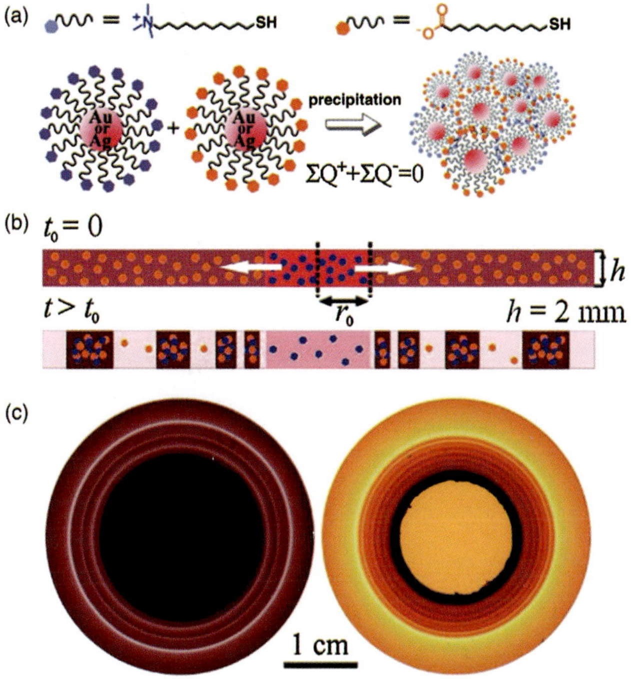

Figure 18.7 Formation of Liesegang rings in colloidal systems. Composition of the capped-nanoparticles (a), schematic cross-section of the distribution of particles in the gel before and after banding (b), and photographs of two different patterns produced by different colloids. The difference in color stems from the plasmonic resonance. (Reprinted with permission from Ref. [93]. Copyright 2010, American Chemical Society.)

interesting consequence then develops that leads to a size sorting of the nanoparticles: When more polydisperse nanoparticles diffuse into a gel that contains less polydisperse nanoparticles, the first precipitation ring contains the largest and most polydisperse particles. As these larger particles are removed from the diffusing ensemble, the consecutive rings contain smaller and increasingly monodisperse nanoparticles. In the opposite case (monodisperse nanoparticles diffuse into a gel that contains polydisperse nanoparticles), the first ring contains the smaller, least polydisperse particles.

The combination of micropatterning technologies with Liesegang patterns has opened the field further. Liesegang himself reported on an apparent "action-at-a-distance" (Fernwirkung), when he placed two silver nitrate drops at certain distances on a gel [108]. He found that the perfect circle that was formed when a single drop of outer electrolyte had been dropped on the gel with the inner electrolyte becomes distorted when two drops were placed at a certain distance from each other. Liesegang explained this with the decrease of the concentration of the inner electrolyte along the precipitation

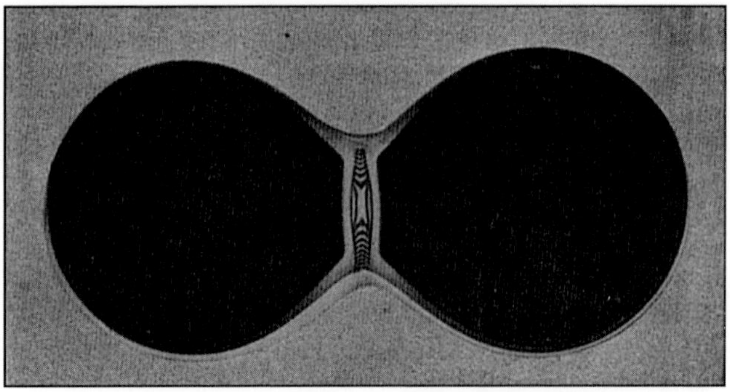

Figure 18.8 Liesegang patterns in spatial patterns. Two drops were placed at a distance of 25 mm and a distortion of the precipitation was observed. (Reprinted with permission from Ref. [108]. Copyright 1906, WILEY-VCH Verlag GmbH.)

front. Along the connection line of the two drops, the concentration of the inner electrolyte is smallest, leading to a faster growth of the precipitate (Figure 18.8).

The recent progress in microfabrication of gels has widened the effort to study the geometry of the reaction field on the Liesegang patterns [109,110].

Liesagang rings also have been used to investigate the diffusion of ions in three dimensions [111] because the local ion concentration can be monitored by the presence of Liesegang rings.

The banding pattern has also been used as a tool to determine the sulfate concentration is single rain droplets [112]. Also, a phase transition in the gel from amorphous to liquid crystalline that led to a stripe pattern has been characterized as a Liesegang pattern [113].

Liesegang patterns are routinely described by the following three laws:

1) The spacing law by Morse and Pierce predicts the spacing between consecutive bands, which follows a geometric series where $X_n/X_{n+1} = P$, with X being the position of a given band (n) from the gel surface and P the so-called spacing coefficient [114].
2) The time law by Jablczynski $X_n \simeq t_n^{1/2}$ gives the relation between the distance of a band and the square root of its time of appearance t_n [115].
3) The Matalon–Packter law shows that the spacing coefficient is not a universal quantity but depends instead on the initial concentrations of the electrolytes:

$$P = F(b_0) + G(b_0)/a_0,$$

where a_0 and b_0 are the initial concentrations of the outer and inner electrolytes, respectively, and F and G are decreasing functions [116].

Even though the Liesegang rings are known since more than a century, the mechanism of their formation is still debated, nonetheless, because there are many different chemical systems that form those rings, and many types of patterns are known. Matalon and Packter found that the pattern obeys a simple relation based on diffusion laws, but the absorption onto the sol formed in the gel medium is the predominant factor for the ring formation. Among the theories proposed to explain the Liesegang phenomenon, that of Dhar and Chatterjee fits the obtained results [116].

Other mechanisms have been proposed, but they can only account for some of the observed patterns. The supersaturation mechanism was published in 1897 by Ostwald in his textbook about general chemistry [117]. According to this mechanism, the solution remains in a supersaturated state until the first crystal seeds are formed. The subsequent growth of those seeds happens very fast in comparison to the diffusion of the ions and a depletion zone is formed around the crystals. This mechanism can account for many of the patterns, especially the ones with increasing distance between the rings, but inverse patterns or more complicated ones are not explicable by this mechanism. Also, the seeding of the gel with nanocrystals of the precipitate should prevent such a supersaturation, but Liesegang rings have been observed in such systems, too.

The spinodal decomposition theory [118] takes into account that a salt precipitate does not form immediately, even though the ion concentration exceeds its solubility product. The precipitate only forms when the supersaturated state becomes unstable. A clear region lies ahead of the diffusion front because the precipitate acts as seeds for further salt crystallization from the surrounding solution. This so-called Ostwald cycle of supersaturation, nucleation, and depletion may be flawed, because seeding the gel with a colloidal dispersion of the precipitate that would prevent any significant region of supersaturation did not prevent the formation of the rings. Later on, Ostwald himself proposed another theory that is based on the diffusion waves of at least three components that propagate along opposite directions [79]. Generally speaking, the outer electrolyte is concentrated and the inner one less concentrated. The precipitation is fast compared to the diffusion, and thus the inner electrolyte becomes depleted around the precipitates. Thus, the threshold concentration cannot be reached in close proximity of a precipitation band. A new precipitation band can only form after a certain distance.

The adsorption theory focuses on the adsorption of one of the precipitating ions onto the colloidal particles of the precipitate [119]. If the particles are small, the absorption is large, diffusion is hindered, and this somehow results in the formation of the rings.

The coagulation theory states that the initial precipitate is in the form of a fine colloidal dispersion that then leads to coagulation because of an excess of the diffusing electrolyte and this results in ring formation [120].

The flocculation theory was formulated in 1970 by Shinohara [121]. He states that diffusion proceeds as the contact surface of the two electrolytes advances. Flocculation occurs suddenly when the total ionic concentration at the end of a sol region exceeds a flocculation value. This flocculation spreads rapidly in the region until it reaches the front of it, and it stops. From there, the next sol region begins to spread.

18.6 Summary

This chapter introduced the formation of a chemical contrast, either on a surface or in bulk. A chemical contrast on a surface can be produced by either top-down (lithography, etc.) or bottom-up (self-assembly, etc.) processes. The chemical contrast on a micrometer scale can be useful to pattern cells or to produce sensors. The bulk process involves the diffusion of two or more components in a hydrogel. Chemical gradients, supersaturation, and related phenomena lead to the periodic precipitation of colloidal particles in a three-dimensional fashion. Such patterns can be used to produce colloidal particles with desired properties.

References

1 Buerk, H., Krampf, G., and Houben, W. (1988) Wet chemical patterning of hafnium boride layers., DE 3708832 A1, filed Mar. 18, 1987 and issued Sep. 29, 1988.

2 Wu, S., Ho, S.T., Xiong, F., and Chang, R.P.H. (1995) Wet chemical patterning of KTP thin films for nonlinear optical waveguide applications. *J. Electrochem. Soc.*, **142** (10), 3556–3557.

3 Curtis, A. and Britland, S. (1996) Surface modification of biomaterials by topographic and chemical patterning, in *Advanced Biomaterials in Biomedical Engineering and Drug Delivery Systems.*, (ed. N. Ogata), Springer, pp. 158–162.

4 Curtis, A. and Riehle, M. (2001) Tissue engineering: the biophysical background. *Phys. Med. Biol.*, **46**, R47–R65.

5 Swain, P.S. and Andelman, D. (2001) Supported membranes on chemically structured and rough surfaces. *Phys. Rev. E*, **63**, 051911.

6 Wade, N., Evans, C., Jo, S.-C., and Cooks, R.G. (2002) Silylation of an OH-terminated self-assembled monolayer surface through low-energy collisions of ions: a novel route to synthesis and patterning of surfaces. *J. Mass Spectrom.*, **37** (6), 591–602.

7 Iqbal, P., Preece, J.A., and Mendes, P.M. (2012) Nanotechnology: the 'Top-Down' and 'Bottom-Up' Approaches, in *Supramolecular Chemistry: From Molecules to Nanomaterials* (eds J.W. Steed and P.A. Gale), John Wiley & Sons, Ltd, Hoboken, NJ, pp. 3589–3978.

8 Allara, D.L. and Nuzzo, R.G. (1985) Spontaneously organized molecular assemblies. 1. Formation, dynamics, and physical properties of n-alkanoic acids adsorbed from solution on an oxidized aluminum surface. *Langmuir*, **1** (1), 45–52.

9 Hofer, R., Textor, M., and Spencer, N.D. (2001) Alkyl phosphate monolayers, self-assembled from aqueous solution onto metal oxide surfaces. *Langmuir*, **17** (13), 4014–4020.

10 Nuzzo, R.G. and Allara, D.L. (1983) Adsorption of bifunctional organic disulfides on gold surfaces. *J. Am. Chem. Soc*, **105**, 4481.

11 Maoz, R. and Sagiv, J.J. (1984) On the formation and structure of self-assembling monolayers. I. A comparative atr-wettability study of Langmuir–Blodgett and adsorbed films on flat substrates and glass microbeads. *Colloid Interface Sci.*, **100**, 465–496.

12 Sagiv, J. (1980) Organized monolayers by adsorption. 1. Formation and structure of oleophobic mixed monolayers on solid surfaces. *J. Am. Chem. Soc.*, **102**, 92–98.

13 Lucy Netzer, L. and Sagiv., J. (1983) A new approach to construction of artificial monolayer assemblies. *J. Am. Chem. Soc.*, **105** (3), 674–676.

14 Linford, M.R., Fenter, P., Eisenberger, P.M., and Chidsey, C.E.D. (1995) Alkyl monolayers on silicon prepared from 1-alkenes and hydrogen-terminated silicon. *J. Am. Chem. Soc.*, **117** (11), 3145–3155.

15 Tillman, N., Ulman, A., Schildkraut, J.S., and Penner, T.L. (1988) Incorporation of phenoxy groups in self-assembled monolayers of trichlorosilane derivatives. Effects on film thickness, wettability, and molecular orientation. *J. Am. Chem. Soc.*, **110**, 6136–6144.

16 Depalma, V. and Tillman, N. (1989) Friction and wear of self-assembled trichlorosilane monolayer films on silicon. *Langmuir*, **5**, 868–872.

17 Al-Abadleh, H.A., Mifflin, A.L., Musorrafiti, M.J., and Geiger, F.M. (2005) Kinetic studies of chromium(VI) binding to carboxylic acid- and methyl ester-functionalized silica/water interfaces important in geochemistry. *J. Phys. Chem. B.*, **109** (35), 16852–16859.

18 Husseini, G.A., Peacock, J., Sathyapalan, A., Zilch, L.W., Asplund, M.C., Sevy, E.T., and Linford, M.R. (2003) Alkyl monolayers on silica surfaces prepared using neat, heated dimethylmonochlorosilanes with low vapor pressures. *Langmuir*, **19**, 5169–5171.

19 Tseng, J.-Y., Lin, M.-H., and Chau, L.-K. (2001) Preparation of colloidal gold multilayers with 3-(mercaptopropyl)-trimethoxysilane as a linker molecule. *Colloids Surf. A*, **182** (1–3), 239–245.

20 Huang, J., Gräter, S.V., Corbellini, F., Rinck, S., Bock, E., Kemkemer, R., Kessler, H., Ding, J., and Spatz, J.P. (2009) Impact of order and disorder in RGD nanopatterns on cell adhesion. *Nano Lett.*, **9** (3), 1111–1116.

21 Barness, Y., Gershevitz, O., Sekar, M., and Sukenik, C.N. (2000) Functionalized silanes for the preparation of siloxane-anchored monolayers. *Langmuir*, **16** (1), 247–251.

22 Kumar, N., Maldarelli, C., Steiner, C., and Couzis, A. (2001) Formation of nanometer domains of one chemical functionality in a continuous matrix of a second chemical functionality by sequential adsorption of silane self-assembled monolayers. *Langmuir*, **17** (25), 7789–7797.

23 Kallury, K.M.R., Macdonald, P.M., and Thompson, M. (1994) Effect of surface-water and base catalysis on the silanization of silica by (aminopropyl) alkoxysilanes studied by X-ray photoelectron-spectroscopy and C-13 cross-polarization magic-angle-spinning nuclear-magnetic-resonance. *Langmuir*, **10**, 492–499.

24 Sugimura, H., Hozumi, A., Kameyama, T., and Takai, O. (2002) Organosilane self-assembled monolayers formed at the vapor/solid interface. *Surf. Interface Anal.*, **34**, 550–554.

25 Ikeda, Y. (1994) Surface modification of polymers for medical applications. *Biomaterials*, **15**, 725–736.

26 Minko, S. (2008) Grafting on solid surfaces: "Grafting to" and "Grafting from" methods, in *Polymer Surfaces and Interfaces* (ed. M. Stamm), Springer, Berlin, pp. 215–234.

27 Kim, M., Schmitt, S.K., Choi, J.W., Krutty, J.D., and Gopalan, P. (2015) From self-assembled monolayers to coatings: advances in the synthesis and nanobio applications of polymer brushes. *Polymers*, **7**, 1346–1378.

28 Kim, J.-K., Shin, D.-S., Chung, W.-J., Jang, K.-H., Lee, K.-N., Kim, Y.-K., and Lee, Y.-S. (2004) Effects of polymer grafting on a glass surface for protein chip applications. *Colloids Surf. B Biointerfaces*, **33**, 67–75.

29 Mévellec, V., Roussel, S., Tessier, L., Chancolon, J., Mayne-L'Hermite, M., Deniau, G., Viel, P., and Palacin, S. (2007) Grafting polymers on surfaces: a new powerful and versatile diazonium salt-based one-step process in aqueous media. *Chem. Mater.*, **19** (25), 6323–6330.

30 Matyjaszewski, K., Dong, H., Jakubowski, W., Pietrasik, J., and Kusumo, A. (2007)

Grafting from surfaces for "Everyone": ARGET ATRP in the presence of air. *Langmuir*, **23** (8), 4528–4531.

31 Minko, S., Müller, M., Luchnikov, V., Motornov, M., Usov, D., Ionov, L., and Stamm, M. (2004) Mixed polymer brushes: switching of surface behavior and chemical patterning at the nanoscale, in *Polymer Brushes* (eds R.C. Advincula, W.J. Brittain, K.C. Caster, and J. Ruehe), Wiley-VCH Verlag GmbH, Weinheim, pp. 403–425.

32 Lee, W.-K., Kaholek, M., Ahn, S.-J., and Zauscher, S. (2006) Nanopatterning of stimuli-responsive polymer brushes by scanning probe and electron beam lithography, in *Responsive Polymer Materials: Design and Applications*, (ed S. Minko), Blackwell Publishing Professional, Ames, pp. 84–100.

33 Lvov, Y., Decher, G., and Sukhorukov, G. (1993) Assembly of thin films by means of successive deposition of alternate layers of DNA and poly(allylamine). *Macromolecules*, **26**, 5396–5399.

34 Jeon, H., Schmidt, R., Barton, J.E., Hwang, D.J., Gamble, L.J., Castner, D.G., Grigoropoulos, C.P., and Healy, K.E. (2011) Chemical patterning of ultrathin polymer films by direct-write multiphoton lithography. *J. Am. Chem. Soc.*, **133** (16), 6138–6141.

35 Okamoto, T., Suzuki, T., and Yamamoto, N. (2000) Microarray fabrication with covalent attachment of DNA using Bubble Jet technology. *Nat. Biotechnol.*, **18**, 438–441.

36 Xu, S. and Liu, G.-Y. (1997) Nanometer-scale fabrication by simultaneous nanoshaving and molecular self-assembly. *Langmuir*, **13** (2), 127–129.

37 Xu, S., Miller, S., Laibinis, P.E., and Liu, G.-Y. (1999) Fabrication of nanometer scale patterns within self-assembled monolayers by nanografting. *Langmuir*, **15** (21), 7244–7251.

38 Piner, R.D., Zhu, J., Xu, F., Hong, S., and Mirkin, C.A. (1999) "Dip-Pen" nanolithography. *Science*, **283**, 661–663.

39 Cai, Y. and Ocko, B.M. (2005) Electro pen nanolithography. *J. Am. Chem. Soc.*, **127** (46), 16287–16291.

40 Wilbur, J.L., Kumar, A., Kim, E., and Whitesides, G.M. (1994) Microfabrication by microcontact printing of self-assembled monolayers. *Adv. Mater.*, **6**, 600–604.

41 Csucs, G., Künzler, T., Feldman, K., Robin, F., and Spencer, N.D. (2003) Microcontact printing of macromolecules with submicrometer resolution by means of polyolefin stamps. *Langmuir*, **19** (15), 6104–6109.

42 Hoff, J.D., Cheng, L.-J., Meyhöfer, E., Guo, L.J., and Hunt, A.J. (2004) Nanoscale protein patterning by imprint lithography. *Nano Lett.*, **4** (5), 853–857.

43 Laibinis, P.E., Hickman, J.J., Wrighton, M.S., and Whitesides, G.M. (1989) Orthogonal systems for self-assembled monolayers: alkanethiols on gold and alkane carboxylic acids on alumina. *Science*, **245**, 845–847.

44 Michel, R., Lussi, J.W., Csucs, G., Reviakine, I., Danuser, G., Ketterer, B., Hubbell, J.A., Textor, M., and Spencer, N.D. (2002) Selective molecular assembly patterning: a new approach to micro- and nanochemical patterning of surfaces for biological applications. *Langmuir*, **18**, 3281–3287.

45 Cai, Y. and Ocko, B.M. (2005) Large-scale fabrication of protein nanoarrays based on nanosphere lithography. *Langmuir*, **21**, 9274–9279.

46 Valsesia, A., Colpo, P., Meziani, T., Lisboa, P., Lejeune, M., and Rossi, F. (2006) Immobilization of antibodies on biosensing devices by nanoarrayed selfassembled monolayers. *Langmuir*, **22**, 1763–1767.

47 Park, C., Yoon, J., and Thomas, E.L. (2003) Enabling nanotechnology with self assembled block copolymer patterns. *Polymer*, **44**, 6725–6760.

48 Steiner, U. (2006) Structure formation in polymer films, in *Nanoscale Assembly: Chemical Techniques* (ed. W.T.S. Huck), Springer, New York.

49 Kajiyama, T., Tanaka, K., and Takahara, A. (1998) Surface Segregation of the higher surface free energy component in symmetric polymer blend films. *Macromolecules*, **31** (11), 3746–3749.

50 Park, M., Harrison, C., Chaikin, P.M., Register, R.A., and Adamson, D.H. (1997) Block copolymer lithography: periodic arrays of $\sim 10^{11}$ holes in 1 square centimeter. *Science*, **276**, 1401–1404.

51 Spatz, J.P., Sheiko, A., and Moller, M. (1996) Ion-stabilized block copolymer micelles: film formation and intermicellar interaction. *Macromolecules*, **29**, 3220–3226.

52 Dupuis, A., Leopoldes, J., Bucknall, D.G., and Yeomans, J.M. (2005) Control of drop positioning using chemical patterning. *Appl. Phys. Lett.*, **87** (2), 024103/1–024103/3.

53 Kusumaatmaja, H. and Yeomans, J.M. (2007) Controlling drop size and polydispersity using chemically patterned surfaces. *Langmuir*, **23**, 956–959.

54 Berejnov, V. and Thorne, R.E. (2005) Enhancing drop stability in protein crystallization by chemical patterning. *Acta Crystallogr. D Biol. Crystallogr.*, **D61**, 1563–1567.

55 Kataoka, D.E. and Troian, S.M. (1999) Patterning liquid flow on the microscopic scale. *Nature*, **402**, 794–797.

56 Wang, J.Z., Zheng, Z.H., Li, H.W., Huck, W.T.S., and Sirringhaus, H. (2004) Dewetting of conducting polymer inkjet droplets on patterned surfaces. *Nat. Mater.*, **3**, 171–176.

57 Krausch, G. (1995) Surface induced self assembly in thin polymer films. *Mater. Sci. Eng. R*, **14**, 1–94.

58 Kim, S., Nealey, P.F., and Bates, F.S. (2014) Directed assembly of lamellae forming block copolymer thin films near the order-disorder transition. *Nano Lett.*, **14** (1), 148–152.

59 Liu, C.-C., Ramírez-Hernández, A., Han, E., Craig, G.S.W., Tada, Y., Yoshida, H., Kang, H., Ji, S., Gopalan, P., de Pablo, J.J., and Nealey, P.F. (2013) Chemical patterns for directed self-assembly of lamellae-forming block copolymers with density multiplication of features. *Macromolecules*, **46**, 1415–1424.

60 Craighead, H.G., James, C.D., and Turner, A.M.P. (2001) Chemical and topographical patterning for directed cell attachment. *Curr. Opin. Solid State Mater. Sci.*, **5** (2–3), 177–184.

61 Lim, J.Y. and Donahue, H.J. (2007) Cell sensing and response to micro- and nanostructured surfaces produced by chemical and topographic patterning. *Tissue Eng.*, **13** (8), 1879–1891.

62 Kane, R.S., Takayama, S., Ostuni, E., Ingber, D.E., and Whitesides, G.M. (1999) Patterning proteins and cells using soft lithography. *Biomaterials*, **20** (23–24), 2363–2376.

63 Healy, K.E., Thomas, C.H., Rezania, A., Kim, J.E., McKeown, P.J., Lom, B., and Hockberger, P.E. (1996) Kinetics of bone cell organization and mineralization on materials with patterned surface chemistry. *Biomaterials*, **17** (2), 195–208.

64 Delacour, C., Bugnicourt, G., Dempsey, N.M., Dumas-Bouchiat, F., and Villard, C. (2014) Combined magnetic and chemical patterning for neural architectures. *J. Phys. D Appl. Phys.*, **47**, 425403/1–425403/6.

65 Wosnick, J.H. and Shoichet, M.S. (2008) Three-dimensional chemical patterning of transparent hydrogels. *Chem. Mater.*, **20** (1), 55–60.

66 Henisch, H.K. (1988) *Crystals in Gels and Liesegang Rings*, Cambridge University Press, Cambridge.

67 Liesegang, R.E. (1896) Über einige eigenschaften von gallerten. *Naturwiss. Wochenschr.*, **11**, 353–362.

68 Liesegang, R.E. (1898) *Chemische Reaktionen in Gallerten* (ed. Liesegang's Verlag) Steinkopff, Düsseldorf.

69 Narwani, C.S. and Gursahani, G.T. (1941) Rhythmic precipitation of silver chloride in gelatin tanned with chromium chloride. *J. Ind. Chem. Soc.*, **18**, 531–534.

70 Sultan, R. and Sadek, S. (1996) Patterning trends and chaotic behavior in Co^{2+}/ NH_4OH Liesegang systems. *J. Phys. Chem.*, **100**, 16912–16920.

71 Volford, A., Izsák, F., Ripszám, M., and Lagzi, L. (2007) Pattern formation and self-organization in a simple precipitation system. *Langmuir*, **23**, 961–964.

72 Al-Ghoul, M., Ammar, M., and Al-Kaysi, R.O. (2012) Band propagation, scaling laws and phase transition in a precipitate system. I: Experimental study. *J. Phys. Chem. A*, **116**, 4427–4437.

73 George, J. and Varghese, G. (2005) Intermediate colloidal formation and the varying width of periodic precipitation bands in reaction-diffusion systems. *J. Colloid Interface Sci.*, **282**, 379–402.
74 Karam, T., El-Rassy, H., and Sultan, R. (2011) Mechanism of revert spacing in a $PbCrO_4$ liesegang system. *J. Phys. Chem. A*, **115** (14), 2994–2998.
75 Xie, A., Zhang, L., Zhu, J., Shen, Y., Xu, A., Zhu, J., Li, C., Chen, L., and Yang, L. (2009) Formation of calcium oxalate concentric precipitate rings in two-dimensional agar gel systems containing $Ca^{2+}RE^{3+}$ (RE=Er, Gd and La)/C_2O_4. *Colloids Surf. A*, **332**, 192–199.
76 Prakash, S.M.D. and Rao, P.M. (1986) Periodic crystallization of barium oxalate in silica hydrogel. *Bull. Mater. Sci.*, **8**, 511–517.
77 Mandalian, L., Fahs, M., Al-Ghoul, M., and Sultan, R. (2004) Morphology, particle size distribution, and composition in one- and two-salt metal oxinate liesegang patterns. *J. Phys. Chem. B*, **108**, 1507–1514.
78 Hausmann, J. (1904) On precipitates in colloids. *Z. Anorg. Chem.*, **40**, 110–145.
79 Ostwald, W. (1925) Zur Theorie der Liesegang'schen Ringe. *Kolloid-Zeitschrift*, **36**, 380–390.
80 Davies, E.C.H. (1922) Liesegang rings. I. Silver chromate in gelatin and colloidal gold in silicic acid gel. *J. Am. Chem. Soc.*, **44**, 2698–2704.
81 Bensemann, I.T., Fialkowski, M., and Gryzybowski, B.A. (2005) Wet stamping of microscale periodic precipitation patterns. *J. Phys. Chem. B*, **109**, 2774–2778.
82 Riegei, E.R. and Reinhard, M.C. (1927) The zone pattern formed by silver dichromate in solid gelatin gel. *J. Phys. Chem.*, **31**, 713–718.
83 Das, I., Pushkarna, A., and Lall, R.S. (1987) Light induced periodic precipitation and crystal growth of PbCrO4: a novel study in two-dimensional gel media. *J. Cryst. Growth*, **82**, 361–366.
84 Badr, L., El-Rassy, H., El-Joubeily, S., and Sultan, R. (2010) Morphology of a 2D Mg/NH_4OH Liesegang pattern in zero, positive and negative radial electric field. *Chem. Phys. Lett.*, **492**, 35–39.
85 Al-Ghoul, M. and Sultan, R. (2003) Front propagation in patterned precipitation. 2. Electric effects in precipitation-dissolution patterning schemes. *J. Phys. Chem.*, **107** (8), 1095–1101.
86 Bena, I., Droz, M., and Rácz, Z. (2005) Formation of Liesegang patterns in the presence of an electric field. *J. Chem. Phys.*, **122** (20), 204502.
87 Bena, I., Droz, M., Lagzi, I., Martens, K., Rácz, Z., and Volford, A. (2008) Designed patterns: flexible control of precipitation through electric currents. *Phys. Rev. Lett.*, **101**, 075701.
88 Lagzi, I. and Ueyama, D. (2009) Pattern transition between periodic Liesegang pattern and crystal growth regime in reaction-diffusion systems. *Chem. Phys. Lett.*, **468**, 188–192.
89 Al-Ghoul, M., El-Rassy, H., Coradin, T., and Mokallad, T. (2010) Reaction-diffusion based co-synthesis of stable alpha- and beta-cobalt hydroxide in bio-organic gels. *J. Cryst. Growth*, **312**, 856–862.
90 Hatschek, E. (1921) Anomalous Liesegang stratifications produced by the action of light. *Proc. R. Soc. Lond. A Math. Phys. Eng. Sci.*, **99**, 496–502.
91 Karam, T., El-Rassy, H., Zaknoun, F., Moussa, Z., and Sultan, R. (2012) Liesegang banding and multiple precipitate formation in cobalt phosphate systems. *Chem. Phys. Lett.*, **525–526**, 54–59.
92 Müller, S.C., Kai, S., and Ross, J. (1982) Mesoscopic structure of pattern formation in initially uniform colloids. *J. Phys. Chem.*, **86** (22), 4294–4297.
93 Lagzi, I., Kowalczyk, B., and Grzybowski, B.A. (2010) Liesegang rings engineered from charged nanoparticles. *J. Am. Chem. Soc.*, **132**, 58–60.
94 Nabika, H., Sato, M., and Unoura, K. (2014) Liesegang patterns engineered by a chemical reaction assisted by complex formation. *Langmuir*, **30**, 5047–5051.
95 Walliser, R.M., Boudoire, F., Orosz, E., Tóth, R., Braun, A., Constable, E.C., Rácz, Z., and Lagzi, I. (2015) Growth of nanoparticles and microparticles by

controlled reaction-diffusion processes. *Langmuir*, **31**, 1828–1834.
96 Wei, Y.H., Han, S.B., Kim, J., Soh, S., and Grzybowski, B.A. (2010) Photoswitchable catalysis mediated by dynamic aggregation of nanoparticles. *J. Am. Chem. Soc.*, **132**, 11018–11020.
97 Narayanan, R. and El-Sayed, M.A. (2004) Shape-dependent catalytic, activity of platinum nanoparticles in colloidal solution. *Nano Lett.*, **4**, 1343–1348.
98 Jiang, W., Kim, B.Y.S., Rutka, J.T., and Chan, W.C.W. (2008) Nanoparticle-mediated cellular response is size-dependent. *Nat. Nanotechnol.*, **3**, 145–150.
99 Euliss, L.E., DuPont, J.A., Gratton, S., and DeSimone, J. (2006) Imparting, size, shape, and composition control of materials for nanomedicine. *Chem. Soc. Rev.*, **35**, 1095–1104.
100 Yu, K., Kelly, K.L., Sakai, N., and Tatsuma, T. (2008) Morphologies and surface plasmon resonance properties of monodisperse bumpy gold nanoparticles. *Langmuir*, **24**, 5849–5854.
101 Boudoire, F., Tóth, R., Heier, J., Braun, A., and Constable, E.C. (2014) Photonic light trapping in self-organized all-oxide microspheroids impacts photoelectrochemical water splitting. *Energy Environ. Sci.* (2014) **7**, 2680–2688.
102 Vestal, C.R. and Zhang, Z.J. (2002) Synthesis of $CoCrFeO_4$ nanoparticles using microemulsion methods and size-dependent studies of their magnetic properties. *Chem. Mater.*, **14**, 3817–3822.
103 Wang, Z.X., Tan, B.E., Hussain, I., Schaeffer, N., Wyatt, M.F., Brust, M., and Cooper, A.I. (2007) Design of polymeric stabilizers for size-controlled synthesis of monodisperse gold nanoparticles in water. *Langmuir*, **23**, 885–895.
104 Taleb, A., Petit, C., and Pileni, M.P. (1997) Synthesis of highly monodisperse silver nanoparticles from AOT reverse micelles: a way to 2D and 3D self-organization. *Chem. Mater.*, **9**, 950–959.
105 Abou-Hassan, A., Sandre, O., and Cabuil, V. (2010) Microfluidics in inorganic chemistry. *Angew. Chem., Int. Ed.*, **49**, 6268–6268.
106 Kalsin, A.M. and Grybowski, B. (2007) Controlling the growth of "Ionic" nanoparticle supracrystals. *Nano Lett.*, **7**, 1018–1021.
107 Bishop, K.J.M. and Grzybowski, B.A. (2007) "Nanoions": fundamental properties and analytical applications of charged nanoparticles. *Chem. Phys. Chem.*, **8**, 2171–2176.
108 Liesegang, R.E. (1906) Eine scheinbar chemische fernwirkung. *Ann. Phys.*, **324**, 395–406.
109 Klajn, R., Fialkowski, M., Bensemann, I.T., Bitner, A., Campbell, C.J., Bishop, K., Smoukov, S., and Grzybowski, B.A. (2004) Multicolour micropatterning of thin films of dry gels. *Nat. Mater.*, **3**, 729–735.
110 Grzybowski, B.A., Bishop, K.J.M., Campbell, C.J., Fialkowski, M., and Smoukov, S.K. (2005) Micro- and nanotechnology via reaction-diffusion. *Soft Matter*, **1**, 114–128.
111 Nell, P. (1905) Studien über diffusionsvorgänge wässeriger lösungen in gelatine. *Ann. Phys.*, **323**, 323–347.
112 Narita, K., Matsumoto, K., and Igawa, M. (2006) Application of Liesegang ring formation on a gelatin film to the determination of sulfate concentration in individual rain droplets. *Anal. Sci.*, **22**, 1559–1563.
113 Narita, T. and Tokita, M. (2006) Liesegang pattern formation in kappa-carrageenan gel. *Langmuir*, **22**, 349–352.
114 Morse, H.W. and Pierce, G.W. (1903) Diffusion and supersaturation in gelatin. *Phys. Rev.*, **17**, 129–150.
115 Jablczynski, K. (1923) Les anneaus de Liesegang. *Bull. Soc. Chim. Fr.*, **33**, 1592–1604.
116 Matalon, R. and Packter, A. (1955) The liesegang phenomenon. I. Sol protection and diffusion. *J. Colloid Sci.*, **10** (1), 46–62.
117 Ostwald, W. (1899) *Lehrbuch der Allgemeinen Chemie*, 2nd edn, vol. **II**, Leipzig, Engelmann, pp. 779.

118 Antal, T., Droz, M., Magnin, J., and Rácz, Z. (1999) Formation of liesegang patterns: a spinodal decomposition scenario. *Phys. Rev. Lett.*, **83**, 2880–2883.
119 Bradford, S.C. (1921) An explanation of liesegang's rings. *Science*, **54**, 463–464.
120 Dhar, N.R. and Chatterji, A.C. (1922) Liesegang'sches phänomenon und niederschlagsbildung. *Kolloid Z.*, **31**, 15–16.
121 Shinohara, S. (1970) A Theory of one-dimensional liesegang phenomena. *J. Phys. Soc. Jpn.*, **29**, 1073–1087.

19
Microcontact Printing

Kiyoshi Yase

National Institute of Advanced Industrial Science and Technology, Department of Materials and Chemistry, Umezono 1-1-1, Tsukuba 305-8565, Japan

19.1
Introduction

Recently organic electroluminescent (EL) display panels have been appearing on the market. In 2007, a Japanese electric company first launched the 11 inch TV and introduced the small EL panels as main displays instead of liquid crystal displays in mobile phones. In 2015, a Korean company started to supply the 55-inch panels to the worldwide. In such TV panels, each pixel was switched by poly-silicon or amorphous silicon thin film transistors (TFTs) on glass substrate, as shown in Figure 19.1.

For flexible and nonfragile displays, the substrate should be changed from glass to plastic sheet. How to fabricate the TFTs on plastic sheet with large and fine pixel size at low temperature process would be explained in this chapter.

19.2
Organic Thin Film Devices

The organic electronic devices such as thin film transistors (TFTs), organic electroluminescent devices, photovoltaic (PV) cells, flexible memories, radiofrequency identification tags, and so on, have been attempted in worldwide. They mainly consist of layered structures of substrate, bottom electrode, active layer, and top electrode, as shown in Figure 19.2.

Now most of the active layers, such as semiconductor, electron/hole transport layer and p, n, or p/n layer, consist of amorphous states, which can be fabricated by wet processes. However, in the case of switching TFT in pixel for flat panel display, there are the complicated in-plane structures as shown in Figure 19.1. Bus lines for source and drain electrodes should overlay on each other for high-resolution display panel. And it is essential for high performance TFT to make

Handbook of Solid State Chemistry, First Edition. Edited by Richard Dronskowski, Shinichi Kikkawa, and Andreas Stein.

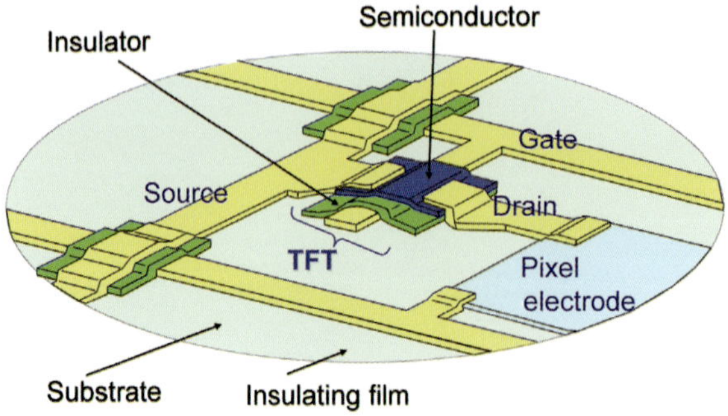

Figure 19.1 Structure of switching TFT of pixel in flat panel display.

fine and small gap between the source and the drain electrodes on the top of the semiconductor.

19.3 Printing Methods

There are a variety of conventional printing techniques such as (a) off-set printing, (b) convex printing, (c) concave printing, (d) gravure printing, (e) screen printing, and (f) ink jet printing, shown in Figure 19.3.

In those printing methods the comparison of coefficient of viscosity of ink, maximum and minimum film thickness, resolution or definition, and printing speed is shown in Table 19.1.

Most of the printing methods represent the resolutions of only 30 μm or larger, which is not enough for high-definition display with active-matrix (AM) liquid crystal display (LCD) and EL panel. And the performance in organic semiconductors have reached only 0.1–0.01 Vs/cm^2 prepared by wet process so that the channel length, which is the distance between the source and drain electrode, should be designed less than 5 μm. Only the microcontact printing is realized with such high resolution, that is, less than micron meter as shown in Table 19.1.

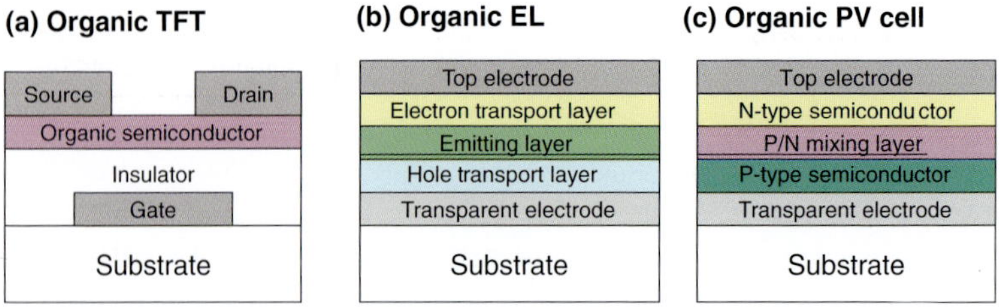

Figure 19.2 Fundamental structures of organic electronic devices, (a) organic thin-film-transistor, (b) organic electroluminescent device, and (c) organic photovoltaic cell.

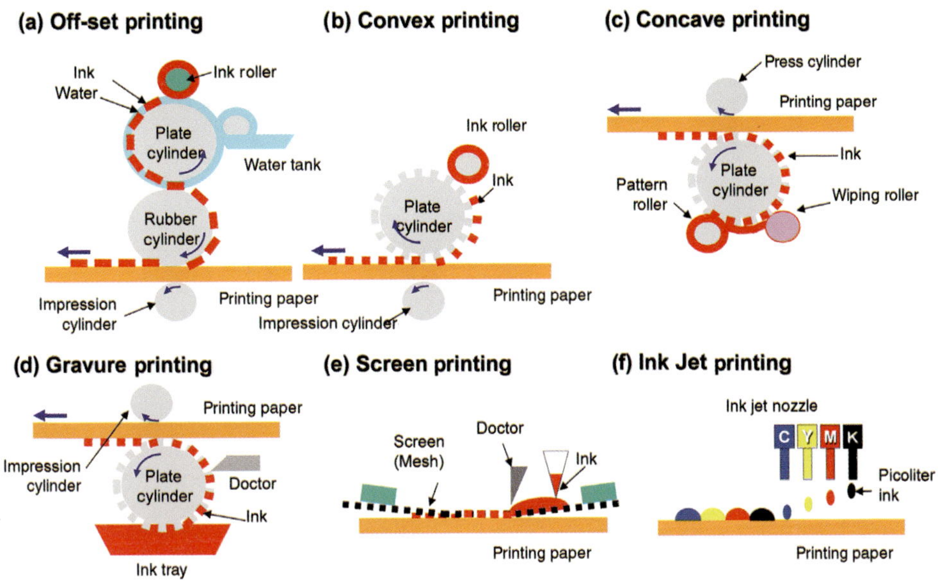

Figure 19.3 Several kinds of printing techniques.

The following sections will present the results of large area and fine patterning of the entirely printed organic thin film transistor array on plastic substrate and driven the polymer network liquid crystal display.

19.4 Microcontact Printing

A microcontact printing (μCP), first demonstrated by Whitesides [1], has an advantage of high resolution printing of self-assembled monolayers in sub-millimeter order. At the early stage of μCP process, one used to demonstrate very small stamper with 1 inch square or less and transfer the self-assembled

Table 19.1 Comparison in several printing techniques.

Printing method	Ink viscosity (mPa s) or (cP)	Maximum/minimum film thickness (μm)	Resolution (definition)	Printing speed (m^2/s)
Screen	$(2–30) \times 10^3$	10–20/100	20 μm	10
Gravur	50–500	2–3/25	30 μm	50
Flexography	100–1000	3–5/15	35 μm	10
Off-Set	$(4–10) \times 10^3$	1/5	10 μm	20
Ink Jet	$1–10^4$	10–20/100	20 μm	0.01
Microcontact	—	1 nm/2 μm	15–20 nm	10^{-5}

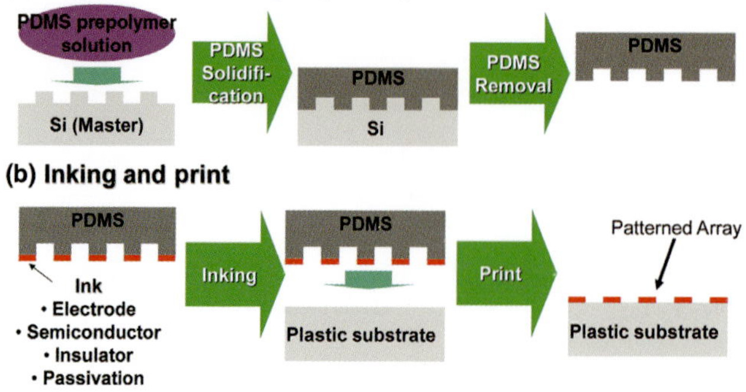

Figure 19.4 Procedures of microcontact printing (μCP).

monolayer of alkane thiol onto the gold thin film on Si or glass substrate. The monomolecular layer was acted as thin mask for further etching process to provide submillimeter scale patterning. So the process was called as "soft lithography."

We have developed up to A4 size from such tiny stamping [2–6]. The procedures are shown in Figure 19.4.

a) The stamper made of poly-dimethyl-siloxane (PDMS) (Figure 19.5) was prepared from Si or glass master, which had been previously patterned by using electron beam and/or photolithographic processes on the substrate. The process is very resemble to "nanoimprint" technique.
b) The conductive, semiconductive, and insulating inks were put on the PDMS stamper with fine pattern and then transferred on the plastic substrate.

There are 1–10 mm line and space (L/S), dots, and holes as test patterns.

Figure 19.6 shows the fine patterns with line and space of 2 μm and line width of 1 μm of silver nanoparticle ink printed on the glass substrate. It is

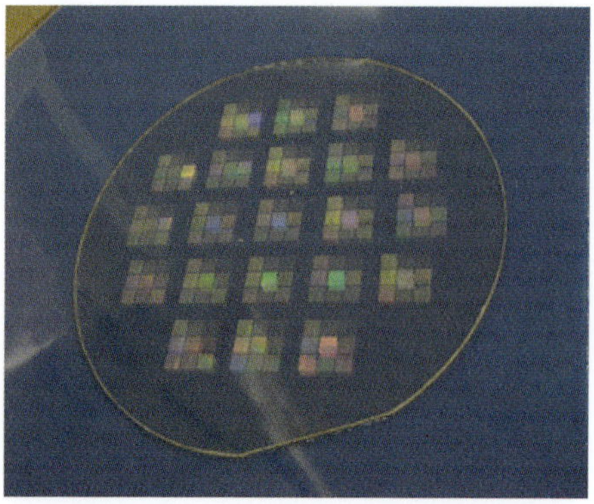

Figure 19.5 Patterned PDMS stamp with a size of 6 inch.

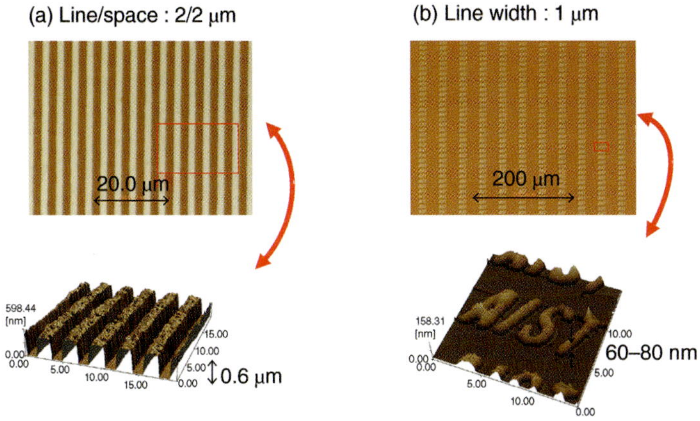

Figure 19.6 Fine patterns of silver nanoparticle ink on glass substrate: (a) line and space of 2 μm and (b) line width of 1 μm.

an important point that the thicknesses of fine lines are 0.6 μm and 60–80 nm, respectively. The former value is enough to get currents as the conductive lines.

In the case of normal printing methods, such as relief, gravure, screen, offset, and inkjet, the ink should be the state of solution with high or moderate viscosity. After the ink was transferred from the mold to substrate, it should change the shape for the fluidity of ink during dry and anneal, as shown in Figure 19.7 (a). However, in the case of microcontact printing, the ink on the stamper had been already semisolidificated and then transferred onto the substrate so that the shape of ink does not change as it was, as shown in Figure 19.7(b).

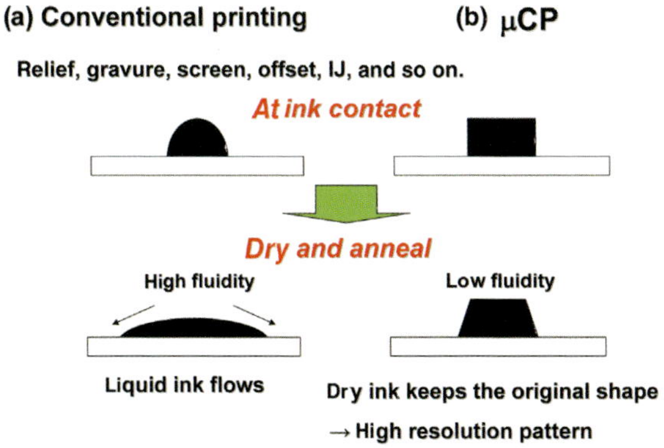

Figure 19.7 Shape changing after contact and transfer from the mold to substrate: (a) conventional and (b) microcontact printing.

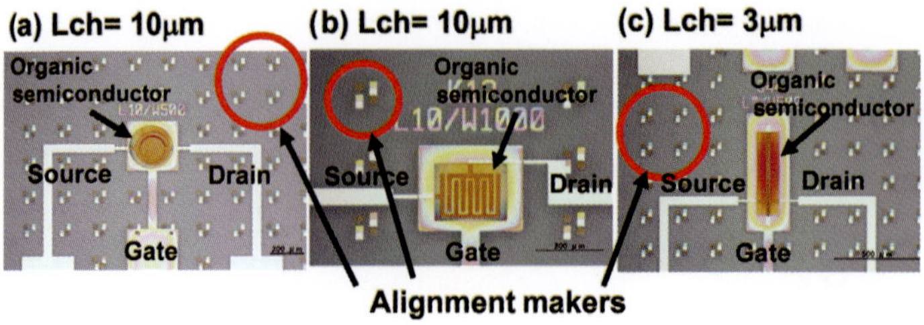

Figure 19.8 Several printed organic TFTs on glass substrate. Channel length (Lch) = (a) 10 μm, (b) 10 μm, and (c) 3 μm.

19.5 Printing of TFT Arrays on Substrate

The glass substrate was chosen as the large area master plate. At first, the photopolymer as resist film was spin-coated on the glass substrate, illuminated by parallel ultraviolet light and then etched to form the fine pattern for each of stampers. Every stamper for gate electrode, source and drain electrode, and organic semiconductor has same alignment makers at the edge of the sheet as shown in Figure 19.8. To optimize the procedures from one layer to other overlayer during annealing and cooling, we have successfully controlled the misalignment less than 5 μm.

The layered structure was fabricated by the procedures given in Table 19.2. Except for insulating layer, every layer was prepared by μCP. The symbols of TSP-μCP and fCT are two step press μCP and flat contact print, in which ink was transferred with unpatterned PDMS, respectively [2–6].

Table 19.2 Fabrication methods and conditions.

Step	Layer	Process	Ink	Bake (°C)
1	Gate electrodes	TSP–μCP	Silver nanoparticle ink (DIC)	180
2	Gate insulator	Spin coating	Poly(4-vinylphenol) copolymer ink (DIC) or polyimide (ADEKA)	180
3	Source, drain, and pixcel electrodes	TSP–μCP	Silver nanoparticle ink (DIC)	180
4	Organic semiconductos	TSP–μCP	Poly(3-hexylthiophene) ink (DIC)	150
5	Passivation layer	fCP	Fluorinated polymer, CYTOP (ASAHI GLASS)	150
6	Light blocking layer	fCP	Pigment dispersed ink (DIC)	150
7	Reflecting floating electrodes	TSP–μCP	Silver nanoparticle ink (DIC)	150

TSP-μCP: Two step press μCP.
fCP: Flat contact print (ink transfer with unpatterned PDMS).

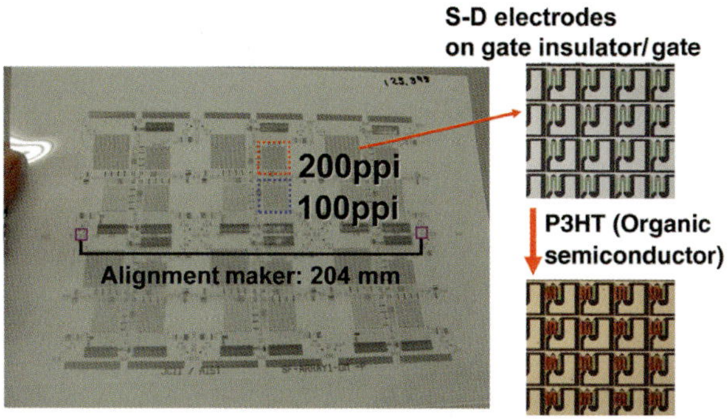

Figure 19.9 Printed organic TFT arrays with 100 and 200 pixel per inch on A4 size plastic sheet.

Figure 19.9 shows the printed organic TFT arrays with several kinds of pixel sizes (100 and 200 pixel per inch) on A4 plastic sheet. In the right hand side, one can see the color change from gray of source and drain electrodes to red of organic semiconductor of poly-3-hexyl-thiophene (P3HT). And there is the alignment maker at the edges of plastic sheet with 204 mm width.

Each of organic TFT shown in Figure 19.10 represents the normal saturation regime in higher gate voltage and TFT parameters are as follows:

$$\text{On/off ratio} : 10^6 - 10^7.$$

$$\text{Mobility} : 3 - 5 \times 10^{-3}\ \text{Vs/cm}^2.$$

19.6 Liquid Crystal Display Packaging

To demonstrate the printed organic TFT arrays as active-matrix back-plane, the polymer network liquid crystal (PNLC) is used as the front panel for printed

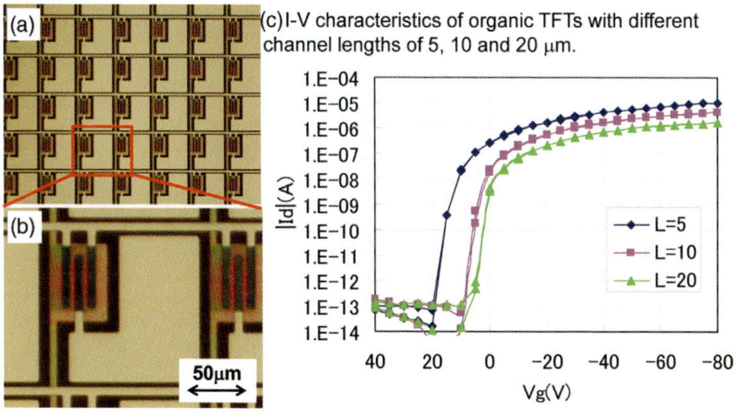

Figure 19.10 Organic TFT array with 200 ppi (a), individual TFT (b), and *I–V* characteristics of TFTs with different channel lengths (c).

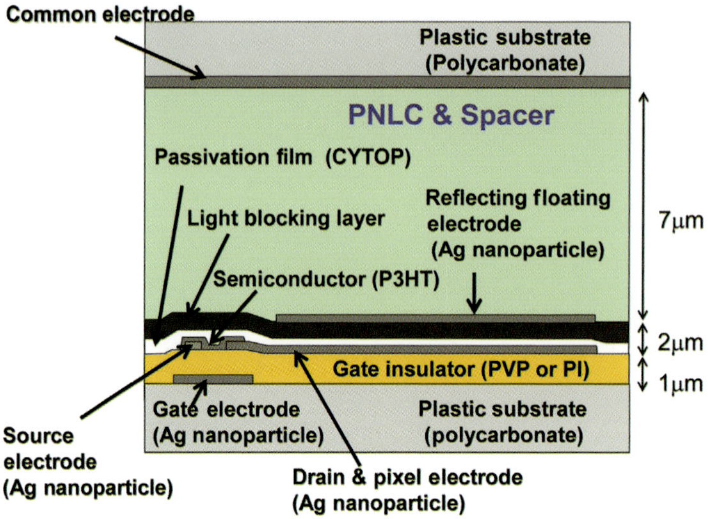

Figure 19.11 Cross section of polymer network liquid crystal panel with printed organic TFT.

organic TFT [6–9]. It does not need any rubbing layer for alignment of liquid crystal but have to be illuminated by UV light to embed the LC molecules in polymer network. To avoid the damage of UV illumination, we covered the organic TFT by light blocking layer consisting of black pigment in addition with the normal passivation layer of fluorinated polymer film (CYTOP) as shown in Figure 19.11.

Figure 19.12 shows the demonstrations of small PNLC panels with 100 and 200 pixel per inch. In them the TFT parameters of line and width (L/W) are 10 μm/160 μm and 5 μm/80 μm, respectively.

The driving condition of 100 ppi panel is represented in Figure 19.13. The contrast of panel is 3.8 with reflectance change of 4–15% when the gate voltage was changed from −20 to + 20 V.

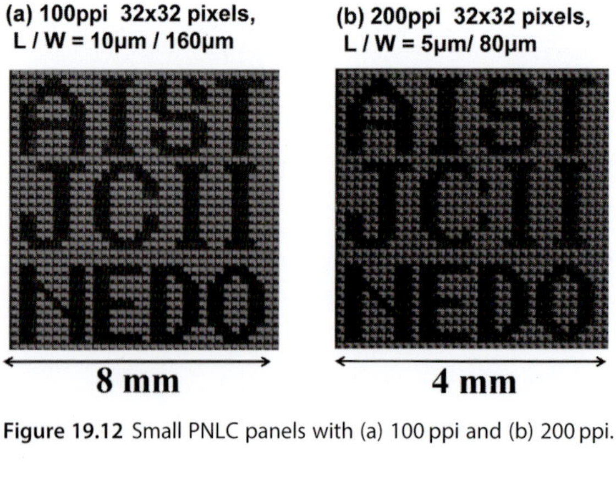

Figure 19.12 Small PNLC panels with (a) 100 ppi and (b) 200 ppi.

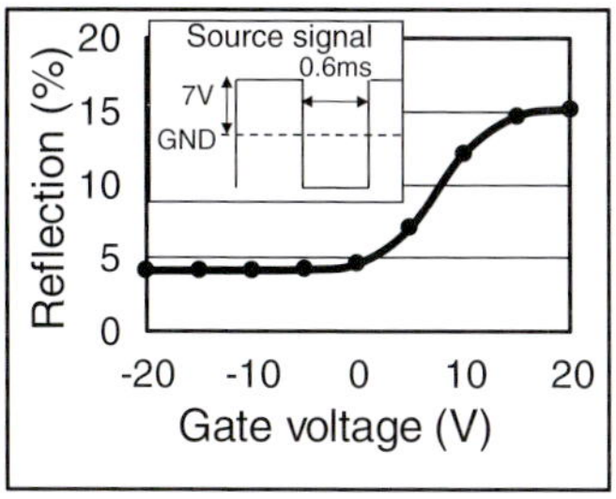

Figure 19.13 Reflectance change of small PNLC panel depending on gate voltage. The driving signal of source electrode was inserted.

Each of the printed devices on plastic sheet with sizes of 6 inch and A4 were cut in 4 and 6 pieces and then packed in the PNLC panel. Several LCD panels were attached by clip probers with 72–200 channels. Figure 19.14 shows two LCDs with 100 and 200 ppi in bending condition with a diameter of 34 mm.

19.7
Conclusion

Figure 19.15 shows full size print of organic TFT arrays on A4 size plastic sheet. In this photograph, there are 1600 × 1200 organic TFT with 200 pixels per inch (ppi). In this photograph there are 1600 × 1200 organic TFTs with 200 pixel per inch (ppi). It corresponds to Ultra Extended Graphics Array (UXGA) panel with 1600 × 1200 pixels.

We have successfully fabricated large area (A4 size) and high precision (200 ppi) organic TFT arrays entirely printed on plastic substrate. And we could also demonstrate the images on flexible displays. In future, we will develop the perfectness of printed TFT arrays without any defects and artifacts.

(a) 100 ppi (200μm):200x200 pixels (b) 200 ppi (125μm): 400x400 pixels

Figure 19.14 Bending PNLCDs with different pixel sizes.

Figure 19.15 UXGA organic TFT array on A4 size plastic sheet.

Acknowledgments

This work was supported by the project "Technological Development of Super-Flexible Display Components" supported by the Ministry of Economy, Trade and Industry (METI) and the New Energy and Industrial Technology Development Organization (NEDO). We have collaborated with the following nine companies and two Institutes: TOPPAN, DNP, DIC, Konica Minolta Technology Center, RICOH, EPSON, ADEKA, Asahi Kasei, Shin-Etsu Chemical, Japan Chemical Innovation Institute (JCII), and AIST.

References

1 Whitesides, G.M. *et al.* (1994) *Langmuir*, **10**, 1498.
2 Takakuwa, A. *et al.* (2007) *Jpn. J. Appl. Phys.*, **46**, 5960–5963.
3 Takakuwa, A. and Azumi, R. (2008) *Jpn. J. Appl. Phys.*, **47**, 1115–1118.
4 Kina, O. *et al.* (2010) *Jpn. J. Appl. Phys.*, **49**, 1AB07.
5 Horii, Y. *et al.* (2009) *Thin Solid Films*, **518**, 642–646.
6 Matsuoka, K. *et al.* (2008) Proceedings of IDW08, AMD1-4L, Niigata, Japan.
7 Yase, K. *et al.* (2009) Proceedings of SID09, 16.3, San Antonio, USA.
8 Matsuoka, K. *et al.* (2009) Proceedings of IDW09, AMD6-2, Miyazaki, Japan.
9 Yase, K. (2010) Proceedings of IDW10, FLX3/EP-1-1, Fukuoka, Japan.

20
Nanolithography Based on Surface Plasmon

Kosei Ueno[1] and Hiroaki Misawa[1,2]

[1]Hokkaido University, Research Institute for Electronic Science, N21, W10, Kita-ku, 001-0021Sapporo, Japan
[2]National Chiao Tung University, Department of Applied Chemistry & Institute of Molecular Science, 1001 Ta Hsueh R., Hsinchu 30010, Taiwan

20.1
Introduction

Photolithography, when used as the basis of semiconductor processing technology, has increased the processing resolution by using an exposure source with a shorter wavelength. The processing resolution has even reached the region from 20 to 30 nm in EUV (extreme ultraviolet light) lithography by using EUV light with an extremely short wavelength of 13.5 nm as an exposure source. A lithography system that employs an ArF excimer laser (wavelength: 193 nm) as an exposure source with a liquid immersion lens in its optical setup has an exposure line width below 38 nm and is already used to manufacture the latest large-scale integration circuits (LSI) in the industry. However, the improvement in the processing resolution realized by an exposure source with a shorter wavelength has already reached its limit. Therefore, a new breakthrough is needed to further improve the performance and integration density of semiconductor processing.

Near-field exposure techniques have received considerable attention as a photolithography technique that can realize processing of a size smaller than the wavelength of light via a simple method. The near field is localized in a domain far smaller than the diffraction limit. That is, nanolithography based on exposure in the near field is a concept that does not employ an exposure source with a short wavelength, but utilizes an exposure source that itself is small. For example, although the near field that stands from a small aperture has been used as the probe light in aperture-type near-field scanning microscopy (NSOM), the principle can also be applied to nanolithography [1].

To evolve the near field effectively, however, localized surface plasmon resonance induced by metallic nanoparticles such as gold and silver has also received attention. When a metallic nanoparticle is irradiated by visible light, collective

Handbook of Solid State Chemistry, First Edition. Edited by Richard Dronskowski, Shinichi Kikkawa, and Andreas Stein.

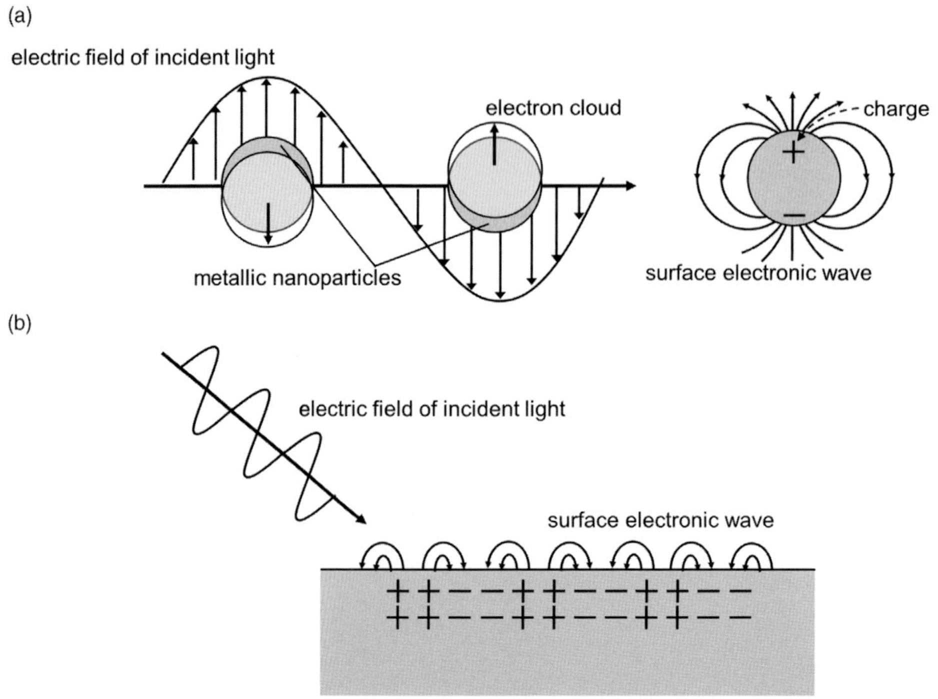

Figure 20.1 (a) A schematic illustration of excitation of LSPR and its charge distribution. (b) A schematic illustration of excitation of the propagating SPR.

oscillations of conduction electrons on the surface of the metallic nanoparticle are induced by the electric field of incident light as shown in Figure 20.1a; thus, condensation and rarefaction structures of electric charges, such as electric dipole, are formed by the induction of electric polarization on the metallic nanoparticle, which is coupled with the evolution of near field near the edges of the metallic nanoparticle (surface plasmon polariton, which is generally called surface plasmon) [2]. This series of phenomena is called localized surface plasmon resonance (LSPR). Because the generated near field exists on the surface until dephasing of LSPR, near-field enhancement is induced. The near field evolved by LSPR induces spatially selective photochemical reactions of the photoresist, and this principle can be applied to nanolithography technology. However, there is also propagating surface plasmon resonance (SPR), in which a surface electronic wave based on the collective oscillations of conduction electrons couples with incident light and propagates on the metal surface in the surface plasmon coupling regime, as shown in Figure 20.1b [3].

The propagating SPR cannot be excited by only the irradiation of visible light onto a metallic flat film vertically because the wave vector of the incident light does not match that of the surface plasmon. That is, the component of the wave vector of the incident light parallel to the metal surface must be equal to the

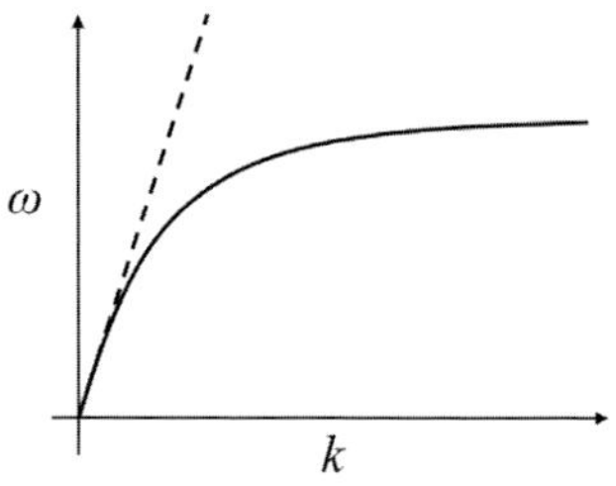

Figure 20.2 Surface plasmon dispersion curve of gold. The broken line indicates a light line in vacuum. ω is the frequency of light and surface plasmon.

wave vector of surface plasmon based on the momentum conservation law. The wave vector of surface plasmon is expressed by the following surface plasmon dispersion relation:

$$k_{sp} = k_0\sqrt{\frac{\varepsilon_d\varepsilon_m}{\varepsilon_d + \varepsilon_m}}, \tag{20.1}$$

where k_0 and k_{sp} are the wave vectors of light in vacuum and surface plasmon and ε_m and ε_d are dielectric constants of a metal and a surrounding dielectric material, respectively. Figure 20.2 shows the surface plasmon dispersion curve of gold. The surface plasmon dispersion curve is asymptotic to the light line in the lower frequency region, whereas the curve settles in a constant value that is decided by $\varepsilon(\omega)$ in the higher frequency region. Importantly, the dispersion curve and the light line do not cross mutually. This means that the propagating SPR is not directly excited by the irradiation of light. Therefore, in general, a prism coupling method such as the Kretschmann configuration (total internal reflection conditions) is employed to match the wave vector of the surface plasmon and the component of the wave vector of the incident light parallel to the metal surface. A grating coupling method is also often utilized for the excitation of the propagating surface plasmon.

In this chapter, we describe the following three characteristic nanolithography techniques based on surface plasmon resonance: (i) the plasmon interference lithography technique, which can form periodic line and space patterns below the diffraction limit on a photoresist substrate utilizing interference of the propagating SPR; (ii) the nanogap-assisted surface plasmon nanolithography technique, which uses ultimate near-field lithography exposed by a local near field at a nanogap with a single nanometer-width induced by LSPR; and (iii) a unique plasmon lithography technique using a scattering light emitted when LSPR couples with a radiation mode as an exposure source. All of these methodologies are nanolithography techniques based on surface plasmon resonance, which transfers the nanopatterns below the diffraction limit of light on a photoresist substrate. Each nanolithography technique has proper applications for use, although each has an advantage and a disadvantage. In this chapter, we describe the principles and features of the nanolithography techniques based on surface plasmon resonance.

20.2 Surface Plasmon Interference Nanolithography

Luo *et al.* proposed a surface plasmon interference nanolithography on a positive-type photoresist film [4]. In this technique, an Ag nanostructured line and space pattern (line width: 240 nm, space width: 60 nm, thickness: 60 nm) was fabricated on a quartz glass by electron beam lithography and lift-off as a photomask. Commercially available g-line positive-type photoresist (TSMR series, Tokyo Ohka Kogyo Co.) was spin-coated on a silicon substrate. An Ag nanostructured photomask was contacted with the photoresist substrate and exposure was performed using a mercury lamp with a wavelength of 436 nm. A cross-sectional near-field intensity profile of these substrates predicted by a finite-difference time-domain (FDTD) simulation is shown in Figure 20.3a. The propagating SPR is excited by a grating coupling and propagates from one end of a silver line to the other end perpendicular to the line and space. Although LSPR may also be induced at the edge of the grating structure, the near-field enhancement factor ($|E|^2$) is not very high, as shown in Figure 20.3a. The interference of the propagating SPR can result in a strongly enhanced nanoscale spatial distribution of the near field on the metal surface. Therefore, the interference pattern of

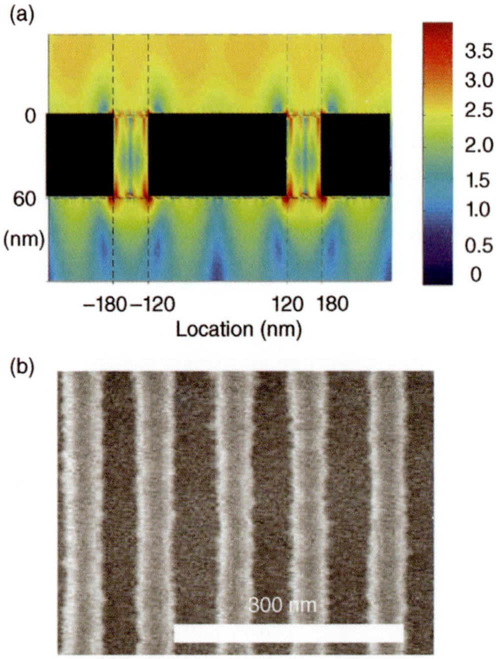

Figure 20.3 (a) Cross-sectional near-field intensity profile of photomask and photoresist substrates predicted by FDTD simulation. (b) SEM image of the photoresist surface after exposure and development by surface plasmon interference nanolithography [4].

the propagating SPR is predicted to be transferred to the photoresist film. In fact, three lines in a 300 nm period were formed almost homogeneously and were strictly reflected on the FDTD simulation result as indicated in the scanning electron microscopy (SEM) image of the photoresist surface after exposure and development as shown in Figure 20.3b. Surface plasmon interference lithography is a very powerful means to expose a periodic line and space pattern whose periodicity is less than a diffraction limit of light [4]. However, the drawback of this method is that the shape of the formed pattern and a certain degree of periodic size are limited because the exposure principle is based on the interference of the propagating SPR. Furthermore, it is necessary to design the structural pitch strictly beforehand for the reason given above. Most importantly, the formed nanopattern is not directly reflected on the photomask design because the pattern is highly dependent on the near-field intensity distribution based on the interference of the propagating SPR.

20.3 Nanogap-Assisted Surface Plasmon Nanolithography

A near field induced by LSPR is also expected as an exposure source to form small nanopatterns on the photoresist substrate because the near field can be localized at a nanometer-sized minute space. A nanogap metallic structure with multiple metallic nanostructures placed several nanometer-sized distances from edge to edge causes the near field to localize in nanogap space and exhibits the near-field enhancement effect as high as 10^5 compared with incident light intensity [5]. Therefore, it is known that nanogap metallic structures can induce nonlinear optical phenomena such as two-photon-induced photoluminescence and higher order harmonic generation. A characteristic of nanolithography utilizing LSPR is the localization of the near field and its enhancement. The nanogap metallic nanostructure was irradiated by a longer wavelength light that does not excite a photoresist directly but excites LSPR, and the plasmonically enhanced near field induces a two-photon-induced photochemical reaction of the photoresist to form a nanopattern. Therefore, the formation of spatially selective nanopatterns such as a structural edge or nanogap where the near field is localized can be achieved on the photoresist substrate.

20.3.1 Verification of Local Photochemical Reaction by LSPR

To demonstrate nanolithography by LSPR, it is necessary to confirm whether a spatially selective photochemical reaction is induced at the nanogap position using nanogap metallic structures as a photomask. On this basis, the spatially selective photopolymerization of a negative-type photoresist on a dimer type of nanogap gold structure, as shown in the SEM image of Figure 20.4a, was explored [6]. The dimer-type nanogap gold structure was fabricated by electron

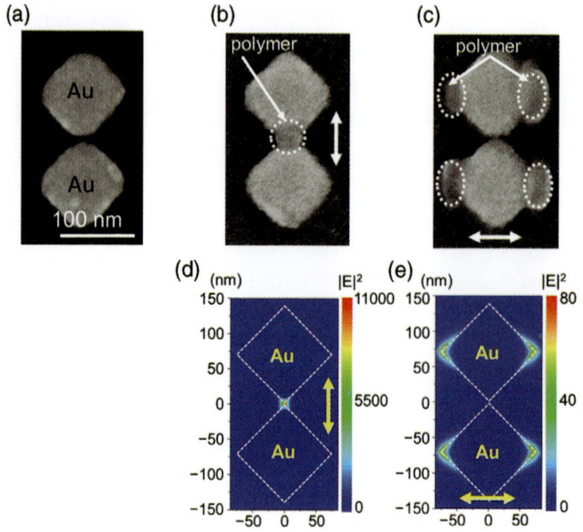

Figure 20.4 (a) SEM image of the nanogap gold structure. (b) SEM image of the dimer structure after exposure for 0.01 s and development under the incident polarization parallel to the dimer structure. (c) SEM image of the dimer structure after exposure for 10 s and development under the incident polarization perpendicular to the dimer structure. (d) Near-field intensity distribution of the dimer structure predicted by FDTD simulation under the incident polarization parallel to the dimer structure. (e) Near-field intensity distribution of the dimer structure predicted by FDTD simulation under the incident polarization perpendicular to the dimer structure. The inset arrows show the incident polarization direction [6].

beam lithography and lift-off processes on a glass substrate. The size of each nanoblock was 100 nm (l) × 100 nm (w) × 40 nm (t), and the gap width was designed as 6 nm. The gap width was defined as the distance from edge to edge.

A commercially available negative-type photoresist (SU-8 2002, Microchem Co.) was spin-coated on the gold nanostructured substrate with a thickness of 1 μm, and a femtosecond (fs) laser pulse (λ_p: 800 nm, τ: 100 fs, f: 82 MHz) was subsequently irradiated onto the substrate to induce LSPR of the nanogap structure. The LSPR band peaked at approximately 800 nm measured after the deposition of the photoresist under the condition of incident polarization parallel to the dimer structure, whose spectrum is almost similar to the LSPR band as shown in Figure 20.5b described later. The photoreaction wavelength of the negative-type photoresist is less than the 400 nm used in this experiment, and therefore, the photochemical reaction proceeds via a nonlinear absorption process with more than two photons when a laser pulse with a wavelength of 800 nm is irradiated. In this experiment, the photopolymerized region was observed by SEM by removing the region where cross-linking was not performed on the substrate via a developer solution (2-methoxy-1-methylethyl acetate) after a laser pulse was irradiated onto the substrate with arbitrary intensity and irradiation time.

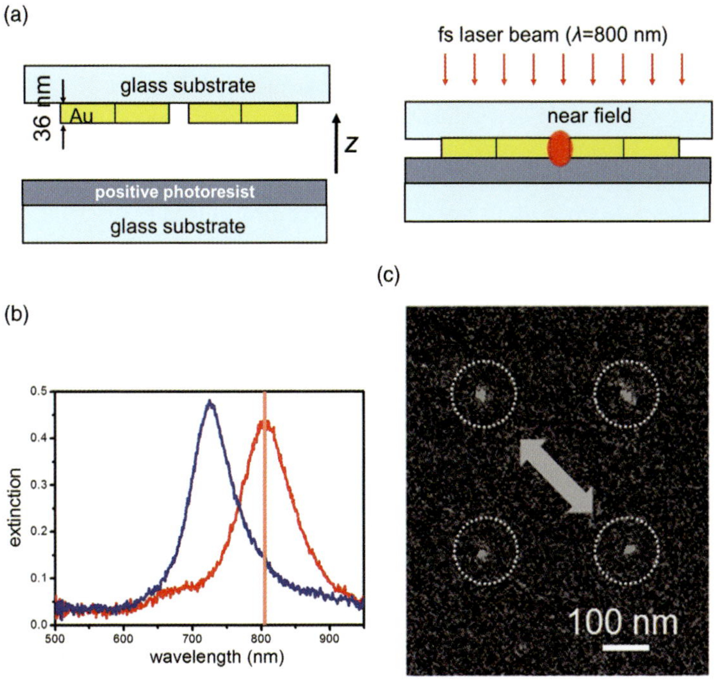

Figure 20.5 (a) A schematic illustration of the exposure system for nanogap-assisted surface plasmon nanolithography. (b) LSPR band of dimer-type nanogap gold structure after deposition of positive-type photoresist film with a thickness of 100 nm measured under incident polarization parallel (red) and perpendicular (blue) to the dimer-type nanogap gold structure. (c) SEM image of the photoresist substrate surface after exposure (0.06 W/cm^2, 10 s) and development. This SEM image achieves a high contrast by using a false color (gray) at the reacted position. The inset arrow shows the incident polarization direction [7].

Figure 20.4b depicts an SEM image of the substrate after exposure for 0.01 s and development under the incident polarization parallel to the dimer structure. It was found that photopolymerization proceeded only at the nanogap position. However, under the incident polarization perpendicular to the dimer structure, photopolymerization proceeded at the left- and right-side corners of each gold nanoblock, as shown in Figure 20.4c, according to the relatively longer irradiation time of 10 s. The photopolymerized region was estimated to be several tens of nanometers in size. From the near-field intensity distribution predicted by the FDTD simulation, a near-field enhancement factor ($|E|^2$) as high as approximately 10^4 was induced at the nanogap position of the dimer-type nanogap gold structures under incident polarization parallel to the dimer structure, as shown in Figure 20.4d. However, the near-field enhancement factor ($|E|^2$) of approximately 100 was estimated at the left- and right-side corners of each nanoblock under the incident polarization perpendicular to the dimer structure, as shown in Figure 20.4e. Therefore, it was considered that the difference in the near-field enhancement factor in the polarizations is related to the difference in irradiation time. It was concluded that the plasmonically enhanced near field promoted the two-photon absorption of the photoinitiator molecule, including the negative-type photoresist and subsequent photopolymerization based on three-dimensional cross-linking [6]. Namely, these results mean that spatially selective two-photon excitation in a nanometer-sized space was realized, and it is expected that the principle can be applied to the nanolithography technique. Thus, it is possible to visualize the near-field intensity distribution localized at a dimer-type nanogap gold structure using nonlinear photopolymerization, and it was confirmed that the spatially selective photochemical reaction of the photoresist was induced according to the near-field enhancement effect based on LSPR.

20.3.2 Demonstration of Nanogap-Assisted Surface Plasmon Nanolithography

To demonstrate nanogap-assisted surface plasmon nanolithography, a dimer-type nanogap gold-structured substrate (24 mm (l) × 24 mm (w) × 0.5 mm (t)) was used as a photomask [7]. Figure 20.5a shows a schematic illustration of the exposure system for nanogap-assisted surface plasmon nanolithography. A photoresist substrate coating with a positive-type photoresist (TSMR-V90LB, Tokyo Ohka Kogyo Co.) with a thickness of 70 nm was contacted with the gold-nanostructured photomask substrate (24 mm (l) × 24 mm (w) × 0.5 mm (t)), and a femtosecond laser pulse (λ_p: 800 nm, τ: 100 fs, f: 82 MHz) was subsequently irradiated onto this substrate from the side of the photomask substrate with arbitrary intensity and irradiation time. After exposure, the photoresist substrate was developed using an alkaline solution (NMD-3, Tokyo Ohka Kogyo Co.), and the formed nanopatterns on the photoresist substrate were observed by SEM.

The size of each gold nanoblock of the dimer-type nanogap gold structure used as a photomask was 80 nm (l) × 80 nm (w) × 35 nm (t), and the gap width was designed as 4 nm from edge to edge of each nanoblock. Figure 20.5b shows the absorption spectrum of the photoresist film and LSPR band of the dimer-type nanogap gold structure after the deposition of a positive-type photoresist film with a thickness of 100 nm via spin coating. The LSPR band measured under incident polarization parallel to the dimer structure exhibits a redshift of approximately 100 nm compared with that measured under the polarization perpendicular to the dimer structure. The redshift of the LSPR band was derived from the near-field interaction, which is a dipole–dipole interaction between two nanoblocks under the incident polarization parallel to the dimer structure, and it was verified that the nanogap was successfully formed. It is noteworthy that the wavelength of the exposure light source overlaps the LSPR band under the incident polarization parallel to the dimer structure. Therefore, it is expected that the near-field enhancement effect is induced at the nanogap position, as described in Section 20.1.

Figure 20.5c shows an SEM image of the photoresist substrate surface after exposure and development. The intensity of the femtosecond laser pulse and the irradiation time were set at 0.06 W/cm^2 and 10 s, respectively, and the incident polarization was set parallel to the dimer block. It was observed that small nanopits with a size less than 10 nm could be formed on the positive-type photoresist surface with a periodic array of 360 nm pitch regularity. Although the size and shape are dispersed, nanopits as small as 5 nm could be formed on the positive-type photoresist film. These formed nanopatterns are very close to the photopolymerized region on the dimer-type nanogap gold structure, as shown in Figure 20.4b, which indicates that the nanogap-assisted surface plasmon nanolithography was demonstrated.

However, in the case of exposure with a relatively higher dose (50 W/cm^2, 20 s), the photochemical reaction of the photoresist proceeded not only at the nanogap position but also at both ends of each nanoblock along the polarization direction as shown in the SEM image of the photoresist surface of Figure 20.6a. Therefore, three nanopatterns were obviously observed on the positive-type photoresist film. The formed nanopatterns are very close to the near-field intensity distribution obtained by the photopolymerization of the negative-type photoresist (2 kW/cm^2, 1 s) and predicted by FDTD simulation represented on a logarithmic scale as shown in Figure 20.6b and c, respectively. It was considered that the photochemical reaction proceeded even with a near-field enhancement factor ($|E|^2$) as small as approximately 100 induced at both ends of gold nanoblocks along the polarization direction.

In summary of nanogap-assisted surface plasmon nanolithography, it is possible to form nanopatterns on the photoresist film as small as the size of the nanogap by utilizing the surface plasmon-assisted photochemical reaction [7]. Namely, the nanogap-assisted surface plasmon nanolithography technique is the ultimate form of nanolithography because it is possible to realize single-nanometer-sized patterning of the photoresist. However, there is a drawback in which the formed nanopattern differs in the exposure dosage. Furthermore, similar to the surface plasmon interference nanolithography introduced in Section 20.2, the formed nanopattern is not directly reflected on the photomask design because the pattern is highly dependent on the near-field intensity distribution. As with most of the main drawbacks described in Sections 20.2 and 20.3, the formed nanopattern is shallow because the exposure source is a near field and the depth direction of the formed nanopattern also depends on the near-field intensity distribution as shown in Figure 20.6c.

20.4
Nanolithography Using Scattering Component of Higher order LSPRs as an Exposure Source

Although a plasmonically enhanced near field based on LSPR excitation can ultimately form small nanopatterns on the photoresist film, the formed nanopattern is not directly reflected on the photomask design. To solve this problem, there is a methodology that employs the scattering component of higher order LSPRs as

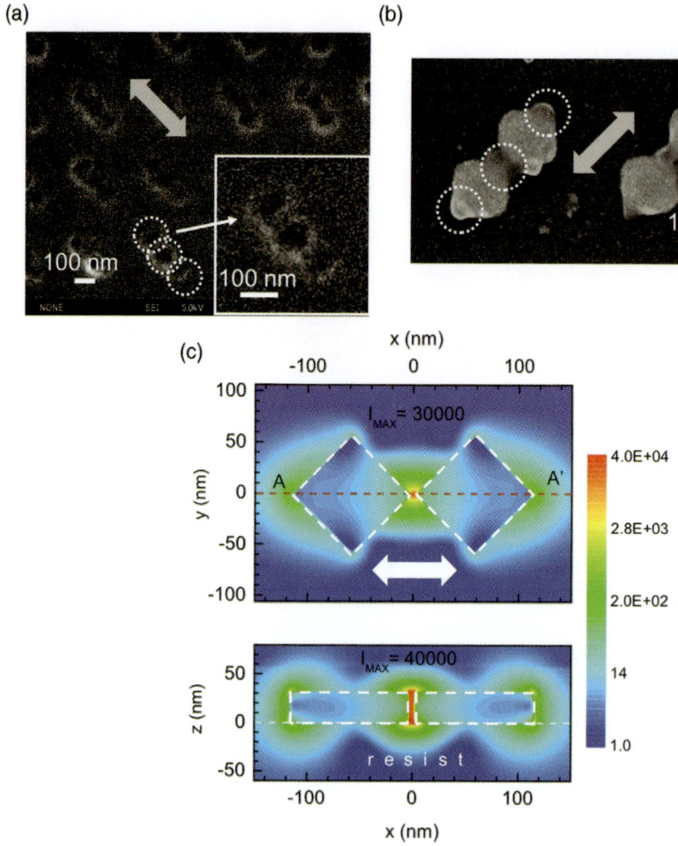

Figure 20.6 (a) SEM image of the photoresist substrate surface after exposure (50 W/cm^2, 20 s) and development. (b) SEM image of a nanogap dimer structure after exposure (2 kW/cm^2, 1 s) and development under incident polarization parallel to the dimer structure. (c) Near-field intensity distribution of nanogap dimer gold structure predicted by FDTD simulation represented on a logarithmic scale. The inset arrows show the incident polarization direction [7].

an exposure source, which can propagate in the far field [8]. In this section, this type of nanolithography is described.

Figure 20.7a shows a schematic illustration of the photomask pattern used in this methodology. The photomask is a glass substrate with a glass nanostructured surface with a cubic shape (100 nm (l) × 100 nm (w) ×100 nm (t)) on which gold thin film with a thickness of 10 nm was deposited. Figure 20.6b depicts the oscillation behavior of the electric field along the x-axis after irradiation by a laser pulse with a wavelength of 800 nm obtained by FDTD simulation. The oscillating cycle of the electric field with a wavelength of 800 nm is 2.66 fs, and Figure 20.7b shows the direction of the electric field along the x-axis at a certain moment with a 1.33 fs cycle corresponding to the pi (π) of the oscillating cycle

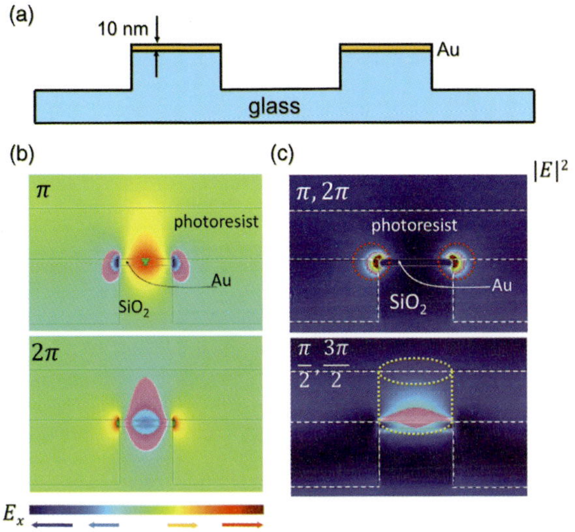

Figure 20.7 (a) A schematic illustration of the photomask pattern used in this methodology. (b) Electric field along the *x*-axis after irradiation by a laser pulse with a wavelength of 800 nm obtained by FDTD simulation. The direction of the electric field along the *x*-axis at a certain moment in a cycle of π and 2π after LSPR excitation is represented by colors. (c) The electromagnetic field intensity distributions with a cycle of π and a phase shift of one-half pi ($\pi/2$) after LSPR excitation predicted by FDTD simulation [8].

after LSPR excitation by colors. The direction of the electric field reverses when the phase changes by 1.33 fs due to the plasma oscillation of conduction electrons based on the LSPR regime. It is noteworthy that the near-field intensity is higher not only at both ends but also at the center of the gold nanostructure, and therefore it was considered that the higher order LSPRs were induced. Electromagnetic field intensity distributions with a cycle of pi (π) and a phase shift of one-half pi ($\pi/2$) after LSPR excitation predicted by FDTD simulation are shown in Figure 20.7c. Most importantly, high near-field intensity is observed only in the vicinity of the gold structure in the cycle of pi (π), and the scattering light of the LSPR modes propagates in the photoresist film in the phase shift of one-half pi ($\pi/2$).

Figure 20.8a shows an SEM image of the photomask substrate. The glass nanostructured photomask was directly fabricated by electron beam lithography using an organic/inorganic hybrid negative-type electron beam resist (XR-1541 (HSQ-based), Dow Corning Co.) on a glass substrate (24 mm (l) × 24 mm (w) × 0.5 mm (t)). It is known that the electron beam resist becomes only a glass component after annealing or oxygen plasma ashing due to the removal of organic components. Gold thin film with a thickness of 10 nm was then deposited on the nanostructured substrate by a sputtering technique. Figure 20.8b exhibits the LSPR band of the fabricated photomask substrate. The LSPR band peaks at a wavelength of approximately 870 nm, with a small shoulder in the

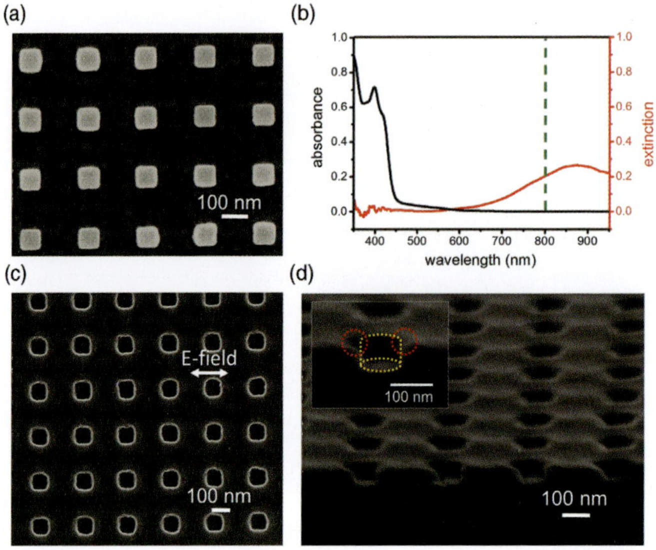

Figure 20.8 (a) SEM image of the photomask substrate. (b) Absorption spectrum of the photoresist film (black line) and LSPR band (red line). The broken line indicates the incident laser wavelength. SEM image of photoresist pattern after exposure (50 W/cm^2, 10 s) and development; top view (c) and cross-sectional view (d). The inset picture shows an enlarged view of the formed nanopattern. Exposure profiles based on the near-field component (red broken line) and scattering component (yellow broken line) are highlighted in the inset picture [8].

shorter wavelength region based on the excitation of higher order LSPRs. When a femtosecond laser pulse with a wavelength of 800 nm is irradiated onto the photomask, higher order LSPRs such as quadrupole mode might be induced because the region with a shorter wavelength than that of the fundamental dipole mode is excited.

A photoresist substrate (24 mm (l) × 24 mm (w) × 0.5 mm (t)) coating with a positive-type photoresist (TSMR-V90LB, Tokyo Ohka Kogyo Co.) and a thickness of 70 nm was contacted with the fabricated photomask substrate, and a femtosecond laser pulse (λ_p: 800 nm, τ: 100 fs, f: 82 MHz) was subsequently irradiated onto this substrate from the side of the photoresist substrate with arbitrary intensity and irradiation time. Figure 20.8c depicts an SEM image of the photoresist pattern after exposure (50 W/cm^2, 10 s) and development. The rectangular-shaped nanohole array, which is highly reflected on the photomask design, was obviously observed on the positive-type photoresist film. The side length of the rectangular-shaped nanohole array was 100 nm. Figure 20.8d depicts an SEM image with a cross-sectional view of the positive-type photoresist film. Two exposure profiles can be seen. One is an exposure profile by the near-field component, and the other is an exposure profile by the scattering component of higher order LSPRs, which correspond to the electromagnetic

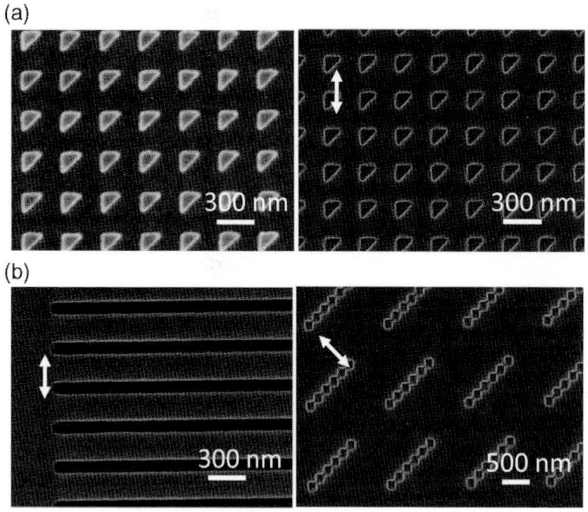

Figure 20.9 (a) SEM images of the photomask with triangular nanostructured photomask (*left*) and the photoresist pattern after exposure ($50\,W/cm^2$, 10 s) and development (*right*). (b) SEM images of the photoresist pattern after exposure ($50\,W/cm^2$, 10 s) and development; line and space (*left*) and nanochain (*right*). The inset arrows indicate the incident polarization direction in the exposure process [8].

field intensity distribution predicted by FDTD simulations as shown in Figure 20.6c with a cycle of pi (π) and a phase shift of one-half pi ($\pi/2$) after LSPR excitation, respectively. The nanopattern was formed only at the surface by the near-field component. However, the photoresist was exposed from the surface to the bottom of the resist film according to the exposure using the scattering component of higher order LSPRs. Thus, nanolithography using the scattering component of higher order LSPRs as an exposure source was clearly demonstrated.

It is possible to form variously shaped nanopatterns on the positive-type photoresist film by using this methodology. As an example, triangular nanopatterns were fabricated on the positive-type photoresist film using a photomask with a triangular nanostructured photomask, as shown in Figure 20.9a. It was demonstrated that this triangle nanopattern could be formed in the entire light irradiation region only at the time of exposure. It is noteworthy that the average size of the formed nanopatterns is estimated to be 141.2 nm, and the variation in size was estimated to be 4.4 nm as a standard deviation from the statistics (100-piece measurement) by SEM measurement. Therefore, it was elucidated that even the triangular nanopattern with an acute angle vertex could be fabricated on the photoresist film with a processing resolution of a single nanometer. Moreover, the nanopatterns of various shapes, such as not only line and space but also nanochain, could be fabricated with high precision by nanolithography using the scattering component of higher order LSPRs as an exposure source as shown in the SEM images of Figure 20.8b.

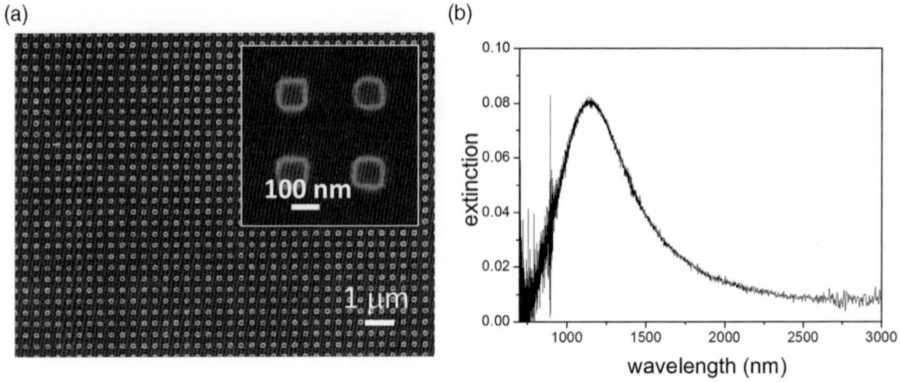

Figure 20.10 (a) SEM image of gold nanostructures on a glass substrate fabricated by nanolithography and lift-off process. (b) LSPR band of the fabricated gold nanostructures measured under nonpolarized conditions [8].

Furthermore, it became possible to fabricate gold nanostructures when titanium and gold film were deposited on the nanopatterns of the positive-type photoresist film formed by the nanolithography and the lift-off process was performed. Figure 20.10a shows the SEM image of the formed gold nanostructures on a glass substrate. The thickness of titanium employed as an adhesion layer and gold film deposited by sputtering was set at 1 and 7 nm, respectively. The variation in size was estimated to be 3.8 nm as a standard deviation. This value is comparable to that of gold nanoblock fabricated by advanced electron beam lithography and lift-off process (standard deviation: 3.2 nm). Moreover, the LSPR band was clearly observed as shown in Figure 20.10b. Therefore, the nanolithography technique can be applied to a lift-off process and is expected to be applied to the process technology for the fabrication of plasmonic solar cells or sensors [8].

20.5 Conclusion

In this chapter, nanolithography techniques based on surface plasmon resonance that can form nanopatterns below the diffraction limit of light on the photoresist substrate by making the exposure source itself small were described. The characteristics of nanolithography techniques introduced in this chapter are summarized in Table 20.1. The depth of the formed nanopattern differs according to whether the exposure source is the near field or scattering component of LSPRs. Because the near field does not propagate into the far field, it is mostly the same as the planer processing resolution in the depth of the formed nanopattern. However, nanogap-assisted surface plasmon nanolithography, which makes the exposure source itself small, has a high planer processing resolution, and

Table 20.1 The characteristics of nanolithography techniques based on surface plasmon resonance.

	Surface plasmon interference nanolithography	Nanogap-assisted surface plasmon nanolithography	Nanolithography using scattering component of higher order LSPRs
Exposure source	Near field based on the propagating SPR	Near field based on LSPR	Scattering component of higher order LSPRs
Processing resolution	~10 nm	~nm	~nm
Depth	~10 nm	~nm	~100 nm

therefore it can be said to be the ultimate near-field nanolithography. Moreover, nanolithography using the scattering component of higher order LSPRs as an exposure source has a high processing resolution, and therefore the fabrication of structures with a high aspect ratio is also somewhat expected. These nanolithography techniques will be expected to realize photolithography technology with a 10 nm node because the methodology for realizing photolithography with a processing resolution higher than 22 nm is not shown.

In recent years, nanoimprint technology has been applied to the fabrication of nanostructures at a large scale [9,10]. However, the nanolithography introduced in this study has an advantage from the viewpoint of mold release between the photomask and photoresist substrate compared with the nanoimprinting method in addition to the high processing resolution. Furthermore, noncontact exposure is theoretically possible in the case of nanolithography using the scattering component of higher order LSPRs as an exposure source. It is possible to reuse a photomask infinitely if gold is deposited after etching the gold and cleaning the photomask substrate surface by oxygen plasma ashing, even if the photomask surface has become dirty. Therefore, nanolithography has different advantages and features than nanoimprint technology, which has problems regarding the mold release and durability of a metal mold. The nanolithography introduced in this chapter and its future development are expected to achieve photolithography technology with high processing resolution.

References

1 Alkaisi, M.M., Blaikie, R.J., McNab, S.J., Cheung, R., and Cumming, D.R.S. (1999) Sub-diffraction-limited patterning using evanescent near-field optical lithography. *Appl. Phys. Lett.*, **75** (22), 3560-1-3.

2 Kelly, K.L., Coronado, E., Zhao, L.L., and Schatz, G.C. (2003) The optical properties of metal nanoparticles: the influence of size, shape, and dielectric environment. *J. Phys. Chem. B*, **107** (3), 668–677.

3 Benson, O. (2011) Assembly of hybrid photonic architectures from nanophotonic constituents. *Nature*, **480** (7376), 193–199.

4 Luo, X. and Ishihara, T. (2004) Surface plasmon resonant interference

nanolithography technique. *Appl. Phys. Lett.*, **84** (23), 4780–4782.

5 Xu, H.X., Aizpurua, J., Käll, M., and Apell, P. (2000) Electromagnetic contributions to single molecule sensitivity in surface-enhanced Raman scattering. *Phys. Rev. E*, **62** (3), 4318–4324.

6 Ueno, K., Juodkazis, S., Shibuya, T., Yokota, Y., Mizeikis, V., Sasaki, K., and Misawa, H. (2008) Nanoparticle plasmon-assisted two-photon polymerization induced by incoherent excitation source. *J. Am. Chem. Soc.*, **130** (22), 6928–6929.

7 Ueno, K., Takabatake, S., Nishijima, Y., Mizeikis, V., Yokota, Y., and Misawa, H. (2010) Nanogap-assisted surface plasmon nanolithography. *J. Phys. Chem. Lett.*, **1** (3), 657–662.

8 Ueno, K., Takabatake, S., Onishi, K., Itoh, H., Nishijima, Y., and Misawa, H. (2011) Homogeneous nano-patterning using plasmon assisted photolithography. *Appl. Phys. Lett.*, **99** (1), 011107-1-3.

9 Chou, S.Y., Krauss, P.R., and Renstrom, P.J. (1995) Imprint of sub-25 nm vias and trenches in polymers. *Appl. Phys. Lett.*, **67** (21), 3114–3116.

10 Jungm, G.-Y., Lim, Z., Wum, W., Chenm, Y., Olynick, D.L., Wang, S.-Y., Tong, W.M., and Williams, R.S. (2005) Vapor-phase self-assembled monolayer for improved mold release in nanoimprint lithography. *Langmuir*, **21** (4), 1158–1161.

Index

a
$AA'B_2O_6$-type compounds
– structure of 83
$AA'_3B_4O_{12}$-type compounds
– high-pressure synthesis of 79
– structure of 77
Abbe number 297
$A_2BB'O_6$-type compounds
– high-pressure synthesis of 75
– structure of 74
ab initio simulation 86
ABO_3 compounds
– Goldschmidt diagram for 68
$A^{2+}B^{4+}O_3$ perovskite
– enthalpy of formation of
– – relationship with tolerance factor 69
– stability of 69
$A^{2+}B^{4+}O_3$ perovskites
– tolerance factors for 69
ABX_3-type compounds 71
– hexagonal- or rhombohedral-layered structure 71
ABX_3 type compounds
– high-pressure synthesis, obtained by 56
accumulative roll bonding (ARB) 340
acetic acid 277
acidic catalysis 182
acidic polyelectrolytes 242
actinide metal 279, 280
active-matrix (AM) 564
adenosine triphosphate (ATP) 253
adsorption theory 555
aerogels 186, 190
AFM. *See* atomic force microscope (AFM)
agar gel matrix 258
$AGeO_3$ perovskites 69
agglomerated particles 409
aging 171, 185
Ag_2O
– as oxidizer 50
– pellet 50
akaganeite crystal growth 168, 174
akaganeite growth 170
Al−Si eutectic droplet 32
Al−Si or Cu−Si eutectic droplet 34
alkali chalcogenide 279
alkali ions 280
alkali metals 276, 277
– carbonates 279
– halides 18
alkaline earth carbonates
– decompose 279
alkaline earth cations 278
alkaline earth-metal halides 277
alkaline (thallium) fluroberyllium borates 114
– hydrothermal growth 114
alkoxysilanes 181, 182, 184, 221
alloys, oxidation resistance 338
AlN films, microstructures of 406
$ALnO_3$ perovskites 69
Al_2O_3 powder 412
Al 2p core-level XPS spectra 21
α-Ga_2O_3
– impurity concentration 462
$AlPO_4$ (berlinite) 113
$AlPO_4$ single crystal, hydrothermal growth 113
aluminosilicate-based oxyfluoride glasses
aluminum halides 362
aluminum nitride with rocksalt structure (rs-AlN) 43
alumosilicate pyrophyllite 26
ammonia 107
ammonium halides
– as hydrogen halide source 368
ammonothermal crystal growth, of GaN 110

Handbook of Solid State Chemistry, First Edition. Edited by Richard Dronskowski, Shinichi Kikkawa, and Andreas Stein.

ammonothermal reaction 110
amorphous precursor 241
1D amorphous SiO_x nanowires 34
amorphous stabilizer 240
amphiphilic surfactants 193
ampoules of glass 358
analytical ultracentrifugation (AUC) 242
anatase 160
– rutile phase transformation 167
anatase-anatase interfaces 167
anatase (TiO_2) crystal 158
anion fugacity 390
anisotropic growth 9
annealing 39
anodic aluminum oxide 218
antimonides transport 370
antimony 276, 281
antimony halides
– stability of 370
antioxidation coating 400
antireflective 196
antireflective coating 191
$A_2O–WO_3$ fluxes 278
$A'A''B_2O_6$-type compounds
– high-pressure synthesis of 82
aprotic condensation reactions 184
ARB processes
– schematic illustrations of 341
arc furnace 17
Archimedean prism 327
arc-melting furnace 9
arc-melt suction casting machine 340
arc-melt suction casting technique 339
arc-melt tilt casting 338, 339
– formation, of centimeter-sized metallic glasses 339
– types of 339
– used to, prepare larger metallic glass samples 339
arc melting 8, 9
armor-piercing shaped-charge 36
Arrhenius-type temperature dependence 292
Arrhenius law 320
arsenic compounds 281
arsenic halides
– stability of 370
artificial bilayer membranes 539
artificial diamond synthesis 43
As-deposited a-C/Ni, Ni film, and grown nanowires 40
As-grown crystal of KBBF 114
$ASiO_3$ perovskites 69
atomic diffusive speed 320
atomic force microscope (AFM) 544
atom transfer radical polymerization (ATRP) 542
ATRP. *See* atom transfer radical polymerization (ATRP)
Au/a-Si/Si wafer 58
Au–Si eutectic droplets 12, 53
Au–Si eutectic temperature 12
Au-catalyzed Si nanowires 7, 12
Au-nanopeapodded nanowires 58
Au nanospheres 57, 58
Au–Si alloy 4, 5, 11
Au–Si eutectic alloy 322
Au–ZnSe interfaces 49
Au particle 4, 12
Au/Si nanoparticles 31
Au_2Si nanopea-podded Si oxide nanowires 31
Au/Si nanospheres 53
Au_2Si nanospheres 53
autoclave 108
– reactors 123
– for vertical reverse temperature gradient 115
Avogadro's number 290
axial nanowire 9
azeotropic point 383
azide decomposition 282

b

back transport 361
$50BaF_2–25Al_2O_3–25B_2O_3$ glass
– optical absorption spectrum for 314
ball-milled Pd–Ni–P amorphous powder 324
ball milling 412
– technique 324
bandgap energy 439, 446, 455
– bond length, relationship between 439
bandgap semiconductors 440
banding pattern 554
$BaO–TiO_2–SiO_2$ glasses 297
base glass, electrical conductivities for 307
batch type reaction system 128
belt-apparatus
– exploded assembly drawing 27
belt module 26
bending PNLCDs with different pixel sizes 571
benzyl alcohol route 192
Bi^{3+} cations 278
bifunctional template 205, 222
binary alloy, schematic illustration of TTT curves of 324
binary compounds 389
binary eutectic systems 323
binary glass-forming systems 324

binary oxides 1
binary phase diagrams 381, 390
– segregation and lever rule for 393
– topology of 383
binary sulfide transport 369
binary system 380
binary transition metal oxides 278
Bi_2O_3-based glasses 300
bio-inspired approaches 256
bio-inspired crystallization process 240
bio-inspired mineralization 258
– silk fibroin-based 258
bio-inspired polymer-controlled crystallization 235, 255, 256
– applications 257
– process 251
bio-inspired polymer-directed routes 246
bio-inspired synthesis 258
bioimaging 192
biological active molecules 195
biological organisms 223
biomacromolecule-ZIF-8 hybrid crystals 257
biomacromolecules 256
biomass conversion 148
biomimetic mineralization 257
– peptide-based 247
biomineralization 155, 233, 234, 547
– eco-friendly process 234
biominerals 213, 233
– formation 234
– systems 242
biomolecule self-assemblies 253
biosensors 1
bismuth 276
bismuth oxides, synthesis 118
bitumen solution 148
block copolymers 220, 544, 545, 547, 548
Boltzmann's constant 128
borate glass
– stability of 305
borates 277
boric acid 278, 279
Born type equation 126
boron nitride 26, 30
borosilicate glass 358
Boson peaks 293
bottom-up approach 180
Boudouard equilibrium 17
bovine serum albumin (BSA) 247
breakdown electric field 440
Bridgman ampoule 377
Bridgman methods 375, 386
Bridgman seals 24
Brownian motion 159
bulk free energy 432
bulk metallic glass 325
– production 343
Bundy's apparatus 26
Burger's vector 448

c
$CaCO_3$
– bio-inspired mineralization 242
cadmium chloride 376
$CaIrO_3$-type structure 86, 87
– high-pressure synthesis of 88
calcinations 193, 212
– temperature 252
calcium carbonate ($CaCO_3$) 233
calcium phosphate (CaP) 233
$CaO–Fe_2O_2$ system
– central part of 387
CaP-coated JEV 260
capillary force 58
capping agents 134, 158, 552
– dissociation of 136
carboloy chamber 26
carbon 214
– sucrose, carbonization of 214
carbonate precursors
– transformation process 252
carbon-coated $FePO_4$ nanotubes 258
carbon sp^2 bonding structure 36
carbothermal reaction 34
carbothermic reduction 17, 18
cartridge heaters 124
casting atmosphere 202, 347
catalyst–semiconductor interface 10
catalysts 407
catalytic CVD 408
catalytic nanoparticles 407, 412
catalytic reaction 402
ccp. *See* cubic closed packing (ccp)
CdSe@ZnS core–shell QDs 255
CdS nanocrystals 254
$CdTiO_3$ compounds
– pressure–temperature phase diagram of 54
cell positioning 547
centrifugation techniques 12
CeO_2 nanoparticles
– oxygen storage capacity 147
– synthesis 140
– with and without organic ligands
– – size and morphology change 146
CeO_2 particle size

– as function of Da 144
– as function of Reynolds number 144
ceramic crucibles 3
– hexagonal boron nitride 3
ceramic oxide materials 16
ceramics 276
– production 14
– synthesis 1, 3
ceria nanoparticles 141
– decanoic acid-capped 148
cerium oxide particles 147
– structure 145
chalcogen 280
chalcogenides 280
– glasses 287, 294
– transport 370
chalcophosphates 281
charge transfer complex 539
Chechevitsa 28
chelating ligands 189
chelation interactions 258
chemical bond formation 136
chemical composition 23, 24
chemical compounds 390
chemical contrast 547
chemical equilibrium 371
chemical inertness 276
chemical modification
– patterned strate 544
– surfaces 540
– – using polymers 543
chemical patterning 539, 544, 546, 547
– biodegradable substrates 539
– micromagnets in 547
– nanoscale 546
– preparation, by conventional lithography 548
– techniques
– – bottom-up 540, 544
– – top-down 540, 542
– in three dimensions 547
– wavelength of 539
chemical reaction-limited process 403
chemical reductants 215
chemical thermodynamics 375
chemical transport 402
chemical vapor deposition (CVD) 2, 399, 429
– basics of 399
– disadvantage, thermal 415
– film formation process in 403
– kinetics of 402
– laser 417
– – laser beam in 419
– – photolytic 417
– – pyrolytic 417
– – CO_2 laser as heat source 417
– – conventional 420
– – deposition rates of films in 418
– – schematics of 417
– plasma 415
– plasma-enhanced (PECVD) 416
– processes
– – in gas phase 402
– – strate surface 402
– rotary 407
– schematics of 401
– thermal equilibrium of 405
chemical vapor deposition materials
– morphology of
– – deposition temperature, effect of 401
– – supersaturation of source gases, effect of 401
chemical vapor transport (CVT) 281, 351, 359
– experiments
– – modeling of 370
– – nonstationary 372
– of halides 366
– reactions 352, 367
– – metal halides obtained by 365
– single equilibrium reaction 352
– zinc sulfide with iodine 352
chemisorption 540
chimie douce 180
chiton teeth 242
Chlorella vulgaris 248
chlorinated silanes 5
chromium(III) chloride, synthesis 366
chromium sublimation source 25
c-plane GaN 454
c-Si/a-SiO_x nanowire 34
classical and nonclassical crystallization pathways
– schematic representation 236
classical crystallization
– morphological control 235
– thermodynamic control-dominated 238
classical ionic compounds 277
classical nucleation theory 289
class II hybrid materials 181
class II type material 194
class I type hybrids 194
coagulation theory 555
coat protein P22 254
coaxial impact, setups 36
coaxial nanowire 40, 54
coaxial Si/Si oxide nanowires 52

coaxial Si/SiO_x nanowires 54
cobalt nanoparticle formation 131
– proposed reaction mechanism 134
cobalt nanoparticles 137
cobalt oxinate system 551
Co-doped $LiMn_2O_4$ 252
co-operative transport 372
coincubation process 258
colloidal crystals 193
– films 220
– template 194, 212
colloidal dispersion 555
colloidal mesoporous silica nanoparticles
colloidal nanocomposite 258
colloidal noble metal nanoparticles 246
– applications 246
colloidal particles 212, 545, 551, 555
colloid gold particles 11
columnar crystals 400
combined use of templates, hierarchical structures 220
complex chalcogen anions 281
complex crystalline superstructures 243
composite materials 14
compound phase diagrams 379
– chemical compounds for 379
compound synthesis 9
compression/impedance method 40
computed-tomography (CT) 139
computer simulations 293
concave bowing 446
concave printing 564
condensation reactions 181, 188, 214
– influencing parameters 184
condensed phase 351
conductivity 2, 155
confocal scanning laser microscope (CSLM) 315
congruent melting 386
conical carboloy pistons 26
conproportionation reactions 362, 365, 366
constitutional supercooling 395
continuous cooling transformation (CCT) 347
continuous flow-type reaction system 124
controlling factors 331
conventional CVD 401
conventional flow-type reactor 138
conventional lithography 548
conventional melt-quenching method 300
conventional pyrolytic CVD
– morphology by 420
conventional pyrolytic laser CVD 419
conventional semiconductors 440
– GaAs as 440
– Si as 440
convex printing 564
cooling agent 36
coordination bonds 205
copolymers
– block 544
– spin-coated diblock 545
copper crucibles 8
copper forms with strontium 386
copper-cold casting method 339
copper mold casting 336
coprecipitation 250
covalent bonds 205
critical free energy 128
critical point 124
critical supersaturation 127
crucible materials 3
cryogenic TEM (cryo-TEM) images 173
cryogenic transmission electron microscopy 168
Cryo-TEM sample preparation 173
crystal growth 155, 156, 161, 166, 167, 172, 235, 275, 276, 375, 383
– conditions 157
– data 162
– modifiers 236
– polymer additive-controlled 237
– process, illustrations of 237
– rate 290
crystalline ice rods 211
crystalline metals 329
crystalline–amorphous phase transition 36
crystalline–crystalline phase transition 36
crystalline phase 346, 375
crystalline porous materials 217
crystalline quartz phases 321
crystallites in secondary intermediate structure, arrangement of 161
crystallites of anatase, tilt boundary between 163
crystallization processes 5, 6, 35, 351, 392
crystallization zone 352
crystallographic theory 235
crystal nucleation 157, 293
crystal structures 117, 118, 119, 163
cubic boron nitride 34
cubic closed packing (ccp) 70
Cu catalyst 6
Cu-catalyzed Si nanowire array 6
cuprates 43
Cu 2p XPS core-level spectra 20
Curie–Weiss law 304

current-activated pressure-assisted
densification 15
Cu_3Si–Si interface 10
Cu_3Si phase 10
CVD. *See* chemical vapor deposition (CVD)
CVT. *See* chemical vapor transport (CVT)
cyclohexane 11
cyclopentadienide 370
cylindrical resistance heaters 30
cytidine triphosphate (CTP) 253
CZ. *See* Czochralski (CZ)
Czochralski (CZ) 456
– growth 383
– methods 386
– process 395

d
DAC. *See* diamond anvil cell (DAC)
Damköhler number 143, 144
Debye temperature 299
decomposition reactions 2, 5
decomposition sublimation 351
– ammonium chloride 352
dehydration 159
delaminated films 416
delocalization 305
dendrimers 234
dendritic crystals 400
dendron rodcoils 253
densification 42
deposition of by-products 438
deposition, of photoresist 579
deposition rates 403
– temperature dependence of 403
– in thickness 419
– in volume 420
derived sol–gel systems 195
Derjaguin, Landau, Verwey, and Overbeek
(DLVO) theory 161
diamond anvil cell (DAC) 23, 32, 43, 52
– inner elements section 33
– technique 32
diamond powder 414
dichlorosilane (Si dichloride, SiH_2Cl_2) 5
dielectric constants 124, 126, 575
differential scanning calorimeter (DSC)
method 294
diffusion 1, 4, 10
– coefficient 11
diffusion-limited process 404
difunctionalized organic molecules 188
dihalides transport 366
diphenylsilane 2, 10, 11, 52
dip-coating solution 546
dip-pen lithography 544
dipole–dipole interaction 581
diquaternary ammonium-type
surfactant 221
directed self-assembly (DSA) 547
direct integrin patterning 547
direct templating process 209
disilane (Si_2H_6) 5
dislocation bending 442
dislocation density 441, 443
dislocation elimination by epitaxial-growth
with inverse-pyramidal Pits (DEEP) 450, 451
dislocation loop formation 443
dispersibilities 213
dissociation energy 292
dissolution reaction 123, 352
dome-shaped liquidus 386
dopant gas 445
– SiH_2Cl_2 as 445
doping, of Si nanowires 4, 13
double hydrophilic block copolymers
(DHBCs) 234
double hydroxide catalysts 413
double-sided laser heating devices 34
double stage diamond anvil cell (ds-DAC) 23
Drumheller source 25
ds-DAC. *See* double stage diamond anvil cell
ductile flow 32
Dulong–Petit heat capacity 310
dynamic compression 35
dynamic high-pressure techniques 37
dynamic light scattering 174
dynamic nanoprecipitation of metastable fcc-
Zr_2Ni
– schematic illustration of 328

e
edge-defined film-fed growth (EFG) 456
EDS spectrum 22
elastic energy 443
electrical conductivity 43, 241
– coatings 191
– effects of pressure 24
electrical insulation 146
electrical insulator 26
electrical resistance 24
electrical resistance furnace 52
electric current heating 401
electrochemical series 130
electrodeposition 215
electro-less deposition 412
electroluminescent (EL) 563

electromagnetic compatibility 10
electromagnetic field intensity 583
electromagnetic radiation 32
electron density 182
electron diffraction (ED) 248
electron holography 241
electron-beam physical vapor deposition (EBPVD) 421
electronic disproportionation 42
electronic polarizability, of glasses 292
electronic transparent packaging 196
electron micrographs 162
electro pen nanolithography 544
electrospinning 218
electrostatic adsorption
– layer-by-layer alternating 542
elemental halogens 370
Ellingham diagrams 131, 390
Ellingham type predominance diagram 391
emission maximum 34
emission wavelength 441
emulsions, assemblies of 211
enantioselective synthesis 205
end-loaded piston-cylinder apparatus 25
endotemplating 201, 202
endothermic equilibrium 355
endothermic reactions 352
endothermic transport 358, 363
energy-consuming process 43
energy transformation 258
enthalpy of relaxation 348
environmental management 258
epitaxial growth 12
– orientation 10
– techniques 429
– – metal-organic chemical vapor deposition (MOCVD) 429
– – molecular beam epitaxy (MBE) 429
epitaxial interfaces 51
epitaxial lateral over-growth (ELO) 442
– facet-controlled (FACELO) 442
– facet-initiated (FIELO) 442
– process flow 442
– sequential processes for 442
epitaxial single-crystal films 400
equilibrium constant (K) 353, 390, 434
equilibrium cooling 394
equilibrium gas phase 371
equilibrium gas pressures 408
equilibrium phase diagram 324
equilibrium states 375
equilibrium thermodynamics 376
– component 379
– system 377
etchants 540
etching 540, 545
ethylenediaminetetraacetic acid 5
eutectic alloy systems
– empirical rules for 325
eutectic melting point 61
eutectic point 11
eutectic precipitation 64
eutectic systems 322
eutectoid reaction 385
excitation of LSPR 574
exotemplating 202
exothermic reaction 18, 352
exothermic transport 363

f
fabrication methods, and conditions 568
face centered solid phases 388
face-centered cubic Si crystal structure 2
faceted crystals
– schematic of 405
facet-controlled epitaxial lateral over-growth (FACELO) 442
– processing sequence of 443
facet-initiated epitaxial lateral over-growth (FIELO) 442
facet-specific peptides 247
fast Fourier transform (FFT) 41
FBCVD
– Rh-precipitated SiO2 powder by 410
– schematic of 410
Fe–O–Si shells 35
Fe–Si eutectic droplet 14
Fe–Si nanodroplets 14
$[(Fe_{0.5}Co_{0.5})_{0.75}Si_{0.05}B_{0.2}]_{96}Nb_4$ metallic glassy particles
– outer appearance of 334
Fe-based heterometallic glasses
– for excellent soft magnetic nanocrystalline alloys 330
Fe-based metallic glasses
– soft magnetic properties 330
Fe-coated Si wafers 34
Fe nanoparticles
– TEM image 136
ferric chloride 159
ferrihydrite 169, 170
ferrihydrite nanoparticles 162
ferroelectric $BaTiO_3$ nanoparticles
– bio-inspired synthesis 256
ferroelectric films 196, 402
ferromolybdenum 17

ferrotungsten alloys 17
field-assisted sintering technology 15
film deposition processes
– deposition rates of 400
film-formation process 402
film preparation 415
films and membranes 219
film thickness
– inhomogeneity of 403
finite-difference time-domain (FDTD) 577
first-order Jahn-Tellar (FOJT) 77
first sharp diffraction peak (FSDP) 293
fishbone-profiled nanowire 56
flexible heat-transfer thin film
– using a polymer-ceramic hybrid material 147
floating zone (FZ) 456
flocculation theory 555
floppy network system 305
flow-type reactor 133, 141
flow-type supercritical hydrothermal reactor 138
flow reactor system 128
flow visualization inside supercritical hydrothermal process 138
fluorapatite–gelatin mesocrystal 240
fluorescence light 138
fluoride glasses 287
fluoride nanocrystals
– size and dispersion state of 313
fluoridic sol–gel process 185
flux growth processes 393
flux-assisted crystal growth 275
flux-assisted synthesis
– metal and salt fluxes 11
flux-melting technique 344
fluxing agents 277
flux treatment 345
flyer plate 38
focused ion beam (FIB) 315
FOJT. *See* first- order Jahn- Tellar (FOJT)
footprint silica gel catalysts, preparation of 207
forced hydrolysis approach 189
formic acid 129
– decomposition 133
– mole fraction 131
– mole fraction plot 133
fragile glasses 292
– schematic illustration, for temperature dependence of viscosity 291
framework composition 222
free energies, schematic illustration for
– glass, stable crystal, super-cooled liquid, and stable liquid 288
– thermodynamically metastable glassy state, metastable crystalline state, and stable crystalline state 289
free solidified side 331
fresnoite 294
frozen-in cybotactic groupings 294
fugacity 389
functional inorganic materials 256
– synthesis via bio-inspired polymer-controlled crystallization 243
functionalized materials 223
furnaces 2
fused quartz tube 23
FZ. *See* floating zone (FZ)

g

$GaAs_{1-x}P_x$ films 429
– epitaxial growth technique for 429
gallium 276
gallium nitride (GaN) 109
– ammonothermal crystal growth 109
– based power devices 441
– crystal growth rate, pressure dependence of 111
– crystal structure of 439
– features and applications of 439
– growth driving force of 435
– growth reactions, equilibrium constants of 434
– growth techniques 456
– – Czochralski (CZ) 456
– – edge-defined film-fed growth (EFG) 456
– – floating zone (FZ) 456
– halide vapor phase epitaxy (HVPE) for 430, 436
– sapphire template 452
– seed crystal 454
– – ammonothermal method 454
– seed crystal Na-flux method 454
– solubility 109
– substrate 441
– thermal conductivity of 446
– thermal decomposition of 444
– thermal expansion coefficients of 444
– wafers 438, 451, 453
– – applications 453
– – base substrate 444
– – thermal expansion coefficients of 444
– – bowing
– – inclination of edge dislocations, mechanism 449

– – island coalescence, mechanism 447
– – strategies for reduction 449
– – concave bending 446
– – concave bowing of 446
– – fabrication of 438, 454
– – by halide vapor phase epitaxy (HVPE) 441
– – strategies and technical issues 440
– – strategies and technical issues, control of electrical properties 445
– – strategies and technical issues, reduction of dislocation density 441
– – strategies and technical issues, reduction of dislocation density, strategies for reduction 441
– – strategies and technical issues, removal of base substrate 443
– – strategies and technical issues, removal of base substrate, strategies 444
– – strategies and technical issues, removal of base substrate, strategies, laser lift-off 444
– – strategies and technical issues, removal of base substrate, strategies, mechanical lapping 444
– – strategies and technical issues, removal of base substrate, strategies, spontaneous separation 444
– – growth methods 440
– – ammonothermal method 440
– – halide vapor phase epitaxy (HVPE) method 440
– – high-pressure high-temperature method 440
– – Na-flux method 440
– – mass production of 430, 451
– – necessity and applications of 440
– – off-angle distribution 446
– – off-angle variation 446
– – reduction of 446
– – production technology of, examples 450
– – DEEP method 450
– – VAS method 451
– – resistivity of 445
gallium/tin self-flux 276
α-Ga_2O_3 456
– crystal structure of 456
– epitaxial growth of, technical issues 460
– growth by halide vapor phase epitaxy (HVPE) 461
– growth rate of 463
– halide vapor phase epitaxy (HVPE) of 460
β-Ga_2O_3 456
– crystal structure of 455
– epitaxial growth of, technical issues 457
– growth by halide vapor phase epitaxy (HVPE) 457
– growth rate of 458
– halide vapor phase epitaxy (HVPE) of 457
– – on foreign strates 459
– homoepitaxy by halide vapor phase epitaxy (HVPE) 458
Ga_2O_3 nanoparticles 253
α-Ga_2O_3 wafers 461
β-Ga_2O_3 wafers 457, 460
$GaPO_4$ single crystal, hydrothermal growth 113
gas atomizer 332
gas complexes 362
– formation of 366
gas constant 126
gaseous transport agent 361
gas flow 14
gas-inlet structure 438
– with separation gas line 438
– without separation gas line 438
gas-phase deposition process
– chemical vapor deposition (CVD) 399
– physical vapor deposition (PVD) 399
gas inlets 437
gas–liquid diffusion method 250, 251
gas phase
– aluminum(III) chloride in
– dissolution of solid in 352
– homogeneous nucleation 407
– nucleation in 407
– solubility 356
– – iron in 357
– – temperature dependence of 356
gas phase solubility of iron
– temperature dependency of 357
gas sensor materials 258
gas/substrate interface 400
$GdFeO_3$-type perovskite 72, 74
$GdFeO_3$-type structure 55, 68, 69, 72, 74
Gd_2O_3-based nanocomposites 259
$21.25Gd_2O_3$–$63.75MoO_3$–$15B_2O_3$ glass
– temperature dependence, of magnetic susceptibility 303
– temperature dependence, of specific heat (C_p), in zero magnetic field 304
Gd_2O_3–MoO_3–B_2O_3 glasses
– composition 303
– crystallization of 303
gel ages 185
gelation 190

gel microfabrication 554
geochemistry 155
germanate (GeO_2) glasses 297
germanides 43
germanium transport 362
Gibb's phase rule 381
Gibbs free energy 53, 130, 381, 387, 404
Gibbs–Thomson effect 12
Gibbs-Thomson relation 157
gigapascal 41
glass compositions and crystalline phases
– in crystallization of off-stoichiometric glasses 296
– in crystallization of stoichiometric and near-stoichiometric glasses 295
glass formation 287, 292
– bond energy criterion for 292
– crystallization, of oxide glasses 295
– crystallization, of super-cooled liquids 289
– fragility concept for 291
– glass fabrication processing 292
– from quenching (cooling) of melt 288
– slow cooling process 288
glass formation
– basic scenario for 288
glass-forming ability
– low Pd content 346
glass-forming alloys 322
glass-forming oxide 305
glass-forming systems 43
glass nanostructured photomask 584
glass strate 563, 568
glass transition 288, 321
Glazer notation 68, 77, 78
goethite nanocrystals 160
gold nanoblock 580
gold nanostructures 586
– SEM image of 586
Goldschmidt diagram 68
– ABO_3 compounds 55
Goldschmidt tolerance factor 55
grafting 540, 542, 543
grain growth 414
graphene sheets 173
graphite heater 26
graphitization 35
graviperception 233
gravure printing 564
grazing angle 9
growth mechanisms 61
growth of Si nanowires under a-C film 501
– composition of nanowire 504–507
– experimental details 501
– morphology of annealed a-C/N film 502
– structural evolution of a-C/N film 501
– structure of nanowire 507
growth orientation 43, 44
– determined by HRTEM image and 2D FFT pattern 44
– determined by TEM image and SAED pattern 44
– origin of 43
– techniques applied to determine 44
growth temperature 43
guanosine triphosphate (GTP) 253
gun method 322

h

Hafnium boride layers
– wet chemical patterning of 539
halide CVD 402
halide melts 279
halide silicates 278
halides, volatilization of 367
halide vapor phase epitaxy (HVPE) 429
– applications 430, 438
– doping techniques in 445
– equipment 435
– – corrosion 438
– – corrosion, indicator of 438
– – fundamentals of design 436
– – HCl, use in 438
– – outline of 435
– high-speed growth 431
– – determinant factors of the growth rate 431
– – spontaneous nucleation by a parasitic gas-phase reaction 432
– hydride gases used in 429
– impurity doping 445
– – problems associated with 445
– metal-organic chemical vapor deposition (MOCVD) and, comparison between 433
– principle of 430
– technical characteristics and functionality of equipment 435
halide vapor phase epitaxy (HVPE) reactor 436
– basic design policies of 436
– counter-measures against corrosion 438
– counter measures against parasitic deposition 437
– depositions on gas inlets 437
– design of source-metal container 437
– film growth
– – thermal stress, effect of 437

– holding of strates 437
– materials for 436
halogen bridges 366
hard and soft templates, combination of 220
hardened steel plates 26
hard templates 210, 220
hcp. *See* hexagonal closed packing (hcp)
heat capacity 126
heat center 17, 19
heat exchanger 128
heating filaments 2
heat of reaction 126
Helbig's apparatus 378
Helgeson–Kirkham–Flowers (HKF) 125
hematite (α-Fe_2O_3) 168
hematite mesocrystals 258
hemoglobin 247
heteroepitaxy 460
heterogeneous interface reaction 403
heterogeneous nucleation 343
heterogeneous precipitation method 412
heterojunction field-effect transistors 109
heterometallic glasses 327
hexadecyltrimethylammonium chloride 221
hexagonal closed packing (hcp) 70
hexane 11
hexanoic acid 136
hexapod building blocks 239
high heat transfer 5
high-energy planetary ball milling techniques
– schematic illustrations of 335
high-melting transition metals 2
high-performance power devices
– metal–oxide–semiconductor field effect transistors (MOSFET) 456
– metal–semiconductor field effect transistors (MESFET) 456
– Schottky barrier diodes (SBD) 456
high-performance power devices 456
high-pressure apparatus
– multianvil-type
– – cell assembly for cubic 53
high-pressure apparatuses 52
– belt type 52
– diamond anvil cell (DAC) 52
– girdle type 52
– multianvil type 52
– – cubic 52
– – octahedral 52
– – tetrahedral 52
– piston cylinder type 52
– shock apparatus 52
high-pressure bomb 50
high-pressure borates 43
high-pressure chemistry principles 41
high-pressure conditions chemistry
– examples 42
high-pressure perovskites 49, 77, 86
– A-site-ordered 73, 77
– apparatus and pressure medium used for synthesis of 52
– B-site-ordered
– – rock salt-type 74, 78
– earth science 49
– engineering 49
– materials science 49
– syntheses of 50
high-pressure polymorphs 42
high-pressure synthesis 49, 54, 69
– under inert condition 55
high-pressure torsion (HPT) 342
high-purity nitrogen gas 282
high-quality plastic materials 15
high-quality prismatic or needle-shaped GaN crystals 282
high-temperature CVD 6
high-temperature synthesis 5
high-temperature techniques 1
high-throughput process 143
high-throughput synthesis 138
– B-site-ordered 74
HIP. *See* hot isostatic pressing (HIP)
homogeneous films 542
homogeneous gas reaction 403
homogeneous nucleation theory 133, 432
horizontal temperature gradient (HTG) 114
horseradish peroxidase (HRP) 257
hot isostatic pressing (HIP) 14, 50
hot isostatic pressing, and reactive sintering 14
hot pressing 341
hot wire method 352
HPT processes
– schematic illustrations of 341
Hugoniot curve 37
Hugoniot states 41
human serum albumin (HSA) 257
HVPE. *See* halide vapor phase epitaxy (HVPE)
HVPE-GaN crystals 441
HVPE of Ga_2O_3
– halide vapor phase epitaxy (HVPE) of 455
HVPE reactor
– carrier gas used in 431

hybrid materials 194
hydraulic jacks 32
hydrazide sol–gel 192
hydrazide sol–gel (HSG) 192
hydride vapor phase epitaxy 429
hydrocarbons 407
hydrochloric acid 5
hydrocodes 38
hydrofluoric acid 3, 44
hydrogen bonds 217
hydrogen deficiency 132
hydrogen halides
– ammonium halides, as source for 361
hydrogen-terminated silicon strate 540
hydrogen–inert gas mixture 5
hydrogen peroxide 277
hydrogen transformation 43
hydrolysis 402
– influencing parameters 184
hydrolytically stable Si–C 186
hydrolytic condensation 181
hydrolytic sol–gel process 181, 184, 185
hydrosilation reactions 187
hydrostatic pressure 14, 23, 346
hydrothermal conditions 160
hydrothermal crystal growth 156
hydrothermal high-pressure autoclaves 24
hydrothermally as-grown KBBF crystal 116
hydrothermal nanoparticle synthesis
– nucleation 127
– particle formation 127
– reaction and solubility in water 125
hydrothermal reactions 107, 109
– reaction rate 140
hydrothermal synthesis 123
– of CeO_2 nanoparticles (NPs)
– – flow-type reactor 124
– method 124
– using $NaBiO_3{\cdot}nH_2O$ 119
hydrothermal system 115
hydrothermal ZnO single crystal growth, conditions for 111
hydroxide fluxes 279
hydroxides 277
hydroxysilanes 542
hyperbolic curves 390
hyperquenched-annealing 294

i

ice templating 212
ideal gas flow 131, 361, 436
IHPP RAS. *See* Institute for High Pressure Physics of the Russian Academy of Sciences (IHPP RAS)
ilmenite, transformation to perovskite 54
impedance modifiers 37
impregnation 412
inÆsitu multianvil technique 31
inÆsitu surface capping method 134, 137
inÆsitu X-ray 29
incongruent, meaning 386
indium 276
indium-rich heavy fermion materials 276
induction heating 10
– of electric polarization 574
industrial scale production 123
inert gas 55
inkjet printing 180, 544, 564
inner diameters (i.d.) 141
In_2O_3 nanocubes 253
inorganic crystals 233
inorganic-biological hybrid materials 194
inorganic ion exchange materials 203
inorganic materials
– bio-inspired synthesis 259
inorganic materials synthesis
– hydrothermal systems, use in 51
inorganic–organic composite materials 234
inorganic–organic hybrid materials 181
inorganic polycondensation reaction 190
inorganic sol–gel network 181
inorganic sol–gel process 190
Institute for High Pressure Physics of the Russian Academy of Sciences (IHPP RAS) 28
insulating materials 16
interfacial energy 290
interfacial structure 48
intermediate chemical species
– formation 403
intermediate compounds 385
intermetallic compounds 282, 363
intermetallic phases
– with wide homogeneity range 363
Intermetallic plutonium compounds 276
intermetallic systems 9, 363
iodide process 352
iodine 6
ion bombardments 416
ion-beam epitaxy (IBE) 400
ionic crystals 42
ionic liquid-type surfactant 220
ionic liquids 1
ionic strength 158

iron nanoparticles
– hexanoic acid modified 136
iron nanoparticle system 137
iron oxyhydroxide materials 250
$IrSb_2$, single crystals of 12
island coalescen 447
isobar case 381
isobaric cooling 382
3-isocyanatopropyltrialkoxysilane 187
isoelectric point 161
iso-oriented crystal 238

j
Jahn-Teller ions 78
Japanese encephalitis vaccine (JEV) 260
Joule heating 10

k
$KClO_3$ as oxidizing agent 283
kinematic viscosity 145
kinetically controlled reactions 179
kinetic energy 58
kinetic models 162, 167, 174
kinetics of homogeneous (steady-state) nucleation rate 290
K_2NiF_4-type structure 83
– stability of 86
$10K_2O–10WO_3–80TeO_2$ sample
– glass transition region for 310
$K_2O–WO_3–TeO_2$ system
– glass-forming region in 309
Kretschmann configuration 575
KTP. *See* potassium titanyl phosphate (KTP)

l
LaMer mechanism 127, 128
lanthanides 280
lanthanoids 367
large-scale integration (LSI) 400, 573
laser ablation 13, 14, 33, 417
laser beam 418
laser chemical vapor deposition (LCVD) 417, 419, 422
– microstructure of 421
laser CVD 419, 421
laser heating 6, 7
– and laser-assisted synthesis 5
laser-assisted high-temperature synthesis 6
laser-induced crystallization technique 302
laser irradiation 402
lattice mismatch 51
lattice spacing 51
layered manganese oxides, transformation of 204
layer-by-layer adsorption 543
LCD panels 571
LCVD. *See* laser chemical vapor deposition (LCVD)
leaching 413
lead 276
lead chloride 13
le Chatelier's principle 355
length scales 203
lever rule 392
levitation 11
Lewis acids 184
lichtbogenofen 9
LiCl flux 279
Liesagang rings 554
Liesegang patterns
– laws 554
– – Matalon–Packter law 554
– – spacing law 554
– – time law 554
– micropatterning and 553
– in spatial patterns 554
Liesegang rings 549, 550, 551
– 2D system 552
– formation 550, 555
– – in colloidal systems 553
– – van der Waals interaction in 552
– nucleation rates of gel, role in 551
Liesegang system 551
light-emitting diode (LED) 196
light-harvesting by periodic mesoporous 223
lightweight alloys 17
Li^+-ion 111
$LiNbO_3$-type oxides
– high-pressure synthesis of 73
line compound 384
$60Li_2O–10Nb_2O_5–30P_2O_5$ glass
– DTA patterns, for bulk and powdered samples of 307
$Li_2O–Nb_2O_5–P_2O_5$ system
– glass formation in 306
liquid aluminum 276
liquid crystal display (LCD) 564
liquid crystal display packaging 569
liquid–solid interface 64
liquid–solid phase transformation 319
liquid sulfur 382
liquidus 383
lithium 282
lithium batteries 1, 2
lithium-ion batteries (LIB) 252, 306
lithium-ion batteries cathode materials 258

lithium nitride fluxes 282
lithographic techniques 14
lithography system 573
Ln^{3+} doped-porous nanoparticles 258
$LnPO_4$ – zebrafish embryo 260
loading–unloading pathways 34
localized surface plasmon resonance (LSPR) 573, 574
local photochemical reaction
– verification by LSPR 578
long-range phenomena 540
– diffusion 540
– electrostatic interaction 540
Lorentz–Lorenz equation 302, 312
low-temperature processes
– requirement for 180
low shrinkage bulk sol–gel materials formation
– silicon alkoxides precursors 191
LSI. *See* large- scale integration (LSI)
LuAG single crystal 113
– harvested from YAG surrogate seeds 114
luminescence materials 282
lysozyme 247
lysozyme-mediated bio-inspired mineralization 258

m
macropores, templates of 211
macroporous
– materials 213
– silica 211, 212
macroscale templates
– biotemplates 213
– colloidal templates 212
– emulsion/reverse micelle templates 211
– water droplets and ice templates 211
Madelung energies 74
magnesium 276
– sponge furnaces 8
magnetic CoPt 248
– properties 283
magnetite nanocrystals 133, 242, 249
magnetotactic bacteria 242
magnetron sputtering technique
– schematic illustration of 331
Maker fringe pattern 299
mass density 436
mass flow controllers (MFC) 435
master alloy ingots 337
materials synthesis 223
MBE. *See* molecular beam epitaxy (MBE)
mean nanoparticle size
– mixing condition effect 141
mechanical alloying (MA) 324, 336
mechanical casting 202
mechanical drilling 33
mechanical grinding (MG) 324, 336
mechanical pressure 16
melted spun $Fe_{83.3}Si_4B_8P_4Cu_{0.7}$ alloy
– high-resolution TEM image of 329
melt-quenched metallic glass 342
melt-solidification 417
melting entropy 290
melting temperature 1, 3, 9
mercury compounds 280
MESFET. *See* metal– semiconductor field effect transistors (MESFET)
mesochannels, vertical alignment of 220
mesocrystal formation 234
– matrix-induced 240
– possible mechanisms 240
mesoporous
– lattice 412
– materials 193, 222
– mesostructured films 219
– metal oxides 209
– solid templates, wall thickness of 211
mesoporous silica
– film 208, 209, 219, 221, 412, 413
– Ni-precipitated 413
mesoscale solid-phase transformation 241
mesoscale templates
– liquid crystalline templates 208
– mesoporous solid templates 210
– microphase-separated block copolymer templates 210
mesoscopically structured crystal 240
mesoscopic self-assembly 235
metal alkoxides 184
– based sol–gel process 188
metal catalysts 4
metal cations 204
metal chalcogenides 241, 253, 255
– mineralization 253
metal fluxes 12
– synthesis 13
metal foil 33
metal halogens 402
metal-assisted chemical etching process 3
metal-covered Si wafers 2
metal-organic chemical vapor deposition (MOCVD) 429
metallic aluminum 47
metallic catalyst 8
metallic conductivity 8

metallic glasses 321
– factors, for preparation 342
– matrix composite 337
– mechanical alloying and mechanical grinding 336
– mechanical properties limits applications of 329, 331
– melt-spinning technique 331
– metallic mold casting method 337
– powder, consolidation techniques 340, 341
– preparation methods for 331
– sputtering 332
– – advantage of 332
– – gas (water) atomization 332
– – pulsated orifice ejection method 334
– – types, of equipment 332
– water quenching 337
metallic impurities 24
metallic materials 275
metallic mold casting method 337
– cooling rate during 337
metallic nanoparticles 129
– formation 124, 129, 130
– powder XRD diffraction patterns 132
– synthesis 134
metallic particles 334
metallic seal 33
metal–metal eutectic system
– atomic clusters of 325
metal–organic framework (MOF) 238, 256
metal–oxide–semiconductor field effect transistors (MOSFET) 456
metal–Si alloy 11
metal–Si droplets 12, 53, 61
metal–Si eutectic droplet 2, 11, 43
metal oxides 123, 214, 215, 248
– bio-inspired synthesis 249
– nanoparticles 138
– solubilities, HKF model, use of 127
– transport 368
– water solubilization 125
metal pnictides 370
– chemical bonding in 370
metal powder additives 36
metal precursors 185, 188
metals, electroless reduction 215
metal/semiconductor interface 10
metal substrate 421
metastable (nonequilibrium) state 289
methanol 278
methyl thiazolyl tetrazolium (MTT) 259
methyltrichlorosilane 404
MFC. *See* mass flow controllers (MFC)
MgO octahedron 30
micelle nanolithography 545
microcontact printing 544, 565, 566
microelectronic applications 195
microelectronics 180, 181, 186
microencapsulation 195
microjets 36
micropatterning 553
microporous zeolite 220
microrings 234
microscale templates
– microporous solid templates 206
– molecular imprinting 205
– molecular templates 205
microwave device 13
microwave heating 13, 14, 282
microwave induction heating 401
mineral bridges 241
mineralogy 155
mirror furnaces 7
mixed alkali silicate glasses 293
mixed crystal 383
mixer channel geometry 145
mixing rate 144
$MnCO_3$
– calcination 252
$MnCO_3$–$CoCO_3$ composite precursors 252
MnO-based porous structures 258
MnO–Co hybrid materials 252
MO. *See* metal- organic (MO)
MOCVD. *See* metal- organic chemical vapor deposition (MOCVD)
modified atomic force microscope 544
moisture-sensitive products 278
molar fractions of iodine
– temperature dependency of 356
molar ratios 43
molar refraction 301
molar volume 290
molecular assemblies 223
molecular beam epitaxy (MBE) 12, 400, 429
molecular crystals 41
molecular educts 179
molecular level interactions 540
– autocatalysis 540
– phase separation 540
– supersaturation 540
molecular orbital (MO) 293
molecular precursors 179, 186
molecular precursors based approach, advantages 186
molybdenum
– compounds 278

– melting point 363
– plates 26, 30
monoalkoxysilanes 183
monodispersed SnO_2 nanospheres 258
monolithic diamond powder 414
monomolecular films 540
monophenylsilane 11
monotectic 388
morphogenesis
– general principles 247
morphological and hierarchical control 217
mosaic microstructures 414, 416
MOSFET. *See* metal– oxide– semiconductor field effect transistors (MOSFET)
multianvil apparatus 29
multianvil devices 23, 29, 31
multianvil experiment, schematic assemblage 30
multianvil techniques 29
multicenter bonding 42
multicomponent bulk metallic glass-forming systems 326
multifunctional coatings 196
multigram scale 44
multiphoton excitation 544
multiple-quantum magic-angle spinning (MQMAS) 293
multiple reaction system 145

n

Na_2CO_3 fluxes 279
nacre 242
$NaGdF_4$
– Ce/Tb@CaP-Apt
– – doxorubicin (DOX)-loaded 259
– Ce/Tb@CaP-Apt nanocomposites
– – fluorescence and magnetic properties 258
Na-K alloys 282
nanocluster-free nanowire tip 38
nanocrystallization 299
nanocrystals 155, 157
– surface 157
nanofibers 218
nanogap gold structure 578
nanogap-assisted surface plasmon nanolithography 576, 577, 581
– demonstration of 580
– a schematic illustration of exposure system 579
nanogap position 579
nano-scale fragility 294
nanoimprinting 544
nanoink for 3D printing 148
nanolaths 250
nanolithography 573, 582
– characteristics of 587
– type of 583
nanomask 452
nanomaterials 7
nanoparticle (NP) 123, 169, 211, 407
– applications 145
– catalysts, use as 147
– formation
– – hydrothermal synthesis 129
– – mechanism of 129
– high- throughput synthesis 124
– industrial scale production 145
– monodisperse 553
– polydisperse 553
– preparation, hydrothermal synthesis 129
– by spatial constraints 241
– superhybrid materials, use as 146
– zeta potential 136
nanoparticles
– organic modified 134
nanoparticulate seeds 219
nanopatterns, by self-assembling 546
nanophase materials 189
nanopores 421
nanoreactors 211
nanorods 218
nanoscale metal fluorides 185
nanosheet building blocks 250
nanospaces, hierarchical integration of 220
nanosphere entrapment, in nanowires 57
nanosphere-entrapped Si nanowires 31
nanostructures
– materials, synthesized by templating methods, applications of 222
– powders 16
– self-construction
– – by oriented attachment 239
– in solid-state materials 223
nanotechnology 201
nanowires 6, 7, 37, 61
– composition 11, 40
– diameter 43
– of grown 39
– growth 8, 9, 12
– – VSS mechanism of 8
– – in work of Li, Si vapor pressure 61
– materials 5
– orientation 10
natural geothermal processes 123
natural magnetite assemblies 250
Nd YAG laser 13, 420, 421

near-field enhancement factor 580
near-field scanning microscopy (NSOM) 573
nearly oriented attachment 163
neutron beam 138
neutron diffraction 8, 29
neutron radiography 138
neutron tomography 140
Newton's law of cooling 339
N-polar GaN 449
n-type doping 446
n-type GaN 449
Ni–Si–O nanoclusters 47, 61, 63
Ni–Si droplet 525, 527
Ni–Si eutectic droplet 503, 515
Ni catalyst 412, 415, 501, 513
Ni catalyst–Si wire spherical interface 515
nickel fine particles 133
nickel-coated silicon wafer 52
nickelocene 411
Ni-catalyzed Si nanowires 29
Ni nanoparticles 411, 412
Ni nanospheres 63
– -entrapped Si nanowire 57
Ni–Al alloy 413
Ni–Si interface 61
niobium(III) chloride transport 366
niobium(IV) chloride
– formation 366
NiO–Si interface 34
Ni particle–Si nanowire interface 54
Ni particles 55, 59, 61, 62, 63
– grain growth of 412
nitridation 402
nitride iodides 279
nitride synthesis via Li_3N and NaN_3 282
nitridometallates 282
nitridophosphates 43
nitridosilicates 282
nitrogenation 283
nitrogen content,in materials $Y_2Fe_{17}N_x$ influence 283
nitrogen diffusion, into intermetallic matrix 283
noble gases 33
noble metal 246
– nanodots 247
– transportation 362, 363
noncentrosymmetry 73
nonclassical crystal growth 158
– mechanisms 159
nonclassical crystallization 238
– particle-based 238
– pathways 238, 242
– processes 235
nonelectrode plasma 416
nonequilibrium films 416
nonhydrolytic hydroxylation reactions 184
nonhydrolytic sol–gel process 184
nonhydrolytic sol–gel reactions 192
non-Si semiconductors 4
nonmetallic particles 334
normal-pressure phase 41
novel array film 250
NP. *See* nanoparticle
nuclear magnetic resonance (NMR) 292
nucleation 49, 402, 432, 433
– frequency 320
– LaMer mechanism 128
– rate 298
nucleophilic attack 182
nucleotide triphosphates
– types 253

o

octahedral edge length (OEL) 31
octahedral-anvil device 29
off-set printing 564
Ohm's law 376
oleylamine 173
optical absorption spectra 298
optical basicity 302
optical devices, efficacy of 441
optical materials 181
optical spectrometer 34
optical transparency 191, 195
optical transparent materials 195
optimal electrode gap 9
optimum transport temperature (Topt) 354
optoelectronics 52, 186
– material 32
3D ordered assembly, of nanoparticles into mesocrystal
– principal possibilities 241
organically functionalized precursors 182, 187
organically functionalized trialkoxysilanes
– selection 187
organic capped nanoparticles 135
organic compounds–water binary systems
– critical loci 135
organic electroluminescent devices 563, 564
organic electronic devices 563, 564
organic field-effect transistor 547
organic matrix, templating effects 240

organic modification method 145
organic modifier 134
organic molecule self-assemblies 253
organic photovoltaic cell 564
organic polymers 205
organic surface modification 147
organic surfactants 193
organic templates 215
organic TFT array 569
organic thin film devices 563
organic thin-film-transistor 564
orientation distribution 46
oriented attachment 159, 238
– between poorly crystalline particles 164
– – major control parameters 164
– – structural recognition 164
– role of attachment
– – in development of microstructure 162
oriented organic matrices 241
original Bridgman anvils 28
orthogonal self-assembly 545
– of alkane thiols 545
orthorhombic sulfur 382
OSC. *See* oxygen storage capacity
Ostwald cycle 555
Ostwald ripening 12, 123, 157
– effect 6
– process 551
oxalic acid 376
oxidation 2, 16
– protection 3
– states 279
oxide 276
– frameworks, synthesis of 214
– materials 278
– synthesis 5
oxide glasses 287
– Boson peaks, in crystallizing glasses 294
– degree of nanoscale heterogeneity 293
– hyperquenching 294
– nanoscale structure in 292
– random and homogeneous structures 293
oxide-assisted growth 15
oxide-assisted growth theory 15
oxidic nanoparticles
– production 191
oxidizing agents 54
– CaO_2 55
– CrO_3 54
– $KClO_3$ 54
– $KClO_4$ 54
oxoacidity 279
oxyfluoride glasses 294, 313
– bulk nanocrstallization 315
– degree of their site asymmetry 315
– density of 313
– effect, of concentration quenching 315
– laser-induced crystallization 315
– microfabrication, of glass materials 315
– synthesis of 313
oxyfluoride-based glass-ceramics 313
oxygen impurities 281
oxygen pressure production 51
– CrO_3 decomposition 51
– using oxygen gas compressor 51
– vaporization of liquid 51
oxygen storage capacity (OSC) 147
oxyhydroxide 169

p

parasitic gas phase reaction 431, 432
Paris-Edinburgh cell 23, 28, 29
partial oxidation 250
partial pressure 367
partial pressures of iodine
– temperature dependency of 353
particle formation reaction 123
– mixing rate, effects of 142
particle-based crystallization 159
particle-mediated growth mechanism 159
particle-particle attachment 164
particle size 161, 165, 166, 171
– mixing rate, effect of 141
– reaction temperature, effect of 142
patterned PDMS stamp 566
pattern multiplication 547
$PbCrO_4$ system 550
$PbO–PbF_2$ flux 278
PDMS. *See* poly(dimethylsiloxane) (PDMS)
Pd–Cu–Si bulk metallic glass 322
Pd–Si metallic glass 322
Pechini method 5
periodically mesoporous organosilica (PMO) 193
periodic band 547
– formation 550
periodicity 210
periodic mesoporous organosilica (PMO) 188
peritectic line 386
– decomposition temperature 387
– meaning 386
perovskite
– in materials science 49
– – materials engineering 49

– – solid-state chemistry 49
– – solid-state physics 49
– structure 49
– synthesis aided by pressure
– – oxygen gas pressure of ~1 MPa 50
– – pressure in GPa range 52
– – pressure of ~100 MPa 50
– type oxides 50
– – $Bi_{1/2}Ag_{1/2}TiO_3$ 50
– – $PrNiO_3$ 50
– – transition metal ions in 51
perovskite-related layered compounds
– Aurivius-type 78
– Dion-Jacobson-type 78
– Ruddelesden–Popper (RP)-type 78
– – high-pressure synthesis of 78, 84
perovskite-type phases 54, 86
perovskite-type structure
– evolution from layered structure 70
perovskite–post-perovskite phase 87
– phase transition 87
peroxides 277
phase
– destabilization 387
– distribution 64
– formation, monitoring of 53
– Gibb's phase rule 380
– purity 155
– separation 189
– stability limits 388
– transitions 24, 54
phase diagram 364, 383
– daltonides in 385
– Pb–Sn 394
– scale of 384
phase transformation 52, 169, 172
– in iron oxides 168
– role of attachment in 166
– tolerance factor for 72
phenomenological formalism 376
phosphides transport 370
phosphorus halides, stability 370
phosphorus nitride imides 43
phosphorus, vapor pressure 389
photocatalysis 222
photodecomposition 402
photolithography 442, 452, 573
photoluminescence (PL) 297
– spectra 298
photolytic laser CVD, schematics 418
photomask
– design 582
– pattern 583
– – schematic illustration of 583
– and photoresist strates 576
– SEM image 585
– substrate, SEM image 584
photonics 52, 181
photopolymerization 578
photoreaction wavelength 579
photoresist
– film 581
– substrate surface 582
photosynthesis 223
phototherapy 259
photovoltaic 563
p-element-rich phases 277
physical adsorption 135
physical confinement 211
physical constants 26
physical vapor deposition (PVD) 399
physical vapor transport (PVT) 375
piezoelectric α-quartz 107
PILP. *See* polymer-induced liquid precursor (PILP)
piston-cylinder apparatus 25, 26
planar impact (flyer plate) 36
planar impact shock wave recovery experiments
– degrees of energy deposition 39
– loading paths 39
Planck formula 6
plasma
– burner 5
– categories 416
– – glow discharge low-temperature 416
– – nonequilibrium low-temperature 416
– – thermal equilibrium high-temperature 416
– formation 416
– -assisted sintering 15, 16
– -enhanced CVD 7
– spray 400, 421
plasma CVD, schematic 417
plasmonic solar cells
– fabrication of 586
plasmon interference lithography technique 575
plastic deformation 329
plate-like crystals 400
platinum 173
platinum group compounds 87
– $BaOsO_3$ 87
– $BaRuO_3$ 87
– $SrIrO_3$ 87
platinum-containing compound 278
plutonium chemistry 276

PMO. *See* periodically mesoporous organosilica (PMO)
pnictide transport 370
pnictogen chalcogenide halide 370
pnictogen halides 370
POEM apparatus, parts of 334
Poisson's ratio (ν) 448
polarized optical microscope (POM) 315, 316
polychalcogenide fluxes 279, 280, 281
polychalcogenides 279, 280
polycrystalline deposition 437
polycrystalline diamonds 29
polycrystallization 455
poly(dimethylsiloxane) (PDMS) 544
polydispersity 547
polyelectrolytes 234
polyhedral oligomeric silsesquioxanes (POSS) 183
polyhedral shrinks 29
poly-dimethyl-siloxane (PDMS) 566
poly-3-hexyl-thiophene (P3HT) 569
poly(l-lysine)-poly(ethylene glycol) (PLL-PEG) 544
polylysine-induced titania 251
polymer additives 236
polymer brushes 542
polymer fillers 192
polymer glasses 287
polymer-controlled crystallization 235, 243
polymer-induced liquid precursor (PILP) 242
polymer-stabilized crystallites 235
polymeric precursor 5
polymer network liquid crystal (PNLC) 565, 569
– panel cross section, with printed organic TFT 570
polymers 214
– pattern, nanometer-scale 545
– phase-separated film 547
– self-assemblies 250
polymer surfaces
– modification of 539
– – embossing 539
– – photolithography 539
polymorphs 42
polyolefine plastics (POP) 544
polypeptides 234
polysiloxanes
– as encapsulants 196
poly(sodium 4-styrenesulfonate) (PSS) 236
polystyrene-poly(vinyl pyrrolidone) (PS-PVP) 545
polysulfides 280
POP. *See* polyolefine plastics (POP)
porous materials
– control of 217
– nanoparticles of 217
porous solids 203
porous templates, fabrication of 452
POSS. *See* polyhedral oligomeric silsesquioxanes (POSS)
post-perovskite
– ABX_3
– – structure of 86
– factors for stabilization of 87
– – covalent character of B–X bond 87
– – tilting of BX_6 octahedron 87
– – tolerance factor 87
– in materials science 87
– $MgSiO_3$ 87
– type phase 87
postshock temperature 40
potassium carbonate 385
potassium sulfate K_2SO_4 385
potassium titanyl phosphate (KTP) 539
– etching of 539
powder metallurgy 341
powder source materials 25
powder synthesis 123
precipitation
– band 555
– behavior 346
– rings 550
precise flow 140
precursors 4, 214
– gallium chloride as 461
predictive Soave-Redlich-Kwong equation of state (PSRK EOS) 131
preformed zeolites 205
prenucleation cluster (PNC) 242
pressure 23
– -homologue rule 24
– -induced crystallization 43
– -induced metallization 42
– -metallization rule 42
– -transmitting medium 26, 33
– –temperature phase diagram 87
– sensor 34
prestructured organic matrix 240
primary nucleation 142
primary shock front 38
printed organic TFT arrays
– with 100 and 200 pixel 569
– on glass strate 568
– with several kinds of pixel sizes 569
printing techniques 564, 565

– comparison in 565
– TFT arrays on substrate 568
process design methodology 140
process design under reaction-controlled condition
– Damköhler numbers, by using 140
– Reynolds number, by using 140
protective coatings 399
protein adsorption 547
protein crystallization 547
pseudobinary complex systems 379
pseudocubic hematite (α-Fe_2O_3)
– crystals 158, 159
pseudomorph 169
$PtCl_2$ 361
Pt single-twinned seeds 247
pulsated orifice ejection (POEM) method 334
pulsed electric current sintering 15
pulsed laser 14
pulverization 1
pumping system 14
purified peptide–CdSe hybrid QDs 255
PVD. *See* physical vapor deposition (PVD)
Pv–pPv phase boundary, Clapeyron slope 87
pyrex sleeve 26
pyrolytic laser CVD
– schematics of 418
– schematics of fiber growth by 419
pyrolytic reactions 417
pyrometer 5
pyrophyllite cylinder 26
P/Zr ratio 116

q
quantum dot (QD) 253
– effect 53
quartz crystal synthetic 156
quartz glass tubes 51
α-quartz, hydrothermal growth of 109
α-quartz single crystal 108, 109
quartz tube 14, 15
quasicrystalline (QC) 347
quasi-isentropic compression 41
quenchable metastable phases 31
quenching rates 292

r
radical polymerization 214
radio frequency (RF) 401
Raman scattering spectra 35, 293, 303
ramp wave 41
rapid prototyping 180
rare earth halide molybdates 278
rare earth-containing platinum metal oxides 279
rare earth transition metal 280
rare earth trihalide 278
rate constant, estimation 140
Rayleigh lines 40
Rayleigh scattering 293
reactant gases 437
reactant materials
– GaCl used as 430
– NH_3 used as 430
reaction enthalpy 354
reaction entropy 354
reaction-controlled condition determination
– dimensionless numbers, use of 143
reaction pressure 43
reactive flux 12, 13
reactive gases 8
reactive metal fluxes 275
reactive salt fluxes 277
receptor-mediated endocytosis 259
reconstructive phase transitions 40
recovered oil
– mixed with 1-methyl naphthalene 148
recrystallization 12, 123, 159, 172, 174, 356
redox reactions 390
reductive supercritical hydrothermal synthesis 130, 131
reflectance change of small PNLC panel, depending on gate voltage 571
refractive indices 302, 311
refractory materials 18
refractory metals 16
relationship, between incubation time and temperature for crystallization 343
relaxation state 348
removal of template
– calcination 215
– chemical etching 216
– solvent extraction 215
RE_2O_3–Bi_2O_3–B_2O_3 glasses 300
resistance furnaces 2
resistive heaters 34
reverse micelle method 134, 211
reverse temperature gradient 361
revert spacing mechanism 550
Reynolds number 143, 145
RF. *See* radio frequency (RF)
RF induction current 402
Rh nanoparticles 409
rhodamine B (RhB) 258
– photocatalytic degradation 251
Rh_2O_3 transport 371

ripening time 123
rock-forming minerals 32
rotary CVD 411, 413, 415
– schematic of 411
rubidium oxide 367
rutile crystals 165

s

SAED pattern 55, 60
salt flux 12
– reactions 277
– synthesis 12
samarium-doped borate 34
saturation-drift velocity 440
SBD. *See* Schottky barrier diodes (SBD)
scaling factor 290
scanning laser 421
scattering techniques 174
Scheil–Gulliver cooling 395
schematic illustration, of the POEM 333
Scherrer's equation 297
Schottky barrier diodes (SBD) 456
scratch and abrasion resistant coatings 195
screening devices 195
screen printing 564
sealed gun barrels 24
sea urchin spine, biomineralization 240
secondary ion mass spectrometry (SIMS) 459
secondary templates 203
second harmonic generation (SHG) 299
segregation 392, 396
– process of 392
selenides 280
self flux 393
self-assembled monolayers 540
– chemical properties 540
self-assembly 203
self-flux synthesis 276
self-flux technique 275, 276
self-organization phenomena 540
self-propagating high-temperature synthesis 18
semiconducting materials 157
semiconductors 282, 400, 429, 439, 457, 564
– crystals 282
– devices 399, 417
– fabrication 415
semimetal alkoxides 186
severe plastic deformation (SPD) 342
SFLS growth 11
SFLS technique 12
shape
– change after contact and transfer from mold to strate 567
– variations, factors affecting 236
shape charge
– armor piercing action 37
shell-type cooler 124
shock compression behavior 40
shock front converges 38
shock-driven reactions 36
shock-loaded metal beads
– microjetting 37
shock-synthesized diamond 37
shock melting 35
shock temperature 40
shock waves 35
– compression 35
– generators 41
– profile 41
– schematic profiles 35
– syntheses 34
– – copper beads, use of 37
– – microjetting, importance of 37
– treatments 43
shrinking process 185
Si atoms 4
$Si-SiO_x$ core–shell nanowires 34, 35
SiC films 404
– morphology of
– – supersaturation, effect of 404
– – temperature, effect of 404
– self-contact of 414
side-by-side biaxial nanowire 54
Si dioxide 1
Si-containing powders 2, 12, 15, 34
silane (SiH_4) 5, 7
– capping agent method 135
– coupling agent 134
– solutions, preparation 542
silica gel 205
silica glass 358
silica-based sol–gel
– materials morphology 183
silicalite-1
– condensation of 208
– crystal symmetry 171
– growth, from amorphous primary particles 170
silica nanoparticles 213
– preparation 191
silica nanotube 58
silica – cyanobacteria 260
silica network formation 187
silica tubes 358

silicides 10
silicon (Si) 1
silicon alkoxides 192
silicon carbide 404
silicon dioxide 109
silicon-based precursors 186
silicon-iron oxide catalyst 148
silicon nanowires 14
– growth, through thermal annealing 34
– thermal evaporation growth of 15
silicon substrate 545, 547
silicon tetra-isopropoxide 188
silicon transport 363
silicon wafers
– silicon nitride-coated 545
silsesquioxanes 183
silver nanoparticle ink on glass substrate, patterns of 567
simple ball milling 335
simple ion packing model 69
Si nanoparticles 15, 47, 53
Si nanowhiskers 12
Si nanowires 1, 2, 3, 6, 8, 9, 11, 12, 13, 14, 15, 32, 33, 34, 36, 41, 43, 45, 47, 52, 61
– diameter-dependent nucleation 7
– experimental condition to grow 16
– flow reactors 11
– growth 11, 14
– – along different orientations 46
– morphology of 38
– statistical distribution of 46
– transition morphology 61
Si nanowires, Al-, Cu-, and Fe-catalyzed growth 33
Si nanowires in alumina tube, growth 30
– alternative growth mechanism
– – Al-catalyzed growth 32
– reduction of alumina tube by H2 30
Si nanowires in quartz tube, growth 15
– alternative growth mechanism 24
– – Al- or Cu-catalyzed growth 24
– – formation of Si oxide sheath 28
– – seeding location, wall of the quartz tube 27
– – trace of metal impurities 29
– metallic Al and Cu in quartz tube 15
– source of Si
– – ejected micrometer-sized Si particles 21
Si–Au interface 46
Si–metal interface 4
Si–Ti–C–Cl–H system
– gas species of 407
– solid species of 407
single crystals 160
– *vs.* mesocrystals 241
2D single crystals 240
single-crystal-like electron diffraction patterns 241
single-crystalline ⟨111⟩-oriented Si nanowires 13
single-crystalline ultralong Ag nanobelts 248
single-phase fields 392
single-roller melt-quenching technique 324
single-roller melt-spinning technique 331
single source precursors 189
single static synthesis 34
sintering 14, 15
sinusoidal nanowire 56
SiO_2 film
– deposition rate of 422
– microstructure of 422
SiO_2-based glasses 297
SiO_2-coated SiC powder 415
– sintering of 416
SiO_2 molecules 47
SiO_2 nanowire 31
SiO_2 substrate 542
– chemical patterns on 547
SiO_2/TiO_2 substrate 544
SiO_2 wettability 414
Si oxide 15
– sheath 32
Si precipitation 43
Si precursor 11
Si/SiO_x coaxial nanowire 42, 54, 59
Si/Si oxide interfacial energy 47
Si/Si oxide nanowires 60, 62, 63, 64
Si/SiO_x nanowires, growth experimental condition 53
$Si/SiO_y/SiO_x$ nanowire 62, 64
$Si/SiO_y/SiO_x$ triple-concentric nanowire 41, 42, 56, 63
Si wafers 2, 3, 4, 12, 35, 37, 61, 64, 416
Si whiskers 4, 13
– VLS growth of 5
Si wire–Ni catalyst 49
Si wires 6
slag phase 391
slow dip coating 545
small angle X-ray scattering 174
small-scale press machine 29
small PNLC panels 570
smooth nanowire 56
smooth nodular films 405
SnO_2 nanodots 252
S_N2 reactions 182
sodium azide 282

sodium flux 282
sodium silicate 293
soft chemistry approaches 180
soft-shell cylindrical nest 30
soft-templating synthesis 214
soft lithography 180, 566
soft templates 180, 220
– combination of 220
solar and mirror furnaces 7
solar cells 43
solar energy conversion 281
solar furnaces 7
solid
– nanoparticles 9
– nanostructural design 201
– phases 409
– Si nanowire 5
– solution 383
– sulfur 382
solid catalyst 9
– particle 12
solid-particle-mediated growth mechanisms 160
solid-state chemistry 23, 155, 159, 180, 201, 275
solid-state materials
– properties 201
– templates 211
solid-state metathesis 18
– reaction 18
solid-state nitride materials 282
solid-state synthesis 3
solidification 33
solid–gas equilibria 389
solid–liquid equilibria 383
solid–liquid interface 43, 275
solid–liquid–solid growth 12
solid–liquid–solid mechanism 34
solid–solid reactions 275
solidus 383
sol–gel chemistry based material
– bulk materials 190
– hybrid materials 194
– nanoparticles 191
– porous materials 193
– thin films and coatings 190
sol–gel coatings 196
sol–gel films 191
sol–gel-based hybrids 196
sol–gel material 185
sol–gel methods 214
– applications 195
sol–gel precursors 181
sol–gel process 179
– silicon-based 182
sol–gel transition 185
solubility 113, 164, 277
– equilibrium 356
– of GaN in supercritical ammonia 111
solution-based chemistry 278
solution-spray plasma method 412
solvent, boiling point of 1
solvothermal crystal growth 172
solvothermal methods 107, 134
spark plasma sintering 15, 16, 341, 414
spherical mesoporous silica nanoparticles 222
spherical Ni–Si interface 61
spin-coated diblock copolymer 545
spinodal curve 388
spinodal decomposition 221
– theory 555
spontaneous phase separation 221
spontaneous separation technique 444
spray coating technology 191
spray drying 218
sputtering 195, 400, 416
square-shaped PbS nanosheets 240
stainless-steel reactor 138
stainless tube reactor 138
standard electrode potentials 276
standard Gibbs free energy 131
standard heating filaments 2
standard hydraulic press 30
statistical thermodynamics 375
steady-state nucleation frequency 319
stoichiometrically composed intermetallic phases 364
stoichiometric coefficient of electron 130
stoichiometric composition 299
stoichiometric compound 384
stoichiometric factors 353
Stöber method 218
stress-binding rings 26
strong glasses
– schematic illustration, for temperature dependence of viscosity 291
structural engineering 193
structure-directing agent (SDA) 171, 202
structure–property relations 277
sublimation 2, 278, 351, 366, 371
substrates 437
– temperature distribution on 421
suction casting 339
– methods 339
sulfide chlorides 278

sulfide–sulfur melts 280
sulfur
– hydride system 43
– pressure–temperature phase diagram 382
superconducting
– films 402
– silicides 43
– synthesis 8
superconductivity 43
supercritical fluid–liquid–solid (SFLS) growth 10
supercritical fluids 11
supercritical hydrothermal synthesis 129, 138, 140, 143, 145, 146
– center of mixing piece
– – neutron CT image 139
supercritical medium 190
supercritical water 139
– mixing behavior 138
supercritical water and room temperature water mixing
– neutron radiography image 139
super-cooled liquid 288
– crystallization 289
– diffusion coefficients 290
– strong and fragile classifications for 291
– value, of viscosity 289
super-cooled liquid state
– crystal nucleation rate and crystal growth rate 290
superoxides 277
supersaturation 58, 61, 64, 142, 157, 400, 404, 555
– approaches 157
super saturation ratio 435
supply partial pressure 435
surface crystallization 299
surface diffusion 9
surface energy 46, 47, 49, 52, 64, 128, 236, 319
surface free energy density (γ) 432
surface functionalization 190
surface hydrophobization 187
surface-modified nanoparticles
– by organic molecules 145
surface-to-volume ratio 148
surface modification 43, 146, 541
surface modified nanoparticles 148
surface modifiers 134, 146
surface morphology 17
surface plasmon coupling regime 575
surface plasmon dispersion curve 575
– of gold 575
surface plasmon dispersion relation 575
surface plasmon interference lithography
– drawback of 577
surface plasmon interference nanolithography 576
surface plasmon resonance (SPR) 574
surface tension 37
surfactant-templated structures, models 209
surfactant-templated syntheses 210
surfactant molecules
– supramolecular self-assembly 193
surfactants 209
– template 208
Swagelok union tee 141
switching TFT of pixel in flat panel display 564
synchrotron sources 32
– X-ray beam 31
syntectic reaction 389
synthetic chemistry 202
synthetic methods 213
– precursors, impregnation of 213
– removal of templates 215
– solidification of wall components 214
– templated synthesis, interfacial control of 216
system CaO–Fe_2O_3 under the high oxygen pressure 380
system entropy 238

t
TBC. *See* thermal barrier coating (TBC)
TEGa. *See* triethylgallium (TEGa)
TEL. *See* truncation edge length (TEL)
tellurium oxide (TeO_2)-based glasses 308
– application 308
– properties 308
temperature control 2
temperature gradient 5, 6, 211, 351, 352, 358
temperature profile for the annealing sequence 276
temperature range 1
templated microparticle formation 195
templated syntheses, general concept of 202
templated synthesis 201
– physical confinement 202
– templates 203
template molecules 205
templates
– atomic-scale templates 203
– macroscale templates 211
– mesoscale templates 207
– microscale templates 204
– replication of 213

– variety of 203
templating 218
tensile strain 447
tensile strength 43
ternary chalcogenide 280
tetraalkoxysilanes 186
tetrachlorosilane (Si tetrachloride, $SiCl_4$) 5
tetraethoxysilane (TEOS) 181
tetraethyl orthosilicate (TEOS) 414
tetrahedral Si nanowires 43
tetramethoxysilane 181
tetramethoxysilane (TMOS) 181
tetrapropylammonium (TPA) hydroxide 171
TFTs with different channel lengths, I–V characteristics of 569
thermal annealing 2, 12, 14, 35, 36, 52, 62, 64
– of Au/amorphous Si bilayers 31
– of the Au/SiN_z/Si wafer 58
– of a piece of Au/SiO 58
thermal barrier coating (TBC) 417, 421
thermal budgets 9
thermal conductive materials 146
thermal conductivity 10
– effects of pressure 24
thermal convection 436
thermal CVD 401
– microstructures in 405
thermal decomposition 4, 186, 214, 361, 402
thermal degradation 2, 3, 52
thermal energy 43
thermal equilibrium 400, 405
thermal evaporation 2, 3, 12, 14, 15, 46, 52
– of Si/SiO_2 powder mixture in 33
– source 25
thermal gradients 377
thermal instability 279
thermal shock resistance 3
thermal treatment 215
thermochemical reactions 399
thermocouples 2, 26, 358
– wires 26
thermodynamic equilibrium 376, 404
– phase diagrams 376
thermodynamics 53, 375
– stability 367
– system, phase 380
thin films preparation 399
thin film transistor (TFT) 563
thioacetamide (TAA) 253
thio sol–gel (TSG) 192
thiourea 283
third nonclassical crystallization mechanism 241
time temperature transformation (TTT) curve 320
tin 276
TiO_2–WO_3
– composite 258
tissue engineering 539
titania nanofibers
– formation 251
titania solubility 168
titanium tetrachloride 18
tolerance factor 83
TOP modified cobalt nanoparticles
– TEM images 137
toroid chamber 29
toroid type 28
toroid type/Paris-Edinburgh cell 28
total energy 43
total quantum yield, for Eusup3+/sup 314
total surface free energy 235, 238
TPA–Si-ZSM-5, structure-directed synthesis of 206
transformation quartz-coesite 28
transitional nanowire 59
transition metal-containing compounds 281
transition metal oxides 18, 277, 278
transition metals 279, 366
transmission electron microscopy (TEM) 172, 238
transparent ceramics 17
transport additives 366
– chlorine as 367
– combination of 368
– hydrogen halides, used as 368
– iodine as 362, 365
– oxygen 367
– tellurium(IV) chloride used as 368
– water 367
transport agents 351, 352, 361, 365, 371
– aluminum(III) chloride 367
– chloride 366
– chlorine 361, 362, 367
– gallium(III) halides 366
– halogens 365, 366
– hydrogen chloride 369
– hydrogen halides 368
– hydrogens 369
– indium(III) halides 366
– iodine 360, 362, 369
– iron(III) halides 366
– oxygen 363, 369
– selection 363
– tellurium(IV) chloride 367
– tellurium(IV) halides 368

– water 368
transport ampoules 360, 361
– preparation 358
transport balance 372
transport effective species 367
transport equation 352, 366, 370
– stoichiometric coefficients 354
transport equilibrium 369
transport furnace 358
transport-effective species 353
– sulfur 353
– zinc iodide 353
transport rates 354, 358, 368, 371
transport reaction
– reaction enthalpy of 352
– reaction entropy of 352
transport reaction molecules
– molar fractions 355
trialkoxysilane precursors 187
triangular truncation 30
triethoxysilane 187
triethylgallium (TEGa) 434
trifluorides 278
trigonal bipyramid (tbp) 308
trigonal pyramid (tp) 308
trimethylgallium (TMGa) 434
trimethylsilyl chloride (Me_3SiCl) 187
trioctyl-phosphine (TOP) 137
triple-concentric nanowire 40, 55
– entrapment of Ni nanospheres in 57
trivalent bismuth oxides 119
truncation edge length (TEL) 31
trypsin, as crystallization modifiers 257
TTT curves, schematic illustration 321
tube furnaces 14, 358
tungsten carbide 25, 33
– core 26
– cubes 30, 31
– pistons 26, 27
tungsten, melting point 363
twinned attachment 162
twinned crystallites 167
two-photon-induced photoluminescence 577
two-zone furnace 358, 359
typical single-roller melt-quenching
– schematic illustration of 330

u
ultra extended graphics array (UXGA) panel 571
undercooled liquid, thermal stability 325
undercooling 319
unheated fused quartz tube wall 23
uniaxial hydraulic press 31
unique plasmon lithography technique 576
$U_4O_4Te_3$ crystals 276
uridine triphosphate (UTP) 253
UV-curing 187
UXGA organic TFT array 572

v
vacuum evaporation 400
van Arkel–de Boer process 362
van der Waals forces 161
van der Waals space 41
van't Hoffs equation 354
vapor deposition 195
vapor-phase transport 214
vapor–liquid–solid (VLS) mechanism 4
vapor–solid–solid growth (VSS) 8
vapor transport
– examples
– – chemical vapor transport of elements 362
– – chemical vapor transport of intermetallics 363
– – halides 365
– – oxides 367
– – selenides 369
– – sulfides 369
– – tellurides 369
– of metal halide 366
VAS. *See* void-assisted separation (VAS)
vertical reverse temperature gradient (VTG) 113
viscosity, effects of pressure 24
visualization method 138
vitrification 173
VLS growth 10
VLS mechanism 5
V_2O_5 flux 278
Vogel–Fulcher–Tammann (VFT) formula 320
void formation 146
void-assisted separation (VAS) 451, 453
voidless zeolite membranes 219
volatile anions 390
volatile components 16
volatile halide 366
volatility 277, 351, 389
Volmer-Weber growth mode 447
volume deposition rate 418
vulcanization 258

w
Walker-type module 30, 43
wall components 216
– precursors of 213

warm extrusion 341
water
– density around critical point 125
– dielectric constant around critical point 125
– properties of 124
– quenching technique, schematic illustrations of 335, 337
water-organic species
– phase behavior 135
water–gas shift reaction 130
wave vector 575
Weiss constant 304
welding 8, 417
well-ordered mesoporous materials formation
– templating approach 194
wet gel 179
whiskers 400
wide bandgap semiconductors 463
wild-type transfer RNA (WT-tRNA) 253
window glasses 296
wire–catalyst interface 49, 52
WO_3–B_2O_3 based glasses 305
Wulff's rule 236
wurtzite-type AlN 43

x
xerogel 190
xerogel powder 186
X-ray beam 31
X-ray diffraction 15, 29, 32, 168, 174
X-ray diffractometers 53
X-ray photoelectron spectroscopy (XPS) 15, 293
XRD curves 18
XRD peak patterns 137
xRO–(100–x)TeO_2 glasses
– refractive index, mean volume thermal expansion coefficient, and electronic polarizability for 312

y
YAG hydrothermal growth 112
$Y_3Fe_5O_{12}$ (YIG) and $Y_3Ga_5O_{12}$ (YGG), hydrothermal growth of 112
yittria-stabilized zirconia (YSZ) 417
– coating 421
– film microstructure 422
Young's modulus 299, 448
YSZ. *See* yittria-stabilized zirconia (YSZ)
yttria-stabilized zirconia 17
yttrium aluminate, $Y_3Al_5O_{12}$ (YAG) 112
YZnPO, single crystals of 11

z
Zeldovich factor 320
zeolites 206, 222
– deposition of 217
– powder, pore size of 413
– structures of 205
– templated carbon 207
zeolitic aluminosilicates 107
zeolitic imidazolate framework
– nucleation 256
zinc 276
zinc oxide (ZnO) 109
– crystals
– – Helbig's apparatus for the growth of 377
– hydrothermal crystal growth, temperature distribution 112
– nanobelts 52
– nanocrystals 252
– nanostructures 52
– nanowire 51
– single crystals 112
– – grown, using KOH mineralizer 113
zirconia sleeve 30
zirconium-iron oxide nanoparticle 148
zirconium(IV) oxide 367
zirconium phosphates 116
– synthesis 115
ZnO–Bi_2O_3–B_2O_3 glasses 300, 301
– densities of 301
– electronic polarizabilities of 301
– refractive indices 301
ZnSe nanowires 49, 50
$Zr_{55}Al_{10}Ni_5Cu_{30}$ alloy 345
$Zr_{65}Al_{7.5}Ni_{10}Cu_{12.5}Pd_5$ glassy alloy
– specific heat curves of 347
$Zr_{65}Al_{7.5}Ni_{10}Pd_{17.5}$ alloy 346
$Zr_{65}Al_{7.5}Ni_{10}Pd_{17.5}$ alloy cast
– cooling curves of 345
$Zr_{65}Al_{7.5}Ni_{10}Pd_{17.5}$ bulk metallic glass
– TEM images of the shear band 328
$Zr_{50}Cu_{40}Al_{10}$ ingots
– cooling curves of 344
$Zr_{60}Cu_{22}Ni_5Al_{10}Au_3$ bulk metallic glass 329
Zr-based metallic glass 338, 348
Zr–Al–Ni–Pd bulk metallic glass
– true compressive stress–strain curve of 327
Zr–Cu–Al metallic glasses
– as bulk glass formers 327